The Cyanobacteria

Preface

When I joined, almost thirty years ago, a small group of enthusiastic botanists in the study of blue-green algae, these photosynthetic microorganisms were considered a curiosity and those engaged in their study were often regarded by others as eccentrics. Over the ensuing thirty years we have witnessed an amazing transformation of many basic concepts in biology. In the course of this transition blue-greens have been renamed cyanobacteria and their remarkable properties have attracted the attention of many other scientists of widely different training and interests, apart from botanists and microbiologists.

Intensive and systematic research during the last five years has provided us not only with a great wealth of new information about cyanobacteria but also with deeper insights into their molecular biology and genetics which require a fundamental revision of earlier views. These, together with recent appreciation of their ecological importance, render a new treatise on cyanobacteria a pressing necessity. We hope that the present volume may serve as a useful source of essential information for specialists in cyanobacteria, as well as for research workers, teachers and students of biology in general.

Not a few of the early explorers have been able to maintain not only their interest but also their undiminished creativity in contributing significantly to the knowledge of blue-greens. Chase Van Baalen was one of them. His tragic death at an early stage of preparations for this volume has deprived me of a dear friend and an invaluable companion in the editorial work. This sad loss could only partly be compensated for by the great enthusiasm of my co-authors and the high scholarly standard of their contributions. I am most grateful for their efforts.

We wish to devote this book to the memory of Chase Van Baalen.

October 1986 Peter Fay

THE CYANOBACTERIA

Editors

P. Fay

Department of Botany and Microbiology,
University College London,
London WC1E 6BT, UK

and

C. Van Baalen

The University of Texas
Marine Science Institute,
Port Aransas, TX 78373, USA

1987
ELSEVIER
Amsterdam, New York, Oxford

ISBN 0-444-80840-X

Published by:
Elsevier Science Publishers B.V. (Biomedical Division)
P.O. Box 211
1000 AE Amsterdam
The Netherlands

Sole distributors for the USA and Canada:
Elsevier Science Publishing Company Inc.
52 Vanderbilt Avenue
New York, NY 10017
USA

Library of Congress Cataloging in Publication Data

The Cyanobacteria.

Includes bibliographies and index.
I. Cyanobacteria. I. Fay, Peter. II. Van Baalen, C. (Chase), 1925–1986.
QR99.63.C9 1987 589,4'6 87-608
ISBN 0-444-80840-X

Acknowledgement
Many illustrations and diagrams in this volume have been obtained from other publications. In all cases reference is made to the original publication. The full source can be found in the reference list. Permission for the reproduction of this material is gratefully acknowledged.

Printed in The Netherlands

List of Contributors

H.J. Bohnert — Department of Biochemistry, Molecular and Cellular Biology and Plant Sciences, University of Arizona, Tucson, AZ 85721, USA

J.J. Boon — FOM Institute for Atomic and Molecular Physics, Kruislaan 407, 1098 SJ Amsterdam, The Netherlands

G.L. Boyer — Department of Chemistry, SUNY College of Environmental Science and Forestry, Syracuse, NY 13210, USA

T. Bricker — Department of Chemistry, University of Southern Mississippi, Southern Station, Hattiesburg, MS 39406, USA

P. Fay — Department of Botany and Microbiology, University College London, London WC1E 6BT, UK

G.E. Fogg — Department of Marine Biology, Marine Science Laboratories, Menai Bridge, Gwynedd LL59 5EH, UK

A. Gillam — Institute of Offshore Engineering, Heriot-Watt University, Edinburgh EH14 4AS, UK

A.N. Glazer — Department of Microbiology and Immunology, University of California, Berkeley, CA 94720, USA

F.K. Gleason — Gray Freshwater Biological Institute, University of Minnesota, Navarre, MN 55392, USA

M.G. Guerrero — Departamento de Bioquimica, Facultad de Biologia, Universidad de Sevilla, 41080 — Sevilla, Spain

J. Guikema — Division of Biology, Kansas State University, Ackert Hall, Manhattan, KS 66506, USA

D.-P. Häder — Fachbereich Biologie-Botanik der Philipps Universität, Lahnberge, D-3550 Marburg, FRG

M. Herdman — Unité de Physiologie Microbienne, Institut Pasteur, 28 rue du Docteur Roux, 75724 Paris Cedex 15, France

J. Houmard — Unité de Physiologie Microbienne, Institut Pasteur, 28 rue du Docteur Roux, 75724 Paris Cedex 15, France

F. Jüttner — Institut für Chemische Pflanzenphysiologie der Universität Tübingen, Correnstrasse 41, D-7400 Tübingen, FRG

C. Lara — Departamento de Bioquimica, Facultad de Biologia, Universidad de Sevilla, 41080 — Sevilla, Spain

J.W. De Leeuw — Organic Geochemistry Unit, Delft University of Technology, De Vries van Heystplantsoen 2, 2628 RZ Delft, The Netherlands

W. Loeffelhardt — Institut für Allgemeine Biochemie und Ludwig-Boltzmann-Forschungsstelle für Biochemie, Universität Wien, A-1090 Wien, Austria

J. Myers — Department of Zoology, University of Texas, Austin, TX 78712, USA

M.L. Narro — Department of Genetics, Stanford University School of Medicine, Stanford, CA 94305, USA
H. Pakrasi — E.I. Dupont de Nemours and Co., Central Research and Development Experimental Station, Wilmington, DE 19898, USA
G.A. Peschek — Institut für Physikalische Chemie der Universität Wien, Wahringerstrasse 42, A-1090 Wien, Austria
L.J. Rogers — Department of Biochemistry, University College of Wales, Aberystwyth, Dyfed, SY23 3DD, UK
L. Sherman — Division of Biological Sciences, University of Missouri, Tucker Hall, Columbia, MO 65211, USA
R.D. Simon — Department of Biology, State University of New York, Geneseo, NY 14454, USA
F.R. Tabita — Center for Applied Microbiology and Department of Microbiology, University of Texas, Austin, TX 78712, USA
N. Tandeau de Marsac — Unité de Physiologie Microbienne, Institut Pasteur, 28 rue du Docteur Roux, 75724 Paris Cedex 15, France
C. Trick — Institute of Environmental Studies, University of Toronto, Toronto, Ontario M5S 1A4, Canada
†C. Van Baalen — The University of Texas Marine Science Institute, Port Aransas, TX 78373, USA
A.E. Walsby — Department of Botany, University of Bristol, Bristol BS8 1UG, UK
C.C. Wasmann — Department of Biochemistry, University of Arizona, Tucson, AZ 85721, USA
B.A. Whitton — Department of Botany, University of Durham, Durham DH1 3LE, UK
J.M. Wood — Gray Freshwater Biological Institute, University of Minnesota, Navarre, MN 55392, USA
M. Wyman — Department of Biological Sciences, University of Warwick, Coventry CV4 7AL, UK

Chase Van Baalen 1925–1986

Chase Van Baalen, one of the editors of this volume, died on January 20, 1986, of complications arising from chemotherapy for a lung cancer. He was a Senior Research Scientist in the Marine Science Institute and a Professor of Botany at The University of Texas.

We remember Chase variously as a convivial drinking buddy, a patient fishing companion, a close-to-the-chest poker player, a quietly competent scientist. In this brief memoir I shall attempt to recount a small sample of Chase's discoveries with something of the flavor of his work and I plead for the pardon of many students and colleagues whose contributions will necessarily go unmentioned. I was a long-time friend, at one time his teacher, in later years his student. Hence this will be a highly personalized account.

Chase once described his undergraduate years as those of a college bum. This must have been too harsh an appraisal because it provided him with a broad background, even including field station work on algae and limnology under G.W. Prescott. In 1950, he came to The University of Texas as a graduate student in microbiology under Jackson Foster. In 1954 he transferred to Zoology, to work under me.

So it was that Chase went to work on the first of the hyperactive blue-greens, *Anacystis nidulans*, then just recently discovered. The problem was to prepare a photochemically active subcellular preparation. He got as far as obtaining high rates of photochemical reduction of indophenol dye. There was a peculiarity that low temperature had to be avoided. In what proved to be an example, often repeated in later years, Chase followed his curiosity: what was bad about low temperature (as 4°C)? He left a cell suspension overnight in the cold room, centrifuged out the cells, and looked at absorption spectra of the supernatant. A component with 270 nm absorption had come out of the cells. All this led to identification and subsequent isolation of several pteridines and, thereafter, several years fruitful work with the collaboration of our colleague, Hugh Forrest.

In 1959, Chase joined the newly formed Kitchawan Research Laboratory at Ossining, New York. He made the best of an unfavorable situation, profited greatly from occasional contact with Luigi Provasoli, stuck to the character-building task of isolating a collection of 22 marine blue-greens, extended the efficacy of the ASP-2 medium, confirmed a widespread need for vitamin B_{12}. Though published results (Botanica Marina 4:129) were not widely appreciated, the experience provided a basis for work of the following 25 years.

In 1961, the timing of a staff vacancy led to an appointment at the Institute of Marine Science in Port Aransas, Texas. Here was a special environment, both for blue-green algae and for people. The shallow and productive inland bays provided a happy home for a wealth of blue-greens in their bottom muds. The sleepy fishing village of Port Aransas became home for Chase and his family. It was a place of beaches, sand, and sun, far removed from the cosmopolitan world.

Most of us engaged in the scientific endeavor take for granted an interdependence among colleagues. This runs from instrumentation to literature review to check-out of new ideas. Self-dependence would impose a scarcely bearable strain. Yet that was the essential feature of the special niche to which Chase adapted. The main university campus at Austin was 200 miles of isolation away. He had to cover the literature by himself, introspectively check out his new ideas, build or repair his own instruments, and solve even trivial logistic problems. One is tempted to suppose that his isolation created a climate of need for invention which nourished his discoveries.

I shall sketchily sample some of his discoveries, mainly to show the many facets of his impact upon our understanding of the cyanobacteria:

— Requirements for quantitative plating
— Conditions for generating and isolating mutants
— Demonstration of iron acquisition systems
— A poorly utilizable N-source can be used as an internal 'chemostat' for growth
— Fructose diphosphate aldolase is indeed a common and required enzyme
— A nickel-requiring *Oscillatoria*
— Ultraviolet killing and photoreactivation
— Ultrastructure of *Trichodesmium* as a key to its fragility
— Marine *Anabaenas* as high-rate N-fixers
— Isolation of psychrophilic diatoms from Arctic ice.

Many may be critical of my sampling though not of the versatility which it displays.

Chase maintained a healthy skepticism on current dogma. It was a long time before he gave up doubts that phycocyanin has no other photochemical function beyond that of an absorber for harvesting light. In somewhat analogous fashion he had a well-nourished distrust of administrative authority. Through seven directors and eight university presidents he observed that the role of the Institute was seldom understood and that it was generally treated as the University's ugly duckling. Chase's continued fight for academic function of the Institute was not always supported. In contrast he enjoyed the appreciation and support of his scientific colleagues in Austin. And it should have been so, for over a span of 25 years he was the Institute's most productive scientist.

The special niche in which Chase developed naturally changed in the course of time. The village grew up into a city. The blue-green algae became cyanobacteria. And Chase reached out. He attracted and worked side-by-side with students and then gloried in their achievements. He developed effective collaborations on the Austin campus and farther afield in Europe, in India, and in Japan.

A fringe benefit of being a scientist is that one achieves some small measure of immortality. His discoveries become a part of the enduring edifice of science. His students go on to fulfill the example of their mentor. His institution is guided onward in the mold which he helped create. In all these ways the mark of Chase Van Baalen has been deeply etched.

Jack Myers

Contents

P. Fay and C. Van Baalen (Eds.) The Cyanobacteria

1

The protein composition of the photosynthetic complexes from the cyanobacterial thylakoid membrane

LOUIS SHERMAN[a], TERRY BRICKER[b], JAMES GUIKEMA[c] and HIMADRI PAKRASI[d]

[a]University of Missouri, Division of Biological Sciences, Tucker Hall, Columbia, MO 65211, [b]University of Southern Mississippi, Department of Chemistry, Southern Station, Hattiesburg, MS 39406, [c]Kansas State University, Division of Biology, Ackert Hall, Manhattan, KS 66506, and [d]E.I. Dupont de Nemours and Co., Central Research and Development, Experimental Station, Wilmington, DE 19898, USA

1. Introduction

The importance of cyanobacteria as model organisms for the study of photosynthesis is increasing due to the confluence of findings from many lines of research. The cyanobacteria are prokaryotes that are capable of oxygenic photosynthesis; as bacteria, they are amenable to a variety of genetic and molecular biological manipulations. While the striking similarity between cyanobacterial and chloroplast photosynthesis has been recognized for some time, the full extent of this relationship is only now being appreciated. The accumulated cytological, physiological, and biochemical evidence over the past century gave rise to the endosymbiont hypothesis for the origin of chloroplasts; this hypothesis states that chloroplasts are descended from cyanobacteria that became endosymbionts in an originally non-photosynthetic host (see Chapter 11 in this volume). This relationship has been refined recently by comparison of ribosomal and transfer RNA sequences, transcription and translation mechanisms, and through the comparison of protein and nucleic acid sequences (Gray and Doolittle, 1982). The cloning and nucleotide sequencing of chloroplast and cyanobacterial genes coding for photosynthesis proteins has now emphasized the profound depth of these similarities. In addition, the ability to readily isolate mutations in cyanobacteria and to exchange genetic information has placed these organisms at the center of current research in photosynthesis.

In this chapter, we will review the protein structure of the cyanobacterial thylakoids, with emphasis on the relationship of structure to function. We will describe the protein composition of the major photosynthetic complexes. For conciseness, we will concentrate on the similarities between cyanobacterial and chloroplast membrane structure to highlight the conserved nature of the membrane components. We will also discuss some of the genetic and mutation studies involving these components to emphasize the reasons why the cyanobacteria are such an interesting organism for research in photosynthesis. A detailed discussion of the cloning of photosynthesis genes from cyanobacteria will be deferred until a later chapter (see Chapter 10).

Our emphasis on the similarities of protein composition between the two types of membranes should not obscure the important differences that distinguish chloroplast and cyanobacterial photosynthesis. Since most strains lack chlorophyll (Chl) *b*, the cyanobacteria contain phycobilins as a major light-harvesting pigment. These pigments are organized in phycobilisomes, which are located on the external surface of the membrane (Gantt, 1981; Glazer, 1984; see Chapter 3 for details). By contrast, the major light-harvesting apparatus of chloroplasts, the Chl *b*-containing light-harvesting protein complex (LHCP), is an integral membrane complex. Additionally, the intracellular organization of the cyanobacterial thylakoids is markedly different from that of chloroplast thylakoids. In most cyanobacteria, the thylakoids are arranged peripherally in three to six concentric layers (Golecki and Drews, 1982). Chloroplast thylakoids can be differentiated into stacked (granal) membranes and unstacked (stromal) membranes; these can be further subdivided into granal portions that are in contact with other membranes (appressed), and those granal or stromal membranes not in contact (non-appressed). It has been determined that the precise composition of the appressed and non-appressed membranes is different and that there is a good deal of lateral heterogeneity between these structural regions. Such lateral heterogeneity has not been determined in cyanobacteria; therefore, the cyanobacteria provide an opportunity to compare the ways in which different membrane organizations function in photosynthesis.

2. The relationship of photosynthesis genes from chloroplasts and cyanobacteria

There are three basic techniques that are currently in use to determine relationships between proteins from different organisms: (1) comparison of amino acid sequences; (2) comparison of nucleic acid sequences and the derived amino acid sequences; and (3) immunological cross-reactivity using antibodies prepared against one of the proteins. All of these procedures have successfully demonstrated that photosynthesis proteins, particularly integral membrane proteins, possess very similar primary sequences. Although amino acid sequencing is the most direct procedure, it is also the most difficult, and relatively few types of cyanobacterial proteins have been so analyzed. The comparison of amino acid sequences for some photosynthesis proteins (e.g., ferredoxins and cyt *f*) isolated from chloroplasts and cyanobacteria have been summarized recently (Schwartz and Dayhoff, 1978, 1981; Gray and Doolittle, 1982; see also Chapter 2). These results indicate a close relationship between these proteins and have been used as evidence in support of the endosymbiont hypothesis for the formation of chloroplasts (Gray and Doolittle, 1982). However, the other two techniques are simpler and more versatile, and have generated far more data to support this contention.

Many of the genes that code for integral membrane proteins of the chloroplast thylakoid have been mapped on the chloroplast genome, and have subsequently been cloned and sequenced (Crouse et al., 1985). These cloned genes have been used as probes to search for homologous sequences in genomes of other plants, algae, or cyanobacteria. This Southern hybridization procedure (Southern, 1975) has permitted the detection and cloning of a number of cyanobacterial photosynthesis genes that are at least partly homologous to their chloroplast counterparts.

The two genes for which the most complete comparative DNA sequence data have been obtained are *rbc*A, which codes for the large subunit of ribulosebisphosphate carboxylase, and *psb*A, which codes for a 32 kDa integral membrane protein (the Q_B protein) associated with Photosystem

II (PSII) (see Section 6.1. of Chapter 10). Curtis and Haselkorn (1983) compared the nucleotide and amino acid sequences of the *Anabaena* 7120 *rbc*A gene with those of corn, spinach, *Chlamydomonas*, and *Synechococcus*. This analysis revealed homology of 71–77 % at the nucleotide level and 80–85 % at the amino acid level; however, this sequence conservation disappeared immediately outside the coding region. These studies demonstrate the close relationship between cyanobacteria and chloroplasts, a relationship that is further emphasized by comparison of the *psb*A genes.

The *psb*A gene codes for a 32 kDa protein which appears to participate in electron transfer to a bound plastoquinone (PQ) molecule in PSII (Arntzen et al., 1982). This protein is sometimes referred to as the Q_B protein, and is the target for photoaffinity labeling with azido-[^{14}C]atrazine (Pfister et al., 1981). It has been demonstrated that herbicides such as atrazine and 3-(3,4-dichlorophenyl)-1,1-dimethylurea (DCMU) compete with PQ binding at this protein, thereby interrupting electron transport and blocking photosynthesis. Since PSII core preparations are DCMU-sensitive, it has been assumed that the 32 kDa protein is closely associated with the PSII reaction center. Additionally, it has been postulated that lesions in *psb*A would give rise to an altered 32 kDa protein, which in turn could generate DCMU or atrazine resistance.

The *psb*A gene has now been cloned and sequenced from about one dozen organisms (Crouse et al., 1985) including spinach, *Amaranthus hybridus*, *Nicotiana debneyi*, *Chlamydomonas reinhardtii*, *Anabaena* 7120, and *Anacystis nidulans* R2 (Golden and Haselkorn, 1985). The nucleotide sequence of *psb*A is remarkably conserved among these organisms, and the spinach and *Nicotiana* sequences are identical (Zurowski et al., 1982). The amino acid sequences of the cyanobacterial *psb*A gene products are 84 % (Golden and Haselkorn, 1985) to 87 % (Curtis and Haselkorn, 1984) homologous relative to the corresponding plant sequences. Both cyanobacterial sequences have a six-amino-acid insertion near the carboxyl terminus, and both contain lysine residues which are not found in the plant sequences. The cyanobacterial genes are about 88 % homologous with each other, which is the same variation found throughout the genes that have been sequenced to date.

One of the more interesting features of cyanobacterial molecular biology is the detection of multiple copies of the *psb*A gene in all strains that have been analyzed (see Section 6.1. of Chapter 10). *Anabaena* 7120 contains at least three copies of this gene (Curtis and Haselkorn, 1984), and *A. nidulans* R2 has three copies (Sherman et al., 1984). The *Anabaena* genes are not identical, but *psb*AI and *psb*AII are 94 % homologous at the amino acid level (Curtis and Haselkorn, 1984). A key question concerns the expression of the different copies of *psb*A. Curtis and Haselkorn (1984) demonstrated that only *psb*AI is expressed in *Anabaena*; on the other hand, at least two *psb*A genes appear to be expressed in *A. nidulans* R2 (Golden and Haselkorn, personal communication). The latter fact is particularly important relative to the matter of atrazine and DCMU resistance in cyanobacteria.

DCMU- and atrazine-resistance mutants have been isolated in *Amaranthus hybridus* (Hirshberg and McIntosh, 1983), *Chlamydomonas reinhardtii* (Erickson et al., 1985), and *Anacystis nidulans* (Golden and Sherman, 1984a), and the corresponding *psb*A genes from the resistant strains have been cloned and sequenced. Resistant mutants from all these strains demonstrate a lesion at amino acid position 264, due to a single nucleotide change. This amino acid is a serine residue in all wild-type strains, and mutates to an alanine in *C. reinhardtii* and *A. nidulans*, but to a glycine in *A. hybridus* and *S. nigrum*. Other lesions that give rise to atrazine resistance have been characterized in *C. reinhardtii* (Erickson et al., 1985), but not yet in the other species. These results indicate that the 32 kDa protein has a similar secondary structure and organization in all photosynthetic membranes studied to date.

Definitive proof that a mutated *psb*A gene is necessary and sufficient for DCMU resistance has only been obtained in *A. nidulans* (Golden and Haselkorn, 1985) by utilizing the high frequency transformation system (Golden and Sherman, 1984b). Golden and Haselkorn (1985) determined that the *psbA* gene giving rise to DCMU resistance was located on the 3.5 kilobase (kb) *XhoI* fragment. This fragment was digested further with restriction enzymes to yield a series of smaller fragments, all of which contained amino acid position 264. DCMU-resistant transformants were obtained with fragments at least 1 kb in size; these fragments also contained the 3′ two-thirds of the *psbA* gene (and position 264 with the ser→ala alteration). No DCMU-resistant transformants were obtained in the absence of added DNA, or when fragments lacking homology to *psbA* were added. These experiments provide proof that a 1 kb segment of the *psbA* gene containing residue 264 is responsible for DCMU resistance in *A. nidulans*. Furthermore, since another copy of *psbA* appears to be expressed in *A. nidulans* (but gave negative results in the transformation experiments), it is tempting to conclude that the resistance trait is dominant in this organism (Golden, personal communication). Although this speculation requires further testing, the transformation systems available in cyanobacteria should prove extremely valuable for the introduction of precise lesions into numerous photosynthesis genes.

A great many unpublished observations indicate that similar homologies exist for other cyanobacterial and chloroplast genes that code for integral thylakoid proteins. Using DNA-DNA hybridization techniques with chloroplast DNA as probes, genes coding for most of the known PSII proteins are now being cloned from cyanobacteria. While this work is proceeding, the alternative technique of Western blotting with antibodies has been used to verify the homologies between these proteins. As shown in Table I, more than one dozen cyanobacterial thylakoid proteins have been identified on Western blots with heterologous antibodies. Such results demonstrate that these proteins have retained antigenic similarities, and extend the nucleotide sequencing data. When coupled with the amino acid sequencing studies that have been performed (Gray and Doolittle, 1982), the experiments provide unequivocal evidence for a great similarity between the chloroplast and cyanobacterial thylakoid protein components. This knowledge will provide the conceptual framework for the remainder of the review. Importantly, a similar analysis comparing chloroplast and photosynthetic bacterial membranes has led Hearst (1985) and Michel and Diesenhofer (1985) to postulate a close relationship between PSII and bacterial reaction center components. We will include this hypothesis in our discussion of PSII in cyanobacteria.

3. Chlorophyll-protein complexes

During the past several years, many studies have been devoted to understanding the pigment/protein substructure of cyanobacterial thylakoids. An important tool in these studies has been polyacrylamide gel electrophoresis (PAGE). Under appropriate conditions of electrophoresis and membrane solubilization, PAGE allows a direct visualization of pigment-protein substructure by resolving Chl-containing protein bands on gels. In a second step, Chl-protein complexes can be excised from gels and examined for protein composition under denaturing conditions. Several objectives have guided these studies: (1) the identification of the apoproteins which bind Chl in cyanobacterial thylakoids; (2) ascertaining the relationship of these apoproteins to the reaction centers of PSI and PSII, (3) determining how the Chl-binding peptides may associate with other proteins to form the PSI and PSII reaction center cores; and, most recently, (4) understanding how the Chl-binding proteins are synthesized and inserted into the photosynthetic membranes. Chl-protein gels have been used extensively in characterizing PSI and PSII, and some of the work discussed here will be directly applicable to sections where each Photosystem is considered independently.

In early work, Thornber and colleagues used sodium dodecyl sulfate (SDS) to solubilize thyla-

TABLE I

Immunological relationship between chloroplast and cyanobacterial photosynthesis proteins

Complex	Protein	Source of chloroplast protein	Cyanobacterial species tested	Reference
PSII	47 kDa	*Chlamydomonas*	*A. nidulans*	1
			Aphanocapsa	2
	43 kDa	*Chlamydomonas*	*A. nidulans*	1
			Aphanocapsa	2
	33 kDa	spinach	*Phormidium laminosum*	3
	cyt b_{559}	spinach	*A. nidulans* and *Aphanocapsa*	2
PSI	65 kDa 20 kDa	Swiss chard	*Mastigocladus*	4
Cyt b_6/f	cyt *f* cyt b_6 Rieske Fe-S Subunit 4	maize	*A. nidulans* and *Aphanocapsa*	2
RuBP carboxylase	large subunit	*Euglena gracilis*	*A. nidulans*	5
ATPase	α-subunit β-subunit	*Chlamydomonas*	*A. nidulans* and *Aphanocapsa*	2
	γ-subunit	spinach	*Spirulina*	6

References: (1) Pakrasi et al., 1985b; (2) Bullerjahn et al., 1985; (3) Stewart et al., 1985; (4) Nechushtai et al., 1983; (5) Bricker and Sherman, 1984; (6) Hicks, 1984.

koids of *Phormidium*, and resolved a major Chl-binding complex by hydroxyapatite chromatography (Thornber, 1975). Subsequently, they examined SDS extracts of *Phormidium* by SDS-PAGE (Hunter and Thornber, 1976; Reinman and Thornber, 1979). The complex which they resolved was approximately 110 kDa, and contained several peptides between 40–45 kDa (Hunter and Thornber, 1976). This work pioneered the study of Chl-protein complexes in cyanobacterial thylakoids, and identified interesting similarities between these membranes and their counterparts in higher plants.

Gel electrophoresis techniques have been augmented by the use of fluorescence spectroscopy. Such analysis provides significant information about the in vivo organization of Chl within membranes, and can serve as a method to identify a specific pool of Chl molecules. When cells or isolated thylakoid membranes are monitored at 77K, three Chl fluorescence emission peaks can be resolved at 686 nm, 696 nm, and 716 nm (Goedheer, 1972; Newman and Sherman, 1978). Of these, the 716 nm peak (F716) has been assigned to PSI, whereas the 686 nm (F686) and the 696 nm (F696) peaks are believed to originate in PSII (Breton, 1983; Guikema and Sherman, 1983a; Pakrasi et al., 1985b). The combination of the fluorescence and gel electrophoresis data has permitted a more definitive functional identification of the numerous Chl-protein bands detected by PAGE.

3.1. PSII Chl-protein complexes

Table II summarizes recent work on the Chl-proteins (CP) associated with PSII and highlights some common features revealed in this work. Chl-protein bands of PSII exhibit 77K fluorescence emission at shorter wavelengths, displaying emission peaks at about 686 or 696 nm. In addition, these bands contain two major Chl-binding proteins at about 43 and 47 kDa. Rusckowski and Zilinskas (1980) observed two Chl-protein bands from *Nostoc* membranes which had a migration rate similar to CPII of spinach thylakoids. When these bands were eluted from gels and monitored for 77K fluorescence emission, F686 was observed. A set of three polypeptides between 43–48 kDa were obtained from these Chl-protein bands. Similarly, Stewart (1980) examined the membranes of *Phormidium luridum* and observed one PSII Chl-containing band, comprised of peptides at 40 and 48 kDa.

TABLE II

PSII Chl-protein complexes resolved by PAGE

Species (Refs.)	Complex	Chl-binding proteins (kDa)	Colorless proteins (kDa)	77 K Chl Fluor. Max. (nm)
Nostoc sp. (1)	CP IIIa	44,48	–	685
	CP IIIb	43,48		
Phormidium luridum (2)	A-2	40,48	–	–
Plectonema boryanum (3,4)	Fraction I	41,46	15,28,52	680
Anacystis nidulans (5,6)	CPVI	45,48	–	–
Synechococcus sp. (7,8)	CP 2a, d	47	–	(686)696
	CP 2b	47	31,66	(686)696
	CP 2c	40	–	686
Anacystis nidulans (9,10)	CPVI-1	42,71	–	685
	CPVI-2	52 (or 48)	–	695
	CPVI-3	45	–	685
	CPVI-4	36	–	685
Aphanocapsa 6714 (11)	CPIII′	64	–	685
	CPIIIa	48,45	–	685
	CPIIIb	36	–	685

References: (1) Rusckowski and Zilinskas, 1980; (2) Stewart 1980; (3) Hladik et al., 1982; (4) Cacka et al., 1981; (5,6) Guikema and Sherman, 1981, 1983a; (7,8) Yamagishi and Katoh, 1983a,b; (9,10) Pakrasi et al., 1985a,b; (11) Bullerjahn et al., 1985.

Yamagishi and Katoh (1983a,b, 1984) used a slightly different strategy to examine the relationship among the PSII Chl-binding proteins. They first isolated a PSII-enriched membrane fraction of thermophilic *Synechococcus* sp. and then used this material to examine the Chl-protein patterns on gels. They observed four Chl-protein bands, each with a slightly different composition of two Chl-binding proteins (40 and 47 kDa), plus two colorless polypeptides (31 and 66 kDa). They attribute these four bands to a successive separation of Chl-proteins from the PSII reaction center complex during treatment with SDS. Their model suggests that the PSII reaction center P680 resides on the 47 kDa Chl-binding polypeptide, and that this peptide is responsible for the fluorescence emission at 696 nm. They suggest that the 40 kDa Chl-binding protein, which emits at 686 nm, serves as an antenna for PSII.

The relationship between PSII proteins was further examined in *A. nidulans* by using dodecyl-β-D-maltoside, a mild non-ionic detergent (Pakrasi et al., 1985b). When membrane extracts were prepared with this detergent and were examined using PAGE, the PSII Chl-protein complex (CPVI) was resolved into four distinct bands (termed CPVI-1 to CPVI-4, see Table II). A majority of the Chl was found in two of the four bands: CPVI-2, which contained the 52 kDa Chl-binding apoprotein and showed a single fluorescence emission peak at 695 nm; and CPVI-3, which contained the 45 kDa Chl-binding apoprotein. Interestingly, the 52 kDa peptide was recognized by an antibody raised against the presumptive PSII reaction center protein of *C. reinhardtii* (Pakrasi et al., 1985b). It is likely, therefore, that CPVI-2 is identical to CP2-a or CP2-d from *Synechococcus* sp. (Yamagishi and Katoh, 1984) and to spinach CPa-1 (Nakatani et al., 1984). On the other hand, CPVI-3, which contained a 45 kDa protein, had a fluorescence emission peak at 686 nm. The 45 kDa peptide was immunologically related to the protein of *C. reinhardtii* which is thought to be the antenna Chl-protein to PSII. Thus, CPVI-3, CP2c from *Synechococcus*, and CPa-2 (CP43) from spinach would appear to be equivalent.

Two novel PSII Chl-protein complexes were also detected in *A. nidulans* R2 using dodecyl-maltoside (Pakrasi et al., 1985b). The largest complex, CPVI-1, contained a 71 kDa and a 42 kDa protein. The 71 kDa protein was determined to be the phycobilisome anchor protein, since it was recognized by an antibody raised against the *Nostoc* anchor (see Section 7.3. of Chapter 3). The 42 kDa was not antigenically-related to any of the PSII proteins, and was determined to be an integral membrane protein. Thus, the 42 kDa protein may act as a binding protein for the 71 kDa anchor polypeptide; at present, it is uncertain which (or both) of the proteins binds chlorophyll. The second new complex, CPVI-4, contained a previously unobserved Chl-binding protein with an apparent molecular mass of 36 kDa. This complex had a fluorescence maximum at 686 nm, and the relative quantity of CPVI-4 increased dramatically in iron-stressed cells (Pakrasi et al., 1985b). Cells grow and divide under iron-deficient conditions, but most of the phycobilisomes disappear and there is an altered organization of the Chl within the thylakoid membrane (Sherman and Sherman, 1983). One possible interpretation is that CPVI-4 acts as an intermediate light-harvesting antennae for PSII and increases in quantity significantly as the level of phycobilisomes diminishes. However, it is still not possible to rule out the involvement of the 36 kDa Chl-binding protein in the primary PSII photochemistry as will be discussed in Section 4.1. of this chapter.

Dodecyl-maltoside has also been useful in examining the membranes from *Aphanocapsa* 6714 (Bullerjahn et al., 1985), and some significant differences in the organization of PSII have been observed in this species. *Aphanocapsa* membranes contain three distinct PSII Chl-protein complexes (termed CPIII′, CPIIIa, and CPIIIb). The largest complex (CPIII′) contains a 64 kDa protein which cross-reacts with an antibody raised against maize light-harvesting complex II (LHCII) (Tables I and II). CPIII′ was isolated with PSII core preparations, but could be easily removed from an active PSII core by chromatography. These results suggest that CPIII′ is an antenna for

PSII. A second PSII Chl-protein complex (CPIIIa) contains two proteins, 48 kDa and 45 kDa, which cross-react to the *C. reinhardtii* PSII antibodies described above. Finally, the third Chl-protein complex (CPIIIb) contains a single 36 kDa polypeptide which was present in very high quantities even in normally-grown cells. The 36 kDa protein from either *Aphanocapsa* or *A. nidulans* R2 did not react with antibodies raised against the other PSII proteins and would appear to be a distinct component of the photosystem. In addition, the *Aphanocapsa* 36 kDa protein cross-reacts with an antibody raised against the *A. nidulans* CPVI-4 apoprotein (Bullerjahn and Sherman, unpublished observations).

3.2. PSI Chl-protein complexes

The studies on the Chl-protein composition of PSI suggest that PSI Chl-proteins emit long wavelength fluorescence at 77K (716–735 nm) and contain a major Chl-binding protein of 58–65 kDa (Table III). Rusckowski and Zilinskas (1980) observed a Chl-containing band from *Nostoc* sp. which had a similar mobility on gels with the major PSI Chl-protein band of spinach. This band had a long wavelength Chl fluorescence emission when monitored at 77K, and contained polypep-

TABLE III

PSI Chl-protein complexes resolved by PAGE

Species (Refs.)	Complex	Chl-binding protein(s) (kDa)	Colorless (minor) proteins (kDa)	77 K Chl Fluor. Max (nm)
Nostoc sp. (1)	CP I	58	74,85	735
Phormidium luridum (2)	A-1	60	12-17	–
Plectonema boryanum (3,4)	Fraction II	60	14,18	680,726
	Fraction III	60	14,16,18	
Anacystis nidulans (5,6)	CP I, CPII	64	10,14.5,16,16.5	680,714
	CP III	64	10,14.5,16,16.5, 35,36	
	CP IV	64	36	
Synechococcus sp. (7–9)	CPIa	60,62	10,13,14	695,728
	CP1b	60,62	10,13,14	723
	CP1c–1d	60,62	10,14	721
	CP1e	60,62	–	721

References: (1) Rusckowski and Zilinskas, 1980; (2) Stewart, 1980; (3) Hladik et al., 1982; (4) Cacka et al., 1981; (5) Guikema and Sherman, 1981; (6) Guikema and Sherman, 1983a; (7) Takahashi et al., 1982; (8) Takahashi and Katoh, 1982; (9) Takahashi et al., 1985.

tides at 58, 75, and 84 kDa. The smallest of these peptides constituted the majority of protein in these bands, and was assumed to be the major PSI Chl-binding protein.

In addition to identifying the PSI Chl-binding proteins, these studies have helped identify the colorless polypeptides of the PSI reaction center. Early studies with cyanobacteria found that, in contrast with higher plants, PSI Chl-binding proteins are not easily separated using SDS-PAGE. Large molecular weight (100–360 kDa) Chl-protein aggregates were resolved instead which contained a mixture of polypeptides (Reinman and Thornber, 1979; Guikema and Sherman, 1981, 1983a). Interestingly, the polypeptides which were observed in the Chl-containing protein aggregates were similar to those found in isolated PSI fractions (Guikema and Sherman, 1982) and this led to the hypothesis that the organization of PSI within the cyanobacterial thylakoid may be responsible for the aggregated Chl-protein forms.

Perhaps the most complete description of the PSI Chl-protein aggregates has come from the work of Katoh and colleagues with the thermophilic strain, *Synechococcus* sp. (Takahashi and Katoh, 1982; Takahashi et al., 1982, 1985). This work resolved six Chl-containing bands which differed in their protein composition, P700 activity, and the PSI acceptors A1 and A2. The major Chl-binding protein in these preparations is a protein of about 60–65 kDa; in addition, colorless polypeptides are observed at 10, 13, and 14 kDa. Takahashi and Katoh (1982) have observed P700 activity, PSI acceptors A1 and A2, and cytochrome c_{553} photooxidation associated with the larger Chl-protein complexes.

The functional identification of the colorless peptides has been attempted by a number of methods. One approach has been to examine ^{59}Fe-labeled membrane proteins on SDS-PAGE. ^{59}Fe can be observed in the larger Chl-protein complexes, suggesting that the Fe-S centers are still associated with these aggregates (Guikema and Sherman, 1982). However, it has been technically difficult to observe ^{59}Fe as an Fe-S ligand with the denatured apoprotein on SDS-PAGE, owing to the lability of the protein-Fe-S association in SDS (Guikema and Sherman, 1982; Lagoutte et al., 1984). In higher plants, Hoyer-Hansen and colleagues have proposed from ^{59}Fe-labeling experiments that the 15 and 18 kDa polypeptides carry the bound Fe-S centers on the reducing side of PSI (Hoyer-Hansen et al., 1985). In contrast, Lagoutte et al. (1984) have described a PSI particle from spinach which contains, in addition to the Chl-binding apoprotein of 65 kDa, major peptides at 10, 17, and 20 kDa, plus minor peptides at 8 and 16.5 kDa. They have chosen an alternative strategy for identifying the Fe-S proteins: carboxymethylation of iodo-^{14}C-acetate to identify available cysteines. Most of the label was observed in the 8 kDa minor peptide, and they concluded that this minor protein, which stains poorly with Coomassie Blue, was a membrane bound acceptor of PSI. Experiments of Guikema and Sherman (1982, 1983a), designed to monitor the topology of these peptides, suggest that most of the small proteins associated with the PSI Chl-binding proteins are only poorly exposed to the membrane surface. In contrast, Bonnerjea et al. (1985) obtained spinach PSI particles and washed these particles with the chaotropic agent, sodium thiocyanate. The only peptides which could not be depleted without influencing the EPR spectra of the PSI Fe-S centers were the 62 kDa and 19 kDa polypeptides.

The reasons for the enhanced resistance of cyanobacterial PSI to SDS solubilization is not entirely clear. Guikema and Sherman (1983b, 1984) showed that PSI from iron-deficient cyanobacterial cells has lost resistance to SDS. They observed that specific proteins were also missing, suggesting that three-dimensional topography might account for some of the stability of the PSI complex. Fork and colleagues (Öquist et al., 1981; Hoshina and Fork, 1983) observed the temperature-dependence of PSI solubility, and correlated this temperature-sensitivity with phase transition temperatures. This suggests a role of specific lipids in the stability of PSI within the fluid phase of the membrane. It is well known that the light harvesting Chl-protein (LHCP) of higher plants often forms oligomers in the presence of SDS. Remy et al. (1984) proposed that LHCP contains a specific lipid (phosphatidyldiacylglycerol with *trans*-C16:1-hexadeconoic acid), and that this lipid-protein combination aids in stabilizing LHCP within the membrane. Cyanobacteria, which do not contain the antenna complexes of chloroplasts, lack this specific lipid. These results suggest that the lipid component of cyanobacterial Chl-protein complexes should be examined in greater detail.

3.3. Chl-protein synthesis and assembly

One exciting use of Chl-protein gels is the examination of membrane biosynthesis. In contrast with higher plants, biosynthesis of cyanobacterial thylakoids does not require the participation of two different genomes within the cell. Unfortunately, membrane biosynthesis is one of the least understood phenomena in cyanobacteria. In ultrastructural studies, the density of the cytoplasmic components prevents a close examination of membrane architecture and development (Guglielmi et al., 1981; Stanier and Cohen-Bazire, 1977). Nonetheless, there is some evidence that synthesis can occur at specific sites on the membrane (Kunkel, 1982).

A second technical problem has interfered with a study of membrane synthesis, or greening: unicellular cyanobacteria are largely photoautotrophs and remain fully greened during periods of darkness (Guikema and Sherman, 1983b). To overcome this problem, it was necessary to find alternative ways to deplete cells of thylakoid membranes. Nutrient deficiency has proven remarkably effective, and most efforts have focused on nitrate (Flores et al., 1983), phosphorus (Stevens et al., 1981), and iron (Guikema and Sherman, 1983b, 1984; Pakrasi et al., 1985a). The most interesting aspects of these studies are the changes which occur when the limited nutrient is restored to deficient cultures. As these cultures recover from nutrient starvation, the repair and assembly of the thylakoid membranes can be monitored by following the Chl-protein patterns on PAGE.

The synthesis of the Chl-proteins has been most extensively examined following iron deficiency (Table IV). Iron-deficient cells are smaller (Sherman and Sherman, 1983), have fewer thylakoid membranes per cell (Sherman and Sherman, 1983; Guikema and Sherman, 1983b), and have an altered ratio of phycocyanin to chlorophyll (Guikema and Sherman, 1983b). In addition, a flavoprotein replaces a soluble ferredoxin in the iron deficient state (Entsch and Smillie, 1972; see Section 2.1. of Chapter 2). Severely stressed cells have a depressed electron transport capacity which can be accounted for, in part, by a depletion of Fe-S proteins (Sandmann and Malkin, 1983) and cytochromes (Guikema and Sherman, 1983b). In addition, an altered Chl-protein pattern observed by PAGE (Guikema and Sherman, 1983b) suggests a changed Chl-protein arrangement in the membrane; this is also exhibited in a shift of the Chl-absorption maxima (Öquist, 1971, 1974a). Furthermore, fluorescence spectroscopy of these cells shows an exaggerated F686 and a greatly diminished cellular emission of F696 and F716 (Öquist, 1974b; Guikema and Sherman, 1983b) (see also Section 3.1. of Chapter 16).

Two phases occur during the recovery of an iron-deprived culture of *Anacystis nidulans* upon readdition of iron (Sherman and Sherman, 1983; Guikema and Sherman, 1983b, 1984; Pakrasi et al., 1985a): an early phase (5–9 h after iron readdition) where membrane components are synthesized and inserted into existing membranes, and a later phase (9–24 h) during which new membranes are synthesized. Most of the interest has focused on the early phase and the changes in membrane structure which occur during that time. Pakrasi et al. (1985a) correlated the recovery of the normal fluorescence emission profile with the assembly of the larger CP bands. In particular, the increase in F696 and F716 was accompanied by an increase in CPI, CPII, and CPIII. It was suggested that the structure of the photosystem was similar to that determined for *C. reinhardtii*

TABLE IV

Summary of the effects which iron deficiency imposes on cyanobacterial photosynthetic physiology

Characteristics	References
General:	
Doubling time lengthened only by severe deficiency	Sandmann and Malkin, 1983
Smaller cells, having fewer thylakoid leaflets per cell, irregular spacing between thylakoid leaflets, and glycogen deposits	Sherman and Sherman, 1983
Reduced number of carboxysomes per cell	Sherman and Sherman, 1983
Electron transport carriers:	
Loss of ferredoxin with a corresponding increase in flavoprotein	Entsch and Smillie, 1972
Depletion of iron/sulfur centers only by severe deficiency	Sandmann and Malkin, 1983
Altered Chl/P700 ratios, a greater sensitivity of PSI to severely iron deficient conditions	Öquist, 1974a Sandmann and Malkin, 1983
Light harvesting:	
Shift of the chlorophyll absorptions maxima 4–5 nm towards the blue	Öquist, 1971
Enhanced chlorophyll fluorescence emission (77 K) at approx. 686 nm	Öquist, 1974b
Increased F_0 upon broad-band blue excitation of iron deficient cells, and a correspondingly decreased F_v	Guikema and Sherman, 1983b
Diminished content of oligomeric forms of chlorophyll-proteins from PSI. An altered banding pattern of chlorophyll-proteins in the 30–40 kDa region of acrylamide gels	Pakrasi et al., 1985a Guikema and Sherman, 1983b

(Gershoni et al., 1982) and has three main functional parts: the reaction center, the main light-harvesting antenna, and an interconnecting or immediate antenna that links the light-harvesting complex with the respective reaction center. The correlation of F696 and F716 with assembly of CPI to CPIII suggests that the interconnecting antennae and their constituent polypeptides are formed during the regreening process. Analysis of the synthesis of proteins in *A. nidulans* R2 cells during iron-induced recovery indicates that certain low molecular weight polypeptides synthesized rapidly between 7 and 13 h after addition of iron may be candidates for the apoproteins of these immediate antennae (Guikema and Sherman, 1984).

At present, the factors which limit membrane synthesis and assembly in the iron-deficient state are not known. A depressed biosynthesis of Chl during iron deficiency has been demonstrated in higher plants, which has been attributed to iron-requiring steps in the protoporphyrin biosynthetic pathway (Spiller et al., 1982; Chereskin and Castelfranco, 1982). Since both iron (as Fe-S centers) and Chl are ligands, a depressed level of these ligands should influence either the synthesis of the ligand-binding apoproteins or the stability of these proteins in the membrane. Alternatively, rRNA and mRNA biosynthesis is depressed in iron-deficient cells (Laulhere and Mache, 1978), which may cause a general depression in the rates of protein biosynthesis (Price, 1968). Guikema (1984, 1985) has presented preliminary evidence on the role of Chl in membrane assembly. Levulinic acid and gabaculine are effective inhibitors of Chl and heme biosynthesis in cyanobacteria (Kipe-Nolt and Stevens, 1979; Guikema, 1985). In the presence of levulinic acid, there is a decreased insertion of Chl-proteins into the membrane of *Anacystis nidulans* during recovery from iron starvation (Guikema, 1984). The 71 kDa polypeptide which presumably links the phycobilisome with PSII is especially sensitive to this inhibitor. Further work with levulinic acid and gabaculine will resolve the differential effects of Chl and iron on the synthesis of specific membrane polypeptides.

Nutrient deficiency will provide an important method for further examination of membrane synthesis in cyanobacteria. Several key areas remain unexplored: (1) the promotors and inhibitors which regulate the synthesis of the mRNA transcripts of integral membrane proteins; (2) the accumulation of any unprocessed precursor apoproteins when the ligands of these proteins are present in limiting concentrations; and, (3) the activation of protease activity to remove unstable, misprocessed peptides which were incorrectly inserted or assembled into the membrane.

Finally, an analysis of Chl-proteins may help us understand the molecular occurrences during state transitions in these organisms. Normal, dark-adapted cyanobacterial cells are often observed in state 2 (where excitation energy distribution primarily occurs from PSII towards PSI) (Fork and Satoh, 1983). Illumination with white light or, more specifically, light absorbed by chlorophyll induces a rearrangement within the membrane towards state 1 (where excitation distribution occurs from PSI towards PSII). These transitions, and the decay to state 2 in the darkness, occur within the range of 0.5–2 min in cyanobacteria (Satoh and Fork, 1983). State transitions have been observed in higher plants as well, and have been attributed, in part, to a phosphorylation/dephosphorylation of LHCP (Kyle et al., 1983). However, the mechanism of state transition in cyanobacteria remains unknown (Biggins et al., 1984; Bruce et al., 1985). The mechanism must account for the quick change from one state to another. It is likely that, if protein modification of a Chl-binding polypeptide is a part of this mechanism, Chl-banding patterns on gels would be a useful approach to the problem.

4. Photosystem II

Photosystem II of cyanobacteria is an integral membrane complex that supports light-mediated electron transfer from water to mobile plastoquinone (PQ) molecules inside the thylakoid. The composition and function of this complex are remarkably similar to the cognate complexes in eukaryotic algae and higher plants. Complexes from all of these organisms contain a reaction center (with the specialized chlorophyll P680) that absorbs light quanta and passes electrons sequentially to a pheophytin molecule, then to a stable primary acceptor quinone (Q_A), and finally to a secondary acceptor quinone (Q_B) through a 2-electron gate (Wraight, 1982). Consequent of the electron transport from P680 to Q_B, a strong oxidant is formed that ultimately oxidizes water with the evolution of oxygen. One of the major goals of this section will be to extend this functional similarity to a structural one, and to establish that the recent analyses of cyanobacterial PSII have led to a detailed understanding of the function of this complex in general. We shall first discuss the composition of the PSII 'core' complex and later describe further details of its O_2-evolving machinery.

4.1. Photosystem II core

The core of PSII has been defined as a complex that can undergo stable light-mediated charge separation in the presence of artificial acceptors and donor molecules but cannot mediate oxidation of water (Kaplan and Arntzen, 1982). Newman and Sherman (1978) isolated a PSII core complex by digitonin treatment of membranes from a mesophilic cyanobacterium *Synechococcus cedrorum.* It contained seven polypeptides (Table V), and could mediate electron transport from semicarbazide to ferricyanide although it could not support O_2-evolution. Table V documents a number of similar preparations derived from other unicellular cyanobacteria. In recent years, Katoh and his colleagues (Takahashi et al., 1982; Yamagishi and Katoh, 1983a,b, 1984, 1985) have done an extensive analysis of the PSII core complexes. Their work, along with work done in other laboratories, has shown that the cyanobacterial PSII core has the same five proteins found in green plant PSII core preparations (Satoh et al., 1983; Bricker et al., 1985). These include: (1) a chlorophyll-binding protein of 47–51 kDa; (2) a second chlorophyll-binding protein of 40–45 kDa; (3) and (4) two proteins of 28–34 kDa and (5) cyt b_{559} ($\sim$ 9 kDa). The complex of Yamagishi and Katoh (1985) contained the 47, 40, 31, 28, and 9 kDa polypeptides at a molar ratio of 1:1:1:1:2. The 31 kDa protein was considered to be equivalent to a recently discovered component of the PSII core, called D2, the exact function of which is not yet known (Delepelaire, 1984). The 28 kDa protein is probably equivalent to the '32' kDa Q_B-protein, a protein that binds the photoaffinity label azidoatrazine (Allen et al., 1983). The Q_B protein does not stain well with Coomassie Blue, but it is readily detected as a ^{35}S-labeled polypeptide or after ^{14}C-azidoatrazine binding.

As indicated in an earlier section, Yamagishi and Katoh (1983a,b, 1984) further treated their isolated PSII reaction center complex with a detergent to obtain a chlorophyll-containing complex (CP2-b) that lacked the Chl-binding 40 kDa polypeptide, but still contained the other four proteins. CP2-b was shown to undergo stable charge separation and could mediate light induced electron transfer from diphenylcarbazide to dichlorophenolindophenol. The particle also contained pheophytin, the intermediate electron acceptor and Q_A, the primary stable acceptor in the PSII reaction center. These experiments demonstrated that the 40 kDa Chl-binding protein does not contain P680, the reaction center chlorophyll. Moreover, the data indicates that the reaction center, pheophytin, and Q_A are on at most four proteins; i.e., the Chl-binding 47 kDa protein, two proteins of 28 and 31 kDa, and cyt b_{559} (9 kDa). This is the first concrete evidence of the nonessential nature of one of the two Chl-binding proteins of PSII in primary photochemistry. The 40 kDa protein probably acts as an immediate antenna for PSII.

The presence of a novel 36 kDa Chl-binding protein in the PSII of *A. nidulans* R2 (Pakrasi et al., 1985b) and *Aphanocapsa* (Bullerjahn et al., 1985) is an interesting finding in the current re-

TABLE V

I. Photosystem II

(A) *Photosystem II: oxygen-evolving preparations*

	References	Protein molecular weight, kDa*	Comments
1.	Stewart and Bendall, 1981	(a) 87, (b) 83, (c) 68, (d) 57, (e) 48, (f) 46, (g) 42, (h) 40, (i) 38, (j) 34, (k) 32 ,(l) 30, (m) 28, (n) 18, (o) 17, (p) 16	*P. laminosum*, LDAO. 1300 μmol O_2/mg Chl/h. (a) and (b) may be heat-aggregated complexes from (f) to (h). (k) cyt *f*; (p) cyt *c*-550.
2.	Bowes et al., 1983	(a) 59, (b) 52, (c) 43, (d) 34, (e) 30, (f) 16, (g) 15, (h) 120, (i) 58, (j) 54, (k) 51, (l) 33	*P. laminosum*; LDAO + dodecyl maltoside. 2000 μmol O_2/mg Chl/h. (f) and (g) may contain allophycocyanin. Proteins (a) to (g) were enriched in preparation. Suggested (b) through (e) required for O_2-evolution.
3.	England and Evans, 1981	(a) 77, (b) 60, (c) 40, (d) 36, (e) 33, (f) 30, (g) 27, (h) 15, (j) 14, (k) 10	*A. nidulans* 625; LDAO. 700 μmol O_2/mg Chl/h. (d) to (f) may be involved in Ca^{2+}-binding. Preparation contains at least 7 more proteins.
4.	Pakrasi and Sherman, 1984a	(a) 97, (b) 73, (c) 68, (d) 48, (e) 45, (f) 42, (g) 36, (h) 34, (k) 32, (j) 26, (k) 25, (l) 8	*A. nidulans* R2; 0.5 M Sucrose, 0.5 M phosphate, 0.3 M citrate buffer, pH 6.8; Triton X-100. 2500 μmol/mg Chl/h. Silver-stained gel. Particles contained attached phycobilisomes. Activity sensitive to calmodulin antagonists; enriched in cyt *b*-559 (l). Fl. max. at 686 and 696 nm. Contains CP IV, VI-2, VI-3, and VI-4. (h) is an intrinsic protein.
5.	Schatz and Witt, 1984	(a) 95, (b) 80, (c) 67, (d) 60, (e) 52, (f) 48, (g) 42, (h) 34, (i) 28	*Synechococcus* sp. Zwittergents (Sulfobetaines). 2500 μmol/mg Chl/h. Contains allophycocyanin.
6.	Satoh et al., 1985	(a) 66, (b) 60, (c) 54, (d) 51, (e) 47, (f) 44, (g) 40, (h) 35, (i) 30, (j) 27, (k) 18, (l) 16, (m) 14, (n) 9	*Synechococcus* sp. Octyl glucoside. 300 μmol O_2/mg Chl/h. 50 Chl/Mn and $Mn/Q_A = 3.2$. Activity requires Cl salt (5mM $MnCl_2$ or $CaCl_2$, or $MgCl_2$). Proteins (e), (g), (h), (i), and (n) are major species.

TABLE V *(continued)*

(B) *Photosystem II: non-oxygen evolving preparations*

	References	Protein molecular weight, kDa	Comments
1.	Newman and Sherman, 1978 Sadewasser and Sherman, 1981	(a) 71, (b) 63, (c) 52, (d) 47, (e) 36, (f) 18, (g) 14	*Synechococcus cedrorum*; digitonin. Semicarbazide to FeCN activity = 2000 μmol/mg Chl/h. 75% DCMU sensitive. (b) is slight PSI contamination. (f) and (g) constitute about 65% total protein. Fl. max = 686 nm. 50 Chl/*b*-559. (a) heavily labeled by lactoperoxidase, linker polypeptide to phycobilisomes.
2.	Guikema and Sherman, 1982; 1983a	(a) 70, (b) 48, (c) 45, (d) 42, (e) 35, (f) 18.5, (g) 18	*A. nidulans* R2; digitonin. Essentially same preparation as above. Particle enriched in CPVI which contains 48 kDa and 45 kDa proteins. These are main PSII chlorophyll-binding proteins equivalent to (c) and (d) above.
3.	Koenig and Vernon, 1981	(a) 50, (b) 48, (c) 38, (d) 31, (e) 27, (f) 14, (g) 54, (h) 35	*A. nidulans* 6301; Miranol. DPC to DPIP activity = 100 mmol/mg Chl/h. (a) and (b) are chlorophyll-binding proteins. Contaminated with CF_1 subunits.
4.	Yamagishi and Katoh, 1983a,b, 1984	(a) 47, (b&c) 28–31, (d) 9, (e) 40	*Synechococcus* sp. octylglucoside. DPC to DPIP activity = 900 μmol/mg Chl/h at pH 6.0. Proteins (a) to (d) in CP complex CP2b (active), and (e) in CP2c (inactive).

TABLE V *(continued)*

II. Photosystem I

(A) *PSI-preparations*

	References	Protein molecular weight kDa	Comments
1.	Newman and Sherman, 1978	(a) 61, (b) 18, (c) 17.5, (d) 16, (e) 10	*Synechococcus cedrorum*; digitonin. 40 chlorophylls/P700; DCPIP to MV = 350 μmol O_2/mg chlorophyll/h; Fl. max = 715 nm.
2.	Guikema and Sherman, 1982	(a) 64, (b) 20, (c) 15.5, (d) 11.5, (e) 10	*Anacystis nidulans* R2; digitonin DCPIP to MV = 380 μmol O_2/mg chlorophyll/h; (d) and (e) are labeled with ^{59}Fe and may be Fe-S proteins. (a) is the Chl-binding protein.
3.	Williams et al., 1983 Lundell et al., 1985	(a) 70, (b) 18, (c) 17.5, (d) 16, (e) 10	*Synechococcus* 6301; Triton X-100; 130 Chl/P700; Fe-S centers A, B, and X are present.
4.	Nechushtai et al., 1983 Muster et al., 1984	(a) 70, (b) 50, (c) 16, (d) 11, (e) 10	*Mastigocladus laminosus*; Triton X-100; 50–100 Chl/P700; active in cytochrome photo-oxidation. Unusually low molecular mass chlorophyll-protein I (a) and associated apoprotein (b) may be due to incomplete denaturation during electrophoresis. (a) and (c) cross-reacted to antibodies prepared against eukaryotic proteins. (Subunits I and II, respectively).
5.	Takahashi and Katoh, 1982 Takahashi et al., 1982; 1985	(a) 62, (b) 60, (c) 14, (d) 10	*Synechococcus* sp.; digitonin; 70 Chl/P700; Fe-S centers P430 and A_2 present.

TABLE V *(continued)*

III. F_1 preparations

	References	Protein molecular weight, kDa	Comments
1.	Binder and Bachofen, 1979 Bohler et al., 1982	α, β, γ, ε; δ isolated separately	*Mastigocladus laminosus*; EDTA and chloroform treatment. 350 kDa molecular mass; reconstitutes photophosphorylation in presence of separately isolated delta subunit. Ca^{2+}-dependent ATPase activity.
2.	Lubberding et al., 1981	no subunit data	*Synechococcus lividus*; EDTA and chloroform treatment. Ca^{2+}-dependent, trypsin-activated ATPase activity.
3.	Hicks and Yocum, 1983	(α) 56, (β) 54, (γ) 36, (δ) 21, (ε) 15	*Spirulina plantensis*; 320 kDa molecular mass; reconstitutes photophosphorylation.
		IV. F_1-F_0 preparations	
1.	Binder et al., 1981	no subunit data	*Mastigocladus laminosus*; active in ^{32}P-ATP exchange; DCCD and oligomycin sensitive; contaminated with phycocyanin.
2.	Lubberding et al., 1983	α, β, γ	*Synechococcus* 6716; active in ^{32}P-ATP exchange; DCCD and oligomycin sensitive; no phycocyanin. No recognizable subunits from F_0 portion of complex.

*Proteins detected by Coomassie Blue staining unless otherwise stated.

search in cyanobacterial photosynthesis. Structural similarities between PSII and the reaction center of purple photosynthetic bacteria have prompted some observers to postulate that the Q_B and D2-proteins of green plant PSII have similar functions as the L and M subunits of bacterial reaction centers (Michel and Diesenhofer, 1985). However, the Q_B and D2 proteins have never been observed to bind chlorophyll. Whether the 36 kDa Chl-binding protein from cyanobacteria is D1 or D2 and, furthermore, can be a reaction center protein is now under intensive investigation. The 36 kDa Chl-binding protein of *A. nidulans* R2 is not the azidoatrazine-binding Q_B protein (Pakrasi and Sherman, unpublished observations), and there is no functional information that links the 36 kDa protein with a reaction center function. However, the X-ray data and the protein homology are impressive, and require that potential similarities be analyzed completely.

The extreme similarity between the core proteins of cyanobacterial and chloroplast PSII has recently been exemplified by the analysis of immunogenic relationship between different proteins (Table I). For example, antibodies raised against each of the two Chl-binding proteins from *Chlamydomonas* (47 kDa and 43 kDa) recognize similar proteins from *Anacystis nidulans* R2 and *Aphanocapsa* 6803. This kind of structural homology indicates that the membrane-integral PSII core has been very well conserved throughout evolution.

4.2. Oxygen-evolving complexes

During recent years, a number of laboratories have successfully isolated PSII preparations that evolve oxygen (Table V). Stewart and Bendell (1979) were able to isolate an O_2-evolving, PSII-enriched fraction from the thylakoids of the thermophilic, filamentous strain, *Phormidium laminosum*. Historically, this was the first O_2-evolving PSII preparation isolated from any photosynthetic organism. Since then, a number of groups have successfully isolated O_2-evolving PSII fractions from different strains of cyanobacteria (Table V). In all these cases, the general feature of the isolation procedure involved solubilization of membranes in a non-ionic detergent followed by sucrose gradient centrifugation and/or column chromatography. As a rule, isolation of O_2-evolving preparations were simpler with thermophilic cyanobacteria (e.g., Stewart and Bendall, 1979; Schatz and Witt, 1984), probably because of the relatively better stability of the complexes at room temperature. All of these different preparations contained many more proteins as compared to the PSII core preparations (Table V). The preparations shared some common features: (1) they were enriched in fluorescence emission at 685 and 695 nm; (2) the activities were sensitive to DCMU and atrazine, inhibitors that act on the reducing side of PSII at the level of Q_B; (3) they contained 3–4 manganese atoms per reaction center; (4) they contained cytochrome b_{559} and almost no other cytochromes. Pakrasi and Sherman (1984a) isolated an O_2-evolving PSII preparation from *A. nidulans* R2 that had the phycobilisomes attached to the PSII complex. The presence of phycobilisomes stabilized the activity of the PSII complex from this non-thermophilic strain. Interestingly, the presence of phycocyanin and allophycocyanin proteins as well as the anchor protein (70–95 kDa between PSII and phycobilisomes have been observed in almost all of the preparations mentioned above. Schatz and Witt (1984) have also suggested that a 110 kDa polypeptide in their preparation may be a stable trimer of allophycocyanin (refer to Chapter 3).

As mentioned above, manganese has been found in all of the O_2-evolving preparations. The presence of manganese has long been implicated to be essential in the process of water-oxidation by PSII. In this context, Okada and Asada (1983) isolated a 13 kDa Mn-binding protein from the thylakoids of *Plectonema boryanum*, and antibodies raised against this protein inhibited the oxidation of water. However, the light-induced flow of electrons from diphenylcarbazide to dichlorophenolindophenol through the reaction center of PSII was not affected by these antibodies. Hence, the 13 kDa protein was thought to function on the oxidizing side of PSII. Unfortunately, a manganese-binding protein of similar molecular weight has not yet been discovered in any other cyanobacterial O_2-evolving PSII preparation.

Four polypeptides that are extrinsic to the membrane and attached to PSII on the lumenal side of thylakoids have recently been discovered in green plant PSII. The nominal molecular masses of these proteins are 33 kDa, 23 kDa, and 16 kDa and 10 kDa, respectively. Although the exact roles of these proteins are as yet unknown, suggestions have been made that they facilitate binding of manganese, calcium, and chloride ions, factors that are essential in O_2-evolution. Similar studies on the cyanobacterial oxygen-evolving complex have recently been done by Stewart et al. (1985). These authors used specific antibodies raised against the above mentioned four extrinsic proteins from spinach thylakoids to search for homologous proteins in the thylakoid membranes of *Phormidium laminosum*. The antibodies raised against the 33 kDa protein recognized a protein of 33 kDa from the cyanobacterium. However, immunoblotting experiments did not show the presence of the 23, 16, or 10 kDa polypeptides. From these data, the authors have concluded that the 23 and the 16 kDa proteins are absent in cyanobacteria. However, a 9 kDa protein was found to have a role on the donor side of PSII of *P. laminosum*. Treatment with Tris, 1 M $CaCl_2$ or 1 M NaCl inhibited activity of O_2-evolving PSII preparations from *P. laminosum* with a concomitant release of this 9 kDa protein. This protein was an extrinsic polypeptide and thus was not equivalent to the cyt b_{559} protein; additionally, it did not bind Mn^{2+}. Hence, it is not similar to the Mn^{2+}-binding 13 kDa protein studied by Okada and Asada (1983). The exact roles of the 33 kDa and 9 kDa proteins in cyanobacterial O_2-evolution remain to be established. However, Koike and Inoue (1985) have recently shown that a 34 kDa protein (equivalent to the 33 kDa protein described above) present in an O_2-evolving PSII complex from the thermophilic cyanobacterium *Synechococcus vulcanus*, can bind to spinach PSII particles that have lost their 33 kDa extrinsic protein and partially restore their O_2-evolving capacity. Similarly, the '33 kDa' protein can partially restore the O_2-evolving activity of cyanobacterial O_2-evolving preparations that have lost their extrinsic '34 kDa' protein. This study again indicates that proteins involved in the essential functions of PSII have been well conserved during evolution from cyanobacteria to green plants.

A major advancement in the isolation of O_2-evolving PSII preparations has recently been achieved by Satoh et al. (1985). These authors have been able to solubilize thylakoid membranes from the thermophilic strain, *Synechococcus* sp., using octylglucoside followed by deoxycholate. They obtained a membrane fraction that had high O_2-evolution activity in the presence of $CaCl_2$. This activity was strongly inhibited by DCMU, indicating that the Q_B protein was still present in the complex. The preparation also contained Q_A. Compared to the non-O_2-evolving PSII core preparation mentioned above, the O_2-evolving preparation was enriched in Mn^{2+}. The most remarkable feature of this preparation is its small size, and the presence of only six to seven proteins, implying that it is a true PSII 'particle' and not merely a fragment of the thylakoid membrane. The particles contained all five proteins (47 kDa, 40 kDa, 31 kDa, 28 kDa, and 9 kDa) present in the core-preparation from the same strain. Additionally, it contained a 35 kDa and possibly a 9 kDa polypeptide. These two proteins most probably correspond to the 33 kDa and 9 kDa extrinsic proteins mentioned above.

4.3. Effect of metal-ions on PSII

Different metals have specific effects on PSII of plants, algae and cyanobacteria. As mentioned above, manganese is intimately associated with the process of water-oxidation. The absence of Mn^{2+} results in severe inhibition on the oxidizing side of PSII. Similarly, copper, lead, zinc, and cadmium inhibit on the oxidizing side of PSII. The role of calcium in PSII has been of particular interest. As in higher plants, Ca^{2+} is a cofactor in the PSII-mediated electron transfer activity in cyanobacteria (Fredericks and Jagendorf, 1964). The addition of calcium increases the rate of PSII-mediated O_2-evolution from cyanobacterial membranes (Piccioni and Mauzerall, 1978). Incubation of cyanobacterial cells in Ca^{2+}-depleted media results in a decrease in O_2-evolution, and, eventually cell death (Becker and Brand, 1982). Brand et al. (1983) have used this calcium depletion system to conclude that the site of action of calcium is at the reaction center of PSII. Recently, England and Evans (1983) have shown that depletion of Ca^{2+} inhibits O_2-evolution activity of PSII enriched fractions of *A. nidulans*. Pakrasi and Sherman (1984b) have shown that calmodulin, a Ca^{2+}-binding protein, can stimulate the O_2-evolving activity of a PSII preparation from *A. nidulans* R2, whereas ethylene glycol-bis(β-amino ethylether) *N*, *N*, *N'*, *N'*-tetraacetic acid (EGTA), a Ca^{2+}-chelator, inhibited the activity. Satoh and Katoh (1985) have also found that treatment of *Synechococcus* sp. O_2-evolving PSII preparations with ethylene diamine tetraacetic acid (EDTA) inhibited the activity. However, the O_2-evolution capacity of the preparation was restored by further addition of $CaCl_2$ or $MnCl_2$. Clearly, both Mn^{2+} and Ca^{2+} are essential for the process of O_2-evolution in cyanobacteria.

4.4. Molecular genetics of Photosystem II of cyanobacteria

One of the major advantages in working with cyanobacteria is that these organisms have a single, relatively simple genome that codes for all its functional components. Moreover, a number of strains are now amenable to genetic manipulation using the techniques of DNA-mediated transformation and conjugation (e.g., Golden and Sherman, 1984b; see Sections 4.1. and 4.2. of Chapter 10). Arntzen and coworkers have recently used the natural transformation system to obtain specific mutations in genes encoding PSII proteins of *Aphanocapsa* 6803. Mutants were created by interrupting functional genes with bacterial antibiotic-resistance genes, a procedure termed interposon mutagenesis. These mutants are produced by cloning the appropriate gene into a bacterial plasmid and inserting a transposon within the gene. The cyanobacteria are then transformed with this construct; antibiotic-resistant transformants are produced when the transposon is inserted into the chromosome after homologous recombination.

Initial characterization of these mutants has indicated that PSII of *Aphanocapsa* 6803 cannot function in the absence of the 47 kDa Chl-binding protein. This analysis also indicated that the Q_B- and D2-proteins are essential for the formation of a functional complex. Such work is in its infancy, but it demonstrates the great potential for the molecular biological analysis of photosynthesis in cyanobacteria. Further creation and analysis of these types of site-directed mutants will give us further insight into the relationship of structure and function for proteins within the PSII complex. These results should be of general importance for both cyanobacterial and higher plant PSII.

5. Photosystem I

5.1. PSI core preparations

Submembrane preparations capable of PSI-dependent electron transport have been isolated from a number of cyanobacteria (Table V) by a variety of detergent solubilization and isolation procedures. These preparations generally exhibit high rates of cytochrome c_{553} photooxidation and will reduce methyl viologen when supplied with artificial electron donors. Additionally, they are highly enriched in the reaction center of PSI, P-700 (Olsen and Thornber, 1979), and spectroscopically detectable iron-sulfur proteins. The similarity between these cyanobacterial PSI preparations and PSI preparations from higher plants and algae is extensive and underlines the general homology which exists between higher plant and cyanobacterial electron transport processes.

P-700 is the reaction center of PSI and is the site of the light-induced charge separation of the photosystem in both cyanobacteria and higher plants. The properties of P-700 have been extensively reviewed (Sauer, 1979; Rutherford and Heathcote, 1985) and will be only briefly summarized here. Optically, the photooxidation of P-700 is characterized by bleaching at 705 nm which is consistent with the formation of Chl *a* cation radical. $P\text{-}700^+$ exhibits an electron paramagnetic resonance (EPR) signal at $g = 2.0026$; this signal could be due to either a dimeric form of chlorophyll or a chlorophyll *a* monomer in a special protein environment. Recent electron nuclear double resonance (ENDOR) evidence supports a monomeric structure for P-700 (O'Malley and Babcock, 1984).

At least three iron-sulfur centers (4Fe-4S), detectable by low temperature EPR spectroscopy, are associated with cyanobacterial and higher plant PSI. Centers A (g = 1.86, 1.94, and 2.05) and B (g = 1.89, 1.92, and 2.05) interact strongly (Evans et al., 1974) and may be located on the same polypeptide. These components are thought to give rise to the signal P-430 detected in optical difference spectra (Hiyama and Ke, 1971; Malkin and Bearden, 1971). Center A2 (also termed 'X', g = 1.78) is probably located on a different polypeptide. The characteristics of these iron-sulfur centers have been reviewed (Evans, 1982). Recently, a new EPR signal which may be associated with another iron-sulfur center (center C; g = 1.97) in PSI has been detected in *A. nidulans* and spinach PSI preparations (Hiyama et al., 1984). This center appears to be closely associated with centers A and B, and has a similar redox potential. It is possible that this iron-sulfur center is located on the same polypeptide as A and B. It has been suggested (Rutherford and Heathcote, 1985) that this signal may arise from an altered form of centers A and B.

In general, the polypeptide compositions of the various cyanobacterial PSI preparations and those isolated from higher plants are quite similar (however, see Klein and Vernon, 1977; Evans and Pullin, 1981). They contain a Chl-binding apoprotein with an apparent molecular mass of 60–70 kDa (50 kDa in *Mastigocladus*), plus several smaller polypeptides. The Chl-binding apoprotein is often observed to migrate as a doublet, and it is unclear as to whether this represents two non-identical (though closely related) gene products or is the result of post-translational modification of one gene product. In spinach, Westhoff et al. (1983) have reported that this polypeptide is the product of one chloroplast-encoded gene which has two conformational variants (Vierling and Alberte, 1983). Recently, however, Fish et al. (1985) have identified two partially homologous genes (45% homology) which may encode the chlorophyll-binding apoproteins in maize. The number of genes and/or gene products for this protein in cyanobacteria has not been determined.

The Chl-binding apoprotein(s) can contain antennae Chl and P-700 (40–100 Chl/P-700) as well as the primary electron acceptor of PSI, Al (see Olsen and Thornber, 1979). Additionally, this component (from *Mastigocladus*) cross-reacts immunologically with polyclonal antisera raised against the analogous polypeptide component isolated from PSI preparations of higher plants (Table I; Nechushtai et al., 1983).

Cyanobacterial and higher plant PSI preparations also contain three to four low molecular mass polypeptides of between 20 and < 10 kDa of uncertain function. Takahashi and Katoh (1982) have suggested that the secondary electron acceptors of the photosystem, A2 and P-430, may reside on two of these smaller polypeptides (10 and 14 kDa in *Synechococcus*) based on indirect photochemical evidence. Guikema and Sherman (1982) observed ^{59}Fe bound to several non-heme proteins in *Anacystis nidulans*, two of which comigrated with subunits of their PSI preparation. It was suggested that these might represent iron-sulfur proteins associated with the photosystem. Recently, Lundell et al. (1985) have shown that both proteinase K treatment and SDS treatment of PSI preparations from *Synechococcus* 6301 remove the low molecular mass subunits of the complex with the concomitant loss of the iron-sulfur centers A, B, and X and bound-iron. Unfortunately, it is not known on which polypeptides the different iron-sulfur centers reside.

Finally, a somewhat larger component with a molecular mass of 16–20 kDa is present in most cyanobacterial and higher plant PSI preparations. It has been suggested that a 20 kDa polypeptide was required for efficient electron donation to PSI in higher plants (Bengis and Nelson, 1977). It is uncertain if this is the case with cyanobacterial PSI, since preparations lacking this polypeptide can photooxidize cytochrome c_{553} at high rates in vitro (Takahashi and Katoh, 1982). This component from *Mastigocladus* (16 kDa) cross-reacts immunologically with antisera raised against the 20 kDa component from Swiss Chard PSI preparations (Table I; Nechushtai et al., 1983).

The stoichiometry of the component polypeptides in cyanobacteria and higher plants has not been fully resolved. Lundell et al. (1985) have proposed a composition of $(70\ \text{kDa})_4$ $(18\ \text{kDa})_1$ $(17.7\ \text{kDa})_1$ $(16\ \text{kDa})_1$ $(10\ \text{kDa})_2$ for their preparation from *Synechococcus* 6301. This differs from determinations made in higher plants (Bengis and Nelson, 1977) and the cyanobacterium *Mastigocladus* (Nechushtai et al., 1983) which suggest a $(70\ \text{kDa})_2$ composition in these organisms. Whether these results are due to species-specific differences, or to differences in experimental procedure has yet to be determined.

As previously mentioned, these PSI preparations contain between 40 and 100 Chl per P-700. These chlorophylls form the interior light-harvesting antennae for the photosystem. It is not known if other Chl-containing light-harvesting proteins are associated with the photosystem as is the case for higher plants (Haworth et al., 1983). A few investigators have reported the association of phycobiliproteins with PSI, although this has not been generally accepted. Evans (1981) reported the presence of allophycocyanin in a PSI preparation from *Chlorogloea fritschii*. Since it was not rigorously demonstrated that the allophycocyanin present was functional in the transfer of excitation energy to PSI, it is possible that the PSI preparation was simply contaminated with the phycobiliproteins. Somewhat more convincingly, Peterson et al. (1981) demonstrated excitation energy transfer from phycobiliproteins to PSI in the heterocysts of *Anabaena variabilis*. Heterocysts do not contain PSII reaction centers, but do contain PSI and significant quantities of phycobiliproteins (see also Section 7 of Chapter 7). 77 K fluorescence emission spectra, 77 K fluorescence excitation spectra, and P-700 photooxidation action spectra seemed to indicate that phycocyanin absorption wavelengths were nearly as efficient in exciting PSI as Chl absorption wavelengths. This suggests that phycocyanin can act (at least in heterocysts) as an exterior antenna for PSI. It is possible, however, that even though PSII reaction centers are not present in heterocysts, the external chlorophyll antenna for PSII is present. If this were the case, then excitation energy transfer from the external PSII antenna to PSI could occur, as during State I-State II transitions. In this model, phycobiliproteins transfer excitation energy to PSI indirectly, with no direct connection between the phycobiliproteins and PSI.

5.2. Donors to PSI

Soluble cytochrome c_{553}, with a molecular mass of 10–12 kDa, is the primary electron donor to PSI in the majority of cyanobacteria which have been studied. Kinetic measurements of absorption changes in the Soret band region induced by short light flashes indicate that cytochrome c_{553} was photooxidized very rapidly by P-700$^+$, with a half time of less than 20 μsec (Nanba and Katoh, 1983). There appears to be about two PSI reaction centers per one cytochrome c_{553}. This stoichiometry, however, is probably quite dependent on the growth conditions of the cyanobacteria (Aoki et al., 1983).

Both basic and acidic forms of this cytochrome have been isolated from a number of cyanobacterial species. Ho and Krogmann (1984) examined 16 species of cyanobacteria and found the pI of cytochrome c_{553} to vary from 3.8 to 9.3. Interestingly, the filamentous cyanobacteria tended to have a basic cytochrome c_{553}, whereas unicellular forms had acidic forms of the cytochrome. Additionally, a number of cyanobacteria studied had multiple electrophoretic forms of cytochrome c_{553}.The functional significance of this heterogeneity has not been determined.

Plastocyanin, the primary electron donor to P-700 in higher plants and most algae, has also been isolated from a number of cyanobacteria. Plastocyanin is a copper-containing protein with a molecular mass of about 10–12 kDa. This protein has also been isolated in both basic and acidic forms (Stewart and Kaethner, 1983; Ho and Krogmann, 1984). The cyanobacteria, which have the genetic ability to produce both plastocyanin and cytochrome c_{553}, apparently regulate the amounts of these two components based on the availability of copper in their environment (Wood, 1978; Sandmann et al., 1983). If these cyanobacterial species possess an acidic cytochrome c_{553} then their plastocyanin is also acidic; species with basic cytochromes possess basic plastocyanins (Ho and Krogmann, 1984). This is as would be expected since it has been shown that the net charge on these proteins is critical with respect to their ability to donate electrons to P-700$^+$ (Davis et al., 1980).

5.3. Acceptors of PSI

Soluble ferredoxin, a low molecular mass (10–12 kDa) iron-sulfur protein (2Fe-2S), is the principal electron acceptor from PSI in both cyanobacteria and higher plants (see Chapter 2). Usually, electrons are then transferred from ferredoxin to ferredoxin-NADP$^+$ reductase. However, analysis of fluorescence induction kinetics, which is a sensitive indicator of the redox state of the plastoquinone pool, indicates that a cyclic pathway from ferredoxin (or NADPH) to the plastoquinone pool may also exist (Aoki and Katoh, 1982). Houchins and Hind (1983) have suggested a similar cyclic pathway around PSI in *Anabaena* heterocysts based on single flash turnover experiments. It is interesting to note that, in higher plants, ferredoxin-NADP$^+$ reductase has been isolated in association with the cytochrome b_6/f complex (Clark and Hind, 1983). If this is the case in cyanobacteria, it would provide a convenient mechanism for the observed cyclic pathway.

Ferredoxin has been isolated from a wide range of cyanobacterial and higher plant species, and primary sequence data are available for many of these (see Section 1.5 of Chapter 2). Sequence analysis of cyanobacterial ferredoxins indicates that the primary sequence has been rather poorly conserved with between 4 and 41 amino acid differences being found among the species examined. Unicellular cyanobacteria can generally be distinguished from filamentous forms by the presence of gaps in the primary sequence (Hase et al., 1982). Despite these differences, the ferredoxin gene from *Anabaena variabilis* has been cloned by using a plant gene as a heterologous probe (van der Plas et al., 1985). When placed under iron-limiting conditions, some cyanobacterial species can replace their ferredoxin complement with flavodoxin, an FMN-containing electron carrier with a molecular mass of 19–22 kDa (Fitzgerald et al., 1977).

The terminal protein of the photosynthetic electron transport chain in higher plants and cyanobacteria is ferredoxin-NADP$^+$ reductase. This flavoprotein, with a molecular mass of 31.5 kDa, was first isolated from a cyanobacterium by Susor and Krogmann (1966). Subsequently, this component has been isolated from a number of other cyanobacterial species (Hutber et al., 1978; Serrano and Rivas, 1982). Most preparations of this enzyme are quite heterogeneous, containing multiple active components of similar molecular masses but which differ slightly in amino acid sequence (Rowell et al., 1981). The functional significance of this heterogeneity, which has also been observed in higher plants, is unknown.

A possible alternative electron acceptor from ferredoxin is thioredoxin. In higher plants, reduced thioredoxin modulates a number of enzymes involved in carbon fixation and carbohydrate metabolism (Buchanan, 1980). A similar control mechanism appears to operate in the cyanobacteria. Schmidt (1980) has isolated two thioredoxins from *Synechococcus* 6301 and has speculated that the higher molecular mass component (> 12 kDa) may be involved in the photosynthetic regulation of soluble proteins. Ye et al. (1981) isolated two thioredoxin-containing fractions from *Anabaena* and demonstrated that these fractions had the ability to activate fructose-1,6-bisphosphatase. Subsequently, Whittaker and Gleason (1984) have purified these thioredoxins to homogeneity. One of these thioredoxins had a molecular mass of 12 kDa and is very similar to higher plant thioredoxins. The other thioredoxin had a molecular mass of 25 kDa and does not appear to be related to other thioredoxins.

6. Cytochrome b_6/f complex

A functional cytochrome b_6/f complex with a polypeptide composition very similar to that of higher plants has been isolated from *A. variabilis* (Krinner et al., 1982). The complex contains cytochrome *f* (31 kDa), cytochrome b_6 (22.5 kDa), the Rieske Fe-S protein (22 kDa), and Subunit 4 (16.5 kDa); in addition, two minor proteins at 9 kDa of unknown functions are associated with the complex (Hauska et al., 1983). The stoichiometry of cyt *f*/cyt b_6/16.5 kDa polypeptides was estimated as 1:2:1, whereas the Rieske Fe-S protein was present in substoichiometric amounts. As a result, the activity of the complex was reduced compared to intact thylakoids. The Rieske Fe-S protein could not be detected separately in an SDS-PAGE system; however, it was purified to homogeneity and identified as a 22 kDa polypeptide. Interestingly, a 21 kDa polypeptide from *Anacystis nidulans* was labeled with ^{59}Fe; the label was lost in the presence of thiol reagents, indicating that this protein might be an Fe-S protein and possibly the Rieske center (Guikema and Sherman, 1982). The mobility of cyt *f* from both *A. variabilis* and *A. nidulans* decreased on heating the sample prior to electrophoresis (Guikema and Sherman., 1982; Hauska et al., 1983). This phenomenon could also be demonstrated by chloroform methanol treatment, implying a close association of lipid to cytochrome *f* (Guikema and Sherman, 1982).

The close relationship of the cytochrome b_6/f complexes from cyanobacteria and chloroplasts has been demonstrated with the aid of heterologous antibodies (Table I). Antibodies prepared against the individual proteins from maize chloroplasts were highly cross-reactive against the respective proteins from *A. nidulans* R2 and *Aphanocapsa* 6714 (Bullerjahn et al., 1985). These experiments verified the tetramethylbenzidine results of Guikema and Sherman (1981) in regard to cytochromes b_6 and *f*, and indicated that the ^{59}Fe-labeled protein at 21 kDa is indeed the Rieske Fe-S center. These relationships have led to the cloning of the cyanobacterial genes for cytochromes b_6 and *f* and Subunit 4 using heterologous DNA probes (T. Kallas and R. Malkin, personal communication).

7. F_1 and F_0-F_1

F_1 (ATPase) and F_0-F_1, the proton-translocating ATP synthetase, have been isolated from only a limited number of cyanobacteria (Table V). These complexes are much more difficult to isolate from the cyanobacteria than from other gram-negative bacteria or higher plants due to the large amount of phycobiliproteins present. The phycobiliproteins form aggregates of similar size, density, and ionic properties as the F_1 and F_0-F_1 complexes.

F_1 isolated from cyanobacteria contains five subunits (α, β, γ, δ and ε) of undetermined stoichiometry. In higher plant chloroplasts and in *E. coli*, the apparent stoichiometry of these five subunits is $(\alpha)_3$, $(\beta)_3$, $(\gamma)_1$, $(\delta)_1$, and $(\varepsilon)_1$ (Amzel and Pedersen, 1983; Moase and Green, 1981; Suss and Schmidt, 1982), and it is likely that the stoichiometry of cyanobacterial F_1 will be found to be similar to that of higher plants. The apparent molecular mass of the holoenzyme is about 320–350 kDa. The molecular mass of the individual subunits of *Spirulina platensis* F_1 have been determined (53.4, 51.6, 36, 21.5, and 14.7 kDa, respectively) from F_1 prepared by chloroform extraction of *Spirulina* photosynthetic membranes (Hicks and Yocum, 1983). These molecular masses are very similar to those obtained from spinach chloroplast F_1 preparations. The similarity of higher plant chloroplast and cyanobacterial F_1 is underlined by the observation that isolated chloroplast F_1 and cyanobacterial F_1 can reconstitute cyclic photophosphorylation reciprocally in cyanobacterial and chloroplast photosynthetic membranes that have been depleted of F_1 with NaBr. Additionally, the alpha and gamma subunits of cyanobacterial F_1 cross-react immunologically with antisera raised against chloroplast alpha and gamma subunits (Hicks, 1984).

F_1 extracted from the photosynthetic membranes of *Mastigocladus laminosus* by chloroform treatment lacks the delta subunit and does not reconstitute photophosphorylation with F_1-depleted membranes (Binder and Bachofen, 1979). The delta subunit can, however, be isolated separately, and combined with delta-deficient F_1 to yield a complex competent in the reconstitution of phosphorylation (Bohler et al., 1982). F_1 isolated by EDTA-extraction of *Mastigocladus* membranes contains a full complement of subunits. Both the *Spirulina* and the *Mastigocladus* preparations exhibit calcium-dependent, trypsin-activated ATPase activity.

Two groups have reported the isolation of F_0-F_1 preparations from cyanobacteria. Binder et al.(1981) isolated a fraction active in ^{32}P-ATP exchange from the thermophilic cyanobacterium *Mastigocladus laminosus* by extraction of the photosynthetic membranes with cholate and octyl glucoside followed by sucrose density centrifugation in the presence of asolectin and detergent. This preparation was only partially purified, since it was contaminated with phycobiliproteins. The ^{32}P-ATP activity of this preparation was markedly inhibited by *N,N'*-dicyclohexylcarbodiimide (DCCD) and the uncoupler carbonyl cyanide chlorophenyl hydrazone (CCCP). Inhibition by CCCP indicated that the F_0-F_1 present in this preparation was incorporated into closed asolectin vesicles.

Lubberding et al. (1983) isolated a similar fraction from the thermophilic cyanobacterium *Synechococcus* 6716. The isolation procedure included solubilization of photosynthetic membranes with octyl glucoside and cholic acid, ammonium sulfate fractionation and gel filtration; this preparation was devoid of phycobiliproteins. However, gel electrophoresis indicated the presence of a number of low molecular mass polypeptides, which could not be attributed to components of the F_0 portion of the complex. The complex exhibited DCCD-sensitive ^{32}P-ATP exchange rates comparable to those observed in chloroplast F_0-F_1 preparations from spinach. When reconstituted into proteoliposomes (Van Walraven et al., 1983) containing cyanobacterial lipid, this complex exhibits an ATP-induced, DCCD-sensitive electrical potential and an ATP-induced, DCCD-sensitive acidification of the internal proteoliposome space. These results suggest that upon ATP hydrolysis an inverted pH gradient and membrane potential are established. It should be cautioned, however, that ATP synthesis has not been rigorously demonstrated in either of these F_0-F_1 preparations.

It is evident from the above studies that little is known of the F_0 portion of the F_0-F_1 complex

TABLE VI

Consensus composition of photosynthetic complexes

Complex	Molecular weight (kDa)	Comment
PSII	(a) 70–95	Phycobilisome linker
	(b) 48	*psb*B gene; PSII reaction center Chl-binding protein (CPVI-2 or CP2a); F696
	(c) 45	*psb*C gene; PSII antennae Chl-binding protein (CPVI-3 or CP2c); F686
	(d) 36	PSII Chl-binding protein (CPVI-4); F686
	(e) 33	extrinsic protein, O_2-evolution
	(f) 32	*psb*A gene; azidoatrazine binding protein (Q_B)
	(g) 8–14	*psb*E gene; cyt b_{559}
PSI	(a) 64–70	PSI Chl-binding protein (present in all PSI aggregates)
	(b) 18	
	(c) 17.5	
	(d) 16	
	(e) 10	
cyt b_6/f	(a) 31	cytochrome *f*
	(b) 22.5	cytochrome b_6
	(c) 22	Rieske Fe-S protein
	(d) 16.5	Subunit 4

from cyanobacteria. No information exists concerning the polypeptide composition or stoichiometry of the F_0 subunits. The amino acid sequence of the *Mastigocladus* DCCD-binding protein, however, has been determined and exhibits strong homology with the chloroplast protein sequence (cited in Binder, 1982).

8. Conclusion

The consensus composition for three of the cyanobacterial photosynthetic complexes is presented in Table VI. It is clear that these structures are remarkably similar to those of chloroplast thylakoids. Thus, there is no doubt that the cyanobacteria can be considered an excellent model system for the analysis of oxygenic photosynthesis. Utilizing the transformation systems now available in the unicellular cyanobacteria (e.g., Golden and Sherman, 1984b) and the mutagenesis procedures described in the PSII section, it will be possible to determine the specific function of each polypeptide. Such information will then be directly applicable to chloroplast proteins, and will provide a new approach to the study of structure and function relationships in photosynthesis.

References

Allen, M.M., Turnburke, A.C., Lagace, E.A. and Steinback, K.E. (1983) Effects of photosystem II herbicides on the photosynthetic membranes of the cyanobacterium *Aphanocapsa* 6308. Pl. Physiol. 71, 388–392.

Amzel, L.M. and Pedersen, P.L. (1983) Proton ATPase: structure and mechanism. Ann. Rev. Biochem. 52, 801–824.

Aoki, M. and Katoh, S. (1982) Oxidation and reduction of plastoquinone by photosynthetic and respiratory electron transport in a cyanobacterium *Synechococcus* sp. Biochim. Biophys. Acta 682, 307–314.

Aoki, M., Hirano, M., Takahashi, R. and Kato, S. (1983) Contents of cytochromes, quinones, and reaction centers of photosystem I and photosystem II in a cyanobacterium *Synechococcus* sp. Plant Cell Physiol. 24, 517–525.

Arntzen, C.J., Darr, S.C., Mullet, J.E., Steinback, K.E. and Pfister, K. (1982) Polypeptide determinants of plastoquinone function in photosystem II of chloroplasts. In B. Trumpower (ed.), Functions of Quinones in Energy Conserving Systems. Academic Press, New York, pp. 443–452.

Becker, D.W. and Brand, J.J. (1982) An in vivo requirement for calcium in Photosystem II of *Anacystis nidulans*. Biochem. Biophys. Res. Commun. 109, 1134–1139.

Bengis, C. and Nelson, N. (1977) Subunit structure of chloroplast PSI reaction center. J. Biol. Chem. 252, 4564–4569.

Biggins, J., Campbell, C.L. and Bruce, D. (1984) Mechanism of the light state transition in photosynthesis. II. Analysis of phosphorylated polypeptides in the red alga, *Porphyridium cruentum*. Biochim. Biophys. Acta 767, 138–144.

Binder, A. (1982) Respiration and photosynthesis in energy-transducing membranes of cyanobacteria. J. Bioenerg. Biomemb. 14, 271–286.

Binder, A. and Bachofen, R. (1979) Isolation and characterization of a coupling factor I ATPase of the thermophilic blue-green algae (cyanobacterium) *Mastigocladus laminosus*. FEBS Lett. 104, 66–70.

Binder, A., Bohler, M.C., Wolf, M. and Bachofen, R. (1981) The proton ATPase of the thermophilic cyanobacterium *Mastigocladus laminosus*. In G. Akoyunoglou (ed.), Photosynthesis II. Electron Transport and Photophosphorylation. Balaban International Science Services, Philadelphia, PA, pp. 771–778.

Bohler, M.-C., Binder, A. and Bachofen, R. (1982) Reconstitution of photophosphorylation with the coupling factor AF_1 of a thermophilic cyanobacterium. FEMS Microbiol. 15, 115–118.

Bonnerjea, J., Ortiz, W. and Malkin, R. (1985) Identification of a 19 kDa polypeptide as an Fe-S center apoprotein in the photosystem I primary electron acceptor complex. Arch. Biochem. Biophys. 240, 15–20.

Bowes J.M., Stewart, A.C. and Bendall, D.S. (1983) Purification of Photosystem II particles from *Phormidium laminosum* using the detergent dodecyl-β-D-maltoside. Properties of the purified complex. Biochim. Biophys. Acta 725, 210–219.

Brand, J.J., Mohanty, P. and Fork, D.C. (1983) Reversible inhibition of photochemistry of Photosystem II by Ca^{+2} removal from intact cells of *Anacystis nidulans*. FEBS Lett. 155, 120–124.

Breton, J. (1983) The 695 nm fluorescence (F695) of chloroplasts at low temperature is emitted from the primary acceptor of photosystem II. FEBS Lett. 159, 1–5.

Bricker, T.M. and Sherman, L.A. (1984) Triton X-114 phase fractionation of membrane proteins of the cyanobacterium *Anacystis nidulans* R2. Arch. Biochem. Biophys. 235, 204–211.

Bricker, T.M., Pakrasi, H.B. and Sherman, L.A. (1985) Characterization of a spinach Photosystem II core preparation isolated by a simplified method. Arch. Biochem. Biophys. 237, 170–176.

Bruce, D., Biggins, J., Steiner, T. and Thewalt, T. (1985) Mechanism of the light state transition in photosynthesis. IV. Picosecond fluorescence spectroscopy of *Anacystis nidulans* and *Porphyridium cruentum* in state 1 and state 2 at 77K. Biochim. Biophys. Acta 806, 237–246.

Buchanan, B.B. (1980) Role of light in the regulation of chloroplast enzymes. Ann. Rev. Plant Physiol. 31, 341–374.

Bullerjahn, G.S., Riethman, H.C. and Sherman, L.A. (1985) Organization of the thylakoid membrane from the heterotrophic cyanobacterium, *Aphanocapsa* 6714. Biochim. Biophys. Acta, 810, 148–157.

Cacka, H., Naus, J., Hladik, J. and Sofrova, D. (1981) Composition and function of thylakoid membranes of the cyanobacterium *Plectonema boryanum* after extraction with n-heptane. Photosynth. 15, 338–346.

Chereskin, B.M. and Castelfranco, P.A. (1982) Effects of iron and oxygen on chlorophyll biosynthesis. Plant Physiol. 68, 112–116.

Clark, R.P. and Hind, G. (1983) Isolation of a five polypeptide cytochrome b_6–f complex from spinach chloroplasts. J. Biol. Chem. 258, 10348–10354.

Crouse, E.J., Schmitt, J.M. and Bohnert, J.-J. (1985) Chloroplast and cyanobacterial genomes, genes and RNAs: a compilation. Plant Mol. Biol. Rep. 3, 43–89.

Curtis, S.E. and Haselkorn, R. (1983) Isolation and sequence of the gene for the large subunit of ribulose -1,5-bisphosphate carboxylase from the cyanobacterium *Anabaena* 7120. Proc. Natl. Acad. Sci. USA 80, 1835–1839.

Curtis, S.E. and Haselkorn, R. (1984) Isolation, sequence and expression of two members of the 32 kDa thylakoid membrane protein gene family from the cyanobacterium *Anabaena* 7120. Plant Mol. Biol. 3, 249–258.

Davis, D.J., Krogmann, D.W. and San Pietro, A. (1980) Electron donation to photosystem I. Plant Physiol. 65, 697–702.

Delepelaire, P. (1984) Partial characterization of the biosynthesis and integration of the Photosystem II reaction centers in the thylakoid membrane of *Chlamydomonas reinhardtii*. EMBO J. 3, 701–706.

England, R.R. and Evans, E.H. (1981) A rapid method for extraction of oxygen-evolving Photosystem II preparations from cyanobacteria. FEBS Lett. 134, 175–177.

England, R.R. and Evans, E.H. (1983) A requirement for Ca^{+2} in the extraction of O_2-evolving Photosystem II preparations from the cyanobacterium *Anacystis nidulans*. Biochem. J. 210, 473–476.

Entsch, B. and Smillie, R.M. (1972) Oxidation reduction properties of phytoflavin, a flavoprotein from blue-green algae. Arch. Biochem. Biophys. 151, 378–386.

Erickson, J.M., Rahire, M., Roachaix, J.-D. and Mets, L. (1985) Herbicide resistance and cross-resistance: changes at three distinct sites in the herbicide-binding protein. Science 228, 204–207.

Evans, E.H. (1981) Photosystem I preparations from dark- and light-grown cells of the cyanobacterium, *Chlorogloea fritschii*. Photosyn. Res. 1, 259–264.

Evans, E.H. and Pullin, C.A. (1981) A requirement for EDTA in the separation of photosystems 1 and 2 from the cyanobacterium *Chlorogloea fritschii*. Biochem. J. 196, 489–493.

Evans, M.C.W. (1982) Photosynthetic iron-sulfur proteins. In: Iron Sulfur Proteins, Vol. 4. J. Wiley & Sons, Inc., New York, pp. 250–284.

Evans, M.C.W., Reeves, W. and Cammack, R. (1974) Determination of the oxidation-reduction potential of the bound iron-sulfur proteins of the primary electron acceptor complex of PSI of spinach chloroplast. FEBS Lett. 49, 111–114.

Fish, L.E., Kuck, U. and Bogorad, L. (1985) Two partially homologous adjacent light-inducible maize chloroplast genes encoding polypeptides of the P700 chlorophyll *a* protein complex of photosystem I. J. Biol. Chem. 260, 1413–1421.

Fitzgerald, M.P., Husain, A., Hutber, G.N. and Roger, L.J. (1977) Studies of the flavodoxins from a cyanobacterium and a red algae. Biochem. Soc. Trans. 5, 1505–1506.

Flores, E., Romero, J.M., Guerrero, M.G. and Losada, M. (1983) Regulatory interaction of photosynthetic nitrate utilization and carbon dioxide fixation in the cyanobacterium *Anacystis nidulans*. Biochim. Biophys. Acta 725, 529–532.

Fork, D.C. and Satoh, K. (1983) State 1-state 2 transition in the thermophilic blue-green alga (cyanobacterium) *Synechococcus cedrorum*. Photochem. Photobiol. 37, 421–427.

Fredericks, W.W. and Jagendorf, A.T. (1964) A soluble component of the Hill Reaction in *Anacystis nidulans*. Arch. Biochem. Biophys. 104, 39–49.

Gantt, E. (1981) Phycobilisomes. Ann. Rev. Plant Physiol. 32, 327–347.

Gershoni, J.M., Shochat, S., Malkin, S. and Ohad, I. (1982) Functional organization of chlorophyll-containing complexes of *Chlamydomonas reinhardtii*. A study of their formation and interconnection with reaction centers in the greening process of the y-1 mutant. Plant Physiol. 70, 637–644.

Glazer, A.N. (1984) Phycobilisomes – a macromolecular complex optimized for light energy transfer. Biochim. Biophys. Acta 748, 29–51.

Goedheer, J.C. (1972) Fluorescence in relation to photosynthesis. Ann. Rev. Plant Physiol. 23, 87–112.

Golden, S.S. and Haselkorn, R. (1985) Mutation to herbicide resistance maps within the *psbA* gene of *Anacystis nidulans* R2. Science 229, 1104–1107.

Golden, S.S. and Sherman, L.A. (1984a) Biochemical and biophysical characterization of herbicide-resistant mutants of the unicellular cyanobacterium, *Anacystis nidulans* R2. Biochim. Biophys. Acta 764, 239–246.

Golden, S.S. and Sherman, L.A. (1984b) Optimal conditions for genetic transformation of the cyanobacterium *Anacystis nidulans* R2. J. Bacteriol. 158, 36–42.

Golecki, J.R. and Drews, G. (1982) Supramolecular organization and composition of membranes. In: N.G. Carr and B.A. Whitton (eds.), The Biology of Cyanobacteria. Blackwell Scientific Publications, Oxford, pp. 125–141.

Gray, M.W. and Doolittle, W.F. (1982) Has the endosymbiont hypothesis been proven? Microbiol. Rev. 46, 1–42.

Guglielmi, G., Cohen-Bazire, G. and Bryant, D.A. (1981) The structure of *Gloebacter violaceus* and its phycobilisomes. Arch. Microbiol. 129, 181–189.

Guikema, J.A. (1984) Membrane organization of *Anacystis nidulans* following iron deficiency and heme deprivation. Proc. VIth Int. Congr. Photosynth. II, 647–650.

Guikema, J.A. (1985) Fluorescence induction characteristics of *Anacystis nidulans* during recovery from iron deficiency. J. Plant Nutr., 8, 891–908.

Guikema, J.A. and Sherman, L.A. (1981) Electrophoretic profiles of cyanobacterial membrane polypeptides showing heme-dependent peroxidase activity. Biochim. Biophys. Acta 637, 189–201.

Guikema, J.A. and Sherman, L.A. (1982) Protein composition and architecture of the photosynthetic membranes from the cyanobacterium, *Anacystis nidulans* R2. Biochim. Biophys. Acta 681, 440–450.

Guikema, J.A. and Sherman, L.A. (1983a) Chlorophyll-protein organization of membranes from the cyanobacterium *Anacystis nidulans*. Arch. Biochem. Biophys. 220, 155–166.

Guikema, J.A. and Sherman, L.A. (1983b) Organization and function of chlorophyll in membranes of cyanobacteria during iron starvation. Plant Physiol. 73, 250–256.

Guikema, J.A. and Sherman, L.A. (1984) Influence of iron deprivation on the membrane composition of *Anacystis nidulans*. Plant Physiol. 74, 90–95.

Hase, T., Inoue, K., Matsubara, H., Williams, M. and Rogers, L. (1982) The amino acid sequence of *Synechocystis* 6714 ferredoxin: A unique structural feature of unicellular blue-green algal ferredoxin. J. Biochem. 92, 1357–1362.

Hauska, G., Hurt, E., Gabellini, N. and Lockau, W. (1983) Comparative aspects of quinol-cytochrome *c*/plastocyanin oxidoreductases. Biochim. Biophys. Acta 726, 97–133.

Haworth, P., Watson, J.L. and Arntzen, C.J. (1983) The detection, isolation and characterization of a light-harvesting complex which is specifically associated with photosystem I. Biochim. Biophys. Acta 724, 151–158.

Hearst, J.E. (1985) Primary structure and function of the reaction center polypeptides of *Rps. capsulata* – The structural and functional analogies with the photosystem II polypeptides of plants. In L.A. Staehelin and C.J. Arntzen (eds.), Encyclopedia of Plant Physiology: Photosynthetic Membranes. Springer-Verlag, Berlin, pp. 382–389.

Hicks, D.B. (1984) The coupling factor ATPase of the cyanobacterium *Spirulina plantensis*: Purification, characterization and relationship to higher plant chloroplast coupling factors. Dissertation, University of Michigan, Ann Arbor.

Hicks, D.B. and Yocum, C.F. (1983) The coupling factor (Ca-ATPase) of the cyanobacterium *Spirulina plantensis*. In: C. Sybesma (ed.), Advances in Photosynthesis, Vol. II, Ch. 6. Martinus Nijhoff/Dr. W. Junk Publishers, The Hague, pp. 599–610.

Hirshberg, J. and McIntosh, L. (1983) Molecular basis of herbicide resistance in *Amaranthus hybridus*. Science 222, 1346–1348.

Hiyama, T. and Ke, B. (1971) A new photosynthetic pigment 'P430'. Its possible role as the primary electron acceptor of photosystem I. Proc. Natl. Acad. Sci. USA 68, 1010–1013.

Hiyama, T., Ohtsuka, T. and Sakurai, H. (1984) A new EPR signal associated with PSI of algal and plant photosynthesis. J. Biochem. 95, 855–860.

Hladik, J., Pancoska, R. and Sofrova, D. (1982) The influence of carotenoids on the conformation of chlorophyll-protein complexes isolated from the cyanobacterium *Plectonema boryanum*. Biochim. Biophys. Acta 681, 263–272.

Ho, K.K. and Krogmann, D.W. (1984) Electron donors to P700 in cyanobacteria and algae and instance of unusual genetic variability. Biochim. Biophys. Acta 766, 310–316.

Hochins, J.P. and Hind, G. (1983) Flash spectroscopic characterization of photosynthetic electron transport in isolated heterocysts. Arch. Biochem. Biophys. 224, 272–282.

Hoshina, S. and Fork, D.C. (1983) Preparation of chlorophyll-protein complexes from the cyanobacterium *Anacystis nidulans* by SDS-sucrose density gradient centrifugation. Carnegie Inst. Wash. Year Book 82, 80–84.

Hoyer-Hansen, G., Honberg, L.S. and Simpson, D.J. (1985) Monoclonal antibodies used for the characterization of the two putative iron-sulfur center proteins associated with photosystem I. Carlsberg. Res. Commun. 50, 23–35.

Hunter, F. and Thornber, J.P. (1976) Further characterization of the P700-Chl-protein from blue-green algae. Plant Physiol. 75S, 95.

Hutber, G.N., Smith, A.J. and Rogers, L.J. (1978a) Isolation of a ferredoxin–$NADP^+$ reductase from a prokaryote. Biochem. Soc. Trans. 6, 1214–1216.

Kaplan, S. and Arntzen, C.J. (1982) Photosynthetic membrane structure and function. In: Govindjee (ed.), Photosynthesis, Vol. 1. Academic Press, New York, pp. 65–152.

Katoh, T. and Gantt, E. (1979) Photosynthetic vesicles with bound phycobilisomes from *Anabaena variabilis*. Biochim. Biophys. Acta 546, 383–393.

Kipe-Nolt, J.A. and Stevens, S.E., Jr. (1979) Effect of levulinic acid on pigment biosynthesis in *Agmenellum quadruplicatum*. J. Bacteriol. 137, 146–152.

Klein, S.M. and Vernon, L.P. (1977) Composition of a photosystem I chlorophyll protein complex from *Anabaena flos-aquae*. Biochim. Biophys. Acta 459, 364–375.

Koenig, G. and Vernon, L.P. (1981) Which polypeptides are characteristic for Photosystem II? Analysis of active Photosystem II particles from the blue-green alga *Anacystis nidulans*. Z. Naturforsch. 36C, 295–304.

Koike, H. and Inoue, Y. (1985) Properties of a peripheral 34 kDa protein in *Synechococcus vulcanus* Photosystem II particles. Its exchangeability with spinach 33 kDa protein in reconstitution of O_2-evolution. Biochim. Biophys. Acta 807, 64–73.

Krinner, M., Hauska, H., Hurt, E. and Lockau, W. (1982) A cytochrome f-b_6 complex with plastoquinol-cytochrome c oxidoreductase activity from *Anabaena variabilis*. Biochim. Biophys. Acta 681, 110–117.

Kunkel, D.D. (1982) Thylakoid centers: Structures associated with the cyanobacterial photosynthetic membrane system. Arch. Microbiol. 133, 97–99.

Kyle, D.J., Staehelin, L.A. and Arntzen, C.J. (1983) Lateral mobility of the light harvesting complex in chloroplast membranes controls excitation energy distribution in higher plants. Arch. Biochem. Biophys. 225, 527–541.

Lagoutte, B., Serif, P. and Duranton, J. (1984) Tentative identification of the apoprotein of iron-sulfur centers of photosystem I. FEBS Lett. 174, 24–29.

Laulhere, J.P. and Mache, R.M. (1978) Induction des synthèses de mRNA et rRNA chloroplastique par le fer lors du verdissement de feuilles chlorosées. Physiol. Veg. 16, 643–656.

Lubberding, H.J., Offerijns, F., Vel, W.A.C. and de Vries, J.R. (1981) Characterization of the ATPase of the thermophilic cyanobacterium *Synechococcus lividus*. In: G. Akoyunoglou. (ed.), Photosynthesis II. Electron Transport and Phosphorylation. Balaban International Science Services, Philadelphia, pp. 779–788.

Lubberding, H.J., Zimmer, G. Van Walraven, H.S., Schrickx, J. and Kraayenhof, R. (1983) Isolation, purification, and characterization of the ATPase complex from the thermophilic cyanobacterium *Synechococcus* 6716. Eur. J. Biochem. 137, 95–99.

Lundell, D.J., Glazer, A.N., Melis, A. and Malkin, R. (1985) Characterization of a cyanobacterial photosystem I complex. J. Biol. Chem. 260, 646–654.

Malkin, R. and Bearden, A.J. (1971) Primary reactions of photosynthesis: photoreduction of a bound chloroplast ferredoxin at low temperature as detected by EPR spectroscopy. Proc. Natl. Acad. Sci. USA 68, 16–19.

Michel, H. and Diesenhofer, J. (1985) X-ray diffraction studies on a crystalline bacterial photosynthetic reaction center: A progress report and conclusions on the structure of photosystem II reaction centers. In: L.A. Staehelin and C.J. Arntzen (eds.), Encyclopedia of Plant Physiology: Photosynthetic Membranes. Springer-Verlag, Berlin, pp. 371–381.

Moase, E.H. and Green, B.R. (1981) Isolation and properties of chloroplast coupling factor from wheat. Eur. J. Biochem. 119, 145–150.

Muster, P., Binder, A. and Bachofen, R. (1984) A single subunit P-700 reaction center of the thermophilic cyanobacterium *Mastigocladus laminosus*. FEBS Lett. 166, 160–164.

Nakatani, H.Y., Ke, B., Dolan, E. and Arntzen, C.J. (1984) Identity of the photosystem II reaction center polypeptide. Biochim. Biophys. Acta 765, 347–352.

Nanba, M. and Katoh, S. (1983) Reaction kinetics of P700, cytochrome C-553, and cytochrome f in the cyanobacterium, *Synechococcus* sp. Biochim. Biophys. Acta 725, 272–279.

Nechushtai, R., Muster, P., Binder, A., Liveanu, V. and Nelson, N. (1983) Photosystem I reaction center from the thermophilic cyanobacterium *Mastigocladus laminosus*. Proc. Natl. Acad. Sci. USA 80, 1179–1183.

Newman, P.J. and Sherman, L.A. (1978) Isolation and characterization of photosystem I and II membrane particles from the blue-green alga, *Synechococcus cedrorum*. Biochim. Biophys. Acta 503, 343–361.

Okada, S. and Asada, K. (1983) A manganese protein from thylakoids of a cyanobacterium, *Plectonema boryanum*; inhibition of oxygen evolution with its antibody. Plant Cell Physiol. 24, 163–172.

Olson, J.M. and Thornber, J.P. (1979) Photosynthetic reaction centers. In: Capaldi, R.A. (ed.) Membrane Proteins, Vol. 2, Membrane Proteins in Energy Transduction. Marcel Dekker, New York, pp. 179–340.

O'Malley, P.J. and Babcock, G.T. (1984) Electron nuclear double resonance evidence supporting a monomeric nature for $P700^+$ in spinach chloroplasts. Proc. Natl. Acad. Sci. USA 81, 1098–1101.

Öquist, G. (1971) Changes in pigment composition and photosynthesis induced by iron deficiency in the blue-green alga, *Anacystis nidulans*. Physiol. Plant 25, 188–191.

Öquist, G. (1974a) Iron deficiency in the blue-green alga, *Anacystis nidulans*. Changes in pigmentation and photosynthesis. Physiol. Plant 30, 30–37.

Öquist, G. (1974b) Iron deficiency in the blue-green alga, *Anacystis nidulans*. Fluorescence and absorption spectra recorded at 77K. Physiol. Plant 31, 55–58.

Öquist, G., Fork, D.C., Schoch, S. and Malmberg, G. (1981) Solubilization and spectral characteristics of chlorophyll-protein complexes isolated from the thermophilic blue-green alga *Synechococcus lividus*. Carnegie Inst. Wash. YearBook 80, 47–49.

Pakrasi, H.B. and Sherman, L.A. (1984a) A highly active oxygen-evolving Photosystem II preparation from the cyanobacterium, *Anacystis nidulans*. Plant Physiol. 74, 742–745.

Pakrasi, H.B. and Sherman, L.A. (1984b) Characterization of an oxygen-evolving membrane fraction from the cyanobacterium, *Anacystis nidulans* R2. In: C. Sybesma (ed.), Advances in Photosynthetic Research, Vol. I. Martinus Nijhoff/Dr. W. Junk Publishers, The Hague, pp. 395–398.

Pakrasi, H., Goldenberg, A. and Sherman, L.A. (1985a) Membrane development in the cyanobacterium, *Anacystis nidulans*, during recovery from iron starvation. Plant Physiol. 79, 290–295.

Pakrasi, H., Riethman, H.C. and Sherman, L.A. (1985b) Organization of pigment-proteins in the photosystem II complex of the cyanobacterium, *Anacystis nidulans* R2. Proc. Natl. Acad. Sci. USA 82, 6903–6907.

Peterson, R., Dolan, E., Calvert, H. and Ke, B. (1981) Energy transfer from phycobiloproteins to PSI in vegetative cells and heterocysts of *Anabaena variabilis*. Biochim. Biophys. Acta 634, 237–248.

Pfister, K., Steinback, K.E., Gardner, G. and Arntzen, C.J. (1981) Photoaffinity labeling of a herbicide receptor protein in chloroplast membranes. Proc. Natl. Acad. Sci. USA 78, 981–985.

Piccioni, R.G. and Mauzerall, D.C. (1978) Calcium and photosynthetic oxygen evolution in cyanobacteria. Biochim. Biophys. Acta 504, 384–397.

Price, C.A. (1968) Iron compounds and plant nutrition. Ann. Rev. Plant Physiol. 19, 239–248.

Reinman, S. and Thornber, J.P. (1979) The electrophoretic isolation and partial characterization of three chlorophyll-protein complexes from blue-green algae. Biochim. Biophys. Acta 547, 188–197.

Remy, R., Tremolieres, A. and Ambard-Bretteville, F. (1984) Formation of oligomeric light-harvesting chlorophyll a/b protein by interaction between its monomeric forms and liposomes. Photochem. Photobiophys. 7, 267–276.

Rowell, P., Diez, J., Apte, S.K. and Stewart, W.D.P. (1981) Molecular heterogeneities of ferredoxin: $NADP^+$ oxidoreductase from the cyanobacterium *Anabaena cylindrica*. Biochim. Biophys. Acta 657, 507–516.

Rusckowski, M. and Zilinskas, B.A. (1980) Chlorophyll-protein complexes of the cyanophyte, *Nostoc* sp. Plant Physiol. 65, 392–396.

Rutherford, A.W. and Heathcote, P. (1985) Primary photochemistry in photosystem I. Photosyn. Res. 6, 295–316.

Sadewasser, D. and Sherman, L.A. (1981) Internal and external membrane proteins of the cyanobacterium *Synechococcus cedrorum*. Biochim. Biophys. Acta 640, 326–340.

Sandmann, G. and Malkin, R. (1983) Iron sulfur centers and activities of the photosynthetic electron transport chain in iron deficient cultures of the blue-green alga *Aphanocapsa*. Plant Physiol. 73, 724–728.

Sandmann, G., Reck, H., Kessler, E. and Boger, P. (1983) Distribution of plastocyanin and soluble plastidic cytochrome c in various classes of algae. Arch. Microbiol. 139, 23–27.

Satoh, K. and Fork, D.C. (1983) State 1-state 2 transitions in the green alga *Scenedesmus obliquus*. Photochem. Photobiol. 37, 428–434.

Satoh, K. and Katoh, S. (1985) Inhibition by ethylenediamine tetraacetate and restoration by Mn^{2+} and Ca^{2+} of oxygen-evolving activity in Photosystem II preparation from the thermophilic cyanobacterium, *Synechococcus* sp. Biochim. Biophys. Acta 806, 221–229.
Satoh, K., Nakatani, H.Y., Steinback, K.E., Watson, J. and Arntzen, C.J. (1983) Polypeptide composition of a Photosystem II core complex, presence of a herbicide-binding protein. Biochim. Biophys. Acta 724, 142–150.
Satoh, K., Ohno, T. and Katoh, S. (1985) An oxygen-evolving complex with a simple subunit structure 'a water-plastoquinone oxidoreductase' from the thermophilic cyanobacterium *Synechococcus* sp. FEBS Lett. 180, 326–330.
Sauer, K. (1979) Photosynthesis – the light reactions. Ann. Rev. Phys. Chem. 30, 155–178.
Schatz, G.H. and Witt, H.T. (1984) Extraction and characterization of oxygen-evolving Photosystem complexes from a thermophilic cyanobacterium *Synechococcus* spec. Photobiochem. Photobiophys. 7, 1–14.
Schmidt, A. (1980) Isolation of two thioredoxins from the cyanobacterium *Synechococcus* 6301. Arch. Microbiol. 127, 259–265.
Schwartz, R.M. and Dayhoff, M.O. (1978) Origins of prokaryotes, eukaryotes, mitochondria, and chloroplasts. Science 199, 395–403.
Schwartz, R.M. and Dayhoff, M.O. (1981) Chloroplast origins: inferences from protein and nucleic acid sequences. Ann NY Acad. Sci. 361, 260–272.
Serrano, A. and Rivas, J. (1982) Purification of ferredoxin-$NADP^+$ oxidoreductase from cyanobacteria by affinity chromatography on 2′, S′-ADP-Sepharose 4B. Anal. Biochem. 126, 109–115.
Sherman, D.A. and Sherman, L.A. (1983) The effects of iron deficiency and iron restoration on the ultrastructure of the cyanobacterium, *Anacystis nidulans*. J. Bacteriol. 156, 393–401.
Sherman, L.A., Golden, S.S. and Vann, C. (1984) Transformation and cloning vectors of the cyanobacterium *Anacystis nidulans*. In: R. Hallick, L.A. Staehelin and J.P. Thornber (eds.), UCLA Symposium on Biosynthesis of the Photosynthetic Apparatus, Allan Liss, Inc., New York, pp. 357–379.
Southern, E.M. (1975) Detection of specific sequences among DNA fragments separated by gel electrophoresis. J. Mol. Biol. 98, 503–517.
Spiller, S.C., Castelfranco, A.M. and Castelfranco, P.A. (1982) Effects of iron and oxygen on chlorophyll biosynthesis. Plant Physiol. 69, 107–111.
Stanier, R.Y. and Cohen-Bazire, G. (1977) Phototrophic prokaryotes: the cyanobacteria. Ann. Rev. Microbiol. 31, 225–274.
Stevens, S.E., Jr., Paone, D.A.M. and Balkwill, D.L. (1981) Accumulation of cyanophycin granules as a result of phosphate limitation in *Agmenellum quadruplicatum*. Plant Physiol. 67, 716–719.
Stewart, A.C. (1980) The chlorophyll-protein complexes of a thermophilic blue-green alga. FEBS Lett. 114, 67–72.
Stewart, A.C. and Bendall, D.S. (1979) Preparation of an active oxygen-evolving Photosystem II particle from a blue-green alga. FEBS Lett. 107, 308–312.
Stewart, A.C. and Bendall, D.S. (1981) Properties of oxygen-evolving Photosystem II particles from *Phormidium laminosum*, a thermophilic blue-green alga. Biochem. J. 194, 877–887.
Stewart, A.C and Kaethner, T.M. (1983) Extraction and partial purification of an acidic plastocyanin from a blue-green alga. Photobiochem. Photobiophys. 6, 67–73.
Stewart, A.C., Ljungberg, U., Akerland, H.-E. and Anderson, B. (1985) Studies on the polypeptide composition of the cyanobacterial oxygen-evolving complex. Biochim. Biophys. Acta 808, 353–362.
Susor, W.A. and Krogmann, D.W. (1966) Triphosphopyridine nucleotide photoreduction with cell-free preparations of *Anabaena variabilis*. Biochim. Biophys. Acta 120, 65–72.
Suss, K.-H. and Schmidt, O. (1982) Evidence for an $alpha_3$, $beta_3$, gamma, delta, I, II, epsilon, III_5 subunit stoichiometry of chloroplast ATP synthetase complex (CF_1-F_0). FEBS Lett. 144, 213–218.
Takahashi, Y. and Katoh, S. (1982) Functional subunit structure of photosystem I reaction center in *Synechococcus* sp. Arch. Biochem. Biophys. 219, 219–227.
Takahashi, Y., Koike, H. and Katoh, S. (1982) Multiple forms of chlorophyll-protein complexes from a thermophilic cyanobacterium *Synechococcus* sp. Arch. Biochem. Biophys. 219, 209–218.
Takahashi, Y., Hirota, Y. and Katoh, S. (1985) Multiple forms of P700-chlorophyll a protein complexes from *Synechococcus* sp. — The iron, quinone and carotenoid contents. Photosynth. Res. 6, 183–192.
Thornber, J.P. (1975) Chlorophyll-proteins: light harvesting and reaction center components of plants. Ann. Rev. Plant Physiol. 26, 127–158.
Van der Plas, J., Smeekens, J.C.M., Cremers, A.F.M., Woortman, M.R., Tuyl, M.J.M., Weisbeek, P.J.and Van Arkel, G.A. (1985) The gene for ferredoxin and its expression in *Anacystis nidulans* R2. V. Inter. Symp. on Photosynth. Prokaryotes. Grindelwald, Switzerland, p. 357.

Van Walraven, H.S., Lubberding, H.J., Marvin, H.J.P. and Kraayenhof, R. (1983) Characterization of reconstituted ATPase complex proteoliposomes prepared from the thermophilic cyanobacterium *Synechococcus* 6716. Eur. J. Biochem. 137, 101–106.

Vierling, E. and Alberte, R.S. (1983) P_{700} chlorophyll a protein: Purification, characterization, and antibody preparation. Plant Physiol. 72, 625–633.

Westhoff, P., Alt, J., Nelson, N., Bottomley, W., Bunemann, H. and Herrmann, R.G. (1983) Genes and transcripts for the P_{700} chlorophyll a apoprotein and subunit 2 of the photosystem I reaction center complex from spinach thylakoid membranes. Plant Mol. Biol. 2, 95–107.

Whittaker, M.M. and Gleason, F.K. (1984) Isolation and characterization of thioredoxin f from the filamentous cyanobacterium, *Anabaena* sp. 7119. J. Biol. Chem. 259, 14088–14093.

Williams, R.C., Glazer, A.N. and Lundell, D.J. (1983) Cyanobacterial Photosystem I: morphology and aggregation behavior. Proc. Natl. Acad. Sci. USA 80, 5923–5926.

Wood, P.M. (1978) Interchangable copper and iron proteins in algal photosynthesis: studies on plastocyanin and cytochrome C-552 in *Chlamydomonas*. Eur. J. Biochem. 87, 9–19.

Wraight, C.A. (1982) Current attitudes in photosynthesis research. In: Govindjee (ed.), Photosynthesis, Vol. I. Academic Press, New York, pp. 17–64.

Yamagishi, A. and Katoh, S. (1983a) Fluorescence emission spectra of photosystem II reaction center complexes and chlorophyll-binding subunits at low temperature. In Y. Inoue, A. Crofts, N. Govindjee, G. Renger and K. Satoh (eds.), The Oxygen Evolving System of Photosynthesis. Academic Press, Tokyo, pp. 39–48.

Yamagishi, A. and Katoh, S. (1983b) Two chlorophyll-binding subunits of the photosystem II reaction center complex isolated from the thermophilic cyanobacterium *Synechococcus* sp. Arch. Biochem. Biophys. 225, 836–846.

Yamagishi, A. and Katoh, S. (1984) A photoactive photosystem II reaction center complex lacking a chlorophyll-binding 40 kDa subunit from the thermophilic cyanobacterium *Synechococcus* sp. Biochim. Biophys. Acta 765, 118–124.

Yamagishi, A. and Katoh, S. (1985) Further characterization of the two Photosystem II reaction center complex preparations from the thermophilic cyanobacterium *Synechococcus* sp. Biochim. Biophys. Acta 807, 74–80.

Ye, B.C., De la Torre, A., Crawford, N., Lara, C., Carlson, D.E. and Buchanan, B.B. (1981) The ferredoxin/thioredoxin system of enzyme regulation in a cyanobacterium. Arch. Microbiol. 130, 14–18.

Zurawski, G., Bohnert, H.-J., Whitfeld, P.R. and Bottomley, R. (1982) Nucleotide sequence of the gene for the M_R 32,000 thylakoid membrane protein from *Spinacia oleracea* and *Nicotiana debneyi* predicts a totally conserved primary translation product of M_R 38,950. Proc. Natl. Acad. Sci. USA 79, 6260–6264.

P. Fay and C. Van Baalen (Eds.) The Cyanobacteria

2

Ferredoxins, flavodoxins and related proteins: structure, function and evolution

LYNDON J. ROGERS

Department of Biochemistry, University College of Wales, Aberystwyth, Dyfed, SY23 3DD, Wales, UK

1. Ferredoxins

1.1 The nature of ferredoxins

The name ferredoxin was given to a non-haem iron-containing protein isolated from the anaerobe, *Clostridium pasteurianum* (Mortenson et al., 1962). A protein with similar characteristics was isolated from *Spinacea oleracea* (spinach) (Tagawa and Arnon, 1962) and proved to be the same as proteins known previously as methaemoglobin-reducing factor, TPN (i.e. $NADP^+$)-reducing factor, and photosynthetic pyridine nucleotide reductase.

Ferredoxins, together with rubredoxins, are the simplest of the iron-sulphur proteins, possessing an iron-sulphur cluster but no additional active group such as flavin or haem. The ferredoxins are characterised by being electron transfer proteins of small molecular weight and usually a negative oxidation-reduction potential. Characteristically they possess either a single [2Fe-2S] cluster or one or two [4Fe-4S] clusters, but recently a [3Fe-3S] cluster, occurring alone or with a [4Fe-4S] cluster, has been recognised in a few ferredoxins and other proteins (Cammack et al., 1981; Beinert and Thompson, 1983). In all these clusters the iron is ligated to an equimolar amount of inorganic sulphur; in contrast, rubredoxins possess one or more [1Fe] clusters ligated to four sulphur-containing ligands, usually cysteine, but inorganic sulphur is absent.

This review is simplified by the fact that photosynthetic oxygen-evolving cells invariably contain a two-iron ferredoxin, except for one organism where no ferredoxin was apparently present. This type of ferredoxin is therefore commonly referred to as a plant-type ferredoxin, though a few non-photosynthetic bacteria and mammalian adrenal cortex cells also possess a ferredoxin with a [2Fe-2S] cluster. Cyanobacteria so far tested have invariably contained this type of ferredoxin. However, it should be noted that membrane-bound iron-sulphur centres also occur in cyanobacteria (Hiyama et al., 1985): three involved with the Photosystem I reaction centre are [4Fe-4S] centres and a Rieske (probably 2Fe-2S) centre is a component of the intermediate electron transport chain (Evans, 1982).

Ferredoxins from a considerable number of cyanobacteria have been isolated and characterised to varying extent (Table I); fourteen have been sequenced. Discussion will be largely confined to studies over the last decade, the earlier work which laid the foundation for these is summarised in a number of earlier comprehensive reviews (e.g.Hall et al., 1973, 1974).

1.2. Chemical properties

The two-iron ferredoxins are strongly acidic proteins with isoelectric points in the range 3.15–3.35 (Dutton and Rogers, 1978). This property has been exploited in their purification by ion-exchange chromatography on DEAE-cellulose, which with a successive molecular sieve chromatography on Sephadex G-50 is usually sufficient to give a ferredoxin of high purity (e.g. Hutson et al., 1978). Their ease of extraction from cyanobacteria is consistent with only a loose association with the photosynthetic lamellae, as with chloroplasts (Bohme, 1977).

In the oxidized state in which they are isolated these ferredoxins are reddish-brown, which facilitates their isolation particularly when two ferredoxins are present. While the absorption spectrum varies slightly depending on source, absorption maxima are typically at 276 sh(shoulder)282, 330, 423 and 465 nm (Fig. 1). The UV absorption is due to contributions from the aromatic amino

TABLE I

Cyanobacteria from which the properties of the ferredoxin have been described

Anabaena flos-aquae	FH	Andrew et al., 1975
*Anabaena variabilis**	FH	Chan et al., 1983a; Schrautemeier and Böhme, 1985
*Aphanothece halophitica**	U	Hase et al., 1983a
Aphanothece sacrum I* and II*	U	Hase et al., 1976a, 1978b
Aphanizomenon flos-aquae I* and II	FH	Lee et al., 1983
*Chlorogloeopsis (Chlorogloea) fritschii**	FH	Takahashi et al., 1982
Gloeothece PCC 6909	U	unpublished data
*Mastigocladus laminosus**	FH	Hase et al., 1978a
Microcystis flos-aquae	U	Rao et al., 1972
Nostoc muscorum I* and II	FH	Hase et al., 1976b
Nostoc strain MAC I* and II*	F	Hase et al., 1982b
Nostoc verrucosum I and II	FH	Shin et al., 1977
Nostoc sp.	FH	Mitsui and Arnon, 1974
Oscillatoria agardhii	F	Daly and Bishop, 1984
Oscillatoria limnetica	F	Cammack et al., 1977
Phormidium foveolarum	F	Mitsui and SanPietro, 1973
Spirulina maxima I* and II	F	Tanaka et al., 1975a,b
*Spirulina platensis**	F	Wada et al., 1975b
Synechococcus lividus	U	Crespi et al., 1972
Synechococcus sp.*	U	Hase et al., 1983b
(*Anacystis nidulans*) *Synechococcus* PCC 6301	U	Yamanaka et al., 1969
Synechococcus PCC 6307	U	unpublished data
(*Aphanocapsa* PCC 6714)* *Synechocystis* PCC 6714	U	Hase et al., 1982a

For some of these the ferredoxin has been characterised only in terms of absorption spectrum and molecular weight. Where the ferredoxin has been sequenced this is indicated by an *; I and II indicate the presence of two ferredoxins.

U, unicellular; F, filamentous; H, heterocystous.

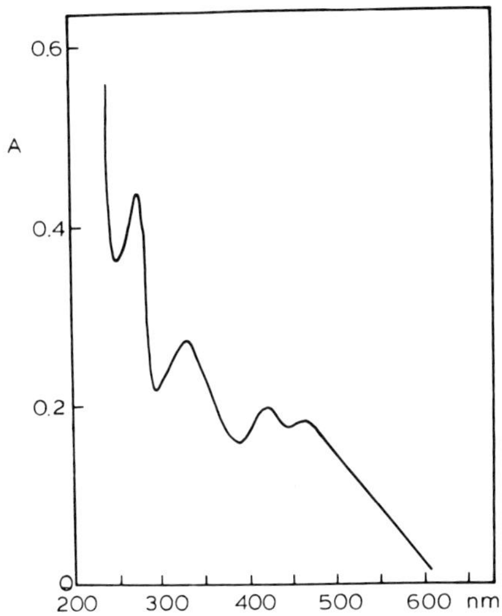

Fig. 1. Absorption spectrum of *Nostoc* MAC ferredoxin I. The ferredoxin (approximately 50 μM) was in the fully oxidized state.

acids and the chromophore. The A_{330}/A_{276}nm ratio is a sensitive index of the purity of a specific ferredoxin, and typically is about 0.65; any loss of the iron-sulphur cluster will decrease this value dramatically. Values for the molar extinction coefficient at 423 nm have been about 6000 $l{\cdot}mol^{-1}{\cdot}cm^{-1}$ for a number of cyanobacterial and algal ferredoxins, but values of about 9000 $l{\cdot}mol^{-1}{\cdot}cm^{-1}$, as for higher plant ferredoxin, are also reported (see Hutson et al., 1978). On reduction the absorption spectrum becomes less well-defined with a decrease of some 50% in the absorption in the visible region. However, the reduced ferredoxin exhibits a characteristic electron paramagnetic reso-

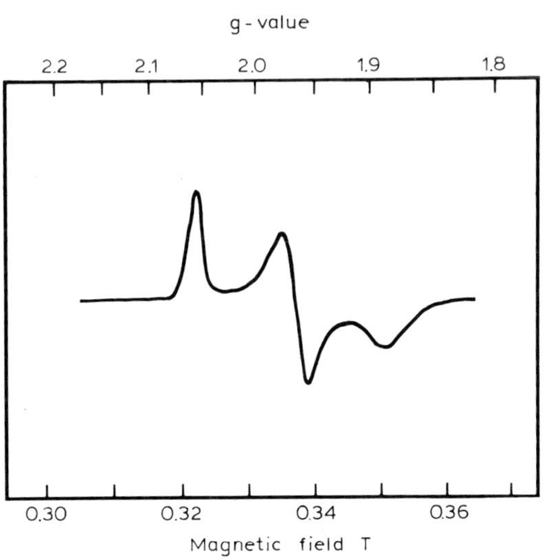

Fig. 2. EPR spectrum at 22K of reduced *Synechocystis* 6714 ferredoxin. Other conditions of measurement were microwave frequency, 9.26 GHz; modulation amplitude, 1.0 mT. (Redrawn from Cammack et al., 1977)

nance (EPR) spectrum arising from the unpaired electron with a signal centred at $g = 1.95$ (Fig. 2). The signal is rather broad with rhombic symmetry because at liquid nitrogen or lower temperatures the ferredoxin molecules are frozen in a range of slightly different conformations and therefore *g*-values (Orme-Johnson and Sands, 1973). Since the oxidized ferredoxin generates no EPR spectrum the 1.95 *g* signal can be exploited to enable the oxidation-reduction potentials of ferredoxins to be determined readily from potentiometric titrations (Cammack et al., 1977). There are also characteristic circular dichroism spectra (Hall et al., 1973) with subtle variations probably reflecting the amino acid environment of the chromophore in particular ferredoxins.

The constraint imposed by the peptide chain on the iron-sulphur cluster can be seen in the quite marked difference in midpoint redox potential (E_m) values for the two *Nostoc* MAC ferredoxins (Hutson et al., 1978). More extreme is the positive E_m value of the membrane-bound Rieske centre which also has the same chromophore. In some other photosynthetic bacteria the peptide chain imposes a considerable difference in E_m through the type of oxidation-reduction reaction the [4Fe-4S] cluster of high potential iron-sulphur protein can undergo compared to this same cluster in a low-potential ferredoxin.

The structure of the [2Fe-2S] cluster, as deduced from spectroscopic and other physicochemical studies is shown in Fig. 3. There is tetrahedral coordination of two iron atoms with four cysteine sulphurs and two bridging atoms of inorganic sulphur (Orme-Johnson and Sands, 1973). The electronic model shows the two iron atoms in the oxidized ferredoxin to be high-spin ferric atoms antiferromagnetically coupled via bridging sulphur ligands, the spins thus cancelling each other; this gives a diamagnetic complex which consequently has no EPR signal. On reduction one of the iron atoms accepts an electron and becomes a high-spin ferrous atom, with a resultant net spin of $s = {}^1/_2$; the characteristic EPR spectrum shown in Fig. 2 is generated (Gibson et al., 1966). Mössbauer studies confirmed the validity of the model and suggested the two iron atoms in the cluster may not be equivalent (Rao et al., 1971), and further information on detail emerged from application of a range of physicochemical techniques to cyanobacterial (Anderson et al., 1975) and plant ferredoxins. The earlier studies are summarised elsewhere (Hall et al., 1973, 1974), with more recent progress being represented by EPR spectroscopy (Gayda et al., 1976), magnetic susceptibility (Petersson et al., 1980), and ^1H nuclear magnetic resonance (NMR) spectroscopy (Nagayama et al., 1983), investigations on *Spirulina* ferredoxins.

The molecular weight of cyanobacterial ferredoxins is about 10 500, including the proteins from *Nostoc verrucosum* (Shin et al., 1980) which were originally thought to be substantially larger (Shin et al., 1977). Their amino acid compositions show a preponderance of (putative) acidic residues

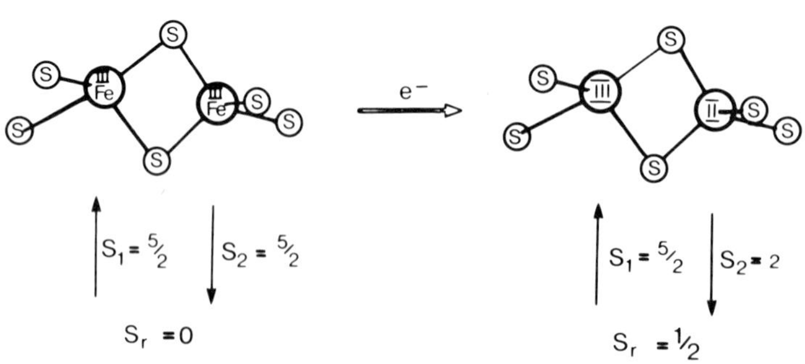

Fig. 3. Model of the active centre of a two-iron two-sulphur ferredoxin in the oxidized and reduced forms (based on references quoted in the text). In formal terms the ferredoxin has a $[2Fe\text{-}2S]^{2+;1+}$ cluster.

(see e.g. Hall et al., 1974) with few basic or aromatic residues; methionine is usually low and tryptophan is absent from cyanobacterial ferredoxins except ferredoxin II from *Aphanothece sacrum*.

The majority of cyanobacteria appear to possess a single soluble ferredoxin. Nevertheless, in recent years it has become clear that a number possess two ferredoxins. and this is the case for *Aphanothece sacrum* (Hase et al., 1975), *Nostoc* MAC (Hutson and Rogers, 1975), *Nostoc muscorum* (Hase et al., 1976b), *Nostoc verrucosum* (Shin et al., 1977), *Spirulina maxima* (Cammack et al., 1977) and most recently *Anabaena variabilis* (Schrautemeier and Böhme, 1985) and *Microcystis aeruginosa* (Cohn et al., 1985). This group includes examples of both unicellular and filamentous cyanobacteria, and within the latter group heterocystous and non-heterocystous cell types. The two proteins are genetically independent isoenzymes (Hase et al., 1978b), and the favoured explanation for their occurrence is a duplication of the ferredoxin gene in a primitive ancestral organism, with divergent evolution leading to the development of proteins with distinctive structural and functional properties (Takaichi et al., 1978). It is possible that the gene for a second ferredoxin is present in a number of cyanobacteria, but is not expressed under the conditions used for maintenance and growth.

Where two ferredoxins are present, their relative amounts differ widely from approaching equal amounts in *Nostoc verrucosum* (Shin et al., 1977); 2:1 for *Nostoc* MAC (Hutson et al., 1978) and 5:1 for *Aphanothece sacrum* (Hase et al., 1975). In the others the second ferredoxin was only present in low quantities, posing the question of contamination by a subspecies where the organism has been isolated from natural sources. However, in heterocystous cyanobacteria the minor ferredoxin may be confined to the specialised nitrogen-fixing cells. Except for *Anabaena variabilis* ferredoxin II (Chan et al., 1983a), ferredoxin I in each case is the major species, though this does not imply coincidence of properties and ferredoxin I from one organism may be more similar to ferredoxin II from another.

Under iron-deficient growth conditions favouring the replacement of *Nostoc* MAC ferredoxins by flavodoxin, the apoproteins of the ferredoxins were not synthesised (Hutber et al., 1978), indicating control is exercised at the transcription stage rather than at the insertion of the iron-sulphur cluster. This is also the case for ferredoxin synthesis in *Chlamydomonas* (Huisman et al., 1979), Neither is ferredoxin synthesis in cyanobacteria dependent on light, as in *Rhodospirillum rubrum*, *Euglena* or higher plants (but see Takahashi et al., 1983), since *Nostoc* MAC grown chemoheterotrophically in the dark possessed both ferredoxins in the same amounts as in photoautotrophic cultures (Hutson and Rogers, 1975).

However, little is known of the biochemical origin of the iron-sulphur cluster and of the method of insertion into the apoprotein. The sulphur transferase activity of rhodanese may be implicated in sulphur utilisation (Bonomi et al., 1977; Pagani et al., 1982, 1983) or, as proposed for *Escherichia coli* ferredoxin, the labile sulphur may originate from cysteine sulphur (White, 1983), possibly through the combined action of a transaminase and 3-mercaptopyruvate sulphur transferase (Taniguchi and Kimura, 1974). Without the need for transport across an intracellular membrane there is no necessity to synthesise an apoprotein bearing a leader sequence, as occurs with ferredoxin (and ferredoxin-$NADP^+$ oxidoreductase) synthesised in the cytoplasm in eukaryotic organisms but destined for the chloroplast (Huisman et al., 1978a,b).

1.3. Oxidation-reduction properties

The first measurement of the oxidation-reduction potential for a cyanobacterial ferredoxin gave an E_m of −406 mV for *Nostoc* sp. ferredoxin (Mitsui and Arnon, 1971), in good agreement with the E_m values for higher plant ferredoxins. However, a survey of a range of ferredoxins (Cammack et al., 1977) showed that E_m values for cyanobacterial ferredoxins varied between −310 mV and

TABLE II

Redox potentials of cyanobacterial ferredoxins

	Potential (mV)
Anabaena variabilis	−355
Aphanothece halophitica	−380[a]
Chlorogloeopsis fritschii	−340
Gloeothece 6909	−342[a]
Mastigocladus laminosus	−325
Nostoc sp.	−406[b]
Nostoc MAC I	−350
Nostoc MAC II	−455
Oscillatoria agardhii	−380[c]
Oscillatoria limnetica	−346
Spirulina maxima I	−390
Spirulina maxima II	−310
Spirulina platensis	−381
Synechococcus 6301	−380[a]
Synechococcus 6307	−395[a]
Synechocystis 6714	−375

Data from Cammack et al., 1977, except: [a], data from authors laboratory; [b], Mitsui and Arnon, 1971; [c], Daly and Bishop, 1984. E_m values are at pH 8.0, except [b,c] which are at pH 7.5.

−455 mV, depending on source (Table II). These data were obtained by redox titrations in the presence of suitable mediators with oxidation state in most cases monitored by EPR spectroscopy. The E_m values of cyanobacterial ferredoxins studied to date are summarised in Table II. Where redox titrations based on EPR spectroscopy or absorbance were used, the plots of signal intensity plotted according to the Nernst equation as a function of redox potential confirmed previous conclusions that the ferredoxins were one-electron-accepting species (Fig. 4).

Of particular interest were the observations that where a cyanobacterium possessed two ferredoxins the E_m values for these differed markedly; this was in contrast to the two ferredoxins from a higher plant where E_m values were very similar. That the difference in E_m values suggests that the ferredoxins may be adapted to specific roles is discussed later.

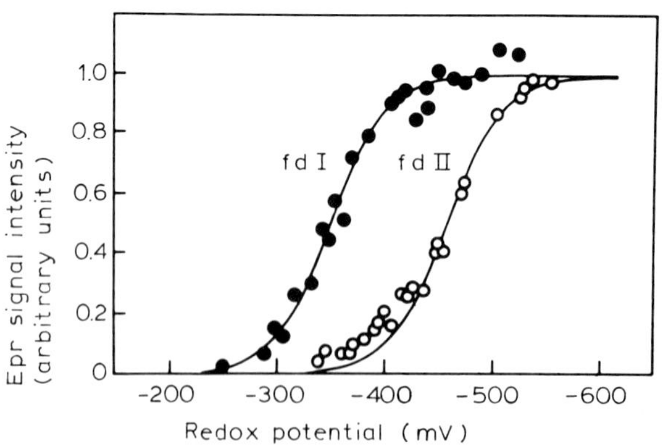

Fig. 4. Potentiometric titrations of *Nostoc* MAC ferredoxins I and II. Experimental points for the intensity of the EPR signal (g_y) at 1.95 for the ferredoxins as a function of redox potential are superimposed on theoretically derived plots for one-electron-transferring species. (From Hutson et al., 1978)

1.4. Biological activity

Yoch and Valentine (1972) were able to list eighteen ferredoxin-dependent reactions in fermentative bacteria, the ferredoxin in many cases interacting with an enzyme also possessing an iron-sulphur cluster. In cyanobacteria the range of compounds to sustain growth is more restricted. However, it is implicit that where a single ferredoxin is present in a cyanobacterium this must be able to fulfil the biochemical roles required of this low potential reductant, as listed in Table III. Of course, depending on E_m and the ability to interact with other proteins, the ferredoxin may be more active in certain of these reactions and less active in others in vivo, or in the in vitro assays, compared to the ferredoxin from a different cyanobacterium. Comparative assays suggest a more negative E_m value ($\leqslant -400$ mV) enhances activity in $NADP^+$ photoreduction, while a less negative E_m (ca. -350 mV) favours activity of the oxidized ferredoxin in a phosphoroclastic assay (Fitzgerald et al., 1980a), and might be expected to enhance effectiveness as a reductant for nitrogenase where the ferredoxin mediates electron flow from NADPH. In its participation in photosynthetic electron flow it is suggested that ferredoxin, loosely attached to the thylakoid membrane, acts as a shuttle picking up electrons from Photosystem I and releasing them to the spatially separate ferredoxin-$NADP^+$ oxidoreductase site (Forti and Grubas, 1985). Evidence for two sites of action of ferredoxin in different Photosystem I-dependent electron flows via either cytochrome b_{563} or ferredoxin-$NADP^+$ oxidoreductase has been presented (Hosler and Yocum, 1985). The involvement of the latter enzyme in cyclic electron flow has also been shown by Shahak et al., (1981), though the $NADP^+$-binding site on the oxidoreductase was not involved. These findings were in contrast to earlier work exploiting antibodies to the proteins which suggested that only ferredoxin was involved in cyclic electron flow (Bohme, 1977).

Environmental and physiological constraints may be reflected in the properties of the ferredoxin(s) in a particular cyanobacterium. This may be seen in the relatively high E_m value for *Gloeothece* 6909 ferredoxin, which would enable a role as a reductant for nitrogenase in this unicellular nitrogen-fixer, while still being effective in $NADP^+$ photoreduction. This may be true also for filamentous nitrogen-fixers such as *Chlorogloeopsis fritschii* and *Mastigocladus laminosus* which apparently possess only a single ferredoxin. Ferredoxins from thermophilic cyanobacteria such as *Mastigocladus laminosus* and *Synechococcus* sp. may be adapted to the environment of these organisms. Even in the isolated state the ferredoxin from the former is stable at 75°C for an hour without destruction of the iron-sulphur cluster (Hase et al., 1978a). The suggestion has been made that the hot springs

TABLE III

Ferredoxin-dependent reactions in cyanobacteria

Fatty acid desaturation	Cammack et al., 1981
Glutamate synthase	Stewart and Rowell, 1975
Hydrogenase	Burris, 1981
Nitrate reductase	Mikami and Ida, 1984
Nitrite reductase	Peschek, 1979
Nitrogenase	Bothe, 1982
Pyruvate-ferredoxin oxidoreductase	Neuer and Bothe, 1982
Sulphite reductase	Cammack et al., 1981
$NADP^+$ photoreduction	Ho and Krogmann, 1982
Thioredoxin-ferredoxin reductase	Yee et al., 1981

The reference cited is to a recent paper or a review which indicates earlier work.

provide a stable environment for ecosystems which may then remain unaltered for long periods of time. On this basis *Mastigocladus* might represent one of the primitive forms of extant algae; supporting this view is its close phylogenetic relationship to *Spirulina*, of which there is fossil evidence. An opposing view is that the thermophilic bacteria have become adapted to their specialised environments through fixing of mutations which increased the thermal stability of their proteins. The reason for the thermostability of *Mastigocladus* ferredoxin is not clear from its sequence, but conclusions from the sequence of *Synechococcus* ferredoxin based on comparisons with the three-dimensional structure for *Spirulina platensis* ferredoxin suggest a role for salt bridges of lysine and arginine side chains with acidic residues lying externally (Hase et al., 1983b).

Of particular interest has been the opportunity in a few cyanobacteria to compare the properties of the two ferredoxins that are present. The activities of the *Nostoc* MAC ferredoxins in four assay systems have been compared (Fitzgerald et al., 1980a). In mediating electron flow from dithionite to hydrogenase from *Clostridium pasteurianum*, or in coupling the reducing potential of spinach chloroplast membranes to this hydrogenase, the two ferredoxins showed similar activity. In $NADP^+$ photoreduction ferredoxin II (E_m: −450 mV) was about twice as active as ferredoxin I (E_m: −350 mV) whereas the latter proved to be the most active in supporting the phosphoroclastic reaction by *Clostridium pasteurianum* extracts. In these latter assays, *Chlorogloeopsis fritschii* and *Synechocystis* 6714 ferredoxins (E_m values of −340 mV and −375 mV, respectively) showed very similar behaviour to *Nostoc* MAC ferredoxin I. In supporting NADPH-cytochrome *c* reductase activity using native ferredoxin-$NADP^+$ oxidoreductase (K_{mFdI} 56 μM; K_{mFdII}16 μM) ferredoxin I was three-fold more active than ferredoxin II (Hutber and Rogers, 1981).

For *Nostoc verrucosum*, ferredoxin II was the more active in $NADP^+$ photoreduction and ferredoxin I more effective in a NADPH-cytochrome *c* reductase system where the ferredoxin accepts electrons from the pyridine nucleotide (Shin et al., 1977). However, quantitative data were not given and the E_m values of these ferredoxins are not known.

Comparative data for the *Aphanothece sacrum* and *Nostoc muscorum* ferredoxins are also available (Wada et al., 1981). In $NADP^+$ photoreduction using higher plant chloroplasts ferredoxin I from both cyanobacteria showed similar activity, approaching twice that shown by ferredoxin II. This distinction was not seen for the *Nostoc muscorum* ferredoxins, however, when the cyanobacterial membranes were used to provide the reducing potential. In a chloroplast-nitrogenase (*Bacillus polymyxa*) system ferredoxin I in each case was more active than ferredoxin II; substantially so in the case of the *Nostoc muscorum* ferredoxins.

More recently has come the suggestion that heterocysts of *Anabaena variabilis* possess a distinctive ferredoxin to that found in vegetative cells (Schrautemeier and Böhme, 1985). Activities of the ferredoxins as mediators in four assay systems were examined: $NADP^+$-photoreduction by heterocyst thylakoids using cytochrome c_{553} as electron donor since heterocysts lack Photosystem II (Tel-Or and Stewart, 1977), and three assays of native nitrogenase activity using as source of reducing potential; (a) NADPH via ferredoxin-$NADP^+$ oxidoreductase, (b) H_2 via hydrogenase (*Clostridium pasteurianum*), or (c) H_2 via the hydrogenase and Photosystem I.

Both ferredoxins were equally active in supporting $NADP^+$-photoreduction, but in assays of nitrogenase activity the heterocyst ferredoxin was more active than the ferredoxin characteristic of vegetative cells. In assays (a) and (b) the latter ferredoxin was inactive while in (c) it showed appreciable activity though significantly less than the heterocyst ferredoxin. Discussion of the basis for the differences in activity must await E_m values and sequence data on the ferredoxins. The close correspondence in supporing $NADP^+$ photoreduction suggests that the E_m values of the ferredoxins are similar, and this is supported by the observation that, though vegetative cell ferredoxin is inactive in the H_2-hydrogenase system, it is reduced by clostridial hydrogenase to the same extent as the heterocyst ferredoxin. It may, therefore, be the ability to interact with the nitrogenase which distinguishes the two ferredoxins rather than their redox potentials.

The parallel with the situation in *Nostoc muscorum* where one of the two ferredoxins is also present in much smaller amounts is relevant, since *Nostoc muscorum* is also a filamentous heterocystous nitrogen-fixer. In this case also, activities in $NADP^+$ photoreduction using cyanobacterial membranes were similar, but here the (major) ferredoxin I was more active in a nitrogenase assay (Wada et al., 1981). However, *Chlorogloeopsis fritschii*, with similar cell characteristics, possesses only a single ferredoxin while *Aphanothece sacrum* which is unicellular has two ferredoxins. The presence of two ferredoxins differing markedly in redox potential in *Nostoc* MAC suggested that these might have different biochemical roles. However, *Nostoc* MAC as now cultured is non-heterocystous and will not fix nitrogen although the original isolate would grow on media free of combined nitrogen (Bowyer and Skerman, 1968); it is recognised that *Nostoc* sp. may lose the ability to fix N_2 after prolonged culture in media containing nitrate (Rippka et al., 1979). On the basis of present knowledge, the occurrence of a specific ferredoxin to support nitrogenase activity in heterocysts is an interesting proposal, but one which needs some further experimental support.

1.5. Amino acid sequence: phylogeny and evolution

The sequences of some forty [2Fe-2S] ferredoxins from oxygenic photosynthetic organisms have been reported, including fourteen from the cyanobacteria. These are shown in Fig. 5; the notable contribution of the group at Osaka University, Japan, is evident. These sequences have proved to be informative in a number of respects, both in detail and for tracing the evolution of the protein (Tsukihara et al., 1982; Matsubara and Hase, 1983).

The cyanobacterial ferredoxins are of 96 to 99 residues with between 4 and 6 cysteines. As shown in the Fig., four of these latter are invariant; these occupy positions 42, 47, 50 and 81 in the sequences aligned for maximum homology though in an individual ferredoxin, because of deletion of residues, the residue numbers may be slightly different. That these cysteines are the ones involved in chelating the iron-sulphur cluster was confirmed unequivocally when *Nostoc muscorum* ferredoxin I (Hase et al., 1976b) and later *Nostoc* MAC ferredoxin I (Hase et al., 1982b) were shown to contain four cysteines only, at positions 41, 46, 49 and 79. The sequences at these two regions of the molecule, 40-53 and 61-88, are remarkably conserved and many residues in these sequences and a few elsewhere are invariant in [2Fe-2S] ferredoxins from oxygenic photosynthetic organisms. These residues, nineteen in all as shown in Fig. 5, but only seventeen with later sequences considered, include Pro-38, Gly-56, Ala-81 and Thr-91 outside the sequences containing the active site cysteines. For details of interest in specific ferredoxins the references given in Table I should be consulted. Microheterogeneity, suggesting the presence of allelic genes, is rare for cyanobacterial ferredoxins but is found in a green-algal ferredoxin (*Dunaliella salina* I) and some plant ferredoxins (Matsubara and Hase, 1983). When arranged as in the Fig. the presence of deletions in specific ferredoxins, or additions in other ferredoxins, is evident. There are deletions at position 3 (except *Nostoc* MAC I) and position 60 (except *Aphanothece sacrum* II) compared to higher plant ferredoxins. The deletions at 11 and 15 in *Synechocystis* 6714 and *Aphanothece sacrum* I ferredoxins are those seen in three green-algal ferredoxins (Hase et al., 1983a) while another unicellular cyanobacterium *Synechococcus* sp. had only the deletion at the first of these positions. These features in particular have influenced the placing of the cyanobacterial and algal ferredoxins in a phylogenetic tree (Hase et al., 1983a) constructed by the method of Fitch and Margoliash (1967) from the amino acid difference matrix.

The amino acid difference matrix for the cyanobacterial ferredoxins (Table IV) also enables

```
      1       10        20        30        40*  *  *     60        70        80*       90
( 1)  AS-YKVTLINEEMGLNETIEVPDDEYILDVAEEEGIDLPYSCRAGACSTCAGKIKEGEI-DQSDQSFLDDDQIEAGYVLTCVAYPASDCTIITHQEEELY
( 2)  AS-YKVTLKT-PDG-DNVITVPDDEYILDVAEEEGLDLPYSCRAGACSTCAGKLVSGPA-PDEDQSFLDDDQIQAGYILTCVAYPTGDCVIETHKEEALY
( 3)  AT-YKVTLINEEEGINAILEVADDQTILDAGEEAGLDLPSSCRAGSCSTCAGKLVSGAAPNQDDQAFLDDDQLAAGWVMTCVAYPTGDCTIMTHQESEVL
( 4)  AT-YKVTLVR-PDGSETTIDVPEDEYILDVAEEQGLDLPFSCRAGACSTCAGKLLEGEV-DQSDQSFLDDDQIEKGFVLTCVAYPRSDCKILTNQEEELY
( 5)  AS-YTVKLIT-PDG-ENSIECSDDTYILDAAEEAGLDLPYSCRAGACSTCAGKITAGSV-DQSDQSFLDDDQIEAGYVLTCVAYPTSDCTIETHKEEDLY
( 6)  AT-YKVTLISEAEGINETIDCDDDTYILDAAEEAGLDLPYSCRAGACSTCAGKITSGSI-DQSDQSFLDDDQIEAGYVLTCVAYPTSDCTIQTHQEEGLY
( 7)  AT-YKVTLINEAEGINETIDCDDDTYILDAAEEAGLDLPYSCRAGACSTCAGTITSGTI-DQSDQSFLDDDQIEAGYVLTCVAYPTSDCTIKTHQEEGLY
( 8)  AT-YKVTLI-DAEGTTTTIDCPDDTYILDAAEEAGLDLPYSCRAGACSTCAGKLVTGTI-DQSDQSFLDDDQVEAGYVLTCVAYPTSDVTIETHKEEDLY
( 9)  ATVYKVTLV-DQEGTETTIDVPDDEYILDIAEDQGLDLPYSCRAGACSTCAGKIVSGTV-DQSDQSFLDDDQIEKGYVLTCVAYPTSDLKIETHKEEDLY
(10)  AT-YKVRLFNAAEGLDETIEVPDDEYILDAAEEAGLDLPFSCRSGSCSSCNGILKKGTV-DQSDQNFLDDDQIAAGNVLTCVAYPTSNCEIETHREDAIA
(11)  AT-FKVTLINEAEGTKHEIEVPDDEYILDAAEEEGYDLPFSCRAGACSTCAGKLVSGTV-DQSDQSFLDDDQIEAGYVLTCVAYPTSDVVIQTHKEEDLY
(12)  AT-YKVTLINDAEGLNQTIEVDDDTYILDAAEEAGLDLPYSCRAGACSTCAGKIKSGTV-DQSDQSFLDDDQIEAGYVLTCVAYPTSDCTIETHKEEELY
(13)  AT-YKVTLINEAEGLNKTIEVPDDQYILDAAEEAGIDLPYSCRAGACSTCAGKLISGTV-DQSDQSFLDDDQIEAGYVLTCVAYPTSDCVIETHKEEELY
```

Fig. 5. Comparison of sequences of [2Fe-2S] ferredoxins from cyanobacteria. Amino acids are written as one letter abbreviations: A. alanine; C. cysteine; D, aspartic acid; E, glutamic acid; F, phenylalanine; G, glycine; H, histidine; I, isoleucine; K, lysine; L, leucine; M, methionine; N, asparagine; P, proline; Q, glutamine, R, arginine; S, serine; T, threonine; V, valine; W, tryptophan; Y, tyrosine. Gaps are inserted to give maximum homology.* indicates a cysteine residue ligated to the iron-sulphur cluster.

The sequences are for (1) *Aphanothece halophitica*, (2) *Aphanothece sacrum* I, (3) *Aphanothece sacrum* II, (4) *Synechococcus* sp., (5) *Synechocystis* 6714, (6) *Spirulina maxima*, (7) *Spirulina platensis*, (8) *Aphanizomenon flos-aquae* I, (9) *Nostoc* MAC I, (10) *Nostoc* MAC II, (11) *Nostoc muscorum* I (shown); *Anabaena variabilis* II**, (12) *Chlorogloeopsis fritschii*, (13) *Mastigocladus laminosus*.

similarities between sequences to be discerned more readily. In this compilation a gap in a sequence was counted as five amino acid differences rather than one on the basis that a deletion or insertion is a rarer event than a point mutation during evolution. A phylogenetic tree on the former basis is shown in Fig. 6; phylogenetic trees based on both methods are compared in Matsubara and Hase (1983).

The phylogenetic tree broadly divides the cyanobacterial ferredoxins into three groups: (1) the ferredoxins of filamentous cyanobacteria; *Aphanizomenon flos-aquae, Chlorogloeopsis fritschii, Mastigocladus laminosus, Nostoc* MAC I, *Nostoc muscorum* I, *Spirulina maxima* I and *Spirulina platensis*, but also including the unicellular *Aphanothece halophitica*; (2) the ferredoxins of the other unicellular cyanobacteria, *Synechococcus* sp., *Synechocystis* 6714 and *Aphanothece sacrum* I; and (3) ferredoxins II from *Aphanothece sacrum* and *Nostoc* MAC. The unicellular *Aphanothece halophitica* thus proved unusual in that its ferredoxin showed more resemblance to those from filamentous cyanobacteria though this organism possesses a 'primitive' ribulose bisphosphate carboxylase compared to other cyanobacteria (Codd et al., 1979). As in plants the differences between the related ferredoxins in different organisms are less than between the two ferredoxins of the same organism. This suggests that gene duplication occurred before species divergence (speciation).

**The sequences as published differ at position 89 (Val and Cys, respectively). Strain histories are uncertain and in view of the confusion in the nomenclature of *Anabaena* strains, which in a recent classification (Rippka et al., 1979) included some isolates designated originally as *Nostoc muscorum*, it is probable that these two sequences are of ferredoxins from closely related strains.

TABLE IV

Amino acid difference matrix of cyanobacterial ferredoxins (one gap was counted as five amino acid differences)

		(1)	(2)	(3)	(4)	(5)	(6)	(7)	(8)	(9)	(10)	(11)	(12)	(13)
Aphanothece halophitica	(1)													
Aphanothece sacrum I	(2)	35												
Aphanothece sacrum II	(3)	38	50											
Synechococcus sp.	(4)	27	33	49										
Synechocystis 6714	(5)	32	24	51	32									
Spirulina maxima	(6)	18	38	35	31	26								
Spirulina platensis	(7)	18	39	35	32	28	4							
Aphanizomenon flos-aquae	(8)	28	40	43	35	32	19	19						
Nostoc MAC I	(9)	34	44	53	33	41	31	31	20					
Nostoc MAC II	(10)	32	45	42	39	43	32	30	35	42				
Nostoc muscorum I	(11)	22	34	38	31	33	21	21	21	28	29			
Chlorogloeopsis fritschii	(12)	16	35	34	30	25	12	11	18	26	25	17		
Mastigocladus laminosus	(13)	16	33	33	29	19	16	15	20	28	25	13	8	

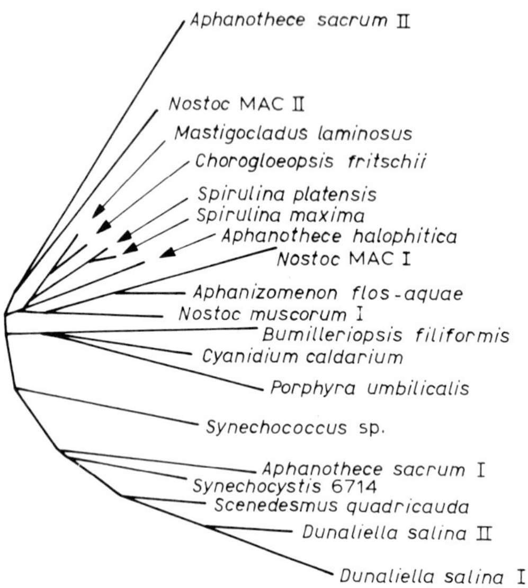

Fig. 6. A phylogenetic tree of cyanobacterial and algal ferredoxins. The calculation of branch lengths was based on the matrix shown in Table IV. (Based on Hase et al., 1983a)

While a phylogenetic tree based on one protein is not necessarily entirely consistent with that deduced from morphological and paleontological analyses it can provide clues about the relationship of organisms. Thus, consideration of the sequences has suggested possible resolution of problems in the taxonomic placement of some organisms. For example, ferredoxin from *Cyanidium caldarium*, an organism whose taxonomic position has been in doubt, shows closest resemblance to those from red algae in the Bangiophyceae rather than those from cyanobacteria (Hase et al., 1978c). Amongst the cyanobacteria the position of *Chlorogloeopsis fritschii* has proved a problem, being first classified in the order Chroococcales, then in the Nostocales, and more recently in the Stigonematales; the resemblance of its ferredoxin to that of *Mastigocladus laminosus* supports the last and its placement in section V of the cyanobacteria as described by Rippka et al. (1979). The relationships are supported by a survey of immunochemical reactivity of a range of cyanobacterial ferredoxins to *Spirulina maxima* ferredoxin antibody (Tel-Or et al., 1977). After *Spirulina platensis* ferredoxin, which was fully reactive, the highest antigenic resemblance was shown by the ferredoxins from *Chlorogloeopsis fritschii* and *Mastigocladus laminosus*.

The belief that ferredoxins are ancient proteins is based on the observation that a fermentative bacterium, *Clostridium butyricum*, possesses a ferredoxin almost entirely composed of nine amino acids which can be synthesized under simulated prebiotic conditions. Some two-thirds of the molecule is of six amino acids which have been found in carbonaceous chondrite meteorites. Moreoever, this ferredoxin has two [4Fe-4S] clusters, and these elements would have been abundant on the primitive earth (Cammack et al., 1981; Matsubara et al., 1980).

A tentative time-scale for evolution of chloroplast-type ferredoxins based on sequence comparisons has been given in Wada et al. (1975a). According to fossil records angiosperms evolved about 1.35×10^8 years ago and higher plants had separated from green algae about 4.2×10^8 years ago. The other photosynthetic prokaryotes diverged from cyanobacteria 1.5×10^9 years or more ago. In the intervening period a unicellular cyanobacterial-like cell may have undergone endosymbiosis with a primitive plant cell precursor to provide ultimately the chloroplast (Craig, 1971; Schwartz and Dayhoff, 1978). The sequences for ferredoxins support closest resemblance of unicellular cyanobacteria to green algae and higher plants. The origin of cyanobacteria is placed in the Archaean era some 3×10^9 years ago. Calculations from differences in the sequences suggest that the evolutionary rate of amino acid replacements in their ferredoxins has not been linear as for haemoglobins, cytochromes *c* and fibrinopeptides (Tanaka et al., 1975c; Matsubara et al., 1978). One possibility is that there was a rapid rate of early evolution of ferredoxins so that mutation numbers were saturated in a relatively short time (Matsubara and Hase, 1983). Comparisons of ferredoxins from remotely related organisms of recent origin would not therefore reflect real phylogenetic distances.

For two cyanobacteria a comparison of the sequences of the two ferredoxins which occur can be made. For both *Aphanothece sacrum* and *Nostoc* MAC the divergence of sequence between ferredoxins I and II is clearly evident. There were 34 differences in the sequences of the two *Nostoc* ferredoxins as calculated in Table IV, which are reflected in the different physicochemical properties of these proteins. *Chlorogloeopsis fritschii* and *Synechocystis* 6714 ferredoxins were similar to ferredoxin I of *Nostoc* MAC in terms of E_m values and biological activities in various assays (Hutson et al., 1980). However, when sequences are compared these ferredoxins are equally divergent from ferredoxins I and II of *Nostoc*, though *Chlorogloeopsis fritschii* ferredoxin is the more closely related. If gaps are counted as a single difference, the other two ferredoxins are more similar to *Nostoc* ferredoxin I than to ferredoxin II (Hase et al., 1982a).

At the present levels of interpretation, sequence information enables regions, and in some instances residues, important to activity or interaction with other proteins to be recognised through homologous sequences, but is unable to explain the differences in biological activity between various ferredoxins. Complementing the use of sequence studies to clarify structure-activity relationships have been investigations of the role of specific amino acids by chemical modification of ferredoxins. For example, functional residues in *Spirulina platensis* ferredoxin have been probed by kinetic studies exploiting inorganic complexes (Adzamli et al., 1983), and NMR techniques have been used to probe the interactions of histidine and tyrosine in *Anabaena variabilis* ferredoxin and how these are influenced by chemical modification of the residues (Chan et al., 1983a; Chan and Markley, 1983). It is clear from studies with bacterial ferredoxins (Lode et al., 1976a,b) that small changes in structure can be responsible for significant alteration in redox potential and biological activity. Clearly, it is a limited number of residues in sequences involved at the active sites that are important in governing these properties, and it is comparisons of tertiary structures that will give the most penetrating insight into structure-function relationships.

1.6. TERTIARY STRUCTURE

The tertiary structures of two cyanobacterial ferredoxins are known; those of *Spirulina platensis* (Fukuyama et al., 1980; Tsukihara et al., 1981) and *Aphanothece sacrum* ferredoxin I (Kunita et al., 1978; Tsutsui et al., 1983), both to 0.25 nm resolution. A representation of the structure of the *Spirulina* ferredoxin is shown in Fig. 7. These studies have confirmed the tetrahedral coordination about each iron atom in the cluster proposed from other studies and also that the [2Fe-2S] core forms a planar 4-atom ring. Distances in nm between atoms in the cluster as estimated for this ferredoxin are: Fe-Fe, 0.272; S-S, 0.294; Fe_1-S_1, 0.211; Fe_1-S_2, 0.209; Fe_2-S_1, 0.193; Fe_2-S_2, 0.189; Fe_1-$S_{\gamma 41}$, 0.305; Fe_1-$S_{\gamma 46}$, 0.220; Fe-$S_{\gamma 49}$, 0.238 and Fe_2-$S_{\gamma 79}$, 0.212. The cluster is near the molecular surface (the iron atoms are within 0.5 nm of the surface) and is mainly surrounded by hydrophobic

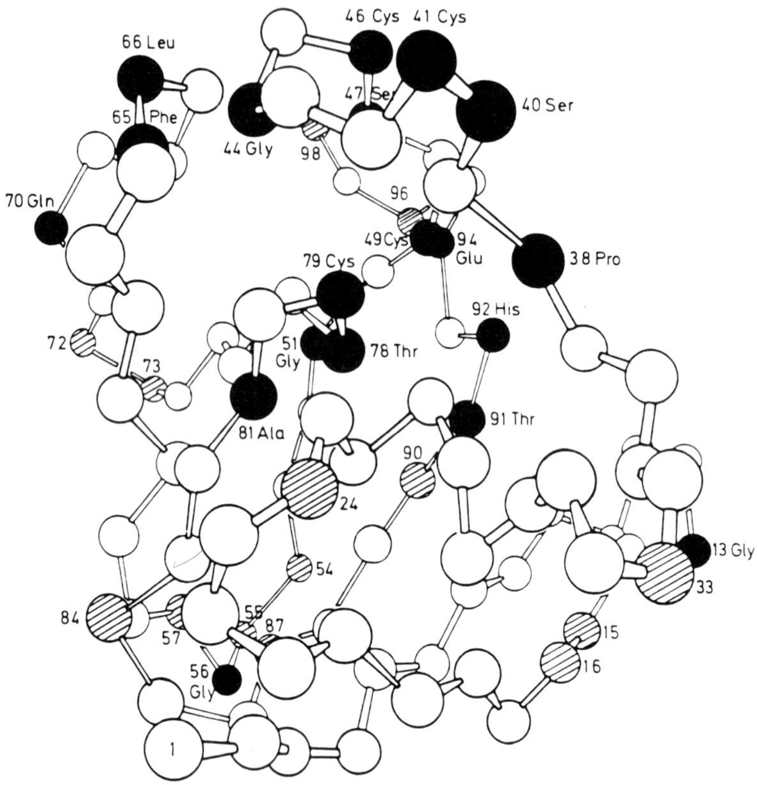

Fig. 7. Representation of the three-dimensional structure of *Spirulina platensis* ferredoxin showing the distribution of invariant (●) and variant (◍) positions in ferredoxins from oxygenic photosynthetic organisms. A variant position is a site occupied by more than six different amino acid residues when the sequences of 24 ferredoxins were compared. Subsequent sequences for a further 15 ferredoxins show one sequence each where Gly-13 and His-92 are replaced by Asn. (From Fukuyama et al., 1980; with permission)

residues, leaving the side having the cysteine side-chains $S_{\gamma41}$ and $S_{\gamma46}$ coordinating Fe_I more exposed.

A possible role of aromatic residues has been suggested for *Peptococcus aerogenes* ferredoxin and *Chromatium* HiPIP but inspection of the structure of the cyanobacterial ferredoxins shows there are no invariant aromatic side-chains in contact with the cluster, suggesting electron transfer may occur via the orbitals of the cysteinyl sulphur atoms. A role for the basic amino acids may be suggested from the observation that their modification in spinach ferredoxin prevented electron transfer between ferredoxin and ferredoxin-$NADP^+$ oxidoreductase but did not prevent complex formation (Sakihama et al., 1983). Amongst some forty sequences of ferredoxins from photosynthetic organisms there are 17 residues which are invariant, including four cysteine (Fig. 7). Most of these surround the cluster cavity, but some residues which appear to be crucial for chain folding elsewhere in the molecule are conserved.

In *Spirulina platensis* ferredoxin about 25% of the residues participate in β-structure, and a further 10% are folded to take the form of a two-turn α-helix. A cylindrical β-structure is at the top of the molecule as shown in the Fig. and segment 41–49, which contains three of the cysteines of the iron-sulphur cluster, form an exposed loop. A region of high homology (residues 65–69) near the cluster may form part of the binding site for ferredoxin-NADP$^+$ oxidoreductase and a region about residue 84 may also play a similar role. *Aphanothece sacrum* ferredoxin has, and by implication other plant-type ferredoxins probably have, the same main chain fold (Tsukihara et al., 1982).

On this basis, a consideration of the sequences of the two *Nostoc* ferredoxins and the ferredoxins from *Chlorogloeopsis fritschii* and *Synechocystis* 6714, proved interesting (Hase et al., 1982a). Compared to these others *Nostoc* ferredoxin II, with its different E_m and biological activities, has fifteen positions where a residue conserved in the other three ferredoxins has been substituted. These are Phe-40, Ser-44, Ser-46, Ser-49, Asn-51, Ile-53, Leu-54, Asn-65, Ala-73, Asn-76, Asn-87, Arg-94, Asp-96, Ile-98 and Ala-99. Whilst distant in the sequence most of these residues are located in the proximity of the [2Fe-2S] cluster in the three-dimensional model. The α-carbons of the first five of these and Asn-65 are within 1.0 nm of the cluster centre. In contrast, when sequences were compared on the same basis for the other ferredoxins the substitutions are: Val-3, Asp-20, Ile-30, Asp-33, Gln-34, Lys-74 and Leu-88 in *Nostoc* ferredoxin I, and Ser-2, Thr-5, Asp-13, Ser-18; Cys-21 and Ser-58 in *Synechocystis* 6714; *Chlorogloeopsis fritschii* ferredoxin had no residues in this category. An inspection of Fig. 7 indicates that these are all residues distant from the active site. Therefore, some at least of the residues unique to *Nostoc* ferredoxin II presumably influence the precise geometry and electronic properties of the cluster and give this ferredoxin its different properties.

With elucidation of the tertiary structures of these ferredoxins the identification of those residues playing a role in electron transfer from the Photosystem I bound iron-sulphur centres to ferredoxin-NADP$^+$ oxidoreductase becomes a realistic goal.

1.7. Ferredoxin-dependent enzymes : ferredoxin-NADP$^+$ oxidoreductase

Ferredoxin-NADP$^+$ oxidoreductase (FNR) (EC 1.6.7.1) is a flavoprotein with one non-covalently bound FAD per molecule and consequently a characteristic absorption spectrum; 273, 389 sh432, 459 sh490 nm for the oxidised flavoprotein (Masaki et al., 1979). The enzyme plays a crucial role as the mediator for electron transfer between ferredoxin (or flavodoxin) and the terminal electron acceptor in photosynthesis. Its evolution must presumably have paralleled those of these electron carriers and it may also be related and show some overlapping activity to other ferredoxin-dependent oxidoreductases in cyanobacteria such as pyruvate-ferredoxin oxidoreductase (Leach and Carr, 1971; Bothe et al., 1971) and ferredoxin-nitrate oxidoreductase (Ortega et al., 1977; Mikami and Ida, 1985). FNR has been purified and characterised to varying extent from five cyanobacteria: *Anabaena* PCC 7119 *(Nostoc muscorum)* (Serrano and Rivas, 1982), *Anabaena cylindrica* (Rowell et al., 1981), *Anabaena variabilis* (Susor and Krogmann, 1966), *Nostoc* MAC (Hutber et al., 1978; Hutber and Rogers, 1981), *Spirulina platensis* (Masaki et al., 1979) and *Synechococcus* sp. (Koike and Katoh, 1980), enzyme from the last only in terms of its thermo-stability.

FNR also possesses NADPH-diaphorase, ferredoxin-dependent cytochrome *c* reductase, and transhydrogenase activities. The physiological reaction represents a difficult assay for routine use and FNR is commonly recognised through its diaphorase activity. However, confidence in the specificity of this assay during purification has sometimes been misplaced since other proteins have diaphorase activity, and the cytochrome *c* reductase reaction is to be preferred.

Despite its stability, FNR has proved difficult to purify, though use of Blue-Sepharose CL-6B (Wada et al., 1983) or 2′,5′-ADP Sepharose 4B (Serrano and Rivas, 1982) has been used advantageously. For the cyanobacterial enzyme, as with that from higher plants, there has been considerable confusion about its molecular properties. For *Nostoc* MAC (Hutber and Rogers, 1981) and *Spirulina platensis* (Masaki et al., 1979) two major isoenzymes were apparent in separations based on ion-exchange chromatography. The two forms of FNR in *Nostoc* MAC did not appear to correspond functionally to the two different ferredoxins present since they showed the same relative activity, with a preference for ferredoxin I, in the cytochrome *c* reductase system (Hutber and Rogers, 1981). In both cyanobacteria the isoenzymes were of closely similar molecular size (M_r about 33 000) with no indication of a higher molecular weight form being also present as in higher plants, the eukaryotic alga *Bumilleriopsis filiformis* (Bookjans and Boger, 1979) or another cyanobacterium *Anabaena cylindrica* (Rowell et al., 1981). For the latter at least three forms of the enzyme were separated on phosphocellulose. In higher plants it has been suggested that the higher molecular weight species (Shin et al., 1981) might be the in situ form of the enzyme, predominantly membrane-bound (Suss, 1979; Zanetti and Arosio, 1980; Forti et al., 1983; Vallejos et al., 1984), and perhaps associated with the cyt *b-f* complex (Clark et al., 1984). The association of two 33 000 FNR molecules to a dimer appears to be stabilised by a specific polypeptide (Shin et al., 1985).

On isoelectric focusing the FNR from *Nostoc* MAC showed multiple ionic forms (pI values 4.6–5.4) the patterns for the two isoenzymes being different. A similar finding for the enzymes from *Spirulina platensis* (Masaki et al., 1979) and *Spirulina* sp. (Wada et al., 1983) was attributed to microheterogeneity at the N-terminus and this has been confirmed subsequently with the elucidation of the complete sequence of the enzyme (Yao et al., 1984). Forms in which up to three of the N-terminal residues were absent could be identified, though whether the differences are due to proteolytic attack during cell disruption or occur in situ remains uncertain. Inclusion of protease inhibitors in extraction media resulted in one of the two major isoenzyme forms increasing at the expense of the other, but both were still present (Wada et al., 1983). In contrast, the presence of protease inhibitors did not affect the multiple ionic forms of the higher plant enzyme (Gozzer et al., 1977) and N-terminal heterogeneity provided only a partial explanation for their occurrence (Hasumi et al., 1983).

It is also interesting that the microheterogeneity is not seen for ferredoxins from *Nostoc* MAC or *Spirulina* sp. though these have the same N-terminal residue as FNR. This might suggest the multiple forms of the latter originate in vivo, maybe because FNR, through its close association with the thylakoid membrane, is particularly susceptible to action of a protease playing a role in turnover of membrane proteins.

The sequence of *Spirulina* sp. FNR is given in Fig. 8; a comparison with the sequence for the spinach enzyme (Karplus et al., 1984) shows that over half the sequence is homologous. However, the higher plant FNR has in addition an extended N-terminus. Absence of this in the cyanobacterial enzyme argues against a role for this region in ferredoxin binding which had been suggested because of its predominantly basic nature (Hasumi et al., 1983). This role may instead be ascribed to the basic sequences at Lys-67–Arg-72 and Lys-284–His-290. The sequence with hydrophobic residues Tyr-172–Tyr-186, may have a function in binding to the thylakoid while that about residue 70 may have a similar role (Yao et al., 1985). By analogy with studies on the spinach enzyme following affinity labelling of a specific lysine with periodate-oxidized $NADP^+$ (Chan et al., 1985), the sequence at residues 222-230 may be part of the nicotinamide nucleotide binding site. The $NADP^+$ binding site is also thought to involve sites for the adenine nucleotide group and the pyrophosphate. X-ray crystallographic studies to 0.37 nm resolution (Sheriff and Herriott, 1981) suggest that the molecule is bi-lobed with the FAD and $NADP^+$ binding sites separate; the latter is associated with the carboxyl half of the sequence (Grossman et al., 1982). Contrary to an earlier report (Karplus et al., 1984) it appears that no disulphide bond is present in spinach FNR, or the cyanobacterial enzyme (Yao et al., 1985). The enzyme shows some sequence homology with erythrocyte glutathione reductase and *Pseudomonas p*-hydroxybenzoate hydroxylase, and this may suggest the derivation of FAD- and $NADP^+$-binding sites through convergent evolution.

1 10 20 30 40
A K T D I P V N I Y K P K N P Y I G K C L S N E E L V R E G G T G T V R H L I F D I S G G D L R Y L

51 60 70 80 90
E G Q S I G I I P P G T D N N G K P H K L R L Y S I A S T R H G D H V D D K T V S L C V R Q L E Y K

101 110 120 130 140
H P E T G E T V Y G V C S T Y L C N L E A G A D V A I T G P V G K E M L L P E D E D A T I I M M A T

151 160 170 180 190
G T G I A P F R A F L W R I F K E Q H E D Y K F K G L A W L F F G I P Y S P N I L Y Q Q E L E E L Q

201 210 220 230 240
E E F P E N F R L T L A I S R E Q Q N P E G G K M Y I Q D R I K E N A D Q L W E L I Q K P N T H T Y

251 260 270 280 290
I C G L K G M E G G I D E G M S A A A G K F D V D W S D Y Q K E L K K K H R W H V E T Y

Fig. 8. Amino acid sequence of *Spirulina* sp. ferredoxin-$NADP^+$ oxidoreductase. Residues underlined show homology with the spinach enzyme. (After Yao et al., 1984)

The mechanism of FNR action is being successfully probed. It forms a one-to-one complex with ferredoxin and with $NADP^+$ (Dykes and Davis, 1982), and the importance of particular amino acids in electron transfer or complex formation is evident, including involvements of lysine and arginine at the nucleotide-binding site (Zanetti et al., 1979; Valle et al., 1982; Bookjans and Böger, 1978), and a role for carboxyl groups in the interaction with ferredoxin (Chan et al., 1983b; Vieira and Davis, 1985).

For the NADPH-diaphorase and ferredoxin-cytochrome *c* reductase activities a ping-pong mechanism involving binary complexes of FNR and ferredoxin or $NADP^+$ has been suggested; i.e. the first product, $NADP^+$ is released from the enzyme complex before the artificial electron-acceptor binds to the fully reduced enzyme (Masaki et al., 1979). However, recent steady-state kinetic and fast reaction studies suggest that the physiological reaction involves formation of a ternary complex with a K_{mFd} of 2.2 μM and a K_{mNADP^+} of 1.7 μM (Masaki et al., 1982), a conclusion supported by work with the enzyme from spinach (Batie and Kamin, 1984). Studies with chemically cross-linked ferredoxin and FNR have also shown that the binding sites for ferredoxin and $NADP^+$ are distinct, allowing formation of a ternary complex (Zanetti et al., 1984).

On the basis of fast-reaction studies, Yoshikawa et al. (1984) have proposed a mechanism for the reduction of $NADP^+$ mediated by FNR (Fig. 9). The reaction scheme is sequential with only one ferredoxin bound at any time; it is therefore not inconsistent with the 1:1 relationship for com-

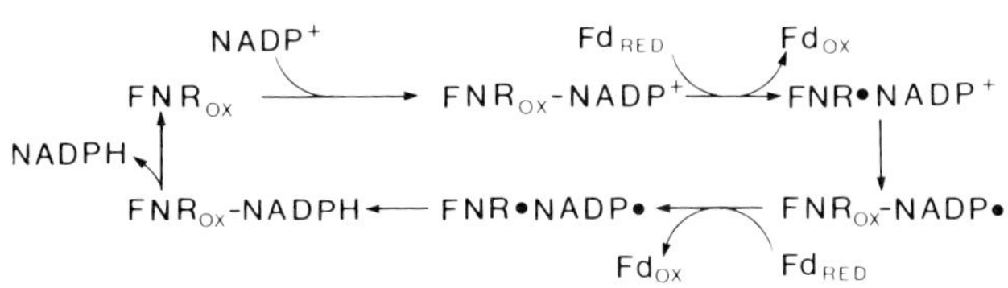

Fig. 9. Proposed reaction mechanism for $NADP^+$ photoreduction by reduced ferredoxin mediated by FNR. The semiquinone forms of the radicals are shown as FNR● and NADP ●. (After Yoshikawa et al., 1984)

plex formation demonstrated by equilibrium studies. Spectral changes accompanying the reaction suggested that at pH 7.4 the oxidized-semiquinone couple of the FAD cofactor was involved. This suggests that the E_m value of this couple bears appropriate relationship to those for the ferredoxin (−390 mV for the closely related *Spirulina maxima* protein) and the $NADP^+$/NADPH couple (−320 mV). While this datum is not available it can be noted that the corresponding E_m for the *Anabaena* 7119 enzyme was −360 mV at pH 8.0 (Serrano et al., 1984).

Comparison with other data is fraught with difficulty not only because of the possibility of differences in the enzymes from different organisms, but also because assays have been at various pH values in the range 6.0 to 10.0 and frequently involved the 'artificial' diaphorase activity of FNR, which utilises NADPH as donor and involves an artificial electron acceptor.

Thus studies with the enzyme from *Anabaena* 7119 (Serrano et al., 1984) showed that the reduction of FNR by NADPH was a two-electron reaction at its optimum near pH 10.0 and generated the hydroquinone. The E_m at this pH was −438 mV. However, at pH values near neutrality, the optimum for two ferredoxin-dependent reactions (NADPH-cyt *c* reductase and oxidation of dithionite-reduced ferredoxin by $NADP^+$), the oxidized-semiquinone couple of FNR (E_m of −256 mV at pH 6.0) was implicated. For the diaphorase activity of *Nostoc muscorum* FNR a two-electron reduction of FNR occurred at high pH values, whereas in the presence of the ferredoxin-FNR complex rather than FNR alone the reaction was enhanced, was optimal at pH values closer to neutrality, but now involved a one-electron reduction with the semiquinone being formed (Melamed-Harel et al., 1985).

In electrochemical titrations at pH 7.3, the FNR from spinach was a two-electron donor shuttling between oxidised and hydroquinone forms without intermediate semiquinone being detected (Smith et al., 1981); however, these workers emphasised the caution necessary in extrapolating data from these experiments to the in vivo situation. The spinach enzyme had an E_m of −350 mV at pH 7.3 when complexed with ferredoxin, compared to −372 mV for free FNR.

As indicated earlier, pH may have a significant effect on reaction mechanism. At physiological pH values the E_m of the oxidized-semiquinone couple may be more appropriate for mediating electron flow from ferredoxin to $NADP^+$, as appears to be the case in vivo for *Chlorella* (Bouges-Bocquet, 1978). However, the E_m of this couple may be pH-dependent, either because of involvement of a proton in the reaction as for FMN in flavodoxins, or because a group(s) in the active site region with a pK of about 8–9 undergo(es) ionisation to produce a negative charge (which would make it more difficult to introduce a reducing electron into the site). If this were so (see e.g. Serrano et al., 1984), the E_m for this couple and the semiquinone-hydroquinone couple could be sufficiently close at higher pH values that both could participate in reactions without intermediate semiquinone being apparent. In these respects, the results for the *Spirulina* sp. FNR provide a substantial base for further investigations of the physiological reaction.

2. Flavodoxins

2.1 Occurrence and regulatory role of iron

In the decade that followed the discovery of ferredoxins a further class of proteins which transferred electrons at low potential was recognised and characterised. These contained flavin, invariably flavin mononucleotide (FMN) as their redox active component, instead of an iron-sulphur cluster. The first flavodoxin* to be isolated was from *Synechococcus* 6301 (*Anacystis nidulans*) (Smillie,

* Alternative names have included azotoflavin, phytoflavin and Shethna flavoprotein.

1965), though the number of organisms from which flavodoxins have been isolated and characterised are limited. Amongst photosynthetic organisms they include an anoxygenic photosynthetic bacterium *Rhodospirillum rubrum* (Cusanovich and Edmondson, 1971; MacKnight et al., 1974), the cyanobacteria *Nostoc* MAC (Hutber et al., 1977, 1981), *Synechococcus* 6301 (Smillie, 1965) and *Synechococcus lividus* (Crespi et al., 1972), and the eukaryotic algae *Chlorella fusca* (Zumft and Spiller, 1971) and *Chondrus crispus* (Fitzgerald et al., 1978). However, the flavodoxins from a few non-photosynthetic bacteria have been more extensively investigated since they represent the simplest models for studying flavoprotein biochemistry; the elegant characterisations of these (summarised in Mayhew and Ludwig, 1975) have set the model for subsequent studies.

In the majority of the non-photosynthetic bacteria which have been studied, the synthesis of significant amounts of flavodoxin occurs only during growth in media low in iron. However, *Escherichia coli* (Vetter and Knappe, 1971) and *Azotobacter vinelandii* (Benemann et al., 1969) produced a flavodoxin in addition to ferredoxin under normal growth conditions, and *Desulfovibrio vulgaris* Miyazaki had a constitutive flavodoxin in place of a ferredoxin (Moura et al., 1980); unlike *Desulfovibrio desulfuricans* Norway where the opposite holds. The photosynthetic organisms, *Rhodospirillum rubrum*, *Synechococcus* 6301, *Nostoc* MAC and *Chlorella fusca* only synthesised flavodoxin under conditions of iron stress. This was probably true also for *Synechococcus lividus* (Crespi et al., 1973), though earlier data were equivocal (Crespi et al., 1972). In the case of the marine red macroalga *Chondrus crispus*, harvested either from the natural environment or grown in culture, the flavodoxin was truly constitutive and there were negligible amounts of ferredoxin (Fitzgerald et al., 1978).

A systemic study of the induction of flavodoxin synthesis during culture in iron-deficient medium has been made using *Nostoc* MAC, a filamentous non-heterocystous cyanobacterium. Here there was the additional interest in the relative balance sustained between the two soluble ferredoxins this organism possesses. Flavodoxin in *Nostoc* MAC became evident when the iron was at, or just below, a critical concentration of 2 μM (Fig. 10) (Hutber et al., 1977). At 1 μM iron the amounts of both ferredoxins had decreased markedly and flavodoxin was present in approaching maximum amount. At even lower iron concentration the level of iron became limiting for cell growth; at this

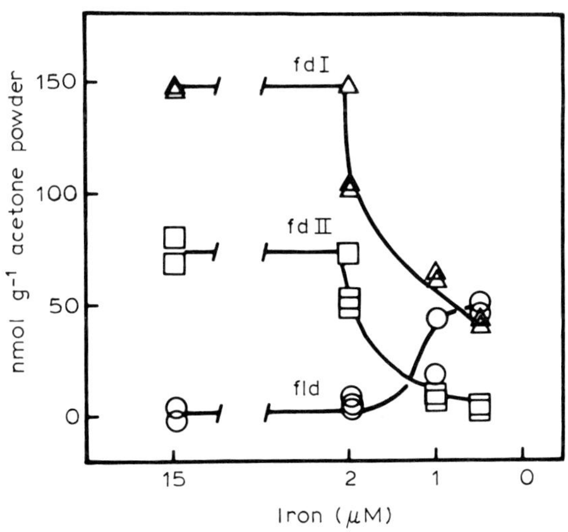

Fig. 10. Variation in flavodoxin and ferredoxins in *Nostoc* MAC with iron level in the medium. Cell material was measured as g of acetone-dried powder from a known volume (32–40 l) of culture. (Based on Hutber et al., 1977)

stage ferredoxin II was not synthesised at all but significant amounts of ferredoxin I were still present. The latter ferredoxin has an E_m of –350 mV at pH 7 which is intermediate between the two E_m values possessed by the flavodoxin. That ferredoxin I has some physiological role which cannot be met by the flavodoxin was an attractive proposal, but any distinction in the in vivo functions of the two ferredoxins is still unclear. In *Synechococcus* 6301 at the lowest iron concentration tested of 2.5 μM the ratio of ferredoxin to flavodoxin was about 3:1 (Bothe, 1969).

Synechocystis 6714 grown in medium containing 3.6 μM iron formed flavodoxin intead of ferredoxin but there was little effect on photosynthetic electron flow and growth. Effects on these occurred when iron was only present at 0.3 μM and now the membrane-bound iron-sulphur centres in Photosystem I were also decreased, though against a background of a general inhibition in formation of all electron transport proteins whether or not they contained iron (Sandmann and Malkin, 1983).

2.2. Chemical properties

From cyanobacteria other than *Synechococcus lividus* the flavodoxins have a molecular weight close to 21 000, similar to the two algal flavodoxins and those from *Rhodospirillum rubrum* and *Azotobacter vinelandii*. A discrepancy in the molecular weight of *Synechococcus* 6301 flavodoxin apparent in reports of 16 000 (Bothe et al., 1971) and 21 000 (Smillie and Entsch, 1971) can be resolved in favour of the higher value (M.P. Fitzgerald and L.J. Rogers,unpublished data). If confirmed, the molecular weight of *Synechococcus lividus* flavodoxin, determined as 17 000 by gel filtration (Crespi et al., 1972), would be an exception. All are a single polypeptide chain, with the possible exception of *Chlorella fusca* flavodoxin (Zumft and Spiller, 1971). Other than for *Azotobacter vinelandii*, flavodoxins from non-photosynthetic bacteria have a molecular weight of 15 000 (Mayhew and Ludwig, 1975). This has made some of this latter group of low molecular weight flavodoxins very attractive as simple models for studying the interactions between flavins and apoproteins.

All flavodoxins have contained a single molecule of non-covalently bound FMN as the redox active component and consequently show a characteristic absorption spectrum (Fig. 11). The flavin is bound tightly to the apoflavodoxin; the association constants for the *Nostoc* MAC and *Synechococcus* 6301 proteins were close to 10^8 M^{-1}, in the same range as those for flavodoxins from other sources. Nevertheless flavin can be dissociated by a number of methods including treatment with ammonium sulphate at acid pH or dialysis against 2M KBr (Husain and Massey, 1978), which enable subsequent recombination of FMN with the apoprotein to be studied. For the cyanobacterial flavodoxins studied (Fitzgerald et al., 1980b) the apoprotein possessed the same molecular characteristics as the protein component of the holoprotein. This was not, however, true for the flavoprotein from *Chondrus crispus* where extensive conformational changes in the apoflavodoxin accompanied dissociation of FMN.

The low isoelectric point for flavodoxins, at about pH 3.5, is slightly higher than that for ferredoxins but likewise signifies a preponderance of acidic residues in their amino acid composition. Consequently, flavodoxins can be readily isolated by use of ion-exchange chromatography.

Application of physiochemical techniques, including ESR and NMR studies on *Synechococcus lividus* flavodoxin (Crespi et al., 1972, 1973), has yielded much information on the chemistry of the flavin-apoprotein interactions in flavodoxins (see Mayhew and Ludwig, 1975).

2.3. Oxidation-reduction properties

The progressive reduction of the oxidized flavodoxin through the semiquinone to the hydroquinone form can be followed from changes in the visible absorption spectrum (see e.g. Fig. 11). Compared to the peaks at 266, 373 and 445 nm in free FMN the corresponding absorption peaks in oxidized flavodoxin are at longer wavelengths and molar absorption coefficients are lower. The peaks in *Nostoc* MAC flavodoxin are at 274 sh283 sh293, 376 sh432, 466 sh488 nm with the molar extinction coefficient at 466 nm being 9500 $mol \cdot l^{-1}cm^{-1}$; the ratio A_{466}/A_{274} is 0.2 (Hutber et al., 1981). The interpretation that the flavin in the holoprotein is in a non-polar environment has been confirmed by structural investigations on some flavodoxins.

Addition of one electron gives a blue semiquinone, the neutral radical form (Entsch and Smillie, 1972). The spectrum is well defined with peaks at 350 sh380, and 576 sh606 nm for *Nostoc* MAC flavodoxin semiquinone. A further one-electron addition yields the hydroquinone. This has a less well-defined spectrum, in *Clostridium* sp. and *Megasphaera elsdenii* (*Peptostreptococcus elsdenii*) flavodoxins having peaks at 315 and 365 nm with a broad shoulder at 450 nm (Mayhew, 1971) but with

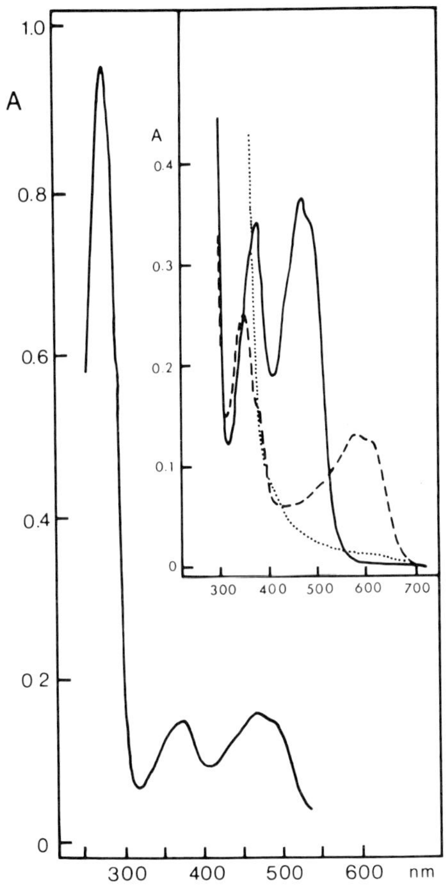

Fig. 11. Absorption spectrum of *Nostoc* MAC flavodoxin. UV and visible spectrum of *Nostoc* strain MAC flavodoxin at approximately 12 μM. The inset shows the visible absorption spectra for the oxidized flavodoxin at approx. 40 μM (full line); semiquinone (dashed line) and hydroquinone (dotted line). The semiquinone was obtained by anaerobic addition of a molar equivalent of dithionite in the presence of 3 μM methyl viologen as mediator, while formation of the hydroquinone required addition of a 50-fold molar excess of the reductant. This obscures absorption maxima for the hydroquinone at low wavelengths. (From Hutber et al., 1981)

the latter feature absent in other flavodoxins, including all from photosynthetic organisms. This suggests that the environment of the flavin chromophore, the latter now in the anionic form, is somewhat different in the two groups of flavodoxins. This division, supported by differences in circular dichroism (CD) spectra, fluorescence properties and flavin binding constants, does not coincide with the division based on molecular weights.

Redox potentials for cyanobacterial and algal flavodoxins have been determined for both the oxidized-semiquinone (E_2) and semiquinone-hydroquinone (E_1) couples. Potentiometric titrations show this sequential reduction clearly (Fig. 12) and this approach has been used to determine the values of E_2 and E_1 for the flavodoxins from *Nostoc* MAC, *Synechococcus* 6301 and the red alga *Chondrus crispus* (Sykes and Rogers, 1984). The values of E_2 and E_1 for the two cyanobacterial flavodoxins were very similar (Table V); those for the algal flavodoxin were significantly different, but this flavoprotein is unusual in a number of other respects also (Fitzgerald et al., 1980a,b). The dominating influence of the apoprotein in influencing the redox potential of the flavin, by stabilising the semiquinone, is shown by the comparison with free FMN where E_2 (−240 mV) is lower than for E_1 (−170 mV). For *Synechococcus* 6301 flavodoxin the value for E_2 agreed with that reported earlier from equilibration with mediator dye (Entsch and Smillie, 1972), and was some 40 mV more negative than the value quoted by VanLin and Bothe (1972). The value for E_1 from potentiometric titration though somewhat more positive was also in better agreement with a value from equilibration with H_2 in the presence of hydrogenase obtained by the former group.

Compared with these data, the flavodoxin from *Synechococcus lividus* appears to be unusual in that E_2 was apparently considerably more negative (Crespi et al., 1972); the molecular weight of 17 000 reported for the flavoprotein would also distinguish it from other cyanobacterial (and algal) flavodoxins.

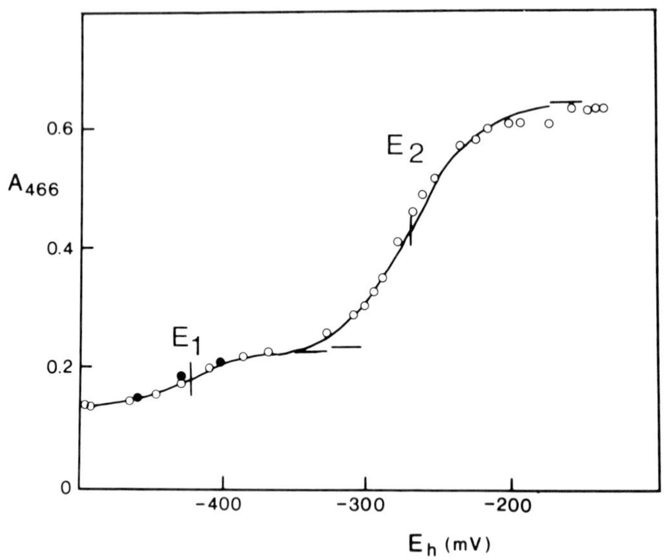

Fig. 12. Midpoint redox potentials of *Nostoc* MAC flavodoxin. Experimental points are of a reductive titration at pH 8.2 with dithionite (○) and a successive oxidative titration with dichlorophenol-indophenol (●), in the presence of mediators. The points are superimposed on theoretically-derived plots for successive one-electron transfers. E_2 and E_1 are the redox potentials of the oxidized-semiquinone and semiquinone-reduced interconversions, respectively. The protein was approximately 60 μM. (From Sykes and Rogers, 1984)

TABLE V

Redox potentials of cyanobacterial flavodoxins. (E_1 and E_2 are the redox potentials for the semiquinone-hydroquinone and oxidized-semiquinone couples; data for E_2 at higher pH have been corrected to pH 7)

	E_2 (mV)	E_1 (mV)	Reference
Nostoc strain MAC	–210	–414	Sykes and Rogers, 1984
Synechococcus PCC 6301	–215	–414	Sykes and Rogers, 1984
	–221	–447	Entsch and Smillie, 1972
	–180	–470	VanLin and Bothe, 1972
Synechococcus lividus	– 50	–450	Crespi et al., 1973

For flavodoxins the value of E_2 changes by –59 mV per pH unit increase which is in accord (Mayhew and Ludwig, 1975) with the reaction sequence:

$$Fld_{ox}H \xrightleftharpoons{e^- + H^+} Fld \cdot H_2 \xrightleftharpoons{e^-} Fld_{red}H_2^-$$

2.4. Biological activity

It is generally accepted that flavodoxins can substitute for ferredoxins in the whole range of reactions in which the latter participate (Mayhew and Ludwig, 1975; Moura et al., 1980; Tollin and Edmondson, 1980). It is also accepted that only the midpoint of the semiquinone-hydroquinone couple is sufficiently negative to enable this substitution (Mayhew et al., 1969). However, it should be noted that for *Chondrus crispus* flavodoxin values of E_1 and E_2 at pH 8, the pH of the chloroplast stroma under conditions of CO_2 assimilation, were only 75 mV apart. For other flavodoxins also, a physiological role for the higher potential of the oxidized-semiquinone couple cannot necessarily be dismissed.

For *Azotobacter vinelandii* the modified flavodoxin from a mutant had E_1 (–435 mV) and E_2 (–190 mV) both more positive by 30 and 45 mV, respectively, than the native flavoprotein. The altered flavodoxin was less effective in supporting nitrogenase activity (Hofsteller and DerVartanian, 1985). Curiously, the flavodoxin from the mutant contained a modified flavin whilst the structure of the apoprotein might have been unchanged. In nitrogen-fixing microorganisms flavodoxin is encoded in the nif region, and its synthesis may be subject to controls governing nitrogenase synthesis (Roberts and Brill, 1981).

For cyanobacteria a crucial role for flavodoxin is to act as an electron transfer protein on the reducing side of Photosystem I, a function well suited to the redox potential of the semiquinone-hydroquinone couple. *Nostoc* MAC flavodoxin was more active than that of ferredoxin I and about as active as ferredoxin II in catalysing $NADP^+$ photoreduction by higher plant chloroplasts or *Nostoc* membranes (Hutber et al., 1978, 1981), or in supporting H_2 evolution in a chloroplast-hydrogenase system. Activity in ferredoxin-cytochrome *c* reduction with ferredoxin-$NADP^+$ reductase was the same as ferredoxin II and substantially less than ferredoxin I (Hutber and Rogers, 1981). However, the flavodoxin was appreciably less active than either ferredoxin in a dithionite-hydrogenase system or in supporting the phosphoroclastic activity of *Clostridium* cell extracts. *Synechococcus* 6301 and *Nostoc* MAC flavodoxins showed very similar activities in the two hydrogenase assays (Fitzgerald et al., 1980a).

Synechococcus 6301 flavodoxin proved effective in substituting for the native ferredoxin in photosynthetic assays, including $NADP^+$ photoreduction, in nitrite reduction, and the phophoroclastic system (Bothe, 1969). Again, the flavoprotein from *Synechococcus lividus* was distinctive and despite a suitable redox potential (E_1) it showed low activity in supporting $NADP^+$ photoreduction (Crespi et al., 1973). Characterisation of the interaction of these flavodoxins with ferredoxin-$NADP^+$ oxidoreductase, such as have been carried out with ferredoxins, are lacking. It would also be of interest to study the interaction of flavodoxin with thioredoxin-ferredoxin oxidoreductase (Schmidt and Christen, 1980; Yee et al., 1981; Ip et al., 1984) in controlling metabolism through modulating the activity of pivotal enzymes.

Finally, roles for flavodoxin other than in electron transport may occur. *Azotobacter vinelandii* flavodoxin is a potent allosteric inhibitor of AMP-nucleosidase in vitro, though a physiological role for this interaction is uncertain (Yoshino et al., 1976).

2.5. Amino acid sequence

The amino acid compositions of cyanobacterial flavodoxins, like those from other sources (summarised in Fox et al., 1972), contain a preponderance of acidic residues since as judged by isoelectric point only a minority of the glutamic acid and aspartic acid in hydrolysates could be in the amide form in the apoprotein. The composition of *Synechococcus* 6301 flavodoxin was Asx_{28}, Glx_{26}, Thr_8, Ser_{13}, Pro_4, Gly_{18}, Ala_{13}, Cys_1, Val_{11}, Met_1, Ile_{13}, Leu_{13}, Tyr_8, Phe_8, Trp_5, His_0, Lys_8, Arg_2 (Fox et al., 1972) and can be regarded as typical. The flavodoxins from *Synechococcus lividus* (Crespi et al., 1972) and *Nostoc* MAC (Hutber et al., 1981) were similar in also possessing appreciable amounts of small non-polar residues, glycine and alanine. Cystine has not been found in flavodoxins and histidine is generally absent or only a single residue; however, *Nostoc* MAC flavodoxin may possess two histidines.

Although the sequences of five of the lower molecular weight flavodoxins from non-photosynthetic bacteria have been reported (summarised in Dubourdieu and Fox, 1977), comparable data for only *Azotobacter vinelandii* flavodoxin in the larger molecular weight group are available (Tanaka et al., 1975a). However, N-terminal amino acid sequences for two cyanobacterial and an algal flavodoxin can be compared (Fig. 13), and the sequencing of one of these, *Synechococcus* 6301, is nearing completion (Ludwig et al., 1984). The conservation of residues in the N-terminal sequence seen in other flavodoxins holds for these flavodoxins also; in particular Gly-13, Thr-15 and Ala-19 are invariant.

A matrix of amino acid differences in the N-terminal sequence for all flavodoxins for which such data are known is shown in Table VI. The differences are quite extensive despite the fact that this portion of the peptide chain probably possesses the highest homology (Fox, 1976). This comparison suggests that the cyanobacterial flavodoxins and that from *Azotobacter vinelandii* are related and distinct from the two other flavodoxins, from *Chondrus crispus* and *Rhodospirillum rubrum* , which are

	1 10 20 30
Nostoc MAC	S K K I G L F Y G T E/Q T G K T E S V A E I I - D E F G D E V V T L D I D
Synechococcus 6301	- A K I G L F Y G T Q T G V T Q T I A E S I Q Q E F G G E S I V D L N D
Chondrus crispus	- - K I G I F F S T S T G N T (T)E V A D F I G K T L G A K A S A P I (S)V

Fig. 13. Comparison of the amino acid sequences of cyanobacterial and algal flavodoxins. Comparison is based on the N-terminal sequences aligned for maximum homology. Sequences are from Smith et al. (1983) for *Synechococcus* 6301 and Takruri et al. (1986) for *Nostoc* strain MAC and *Chondrus crispus*.

TABLE VI

Amino acid difference matrix of flavodoxins

		1	2	3	4	5	6	7	8	9
(1)	*Nostoc* strain MAC		0.72	1.00	1.03	0.89	1.08	1.06	1.00	1.08
(2)	*Synechococcus* 6301	15		0.97	1.25	0.78	1.14	1.22	1.08	1.19
(3)	*Chondrus crispus*	24	25		1.11	0.92	1.00	1.08	1.17	1.00
(4)	*Rhodospirillum rubrum*	26	29	29		1.22	1.11	0.83	1.19	1.03
(5)	*Azotobacter vinelandii*	19	20	24	28		1.11	1.22	1.14	1.22
(6)	*Clostridium* MP	27	28	27	27	27		0.61	0.83	0.72
(7)	*Clostridium pasteurianum*	28	29	28	23	27	14		0.92	0.56
(8)	*Desulfovibrio vulgaris*	24	26	26	26	27	19	20		0.89
(9)	*Megasphaera elsdenii*	29	30	26	26	30	18	16	20	

The matrix is based on the N-terminal sequences (first 36 positions) given in Fig. 13 and on sequences summarised in Fox (1976). Figures above the diagonal indicate the minimum base changes per codon between pairs of sequences, with gaps counted as 2. Differences in residues, with gaps counted as 1, are shown below the diagonal. Sequences were aligned for maximum homology.

also in the larger molecular weight group. The smaller flavodoxins form a separate group. These relationships are supported by the consideration of minimum base changes for differing amino acids at each position in the sequences.

2.6. Tertiary structure and evolution

Considerably more information on the structure of the flavodoxin from *Synechococcus* 6301 is now emerging following the recent elucidation of the structure of the oxidized flavodoxin, finally to 0.165 nm resolution (Smith et al., 1983; Ludwig et al., 1984), and the semiquinone to 0.3 nm resolution.

Despite uncertainties in detailed interpretation because only two-thirds of the sequence is known, it is clear that the polypeptide backbone follows the β-α-β conformation seen in other flavodoxins, and which characterises the nucleotide-binding domains of the pyridine nucleotide dehydrogenases. Most of the extra residues compared to the smaller flavodoxin from *Clostridium* MP have been added in the form of a small lobe (Fig. 14) to the underlying structural base of the five parallel strands in the central sheet bounded by two helical regions. Residues 9-14, 55-61 and 88-99 comprise the FMN-binding site and a number of the interactions involved have been elucidated. For the first of these segments, the sequence has been given in Fig. 13. In *Clostridium* MP flavodoxin Ser-10, Thr-12 and Thr-15 (also Ser-58) hydrogen-bond to the FMN phosphate and Asn-14 (Thr-59, Asp-129) to the 4′hydroxyl of the ribityl side-chain. As a consequence of these interactions the flavin isoalloxazine ring in *Clostridium* MP (and *Desulfovibrio vulgaris*) flavodoxin is planar and lies at the periphery of the holoenzyme between Trp-58 and Tyr-98 with the dimethyl benzene portion accessible. The pyrimidine portion is enclosed by the polypeptide chain and the ribityl side-chain extends towards the interior (Tollin and Edmondson, 1980). Work with flavin analogues suggests that the ribityl phosphate is essential for semiquinone stabilization. In the cyanobacterial and algal flavodoxins there is a Thr at position 10 and Asn-14 is replaced in the former by Val or Lys. Differences in binding of the flavin due to these amino acid replacements in the sequences will give subtle differences in flavin environments and could explain the range of redox potentials seen in the flavodoxins from different sources.

In the cyanobacterial flavodoxin, protein-isoalloxazine orientation is altered and the FMN-apoprotein interaction appears to be weaker than in the *Clostridium* flavoprotein, and this might have some bearing on the significantly more negative E_m for the oxidized-semiquinone couple for this flavodoxin, this interconversion being less favourable. There is no insight as yet into the structural changes involved in the semiquinone-hydroquinone interconversion, but studies of the *Clostridium* flavodoxin suggests these may be slight.

However, these partial sequences would be an insecure basis on which to deduce phylogenetic relationships and evolutionary progressions. Analysis of sequences of nucleotide binding proteins have led to the speculation that flavodoxins are ancient proteins originating 3.2 to 4.5 billion years ago, and may be ancestors of the pyridine nucleotide dehydrogenases (Rossmann et al., 1974). They may even have preceded the ferredoxins in evolution (Fox, 1976), a view based on the argument that flavins could have been formed in the primordial period by polymerisation of amino acids while what little iron was present would be largely as the silicate, and hence inaccessible. The ferredoxins may have evolved subsequently as oxygen became available as an evolutionary response to the oxygen sensitivity of reduced flavin. If this view was accepted, the high homology between ferredoxin sequences would be a consequence of their more recent origin and less evolution of sequence. The absense of sequence homology between flavodoxins and ferredoxins from the same organisms, lack of conformational homologies, and different cofactor requirements argue that evolution of these proteins for the same function have been convergent not divergent (Dubourdieu and Fox, 1977).

Like the ferredoxins, repetitious gene duplications may have occurred in the early stage of flavodoxin evolution (Kobayashi and Fox, 1978), and a repeating pentapeptide sequence in (only) *Clos-*

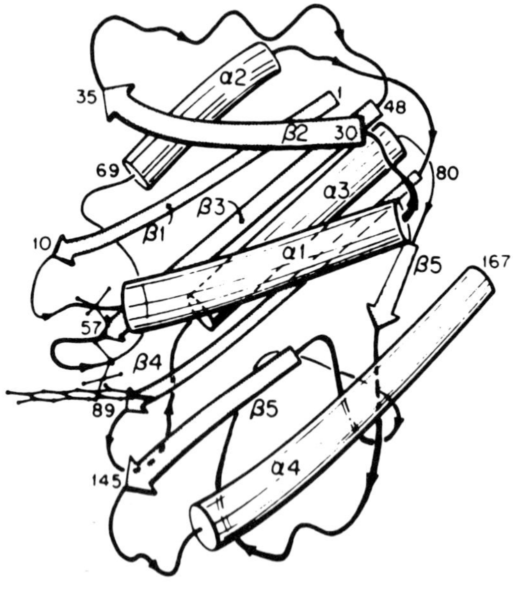

Fig. 14. The polypeptide backbone of the flavodoxin from *Synechococcus* 6301 with bound FMN emphasising the major secondary structures as viewed down the *c*-axis, approximately perpendicular to the parallel sheet. The lobe accounting for the majority of extra residues compared to smaller flavodoxins is at the lower right. (From Smith et al., 1983; with permission)

tridium pasteurianum flavodoxin suggests less evolutionary change for this organism than the others listed in Table VI for which sequences are available.

For the smaller flavodoxins a sequence homology of about 25 to 35% is present (Dubourdieu and Fox, 1977), yet studies of 3d-structure show they are structurally very similar, suggesting that their origin by convergent evolution through the constraints imposed by a flavin binding site remains a possibility. This speculation apart, the idea of evolution selection processes occurring at the topographical rather than the sequence level should prove another incentive to continuing studies of these intriguing proteins, the simplest of the flavoproteins.

3. Acknowledgements

The help of Mrs. P. Jones in preparation of the typescript and of Miss M.M. Williams in preparation of the Figures is gratefully acknowledged.

References

Adzamli, I.K., Petrou, A., Sykes, A.G., Rao, K.K. and Hall, D.O. (1983) Kinetic studies on reactions of iron-sulphur proteins: oxidation of the reduced form of *Spirulina platensis* two-iron two-sulphur ferredoxin with inorganic complexes. Biochem. J. 211, 219–226.

Anderson, R.E., Dunham, W.R., Sands, R.H., Bearden, A.J. and Crespi, H.L. (1975) On the nature of the iron-sulphur cluster in a deuterated algal ferredoxin. Biochim. Biophys. Acta 408, 306–318.

Andrew, P.W., Delaney, M.E., Rogers, L.J. and Smith, A.J. (1975) Isolation and properties of ferredoxin from *Anabaena flos-aquae*. Phytochemistry 14, 931–935.

Batie, C.J. and Kamin, H. (1984) Electron transfer by ferredoxin-NADP$^+$ reductase – rapid reaction evidence for participation of a ternary complex. J. Biol. Chem. 259, 11976–11985.

Beinert, H. and Thomson, A.J. (1983) Three iron clusters in iron-sulfur proteins. Arch. Biochem. Biophys. 222, 333–361.

Benemann, J.R., Yoch, D.C., Valentine, R.C. and Arnon, D.I. (1969) The electron transport system in nitrogen fixation by *Azotobacter*: azotoflavin as an electron carrier. Proc. Natl. Acad. Sci. USA 64, 1079–1086.

Bohme, H. (1977) On the role of ferredoxin and ferredoxin-NADP$^+$ reductase in cyclic electron transport of spinach chloroplasts. Eur. J. Biochem. 72, 283–289.

Bonomi, F., Pagani, S. and Cerletti, P. (1977) Insertion of sulphide into ferredoxins catalysed by rhodanese. FEBS Lett. 84, 149–152.

Bookjans, G. and Böger, P. (1978) Complex-forming properties of butanedione-modified ferredoxin-NADP$^+$ reductase with NADP$^+$ and ferredoxin. Arch. Biochem. Biophys. 194, 387–393.

Bookjans, G. and Böger, P. (1979) Algal ferredoxin-NADP$^+$ reductase with different molecular weight forms. Z. Naturforsch. 34C, 637–640.

Bothe, H. (1969) The role of phytoflavin in photosynthetic reactions. Prog. Photosynth. Res. Proc. Int. Congr. 1st. 3, 1483–1491.

Bothe, H. (1982) Nitrogen fixation. In N.G. Carr and B.A. Whitton (eds.), The Biology of Cyanobacteria. Blackwell, London, pp. 87–104.

Bothe, H., Hemmerich, P. and Sund, H. (1971) Some properties of phytoflavin isolated from the blue-green alga *Anacystis nidulans*. In D. Kamin (ed.), Flavins and Flavoproteins. Proc. 3rd Int. Symp., Butterworths, London, pp. 211–226.

Bouges-Bocquet, B. (1978) Reactions leading to the photoinduced reaction of ferredoxin-NADP$^+$ reductase in *Chlorella* cells. FEBS Lett. 94, 95–99.

Bowyer, J.W. and Skerman, V.B.D. (1968) Production of axenic cultures of soil-borne and endophytic blue-green algae. J. Gen. Microbiol. 54, 299–306.

Burris, R.H. (1981) Two hydrogenases in *Anabaena*. J. Bacteriol. 146, 209–214.

Cammack, R., Rao, K.K., Bargeron, C.P., Hutson, K.G., Andrew, P.W. and Rogers, L.J. (1977) Midpoint redox potentials of plant and algal ferredoxins. Biochem. J. 168, 205–209.

Cammack, R., Rao, K.K. and Hall, D.O. (1981) Metalloproteins in the evolution of photosynthesis. Bio-Systems 14, 57–80.
Chan, T-M., and Markley, J.L. (1983) NMR studies of two-iron two-sulphur ferredoxins: properties of the histidine residues. Biochemistry 22, 5982–5987.
Chan, T-M., Hermodson, M.A., Ulrich, E.L. and Markley, J.L. (1983a) NMR studies of 2Fe-2S ferredoxins II Determination of the sequence of *Anabaena variabilis* ferredoxin II, assignment of aromatic resonances in proton spectra and effects of chemical modifications. Biochemistry 22, 5988–5995.
Chan, T-M., Ulrich, E.L. and Markley, J.L. (1983b) NMR studies of two-iron two-sulfur ferredoxins : interaction with redox partners. Biochemistry 22, 6002–6007.
Chan, R.L., Carrillo, N. and Vallejos, R.H. (1985) Isolation and sequencing of an active site peptide from spinach ferredoxin-NADP$^+$ oxidoreductase after affinity labelling with periodate-oxidized NADP$^+$. Arch. Biochem. Biophys. 240, 172–177.
Clark, R.D., Hawkesford, M.J., Coughlan, S.J., Bennett, J. and Hind, G. (1984) Association of ferredoxin-NADP$^+$ reductase with the chloroplast cytochrome *b-f* complex. FEBS Lett. 174, 137–142.
Codd, G.A., Cook, C.M. and Stewart, W.D.P. (1979) Purification and subunit structure of D-ribulose 1,5-bisphosphate carboxylase from the cyanobacterium *Aphanothece halophytica*. FEMS Microbiol. Lett. 6, 81–86.
Cohn, C.L., Alam, J. and Krogmann, D.W. (1985) Multiple ferredoxins from cyanobacteria. Physiol. Veg. 23, 659–667.
Craig, I. (1971) Chloroplasts: little green slaves. New Sci. & Sci. J. 51, 313–314.
Crespi, H.L., Smith, U., Gajda, L., Tisue, T. and Ammeraal, R.M. (1972) Extraction and purification of ^{1}H, ^{2}H, and isotope hybrid algal cytochrome, ferredoxin and flavodoxin. Biochim. Biophys. Acta 256, 611–618.
Crespi, H.L., Norris, J.R., Bays, J.P. and Katz, J.J. (1973) ESR and NMR studies with deuterated flavodoxin. Ann. NY Acad. Sci. 222, 800–815.
Cusanovich, M.A. and Edmondson, D.E. (1971) The isolation and characterization of *Rhodospirillum rubrum* flavodoxin. Biochem. Biophys. Res. Commun. 45, 327–335.
Daly, M.G. and Bishop, R.H. (1984) Ferredoxin from the cyanobacterium *Oscillatoria agardhii*. Phytochemistry, 23, 2669–2670.
Dubourdieu, M. and Fox, J.L. (1977) Amino acid sequence of *Desulfovibrio vulgaris* flavodoxin. J. Biol. Chem. 252, 1453–1463.
Dutton, J.E. and Rogers, L.J. (1978) Iso-electric focusing of ferredoxins, flavodoxins and a rubredoxin. Biochim. Biophys. Acta 537, 501–506.
Dykes, J. and Davis, D.J. (1982) Re-examination of the interaction between ferredoxin-NADP$^+$ reductase and NADP$^+$. Arch. Biochem. Biophys. 218, 46–50.
Entsch, B. and Smillie, R.M. (1972) Oxidation-reduction properties of phytoflavin, a flavoprotein from blue-geen algae. Arch. Biochem. Biophys. 151, 378–386.
Evans, M.C.W. (1982) Iron-sulphur centres in photosynthetic electron transport. In: T.G. Spiro (ed.), Iron-sulphur Proteins, Vol. 4. Wiley Interscience, New York, pp. 250–284.
Fitch, W.M. and Margoliash, E. (1967) Construction of phylogenetic trees. Science 155, 279–284.
Fitzgerald, M.P., Husain, A. and Rogers, L.J. (1978) A constitutive flavodoxin from a eukaryotic alga. Biochem. Biophys. Res. Commun. 81, 630–635.
Fitzgerald, M.P., Rogers, L.J., Rao, K.K. and Hall, D.O. (1980a) Efficiency of ferredoxins and flavodoxins as mediators in systems for hydrogen evolution. Biochem. J. 192, 665–672.
Fitzgerald, M.P., Sykes, G.A. and Rogers, L.J. (1980b) Apoflavodoxin aggregation following dissociation of flavin. Biochim. Biophys. Acta 625, 127–132.
Forti, G. and Grubas, P.M.G. (1985) Two sites of action of ferredoxin with thylakoids. FEBS Lett. 186, 149–152.
Forti, G., Cappelletti, A., Nobile, R.L., Garlaschi, F.M., Gerola, P.D. and Jennings, R.C. (1983) Interaction of ferredoxin and ferredoxin-NADP$^+$ reductase with thylakoids. Arch. Biochem. Biophys. 221, 507–513.
Fox, J.L. (1976) Evolution of the flavodoxins. In: T.P. Singer (ed.), Flavins and Flavoproteins, Elsevier, Amsterdam, pp. 439–446.
Fox, J.L., Smith, S.S. and Brown, J.R. (1972) Amino acid sequences of *Clostridium pasteurianum* flavodoxin. Z. Naturforsch. 27b, 1096–1100.
Fukuyama, K., Hase, T., Matsumoto, S., Tsukihara, T., Katsube, Y., Tanaka, N., Kakudo, M., Wada, K. and Matsubara, H. (1980) Nature 286, 522–524.
Gayda, J-P., Gibson, J.F., Cammack, R., Hall, D.O. and Mullinger, R. (1976) Spin lattice relaxation and exchange interaction in a 2-iron, 2-sulphur protein. Biochim. Biophys. Acta 434, 153–164.

Gibson, J.F., Hall, D.O., Thornley, J.H.M. and Whatley, F.R. (1966) The iron-complex in spinach ferredoxin. Proc. Natl. Acad. Sci. USA 56, 987–990.

Gozzer, C., Zanetti, G., Galliano, M., Sacchi, G.A., Minchiotti, L. and Curti, B. (1977) Molecular heterogeneity of ferredoxin-$NADP^+$ reductase from spinach leaves. Biochim. Biophys. Acta 485, 278–290.

Grossman, A.R., Bartlett, S.G., Schmidt, G.W., Mullet, J.E. and Chua, N-H. (1982) Optimal conditions for post-translational uptake of proteins by isolated chloroplasts. J. Biol. Chem. 257, 1558–1563.

Hall, D.O., Cammack, R. and Rao, K.K. (1973) The plant ferredoxins and their relationship to the evolution of ferredoxins from primitive life. Pure Appl. Chem. 34, 553–577.

Hall, D.O., Cammack, R. and Rao, K.K. (1974) Non-haem iron proteins. In A. Jacobs and M. Worwood (eds.), Iron in Biochemistry and Medicine. Academic Press, London, pp. 279–334.

Hase, T., Wada, K. and Matsubara, H. (1975) A minor component of ferredoxin from *Aphanothece sacrum* cells. J. Biochem. (Tokyo) 78, 605–610.

Hase, T., Wada, K. and Matsubara, H. (1976a) Amino acid sequence of the major component of *Aphanothece sacrum* ferredoxin. J. Biochem. (Tokyo) 79, 329–343.

Hase, T., Wada, K., Ohmiya, M. and Matsubara, H. (1976b) Amino acid sequence of the major component of *Nostoc muscorum* ferredoxin. J. Biochem. (Tokyo) 80, 993–999.

Hase, T., Wakabayashi, S., Matsubara, H., Rao, K.K., Hall, D.O., Widmer, H., Gysi, J. and Zuber, H. (1978a) The amino acid sequence of ferredoxin from the alga *Mastigocladus laminosus*. Phytochemistry 17, 1863–1867.

Hase, T., Wakabayashi, S., Wada, K. and Matsubara, H. (1978b) Amino acid sequence of *Aphanothece sacrum* ferredoxin II (minor component). J. Biochem. (Tokyo) 83, 761–770.

Hase, T., Wakabayashi, S., Wada, K., Matsubara, H., Juttner, F., Rao, K.K., Fry, I. and Hall, D.O. (1978c) *Cyanidium caldarium* ferredoxin: a red algal type? FEBS Lett. 96, 41–44.

Hase, T., Inoue, K., Matsubara, H., Williams, M.M. and Rogers, L.J. (1982a) Amino acid sequence of *Synechocystis* 6714 ferredoxin: a unique structural feature of unicellular blue-green algal ferredoxins. J. Biochem. (Tokyo) 92, 1357–1362.

Hase, T., Matsubara, H., Hutber, G.N. and Rogers, L.J. (1982b) Amino acid sequences of *Nostoc* strain MAC ferredoxins I and II. J. Biochem. (Tokyo) 92, 1347–1355.

Hase, T., Inoue, K., Hagihara, N., Matsubara, H., Williams, M.M. and Rogers, L.J. (1983a) Ferredoxin from *Aphanothece halophitica*, a unicellular blue-green alga: close relationship to ferredoxins from filamentous blue-green algae and phylogenetic implications. J. Biochem. (Tokyo) 94, 1457–1464.

Hase, T., Matsubara, H., Koibe, H. and Katoh, S. (1983b) Amino acid sequence of ferredoxin from the thermophilic blue-green alga *Synechococcus* sp.: salt bridges responsible for thermostability. Biochim. Biophys. Acta 744, 46–52.

Hasumi, H., Nagata, E. and Nakamura, S. (1983) Molecular heterogeneity of ferredoxin-$NADP^+$ reductase from spinach leaves. Biochem. Biophys. Res. Commun. 110, 280–286.

Hiyama, T., Murakami, A., Itoh, S., Fujita, Y. and Sakurai, H. (1985) A low potential iron-sulphur cluster bound to membranes in blue-green algae. J. Biochem. (Tokyo) 97, 89–95.

Ho, K.K. and Krogmann, D.W. (1982) Photosynthesis. In N.G. Carr and B.A. Whitton (eds.), The Biology of Cyanobacteria. Blackwell, London, pp. 191–214.

Hofsteller, W. and der Vartanian, D.V. (1985) A modified flavodoxin with altered E_m is less efficient in electron transfer to nitrogenase. Biochem. Biophys. Res. Commun. 128, 643–649.

Hosler, J.P. and Yocum, C.F. (1985) Evidence for two cyclic photophosphorylation reactions concurrent with ferredoxin-catalysed non-cyclic electron transport. Biochim. Biophys. Acta 808, 21–31.

Huisman, J.G., Stapel, S. and Muijsers, A.O. (1978a) Two different plant-type ferredoxins in each of two *Petunia* species. FEBS Lett. 85, 198–202.

Huisman, J.G., Moorman, A.F.M. and Verkley, F.N. (1978b) In vitro synthesis of chloroplast ferredoxin as a high molecular weight precursor in a cell-free protein synthesizing system from wheat germ. Biochem. Biophys. Res. Commun. 82, 1121–1131.

Huisman, J.G., Touw, I., Liebregts, P. and Bernards, A. (1979) Biosynthesis of ferredoxin in *Chlamydomonas reinhardii*. Planta 145, 351–356.

Husain, A. and Massey, V. (1978) Reversible resolution of flavoproteins into apoproteins and free flavins. Methods Enzymol. 53, 429–437.

Hutber, G.N. and Rogers, L.J. (1981) Isoenzymes of ferredoxin-$NADP^+$ oxidoreductase from the cyanobacterium *Nostoc* strain MAC. Photosyn. Res. 2, 269–280.

Hutber, G.N., Hutson, K.G. and Rogers, L.J. (1977) Effect of iron deficiency on levels of two ferredoxins and flavodoxin in a cyanobacterium. FEMS Microbiol. Lett. 1, 193–196.

Hutber, G.N., Smith, A.J. and Rogers, L.J. (1978) Isolation of ferredoxin-NADP$^+$ reductase from a prokaryote. Biochem. Soc. Trans. 6, 1214–1216.

Hutber, G.N., Smith, A.J. and Rogers, L.J. (1981) Flavodoxin from the blue-green alga *Nostoc* strain MAC. Phytochemistry 20, 383–387.

Hutson, K.G. and Rogers, L.J. (1975) Two plant-type ferredoxins in light-grown and dark-grown cells of the blue-green bacterium *Nostoc* strain MAC. Biochem. Soc. Trans. 3, 377–379.

Hutson, K.G., Rogers, L.J., Haslett, B.G., Boulter, D. and Cammack, R. (1978) Comparative studies on two ferredoxins from the cyanobacterium *Nostoc* strain MAC. Biochem. J. 172, 465–477.

Hutson, K.G., Rogers, L.J., Haslett, B.G. and Boulter, D. (1980) Ferredoxins from two cyanobacteria capable of heterotrophic growth in the dark. FEMS Microbiol. Lett. 7, 279–284.

Ip, S-M., Rowell, P., Aitken, A. and Stewart, W.D.P. (1984) Purification and characterisation of thioredoxin from the N_2-fixing cyanobacteria *Anabaena cylindrica*. Eur. J. Biochem. 141, 497–504.

Karplus, P.A., Walsh, K.A. and Herriott, J.R. (1984) Amino acid sequence of spinach ferredoxin-NADP$^+$ oxidoreductase. Biochemistry 23, 6576–6583.

Kobayashi, K. and Fox, J.L. (1978) The evolution of protein sequences by repetitious gene duplication: *Clostridium* flavodoxin. J. Mol. Evol. 11, 233–243.

Koike, H. and Katoh, S. (1980) Thermal properties of NADP : ferredoxin oxidoreductase and ferredoxin isolated from a thermophilic blue-green alga. Photosyn. Res. 1, 163–170.

Kunita, A., Koshibe, M., Nishikawa, Y., Fukuyama, K., Tsukihara, T., Katsube, Y., Matsu-ura, Y., Tanaka, N., Kakudo, M., Hase, T. and Matsubara, H. (1978) Crystallisation and a 5A X-ray diffraction study of *Aphanothece sacrum* ferredoxin. J. Biochem. (Tokyo) 84, 989–992.

Leach, C.K. and Carr, N.G. (1971) Pyruvate: ferredoxin oxidoreductase and its activation by ATP in the blue-green alga, *Anabaena variabilis*. Biochim. Biophys. Acta 245, 165–171.

Lee, I.S., Hase, T., Matsubara, H., Ho, K.K. and Krogmann, D. (1983) Amino acid sequence of ferredoxin I from *Aphanizomenon flos-aquae*. Biochim. Biophys. Acta 744, 53–56.

Lode, E.T., Murray, C.L. and Rabinowitz, J.C. (1976a) Derivatives of *Clostridium acidi-urici* ferredoxin containing altered amino acid sequences. J. Biol. Chem. 251, 1675–1682.

Lode, E.T., Murray, C.L. and Rabinowitz, J.C. (1976b) Apparent oxidation-reduction potential of *Clostridium acidi-urici* ferredoxin. J. Biol. Chem. 251, 1683–1687.

Ludwig, M.L., Pattridge, K.A. and Tarr, G. (1984) FMN: Protein interactions in flavodoxin from *A. nidulans*. In: R.C. Bray, P.C. Engel and S.G. Mayhew (eds.), Flavins and Flavoproteins. Walter de Gruyter, Berlin-New York, pp. 253–259.

MacKnight, M.L., Gray, W.R. and Tollin, G. (1974) N-Terminal amino acid sequences of *Azotobacter vinelandii* and *Rhodospirillum rubrum* flavodoxins. Biochem. Biophys. Acta 59, 630–637.

Masaki, R., Wada, K. and Matsubara, H. (1979) Isolation and characterisation of two ferredoxin-NADP$^+$ reductases from *Spirulina platensis*. J. Biochem. (Tokyo) 86, 951–962.

Masaki, R., Yoshikawa, S. and Matsubara, H. (1982) Steady state kinetics of oxidation of reduced ferredoxin with ferredoxin-NADP$^+$ reductase. Biochim. Biophys. Acta 700, 101–109.

Matsubara, H. and Hase, T. (1983) Phylogenetic consideration of ferredoxin sequences in plants, particularly algae. In U. Jensen and D.E. Fairbrothers (eds.), Proteins and Nucleic acids in Plant Systematics. Springer-Verlag, Berlin, pp. 168–181.

Matsubara, H., Hase, T., Wakabayashi, S. and Wada, K. (1978) Gene duplication during evolution of chloroplast-type ferredoxins. In: H. Matsubara and T. Yamanaka (eds.), Evolution of Protein Molecules. Japan Sci. Soc. Press, Tokyo, pp. 209–219.

Matsubara, H., Hase, T., Wakabayashi, S. and Wada, K. (1980) Structure and evolution of chloroplast- and bacterial-type ferredoxins. In D.S. Sigman and D.S. Brazier (eds.), The Evolution of Protein Structure and Function. Academic Press, New York, pp. 245–266.

Mayhew, S.G. (1971) Properties of two clostridial flavodoxins. Biochim. Biophys. Acta 235, 276–288.

Mayhew, S.G. and Ludwig, M.L. (1975) Flavodoxins and electron-transferring flavoproteins. In: Enzymes, 3rd Edn. 12, 57–118, Academic Press, London.

Mayhew, S.G., Foust, G.P. and Massey, V. (1969) Oxidation-reduction properties of flavodoxin from *Peptostreptococcus elsdenii*. J. Biol. Chem. 244, 803–810.

Melamed-Harel, H., Tel-Or, E. and SanPietro, A. (1985) Effect of ferredoxin on the diaphorase activity of cyanobacterial ferredoxin-NADP$^+$ oxidoreductase. Plant Physiol. 77, 229–231.

Mikami, B. and Ida, S. (1984) Purification and properties of ferredoxin-nitrate reductase from the cyanobacterium *Plectonema boryanum*. Biochim. Biophys. Acta 791, 294–304.

Mitsui, A. and Arnon, D.I. (1971) Crystalline ferredoxin from a blue-green alga, *Nostoc* sp. Physiol. Plant 25, 135–140.

Mitsui, A. and SanPietro, A. (1973) Large ferredoxin crystals from a blue-geen alga, *Phormidium foveolarum*. Plant Sci. Lett. 1, 157–163.

Mortenson, L.E., Valentine, R.C. and Carnahan, J.E. (1962) Biochem. Biophys. Res. Commun. 7, 448–452.

Moura, I., Moura, J.J.G., Bruschi, M. and LeGall, J. (1980) Flavodoxin and rubredoxin from *Desulfovibrio salexigens*. Biochim. Biophys. Acta 591, 1–8.

Nagayama, K., Ozaki, Y., Kyogoku, Y., Hase, T. and Matsubara, H. (1983) Classification of iron-sulphur cores in ferredoxins by ^{1}H NMR spectroscopy. J. Biochem. (Tokyo) 94, 893–902.

Neuer, G. and Bothe, H. (1982) The pyruvate-ferredoxin oxidoreductase in heterocysts of the cyanobacterium *Anabaena cylindrica*. Biochim. Biophys. Acta 716, 358–366.

Orme-Johnson, W.H. and Sands, R.H. (1973) Probing iron-sulphur proteins with EPR and ENDOR spectroscopy. In W. Lovenberg (ed.), Iron-sulphur Proteins, Vol II. Academic Press, London, pp. 195–238.

Ortega, T., Rivas, J., Cardenas, J. and Losada, M. (1977) Metabolic inter- conversion of ferredoxin-nitrate reductase and $NADP^+$ reductase of *Nostoc muscorum*. Biochem. Biophys. Res. Commun. 78, 185–193.

Pagani, S., Bonomi, F. and Cerletti, P. (1982) Sulfide insertion into spinach ferredoxin by rhodanese. Biochim. Biophys. Acta 700, 154–164.

Pagani, S., Bonomi, F. and Cerletti, P. (1983) The inhibition of rhodanese by lipoate and iron-sulphur proteins. Biochim. Biophys. Acta 742, 116–121.

Peschek, G.A. (1979) Nitrate and nitrite reductase and hydrogenase in *Anacystis nidulans* grown in iron and molybdenum deficient media. FEMS Microbiol. Lett. 6, 371–375.

Petersson, L., Cammack, R. and Rao, K.K. (1980) Anti-ferromagnetic exchange reactions in the two-iron two-sulphur ferredoxin from *Spirulina maxima* studied with a highly sensitive magnetic balance. Biochim. Biophys. Acta 622, 18–24.

Rao, K.K., Cammack, R., Hall, D.O. and Johnson, C.E. (1971) Mossbauer effect in *Scenedesmus* and spinach ferredoxins. Biochem. J. 122, 257–265.

Rao, K.K., Smith, R.V., Cammack, R., Evans, M.C.W., Hall, D.O. and Johnson, C.E. (1972) Spectroscopic characterization of ferredoxin from the blue-green alga *Microcystis flos-aquae*. Biochem. J. 129, 1159–1162.

Rippka, R., Deruelles, J., Waterbury, J.B., Herdman, M. and Stanier, R.Y. (1979) Generic assignments, strain histories and properties of pure cultures of cyanobacteria. J. Gen. Microbiol. 111, 1–61.

Roberts, G.P. and Brill, W.J. (1981) Genetics and regulation of nitrogen fixation. Ann. Rev. Microbiol. 35, 207–235.

Rossmann, M.G., Moras, D. and Olsen, K.W. (1974) Chemical and biological evolution of a nucleotide-binding protein. Nature 250, 194–199.

Rowell, P., Diez, J., Apte, S.K. and Stewart, W.D.P. (1981) Molecular heterogeneity of ferredoxin-$NADP^+$ oxidoreductase from the cyanobacterium *Anabaena cylindrica*. Biochim. Biophys. Acta 657, 507–516.

Sakihama, N., Nimi, K., Nakamura, N. and Shin, M. (1983) Trinitrophenylation of spinach ferredoxin and its effects on function. J. Biochem. (Tokyo) 93, 1685–1690.

Sandmann, G. and Malkin, R. (1983) Iron-sulphur centres and activities of the photosynthetic electron transport chain in iron-deficient cultures of the blue-green alga *Aphanocapsa* 6714. Plant Physiol. 73, 724–728.

Schmidt, A. and Christen, U. (1980) Distribution of thioredoxins in cyanobacteria. Z. Naturforsch. 34C, 1272–1275.

Schrautemeier, B. and Böhme, H. (1985) A distinct ferredoxin for nitrogen fixation isolated from heterocysts of the cyanobacterium *Anabaena variabilis*. FEBS Lett. 184, 304–308.

Schwartz, R.M. and Dayhoff, M.O. (1978) Origins of prokaryotes, eukaryotes, mitochondria and chloroplasts. Science 199, 395–403.

Serrano, A. and Rivas, J. (1982) Purification of ferredoxin-$NADP^+$ oxidoreductase from cyanobacteria by affinity chromatography on 2′,5′-ADP Sepharose 4B. Anal. Biochem. 126, 109–115.

Serrano, A., Rivas, J. and Losado, M. (1984) Studies on the one- and two-electron FAD-mediated reactions by ferredoxin-$NADP^+$ oxidoreductase from *Anabaena* 7119. FEBS Lett. 170, 85–88.

Shahak, Y., Crowther, D. and Hind, G. (1981) The involvement of ferredoxin-$NADP^+$ oxidoreductase in cyclic electron transport in chloroplasts. Biochim. Biophys. Acta 636, 234–243.

Sheriff, S. and Herriott, J.R. (1981) Structure of ferredoxin-$NADP^+$ oxidoreductase and the location of the $NADP^+$ binding site. J. Mol. Biol. 145, 441–451.

Shin, M., Sukenobu, M., Oshino, R. and Kitazume, Y. (1977) Two plant-type ferredoxins from a blue-green alga, *Nostoc verrucosum*. Biochim. Biophys. Acta 460, 85–94.

Shin, M., Yokoyama, Z., Oshino, R., Sukenobu, M. and Kitazume, Y. (1980) Further characterisation of *Nostoc verrucosum* ferredoxins. Plant Cell Physiol. 21, 1681–1684.

Shin, M., Wakita, R., Yamasaki, Y. and Oshino, R. (1981) Interrelation of two forms of ferredoxin-NADP$^+$ reductase with different molecular weights. Plant Cell Physiol. 22, 343–346.
Shin, M., Ishida, H. and Nozaki, Y. (1985) A new protein factor, connectein, as a constituent of the large form of ferredoxin-NADP$^+$ oxidoreductase. Plant Cell Physiol. 26, 559–563.
Smillie, R.M. (1965) Isolation of phytoflavin, a flavoprotein with chloroplast ferredoxin activity. Plant Physiol. 40, 1124–1128.
Smillie, R.M. and Entsch, B. (1971) Phytoflavin. Methods Enzymol. 23A, 504–515.
Smith, J.M., Smith, W.H. and Knaff, D.B. (1981) Electrochemical titrations of a ferredoxin-ferredoxin-NADP$^+$ oxidoreductase complex. Biochim. Biophys. Acta 635, 405–411.
Smith, W.W., Pattridge, K.A., Ludwig, M.L., Petsko, G.A., Tsernoglou, D., Tanaka, M. and Yasunobu, K.T. (1983) Structure of oxidized flavodoxin from *Anacystis nidulans*. J. Mol. Biol. 165, 737–755.
Stewart, W.D.P. and Rowell, P. (1975) Effects of L-methionine DL-sulphoximine on the assimilation of newly fixed NH_3, acetylene reduction and heterocyst production in *Anabaena cylindrica*. Biochem. Biophys. Res. Commun. 65, 846–856.
Susor, W.A. and Krogmann, D.W. (1966) Triosephosphopyridine nucleotide photoreduction with cell free preparations of *Anabaena variabilis*. Biochim. Biophys. Acta 120, 65–72.
Suss, K-H. (1979) Isolation and partial characterisation of membrane-bound ferredoxin-NADP$^+$ reductase from chloroplasts. FEBS Lett. 101, 305–310.
Sykes, G.A. and Rogers, L.J. (1984) Redox potentials of algal and cyanobacterial flavodoxins. Biochem. J. 217, 845–850.
Tagawa, K. and Arnon, D.I. (1962) Ferredoxins as electron carriers in photosynthesis and in the biological production and consumption of hydrogen gas. Nature (Lond.) 195, 537–543.
Takahashi, Y., Hase, T., Matsubara, H., Hutber, G.N. and Rogers, L.J. (1982) Amino acid sequence of *Chlorogloeopsis fritschii* ferredoxin : Taxonomic and evolutionary aspects. J. Biochem. (Tokyo) 92, 1363–1368.
Takahashi, Y., Hase, T., Wada, K. and Matsubara, H. (1983) Ferredoxins in developing spinach cotyledons : the presence of two molecular species. Plant Cell Physiol. 24, 189–198.
Takaichi, S., Makino, K., Iwasaki, A. and Suzuki, K. (1978) Phytochemical reaction systems of photosynthesis in *Phytolacca americana*: two ferredoxins from pokeweed; isolation and characterization. J. Biochem. (Tokyo) 83, 1151–1158.
Takruri, I., Boulter, D., Fitzgerald, M.P., Hutber, G.N. and Rogers L.J. (1986) N-Terminal amino acid sequences of flavodoxins from *Chondrus crispus* and *Nostoc* strain MAC. Phytochemistry, 25, 2113–2115.
Tanaka, M., Haniu, M. and Yasunobu, K.T. (1975a) The amino acid sequence of the *Azotobacter vinelandii* flavodoxin. Biochem. Biophys. Res. Commun. 66, 639–644.
Tanaka, M., Haniu, M., Yasunobu, K.T., Rao, K.K. and Hall, D.O. (1975b) Modification of the automated sequence determination as applied to the sequence determination of the *Spirulina maxima* ferredoxin. Biochemistry 14, 5535–5540.
Tanaka, M., Haniu, M., Zeitlin, S., Yasunobu, K.T., Evans, M.C.W., Rao, K.K. and Hall, D.O. (1975c) Amino acid sequence of the *Spirulina maxima* ferredoxin, a ferredoxin from a prokaryote. Biochem. Biophys. Res. Commun. 64, 399–407.
Taniguchi, T. and Kimura, T. (1974) Role of 3-mercaptopyruvate sulfurtransferase in the formation of the iron-sulfur chromophore of adrenal ferredoxin. Biochim. Biophys. Acta 364, 284–295.
Tel-Or, E. and Stewart, W.D.P. (1977) Photosynthetic components and activities of nitrogen-fixing isolated heterocysts of *Anabaena cylindrica*. Proc. Roy. Soc. Lond. B 198, 61–86.
Tel-Or, E., Cammack, R., Rao, K.K., Rogers, L.J., Stewart, W.D.P. and Hall, D.O. (1977) Comparative immunochemistry of bacterial, algal and plant ferredoxins. Biochim. Biophys. Acta 490, 120–131.
Tollin, G. and Edmondson, D.E. (1980) Purification and properties of flavodoxins. Methods Enzymol. 69, 392–406.
Tsukihara, T., Fukuyama, K., Nakamura, M., Katsube, Y., Tanaka, N., Kakudo, M., Wada, K., Hase, T. and Matsubara, H. (1981) X-ray analysis of a [2Fe-2S] ferredoxin from *Spirulina platensis* – main chain folding and location of side-chains at 2.5A resolution. J. Biochem. (Tokyo) 90, 1763–1773.
Tsukihara, T., Kobayashi, M., Nakamura, M., Katsube, Y., Fukuyama, K., Hase, T., Wada, K. and Matsubara, H. (1982) Structure-function relationships of [2Fe-2S] ferredoxins and design of a model molecule. Biosystems 15, 243–257.
Tsutsui, T., Tsukihara, T., Fukuyama, K., Katsube, Y., Hase, T., Matsubara, H., Nishikawa, Y. and Tanaka, N. (1983) Main chain fold of a [2Fe-2S] ferredoxin I from *Aphanothece sacrum* at 2.5A resolution. J. Biochem. (Tokyo) 94, 299–302.
Valle, E.M., Carrillo, N. and Vallejos, R.H. (1982) Functional sulphydryl groups of ferredoxin-NADP$^+$ reductase. Biochim. Biophys. Acta 681, 412–418.

Vallejos, R.H., Ceccarelli, E. and Chan, R. (1984) Evidence for the existence of a thylakoid intrinsic protein that binds ferredoxin-NADP$^+$ oxidoreductase. J. Biol. Chem. 259, 8048–8052.

VanLin, B. and Bothe, H. (1972) Flavodoxin from *Azotobacter vinelandii*. Arch. Mikrobiol. 82, 155–172.

Vetter, H. and Knappe, J. (1971) Flavodoxin and ferredoxin of *Escherichia coli*. Hoppe-Seyler's Z. Physiol. Chem. 352, 433–466.

Vieira, B.J. and Davis, D.J. (1985) Chemical modification of carboxyl groups in spinach ferredoxin alters its ability to interact with ferredoxin-NADP$^+$ oxidoreductase. Biochem. Biophys. Res. Commun. 129, 467–471.

Wada, K., Hase, T. and Matsubara, H. (1975a) Evolutionary information involved in primary structures of chloroplast-type ferredoxins. J. Biochem. (Tokyo) 78, 637–639.

Wada, K., Hase, T., Tokunaga, H. and Matsubara, H. (1975b) Amino acid sequence of *Spirulina platensis* ferredoxin: a far divergency of blue-green algal ferredoxins. FEBS Lett. 55, 102–104.

Wada, K., Matsubara, H., Chain, R.K. and Arnon, D.I. (1981) A comparative study of the biological activities of two molecular species of chloroplast-type ferredoxins. Plant Cell Physiol. 22, 275–281.

Wada, K., Tamura, T., Matsubara, H. and Kodo, K. (1983) Spirulina ferredoxin-NADP$^+$ reductase. Further characterisation with an improved preparation. J. Biochem. (Tokyo) 94, 387–393.

White, R.H. (1983) Origin of the labile sulphur in the iron-sulphur proteins of *E. coli*. Biochem. Biophys. Res. Commun. 112, 66–72.

Yamanaka, T., Takenami, S., Wada, K. and Okunuki, K. (1969) Purification and properties of ferredoxin derived from the blue-green alga, *Anacystis nidulans*. Biochim. Biophys. Acta 180, 196–198.

Yao, Y., Tamura, T., Wada, K., Matsubara, H. and Kodo, K. (1984) Spirulina ferredoxin-NADP$^+$ reductase. The complete amino acid sequence. J. Biochem. (Tokyo) 95, 1513–1516.

Yao, Y., Wada, K., Takahashi, Y., Katoh, S. and Matsubara, H. (1985) The sulphydryl groups of ferredoxin-NADP$^+$ oxidoreductase: is a disulphide bond present? J. Biochem. (Tokyo) 98, 1079–1082.

Yee, B.C., De la Torre, A., Crawford, N.A., Lara, C., Carlson. D.E. and Buchanan, B.B. (1981) The ferredoxin-thioredoxin system of enzyme regulation in a cyanobacterium *Nostoc muscorum*. Arch. Microbiol. 130, 14–18.

Yoch, D.C. and Valentine, R.C. (1972) Ferredoxins and flavodoxins of bacteria. Ann. Rev. Microbiol. 26, 139–162.

Yoshikawa, S., Ohnishi, N., Morigiwa, A., Takeshima, K., Matsumoto, M., Nishiyama, K., Matsubara, H. and Kodo, K. (1984) Transient kinetics of the ferredoxin-NADP$^+$ reductase reaction. In R.C. Bray, P.C. Engel and S.G. Mayhew (eds.), Flavins and Flavoproteins. De Gruyter, Berlin, pp. 489–492.

Yoshino, M., Murakami, K. and Tsushima, K. (1976) Flavodoxin – an allosteric inhibitor of AMP nucleosidase from *Azotobacter vinelandii*. J. Biochem. (Tokyo) 80, 839–843.

Zanetti, G. and Arosio, P. (1980) Solubilization from spinach thylakoids of a higher molecular weight form of ferredoxin NADP$^+$ reductase. FEBS Lett. 111, 373–376.

Zanetti, G., Gozzer, C., Sacchi, G. and Curti, B. (1979) Modification of arginyl residues in ferredoxin-NADP$^+$ reductase from spinach leaves. Biochim. Biophys. Acta 569, 127–134.

Zanetti, G., Aliverti, A. and Curti, B. (1984) A cross-linked complex between ferredoxin and ferredoxin-NADP$^+$ oxidoreductase. J. Biol. Chem. 259, 6153–6157.

Zumft, W.G. and Spiller, H. (1971) Characterization of a flavodoxin from the green alga *Chlorella*. Biochem. Biophys. Res. Commun. 45, 112–118.

P. Fay and C. Van Baalen (Eds.) The Cyanobacteria

3

Phycobilisomes: assembly and attachment

ALEXANDER N. GLAZER

Department of Microbiology and Immunology, University of California, Berkeley, CA 94720, USA

1. Introduction

Oxygen-evolving photosynthesis is believed to have arisen some 2.5 billion years ago. Geological evidence indicates that the particular Precambrian fossils found in conjunction with products of oxidation bear a remarkable morphological resemblance to modern cyanobacteria (Schopf, 1978; Schopf, 1983). The high concentration of oxygen in the atmosphere which resulted from the activities of these early photosynthesizers provided both problems and opportunities for the ancestral cyanobacteria and influenced their further biochemical evolution. For example, many filamentous cyanobacteria are nitrogen fixers. Since the nitrogenase enzyme system is exquisitely oxygen-sensitive, retention of the ability to fix nitrogen required the elaboration of a protection mechanism for the enzyme. Such a protection was achieved by the differentiation of a specialized cell type, the heterocyst. It is interesting to note that heterocyst development passes through an obligatory oxygen-requiring step (Rippka and Stanier, 1978). Another innovation, central to the evolution of phycobilisomes, the subject of this chapter, was the synthesis of bilins, the visible light-absorbing chromophores of the phycobiliproteins. The precursor of these open-chain tetrapyrroles, biliverdin IXα, is produced by the cleavage of heme by the enzyme heme oxygenase in a reaction which requires molecular oxygen (Troxler et al., 1979).

The chemistry of the reaction centers of Photosystems I and II (PSI and PSII) of cyanobacteria, as judged from their redox potentials and spectroscopic signatures, does not differ from that of the corresponding components of higher plant chloroplasts (see Chapter 1). It is with respect to the light-harvesting antenna composition and structure that cyanobacteria are seen to be distinctive. Whereas higher plants utilize both chlorophylls *a* and *b* as light-harvesting pigments, chlorophyll *a* only is present in cyanobacteria. In cyanobacteria, a central role in light-harvesting is played by phycobiliproteins (Glazer, 1981). The intense colors of these proteins reflect the presence of numerous, chemically diverse, bilin prosthetic groups. The phycobiliproteins are present at very high concentrations in cyanobacterial cells. The obvious contribution to the color of these cells made by the phycobiliproteins was responsible for the common name for these organisms of 'blue-green algae', which has persisted in the literature to the present day.

The cyanobacterial photosynthetic apparatus consists of three macromolecular complexes. Two of these, PSI and PSII, are intrinsic complexes of thylakoid membranes. The third complex, the phycobilisome, is attached to the external surface of these membranes. All three of these complexes have been isolated and partially characterized biochemically. Cyanobacterial PSI, solubilized with Triton X-100, is a particle of ~460 000 daltons, and contains 130–140 antenna chlorophyll *a* molecules per P700 (Lundell et al., 1985). A PSII preparation obtained from a thermophilic *Synechococcus* sp. by solubilization with *β*-octylglucoside, followed by polyacrylamide gel electrophoresis in the presence of digitonin, has been particularly well characterized (Yamagishi and Katoh, 1983, 1984, 1985). This complex contains 32 chlorophyll *a* molecules per P680, and has a minimum molecular mass of about 200 000. A similar ratio of chlorophyll *a* per reaction center has been reported for an oxygen-evolving PSII preparation isolated from the thermophilic filamentous cyanobacterium *Phormidium laminosum* (Ke et al., 1982) (for details see Chapter 1). The phycobiliprotein-containing complexes, phycobilisomes, have been isolated from numerous cyanobacteria and extensively characterized. Phycobilisomes contain between 300 and 700 bilins per particle and are physically associated with one or more PSII complexes. In relation to its effectiveness in the absorption of visible light, the absorption cross-section of a bilin chromophore within a phycobilisome is approximately equal to that of chlorophyll *a* in a PSI or PSII protein-chlorophyll complex.

The cyanobacterial photosynthetic apparatus responds to changes in both light intensity and light quality. The responses include change in the relative ratio of PSII to PSI complexes (Myers et al., 1980; Manodori and Melis, 1986), change in proportions of the major phycobiliproteins of the phycobilisome (Yamanaka and Glazer, 1981; Lönneborg et al., 1985), and change in the ratio and nature of the phycobiliproteins absorbing in the green and red parts of the spectrum (chromatic adaptation) (Tandeau de Marsac, 1983). The complementary chromatic adaptation response is seen in some but not all cyanobacteria (see Section 4.2.1. of Chapter 13).

This chapter gives a brief account of the current understanding of the structure and function of the phycobilisome, the relationship of this particle to PSII, and the changes in phycobilisome structure that take place in response to changes in environmental conditions. Zuber (1985) has recently reviewed the literature on the structure and function of light-harvesting complexes in general.

2. Morphology of phycobilisomes

Cyanobacterial phycobilisomes are supramolecular assemblies of $6–8 \times 10^6$ daltons. Electron microscopic studies of thin sections have established that these particles are attached in regular arrays to the stromal surfaces of the thylakoid membranes. The ultrastructural studies are discussed in several excellent reviews (Gantt, 1980; Cohen-Bazire and Bryant, 1982; Wehrmeyer, 1983). The gross dimensions of the phycobilisomes in various organisms have been determined in this manner. The development of procedures for the isolation of intact phycobilisomes (Gantt and Lipschultz, 1972; Gantt et al., 1979) provided the opportunity to examine the fine structure of these particles in detail.

Electron microscopy of purified phycobilisomes of the unicellular red alga *Porphyridium cruentum* showed these particles to approximate prolate ellipsoids 47.5 nm in diameter, 33 nm in height, and 32 nm in width (Gantt and Lipschultz, 1972; however, see Wanner and Köst, 1980; Wehrmeyer, 1983). The structure could be seen to be made up of uniformly sized building blocks, described as 'globular' by Wanner and Köst (1980), but the precise arrangement of these building blocks could not be discerned because of the superimposition of structural elements. A much different situation was found with the phycobilisomes of the unicellular red alga *Rhodella violacea* (Wehrmeyer,

1971). These phycobilisomes are hemidiscoidal with a base 36 nm long, a height of 29 nm, and a thickness of 8 nm (Mörschel et al., 1977). Electron micrographs of purified and glutaraldehyde-fixed preparations of these phycobilisomes showed these particles almost exclusively in 'face-view' (i.e., at right angles to their orientation in situ with respect to the thylakoid membrane). *Rhodella violacea* phycobilisomes were seen to consist of a triangular core composed of three cylindrical structures each with an average diameter of 11–12 nm. From two sides of the triangle six rods, each made up of three stacked disks, 10–12 nm in diameter and 6 nm thick, radiated in a hemidiscoidal array (Mörschel et al., 1977). The hemidiscoidal morphology described above was found to be typical of phycobilisomes from numerous cyanobacteria (e.g., Bryant et al., 1979; Glazer et al., 1979; Williams et al., 1980; Nies and Wehrmeyer, 1980; Rosinski et al., 1981). The only notable departure was noted for the hemidiscoidal phycobilisomes of *Synechococcus* 6301, where the core is made up of only two cylindrical elements (Glazer et al., 1979).

Phycobilisomes with ultrastructure somewhat different from the types discussed above have been described. Those from *Griffithsia pacifica* (Gantt and Lipschultz, 1980) are block-shaped. The unicellular cyanobacterium *Gloeobacter violaceus* has the most unusual phycobilisomes. This cyanobacterium is devoid of thylakoids and its photosynthetic apparatus is associated with the cytoplasmic membrane (Rippka et al., 1974). The phycobilisomes isolated from this organism are seen to be bundles of rods (Guglielmi et al., 1981; Fig. 1) and in situ the rods project towards the interior of the cell at right angles to the cell membrane (Guglielmi et al., 1981).

In certain cyanobacterial mutants, it has been possible to obtain 'top' views of the cylindrical core elements. In *Synechococcus* 6301 mutant AN112, each core cylinder is seen to be a stack of four disks 30 Å thick (Yamanaka et al., 1980), and the same substructure was detected in the core cylinders of *Synechocystis* 6701 mutant UV16 phycobilisomes (Anderson et al., 1984; see also Bryant et al., 1979).

3. Composition of phycobilisomes

Phycobilisomes are made up entirely of proteins. In cyanobacterial phycobilisomes, phycobiliproteins make up some 85% of the mass. The balance is made up of 'linker' polypeptides, some of which also carry bilin prosthetic groups (Tandeau de Marsac and Cohen-Bazire, 1977; Yamanaka et al., 1978; Lundell et al., 1981a). Phycobilisomes are stable only under rather unusual conditions, either in the presence of high concentrations ($\sim$0.65 M) of Na-K phosphate buffers at pH 7–8 (Gantt and Lipschultz, 1972; Gantt et al., 1979; Yamanaka et al., 1978), or those containing high concentrations of Na-sulfate at pH 8.0 (Yamanaka et al., 1978). At lower salt concentrations, phycobilisomes dissociate releasing their building blocks, the phycobiliproteins, as water-soluble protein assemblies whose complexity and degree of aggregation depend on several factors such as pH, temperature, ionic strength, protein concentration, retention or loss of linker polypeptides. 'Pure' phycobiliproteins are then obtained by fractionation of the products of phycobilisome dissociation by chromatography on ion-exchange resins, hydroxylapatite, and gel filtration.

An arbitrary, but widely accepted definition of a phycobiliprotein is that it is a protein whose monomer unit is made up of two dissimilar polypeptide chains, α and β, each of $<22\,000$ daltons. Certain phycoerythrins are isolated as monodisperse 'hexameric' aggregates, $(\alpha\beta)_6\gamma$, in which the γ subunit has a molecular weight of $\sim 30\,000$ daltons and carries bilins (Gantt and Lipschultz, 1974; Glazer and Hixson, 1977). The definition of a phycobiliprotein has been applied in a manner sufficiently elastic to permit the inclusion of these more complex aggregates under the general rubric of phycobiliproteins even though, as will become evident from further discussion, these are

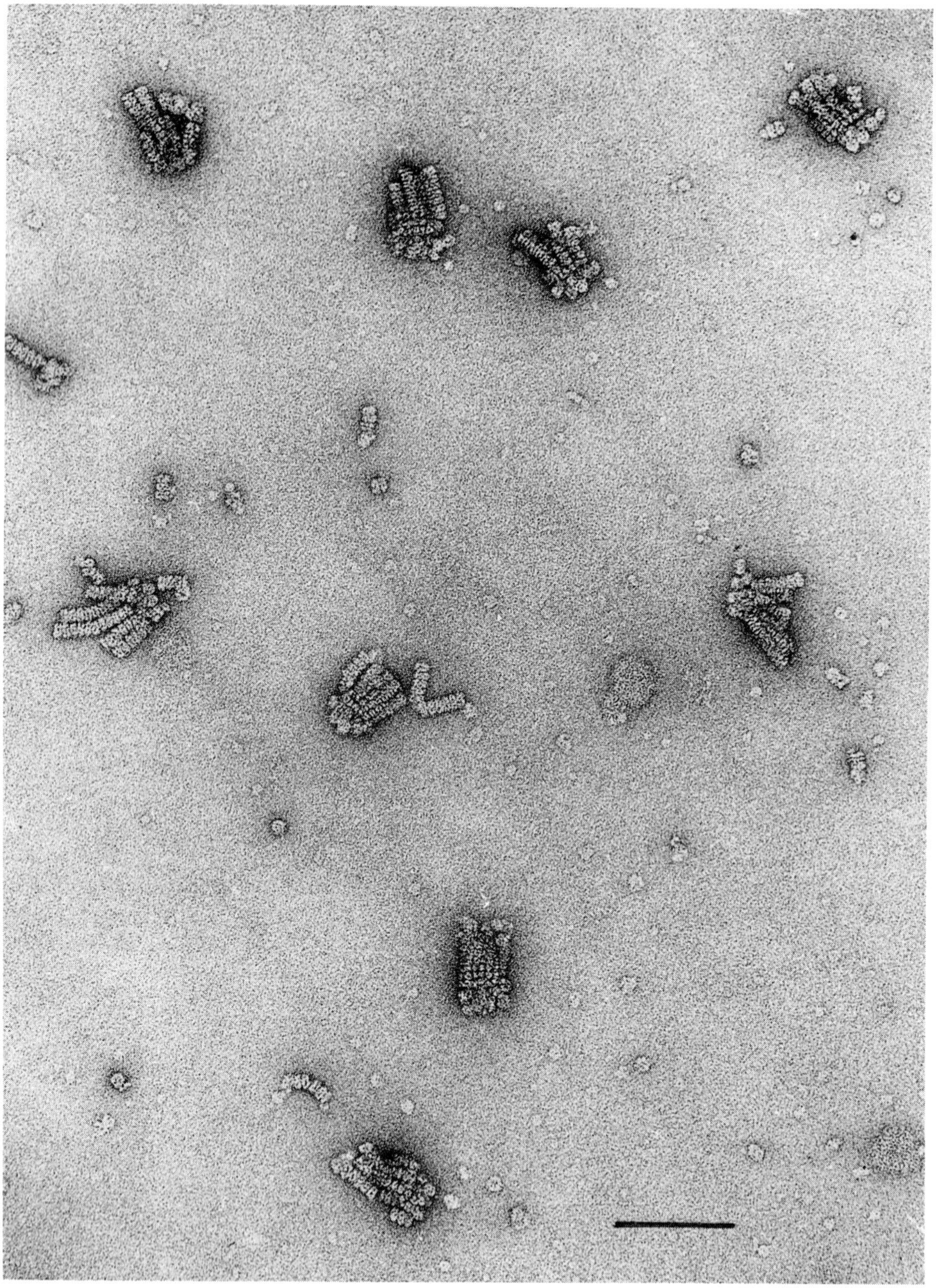

Fig. 1. *Gloeobacter violaceus* phycobilisomes in 0.04% (w/v) lauryldimethylamine oxide – 0.05 M K-phosphate – 0.26 M Na-citrate – 10 mM $MgCl_2$, pH 7.0. Fixed for 5 min on the grid with 0.3% glutaraldehyde in the same buffer at a protein concentration of 30 μg/ml. Uranyl formate negative stain. Bar represents 1 μm. (L.J. Ong, R.C. Williams and A.N. Glazer, unpublished data)

phycobiliprotein-linker polypeptide complexes. Biliprotein subunits, linker polypeptides, bilins, and their abbreviations are given in Table I.

4. Chemistry of the phycobiliproteins

A list of phycobiliproteins is provided in Table II together with details of their prosthetic group composition. Information on their spectroscopic properties is given in Fig. 2. The prosthetic groups of the phycobiliproteins are a family of open-chain tetrapyrroles (bilins) formed from heme, presumably via biliverdin (Troxler et al., 1978, 1979; Beale and Cornejo, 1984) by as yet largely unknown biosynthetic pathways, and are all isomers of one another (Fig. 3). The prosthetic groups are attached to the polypeptides through either one or two thioether linkages (Figs. 3 and 4). The α and β subunits of the phycobiliproteins are likewise a family of closely related homologous polypeptides. Inspection of their amino acid sequences (for examples, see Fig. 5) reveals that this family of polypeptides originated from a common ancestral gene.

5. X-ray crystallographic structures of phycobiliproteins

The structures of the C-phycocyanins of *Mastigocladus laminosus* and *Agmenellum quadruplicatum* (*Synechococcus* 7002) have been determined to about 3 Å resolution by X-ray diffraction methods (Schirmer et al., 1985; Deisenhofer et al., 1985). Preliminary data on the crystal structure of phycoerythrocyanin have also been reported (Rümbeli et al., 1985). The conformation of both phyco-

TABLE I

Abbreviations for biliprotein subunits, linker polypeptides and bilins (Glazer, 1985)

Type of polypeptide	Abbreviations
Phycoerythrin subunits	α^{PE}, β^{PE}, γ^{PE}
Phycoerythrocyanin subunits	α^{PEC}, β^{PEC}
R-Phycocyanin subunits	α^{RPC}, β^{RPC}
C-Phycocyanin subunits	α^{PC}, β^{PC}
Allophycocyanin subunits	α^{AP}, β^{AP}
Allophycocyanin B α subunit	α^{APB}
β-Type core biliprotein subunits	β^{MW^a}
Rod linker polypeptides	L_R^{MW}
Linker attaching rod elements to core	L_{RC}^{MW}
Core linker polypeptides	L_C^{MW}
Linker attaching core to membrane	L_{CM}^{MW}

Bilin	Abbreviation
Phycourobilin	PUB
Phycoerythrobilin	PEB
Cryptoviolin	PXB
Phycocyanobilin	PCB

TABLE II

Some properties of the major cyanobacterial and red algal phycobiliproteins

Protein	Aggregation state[a]	Bilin content per subunit[b]	MW × 10^{-3}	References
Allophycocyanin	$(\alpha\beta)_3$	α 1 PCB; β 1 PCB	100	c
C-Phycocyanin	$(\alpha\beta)_n$ ($n=1$–6)	α 1 PCB; β 2 PCB	36.5–220	c
R-Phycocyanin	$(\alpha\beta)_3$	α 1 PCB; β 1 PCB, 1 PEB	120	c
R-Phycocyanin II	$(\alpha\beta)_n$ ($n=1$–3)	α 1 PEB; β 1 PCB, 1 PEB	36.5–120	d
Phycoerythrocyanin	$(\alpha\beta)_3$	α 1 PXB; β 2 PCB	120	c
C-Phycoerythrin	$(\alpha\beta)_n$ ($n=1$–6)	α 2 PEB; β 3 PEB	40–240	Klotz and Glazer, 1985; Sidler et al., 1985
B-Phycoerythrin[e]	$(\alpha\beta)_6\gamma$	α 2 PEB; β 3 PEB; γ 2 PEB, 2 PUB		Lundell et al., 1984
R-Phycoerythrin[e]	$(\alpha\beta)_6\gamma$	α 2 PEB; β 2 PEB, 1 PUB; γ 1 PEB 3 PUB		Klotz and Glazer, 1985

[a]The molecular weights are those of the aggregates specified in this column.
[b]For abbreviations, Table I.
[c]For original references, see Glazer, 1981, 1982, 1984, 1985.
[d]Obtained from the marine *Synechococcus* strain WH8103. (Ong, L.J., and Glazer, A.N., unpublished results)
[e]The data given are for *P. cruentum* B-phycoerythrin and *Gastroclonium coulteri* R-phycoerythrin. B- and R-phycoerythrins from different cyanobacteria and red algae vary widely in their PEB/PUB ratio. For references, see legend to Fig. 2.

biliproteins appears to be very similar. These studies have provided much important information relevant to the structure and function of phycobilisomes.

In accord with numerous biophysical and electron microscopic studies, the phycocyanin trimer was found to consist of three identical $\alpha\beta$ units which are arranged around a threefold symmetry axis to form a disk of approximately 110 Å x 30 Å with a central channel of 35 Å diameter (Schirmer et al., 1985; Fig. 6). Such a channel is of adequate size to contain the bulk of a rod linker polypeptide (Schirmer et al., 1985; Yu and Glazer, 1982).

The C-phycocyanin α and β subunits exhibit a very similar tertiary structure. Each subunit is folded into eight α-helices and irregular loops. The helical content of the secondary structure is about 65% for the α subunit and 62% for the β subunit. In each subunit, six helices are arranged to form a globular domain whereas two helices protrude from this domain and are involved in extensive contacts between the subunits. A particularly noteworthy observation is that the arrangement of the helices within the globular domains closely resembles that seen in myoglobin. Moreover, the residue that corresponds in location to Cys-84 in the myoglobin structure is His E7 which acts as the distal ligand for the heme-iron. The amino acid sequence homologies between myoglobin and the α and β subunits of C-phycocyanin are low. Schirmer et al. (1985) conclude: "Phycocyanin and the globins may have evolved from different ancestors, but acquired a similar fold which represents a particularly stable form of α-helix association. However, during evolution, the tertiary structure of proteins seems to be more conserved than the primary structure (Bajaj and Blundell, 1984). Therefore, it is also possible that despite their low sequence homology the strong resemblance of the C-phycocyanin and globin structures points to a very distant evolutionary relationship."

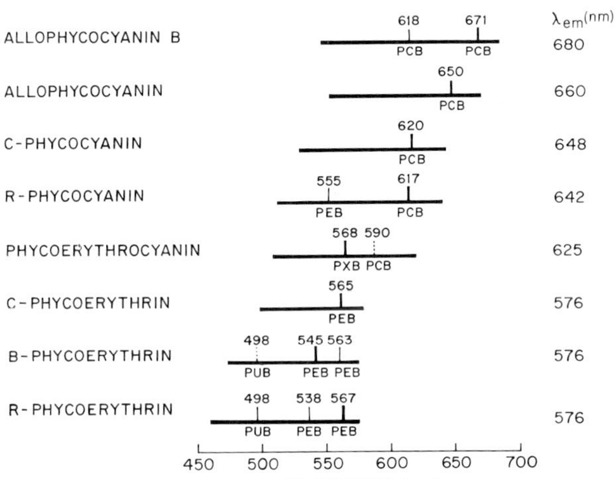

Fig. 2. Spectroscopic characteristics of native cyanobacterial and rhodophytan phycobiliproteins at neutral pH. The main absorption peak in the visible region of the room temperature spectrum is indicated for each protein with a thick vertical line, subsidiary maxima are indicated by thinner vertical lines. Broken vertical lines represent shoulders. The horizontal lines cover the portion of the spectrum over which the absorption of the particular protein is >20% of its absorption maximum. The spectra of B- and R-phycoerythrins vary widely (Glazer et al., 1982; Alberte et al., 1984; Honsell et al., 1984; Ong et al., 1984). The data given here are for *Porphyridium cruentum* B-phycoerythrin and for *Cryptonemia obovata* R-phycoerythrin (Glazer and Hixson, 1977). The values for the fluorescence emission maximum (λ_{em}) for each protein are independent of excitation wavelength. The abbreviation (defined in Table I) indicates the bilin which makes the major contribution at each maximum or shoulder in the absorption spectrum.

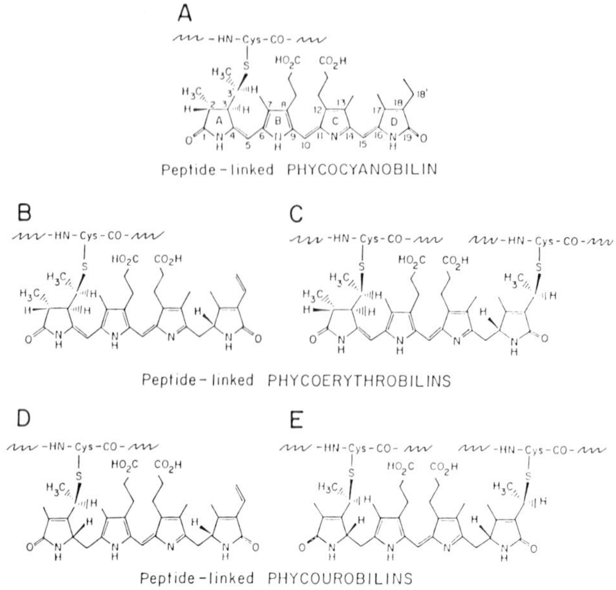

Fig. 3. Bilin-peptide linkages in phycobiliproteins (for locations of attachments to the α and β subunits, see Fig. 4). (A) PCB linked through ring A (Lagarias et al., 1979; Rapoport and Glazer, 1985); (B) PEB linked through ring A (Lundell et al., 1984; Schoenleber et al., 1983; Schoenleber et al., 1984b; Klotz and Glazer, 1985); (C) doubly-linked PEB (Lundell et al., 1984; Schoenleber et al., 1984a); (D) singly-linked PUB (the structure shown is that of the A-ring linked bilin, but the possibility that PUB is D-ring linked has not been rigorously excluded); (E) doubly-linked PUB (Klotz and Glazer, 1985; Nagy et al., 1985). For abbreviations, see Table I.

The crystals of *A. quadruplicatum* C-phycocyanin (Deisenhofer et al., 1985; Schirmer et al., 1985) consist of hexameric units that are composed of two trimers stacked upon one another face-to-face. Earlier studies of *P. cruentum* B-phycoerythrin, $(\alpha\beta)_6\gamma$, showed the presence of a local twofold axis perpendicular to the trimer axis, indicating that in this phycobiliprotein the trimers are also stacked face-to-face (Fisher et al., 1980).

Detailed data on the three-dimensional structure of the three phycocyanobilin chromophores is yet to be obtained. However, all three of the bilins in C-phycocyanin are in an extended conformation (Schirmer et al., 1985). This is consistent with predictions based on spectroscopic properties of native phycocyanin (Scheer and Kufer, 1977; see also Scheer, 1982; and Glazer, 1985, for reviews). The pairwise distances between the chromophores in $\alpha\beta$ monomers or $(\alpha\beta)_3$ trimers appear to exceed those which would permit exciton interactions but are clearly short enough for efficient radiationless energy transfer by the Förster mechanism (Förster, 1965). The arrangement of bilins within the hexameric stacks of *A. quadruplicatum* C-phycocyanin is illustrated in Fig. 7. This arrangement suggests that the bilin linked to β Cys-84 mediates the transfer of energy from trimer to trimer within stacks (Schirmer et al., 1985).

→

Fig. 4. Attachment positions and modes of linkage of the bilin prosthetic groups. Residue numbers are taken from the sequences shown in Fig. 5. The superscripts, A or D, indicate mode of linkage, either through ring A or through rings A and D (see Fig. 3) for the particular bilin. References to the sources of the data are given in the legends to Figs. 3 and 5. For abbreviations, see Table I.

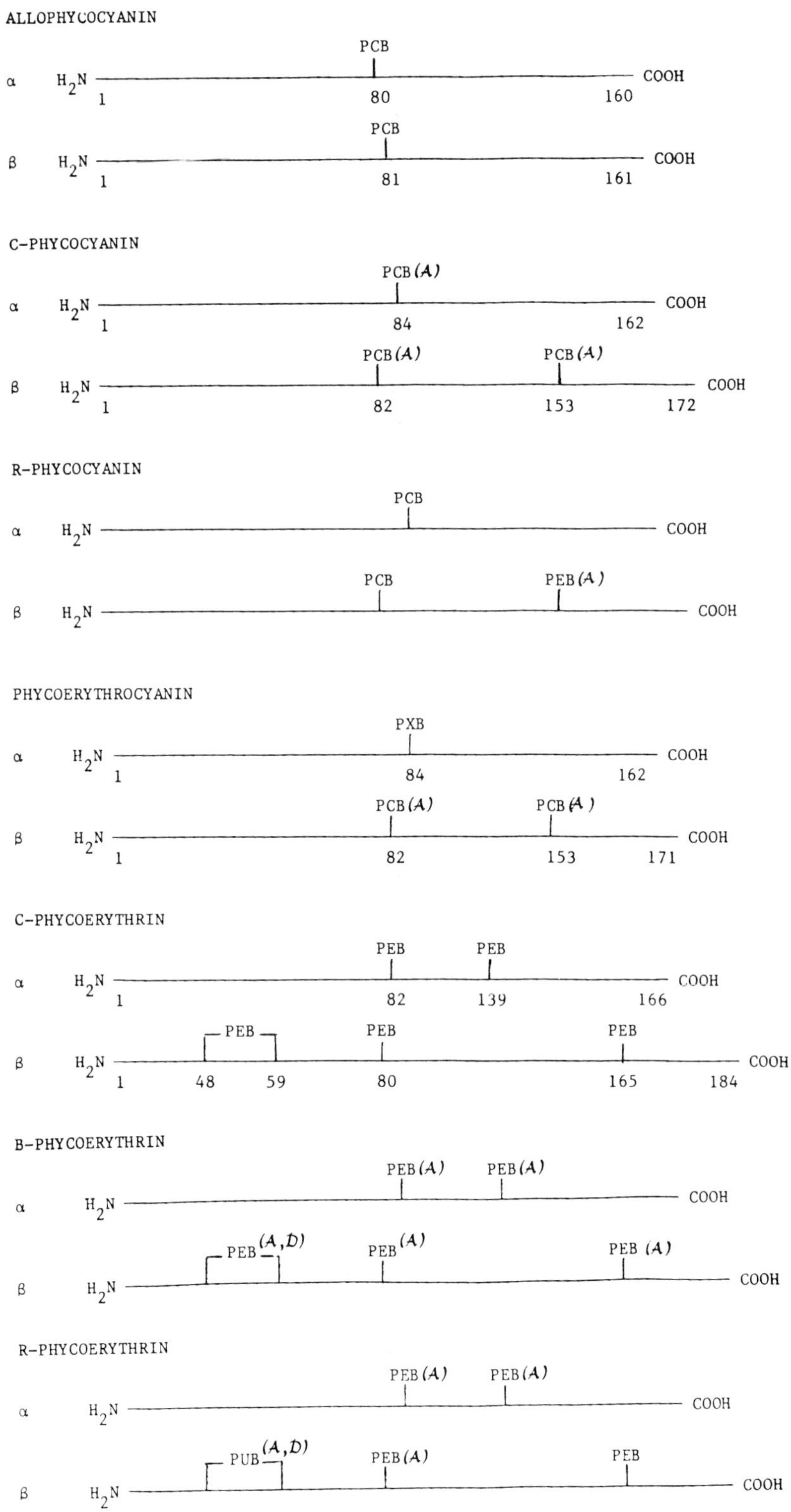
ALLOPHYCOCYANIN
PCB
α H2N COOH
1 80 160
PCB
β H2N COOH
1 81 161
C-PHYCOCYANIN
PCB(A)
α H2N COOH
1 84 162
PCB(A) PCB(A)
β H2N COOH
1 82 153 172
R-PHYCOCYANIN
PCB
α H2N COOH
PCB PEB(A)
β H2N COOH
PHYCOERYTHROCYANIN
PXB
α H2N COOH
1 84 162
PCB(A) PCB(A)
β H2N COOH
1 82 153 171
C-PHYCOERYTHRIN
PEB PEB
α H2N COOH
1 82 139 166
PEB PEB PEB
β H2N COOH
1 48 59 80 165 184
B-PHYCOERYTHRIN
PEB(A) PEB(A)
α H2N COOH
PEB(A,D) PEB(A) PEB (A)
β H2N COOH
R-PHYCOERYTHRIN
PEB(A) PEB(A)
α H2N COOH
PUB(A,D) PEB(A) PEB
β H2N COOH

```
α-AP     SIVTKSIVNADAEARYLSPGELDRIKSFVSSGEKRLRIAQILTDNRERIVKQAGDQLF
α-PC    VKTPITDAIAAADTQGRFLSNTELQAVNGRYQRAAASLEAARALTANAQRLIDGAAQAVY
α-PEC   MKTPLTEAIAAADLRGSYLSNTELQAVFGRFNRARAGLEAARAFANNGKKWAEAAANHVY

α-PE    MKSVVTTVIAAADAAGRFPSTSDLESVQGSIQRAAARLEAAEKLANNIDAVATEAYNACI


α-AP    QKRPDVVSPGGNAYGQEM--TATCLRDLDYYLRLITYGIVAGDVTPIEEIGIVGVREMYK
α-PC    QKFPYLIQTSGPNYAADARGKSKCARDIGHYLRIITYSLVAGGTGPLDEYLIAGLNEIND
α-PEC   QKFPYTTQMQGPQYASTPEGKAKCVRDIDHYLRTISYCCVVGGTGPLDDYVVAGLKEFNS

α-PE    KKYPYLNN-SGEANSTDTF-KAKCARDIKHYLRLIQYSLVVGGTGPLDEWGIAGQREVYR


α-AP    SLGTPIDAVAAGVSAMKNVASSILSAEDAA----------EAGAYFDYVAGALA
α-PC    AFELSPSWYIEALKYIKANHG--LSGQAAN----------EANTYIDYVINALS
α-PEC   ALGLSPSWYIAALEFVRDNHG--LTGDVAG----------EANTYINYAINALS

α-PE    ALGLPTAPYVEALSFAR-NRG  MSAQALT----------EYNALLDYAINSLS
                           /   \
                          CAPRD


β-AP    MQDAITAVINSSDVQGKYLDTAALEKLKSYFSTGELRVRAATTIAANAAAIVKEAVAKLT
β-PC    AYDVFTKVVSQADSRGEFLSNEQLDALANVVKEGNKRLDVVNRITSNASTIVTNAARALF
β-PEC   MLDAFSRVVEQADKKGAYLSNDEINALQAIVADSNKRLDVVNRLTSNASSIVANAYRALV

β-PE    MLDAFSRAVVSADASGSTVS--DIAALRAFVASGNRRLDAVNAIASNASCMVSDAVAGMI


β-AP    LYS-DITLPGGDMYTTRR--YAACIRDLDYYLRYATYAMLAGDPSILDERVLNGLKETYN
β-PC    EEQPQLIAPGGS--ATRNGTMAACLRDMEIILRYITYAILAGDASILDDRCLNGLRETYQ
β-PEC   AERPQVFNPGGPCFHHNR--QAACIRDLGFILRYVTYSVLAGDTSVMDDRCLNGLRETYQ

β-PE    CENQGLIQAGGSCYPNRR--MAACLRDAEIVLRYVTYALLAGDASVLDDRCLNGLKETYA


β-AP    SLGVPISATVQAIQAMKEVTA SLVGPDAGK----------EMGVYFDYICSGLS
β-PC    ALGTPGSSVAVGIQKMKEAAI NIANDPNGITKGDCSALISEVASYFDRAAAAVA
β-PEC   ALGTPGDAVASGIKKMKEAAL KIANDPNGITKGDCSQLMSELASYFDRAAAAV

β-PE    ALGVPTTSTVRAVQIMKAQAA KLRKMGTPVVEDRCASLVAEASRYFDRVISALS
                            /  \
                    AHIQDTPSEARAGA
```

Fig. 5. Amino acid sequences of the α and β subunits of *Mastigocladus laminosus* allophycocyanin (AP) (Sidler et al., 1981), C-phycocyanin (PC) (Frank et al., 1978), phycoerythrocyanin (PEC) (Füglistaller et al., 1983), and *Fremyella diplosiphon* C-phycoerythrin (PE) (Sidler et al., 1985).

6. Assembly of phycobiliproteins into phycobilisomes

As noted earlier, phycobilisomes consist of two morphologically distinct assemblies, rods and a core. A schematic representation of an extensively studied hemidiscoidal cyanobacterial phycobilisome is shown in Fig. 8. The structure of the rods has been examined in some detail in numerous organisms whereas the range of studies on the structure of the core is much more limited.

6.1. Rods

The data on the rod organization in seven cyanobacterial and two red algal phycobilisomes presented in Fig. 9, as well as much additional information of a less complete nature, lead to the following general conclusions. (a) Each disk in the rods is a complex of a phycobiliprotein hexamer with a position-specific linker polypeptide. (b) When different phycobiliproteins are present in a

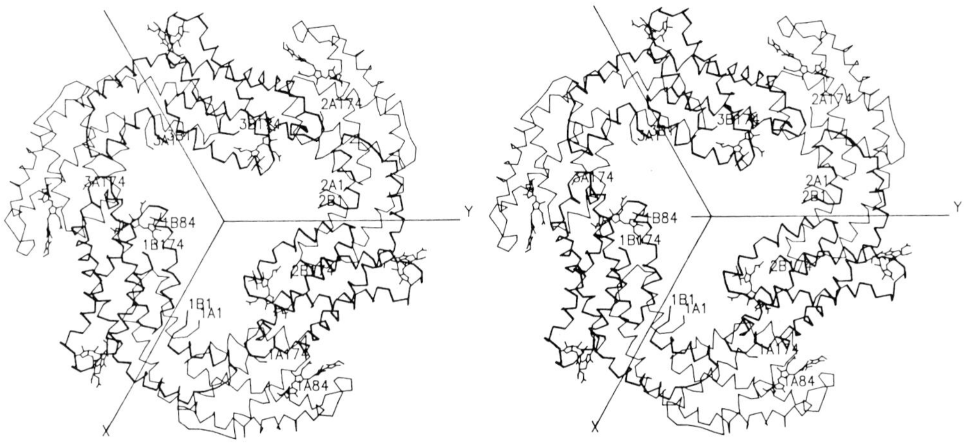

Fig. 6. *Agmenellum quadruplicatum* C-phycocyanin $(\alpha\beta)_3$ trimer at 2.5 A resolution projected down the crystallographic 3-fold axis. C_α-backbone (β subunit, heavy lines; α subunit, light lines) and chromophores are shown. Chromophore β84 is stretching into the central channel, α84 is near it and β155 is at the periphery. (Huber, R., personal communication)

rod, they are arranged in order of their absorption maxima with those absorbing in the green (C-, B-, or R-phycoerythrin, phycoerythrocyanin) forming the end of the rod distal to the core. (c) Phycobilisomes with more than one type of rod, but as far as is known identical cores, are present in certain cyanobacteria (see examples 5 and 6 in Fig. 9). (d) Complementary chromatic adaptation (Tandeau de Marsac, 1983) is accomplished by changing the composition of the rods. In type III chromatic adapters, where the synthesis of both phycoerythrin and phycocyanin is controlled by light quality, the rod length is the same in red and green light, but the distal portion of the rod

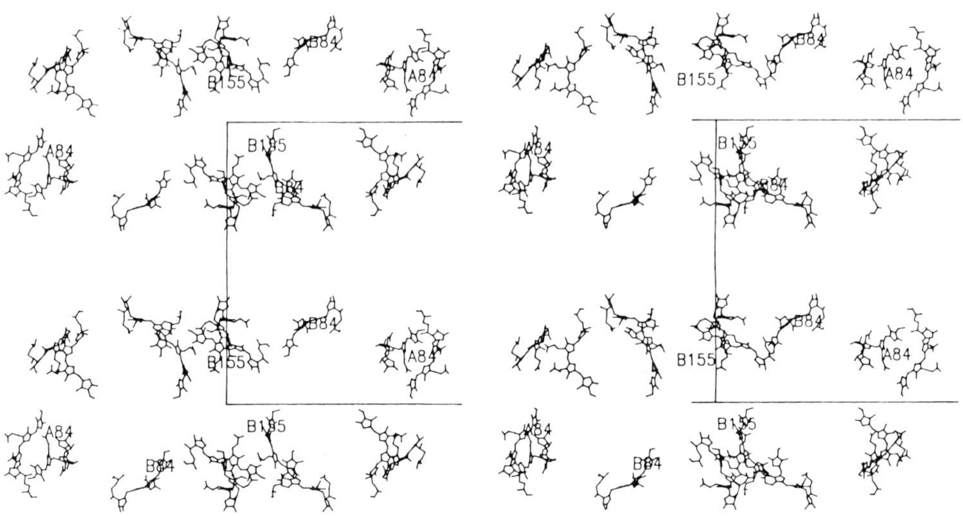

Fig. 7. The arrangement of the phycocyanobilins in stacked hexameric disks in crystals of *Agmenellum quadruplicatum* C-phycocyanin. Bilin attached to the α subunit is identified as A84, those attached to the β subunit as B84 and B155. (Huber, R., personal communication)

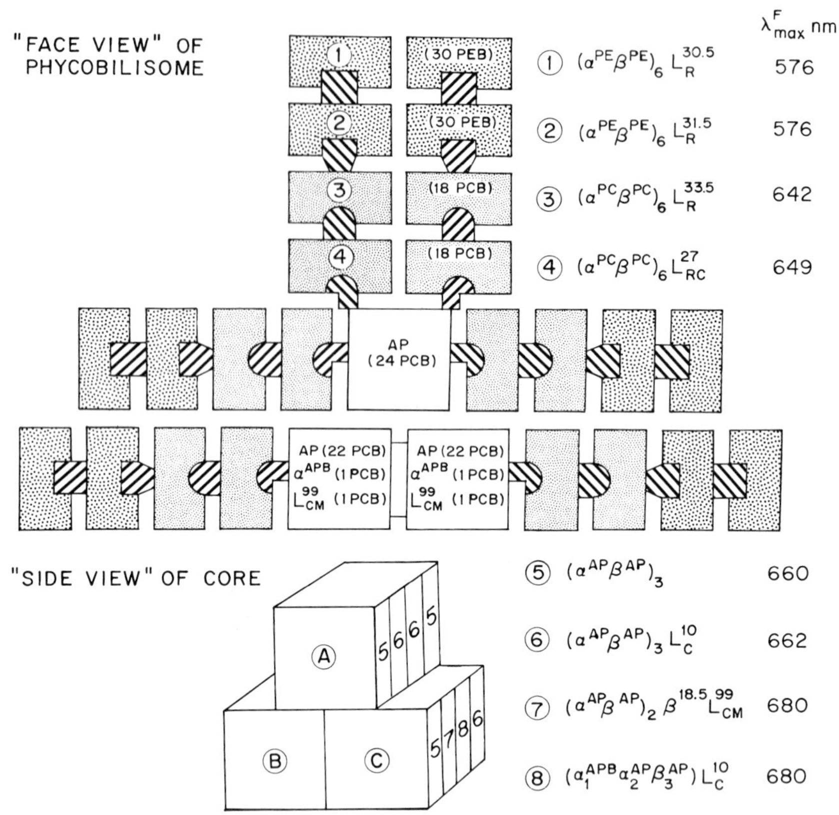

Fig. 8. Schematic representation of the hemidiscoidal *Synechocystis* 6701 phycobilisome. A rod is made up of four hexameric biliprotein complexes, each of which is attached through its specific linker polypeptide to the component adjacent to it in the phycobilisome. Core cylinders B and C are believed to be identical (Gingrich et al., 1983). The number and type of bilins present in each domain of the structure is indicated. For each complex, λ^{F}_{max} is the fluorescence emission maximum. Abbreviations are defined in Table I.

is made up of phycoerythrin disks in 'green light particles', and of phycocyanin disks in 'red light particles' (Fig. 9, examples 8 and 9). In type II chromatic adapters (e.g., Fig. 9, example 7), where only phycoerythrin synthesis is controlled by light quality, the rods of red light phycobilisomes are shorter than those of green light phycobilisomes. A parallel observation has been reported for the phycobilisomes of *Oscillatoria splendida* L_3, which synthesizes phycoerythrocyanin in white light but not in red light (Thomas et al., 1984). (e) Alterations in phycobilisome structure in response to variation in light intensity or to nitrogen starvation likewise involve changes in rod length. For example, the rods of *Synechococcus* 6301 (Fig. 9, example 1) isolated from cells grown at high light intensity, are reduced in length to a single disk (Lönneborg et al., 1985), and a decrease in rod length, as well as in phycobilisome number, is seen in response to nitrogen starvation (Yamanaka and Glazer, 1980). In *Rhodella violacea* phycobilisomes, the B-phycoerythrin content varies inversely with light intensity, and is also reduced when the cells are in a medium depleted of a nitrogen source (Koller, 1979). (f) Whereas in the majority of phycobilisomes each disk contains a hexamer of a single phycobiliprotein, there appear to be exceptions to this structural motif. In *Mastigocladus laminosus*, the terminal disk has been reported to have the composition $(\alpha^{PC}\beta^{PC})_3(\alpha^{PEC}\beta^{PEC})_3L_R$ (Nies and Wehrmeyer, 1981).

6.2. Rod linker polypeptides

Each disk in a phycobilisome rod is associated with a specific linker polypeptide in the 27 – 33 kDa range. The linker polypeptides play the following roles (Lundell et al., 1981a; Yu et al., 1981; Yu and Glazer, 1982; Gingrich et al., 1982a,b; Lundell and Glazer, 1983a,b): (a) stabilization of a specific assembly form of a structural element of the phycobilisome such as a phycocyanin hexamer disk in the rod substructure, (b) determination of the relative positions of the various building blocks of the phycobilisome, (c) modulation of the spectroscopic properties of the phycobiliproteins to generate polar pathways of energy transfer within the phycobilisome.

The *Synechococcus* 7002 gene encoding the 33 kDa polypeptide associated with phycocyanin has

	ORGANISM	LIGHT QUALITY	CORE-PROXIMAL TO CORE-DISTAL ROD DISKS
1	*Synechococcus* 6301	white	$[(\alpha^{PC}\beta^{PC})_6L^{27}_{RC}][(\alpha^{PC}\beta^{PC})_6L^{33}_{C}][(\alpha^{PC}\beta^{PC})_6L^{30}_{C}]$
2	*Rhodella violacea*	"	$[(\alpha^{PC}\beta^{PC})_6L^{a}_{RC}][(\alpha^{PE}\beta^{PE})_6L^{b}_{C}][(\alpha^{PE}\beta^{PE})_6L^{c}_{C}]$
3	*Porphyridium cruentum*	"	$[(\alpha^{PC}\beta^{PC})_6L^{a}_{RC}][(\alpha^{PE}\beta^{PE})_6L^{b}_{C}][(\alpha^{PE}\beta^{PE})_6L^{c}_{C}]$
4	*Mastigocladus laminosus*	"	$[(\alpha^{PC}\beta^{PC})_6L^{a}_{RC}][(\alpha^{PC}\beta^{PC})_6L^{b}_{C}][(\alpha^{PC}\beta^{PC})_3(\alpha^{PEC}\beta^{PEC})_3L^{c}_{C}]$
5	*Anabaena variabilis*	"	$[(\alpha^{PC}\beta^{PC})_6L^{27}_{RC}][(\alpha^{PC}\beta^{PC})_6L^{32.5}_{C}][(\alpha^{PC}\beta^{PC})_6L^{29}_{C}]$
			$[(\alpha^{PC}\beta^{PC})_6L^{27}_{RC}][(\alpha^{PC}\beta^{PC})_6L^{32.5}_{C}][(\alpha^{PEC}\beta^{PEC})_6L^{30.5}_{C}]$
6	*Nostoc* MAC	"	$[(\alpha^{PC}\beta^{PC})_6L^{29}_{RC}][(\alpha^{PC}\beta^{PC})_6L^{34.5}_{C}]$
			$[(\alpha^{PC}\beta^{PC})_6L^{29}_{RC}][(\alpha^{PE}\beta^{PE})_6L^{34}_{C}][(\alpha^{PE}\beta^{PE})L^{32}_{C}]$
7	*Synechocystis* 6701	green	$[(\alpha^{PC}\beta^{PC})_6L^{27}_{RC}][(\alpha^{PC}\beta^{PC})_6L^{33.5}_{C}][(\alpha^{PE}\beta^{PE})_6L^{31.5}_{C}][(\alpha^{PE}\beta^{PE})_6L^{30.5}_{C}]$
		red	$[(\alpha^{PC}\beta^{PC})_6L^{27}_{RC}][(\alpha^{PC}\beta^{PC})_6L^{33.5}_{C}][(\alpha^{PE}\beta^{PE})_6L^{31.5}_{C}]_{trace}$
8	*Pseudoanabaena* 7409	green	$[(\alpha^{PC}\beta^{PC})_6L^{a}_{RC}][(\alpha^{PE}\beta^{PE})_6L^{b}_{C}][(\alpha^{PE}\beta^{PE})_6L^{c}_{C}][(\alpha^{PE}\beta^{PE})_6L^{d}_{C}]$
		red	$[(\alpha^{PC}\beta^{PC})_6L^{a}_{RC}][(\alpha^{PC}\beta^{PC})_6L^{e}_{C}][(\alpha^{PC}\beta^{PC})_6L^{f}_{C}][(\alpha^{PC}\beta^{PC})_6L^{g}_{C}]$
9	*Fremyella diplosiphon*	green	$[(\alpha^{PC}\beta^{PC})_6L^{a}_{RC}][(\alpha^{PE}\beta^{PE})_6L^{b}_{C}][(\alpha^{PE}\beta^{PE})_6L^{c}_{C}][(\alpha^{PE}\beta^{PE})_6L^{d}_{C}]$
		red	$[(\alpha^{PC}\beta^{PC})_6L^{a}_{RC}][(\alpha^{PC}\beta^{PC})_6L^{d}_{C}][(\alpha^{PC}\beta^{PC})_6L^{e}_{C}]$

Fig. 9. Composition of the rods of cyanobacterial and red algal phycobilisomes. In cases where the apparent size of the linker polypeptide associated with a particular complex has not been established it is designated as L^a_R, L^b_R, etc. The sources of the data are (1) Glazer et al., 1983; (2) Koller et al., 1978; Mörschel et al., 1980; (3) Wehrmeyer et al., 1985; (4) Nies and Wehrmeyer, 1981; (5) Yu et al., 1981; Yu, 1981; Yu and Glazer, 1982; (6) Anderson et al., 1983; Zilinskas and Howell, 1983; (7) Gingrich et al., 1982a,b; (8) Bryant and Cohen-Bazire, 1981; (9) Rosinski et al., 1981.

been cloned and sequenced (Bryant et al., 1985; Fig. 10). Two different linker polypeptides, both 34.5 kDa, from *Mastigocladus laminosus*, one associated with phycocyanin, the other with phycoerythrocyanin, have been purified and partially sequenced (Fig. 10). These linker polypeptides are basic at physiological pH whereas the phycobiliproteins are acidic. Consequently, electrostatic in-

Synechococcus 7002 $L_R^{33,PC}$

M P V T V A A S **R** L G T A A F D Q S P V E L **R** A N Y S **R** D D A Q T V I **R**

A V Y **R** Q V L G N D Y V M S S E **R** L T A A E S L F T N G F I S V **R** D F V

R A V A Q S E L Y **K** E **K** F L Y N N F Q T **R** V I E L N F **K** **H** L L G **R** A P Y

D E A E V I E **H** L D **R** Y Q N E G F E A D I N S Y I D S A E Y T E N F G D

N I V P Y I **R** S Y V V Q T G **H** **R** T V G F T **R** M F S L Q **R** G Y A N S D **R** A

Q I A G N A S **R** L A Q E L A **R** N T T S A V V G P S G V N E G W A F **R** S A

A D D Y **H** P G Q S L G G S T G L S A D D Q V V **R** V E V A A L S T P **R** Y P

R I **R** **R** S S **R** V F F V P V S **R** L S Q **K** L Q E I Q **R** M G G **R** V A S I S P A

G Q

Mastigocladus laminosus $L_R^{34.5,PEC}$

N-terminal sequence S T S V A E R L A I K D E V D K K I E L R P N W (S)

E D E L Q I V F K (T) A Y E Q V F G R Q

C-terminal sequence (M) K T V G F N R I F E L Y R G R A N S D N A Q F G

G K S A R L R S K I (S) M N L A N T I V P P T S P I A A S T S S A R T L

V V (G M) F I V E A I A G T L N T N V A V R R S R Q V Y T V P Y D R L S

A T Y Q E I H K R G G K I V K I T P A S

Mastigocladus laminosus $L_R^{34.5,PC}$

N-terminal sequence A I T A A A S R L G T E P F S N A A K I E L R S D

A S R E E V E A V I N A V Y (R) E V L G

Fig. 10. Amino acid sequence data on rod linker polypeptides. $L_R^{33,PC}$: linker polypeptide which bridges the two phycocyanin disks of the rods of *Synechococcus* 7002 (*Agmenellum quadruplicatum*) phycobilisome (Bryant et al., 1985; Bryant, personal communication). $L_R^{34.5,PEC}$: partial amino acid sequence of the phycoerythrocyanin-associated linker polypeptide, and $L_R^{34.5,PC}$: amino-terminal sequence of phycocyanin-associated linker polypeptide (Füglistaller et al., 1985). Basic residues are shown in bold face, acidic residues are underlined.

teractions are likely to make a significant contribution to the stability of phycobiliprotein-linker polypeptide complexes. A low molecular weight polypeptide of ~9 kDa is also present in the rods of cyanobacterial phycobilisomes (Glazer et al., 1983). This polypeptide has been purified from the phycobilisomes of *M. laminosus* and sequenced (Fig. 11). The gene encoding this polypeptide in *Synechococcus* 7002 has also been sequenced (Fig. 11).

Phycobilisome preparations invariably contain very low amounts of a 45 kDa polypeptide. In *Synechococcus* 6301, it has been shown that this polypeptide is present at less than one copy per phycobilisome (Yamanaka et al., 1982). Anderson (1985) has compared the compositions of phycobilisomes isolated from *Synechocystis* 6701 wild type and various mutants with respect to the presence or absence of the 45 kDa polypeptide. He also fractionated the products of partial dissociation of these phycobilisomes. From these studies, he has suggested that the 45 kDa polypeptide is associated with the core-proximal phycocyanin disk and that it plays some role (catalytic?) in phycobilisome assembly. Zilinskas (1982) described the isolation from *Nostoc* sp. phycobilisomes of a 10.3 S fraction containing 46 kDa and 44 kDa polypeptides in addition to the α and β subunits of allophycocyanin, and speculated that these polypeptides may function in the assembly of aggregates of allophycocyanin higher than a trimer.

6.3. Core organization

In contrast to the extensive body of information on the rod structure in numerous phycobilisomes, the detailed data on the core is available only for three hemidiscoidal phycobilisomes, those of *Syne-*

Mastigocladus laminosus $L_R^{8.9}$

M F G Q T T L G I D S V S S S A S **R** V F **R** F E V V G M **R** Q N E E N D **K** N

K Y N I **R** **R** S G S V Y I T V P Y N **R** M S E E M Q **R** I **H** **R** L G G **K** I V **K** I

E P L T **R** A A G

Synechococcus 7002 L_R^{9}

M L S Q F A N G T E A A S **R** V F T Y E V Q G L **R** Q T E E T D N Q E Y A F

R **R** S G S V F I N V P Y A **R** M N G E M Q **R** I L **R** L G G **K** I V S I **K** P Y T

G A T A S D E E

Mastigocladus laminosus L_C^{9}

G **R** L F **K** I T A C/S V P S Q T **R** I **R** T Q **R** E L Q N T Y F T **K** L V P Y E N

W F **R** E Q Q **R** I Q **K** M G G **K** I V **K** V E L A T G **K** Q G I N T G L A

Fig. 11. Low molecular weight linker polypeptides associated with the rods of *M. laminosus* and *Synechococcus* 7002 phycobilisomes (Füglistaller et al., 1985; Bryant, personal communication), and the core of *M. laminosus* phycobilisomes (Füglistaller et al., 1985). Basic residues are shown in bold face, acidic residues are underlined.

chococcus 6301, *Synechocystis* 6701, and *Nostoc* sp. The quantitatively major components of the core are the α and β subunits of allophycocyanin, but four other types of polypeptides are also present. The core contains the largest polypeptide of the phycobilisome (L_{CM}) whose apparent molecular weight ranges from 75 to 124 kDa depending on the source of the phycobilisome (Tandeau de Marsac and Cohen-Bazire, 1977). This polypeptide carries a single PCB group (Lundell et al., 1981b; Redlinger and Gantt, 1982). Two polypeptides homologous to allophycocyanin subunits are present. α^{APB} is a polypeptide homologous to α^{AP} (Lundell and Glazer, 1981), and $\beta^{18.3}$ which is homologous to β^{AP} (Lundell and Glazer, 1983a; Fuglistaller et al., 1984). α^{APB} and $\beta^{18.3}$ each carry a single PCB group. The core also contains a small linker polypeptide, L_C^{10} (Lundell and Glazer, 1983b). This polypeptide has been purified from the phycobilisome of *M. laminosus* and sequenced (Fig. 11). These polypeptides are distributed within four different allophycocyanin complexes which make up the core cylinders of the phycobilisome.

Each of the two core cylinders of the *Synechococcus* 6301 phycobilisome contains the following complexes: *A*, $(\alpha^{AP}\beta^{AP})_3$; *B*, $(\alpha^{AP}\beta^{AP})_2\beta^{18.3}L_{CM}^{75}$; *C*, $(\alpha_1^{APB}\alpha_2^{AP}\beta_3^{AP})\ L_C^{10.5}$; and *D*, $(\alpha^{AP}\beta^{AP})_3L_C^{10.5}$. From a variety of studies (reviewed in Glazer et al., 1983; Lundell and Glazer, 1983c) it has been concluded that these complexes are arranged in the order A-B-C-D in each core cylinder, although the arrangements D-A-B-C or C-A-B-D have not been rigorously excluded. The allophycocyanin complexes that make up the three-cylinder core of the *Synechocystis* 6701 phycobilisome have been isolated and their relative amounts determined (Gingrich et al., 1983). From these data and by analogy to the results obtained with *Synechococcus* 6301 phycobilisomes, it has been concluded that the two basal cylinders have the composition *A*, $(\alpha^{AP}\beta^{AP})_3$; *B*, $(\alpha^{AP}\beta^{AP})_2\beta^{18.5}L_{CM}^{99}$; *C*, $(\alpha_1^{APB}\alpha_2^{AP}\beta_3^{AP})L_C^{10}$; *D*, $(\alpha^{AP}\beta^{AP})_3L_C^{10}$, and that the upper cylinder contains two copies each of complexes *A* and *D* (Fig. 8).

Comparison of the properties of the allophycocyanin complexes isolated from *Synechocystis* phycobilisomes with those isolated from the three-cylinder core-containing *Nostoc* sp. phycobilisomes (Zilinskas et al., 1978; Canaani and Gantt, 1980; Zilinskas, 1982) strongly suggests the same overall organization for the phycobilisome cores from these very different types of cyanobacteria.

Anderson et al. (1984) have described UV-induced mutants of *Synechocystis* 6701 which can be exploited as a source of the three-cylinder phycobilisome cores. From partial dissociation studies on such core preparations, Anderson (1985) has concluded that the two basal core cylinders lie in parallel (i.e., that the two A-B-C-D stacks of allophycocyanin complexes point in the same direction) and that the two basal rods of the *Synechocystis* 6701 phycobilisome are each attached to the half of a basal core cylinder made up of the $(\alpha^{AP}\beta^{AP})_3$ and $(\alpha^{AP}\beta^{AP})_2\beta^{18.5}L_C^{99}$ complexes.

7. Relationship of phycobilisomes to Photosystem II

7.1. Pathway of energy flow in intact cells and phycobilisomes

There is a wealth of information on phycobilisome structure, but little clear detail on how these particles interact with the thylakoid membrane. Likewise, there is ample speculation but little convincing evidence for the way in which excitation energy initially resident in phycobilisomes is partitioned between Photosystems I and II. The second question will be discussed first.

Early studies of the action spectra for photosynthetic oxygen production showed that in cyanobacteria and red algae phycobiliproteins make the major contribution to light harvesting for Photosystem II whereas chlorophyll *a* makes a relatively minor contribution (Haxo and Blinks, 1950; Blinks, 1954, 1960; Duysens, 1952, 1964; Haxo, 1960; Fork and Amesz, 1969; Halldal, 1970). For Photosystem I, on the other hand, the action spectra showed chlorophyll *a* to be the major contributor (Mimuro and Fujita, 1977; Wang et al., 1977). The early inferences concerning the distribution of chlorophyll *a* between the two photosystems were fully supported by later in vitro studies of protein-chlorophyll complexes. As noted earlier, cyanobacterial PSII complexes contain 30–40 chlorophyll *a* per reaction center whereas the number of chlorophyll *a* molecules per P700 in PSI is 130–150. Moreover, the number of PSI complexes exceeds that of PSII complexes under most environmental conditions (Kawamura et al., 1979; Myers et al., 1980; Manodori and Melis, 1984; Manodori et al., 1984) (see Chapter I).

The advent of time-resolved fluorescence spectroscopy with picosecond resolution has made it possible to trace the path of energy transfer in intact cells. In both cyanobacteria and red algae, this path has been shown to be phycoerythrin → phycocyanin → allophycocyanin → chlorophyll *a*. Less than 200 ps are required for the energy delivered to the phycobiliprotein at the distal end of phycobilisome rod (e.g., phycoerythrin) to reach chlorophyll *a*. Losses of energy during this process due to fluorescence emission account for less than 5% of the excitation energy (Porter et al., 1978; Brody et al., 1981; Mimuro et al., 1984; Yamazaki et al., 1984). Time-resolved fluorescence emission studies on phycobilisomes show that the kinetic parameters for the pathway phycoerythrin → phycocyanin → allophycocyanin within the isolated particles are similar to those for the corresponding transfer steps in intact cells (e.g., Searle et al., 1978; Wendler et al., 1984).

There is consensus on the conclusion that the chlorophyll *a* fluorescence sensitized by the phycobiliproteins originates from PSII (Karukstis and Sauer, 1983). Consequently, PSII must function as an energy acceptor, but the various interpretations of the numerous studies on the fate of the energy harvested by phycobiliproteins in intact cells lead to three different scenarios: (a) From direct measurements of the absorption cross-sections for PSI and PSII, it was concluded that phycobilisomes deliver energy directly to PSII. This energy is either utilized for chemistry within PSII or lost through radiationless decay pathways and fluorescence. Little, if any, of this energy is transferred ('spilled over') to PSI (Manodori and Melis, 1984, 1986; Manodori et al., 1984). (b) Phycobiliproteins deliver excitation energy directly to PSII. However, a portion of this energy is transferred from the chlorophyll *a* pigment bed of PSII to that of PSI. The 'spillover' is controlled by the proportion of PSII reaction centers that are closed (Ley and Butler, 1976, 1977, 1980).

In the context of scenario (b), the phenomenon of State 1 to State 2 transition has received considerable attention (Bonaventura and Myers, 1969; Murata, 1970; Ley, 1984). Illumination of cyanobacteria or red algae with light absorbed primarily by PSII results in a rapid (light intensity dependent rate) increase in fluorescence emanating from PSI. This state has been designated State 2. Conversely, when the excitation is by light absorbed primarily by PSI, an increase in PSII fluorescence is observed and the cells are said to be in State 1. From these changes in the steady state fluorescence emission, it has been inferred that the transition to State 2 is accompanied by an increase in the absorption cross-section of PSI (Bonaventura and Myers, 1969). Recent studies by Biggins and co-workers (Biggins, 1983; Biggins et al., 1984; Biggins and Bruce, 1985; Bruce et al., 1985) on the mechanism of the light state transition in the cyanobacterium *A. nidulans* and the red alga *P. cruentum* have led to the following model. In both, State 1 and State 2 phycobilisomes are attached to PSII complexes. It is postulated that the distance between PSI and PSII complexes differs in State 1 and State 2. In State 2, these complexes are envisaged to be in close contact allowing rapid energy transfer between the pigment beds of the two photosystems; in State 1, these complexes are somewhat further apart. Time-resolved fluorescence emission spectra were determined for *A. nidulans* and *P. cruentum* in State 1 and State 2 (Bruce et al., 1985). The values obtained for the initial decay lifetime of fluorescence emission from PSII chlorophyll *a* were 2.8 times larger for cells in State 1 than those in State 2 in *A. nidulans* and 1.6 times larger in *P. cruentum*. The direction of these changes is consistent with the model proposed by Biggins et al. (1984), but their magnitude is modest.

Scenarios (a) and (b) both imply attachment of phycobilisomes to PSII complexes. (c) In (b), the pathway of energy transfer is proposed to be phycobiliproteins → PSII → PSI. However, it has been claimed that direct energy transfer from phycobiliproteins to PSI (by-passing PSII) can be observed. For example, Kramer et al. (1985) concluded from studies of low temperature spectra of intact *A. nidulans* and *C. caldarium* cells that part of the energy absorbed by phycocyanin is directly transferred to PSI chlorophyll by-passing allophycocyanin. This conclusion is clearly at odds with scenario (a). Additional experiments are needed to interpret these seemingly incompatible findings.

7.2 Ultrastructural studies

Freeze-fracture electron micrographs of thylakoids of cyanobacteria, red algae, and cyanelles reveal long parallel rows of particles covering most of the exoplasmic (E) face. These particles have a mean diameter of ~10 nm and a uniform center-to-center spacing. Examination of thin sections of thylakoids which have retained nearly their full complement of phycobilisomes, shows that the spacing of the rows of phycobilisomes corresponds to that of the rows of E-face particles (Lefort-Tran et al., 1973; Lichtlé and Thomas, 1976; Giddings et al., 1983). Moreover, the level of PSII activity correlates directly with the density of the 10 nm particles (Wollman, 1979). From these observations, it has been conluded that each 10 nm E-face particle represents a PSII complex and that one phycobilisome is bound per PSII particle (Giddings et al., 1983).

However, spectroscopic studies indicate that in *Synechococcus* 6301, two PSII reaction centers are associated with each phycobilisome (Manodori and Melis, 1985), and that in red algae there may be as many as four PSII reaction centers per phycobilisome (Kursar and Alberte, 1983; Ley, 1984). Giddings et al. (1983) noted that the 10 nm E-face particles on the thylakoids of *Cyanophora paradoxa* appear to be made up of two subunits. It is noteworthy that the phycobilisomes of *C. paradoxa* are hemidiscoidal and presumably possess a three-cylinder triangular core. It is possible that the two-subunit particle may be a dimer of PSII reaction center complexes each unit of which is bound to one of the equivalent basal cylinders of the phycobilisome core. This arrangement corresponds to the diagrammatic representation of the putative phycobilisome-PSII complex illustrated in Fig. 19 of Giddings et al. (1983). As pointed out by Gantt et al. (1981), the structural correlations detailed above provide only circumstantial evidence for the relationship of phycobilisomes to PSII.

7.3. Role of L_{CM}, the putative 'anchor' polypeptide

All phycobilisomes contain a high molecular weight polypeptide (L_{CM}) located in the core (see Section 6.3.). This polypeptide is found in two copies per phycobilisome in *Synechococcus* 6301 and *Synechocystis* 6701 (Glazer et al., 1983; Gingrich et al., 1983), and in a similar amount per phycobilisome in *P. cruentum* (Redlinger and Gantt, 1982). When phycobiliproteins are removed from thylakoid membranes by repeated washing with low ionic strength buffer, L_{CM} is recovered in the thy-

→

Fig. 12. Photosystem II-phycobilisome preparation (A and C–I) and unfractionated thylakoids (B) negatively stained with uranyl sulfate. (A) Phycobilisomes are frequently in clusters. For easy identification individual phycobilisomes have been numbered in one of the clusters. Arrows indicate possible thylakoid components at basal (flattened) portions of phycobilisomes. (B) Unfractionated photosynthetically active thylakoid-phycobilisome preparation. (C–I) Selected areas with possible thylakoid fragments. (A) Bar equals 100 nm. (B–I) Bar equals 50 nm. (From Clement-Metral et al., 1985)

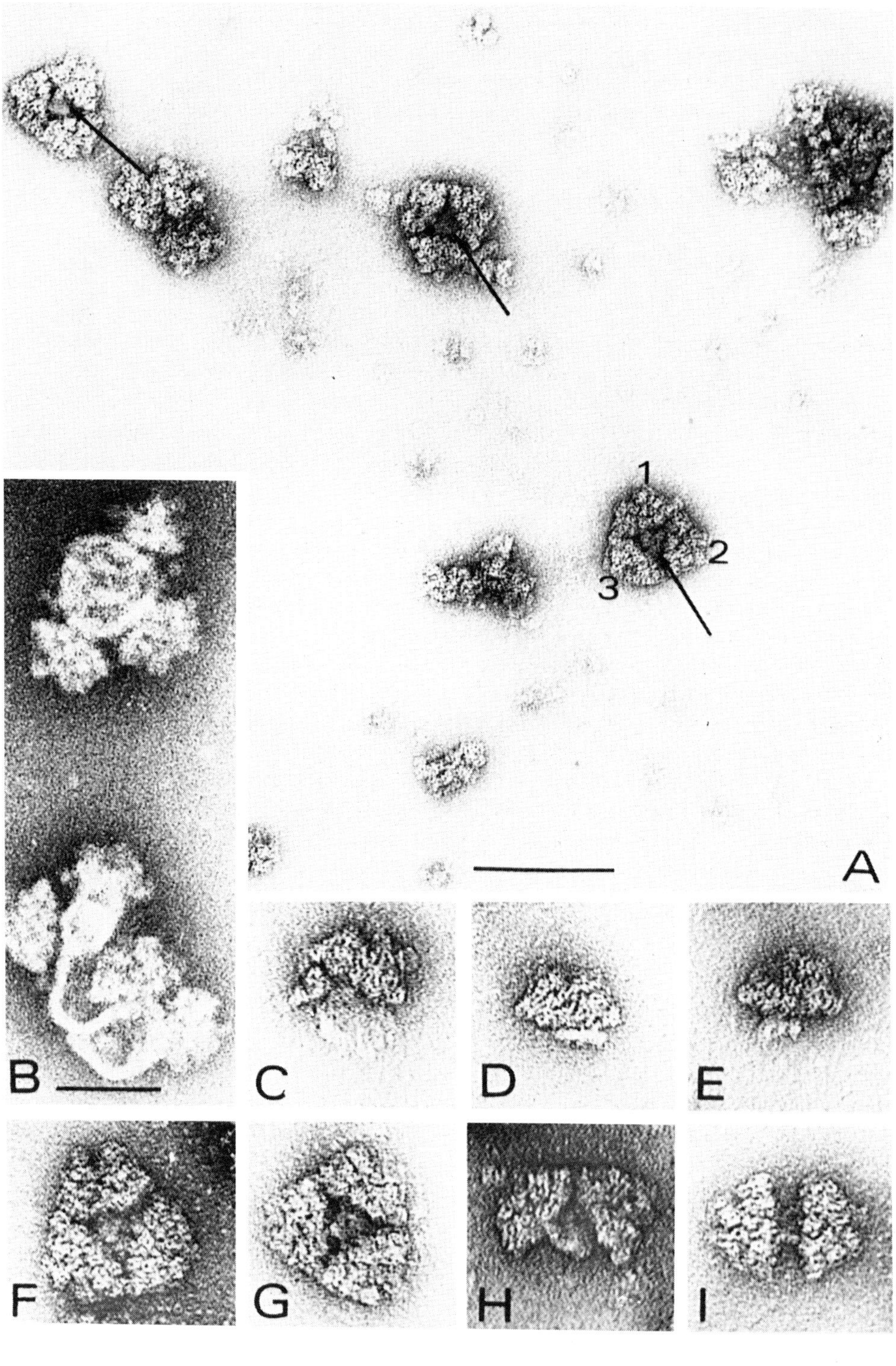
1
2
3
A
B
C
D
E
F
G
H
I

lakoid fraction (Tandeau de Marsac and Cohen-Bazire, 1977; Redlinger and Gantt, 1982; Ruskowski and Zilinskas, 1982). Redlinger and Gantt (1982) showed that the 95 kDa polypeptide recovered from *P. cruentum* phycobilisomes, or from washed thylakoid membranes, exhibited identical molecular weight, identical gel electrophoresis pattern after limited protease digestion with *S. aureus* V8 protease, and immunological cross-reactivity with an immunoglobulin G antibody fraction directed against the 95 kDa polypeptide from phycobilisomes. Kirilovsky et al. (1984) reported that the loss of the ability of trypsin-treated phycobilisomes to re-associate with cyanobacterial thylakoid membranes parallelled the degradation of the L_{CM} polypeptide.

These results suggest that a domain of L_{CM} is involved in the attachment of phycobilisomes to thylakoid membranes. However, the identity of the component(s) within the membranes that serves as the recognition site for phycobilisome attachment is not known. From studies of the release of phycobilisomes from the thylakoid membranes of *Griffithsia monilis* by trypsin, Hiller et al. (1983) suggested that a trypsin-sensitive peptide associated with the thylakoid membrane is required for attachment in addition to intact L_{CM}. However, Kirilovsky et al. (1984) found that trypsin treatment of isolated thylakoid membranes did not prevent them from binding phycobilisomes.

7.4. Phycobilisome-PSII complexes

Katoh and Gantt (1979) showed that disruption of cyanobacterial cells in sucrose/phosphate/citrate (0.3:0.5:0.3 M, respectively) permitted isolation of photosynthetically active (oxygen-evolving) vesicles with attached phycobilisomes. In such preparations, excitation of phycobiliproteins sensitized the fluorescence of PSII chlorophyll. This finding provided the basis for a procedure for the isolation of phycobilisome-PSII complexes. Disruption of *P. cruentum* cells in sucrose/phosphate/citrate (0.5:0.5:0.26 M, respectively) followed by exposure to low concentrations of the detergent lauryldimethylamine oxide and sucrose density centrifugation led to the isolation of phycobilisome-PSII particles (Fig. 12). These particles contained ~90 chlorophyll *a*/phycobilisome as compared to a ratio of ~1200 in unfractionated thylakoids and contained no detectable P700 (Clement-Metral and Gantt, 1983; Clement-Metral et al., 1985). In these particles, green light absorbed almost exclusively by the phycobilisomes was effective in promoting both oxygen evolution and 2,6-dichlorophenol-indophenol reduction. Moreover, these activities were inhibited by PSII-specific herbicides such as 3-(3,4-dichlorophenyl)-1, 1-dimethylurea. These observations demonstrate a functional coupling between phycobilisomes and PSII within the isolated complexes (Chereskin et al., 1985). The medium used for the isolation of the complexes from *P. cruentum* may also permit isolation of such particles from cyanobacteria (Pakrasi and Sherman, 1984). Detailed biochemical studies of these complexes should reveal the structural basis of phycobilisome-PSII interaction and lead to the definition of the pathway(s) of energy flow from the phycobilisome to the reaction center of PSII.

References

Alberte, R.S., Wood, A.M., Kursar, T.A. and Guillard, R.R.L. (1984) Novel phycoerythrins in marine *Synechococcus* spp. Characterization and evolutionary and ecological implications. Plant Physiol. 75, 732–739.

Anderson, L.K. (1985) Cyanobacterial phycobilisomes: mutant analysis and core structure. Ph.D. Thesis, University of California, Los Angeles, CA.

Anderson, L.K., Rayner, M.C., Sweet, R.M. and Eiserling, F.A. (1983) Regulation of *Nostoc* sp. phycobilisome structure by light and temperature. J. Bacteriol. 155, 1407–1416.

Anderson, L.K., Rayner, M.C. and Eiserling, F.A. (1984) Ultraviolet mutagenesis of *Synechocystis* sp. 6701: mutations in chromatic adaptation and phycobilisome assembly. Arch. Microbiol. 138, 237–243.

Bajaj, M. and Blundell, T. (1984) Evolution and the tertiary structure of proteins. Ann. Rev. Biophys. Bioeng. 13, 453–492.

Beale, S.I. and Cornejo, J. (1984) Enzymic transformation of biliverdin to phycocyanobilin by extracts of the unicellular red alga *Cyanidium caldarium*. Plant Physiol. 76, 7–15.

Biggins, J. (1983) Mechanism of the light state transition in photosynthesis. I. Analysis of the kinetics of cytochrome *f* oxidation in state 1 and state 2 in the red alga *Porphyridium cruentum*. Biochim. Biophys. Acta 724, 111–117.

Biggins, J. and Bruce, D. (1985) Mechanism of the light state transition in photosynthesis. III. Kinetics of the state transition in *Porphyridium cruentum*. Biochim. Biophys. Acta 806, 230–236.

Biggins, J., Campbell, C.L. and Bruce, D. (1984) Mechanism of the light state transition in photosynthesis. II. Analysis of phosphorylated polypeptides in the red alga *Porphyridium cruentum*. Biochim. Biophys. Acta 767, 138–144.

Blinks, L.R. (1954) The photosynthetic function of pigments other than chlorophyll. Ann. Rev. Plant Physiol. 5, 93–114.

Blinks, L.R. (1960) Action spectra of chromatic transients and the Emerson effect in marine algae. Proc. Natl. Acad. Sci. USA 46, 327–373.

Bonaventura, C. and Myers, J. (1969) Fluorescence and oxygen evolution from *Chlorella pyrenoidosa*. Biochim. Biophys. Acta 189, 366–383.

Brody, S.S., Treadwell, C.J. and Barber, J. (1981) Picosecond energy transfer in *Porphyridium cruentum* and *Anacystis nidulans*. Biophys. J. 34, 439–449.

Bruce, D., Biggins, J., Steiner, T. and Thewalt, M. (1985) Mechanism of the light state transition in photosynthesis. IV. Picosecond fluorescence spectroscopy of *Anacystis nidulans* and *Porphyridium cruentum* in state 1 and state 2 at 77K. Biochim. Biophys. Acta 806, 237–246.

Bryant, D.A. and Cohen-Bazire, G. (1981) Effects of chromatic illumination on cyanobacterial phycobilisomes. Evidence for the specific induction of a second pair of phycocyanin subunits in *Pseudanabaena* 7409 grown in red light. Eur. J. Biochem. 119, 415–424.

Bryant, D.A., Guglielmi, G., Tandeau de Marsac, N., Castets, A.M. and Cohen-Bazire, G. (1979) The structure of cyanobacterial phycobilisomes: a model. Arch. Microbiol. 123, 113–127.

Bryant, D.A., DeLorimier, R., Guglielmi, G., Stirewalt, V.L., Dubbs, J.M., Illman, B., Porter, R.D. and Stevens, S.E., Jr. (1985) Genes for phycobilisome components in *Synechococcus* 7002, *Pseudanabaena* 7409, and *Mastigocladus laminosus*. Vth International Symp. on Photosynthetic Prokaryotes. Grindelwald, Switzerland, Abstracts, p. 103.

Canaani, O.D. and Gantt, E. (1980) Circular dichroism and polarized fluorescence characteristics of blue-green algal allophycocyanins. Biochemistry 19, 2950–2956.

Chereskin, B.M., Clement-Metral, J.D. and Gantt, E. (1985) Characterization of a purified photosystem II-phycobilisome particle preparation from *Porphyridium cruentum*. Plant Physiol. 77, 626–629.

Clement-Metral, D. and Gantt, E. (1983) Isolation of oxygen-evolving phycobilisome-photosystem II particles from *Porphyridium cruentum*. FEBS Lett. 156, 185–188.

Clement-Metral, J.D., Gantt, E. and Redlinger, T. (1985) A photosystem II-phycobilisome preparation from the red alga *Porphyridium cruentum*: oxygen evolution, ultrastructure, and polypeptide resolution. Arch. Biochem. Biophys. 238, 10–17.

Cohen-Bazire, G. and Bryant, D.A. (1982) Phycobilisomes: composition and structure. In N.G. Carr and B.A. Whitton (eds.), The Biology of the Cyanobacteria, Vol. 19. Blackwell, Oxford, pp. 143–189.

Deisenhofer, J., Michel, H. and Huber, R. (1985) The structural basis of photosynthetic light reactions in bacteria. Trends Biochem. Sci. 10, 243–248.

Duysens, L.M.N. (1952) Transfer of excitation energy in photosynthesis. Doctoral thesis, University of Utrecht, The Netherlands, 96 pp.

Duysens, L.M.N. (1964) Photosynthesis. Prog. Biophys. Mol. Biol. 14, 3–104.

Fisher, R.G., Woods, N.E., Fuchs, H.E. and Sweet, R.M. (1980) Three-dimensional structures of C-phycocyanin and B-phycoerythrin at 5A resolution. J. Biol. Chem. 255, 5082–5089.

Fork, D.C. and Amesz, J. (1969) Action spectra and energy transfer in photosynthesis. Ann. Rev. Plant Physiol. 20, 305–328.

Förster, Th. W. (1965) Delocalized excitation and excitation transfer. In O. Sinanoğlou (ed.), Modern Quantum Chemistry — Istanbul Lectures, Ser. IIIB. Academic Press, London, pp. 93–137.

Frank, G., Sidler, W., Widmer, H. and Zuber, H. (1978) The complete amino acid sequence of both subunits of C-phycocyanin from the cyanobacterium *Mastigocladus laminosus*. Hoppe-Seyler's Z. Physiol. Chem. 359, 1491–1507.

Füglistaller, P., Suter, F. and Zuber, H. (1983) The complete amino acid sequence of both subunits of phycoerythrocyanin from the thermophilic cyanobacterium *Mastigocladus laminosus*. Hoppe-Seyler's Z. Physiol. Chem. 364, 691–712.

Füglistaller, P., Rümbeli, R., Suter, F. and Zuber, H. (1984) Minor polypeptides from the phycobilisome of the cyanobacterium *Mastigocladus laminosus*. Hoppe-Seyler's Z. Physiol. Chem. 365, 1085–1096.

Füglistaller, P., Suter, F. and Zuber, H. (1985) Linker polypeptides of the phycobilisome from the cyanobacterium *Mastigocladus laminosus*: amino acid sequences and relationships. Hoppe-Seyler's Z. Physiol. Chem. 366, 993–1001.

Gantt, E. (1980) Structure and function of phycobilisomes: light-harvesting pigment complexes in red and blue-green algae. Int. Rev. Cytol. 66, 45–80.

Gantt, E. and Lipschultz, C.A. (1972) Phycobilisomes of *Porphyridium cruentum*. I. Isolation. J. Cell Biol. 54, 313–324.

Gantt, E. and Lipschultz, C.A. (1974) Phycobilisomes of *Porphyridium cruentum*: pigment analysis. Biochemistry 13, 2960–2966.

Gantt, E. and Lipschultz, C.A. (1980) Structure and phycobiliprotein composition of phycobilisomes from *Griffithsia pacifica* (Rhodophyceae). J. Phycol. 16, 394–398.

Gantt, E., Lipschultz, C.A., Grabowski, J. and Zimmerman, B.K. (1979) Phycobilisomes from blue-green and red algae. Isolation criteria and dissociation characteristics. Plant Physiol. 63, 615–620.

Gantt, E., Canaani, O., Lipschultz, C.A. and Redlinger, T. (1981) Structure of phycobilisomes and their relationship to photosystem II. In G. Akoyunoglou (ed.), Photosynthesis III. Structure and Molecular Organization of the Photosynthetic Apparatus. Balaban International Science Services, Philadelphia, pp. 143–153.

Giddings, T.H., Jr., Wasmann, C. and Staehelin, L.A. (1983) Structure of the thylakoids and envelope membranes of the cyanelles of *Cyanophora paradoxa*. Plant Physiol. 71, 409–419.

Gingrich, J.C., Blaha, L.K. and Glazer, A.N. (1982a) Rod substructure in cyanobacterial phycobilisomes: analysis of *Synechocystis* 6701 mutants low in phycoerythrin. J. Cell. Biol. 92, 261–268.

Gingrich, J.C., Williams, R.C. and Glazer, A.N. (1982b) Rod substructure in cyanobacterial phycobilisomes: phycoerythrin assembly in *Synechocystis* 6701 phycobilisomes. J. Cell Biol. 95, 170–178.

Gingrich, J.C., Lundell, D.J. and Glazer, A.N. (1983) Core substructure in cyanobacterial phycobilisomes. J. Cell. Biochem. 22, 1–14.

Glazer, A.N. (1981) Photosynthetic accessory proteins with bilin prosthetic groups. In M.D. Hatch and N.K. Boardman (eds.), The Biochemistry of Plants, Vol. 8, Academic Press, New York, pp. 51–96.

Glazer, A.N. (1982) Phycobilisomes: structure and dynamics. Ann. Rev. Microbiol. 36, 171–196.

Glazer, A.N. (1984) Phycobilisome. A macromolecular complex optimized for light energy transfer. Biochim. Biophys. Acta 768, 29–51.

Glazer, A.N. (1985) Light harvesting by phycobilisomes. Ann. Rev. Biophys. Biophys. Chem. 14, 47–77.

Glazer, A.N. and Hixson, C.S. (1977) Subunit structure and chromophore composition of rhodophytan phycoerythrins. J. Biol. Chem. 252, 32–42.

Glazer, A.N., Williams, R.C., Yamanaka, G. and Schachman, H.K. (1979) Characterization of cyanobacterial phycobilisomes in zwitterionic detergents. Proc. Natl. Acad. Sci. USA 76, 6162–6166.

Glazer, A.N., West, J.A. and Chan, C. (1982) Phycoerythrins as chemotaxonomic markers in red algae: a survey. Biochem. Syst. Ecol. 10, 203–215.

Glazer, A.N., Lundell, D.J., Yamanaka, G. and Williams, R.C. (1983) The structure of a 'simple' phycobilisome. Ann. Microbiol. Inst. Pasteur 134B, 159–180.

Guglielmi, G., Cohen-Bazire, G. and Bryant, D.A. (1981) The structure of *Gloeobacter violaceus* and its phycobilisomes. Arch. Microbiol. 129, 181–189.

Halldal, P. (1970) The photosynthetic apparatus of microalgae and its adaptation to environmental factors. In P. Halldal (ed.), Photobiology of Microorganisms. Wiley, London, pp. 17–55.

Haxo, F.T. (1960) The wavelength dependence of photosynthesis and the role of accessory pigments. In M.B. Allen (ed.), Comparative Biochemistry of Photoreactive Systems, Academic Press, New York, pp. 339–360.

Haxo, F.T. and Blinks, L.R. (1950) Photosynthetic action spectra of marine algae. J. Gen. Physiol. 33, 389–422.

Hiller, R.G., Post, A. and Stewart, A.C. (1983) Isolation of intact detergent-free phycobilisomes by trypsin. FEBS Lett. 156, 180–184.

Honsell, E., Kosovel, V. and Talarico, L. (1984) Phycobiliprotein distribution in Rhodophyta: studies and interpretation on the basis of their absorption spectra. Bot. Mar. 27, 1–16.

Karukstis, K.K. and Sauer, K. (1983) Fluorescence decay kinetics of chlorophyll in photosynthetic membranes. J. Cell. Biochem. 23, 131–158.

Katoh, T. and Gantt, E. (1979) Photosynthetic vesicles with bound phycobilisomes from *Anabaena variabilis*. Biochim. Biophys. Acta 546, 383–393.

Kawamura, M., Mimuro, M. and Fujita, Y. (1979) Quantitative relationship between two reaction centers in the photosynthetic system of blue-green algae. Plant Cell Physiol. 20, 697–705.

Ke, B., Inoue, H., Babcock, G.T., Fang, Z.-X. and Dolan, E. (1982) Optical and EPR characterizations of oxygen-evolving photosystem II subchloroplast fragments isolated from the thermophilic blue-green alga *Phormidium laminosum*. Biochim. Biophys. Acta 682, 297–306.

Kirilovsky, D., Lavintman, N., Ish-Shalom, D. and Ohad, I. (1984) Specificity of energy transfer to photosystem II by in vitro reassociated homologous and heterologous membrane-bound phycobilisomes. Biochim. Biophys. Acta 767, 451–459.

Klotz, A.V. and Glazer, A.N. (1985) Characterization of the bilin attachment sites in R-phycoerythrin. J. Biol. Chem. 260, 4856–4863.

Koller, K.-P. (1979) Untersuchungen zur Biliproteinkomposition und Lokalisation in isolierten Phycobilisomen von *Rhodella violacea*. Doctoral dissertation. Philipps-Universitat Marburg/Lahn.

Koller, K.-P., Wehrmeyer, W. and Mörschel, E. (1978) Biliprotein assembly in the disc-shaped phycobilisomes of *Rhodella violacea*. On the molecular composition of energy-transferring complexes (tripartite units) forming the periphery of the phycobilisome. Eur. J. Biochem. 91, 57–63.

Kramer, H.J.M., Westerhuis, W.H.J. and Amesz, J. (1985) Low temperature spectroscopy of intact algae. Physiol. Veg. 23, 535–543.

Kursar, T.A. and Alberte, R.S. (1983) Photosynthetic unit organization in a red alga. Relationships between light-harvesting pigments and reaction centers. Plant Physiol. 72, 409–414.

Lagarias, J.C., Glazer, A.N. and Rapoport, H. (1979) Chromopeptides from C-phycocyanin. Structure and linkage of a phycocyanobilin bound to the β subunit. J. Am. Chem. Soc. 101, 5030–5037.

Lefort-Tran, M., Cohen-Bazire, G. and Pouphile, M. (1973) Les membranes photosynthétiques des algues à biliprotéines observées par cryodécapage. J. Ultrastruct. Res. 44, 199–209.

Ley, A.C. (1984) Effective absorption cross-sections in *Porphyridium cruentum*. Implications for energy transfer between phycobilisomes and photosystem II reaction centers. Plant Physiol. 74, 451–454.

Ley, A.C. and Butler, W.L. (1976) The efficiency of energy transfer from Photosystem II to Photosystem I in *Porphyridium cruentum*. Proc. Natl. Acad. Sci. USA 73, 3957–3960.

Ley, A.C. and Butler, W.L. (1977) The distribution of excitation energy between photosystem I and photosystem II in *Porphyridium cruentum*. In S. Miyachi, K. Katoh, Y. Fujita and K. Shibata (eds.), Special Edition of Plant and Cell Physiology. Japanese Society of Plant Physiologists, Tokyo, pp. 33–46.

Ley, A.C. and Butler, W.L. (1980) Energy distribution in the photochemical apparatus of *Porphyridium cruentum* in state I and state II. Biochim. Biophys. Acta 592, 349–363.

Lichtlé, C. and Thomas, J.C. (1976) Étude ultrastructurale des thylacoides des algues à phycobiliprotéines, comparaison des résultats obtenus par fixation classique et cryodécapage. Phycologia 15, 393–404.

Lönneborg, A., Lind, L.K., Kalla, S.R., Gustaffson, P. and Öquist, G. (1985) Acclimation processes in the light-harvesting system of the cyanobacterium *Anacystis nidulans* following a light shift from white to red light. Plant Physiol. 78, 110–114.

Lundell, D.J. and Glazer, A.N. (1981) Allophycocyanin B. A common β subunit in *Synechococcus* allophycocyanin B (λ_{max} 670 nm) and allophycocyanin (λ_{max} 650 nm). J. Biol. Chem. 12600–12606.

Lundell, D.J. and Glazer, A.N. (1983a) Molecular architecture of a light-harvesting antenna. Structure of the 18S rod-core subassembly of the *Synechococcus* 6301 phycobilisome. J. Biol. Chem. 258, 894–901.

Lundell, D.J. and Glazer, A.N. (1983b) Molecular architecture of a light-harvesting antenna. Core substructure in *Synechococcus* 6301 phycobilisomes: two new allophycocyanin and allophycocyanin B complexes. J. Biol. Chem. 258, 902–908.

Lundell, D.J. and Glazer, A.N. (1983c) Molecular architecture of a light-harvesting antenna. Quaternary interactions in the *Synechococcus* 6301 phycobilisome core as revealed by partial tryptic digestion and circular dichroism. J. Biol. Chem. 258, 8708–8713.

Lundell, D.J., Williams, R.C. and Glazer, A.N. (1981a) Molecular architecture of a light-harvesting antenna. In vitro assembly of the rod substructures of *Synechococcus* 6301 phycobilisomes. J. Biol. Chem. 256, 3580–3592.

Lundell, D.J., Yamanaka, G. and Glazer, A.N. (1981b) A terminal energy acceptor of the phycobilisome: the 75,000-dalton polypeptide of *Synechococcus* 6301 phycobilisomes – A new biliprotein. J. Cell Biol. 91, 315–319.

Lundell, D.J., Glazer, A.N., DeLange, R.J. and Brown, D.M. (1984) Bilin attachment sites in the α and β subunits of B-phycoerythrin. Amino acid sequence studies. J. Biol. Chem. 259, 5472–5480.

Lundell, D.J., Glazer, A.N., Melis, A. and Malkin, R. (1985) Characterization of a cyanobacterial photosystem I complex. J. Biol. Chem. 260, 646–654.

Manodori, A. and Melis, A. (1984) Photochemical apparatus organization in *Anacystis nidulans* (Cyanophyceae). Effect of CO_2 concentration during cell growth. Plant Physiol. 74, 67–71.

Manodori, A. and Melis, A. (1985) Phycobilisome-photosystem II association in *Synechococcus* 6301 (Cyanophyceae). FEBS Lett. 181, 79–82.

Manodori, A. and Melis, A. (1986) Light quality regulates photosystem stoichiometry in cyanobacteria. In G. Akoyunoglou (ed.), Regulation of Chloroplast Differentiation. Alan R. Liss, Inc., New York, pp. 653–662.

Manodori, A., Alhadeff, M., Glazer, A.N. and Melis, A. (1984) Photochemical apparatus organization in *Synechococcus* 6301 (*Anacystis nidulans*). Effect of phycobilisome mutation. Arch. Microbiol. 139, 117–123.

Mimuro, M. and Fujita, Y. (1977) Estimation of chlorophyll *a* distribution in the photosynthetic pigment systems I and II of the blue-green alga *Anabaena variabilis*. Biochim. Biophys. Acta 459, 376–389.

Mimuro, M., Yamazaki, I., Murao, T., Yamazaki, T., Yoshihara, K. and Fujita, Y. (1984) Excitation transfer in phycobilin-chlorophyll *a* system in blue-green and red algae. In C. Sybesma (ed.), Advances in Photosynthesis Research, Vol. I. Martinus Nijhoff/Dr.W. Junk Publishers, The Hague/Boston/Lancaster, pp. 21–28.

Mörschel, E., Koller, K.-P., Wehrmeyer, W. and Schneider, H. (1977) Biliprotein assembly in the disc-shaped phycobilisomes of *Rhodella violacea*. I. Electron microscopy of phycobilisomes in situ and analysis of their architecture after isolation and negative staining. Cytobiologie 16, 118–129.

Mörschel, E., Wehrmeyer, W. and Koller, K.-P. (1980) Biliprotein assembly in the disc-shaped phycobilisomes of *Rhodella violacea*. Electron microscopical and biochemical analysis of B-phycoerythrin and B-phycoerythrin-C-phycocyanin aggregates. Eur. J. Cell Biol. 21, 319–327.

Murata, N. (1970) Control of excitation transfer in photosynthesis. IV. Kinetics of chlorophyll *a* fluorescence in *Porphyra yezoensis*. Biochim. Biophys. Acta 205, 379–389.

Myers, J., Graham, J.R. and Wang, R.T.K. (1980) Light harvesting in *Anacystis nidulans* studied in pigment mutants. Plant Physiol. 66, 1144–1149.

Nagy, J.O., Bishop, J.E., Klotz, A.V., Glazer, A.N. and Rapoport, H. (1985) Bilin attachment sites in the α, β, and γ subunits of R-phycoerythrin. Structural studies on singly and doubly linked phycourobilins. J. Biol. Chem. 260, 4864–4868.

Nies, M. and Wehrmeyer, W. (1980) Isolation and biliprotein characterization of phycobilisomes from the thermophilic cyanobacterium *Mastigocladus laminosus* Cohn. Planta 150, 330–337.

Nies, M. and Wehrmeyer, W. (1981) Biliprotein assembly in the hemidiscoidal phycobilisomes of the thermophilic cyanobacterium *Mastigocladus laminosus* Cohn. Characterization of the dissociation products with special reference to the peripheral phycoerythrocyanin-phycocyanin complexes. Arch. Microbiol. 129, 374–379.

Ong, L.J., Glazer, A.N. and Waterbury, J.B. (1984) An unusual phycoerythrin from a marine cyanobacterium. Science 224, 80–83.

Pakrasi, H.B. and Sherman, L.A. (1984) A highly active oxygen-evolving photosystem II preparation from the cyanobacterium *Anacystis nidulans*. Plant Physiol. 74, 742–745.

Porter, G., Tredwell, C.J., Searle, G.F.W. and Barber, J. (1978) Picosecond time-resolved energy transfer in *Porphyridium cruentum*. I. In the intact alga. Biochim. Biophys. Acta 501, 232–245.

Rapoport, H. and Glazer, A.N. (1985) Bilins and bilin-protein linkages in phycobiliproteins: structural and spectroscopic studies. In G. Blauer and H. Sund (eds.), Optical Properties and Structure of Tetrapyrroles. Walter de Gruyter & Co., Berlin/New York, pp. 411–423.

Redlinger, T. and Gantt, E. (1982) A Mr 95,000 polypeptide in *Porphyridium cruentum* phycobilisomes and thylakoids: possible function in linkage of phycobilisomes to thylakoids and in energy transfer. Proc. Natl. Acad. Sci. USA 79, 5542–5546.

Rippka, R. and Stanier, R.Y. (1978) The effects of anaerobiosis on nitrogenase synthesis and heterocyst development by nostocacean cyanobacteria. J. Gen. Microbiol. 105, 83–94.

Rippka, R., Waterbury, J.B. and Cohen-Bazire, G. (1974) A cyanobacterium which lacks thylakoids. Arch. Microbiol. 100, 419–436.

Rosinski, J., Hainfeld, J.F., Rigbi, M. and Siegelman, W.H. (1981) Phycobilisome ultrastructure and chromatic adaptation in *Fremyella diplosiphon*. Ann. Bot. 47, 1–12.
Rümbeli, R., Schirmer, T., Bode, W., Sidler, W. and Zuber, H. (1985) Crystallization of phycoerythrocyanin from the cyanobacterium *Mastigocladus laminosus* and preliminary characterization of two crystal forms. J. Mol. Biol. 186, 197–200.
Ruskowski, M. and Zilinskas, B.A. (1982) Allophycocyanin I and the 95 kilodalton polypeptide. The bridge between phycobilisomes and membranes. Plant Physiol. 70, 1055–1059.
Scheer, H. (1982) Phycobiliproteins: molecular aspects on photosynthetic antenna system. In F.K. Fong (ed.), Light Reaction Path of Photosynthesis. Springer-Verlag, Berlin, pp. 7–45.
Scheer, H. and Kufer, W. (1977) Conformational studies on C-phycocyanin from *Spirulina platensis*. Z. Naturforsch. 32c, 513–519.
Schirmer, T., Bode, W., Huber, R., Sidler, W. and Zuber, H. (1985) X-Ray crystallographic structure of the light-harvesting biliprotein C-phycocyanin from the thermophilic cyanobacterium *Mastigocladus laminosus* and its resemblance to globin structures. J. Mol. Biol. 184, 257–277.
Schoenleber, R.W., Leung, S.-L., Lundell, D.J., Glazer, A.N. and Rapoport, H. (1983) Chromopeptides from phycoerythrins. Structure and linkage of a phycoerythrobilin tryptic tripeptide derived from B-phycoerythrin. J. Am. Chem. Soc. 105, 4072–4076.
Schoenleber, R.W., Lundell, D.J., Glazer, A.N. and Rapoport, H. (1984a) Bilin attachment sites in the α and β subunits of B-phycoerythrin. Structural studies on a doubly-linked phycoerythrobilin. J. Biol. Chem. 259, 5481–5484.
Schoenleber, R.W., Lundell, D.J., Glazer, A.N. and Rapoport, H. (1984b) Bilin attachment sites in the α and β subunits of B-phycoerythrin. Structural studies on the singly linked phycoerythrobilins. J. Biol. Chem. 259, 5485–5489.
Schopf, J.W. (1978) The evolution of the earliest cells. Sci. Am. 239 (Sept.), 111–138.
Schopf, J.W., ed. (1983) Earth's Earliest Biosphere. Its Origin and Evolution. Princeton University Press, Princeton, 543 pp.
Searle, G.F.W., Barber, J., Porter, G. and Tredwell, C.J. (1978) Picosecond time-resolved energy transfer in *Porphyridium cruentum*. Part II. In the isolated light harvesting complex (phycobilisomes). Biochim. Biophys. Acta 501, 246–256.
Sidler, W., Gysi, J., Isker, E. and Zuber, H. (1981) The complete amino acid sequence of both subunits of allophycocyanin, a light harvesting protein-pigment complex from the cyanobacterium *Mastigocladus laminosus*. Hoppe-Seyler's Z. Physiol. Chem. 362, 611–628.
Sidler, W., Kumpf, B., Zuber, H. and Rüdiger, W. (1985) The complete amino acid sequence of C-phycoerythrin from *Fremyella diplosiphon*. Vth International Symposium on Photosynthetic Prokaryotes, Grindelwald, Switzerland, Abstracts, p. 303.
Tandeau de Marsac, N. (1983) Phycobilisomes and complementary chromatic adaptation in cyanobacteria. Bull. Inst. Pasteur 81, 201–254.
Tandeau de Marsac, N. and Cohen-Bazire, G. (1977) Molecular composition of cyanobacterial phycobilisomes. Proc. Natl. Acad. Sci. USA 74, 1635–1639.
Thomas, J.C., Mousseau, A. and Hauswirth, N. (1984) Photocontrol of phycoerythrocyanin synthesis in an *Oscillatoria* strain (Cyanobacteria). In C. Sybesma (ed.), Advances in Photosynthesis Research, Vol. II. Martinus Nijhoff/Dr. W. Junk Publishers, The Hague/Boston/Lancaster, pp. 691–694.
Troxler, R.F., Kelly, P. and Brown, S.B. (1978) Phycocyanobilin synthesis in the unicellular rhodophyte *Cyanidium caldarium*. Biochem. J. 172, 569–576.
Troxler, R.F., Brown, A.S. and Brown, S.B. (1979) Bile pigment synthesis in plants. Mechanism of ^{18}O incorporation into phycocyanobilin in the unicellular rhodophyte *Cyanidium caldarium*. J. Biol. Chem. 254, 3411–3418.
Wang, R.T., Stevens, C.L.R. and Myers, J. (1977) Action spectra for photoreactions I and II of photosynthesis in the blue-green alga *Anacystis nidulans*. Photochem. Photobiol. 25, 103–108.
Wanner, G. and Köst, H.-P. (1980) Investigations on the arrangement and fine structure of *Porphyridium cruentum* phycobilisomes. Protoplasma 102, 97–109.
Wehrmeyer, W. (1971) Elektronenmikroskopische Untersuchung zur Feinstruktur von *Porphyridium violaceum* (Kornmann) mit Bemerkungen über seine taxonomische Stellung. Arch. Mikrobiol. 75, 121–139.
Wehrmeyer, W. (1983) Organization and composition of cyanobacterial and rhodophytan phycobilisomes. In G.C. Papageorgiou and L. Packer (eds.), Photosynthetic Prokaryotes: Cell Differentiation and Function, Elsevier, Amsterdam, pp. 1–22.

Wehrmeyer, W., Wendler, J. and Holzwarth, A.R. (1985) Biochemical and functional characterization of a peripheral phycobilisome unit from *Porphyridium cruentum*. Measurement of picosecond energy transfer kinetics. Eur. J. Cell Biol. 36, 17–23.

Wendler, J., Holzwarth, A.R. and Wehrmeyer, W. (1984) Picosecond time-resolved energy transfer in phycobilisomes isolated from the red alga *Porphyridium cruentum*. Biochim. Biophys. Acta 765, 58–67.

Williams, R.C., Gingrich, J.C. and Glazer, A.N. (1980) Cyanobacterial phycobilisomes. Particles from *Synechocystis* 6701 and two pigment mutants. J. Cell Biol. 85, 558–566.

Wollman, F.-A. (1979) Ultrastructural comparison of *Cyanidium caldarium* wild type and IIIC mutant lacking phycobilisomes. Plant Physiol. 63, 375–381.

Yamagishi, A. and Katoh, S. (1983) Two chlorophyll-binding subunits of the photosystem 2 reaction center complex isolated from the thermophilic cyanobacterium *Synechococcus* sp. Arch. Biochem. Biophys. 225, 836–846.

Yamagishi, A. and Katoh, S. (1984) A photoactive photosystem II reaction center complex lacking a chlorophyll-binding 40 kilodalton subunit from the thermophilic cyanobacterium *Synechococcus* sp. Biochim. Biophys. Acta 765, 118–124.

Yamagishi, A. and Katoh, S. (1985) Further characterization of the two photosystem II reaction center complex preparations from the thermophilic cyanobacterium *Synechococcus* sp. Biochim. Biophys. Acta 807, 74–80.

Yamanaka, G. and Glazer, A.N. (1980) Dynamic aspects of phycobilisome structure. Phycobilisome turnover during nitrogen starvation in *Synechococcus* sp. Arch. Microbiol. 124, 39–47.

Yamanaka, G. and Glazer, A.N. (1981) Dynamic aspects of phycobilisome structure: modulation of phycocyanin content of *Synechococcus* phycobilisomes. Arch. Microbiol. 130, 23–30.

Yamanaka, G., Glazer, A.N. and Williams, R.C. (1978) Cyanobacterial phycobilisomes. Characterization of the phycobilisomes of *Synechococcus* sp. 6301. J. Biol. Chem. 253, 8303–8310.

Yamanaka, G., Glazer, A.N. and Williams, R.C. (1980) Molecular architecture of a light-harvesting antenna. Comparison of wild type and mutant *Synechococcus* 6301 phycobilisomes. J. Biol. Chem. 255, 11004–11010.

Yamanaka, G., Lundell, D.J. and Glazer, A.N. (1982) Molecular architecture of a light-harvesting antenna. Isolation and characterization of phycobilisome subassembly particles. J. Biol. Chem. 257, 4077–4086.

Yamazaki, I., Mimuro, M., Murao, T., Yamazaki, T., Yoshihara, K. and Fujita, Y. (1984) Excitation energy transfer in the light harvesting antenna system of the red alga *Porphyridium cruentum* and the blue-green alga *Anacystis nidulans*: analysis of time-resolved fluorescence spectra. Photochem. Photobiol. 39, 233–240.

Yu, M.-H. (1981) Biliproteins of cyanobacteria and rhodophyta: studies of structure and assembly. Ph.D. Thesis, University of California, Berkeley, CA.

Yu, M.-H. and Glazer, A.N. (1982) Cyanobacterial phycobilisomes. Role of the linker polypeptides in the assembly of phycocyanin. J. Biol. Chem. 257, 3429–3433.

Yu, M.-H., Glazer, A.N. and Williams, R.C. (1981) Cyanobacterial phycobilisomes. Phycocyanin assembly in the rod substructures of *Anabaena variabilis* phycobilisomes. J. Biol. Chem. 256, 13130–13136.

Zilinskas, B.A. (1982) Isolation and characterization of the central component of the phycobilisome core. Plant Physiol. 70, 1060–1065.

Zilinskas, B.A. and Howell, D.A. (1983) Role of the colorless polypeptides in phycobilisome assembly in *Nostoc* sp. Plant Physiol. 71, 379–387.

Zilinskas, B.A., Zimmerman, E. and Gantt, E. (1978) Allophycocyanin forms isolated from *Nostoc* sp. Photochem. Photobiol. 27, 587–595.

Zuber, H. (1985) Structure and function of light-harvesting complexes and their polypeptides. Photochem. Photobiol. 42, 821–844.

P. Fay and C. Van Baalen (Eds.) The Cyanobacteria

4

Carbon dioxide fixation and its regulation in cyanobacteria

F. ROBERT TABITA

Center for Applied Microbiology, and Department of Microbiology, The University of Texas at Austin, Austin, Texas, 78712-1095, USA

1. Introduction

Carbon metabolism in cyanobacteria, of necessity, is concerned with the assimilation of carbon dioxide, since without exception, this gas is the preferred source of carbon. Those few cyanobacteria capable of dark chemoheterotrophic growth seem to tolerate this mode of metabolism at best. For example, *Nostoc* sp. strain MAC, perhaps the most rapidly metabolizing chemoheterotroph, is poised for immediate and full photosynthetic metabolism when supplied with light (Bottomley and Van Baalen, 1978a, b). The preference for photoautotrophic metabolism is also apparent when the maximum autotrophic growth rates are compared to heterotrophic growth (Bottomley and Van Baalen, 1978a). From these results, it is obvious that carbon dioxide fixation is of paramount importance to these organisms and should be fully considered in any discussion of metabolism. It will thus be the purpose of this contribution to consider the various factors which are involved in the control of carbon dioxide fixation in cyanobacteria. Heterotrophic metabolism has been considered in several reviews (Smith, 1973; Smith, 1982; Stanier, 1973; Van Baalen and Pulich, 1973) and the reader is encouraged to consult these articles and others for further information on this subject. (See also Sections 1.1. and 1.2. of Chapter 5.)

2. The reductive assimilation of carbon dioxide in cyanobacteria

The cyanobacteria, along with several diverse groups of organisms found in the microbial world, including purple and green sulfur and non-sulfur photosynthetic bacteria, chemolithotrophic bacteria, the methanogens, various *Pseudomonas* species, *Rhizobium* species, Actinomycetes, and perhaps several other prokaryotes, as well as eukaryotic algae, are all capable of utilizing carbon dioxide as their sole source of carbon. With the exception of the methanogens and the green sulfur photosynthetic bacteria, the primary route by which carbon is assimilated in these organisms involves the reactions of the Calvin-Benson reductive pentose phosphate pathway. There are two enzymes unique to this metabolic cycle. These are phosphoribulokinase (ATP: D-ribulose 5-phosphate 1-phosphotransferase; EC 2.7.1.19) and ribulose 1,5-bisphosphate (RuBP) carboxylase/oxygenase (3-phospho-D-glycerate carboxylase [dimerizing]; EC 4.1.1.39) (RuBPC/O). RuBPC/O catalyzes the actual fixation of CO_2, with the gaseous substrate fixed into the enediol of RuBP to form a six-carbon carboxylated intermediate which then splits to yield two molecules of 3-phosphoglyceric acid. Phosphoglyceric acid is then subsequently metabolized by enzymes common to intermediary metabolism, with the CO_2 acceptor molecule, RuBP, regenerated through the action of phosphoribulokinase. The enzymes thus catalyze sequential reactions (Fig. 1). The net effect of this metabolism is to produce one molecule of glyceraldehyde-3-phosphate from three molecules of CO_2 (Gottschalk, 1979). This conversion of inorganic gaseous CO_2 into utilizable organic carbon represents a huge energy commitment by the cell, one that is obviously necessary in order to ensure further growth.

There is evidence to indicate that the fixation of $^{14}CO_2$ into 3-phosphoglyceric acid may not be the only mechanism by which cyanobacteria assimilate carbon. Early results with *Anacystis nidulans* indicated that significant amounts of label were incorporated into glutamate, aspartate, alanine and phosphoenolpyruvate as well as 3-phosphoglycerate (Richter, 1961). Similar results were obtained by Dohler (1974, 1976a, b) who showed that light intensity, light quality and the gas phase markedly affected the pattern of isotope distribution. These results suggest that the fixation of carbon dioxide onto 3-carbon acceptors, i.e., via phosphoenolpyruvate carboxylase (PEPC) or pyruvate carboxylase may be more important than previously realized. Certainly, with the recent isolation and cloning of the PEPC gene from *Anacystis nidulans* (Izui et al, 1985), a more definitive role for this enzyme and its control may be forthcoming. Previous studies reporting higher levels of PEPC compared to RuBPC/O are tempered by the fact that, apparently, there was no activation of the latter enzyme prior to initiation of the reaction (Colman et al., 1976). The curious finding that up to 17% of the total isotope incorporation (20.9% of the label found in an ethanol extract) is present in an unidentified compound (Norris et al., 1955), subsequently identified as citrulline (Linko et al., 1957) was obtained with *Nostoc muscorum*. Approximately 7% of the label in an ethanol extract was found in this compound in an unidentified *Synechococcus* as well. No explanation for this unusual labeling pattern was offered, but since most of the label was incorporated into the carbamyl portion of the molecule, it now seems evident that carbamyl phosphate synthetase may be the enzyme responsible for this labeling pattern. We have recently found (G.H. Chen and F.R. Tabita, manuscript in preparation) that a marine nitrogen fixing cyanobacterium, *Anabaena* sp. 1F (Gotto et al., 1979), incorporates bicarbonate into large intracellular pools of citrulline. Furthermore, significant levels of glutamine-dependent carbamyl phosphate synthetase activity were readily demonstrated, as was the incorporation of ornithine and CO_2 into citrulline in vivo and in vitro. Thus, at least a portion of the required carbon and nitrogen is assimilated through the pathway

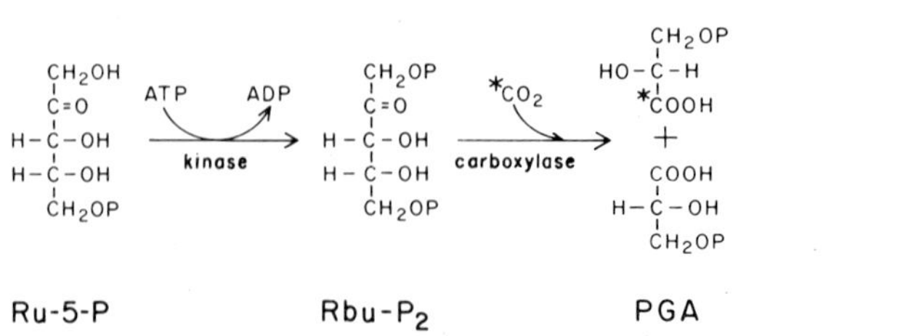

Fig. 1. Enzymatic reactions unique to the Calvin reductive pentose phosphate pathway.

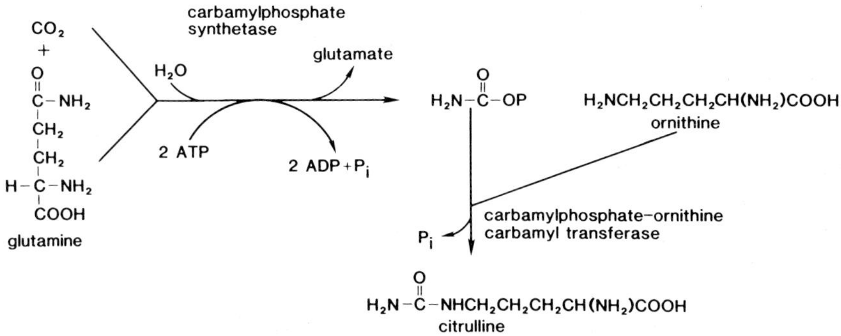

Fig. 2. An alternative CO_2 assimilatory pathway in cyanobacteria.

represented in Fig. 2. Current studies are directed at elucidating the significance, control and distribution of this alternative CO_2 assimilatory mechanism. It has been suggested that the fixation of CO_2 into citrulline may be related to the need for arginine in the synthesis of cyanophycin granules (Weathers and Allen, 1978).

3. Transport of inorganic carbon in cyanobacteria

It has been shown that the affinity of green algae for CO_2 greatly depends on whether cells are grown on high or low levels of CO_2 (Berry and Bowes, 1976). Low CO_2-grown *Chlamydomonas reinhardtii* exhibit a $K_m(CO_2)$ which is 10-fold lower than high CO_2-grown cells and thus low-CO_2 grown cells have a much greater capacity to concentrate CO_2 from the medium. The consequence of this increased ability to concentrate CO_2 is probably related to a decrease in the amount of glycolate excreted into the medium, since the oxygenase activity of RuBPC/O will be inhibited at high intracellular levels of CO_2 (section 4 of this chapter). Similar experiments with the cyanobacterium *Anabaena variabilis* indicated that the half-maximum rate of photosynthesis was obtained at 0.25 and 2.5 μM CO_2 for low and high CO_2-grown cells, respectively, at pH 8.0. The level of fully activated RuBPC/O was about 10-fold higher than the level of PEPC and neither of these carboxylation enzymes were affected by the CO_2 concentration used for growth (Kaplan et al., 1980). Thus, in contrast to previous findings where enzymes were not assayed properly (Dohler, 1974; 1976a and b; Colman et al., 1976), RuBPC/O is obviously the major CO_2 fixation enzyme. In addition, intact cells exhibited a substantially higher affinity for CO_2 than the enzyme in vitro. Since there is enough RuBPC/O present within the cells to account for the rate of CO_2 fixation, it was concluded that the concentration of CO_2 at or near RuBPC/O within the cells was high. Such studies stimulated further investigations of the transport of inorganic carbon (C_{inorg}) by cyanobacteria and the initial results indicated that C_{inorg} transport within *A. variabilis* is an active process (Kaplan et al., 1980). Similar results were obtained with *Coccochloris peniocystis (Synechococcus)* (Miller and Colman, 1980). In this investigation, the authors point out that cyanobacteria are particularly useful for C_{inorg} transport studies because growth and photosynthesis is optimized at alkaline pH, suggesting that HCO_3^- is the species of carbon transported. Since there was no detectable extracellular carbonic anhydrase, it also follows that HCO_3^- is the species of C_{inorg} that is transported; its subsequent dehydration within the cells serves as the source of CO_2 required for the carboxylation of RuBP. The spontaneous dissociation of HCO_3^- in the medium could not account for the photosynthetic rates obtained, again pointing to the active transport of HCO_3^- ions, balanced by a corresponding efflux of OH^- ions in the medium (Miller and Colman, 1980). More recently, it was shown that a primary electrogenic pump for HCO_3^- transport may be operable (Kaplan et al., 1982); however, Badger and Andrews (1982) report that CO_2 and HCO_3^- can both be taken up by a marine *Synechococcus* species, a result recently corroborated by Volokita et al. (1984) with *Anabaena variabilis*. This latter study provides evidence that HCO_3^- reaches the inside of the cell no matter which form of C_{inorg} is supplied. A model for C_{inorg} uptake was proposed that apparently accounts for all the observed results. In this model, Volokita et al. (1984) propose that a subunit of the transport apparatus in the cytoplasmic membrane binds CO_2, similar to carbonic anhydrase, and this CO_2 is then converted to HCO_3^-. A HCO_3^- porter then transfers the HCO_3^- to the inner membrane surface and subsequently to the inside of the cell. This model is certainly intriguing and should provide the basis for more detailed experimentation. Obviously, the identification and isolation of the transport apparatus is awaited with great interest.

In summary, all the available evidence indicates that HCO_3^- is the predominant form of C_{inorg} transported into cyanobacterial cells. The transport process is dynamic, requiring energy, and appears to be mediated by specific carriers in the cytoplasmic membrane.

4. Ribulose bisphosphate carboxylase/oxygenase in cyanobacteria

It has been stressed that RuBPC/O is probably the most abundant protein found on earth (Ellis, 1979; Miziorko and Lorimer, 1983), the rationale for this being the large amounts of enzyme found in extracts of photosynthetic tissue. Early studies indicated that the enzyme from cyanobacteria approximated the large size ($M_r \cong 500\,000$) reported for higher plant RuBPC/O (Kieras and Haselkorn, 1968; Anderson et al., 1968). The large size of this enzyme suggested that the facile sucrose density gradient procedure developed for the purification of plant RuBPC/O (Goldthwaite and Bogorad, 1971) might be employed to purify the cyanobacterial enzyme. Accordingly, this procedure, adapted for the purification of bacterial RuBPC/O (Tabita and McFadden, 1974a), was employed to purify the first electrophoretically homogeneous cyanobacterial RuBPC/O, from *Agmenellum quadruplicatum* strain PR-6 (Tabita et al., 1974). In these studies, the enzyme was determined by gel filtration and sucrose density gradient centrifugation to possess a molecular weight of about 450 000 (Tabita et al., 1974, 1976). Similarly, a molecular weight of 440 000 was determined by pore-gradient electrophoresis and gel filtration for the enzyme from *Synechococcus* strain RRIMPN1 (now strain ACMM 323), a marine organism related to *Agmenellum* (Andrews et al., 1981). However, when the molecular weight was determined by sedimentation equilibrium, considerably different results were obtained and a value of 530 000 was found. Andrews et al. (1981) attributed the atypical pore penetration behavior of the enzyme to an asymmetry suggested by the unusual appearance of the enzyme in the electron microscope. They suggest that erroneous molecular weight estimations obtained by gel filtration and pore-gradient electrophoresis may be due to this asymmetry, which is not exhibited by the spinach enzyme.

It was originally reported that the *Agmenellum* and *Anabaena* enzymes were composed only of large subunits (Tabita et al., 1974, 1976). However, subsequent studies in the author's laboratory (Tabita and Colletti, 1979) and in several other laboratories (Takabe et al., 1976; Codd and Stewart, 1977a; Andrews and Abel, 1981) show conclusively that both large and small subunits are found in RuBPC/O from cyanobacteria. In addition, the previous report that RuBPC/O isolated from the extreme halophile *Aphanothece* is a tetramer of large subunits (Codd et al., 1979) has been corrected and it is now apparent that this enzyme is also comprised of both large and small subunits (Asami et al., 1983). The failure to demonstrate small subunits in the case of *Agmenellum* and *Anabaena* (Tabita et al., 1974, 1976) was probably due to the poor sensitivity of the SDS tube gel electrophoresis compared to the SDS-slab gel electrophoresis procedures currently employed and also the general difficulty often experienced in staining small subunits of RuBPC/O (Andrews and Ballment, 1983). Moreover, Tabita et al. (1976) suggested that the failure to observe small subunits might have been due to the isolation procedure employed to obtain the *Anabaena* enzyme, which included an acid precipitation step. Subsequent studies showed that acid precipitation removed small subunits from the *Aphanocapsa* 6308 enzyme (Codd and Stewart, 1977a) and mild acid treatment has become the technique of choice for dissociation-reconstitution, structure-function studies (Andrews and Abel, 1981; Andrews and Ballment, 1983; Andrews et al., 1984; Andrews and Lorimer, 1985). In addition, the *Aphanothece halophytica* enzyme dissociates into aggregates of large and small subunits after lowering the ionic strength (Takabe et al., 1984); presumably a high ionic strength is required to maintain the enzyme from this extreme halophile in its native conformation.

4.1. Regulation of RuBPC/O synthesis

Several studies indicate that the level of RuBPC/O synthesis is essentially invariant, even when cells are cultured under photoheterotrophic or dark heterotrophic-growth conditions. This lack of control over RuBPC/O synthesis, which is decidedly different from the situation with other photosynthetic prokaryotes (Tabita, 1981; Tabita et al., 1984), is also true for other enzymes of carbon metabolism (Carr, 1973). Variations in the specific activity of RuBPC/O which have been reported for various heterotrophic cyanobacteria (Raboy et al., 1976; Joset-Espardellier et al., 1978), may be misleading due to changes in the content of phycocyanin, the major protein of phototrophically-grown cells. Anoxygenic photosynthesis also seems to preserve the Calvin cycle as the major CO_2 assimilatory path (Peschek, 1979). Such findings have led Carr (1973) to postulate that there is a lack of transcriptional control over the enzymes of carbon metabolism in cyanobacteria. With recent advances in the molecular biology of cyanobacteria, experiments similar to those described to study the transcriptional control of phycocyanin synthesis (Conley et al., 1985) may now be attempted to examine the transcriptional control, or lack thereof, over RuBPC/O synthesis (see Section 6.1. of Chapter 10).

The recent isolation, cloning, and sequencing of the structural genes for the cyanobacterial RuBPC/O have been reported from several laboratories (Curtis and Haselkorn, 1983; Heinhorst and Shively, 1983; Reichelt and Delaney, 1983; Shinozaki et al., 1983; Shinozaki and Sugiura, 1983; Nierzwicki-Bauer et al., 1984; Tiboni et al., 1984; Shinozaki and Sugiura, 1985; see also Section 5.2. of Chapter 10). In all cases thus far examined, both the large and small subunit genes are contained within a single transcriptional unit, indicating that the large and small subunit genes constitute a single operon. Unlike the situation in higher plants, cyanobacterial small subunit genes do not contain a transit sequence (Shinozaki and Sugiura, 1983; Nierzwicki-Bauer et al., 1984), obviously because there is no need for such a sequence in the absence of compartmentation of large and small subunit synthesis in prokaryotes. In addition, there are no intervening sequences present in the prokaryotic small subunit genes examined to date. The co-transcription of the large and small subunit genes in cyanobacteria seems to insure that stoichiometric amounts of large and small subunits are synthesized such that a holoenzyme of eight large and eight small subunits is assembled (L_8S_8 structure). It is well established that the catalytic site resides on the large subunit of RuBPC/O (Miziorko and Lorimer, 1983) and the very fact that the small subunit is co-transcribed with the large subunit implies some functional role. It is interesting that a spacer sequence of 93 base pairs exists between the *Anacystis* large and small subunit genes (Shinozaki and Sugiura, 1983) whereas in *Anabaena* 7120, there is a much larger spacer region of 545 base pairs (Nierzwicki-Bauer et al., 1984). The significance of this variable spacer region is not yet apparent. The deduced amino acid sequence of the *Anacystis (Synechococcus)* large subunit shows 81% homology to the spinach large subunit and 78% homology to the maize large subunit; at the nucleotide level, there is 70% homology with the plant large subunit gene (Shinozaki and Sugiura, 1983). The deduced amino acid sequence of the *Anacystis* small subunit protein shows 40% homology to the pea small subunit protein and the *Anacystis* protein is 12 amino acids shorter than the pea protein (Bedbrook et al., 1980). The *Anabaena* small-subunit protein is 2 amino acids shorter than the *Anacystis* small subunit and exhibits 61% homology with the plant small subunit protein (Nierzwicki-Bauer et al., 1984). The large subunit of *Anabaena* RuBPC/O shows 84% homology to both *Anacystis* and higher plant large subunit RuBPC/O. At the nucleotide level, the *Anabaena* small subunit gene is 60% homologous with that of *Anacystis* and 53–56% homologous with the plant small subunit genes (Nierzwicki-Bauer et al., 1984).

4.2. Regulation of RuBPC/O activity

As discussed previously, gentle procedures, such as manipulation of the pH or ionic strength, serve to dissociate the bulk of small subunits from the native cyanobacterial L_8S_8 RuBPC/O. This finding has led to investigations on the structure-function relationships of small subunit-depleted and reconstituted cyanobacterial RuBPC/O. Andrews and Abel (1981) noted that repeated isoelectric precipitation of RuBPC/O from a marine *Synechococcus* (strain ACMM 323) yielded a catalytic core, containing only 13% of the original complement of small subunits, which retained 41% of the activity of the native enzyme. These studies, although indicating that activity was not correlated with the degree of saturation of small subunits, pointed to a role of the small subunit in catalysis. Subsequent studies, which optimized the removal and reconsitution of small subunits, indicated, however, that the activity of small subunit-depleted enzyme (catalytic core) was proportional to the amount of small subunit that remained associated with the enzyme. Moreover, reconstitution of enzyme activity correlated with the reassociation of small subunits to the catalytic core (Andrews and Ballment, 1983). Similar conclusions were reached with the small subunit-depleted *Aphanothece* enzyme (Takabe et al., 1984), and the alkali-dissociated enzyme from the purple sulfur photosynthetic bacterium *Chromatium* (Incharoensakdi et al., 1985). In toto, these studies point to a role of small subunits in the activity of RuBPC/O, despite the fact that the large subunit contains the catalytic and activation sites and the fact that dissociated catalytic core also retains unaffected activator and catalytic sites (Andrews and Ballment, 1984; Jordan and Chollet, 1985).

If the above scenario for the role of small subunits is correct, there remain some unanswered and important anomalies. First of all, the postulated role does not explain the existence of RuBPC/O which is fully active in the absence of small subunits; i.e. the L_2 enzyme from *Rhodospirillum rubrum* (Tabita and McFadden, 1974b; Schloss et al., 1979) and the form II RuBPC/O from *Rhodopseudomonas sphaeroides* (Gibson and Tabita, 1977a, 1985) and *R. capsulata* (Gibson and Tabita, 1977b; Shively et al., 1984). The form II *Rhodopseudomonas* enzyme is particularly interesting since the subunit molecular weight is 60 000 (Gibson and Tabita, 1985; Shively et al., 1984), somewhat larger than the usual 52–57 000, perhaps suggesting that amino acid residues normally required for activity and supplied by small subunits may have been incorporated into the large subunit of the *Rhodopseudomonas* form II and/or *Rs. rubrum* RuBPC/O during evolution. Secondly, Torres-Ruiz and McFadden (1985) have recently shown that extracts of *Chromatium vinosum* contain highly active L_8S_8 and L_8 RuBPC/O, which are separately isolated by differential centrifugation of polyethylene glycol treated crude extracts. These authors postulate that an equilibrium exists between the two forms

$$L_8S_8m \rightleftharpoons L_8 + S_8m$$

where S_8m represents S_8 that is associated with membrane fragments that are removed by high speed centrifugation in the highly viscous polyethyleneglycol medium. If subsequent studies show these two forms of RuBPC/O are derived from the same L_8S_8 complex and the large subunits are not separate gene products as in *R. sphaeroides* (Gibson and Tabita, 1977a, 1985; Quivey and Tabita, 1984), the role of small subunits may merely be to facilitate renaturation of L_8S_8 with a consequent increase in activity (Torres-Ruiz and McFadden, 1985). Thus, the small subunits may function primarily to place the large subunits in the proper conformation for catalysis (Gibson and Tabita, 1979).

Perhaps, a clue to the function of small subunits in catalysis and the assembly of intact RuBPC/O may come from a different method of approach. Shinozaki and Sugiura (1983) first noted that there are two major portions of the sequence of the *Anacystis* small subunit which show substantial homology with the small subunits of plants. These authors note that the first region (amino acid positions 12–21) (Fig. 3) is quite hydrophobic and may be involved in binding small subunits to large subunits. The second region (amino acid positions 54–63) has an α-helical structure and perhaps may be somehow involved in catalysis. Nierzwicki-Bauer et al. (1984) observed that the *Anabaena* small subunit sequence is quite homologous to the *Anacystis* sequence and further indicate that there is a third region of substantial homology to plant small subunits. These very interesting correlations of sequence homology suggest that the sequence at these regions might be altered by site-specific oligonucleotide mutagenesis in order to detect a functional role for residue(s) in these homologous regions. Before such potentially rewarding experiments might be attempted, however, it is necessary to express active and fully assembled cyanobacterial RuBPC/O in *Escherichia coli* or some other host where the gene can be easily manipulated. Mutated RuBPC/O genes may then be reassembled in the host, the enzyme isolated, and any effect of the specific mutation on either catalysis or assembly noted. In this way, using recombinant DNA procedures, the role of small subunits in catalysis may be determined as well as the specific sites on the small subunit (and the large subunit) required for assembly of the holoenzyme.

Several groups have attempted to express the active hexadecameric cyanobacterial RuBPC/O protein in *E. coli*. The cyanobacterial system seems particularly advantageous for these experiments

		10 20		60		100 110	
Anabaena	. . .	RRYETLSYLPPLT	. . .	YWTLWKLPLFG	. . .	YIRVVGFDNIKQCQILSFI	. . .
Anacystis	. . .	RRFETFSYLPPLS	. . .	YWTMWKLPLFD	. . .	YIRVAGFDNIKQCQTVSFI	. . .
Tobacco	. . .	KKYETLSYLPDLS	. . .	YWTMWKLPMFG	. . .	WIRIIGFDNVRQVQCISFI	. . .
Wheat	. . .	KKFETLSYLPPLS	. . .	YWTMWKLPMFG	. . .	YVRVIGFDN[M/L]RQVQCVSFI	. . .
Spinach	. . .	KKFETLSYLPPLT	. . .	YW---KLPMFG	. . .	FVRFIGFDNKREVQCISFI	. . .
Soybean	. . .	KKFETLSYLPDLD	. . .	YWTMWKLPMFG	. . .	FIRIIGFDNVRQVQCISFI	. . .
Pea pSS15	. . .	KKFETLSYLPPLT	. . .	YWTMWKLPMFG	. . .	FVRIIGFDNVRQVQCISFI	. . .
Pea pSSU1	. . .	KKFETLSYLPPLT	. . .	YWTMWKLPMFG	. . .	FVRVIGFNNVRQVQCISFI	. . .

Fig. 3. Regions of sequence homology in the small subunits of RuBPC/O from cyanobacteria and higher plants. (From Nierzwicki-Bauer et al., 1984, with permission)

since the large and small subunit genes are localized in a single transcriptional unit (Nierzwicki-Bauer et al., 1984; Shinozaki and Sugiura, 1985). Gatenby et al. (1985) cloned the 2.17-kbp *Pst* I fragment of *Synechococcus (Anacystis)* 6301 DNA containing the large and small subunit genes into the *Pst* I site of plasmid pLa2311, which is under control of the P_L promoter. Using anti-wheat RuBPC/O, they were able to show that both large and small subunits were synthesized in *E. coli*, as well as a 72 000 M_r fusion protein, after temperature induction of cells containing the hybrid pLa2311 plasmid. However, there appeared to be a lack of completely assembled enzyme of high specific activity in this investigation. Cloned RuBPC/O genes from *Spirulina platensis* were recently used to express large subunits in *E. coli* minicells, but the amount of small subunits produced was very low, which appears to correlate with the lack of enzyme activity produced in this system (Tiboni et al., 1984). Similarly, *Anabaena* 7120 RuBPC/O genes were placed under *lac* transcriptional control in plasmid pUC19 (Gurevitz et al., 1985) with the result that small amounts of active, hexadecameric enzyme (0.6% of the soluble protein) were produced in extracts of *E. coli*. In this study also, the synthesis of large subunits far exceeds the synthesis of small subunits, and excess large subunits accumulate in an inactive, insoluble form.

There has been only a single report of the expression of large quantities of active, fully assembled, hexadecameric RuBPC/O in *E. coli* (Tabita and Small, 1985). In this investigation, plasmid pANP1155 [a hybrid pBR322 molecule (Shinozaki and Sugiura, 1983) containing a 2.3-kbp DNA fragment from *Anacystis* which codes for large and small subunits of RuBPC/O] was digested with *Pst* I to liberate the cloned *Anacystis* DNA. This material was then ligated into the expression vector pUC9 (Vieira and Messing, 1982) in order to study the possible translation products while under

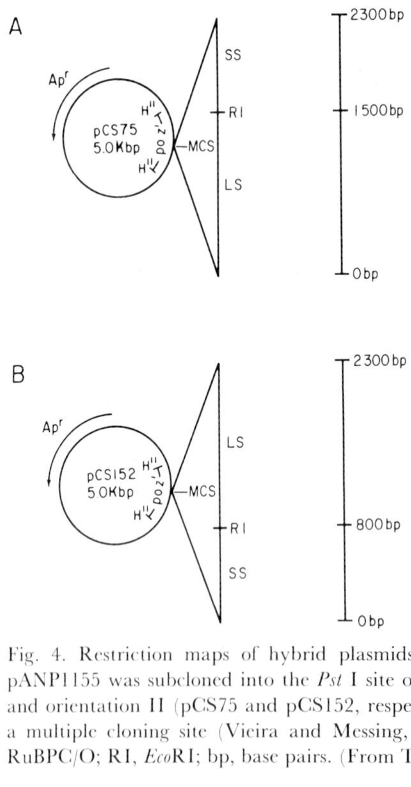

Fig. 4. Restriction maps of hybrid plasmids pCS75 and pCS152. The 2.3-kbp insert DNA in plasmid pANP1155 was subcloned into the *Pst* I site of pUC9. The pUC9 hybrids, with insert DNA in orientation I and orientation II (pCS75 and pCS152, respectively) are shown with insert DNA offset in A and B. MCS is a multiple cloning site (Vieira and Messing, 1982). LS, large subunit of RuBPC/O; SS, small subunit of RuBPC/O; RI, *Eco*RI; bp, base pairs. (From Tabita and Small, 1985, with permission)

TABLE I

Expression of cloned *Anacystis nidulans* RuBPC/O genes in *E. coli*

Hybrid plasmid	Orientation of insert	Specific activity of RuBPC/O (U/l/OD_{650})
pCS75	I, induced[a]	54.0
pCS75	I, uninduced	2.0
pCS152	II, induced	0.02
pCS152	II, uninduced	0.02

[a]Cells were induced with 1mM isopropyl-β-D-thiogalactopyranoside and harvested after reaching the logarithmic phase of growth. (From Tabita and Small, 1985, with permission)

lac promoter control. Restriction endonuclease analyses revealed subclones with the *Anacystis* DNA inserted in opposite orientations (Fig. 4). Orientation I (pCS75) was found to contain the fragment with the large subunit proximal to the *lac* promoter, while in orientation II (pCS152) the small subunit coding region was proximal to the *lac* promoter. Since the genes are transcribed in a large subunit to small subunit direction (Shinozaki and Sugiura, 1983; Nierzwicki-Bauer et al., 1984), orientation I was presumed to be properly positioned for proper transcription from the *lac* promoter. This was borne out by examining the two plasmids after induction with the gratuitous inducer isopropyl β-D-thiogalactopyranoside (IPTG) (Table I). Obviously, plasmid pCS75, with the fragment in orientation I, showed substantial enzymatic activity. To determine if the enzyme was assembled in a proper stoichiometry of large and small subunits in *E. coli*, the recombinant protein was purified and its subunit structure analyzed. When purified, the enzyme exhibited a molecular weight similar to spinach RuBPC/O, since it migrated only slightly slower in sucrose gradients. The specific activity ranged from 3.6–4.5 and showed a single stained band on non-denaturing polyacrylamide gels which co-migrated with authentic *Anacystis* RuBPC/O (F.R. Tabita and C.S. Small, unpublished results). Since the specific activity of this recombinant protein is somewhat higher than that obtained for previously isolated cyanobacterial RuBPC/O (Andrews and Ballment, 1983), proper assembly to the hexadecameric structure was indicated. This was shown directly by sodium dodecyl sulfate electrophoresis of the *E. coli* (pCS75) enzyme which revealed the presence of both large and small subunits (Fig. 5) with molecular weights of 56–57 000 and 12 000, respectively. The ratio of the molecular weights of the *E. coli* (pCS75) RuBPC/O subunits, as well as the ratio of the masses of the subunits, indicated that assembly to the holoenzyme proceeded normally in *E. coli*. We conservatively estimate that RuBPC/O represents about 10 % of the soluble protein in *E. coli*, which may be somewhat higher when plasmid pCS75 is placed in *E. coli* HB101 (Gibson, J.L., and Tabita, F.R., unpublished observations).

The successful expression of high levels of active hexadecameric *Anacystis* RuBPC/O in *E. coli* provides a basic system for site-specific mutagenesis studies, followed by structure-function analysis of specific mutations. These studies will also provide information relative to the mechanism of enzyme assembly. Certainly the necessity for an assembly specific protein, as hypothesized to function in the chloroplast (Ellis, 1981; Roy et al., 1982), does not appear to be required to assemble the cyanobacterial enzyme, unless *E. coli* contains such a protein. Such an assembly protein may not be required because the two cyanobacterial subunits are cotranslated and may interact immediately (Gurevitz et al., 1985).

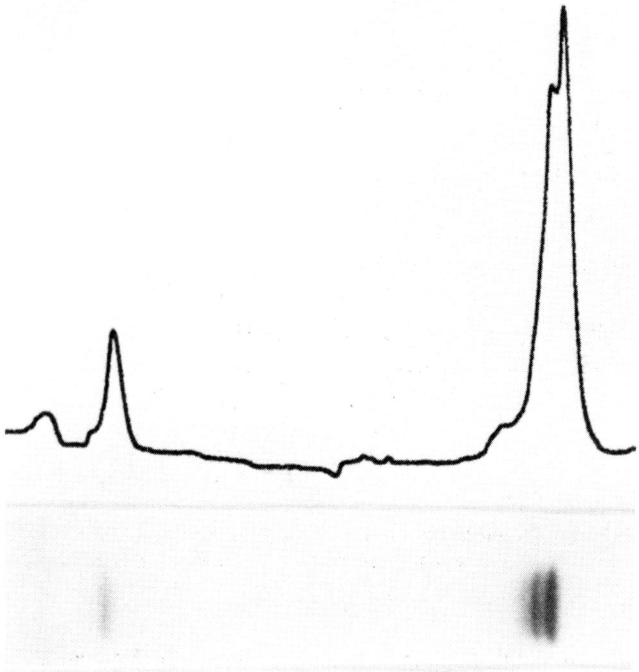

Fig. 5. Sodium dodecyl sulfate polyacrylamide slab gel of purified recombinant *E. coli* (pCS75) RuBPC/O. Protein was stained with Coomassie Blue. A densitometric trace is shown above the gel. (From Tabita and Small, 1985, with permission)

4.3. Kinetic factors and the regulation of RuBPC/O in cyanobacteria

In addition to catalyzing the carboxylation of RuBP, it has been shown that RuBPC/O is also capable of catalyzing the RuBP-dependent oxygenolysis of RuBP, with the resultant formation of one molecule each of 3-phosphoglycerate and 2-phosphoglycolate (Miziorko and Lorimer, 1983). The same polypeptide chain catalyzes both reactions and thus functions as the first reaction of the two competing pathways of CO_2 assimilation and the oxidative metabolism of glycolate (Fig. 6). Convincing evidence for the oxygenase activity of cyanobacterial RuBPC/O has been provided (Andrews and Abel, 1981; Andrews and Lorimer, 1985; Codd and Stewart, 1977a; Jordan and Ogren, 1981) and it has been demonstrated that cyanobacteria will excrete glycolate (Dohler and Braun, 1971; Cheng et al., 1972; Han and Eley, 1973; Bergman et al., 1984) as well as metabolize this 2-carbon compound intracellularly (Codd and Stewart, 1973). Apparently, however, glycolate does not serve as a substrate for dark heterotrophic or photoheterotrophic growth (Codd and Stewart, 1974).

If one could selectively shut off the oxygenase function by either mutation or chemical means, CO_2 assimilation and, ultimately, plant productivity would be greatly enhanced. Thus, a great deal of effort has been directed to understand all aspects of the biochemistry of RuBPC/O. Every RuBPC/O that has been isolated is capable of catalyzing the oxygenase reaction, including the enzyme from anaerobic bacteria, and this has led to the general belief that the oxygenase function is an inherent property of the protein and proceeds as a consequence of the enzymatic formation

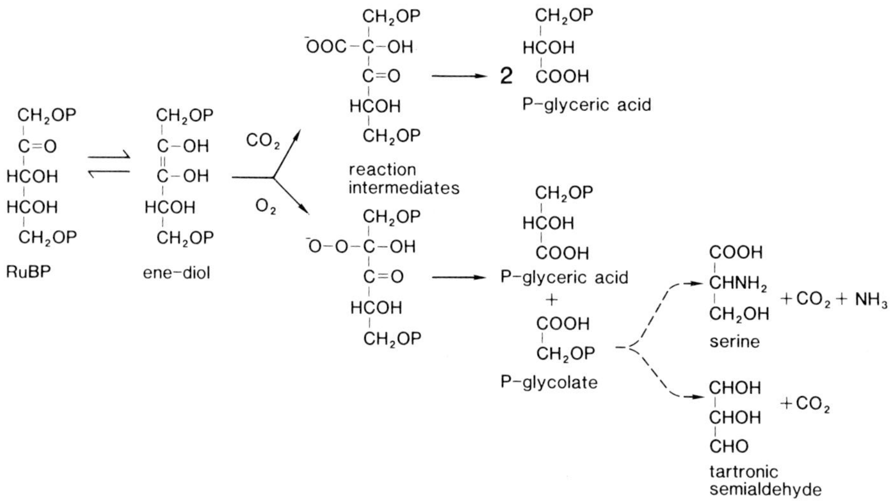

Fig. 6. Reactions catalyzed by RuBPC/O and their metabolic significance. Dashed arrows refer to proposed routes for glycolate metabolism in *Anabaena cylindrica* (Codd and Stewart, 1973) and *Anacystis nidulans* (Dohler and Przybylla, 1973).

of the enediol of RuBP (Lorimer and Andrews, 1973). On the basis of studies with effector metabolites, and limited experiments with active-site probes, it has been assumed that the same active site is responsible for both activities. However, there is no compelling evidence to indicate that there are no differences in the two activities that might be exploited. For example, recent studies indicate that there are distinct differences in the metal requirements of the carboxylase and oxygenase activities of RuBPC/O, with Mg^{2+} the preferred divalent cation for the carboxylase activity and Mn^{2+} for the oxygenase activity (Christeller and Laing, 1979; Wildner and Henkel, 1978; Robison et al., 1979). These results were obtained with both the *Rs. rubrum* and plant RuBPC/O. Furthermore, Co^{2+} was found to support only the oxygenase activity of the *Rs. rubrum* enzyme, the first demonstration for a differential effect on one activity and not the other (Robison et al., 1979). Following these studies it was found that the oxygenase activity of both the bacterial (*Rs. rubrum*) and plant enzymes exhibited a significantly greater affinity for the substrate RuBP than did the carboxylase (Martin and Tabita, 1981). Similar results were obtained with effectors such as fructose 1,6-bisphosphate. The K_m (RuBP) for the oxygenase activity of the cyanobacterial RuBPC/O is also lower than the K_m (RuBP) for the carboxylase activity (Andrews and Abel, 1981; Badger et al., 1980). It was suggested that these results might be significant when the intracellular level of RuBP or fructose bisphosphate is low, such that the oxygenase activity might be favored (Martin and Tabita, 1981; Tabita et al., 1984); the actual physiological significance of these findings remains to be established. These results do suggest, however, that the carboxylase/oxygenase ratio may be changed. Further support for the fact that the ratio of the two activities is not immutable came from a comparison of the specificities of RuBPC/O from several species toward the gaseous substrates CO_2 and O_2 (Jordan and Ogren, 1981). In this investigation, significant differences in specificity towards the gaseous substrates CO_2 and O_2 were found, such that RuBPC/O from evolutionarily advanced organisms possesses a higher affinity for CO_2 and thus a higher efficiency for CO_2 fixation, compared to various bacterial enzymes. These studies certainly suggest that the balance between RuBP carboxylation and oxygenation has changed during evolution, providing a rationale for further alteration of the enzyme by suitable selection pressure or mutagenesis procedures (Jordan and Ogren, 1981). In addition, recent studies of hybrid enzyme prepared from large subunits of *Synechococcus* RuBPC/O and small subunits from spinach RuBPC/O (Andrews and Lorimer, 1985) indicate that the degree of partitioning between carboxylation and oxygenation by the hybrid is identical to that of the native and reassembled *Synechococcus* enzymes and different from spinach RuBPC/O. These results seemingly provide evidence that the small subunit does not influence the specificity for CO_2 and O_2, a result which is extremely interesting since RuBPC/O from *Rs. rubrum* and *R. sphaeroides* (form II), enzymes which do not have small subunits, possess significantly different specificity factors [partition coefficients (Andrews and Lorimer, 1985)] compared to enzymes which do possess small units (Jordan and Ogren, 1981). However, the hybrid enzyme does exhibit a higher K_m (CO_2) (Andrews and Lorimer, 1985).

In order to understand how RuBPC/O activity might be regulated in the cell, it is necessary to appreciate that this enzyme must first be activated before the enzyme is capable of catalyzing either the carboxylation or oxygenation of RuBP (Badger and Lorimer, 1976; Lorimer et al., 1976). Both CO_2 and divalent cations are involved in converting the unactivated enzyme to a fully active ternary complex ($E \cdot CO_2 \cdot Mg^{2+}$). The rate-determining step in this activation process is the formation of a carbamate with the activator CO_2, and the ε-amino group of a specific lysine residue, followed by the rapid binding of the metal (Lorimer and Miziorko, 1980). Several studies have shown that phosphorylated metabolites may cause full activation of partially active enzyme (i.e., enzyme incubated with suboptimal concentrations of activator CO_2), while the same metabolites are competitive inhibitors with respect to RuBP for fully activated enzyme (Jensen and Bahr, 1977). With regard to the cyanobacteria, kinetic studies of $^{14}CO_2$ fixation and ^{32}P incorporation into *Aphanocapsa* strains 6308 and 6714 showed a rapid increase in the concentration of phosphogluconate and changes in the concentrations of sugar mono- and diphosphates when cells were transferred from the light to the dark (Pelroy and Bassham, 1972, 1976). This transition also caused an immediate cessation of biosynthetic CO_2 fixation. Since, phosphogluconate was found to be a potent competitive inhibitor of fully activated RuBPC/O, and since the levels increase dramatically in chloroplasts in the dark as well (Bassham and Kirk, 1968), it was proposed that phosphogluconate might play a role in the physiological regulation of CO_2 fixation in the dark (Tabita and McFadden, 1972). Subsequent studies were performed with various species of cyanobacteria, including *Anabaena* strain CA, *Agmenellum quadruplicatum* strain PR-6 and *Aphanocapsa* strain 6714 (Tabita and Colletti, 1979). The properties of the enzyme within the cell, using permeabilized cells (Tabita et al., 1978), as well as the properties of crude, partially purified, and homogeneous RuBPC/O were examined (Tabita and Colletti, 1979). The whole-cell assay is particularly significant because the activity measured is free from potential artifacts of enzyme isolation. With all three organisms studied, carboxylase activity was enhanced by various phosphorylated metabolites. In particular, phosphogluconate and NADPH were the most effective regulator metabolites, with the partially activated *Anabaena* CA enzyme under the greatest metabolic control (Figs. 7 and 8). Interestingly, the in situ RuBP carboxylase activity is significantly more sensitive to effectors than is the homogeneous or freshly prepared crude enzyme (Tabita and Colletti, 1979), suggesting that the microenvironment of the enzyme might contribute to its regulation. Recent studies with higher plants indicate that there is a factor(s) in the chloroplast which promotes activation of the enzyme in the light, a factor(s) which is separate from and exceeds any control exerted by the intracellular level of CO_2 and divalent cation (Perchorowicz and Jensen, 1983; Servaites and Torisky, 1984; Torisky and Servaites, 1984; Perchorowicz et al., 1981; Vu et al., 1983, 1984; Servaites et al., 1984). To date, the basis for this light-mediated regulation of RuBPC/O activity is unknown and similar experiments have not been performed in cyanobacteria. The intracellular level of RuBPC/

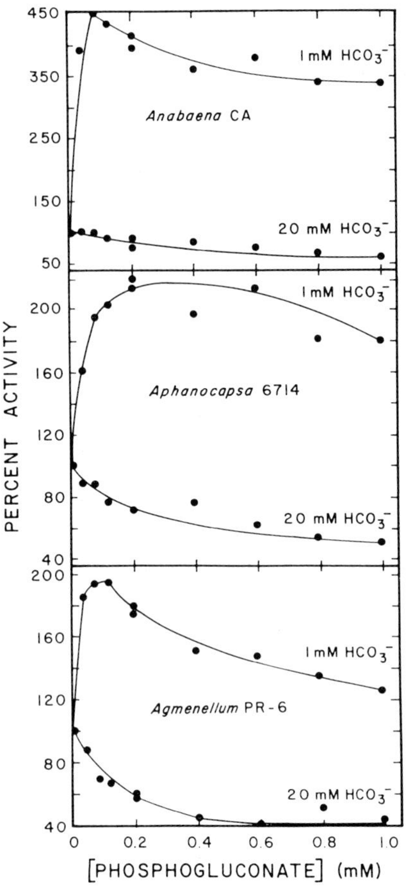

Fig. 7. Regulation of RuBP carboxylase activity by 6-phosphogluconate in permeabilized cyanobacteria at 1 and 20 mM HCO_3^- (partially activated and fully activated enzyme). (From Tabita and Colletti, 1979, with permission)

O active sites in cyanobacteria is at least an order of magnitude lower than the level found in plant chloroplasts [where it is about 4 mM (Miziorko and Lorimer, 1983)]; the flux of various phosphorylated metabolites might thus have an effect on the carboxylase activity in vivo, since these metabolites are present at millimolar levels (Pelroy and Bassham, 1972, 1976). Recent results indicate that effectors promote activation of inactive enzyme by altering the relative rates of activation and deactivation to favor active enzyme (Jordan et al., 1983). These results contradict the alternative theory that the active enzyme has a greater binding affinity than the inactive enzyme for positive effectors (Badger and Lorimer, 1981), since equilibrium binding studies indicate that the active and inactive enzyme have closely similar binding constants for positive effectors (Jordan and Ogren, 1983). In any case, it seems likely that fluctuations in the cellular level of certain key metabolites may exert a fine control over CO_2 assimilation in the cyanobacteria.

4.4. Localization of RuBPC/O

Aerobic heterocystous cyanobacteria compartmentalize the processes of nitrogen fixation and photosynthetic CO_2 fixation (Fogg et al., 1973). The assimilation of CO_2 was shown by Wolk (1968) to occur primarily in the vegetative cells of *Anabaena cylindrica*, followed by the transfer of labeled carbon to the heterocysts. Subsequent experiments showed that RuBPC/O activity was greatly enriched in vegetative cells compared to isolated heterocysts (Winkenbach and Wolk, 1973; Stewart and Codd, 1975) and Codd and Stewart (1977b) showed that the lack of activity was due to a lack of immunologically detected RuBPC/O protein in the heterocysts. More recently, anti-*Microcystis* RuBPC/O was used, in conjunction with colloidal gold, to confirm the fact that the protein is not present in heterocysts of *Anabaena cylindrica* (Cossar et al., 1985).

All cyanobacteria contain cellular inclusions, polyhedral bodies, that resemble structures found in thiobacilli and nitrifying bacteria (Shively, 1974; see Section 4 of Chapter 8). Shively and colleagues have shown that these structures are greatly enriched with large concentrations of a para-

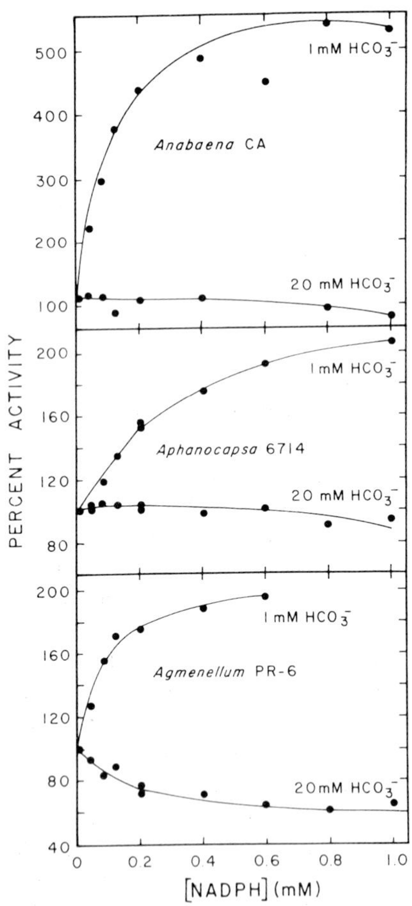

Fig. 8. Regulation of RuBP carboxylase activity by NADPH in permeabilized cyanobacteria at 1 and 20 mM HCO_3^-. (From Tabita and Colletti, 1979, with permission)

crystalline array of RuBPC/O protein (Shively et al., 1973a, b). These polyhedral bodies or 'carboxysomes' (Shively et al., 1973a) are not present in the heterocysts of cyanobacteria, but they are present in vegetative cells of filamentous cyanobacteria, suggesting that the carboxysomes may somehow be involved with CO_2 fixation (Stewart and Codd, 1975). Subsequent studies have shown that crude extracts contain both soluble and particulate RuBPC/O (Badger, 1980; Codd and Stewart, 1976) and the particulate enzyme appears associated with the carboxysomes (Codd and Stewart, 1976; Lanaras and Codd, 1981a, b). Most of the RuBPC/O appeared to be soluble in the early phases of growth while the distribution changed, and was predominantly particulate, in the latter stages of growth (Lanaras and Codd, 1982). Although, both enzymes were capable of catalyzing CO_2 fixation in vitro, there seemed to be less activity per mg of carboxylase protein (as determined immunologically) for the carboxysome RuBPC/O compared to the soluble enzyme. These results suggest that the carboxysomal enzyme might be under some form of posttranslational regulation and further results on this interesting system are awaited with great interest as are studies of the physiological significance of carboxysomes (Beudeker et al., 1981). It should be stressed that considerable variation in the level of RuBPC/O activity of CO_2-limited chemostat cultures of *Anacystis nidulans* has been reported, a result which also suggests the possibility of translational, or perhaps even transcriptional control under CO_2 limitation (Karagouni and Slater, 1979).

5. Phosphoribulokinase and other enzymes important for CO_2 fixation

Unlike RuBPC/O, comparatively little is known of the structure, function and regulation of phosphoribulokinase in cyanobacteria. The enzyme was first purified to homogeneity as a 200–232 000 molecular weight protein from *Rhodopseudomonas capsulata* and shown to contain a single polypeptide chain with a molecular weight of 36 000 (Tabita, 1980). Subsequently, the enzyme was purified from the chemolithotroph *Alcaligenes eutrophus* and was found to have a native molecular weight of 256 000 and a subunit molecular weight of 33 000 (Siebert et al., 1981); the enzyme from *R. acidophila* was similarly shown to have a native molecular weight of 248 000 and a subunit molecular weight of 32 000 (Rippel and Bowien, 1984). To date, phosphoribulokinase has been purified from only one cyanobacterium, *Chlorogloeopsis fritschii*, where it exhibits a molecular weight of 230 000 and is composed of subunits of 40 000 molecular weight (Marsden and Codd, 1984). By contrast, the tobacco and wheat enzymes had native molecular weights of 90 000 and 83 000 and subunit molecular weights of 46 000 and 42 000, respectively (Kagawa, 1982; Surek et al., 1985). Early studies (Rindt and Ohmann, 1969) indicated that various effectors modified the bacterial enzyme; of particular interest was NADH, which was found to be required to fully activate the enzyme from many diverse bacteria (Rindt and Ohmann, 1969; Rippel and Bowien, 1984; Tabita, 1980, 1981; MacElroy et al., 1969; Kiesow et al., 1977; Siebert et al., 1981). Phosphoribulokinase from oxygenic photosynthetic organisms, including the cyanobacteria, was not regulated by NADH, nor was the enzyme affected by NADPH, NAD, or NADP (Kiesow et al., 1977; Marsden and Codd, 1984; Tabita, 1980, 1981). However, plant phosphoribulokinase, and other enzymes important for CO_2 fixation, including fructose bisphosphatase and sedoheptulose bisphosphatase, were shown to be subject to control by a unique light-dependent ferredoxin/thioredoxin system such that the redox status of key sulfhydryl groups on these enzymes was affected in a reversible manner (Buchanan, 1980; Wolosiuk and Buchanan, 1978). In cyanobacteria, light-dependent regulation of phosphoribulokinase was first noted in *Anacystis nidulans* (Duggan and Anderson, 1975), a result which corroborated previous steady-state labeling experiments with whole cells of *Aphanocapsa* strain 6714 (Pelroy and Bassham, 1972). The effects of light can be mimicked with dithio-

TABLE II

Summary of structure and regulation of homogeneous phosphoribulokinase from various sources

Organism	Native (M_r)	Subunit (M_r)	NADH activation	Thioredoxin or dithiothreitol activation
Rhodopseudomonas capsulata	200–232 000	36 000	Yes	No
Rhodopseudomonas acidophila	248 000	32 000	Yes	No
Alcaligenes eutrophus	256 000	33 000	Yes	No
Chlorogloeopsis fritschii	230 000	40 000	No	Yes
Nicotiana (tobacco)	90 000	46 000	No	Yes
Triticum (wheat)	83 000	42 000	No	Yes

threitol (Duggan and Anderson, 1975; Pelroy and Bassham, 1976; Crawford et al., 1984), which regulated fructose bisphosphatase and sedoheptulose bisphosphatase as well as phosphoribulokinase. Two forms of phosphoribulokinase were found in *Nostoc muscorum* which were similar in regulatory properties (Crawford et al., 1984), unlike the situation in spinach chloroplasts, where there exist regulatory and nonregulatory isoenzymes (Wolosiuk and Buchanan, 1978). In addition, the two forms of *N. muscorum* phosphoribulokinase differed from the regulatory enzyme from spinach (Wolosiuk et al., 1979) in that both thioredoxin *f* and thioredoxin *m* were effective (Crawford et al., 1984). The structural properties and regulation of purified phosphoribulokinase from various sources is summarized in Table II. From these studies, it is apparent that cyanobacteria, similar to chloroplasts from higher plants, contain all the components of the ferredoxin/thioredoxin system, which functions to regulate key enzymes of CO_2 assimilation as well as other enzymes of carbon and nitrogen metabolism (Papen and Bothe, 1974; Papen et al., 1983; Cossar et al., 1984).

6. Conclusion

It is apparent from the almost exponential increase in the use of cyanobacteria as experimental tools that forthcoming advances in the biochemistry and molecular regulation of CO_2 fixation can be expected. In this review, I have attempted to bring together the various approaches that have been taken as well as collect the relevant information. Studies performed to date certainly indicate that cyanobacteria employ several mechanisms to regulate the photoassimilation of CO_2. These organisms should continue to serve as a viable model system for these various approaches; the close evolutionary relationship to the chloroplasts of higher plants is well appreciated (Carr and Whitton, 1982).

Acknowledgements

The contributions of my colleagues, students, and postdoctoral associates are gratefully acknowledged. Recent studies were supported by grants from the National Institutes of Health, Robert A. Welch Foundation, and U.S. Department of Agriculture.

References

Anderson, L.E., Price, G.B. and Fuller, R.C. (1968) Molecular diversity of the ribulose diphosphate carboxylase from photosynthetic microorganisms. Science 161, 482–484.

Andrews, T.J. and Abel, K.M. (1981) Kinetics and subunit interactions of ribulose bisphosphate carboxylase-oxygenase from the cyanobacterium, *Synechococcus* sp. J. Biol. Chem. 256, 8445–8451.

Andrews, T.J. and Ballment, B. (1983) The function of the small subunits of ribulose bisphosphate carboxylase-oxygenase. J. Biol. Chem. 258, 7514–7518.

Andrews, T.J. and Ballment, B. (1984) Active-site carbamate formation and reaction-intermediate-analog binding by ribulose bisphosphate carboxylase/oxygenase in the absence of its small subunits. Proc. Natl. Acad. Sci. USA 81, 3660–3664.

Andrews, T.J. and Lorimer, G.H. (1985) Catalytic properties of a hybrid between cyanobacterial large subunits and higher plant small subunits of ribulose bisphosphate carboxylase-oxygenase. J. Biol. Chem. 260, 4632–4636.

Andrews, T.J., Abel, K.M., Menzel, D. and Badger, M.R. (1981) Molecular weight and quaternary structure of ribulose bisphosphate carboxylase from the cyanobacterium, *Synechococcus* sp. Arch. Microbiol. 130, 344–348.

Andrews, T.J., Greenwood, D.M. and Yellowlees, D. (1984) Catalytically active hybrids formed in vitro between large and small subunits of different procaryotic ribulose bisphosphate carboxylases. Arch. Biochem. Biophys. 234, 313–317.

Asami, S., Takabe, T., Akazawa, T.T. and Codd, G.A. (1983) Ribulose 1,5-bisphosphate carboxylase from the halophilic cyanobacterium *Aphanothece halophytica*. Arch. Biochem. Biophys. 225, 713–721.

Badger, M.R. (1980) Kinetic properties of ribulose 1,5-bisphosphate carboxylase/oxygenase from *Anabaena variabilis*. Arch. Biochem. Biophys. 201, 247–254.

Badger, M.R. and Andrews, T.J. (1982) Photosynthesis and inorganic carbon usage by the marine cyanobacterium, *Synechococcus* sp. Pl. Physiol. 70, 517–523.

Badger, M.R. and Lorimer, G.H. (1976) Activation of ribulose-1,5-bisphosphate oxygenase. The role of Mg^{2+}, CO_2 and pH. Arch. Biochem. Biophys. 175, 723–729.

Badger, M.R. and Lorimer, G.H. (1981) Interaction of sugar phosphates with the catalytic site of ribulose-1,5-bisphosphate carboxylase. Biochemistry 20, 2219–2225.

Badger, M.R., Andrews, T.J., Canvin, D.T. and Lorimer, G.H. (1980) Interactions of hydrogen peroxide with ribulose bisphosphate carboxylase oxygenase. J. Biol. Chem. 255, 7870–7875.

Bassham, J.A. and Kirk, M. (1968) Dynamic metabolic regulation of the photosynthetic carbon reduction cycle. In K. Shibata, A. Takamiya, A.T. Jagendorf and R.C. Fuller (eds.), Comparative Biochemistry and Biophysics of Photosynthesis. University of Tokyo Press, Tokyo, pp. 365–378.

Bedbrook, J.R., Smith, S.M. and Ellis, R.J. (1980) Molecular cloning and sequencing of cDNA encoding the precursor to the small subunit of chloroplast ribulose 1,5-bisphosphate carboxylase. Nature 287, 692–697.

Bergman, B. Codd, G.A. and Hallbom, L. (1984) Glycolate excretion by N_2-fixing cyanobacteria treated with photorespiratory inhibitors. Z. Pflanzenphysiol. 113, 457–460.

Berry, J.A. and Bowes, G.W. (1976) Oxygen uptake in vitro by RuBP carboxylase of *Chlamydomonas reinhardtii* as a function of CO_2 concentration. Carnegie Inst. Washington Yearb. 75, 423–432.

Beudeker, R.F., Codd, G.A. and Kuenen, J.G. (1981) Quantitation and intracellular distribution of ribulose 1,5-bisphosphate carboxylase in *Thiobacillus neapolitanus*, as related to possible functions of carboxysomes. Arch. Microbiol. 129, 361–367.

Bottomley, P.J. and Van Baalen, C. (1978a) Characteristics of heterotrophic growth in the blue-green alga *Nostoc* sp. strain Mac. J. Gen. Microbiol. 107, 309–318.

Bottomley, P.J. and Van Baalen, C. (1978b) Dark hexose metabolism by photoautotrophically and heterotrophically grown cells of the blue-green alga (cyanobacterium) *Nostoc* sp. strain Mac. J. Bacteriol. 135, 888–894.

Buchanan, B.B. (1980) Role of the light on the regulation of chloroplast enzymes. Ann. Rev. Plant Physiol. 31, 341–374.

Carr, N.G. (1973) Metabolic control and autotrophic physiology. In N.G. Carr and B.A. Whitton (eds.), The Biology of Blue-Green Algae. Blackwell, Oxford, UK, pp. 39–65.

Carr, N.G. and Whitton, B.A., Eds. (1982) The Biology of Cyanobacteria, Blackwell, Oxford, UK.

Cheng, K.H., Miller, A.G. and Colman, B. (1972) An investigation of glycolate excretion in two species of blue-green algae. Planta 103, 110–116.

Christeller, J.T. and Laing, W.A. (1979) Effects of manganese ions and magnesium ions on the activity of soya-bean ribulose bisphosphate carboxylase/oxygenase. Biochem. J. 183, 747–750.

Codd, G.A. and Stewart, W.D.P. (1973) Pathways of glycollate metabolism in the blue-green alga *Anabaena cylindrica*. Arch. Mikrobiol. 94, 11–28.

Codd, G.A. and Stewart, W.D.P. (1974) Glycollate oxidation and utilization by *Anabaena cylindrica*. Plant Sci. Lett. 3, 199–205.

Codd, G.A. and Stewart, W.D.P. (1976) Polyhedral bodies and ribulose 1,5-diphosphate carboxylase of the blue-green alga *Anabaena cylindrica*. Planta 130, 323–326.

Codd, G.A. and Stewart, W.D.P. (1977a) D-Ribulose 1,5-diphosphate carboxylase from the blue-green alga *Aphanocapsa* 6308. Arch. Microbiol. 113, 105–110.

Codd, G.A. and Stewart, W.D.P. (1977b) Ribulose-1,5-diphosphate carboxylase in heterocysts and vegetative cells of *Anabaena cylindrica*. FEMS Microbiol. Lett. 2, 247–249.

Codd, G.A., Cook, G.M. and Stewart, W.D.P. (1979) Purification and subunit structure of D-ribulose 1,5-bisphosphate carboxylase from the cyanobacterium *Aphanothece halophytica*. FEMS Microbiol. Lett. 6, 81–86.

Colman, B., Cheng, K.H. and Ingle, R.K. (1976) The relative activities of PEP carboxylase and RuDP carboxylase in blue-green algae. Plant Sci. Lett. 6, 123–127.

Conley, P.B., Lemaux, P.G. and Grossman, A.R. (1985) Cyanobacterial light-harvesting complex subunits encoded in two red light-induced transcripts. Science 230, 550–553.

Cossar, J.D., Rowell, P., Darling, A.J., Murray, S., Codd, G.A. and Stewart, W.D.P. (1985) Localization of ribulose 1,5-bisphosphate carboxylase-oxygenase in the N_2-fixing cyanobacterium *Anabaena cylindrica*. FEMS Microbiol. Lett. 28, 65–68.

Cossar, J.D., Rowell, P. and Stewart, W.D.P. (1984) Thioredoxin as a modulator of glucose-6-phosphate dehydrogenase in a N_2-fixing cyanobacterium. J. Gen. Microbiol. 130, 991–998.

Crawford, N.A., Sutton, C.W., Yee, B.C., Johnson, T.C., Carlson, D.C. and Buchanan, B.B. (1984) Contrasting modes of photosynthetic enzyme regulation in oxygenic and anoxygenic prokaryotes. Arch. Microbiol. 139, 124–129.

Curtis, S.E. and Haselkorn, R. (1983) Isolation and sequence of the gene for the large subunit of ribulose-1,5-bisphosphate carboxylase from the cyanobacterium *Anabaena* 7120. Proc. Natl. Acad. Sci. USA 80, 1835–1839.

Dohler, G. (1974) C_4-Weg der Photosynthese in der Blaualge *Anacystis nidulans*. Planta 118, 259–269.

Dohler, G. (1976a) Wirkung anaerober Bedingungen auf die photosynthetischen Carboxylierungsreaktionen von *Anacystis* und *Chlorella*. Z. Pflanzenphysiol. 78, 416–420.

Dohler, G. (1976b) Photosynthetische Carboxylierungsreaktionen verschieden pigmentierter *Anacystis*-Zellen. Planta 131, 129–133.

Dohler, G., and Braun, F. (1971) Untersuchung der Beziehung zwischen extracellularer Glycolsaure-Ausscheidung und der photosynthetischen CO_2-Aufnahme bei der Blaualge *Anacystis nidulans*. Planta 98, 357–361.

Dohler, G. and Przybylla, K.-R. (1973) Einfluss der Temperatur auf die Lichtatmung der Blaualge *Anacystis nidulans*. Planta 110, 153–158.

Duggan, J.X. and Anderson, L.E. (1975) Light regulation of enzyme activity in *Anacystis nidulans* (Richt.) Planta 122, 293–297.

Ellis, R.J. (1979) The most abundant protein in the world. Trends Biochem. Sci. 4, 241–244.

Ellis, R.J. (1981) Chloroplast proteins: synthesis, transport, and assembly. Ann. Rev. Plant Physiol. 32, 111–137.

Fogg, G.E., Stewart, W.D.P., Fay, P. and Walsby, A.E. (1973) The Blue-Green Algae. Academic Press, London and New York, pp. 180–213.

Gatenby, A.A., van der Vies, S. and Bradley, D. (1985) Assembly in *E. coli* of a functional multi-subunit ribulose bisphosphate carboxylase from a blue-green alga. Nature 314, 617–620.

Gibson, J.L. and Tabita, F.R. (1977a) Different molecular forms of ribulose 1,5-bisphosphate carboxylase from *Rhodopseudomonas sphaeroides*. J. Biol. Chem. 252, 943–949.

Gibson, J.L. and Tabita, F.R. (1977b) Isolation and preliminary characterization of two forms of ribulose 1,5-bisphosphate carboxylase from *Rhodopseudomonas sphaeroides*. J. Bacteriol. 132, 818–823.

Gibson, J.L. and Tabita, F.R. (1979) Activation of ribulose 1,5-bisphosphate carboxylase from *Rhodopseudomonas sphaeroides:* probable role of the small subunit. J. Bacteriol. 140, 1023–1027.

Gibson, J.L. and Tabita, F.R. (1985) Structural differences in the catalytic subunits of form I and form II ribulose 1,5-bisphosphate carboxylase/oxygenase from *Rhodopseudomonas capsulata*. J. Bacteriol. 164, 1188–1193.

Goldthwaite, J.J. and Bogorad, L. (1971) A one-step method for the isolation and determination of leaf ribulose-1,5-diphosphate carboxylase. Anal. Biochem. 41, 57–66.

Gottschalk, G. (1979) Bacterial Metabolism. Springer-Verlag, New York, pp. 233–237.

Gotto, J.W., Tabita, F.R. and Van Baalen, C. (1979) Isolation and characterization of rapidly-growing marine, nitrogen-fixing strains of blue-green algae. Arch. Microbiol. 121, 155–159.

Gurevitz, M., Somerville, C.R. and McIntosh, L. (1985) Pathway of assembly of ribulose bisphosphate carboxylase/oxygenase from *Anabaena* 7120 expressed in *Escherichia coli*. Proc. Natl. Acad. Sci. USA 82, 6546–6550.

Han, T.-W. and Eley, J.H. (1973) Glycolate excretion by *Anacystis nidulans*: effect of HCO_3^- concentration, oxygen concentration and light intensity. Plant Cell Physiol. 14, 285–291.

Heinhorst, S. and Shively, J.M. (1983) Encoding of both subunits of ribulose-1,5-bisphosphate carboxylase by organelle genome of *Cyanophora paradoxa*. Nature 304, 373–374.

Incharoensakdi, A., Takabe, T. and Akazawa, T. (1985) Factors affecting the dissociation and reconstitution of ribulose-1,5-bisphosphate carboxylase/oxygenase from *Aphanothece halophytica*. Arch. Biochem. Biophys. 237, 445–453.

Izui, K., Kodaki, T., Katagiri, F., Fujita, N., Miwa, T., Ishijima, S., Hirai, K., Murata, T., Kuriyama, H. and Katsuki, H. (1985) PEP carboxylase of *Synechococcus* 6301 (*A. nidulans*): cloning and sequencing of the *PPC* gene, and comparison with the *E. coli PPC* gene. Abst. Vth Int. Symp. on Photosyn. Proc., p. 31.

Jensen, R.G. and Bahr, J.T. (1977) Ribulose-1,5-bisphosphate carboxylase/oxygenase. Ann. Rev. Plant Physiol. 28, 379–400.

Jordan, D.B. and Chollet, R. (1985) Subunit dissociation and reconstitution of ribulose-1,5-bisphosphate carboxylase from *Chromatium*. Arch. Biochem. Biophys. 236, 487–496.

Jordan, D.B. and Ogren, W.L. (1981) Species variation in the specificity of ribulose bisphosphate carboxylase/oxygenase. Nature, 291, 513–515.

Jordan, D.B., Chollet, R. and Ogren, W.L. (1983) Binding of phosphorylated effectors by active and inactive forms of ribulose-1,5-bisphosphate carboxylase. Biochemistry 22, 3410–3418.

Joset-Espardellier, F., Astier, C., Evans, E.H. and Carr, N.G. (1978) Cyanobacteria grown under photoautotrophic, photoheterotrophic, and heterotrophic regimes: sugar metabolism and carbon dioxide fixation. FEMS Microbiol. Lett. 4, 261–264.

Kagawa, T. (1982) Isolation and purification of ribulose-5-phosphate kinase from *Nicotiana glutinosa*. In M. Edelmann, R.B. Hallick and N.-H. Chua (eds.), Methods in Chloroplast Molecular Biology. Elsevier Biomedical Press, Amsterdam, pp. 695–705.

Kaplan, A., Badger, M.R. and Berry, J.A. (1980) Photosynthesis and the intracellular inorganic carbon pool in the bluegreen alga *Anabaena variabilis:* response to external CO_2 concentration. Planta 149, 219–226.

Kaplan, A., Zenvirth, D., Reinhold, L. and Berry, J.A. (1982) Involvement of a primary electrogenic pump in the mechanism for HCO_3^- uptake by the cyanobacterium *Anabaena variabilis*. Pl. Physiol. 69, 978–982.

Karagouni, A.D. and Slater, J.H. (1980) Enzymes of the Calvin cycle and intermediary metabolism in the cyanobacterium *Anacystis nidulans* grown in chemostat culture. J. Gen. Microbiol. 115, 369–376.

Kieras, F.J. and Haselkorn, R. (1968) Properties of ribulose-1,5-diphosphate carboxylase (carboxydismutase) from Chinese cabbage and photosynthetic microorganisms. Pl. Physiol. 43, 1264–1270.

Kiesow, L.A., Lindsley, B.F. and Bless, J.W. (1977) Phosphoribulokinase from *Nitrobacter winogradskyi:* activation by reduced nicotinamide dinucleotide and inhibition by pyridoxal phosphate. J. Bacteriol. 130, 20–25.

Lanaras, T. and Codd, G.A. (1981a) Ribulose 1,5-bisphosphate carboxylase and polyhedral bodies of *Chlorogloeopsis fritschii*. Planta 153, 279–285.

Lanaras, T. and Codd, G.A. (1981b) Structural and immunoelectrophoretic comparison of soluble and particulate ribulose bisphosphate carboxylases from the cyanobacterium *Chlorogloeopsis fritschii*. Arch. Microbiol. 130, 213–217.

Lanaras, T. and Codd, G.A. (1982) Variations in ribulose-1,5-bisphosphate carboxylase protein levels, activities and subcellular distribution during photoautotrophic batch culture of *Chlorogloeopsis fritschii*. Planta 154, 284–288.

Linko, P., Holm-Hansen, O., Bassham, J.A. and Calvin, M. (1957) Formation of radioactive citrulline during photosynthetic $^{14}CO_2$-fixation by blue-green algae. J. Exp. Bot. 8, 147–156.

Lorimer, G.H. and Andrews, T.J. (1973) Plant photorespiration – an inevitable consequence of the existence of atmospheric oxygen. Nature 243, 359–360.

Lorimer, G.H. and Miziorko, H.M. (1980) Carbamate formation on the ε-amino group of a lysyl residue as the basis for activation of ribulose bisphosphate carboxylase by CO_2 and Mg^{2+}. Biochemistry 19, 5321–5328.

Lorimer, G.H., Badger, M.R. and Andrews, T.J. (1976) The activation of ribulose-1,5-bisphosphate carboxylase by carbon dioxide and magnesium ions. Equilibria, kinetics, a suggested mechanism, and physiological implications. Biochemistry 15, 529–536.

MacElroy, R.D., Johnson, E.J. and Johnson, M.K. (1969) Control of ATP-dependent CO_2 fixation in extracts of *Hydrogenomonas facilis:* NADH regulation of phosphoribulokinase. Arch. Biochem. Biophys. 131, 272–275.

Marsden, W.J.N. and Codd, G.A. (1984) Purification and molecular and catalytic properties of phosphoribulokinase from the cyanobacterium *Chlorogloeopsis fritschii*. J. Gen. Microbiol. 130, 999–1006.

Martin, M.N. and Tabita, F.R. (1981) Differences in the kinetic properties of the carboxylase and oxygenase activities of ribulose bisphosphate carboxylase/oxygenase. FEBS Lett. 129, 39–43.

Miller, A.G. and Colman, B. (1980) Evidence for HCO_3^- transport by the blue-green alga (cyanobacterium) *Coccochloris peniocystis*. Pl. Physiol. 65, 397–402.

Miziorko, H.M. and Lorimer, G.H. (1983) Ribulose-1,5-bisphosphate carboxylase-oxygenase. Ann. Rev. Biochem. 52, 507–535.

Nierzwicki-Bauer, S.A., Curtis, S.E. and Haselkorn, R. (1984) Cotranscription of genes encoding the small and large subunits of ribulose-1,5-bisphosphate carboxylase in the cyanobacterium *Anabaena* 7120. Proc. Natl. Acad. Sci. USA 81, 5961–5965.

Norris, L, Norris, R.E. and Calvin, M. (1955) A survey of the rates and products of short-term photosynthesis in plants of nine phyla. J. Exp. Bot. 6, 64–74.

Papen, H. and Bothe, H. (1984) The activation of glutamine synthetase from the cyanobacterium *Anabaena cylindrica* by thioredoxin. FEMS Microbiol. Lett. 23, 41–46.

Papen, H., Neuer, G., Refaian, M. and Bothe, H. (1983) The isocitrate dehydrogenase from cyanobacteria. Arch. Microbiol. 134, 73–79.

Pelroy, R.A. and Bassham, J.A. (1972) Photosynthetic and dark carbon metabolism in unicellular blue-green algae. Arch. Mikrobiol. 86, 25–38.

Pelroy, R.A. and Bassham, J.A. (1976) Kinetics of light-dark CO_2 fixation and glucose assimilation by *Aphanocapsa* 6714. J. Bacteriol. 128, 633–643.

Perchorowicz, J.T. and Jensen, R.G. (1983) Photosynthesis and activation of ribulose bisphosphate carboxylase in wheat seedlings. Plant Physiol. 71, 955–960.

Perchorowicz, J.T., Raynes, D.A. and Jensen, R.G. (1981) Light limitation of photosynthesis and activation of ribulose bisphosphate carboxylase in wheat seedlings. Proc. Natl. Acad. Sci. USA 78, 2985–2989.

Peschek, G.A. (1979) The role of the Calvin cycle for anoxygenic CO_2 photoassimilation in *Anacystis nidulans*. FEBS Lett. 106, 34–38.

Quivey, R.G., Jr. and Tabita, F.R. (1984) Cloning and expression in *Escherichia coli* of the form II ribulose 1,5-bisphosphate carboxylase/oxygenase gene from *Rhodopseudomonas sphaeroides*. Gene 31, 91–101.

Raboy, B., Padan, E. and Shilo, M. (1976) Heterotrophic capacities of *Plectonema boryanum*. Arch. Microbiol. 110, 77–85.

Reichelt, B.Y. and Delaney, S.F. (1983) The nucleotide sequence for the large subunit of ribulose 1,5-bisphosphate carboxylase from a unicellular cyanobacterium, *Synechococcus* PCC 6301. DNA 2, 121–129.

Richter, G. (1961) The lack of diphosphofructose aldolase in two photosynthetic organisms, *Anacystis nidulans* and *Rhodopseudomonas sphaeroides*. Biochim. Biophys. Acta 48, 606–608.

Rindt, K.-P. and Ohmann, E. (1969) NADH and AMP as allosteric effectors of ribulose-5-phosphate kinase in *Rhodopseudomonas sphaeroides*. Biochem. Biophys. Res. Commun. 35, 357–364.

Rippel, S. and Bowien, B. (1984) Phosphoribulokinase from *Rhodopseudomonas acidophila*. Arch. Microbiol. 139, 207–212.

Robison, P.D., Martin, M.N. and Tabita, F.R. (1979) Differential effects of metal ions on *Rhodospirillum rubrum* ribulose bisphosphate carboxylase/oxygenase and stoichiometric incorporation of HCO_3^- into a cobalt (III)-enzyme complex. Biochemistry 18, 4453–4458.

Roy, H., Bloom, M. and Monroe, M. (1982) Studies on the assembly of large subunits of ribulose bisphosphate carboxylase in isolated pea chloroplasts. J. Cell Biol. 94, 20–27.

Schloss, J.V., Phares, E.F., Long, M.V., Norton, I.L., Stringer, C.D. and Hartman, F.C. (1979) Isolation, characterization, and crystallization of ribulose bisphosphate carboxylase from autotrophically grown *Rhodospirillum rubrum*. J. Bacteriol. 137, 490–501.

Servaites, J.C. and Torisky, R.S. (1984) Activation state of ribulose bisphosphate carboxylase in soybean leaves. Plant Physiol. 74, 681–686.

Servaites, J.C., Torisky, R.S. and Chao, S.F. (1984) Diurnal changes in ribulose 1,5-bisphosphate carboxylase activity and activation state in leaves of field-grown soybeans. Plant Sci. Lett. 35, 115–121.

Shinozaki, K. and Sugiura, M. (1983) The gene for the small subunit of ribulose-1,5-bisphosphate carboxylase/oxygenase is located close to the gene for the large subunit in the cyanobacterium *Anacystis nidulans* 6301. Nucl. Acids Res. 11, 6957–6964.

Shinozaki, K. and Sugiura, M. (1985) Genes for the large and small subunits of ribulose-1,5-bisphosphate carboxylase/oxygenase constitute a single operon in a cyanobacterium *Anacystis nidulans*. Mol. Gen. Genet. 200, 27–32.

Shinozaki, K., Yamada, C., Takahata, N. and Sugiura, M. (1983) Molecular cloning and sequence analysis of the cyanobacterial gene for the large subunit of ribulose-1,5-bisphosphate carboxylase/oxygenase. Proc. Natl. Acad. Sci. USA 80, 4050–4054.

Shively, J.M. (1974) Inclusion bodies of prokaryotes. Ann. Rev. Microbiol. 28, 167–187.

Shively, J.M., Ball, F., Brown, D.H. and Saunders, R.E. (1973a) Functional organelles in prokaryotes: polyhedral bodies (carboxysomes) of *Thiobacillus neapolitanus*. Science 182, 584–586.

Shively, J.M., Ball, F.L. and Kline, B.W. (1973b) Electron microscopy of the carboxysomes (polyhedral bodies) of *Thiobacillus neapolitanus*. J. Bacteriol. 116, 1405–1411.

Shively, J.M., Davidson, T. and Marrs, B.L. (1984) Derepression in the synthesis of the intermediate and large forms of ribulose-1,5-bisphosphate carboxylase/oxygenase in *Rhodopseudomonas capsulata*. Arch. Microbiol. 138, 233–236.

Siebert, K., Schobert, P. and Bowien, B. (1981) Purification, some catalytic and molecular properties of phosphoribulokinase from *Alcaligenes eutrophus*. Biochim. Biophys. Acta 658, 35–44.

Smith, A.J. (1973) Synthesis of metabolic intermediates. In N.G. Carr and B.A. Whitton (eds.), The Biology of Blue-Green Algae. Blackwell, Oxford, UK, pp. 1–38.

Smith, A.J. (1982) Modes of cyanobacterial carbon metabolism. In N.G. Carr and B.A. Whitton (eds.), The Biology of Cyanobacteria. Blackwell, Oxford, UK, pp. 47–85.

Stanier, R.Y. (1973) Autotrophy and heterotrophy in unicellular blue-green algae. In N.G. Carr and B.A. Whitton (eds.), The Biology of Blue-Green Algae. Blackwell, Oxford, UK, pp. 501–518.

Stewart, W.D.P. and Codd, G.A. (1977) Polyhedral bodies (carboxysomes) of nitrogen-fixing blue-green algae. Br. Phycol. J. 10, 273–278.

Surek, B., Heilbronn, A., Austen, A. and Latzko, E. (1985) Purification and characterization of phosphoribulokinase from wheat leaves. Planta 165, 507–515.

Tabita, F.R. (1980) Pyridine nucleotide control and subunit structure of phosphoribulokinase from photosynthetic bacteria. J. Bacteriol. 143, 1275–1280.

Tabita, F.R. (1981) Molecular regulation of carbon dioxide assimilation in autotrophic microorganisms. In H. Dalton (ed.), Microbial Growth on C_1 Compounds. Heydon & Sons, London, pp. 70–81.

Tabita, F.R. and Colletti, C. (1979) Carbon dioxide assimilation in cyanobacteria: regulation of ribulose 1,5-bisphosphate carboxylase. J. Bacteriol. 140, 452–458.

Tabita, F.R. and McFadden, B.A. (1972) Regulation of ribulose-1,5-diphosphate carboxylase by 6-phospho-D-gluconate. Biochem. Biophys. Res. Commun. 48, 1153–1160.

Tabita, F.R. and McFadden, B.A. (1974a) One-step isolation of microbial ribulose-1,5-diphosphate carboxylase. Arch. Microbiol. 99, 231–240.

Tabita, F.R. and McFadden, B.A. (1974b) D-Ribulose 1,5-diphosphate carboxylase from *Rhodospirillum rubrum*. II. Quaternary structure, composition, catalytic and immunological properties. J. Biol. Chem. 249, 3459–3464.

Tabita, F.R. and Small, C.L. (1985) Expression and assembly of active cyanobacterial ribulose-1,5-bisphosphate carboxylase/oxygenase in *Escherichia coli* containing stoichiometric amounts of large and small subunits. Proc. Natl. Acad. Sci. USA 82, 6100–6103.

Tabita, F.R., Stevens, S.E., Jr. and Quijano, R. (1974) D-Ribulose 1,5-diphosphate carboxylase from blue-green algae. Biochem. Biophys. Res. Commun. 61, 45–52.

Tabita, F.R., Stevens, S.E., Jr. and Gibson, J.L. (1976) Carbon dioxide assimilation in blue-green algae: initial studies on the structure of ribulose 1,5-bisphosphate carboxylase. J. Bacteriol. 125, 531–539.

Tabita, F.R., Caruso, P. and Whitman, W. (1978) Facile assay of enzymes unique to the Calvin cycle in intact cells, with special reference to ribulose 1,5-bisphosphate carboxylase. Anal. Biochem. 84, 462–472.

Tabita, F.R., Martin, M.N., Beudeker, R.F., Quivey, R.G., Jr., Sarles, L.S. and Weaver, K.E. (1984) Control of carbon dioxide fixation. In R.L. Crawford and R.S. Hansen (eds.), Microbial Growth on C_1 Compounds. American Soc. for Microbiol., Washington, D.C., pp. 3–8.

Takabe, T., Nishimura, M. and Akazawa, T. (1976) Presence of two subunit types in ribulose-1,5-bisphosphate carboxylase from blue-green algae. Biochem. Biophys. Res. Commun. 68, 537–544.

Takabe, T., Rai, A.K. and Akazawa, T. (1984) Interaction of constituent subunits in ribulose 1,5-bisphosphate carboxylase from *Aphanothece halophytica*. Arch. Biochem. Biophys. 229, 202–211.

Tiboni, O., DiPasquale, G. and Ciferri, O. (1984) Cloning and expression of the genes for ribulose-1,5-bisphosphate carboxylase from *Spirulina platensis*. Biochim. Biophys. Acta 783, 258–264.

Torisky, R.S. and Servaites, J.C. (1984) Effect of irradiance during growth of *Glycine max* on photosynthetic capacity and percent activation of ribulose 1,5-bisphosphate carboxylase. Photosyn. Res. 5, 251–261.

Torres-Ruiz, J.A. and McFadden, B.A. (1985) Isolation of L_8S_8 forms of ribulose bisphosphate carboxylase/oxygenase from *Chromatium vinosum*. Arch. Microbiol. 142, 55–60.

Van Baalen, C. and Pulich, W.M. (1973) Heterotrophic growth of microalgae. CRC Crit. Rev. Microbiol., Jan. issue, 229–255.

Vieira, J. and Messing, J. (1982) The pUC plasmids, an M13 mp7-derived system for insertion mutagenesis and sequencing with synthetic universal primers. Gene 19, 259–268.

Volokita, M., Zenvirth, D., Kaplan, A. and Reinhold, L. (1984) Nature of the inorganic carbon species actively taken up by the cyanobacterium *Anabaena variabilis*. Pl. Physiol. 76, 599–602.

Vu, C.V., Allen, L.H., Jr. and Bowes, G. (1983) Effects of light and elevated atmospheric CO_2 on the ribulose bisphosphate carboxylase activity and ribulose bisphosphate level of soybean leaves. Plant Physiol. 73, 729–734.

Vu, C.V., Allen, L.H., Jr. and Bowes, G. (1984) Dark/light modulation of ribulose bisphosphate carboxylase activity in plants from different photosynthetic categories. Plant Physiol. 76, 843–845.

Weathers, P.J. and Allen, M.M. (1978) Variations in short term products of inorganic carbon fixation in exponential and stationary phase cultures of *Aphanocapsa* 6308. Arch. Microbiol. 116, 231–234.

Wildner, G.F. and Henkel, J. (1978) Differential reactivation of ribulose 1,5-bisphosphate oxygenase with low carboxylase activity with Mn^{2+}. FEBS Lett. 91, 99–103.

Winkenbach, F. and Wolk, C.P. (1973) Activities of enzymes of the oxidative and the reductive pentose phosphate pathways in heterocysts of a blue-green alga. Pl. Physiol. 52, 480–483.

Wolk, C.P. (1968) Movement of carbon from vegetative cells to heterocysts in *Anabaena cylindrica*. J. Bacteriol. 962, 2138–2143.

Wolosiuk, R.A. and Buchanan, B.B. (1978) Regulation of chloroplast phosphoribulokinase by the ferredoxin/thioredoxin system. Arch. Biochem. Biophys. 189, 97–101.

Wolosiuk, R.A., Crawford, N.A., Yee, B.C. and Buchanan, B.B. (1979) Isolation of three thioredoxins from spinach leaves. J. Biol. Chem. 254, 1627–1632.

P. Fay and C. Van Baalen (Eds.) The Cyanobacteria

5

Respiratory electron transport

GÜNTER A. PESCHEK

Biophysical Chemistry Group, Institute of Physical Chemistry, University of Vienna, Währingerstraße 42, A-1090 Vienna, Austria

1. Introduction

This review will focus on respiratory electron transport in cyanobacteria including components, energy-transducing function, localization, and interaction with photosynthetic electron transport. Selected aspects of respiratory carbon metabolism will also be discussed. Several comprehensive reviews and monographs are available on cyanobacterial ultrastructure, ecology, physiology, biochemistry, and energy-transduction in general (Lang, 1968; Holm-Hansen, 1968; Wolk, 1973; Fogg et al., 1973; Carr and Whitton, 1973, 1982; Stanier and Cohen-Bazire, 1977; Doolittle, 1979; Padan, 1979; Stewart, 1980; Binder, 1982; Peschek, 1984a). However, cyanobacterial respiration is among the more neglected fields.

1.1. Chemoheterotrophic growth

Only a minority of cyanobacterial species so far tested are capable of dark aerobic growth on organic substrates (Rippka, 1972; Stanier, 1973). Although some possess photoheterotrophic potential (Rippka et al., 1979), quite a few species have never been successfully cultured in the dark under air and in the presence of substrates (Smith, 1982), and not a single case of dark anaerobic heterotrophy (fermentative growth) is known. Thus, many cyanobacteria are obligate photo(auto)trophs relying entirely on energy from light (and carbon from CO_2) for growth and prolifera-

Abbreviations: cyt, cytochrome; DBMIB, dibromothymoquinone (2,5-dibromo-3-methyl-6-isopropyl-*p*-benzoquinone); DAB, diaminobenzidine; DCPIP, 2,6-dichlorophenol indophenol; DMO, 5,5-dimethyl-oxazolidine-2,4-dione; fd, ferredoxin; HOQNO, 2-*n*-heptyl-8-hydroxyquinoline-*N*-oxide; INT-DNP, 2-iodo-6-isopropyl-3-methyl-2′,4,4′-trinitrophenylether; PMS, phenazine methosulfate (*N*-methyl-phenazonium sulfate); PS, photosystem; SDS, sodium dodecylsulfate; TMPD, *N*,*N*,*N*′,*N*′-tetramethyl-*p*-phenylenediamine; TPP^+, tetraphenylphosphonium; DCMU, 3-(3′,4′-dichlorophenyl)-1,1-dimethylurea; NBD, 7-chloro-4-nitrobenz-2-oxa-1,3-diazole; TTFA, thenoyltrifluoroacetone; FeCy, ferricyanide; CM, plasma membrane; DCCD, *N*,*N*′-dicyclohexylcarbodiimide.

tion; they share this property with a variety of chemolithotrophic and methylotrophic bacteria (Kelly, 1971; Whittenbury and Kelly, 1977; Smith and Hoare, 1977; Matin, 1978; Benedict, 1978; Higgins et al., 1981; Peschek, 1984a). Apparently, there is no singular property or lack of property, like the lack of a functional tricarboxylic acid cycle (Pearce et al., 1969; Pearce and Carr, 1969; Smith et al., 1967), and of NADH oxidase (Smith et al., 1967; Horton, 1968; Leach and Carr, 1970), or an impermeability of the cell to organic substrates (Stanier, 1973), an insufficient ATP supply from oxidative phosphorylation (Pelroy and Bassham, 1973; Nitschmann and Peschek, 1982; Nitschmann et al., 1982), or a general lack of transcriptional control (Carr, 1973), responsible for the inability to grow aerobically in the dark on organic substrates. We are still left with the enigma of obligate (photo- or chemo-)autotrophic growth behavior which, though sometimes

TABLE I

Growth rates of cyanobacteria; values shown are generation times (h) estimated from exponential growth rates

Strain	Growth conditions			Substrate for heterotrophic growth (mM)	Temperature (°C)	References
	photo-autotrophic	photo-heterotrophic[a]	chemo-heterotrophic			
Synechocystis 6714	15	30	50	Glc (55)	30	Rippka, 1972
	6	6	27	Glc (10)	34	Joset-Espardellier et al., 1978
Nostoc sp. strain MAC	6.5	–	48	Glc (28)	39	Ingram et al., 1973; Bottomley and Van Baalen, 1978
	6.5	–	144	Fru (28)	39	Bottomley and Van Baalen, 1978
	–	–	22	Fru + 0.1% casamino acids	39	Bottomley and Van Baalen, 1978
	30	–	96	Glc (10)	30	Peschek and Sleytr, 1983
Plectonema boryanum	24	–	70–100[b]	Glc (30)	30	Raboy et al., 1976
	–	–	40[c]	Glc (30)	30	Raboy et al., 1976
Anabaena ATCC 29413	(9)[d]	–	36	Fru (5)	30	Wolk and Shaffer, 1976
	14	25	40	Fru (?)	?	Smith, 1982
	20	–	38	Fru (10)	30	Peschek and Sleytr, 1983
Chlorogloeopsis fritschii	30	25	80	Glc (10)	34	Joset-Espardellier et al., 1978

[a] In the presence of DCMU; [b] after adaptation period of 2 generations; [c] after 1 year of dark growth; [d] estimated from growth on agar; Glc, glucose; Fru, fructose; dash indicates 'not determined'.

disputed (Rittenberg, 1969, 1972; Whittenbury and Kelly, 1977), all the same does exist in several groups of microorganisms.

1.2. Substrate catabolism

Chemoheterotrophic growth is generally much slower than photoautotrophic growth (Table I). Substrates which support dark growth of cyanobacteria are invariably carbohydrates or compounds readily convertible to intermediates of the oxidative pentose phosphate pathway, the main route of substrate breakdown in cyanobacteria. Accordingly, the main endogenous reserve substrate in cyanobacteria is glycogen (Pelroy et al., 1972; Lehmann and Wöber, 1976; Van Liere et al., 1979). The first two enzymes of the oxidative pentose phosphate pathway, glucose 6-phosphate dehydrogenase and 6-phosphogluconate dehydrogenase, have been shown to be specific for $NADP^+$ rather than NAD^+ (Smith et al., 1967; Pelroy et al., 1972). NADPH thus would be expected to be the native electron donor to the cyanobacterial respiratory chain. This was first pointed out by Biggins (1969) who demonstrated that on transition of whole cells of *Anacystis nidulans* from aerobiosis to anaerobiosis (in the dark), the intracellular level of NADH remained constant, while that of NADPH increased in proportion to the decrease in $NADP^+$ (Peschek, 1979d). The pivotal importance of the oxidative pentose phosphate pathway for the heterotrophic carbon metabolism in respiring cyanobacteria became evident from radiorespirometry using specifically labeled ^{14}C-glucose (Table II) and from enzymology (Table III). Using these techniques the key enzymes of the Entner-Doudoroff pathway could not be detected (Pelroy et al., 1972). On the other hand, variable and inconclusive results were obtained with respect to 6-phosphofructokinase (PFK) (Fewson et al., 1962; Pearce and Carr, 1969; Neuer and Bothe, 1982; Pelroy et al., 1972). It may be mentioned that PFK, the key enzyme of the Embden-Meyerhof-Parnas pathway, is present in chloroplasts (Kelly and Latzko, 1977a,b; Croxdale, 1983). Its possible function in glucose (but not fructose?) catabolism in three species of cyanobacteria is indicated from studies with specifically labeled fructose and glucose (I. Jansen, V. Molitor, L. Kindslehner and G.A. Peschek, unpublished; see Table II).

1.3. Respiratory oxygen uptake

Cyanobacteria are often referred to as exhibiting very low rates of endogenous respiration. Table IV shows that this assumption is not justified if we compare cyanobacteria with, e.g., eukaryotic algae. Instead, what appears to be characteristic of cyanobacteria is the poor response of respiration rates to exogenous substrates. Respiratory oxygen uptake by whole cells of cyanobacteria was found to be stimulated by uncouplers of electron transport phosphorylation (Biggins, 1969) and inhibited, to varying degrees, by poisons of hemoprotein cytochrome oxidase (Webster and Frenkel, 1953; Kratz and Myers, 1955; Biggins, 1969). Cyanobacterial respiration also is inhibited by light (Brown, 1953; Brown and Webster, 1953; Peschek, 1975; Scherer and Böger, 1982). This photoinhibition of respiratory oxygen uptake, characteristic of all phototrophic prokaryotes (Keister, 1978), reflects the intimate spatial relationship between photosynthetic and respiratory systems in a 'non-compartmentalized' prokaryotic cell (see Section 6 of this chapter).

TABLE II

Relative rates of $^{14}CO_2$ evolution from positionally labeled ^{14}C-glucose and ^{14}C-fructose by cyanobacteria incubated in dark aerobic conditions; for comparison, results with *Rhodopseudomonas capsulata* are also shown (maximum rates of CO_2 evolution within each series are taken as 100)

Organism	Glucose				Fructose				Incubation time (min)	References	Comments
	C-1	C-2	C-3,4	C-6	C-1	C-2	C-3,4	C-6			
Synechocystis 6714	100	–	40.6	6.8	–	–	–	–	30	Pelroy et al., 1972	Facultative chemoheterotroph. Photoautotrophically grown cells were used
Plectonema boryanum	100	–	–	43.5	–	–	–	–	40	Raboy et al., 1976	After photoautotrophic growth
	100	–	–	41.7	–	–	–	–	40	Raboy et al., 1976	After chemoheterotrophic growth on glucose
Anacystis nidulans	100	22.8	59.5	33.8	100	62.8	57.0	43.7	20	Molitor, 1984	
Anabaena ATCC 29413	100	41.3	63.0	17.4	100	87.8	60.0	34.4	20	Molitor, 1984	After photoautotrophic growth
	100	15.1	14.1	5.2	100	24.6	22.0	20.2	15	Molitor, 1984	After chemoheterotrophic growth on fructose
Nostoc sp. strain MAC	100	21.9	39.0	15.2	100	73.3	56.3	29.5	20	Molitor, 1984	After photoautotrophic growth
	100	24.4	48.8	13.4	100	50.0	35.0	30.0	15	Molitor, 1984	After chemoheterotrophic growth on glucose
Rhodopseudomonas capsulata	100	35.1	56.7	21.6	37.2	31.5	100	22.9	30	Conrad and Schlegel, 1977	After chemoheterotrophic growth on glucose (Entner-Doudoroff-pathway) and fructose (Embden-Meyerhof-pathway), resp.

Dash indicates 'not determined'.

TABLE III

Activities of key enzymes of catabolic pathways in cell-free extracts of cyanobacteria. Activities expressed as nmol substrate transformed/min per mg protein. In case of facultative heterotrophs the mode of growth preceding the assay is indicated. GAP-DH, glyceraldehyde 3-phosphate dehydrogenase; G6P-DH, glucose 6-phosphate dehydrogenase; 6PG-DH, 6-phosphogluconate dehydrogenase; PFK, phosphofructokinase; KDPG-aldolase, 2-keto-3-deoxy-6-phosphogluconate aldolase

Strain	Hexokinase	GAP-DH		G6P-DH	6PG-DH	PFK	KDPG-aldolase	References
		NAD^+	$NADP^+$					
Anabaena variabilis	1.2	8.3	7.5	8.7	12.0	8.1	–	Pearce et al., 1969; Pearce and Carr, 1967, 1969
	–	120	60	184	284	16.7	–	Fewson et al., 1962
Anacystis nidulans	251	43.4	80.2	184	517	13.4	–	Fewson et al., 1962; Pelroy et al., 1972
Synechocystis 6714								
photoautotrophic	9	13	–	5	7	0.05	0.02	Pelroy et al., 1972
photoautotrophic	–	1000[a]		298	250	–	–	Joset-Espardellier et al., 1978
chemoheterotrophic	–	230[a]		250	220	–	–	Joset-Espardellier et al., 1978
Anabaena cylindrica								Neuer and Bothe, 1982)
isolated heterocysts	9.5	11.1	–	15.3	–	1.5	–	
vegetative cells	7.4	24.5	–	1.1	–	1.6	–	
Plectonema boryanum								Raboy et al., 1976
photoautotrophic	–	–	–	6.7	30	–	–	
chemoheterotrophic	–	–	–	113.3	50	–	–	
Chlorogloeopsis fritschii								Joset-Espardellier et al., 1978
photoautotrophic	–	108[a]		16	–	–	–	
chemoheterotrophic	–	68[a]		52	–	–	–	

[a] Not indicated if assayed with NAD^+ or $NADP^+$. Dash indicates 'not determined'.

TABLE IV

Rates of endogenous and exogenous respiration of cyanobacteria and other microorganisms (nmol O_2/min per mg dry weight)[a]

Organism	Endogenous	In presence of substrate	References	Comments
Cyanobacteria[a]				
Anacystis nidulans	0.2–1.2	–	Kratz and Myers, 1955; Peschek and Broda, 1973	
		3.6 (Fru, 50 mM) 1.5 (Man, 50 mM) 1.1 (Glc, 50 mM)	Peschek and Broda, 1973	35°C, pH 7–8
Anabaena ATCC 27892	6.2–6.3	–	Webster and Frenkel, 1953	
Anabaena ATCC 29413	6.3–7.9	–	Nitschmann and Peschek, 1982	35°C, pH 7.4
Nostoc sp.strain MAC	7.8–9.2	–	Nitschmann and Peschek, 1985	35°C, pH 7.4
Dermocarpa ATCC 29372	0.9	–	Nitschmann and Peschek, 1985	35°C, pH 7.4
Plectonema boryanum	9.7–11.3	–	Raboy et al., 1976	Photoautotrophically grown cells
	12.3–15.1	–		Chemoheterotrophically grown cells
Eukaryotic algae				
Chlorella pyrenoidosa	0.74–1.04	3.7 (Glc)	Genevois, 1927	Photoautotrophically grown cells
	–	2.1 (Glc)		Chemoheterotrophically grown cells (on glucose)
Scenedesmus sp.	1.33	2.83	Gaffron, 1939	Chemoheterotrophically grown cells (on glucose)
Heterotrophic bacteria				
Azotobacter sp.	–	1500	Schlegel, 1981	Assayed in the presence of growth substrate
Acetobacter sp.	–	1350		
Pseudomonas sp.	–	900		
Yeasts				
Saccharomyces cerevisiae	7.44	70.7 (Glc)	Meyerhof, 1925	

Dash indicates 'not determined'; Glc, glucose; Fru, fructose; Man, mannose.

[a] For comparison, typical rates of photosynthetic oxygen evolution by both prokaryotic and eukaryotic algae are between 50 and 200 nmol O_2/min per mg dry weight.

2. Components of respiratory electron transport

2.1. Reduced pyridine nucleotide dehydrogenases

NADPH may be the natural electron donor to the respiratory chain of intact cyanobacteria (Biggins, 1969), yet, isolated membranes from cyanobacteria have invariably been found to oxidize both NADPH and NADH, photosynthetically via Photosystem I (PSI) and also in the dark via respiratory electron transport.

In cell-free preparations of *Plectonema boryanum* NADH and NADPH were oxidized at the thylakoid membrane, with different dehydrogenases for either nucleotide (Matthijs et al., 1984a). Oxidation of NADH was inhibited 70% by 10 μM antimycin A, while NADPH oxidation was rather insensitive to this inhibitor, suggesting different sites of electron transfer. Both types of oxidation were equally sensitive to plastoquinone antagonists such as INT-DNP or DBMIB, and to HOQNO. Comparing their results with membranes derived from sonicated cells ('scrambled membranes') and from osmotically shocked spheroplasts, the authors arrived at the following scheme of electron transport from NAD(P)H to P700 (photosynthetic pathway) and/or oxygen (respiratory pathway) (Matthijs, 1984; Matthijs et al., 1984b):

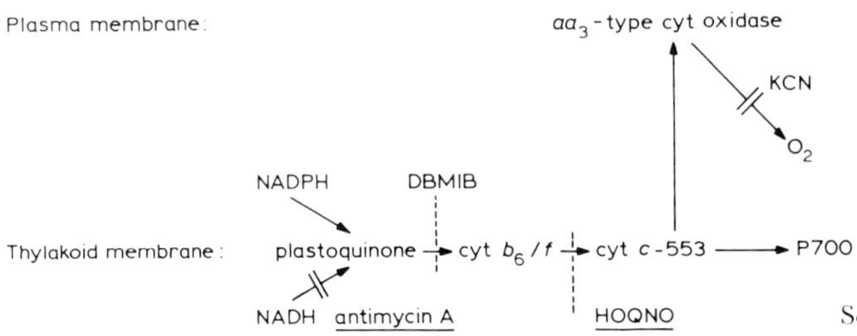

Scheme 1.

Crude membrane preparations of the thermophilic cyanobacterium *Mastigocladus laminosus* have shown optimum rates of dark respiratory electron transport when isolated from cells grown with limiting CO_2 concentrations. Photosystem II (PSII) activity was low in these preparations. The opposite effect was noted when the cells were grown at high CO_2 (Binder et al., 1984). Again, both NADH and NADPH could serve as substrates to both PSI-dependent and respiratory electron transport in isolated membranes, with almost equal overall efficiency. 1 mM KCN inhibited dark oxygen uptake supported by NADH and NADPH by 70 and 20%, respectively. NADH-dependent O_2 uptake, in addition, was markedly inhibited by azide, antimycin A, rotenon, and the plastoquinone antagonist INT-DNP.

Different kinetic parameters (V_{max},K_m) were measured for the oxidation of NADH and NADPH, respectively, in membranes from *Anabaena* ATCC 29413, in the light and in the dark (Stürzl et al., 1984), suggesting some light modulation of the activities of both NADH and NADPH dehydrogenases, but in similar experiments with NAD(P)H dehydrogenase preparations from *Synechocystis* 6714 (Sandmann and Malkin, 1983a) and *Microcystis aeruginosa* (Viljoen et al., 1985) no effect of light on the dehydrogenation of NAD(P)H was found. NAD(P)H-dependent O_2 uptake, on the other hand, was inhibited by light. Photoinhibition of NAD(P)H-supported O_2 uptake in isolated membranes of cyanobacteria was first observed by Peschek and Schmetterer (1982) who proposed that the dehydrogenase-plastoquinone-cytochrome b_6/f sequence of the electron transport chain might be a common donor to both P700 and cytochrome oxidase. A similar conclusion was reached by Sandmann and Malkin (1984) from results with *Synechocystis* 6714 (see Section 6).

NADH and NADPH (and H_2, via a membrane-bound uptake hydrogenase) donate electrons to photosynthetic and respiratory electron transport systems in isolated heterocysts of *Anabaena* ATCC 27893 (Houchins and Hind, 1982). NADH-supported O_2 uptake is strongly inhibited by cyanide and unaffected by methyl viologen while NADPH-supported O_2 uptake is only weakly cyanide-sensitive but markedly stimulated by methyl viologen. This would suggest that *Anabaena* heterocysts can oxidize NADH through a membrane-bound dehydrogenase, while NADPH-dependent O_2 uptake proceeds via ferredoxin:$NADP^+$ oxidoreductase; the latter, by virtue of its well-known diaphorase activity (Rowell et al., 1982), would reduce some cytochrome(s) of the respiratory/photosynthetic electron flow chain (Houchins and Hind, 1982). In heterocysts the ultimate purpose of electron donation from NAD(P)H (and H_2) is the supply of reducing power and energy to the nitrogenase (Haselkorn, 1978; Stewart, 1980). For isolated heterocysts of *Anabaena* ATCC 29413 it has been suggested that the tandem operation of NADH dehydrogenase, PSI, ferredoxin, and fd:$NADP^+$ oxidoreductase constitutes an efficient light-dependent transhydrogenase system (Schrautemeier and Böhme, 1984). Table V summarizes electron transport rates supported by NAD(P)H in particle preparations of several cyanobacteria, with O_2 as the terminal electron acceptor.

TABLE V

NAD(P)H:O_2 oxidoreductase activities of cyanobacteria

Organism	Electron donor		References	Comments
	NADH	NADPH		
Phormidium luridum[a]	0	30	Biggins, 1969	Isolated membranes (protein/chlorophyll = 6.9)
Anabaena variabilis ATCC 27892[b]	2.4	4.2	Leach and Carr, 1970	Isolated membranes
Anacystis nidulans[b]	20	54	Peschek, 1980a	Isolated membranes (protein/chlorophyll = 10)
Anabaena ATCC 27893[a]	4800	12670	Houchins and Hind, 1982	Freeze-thawed heterocysts (protein/chlorophyll = 71.4)
Synechocystis 6714[a]	1340	1800	Sandmann and Malkin, 1983a	Osmotically shocked spheroplasts
Mastigocladus laminosus[b]	172	64	Binder et al., 1984	Isolated membranes
Plectonema boryanum[a]	70	180	Matthijs et al., 1984a	Osmotically shocked spheroplasts
Anabaena ATCC 29413[a,c]	Dark: 133	400	Stürzl et al., 1984	K_m toward NADH was 110 μM in the light and 40 μM in the dark; with NADPH the values were 48 and 18, resp.
	Light: 457	533		

Rates expressed as nequiv. e^-/min per mg chlorophyll (a) or protein (b). Corrected for O_2 uptake by DCMU-poisoned membranes in the absence of NAD(P)H (c).

Recently, isolation and characterization of NAD(P)H dehydrogenases from *Synechocystis* 6714 (Sandmann and Malkin, 1983b; Sandmann, 1985) and *Microcystis aeruginosa* (Viljoen et al., 1985) were reported. The soluble fraction of extracts from *Synechocystis* contained a dehydrogenase which is inhibited by rotenone, *o*-phenanthroline and TTFA, and shows high specificity for NADH when assayed with benzoquinone as electron acceptor (Sandmann and Malkin, 1983b). EPR spectrometry suggests that this dehydrogenase is involved in the NADH-supported reduction of ferredoxin via some soluble NAD(P)H:fd oxidoreductase. The particulate fraction of the extract, on the other hand, revealed the presence of a dehydrogenase whose activity with NADPH was 2.5 times the activity with NADH (K_m toward NADH = 220 μM, toward NADPH = 77 μM). Inhibition of this enzyme by rotenone, *o*-phenanthroline and TTFA was much less pronounced. The membrane-bound NAD(P)H dehydrogenase was solubilized and purified to some extent (Sandmann and Malkin, 1983b). The partly purified enzyme reduced various quinones with NADH or NADPH as electron donor, K_m values (nM) being 690 toward duroquinone, 53 toward benzoquinone, and 13 for 1,4-naphthoquinone. The high affinity of the enzyme for naphthoquinone is remarkable in view of a possible participation of native phylloquinone in the respiratory electron transport (Peschek, 1980b; Section 2.4). Another peculiar feature of the dehydrogenase is the presence of only one redox active iron-sulfur center (Sandmann, 1985; Section 2.3.).

A soluble NAD(P)H dehydrogenase was isolated and extensively purified from *Microcystis aeruginosa* (Viljoen et al., 1985). The enzyme was described as a globular flavoprotein with molecular mass of 41 kD, and comprising 8 identical subunits. Though soluble, the enzyme transfers electrons from NAD(P)H to the membrane particles with plastoquinone possibly acting as the immediate acceptor. In view of similar findings with NAD(P)H:quinone oxidoreductase activities in *Synechocystis* 6714 (Sandmann and Malkin, 1983b) and *Anacystis nidulans* (Peschek, 1980b) the markedly higher affinity of the *Microcystis* enzyme for naphthoquinone, when compared to benzoquinone, is noteworthy (also see Section 2.4.). However, in contrast to *Anabaena variabilis* (Stürzl et al., 1984) and *Synechocystis* 6714 (Sandmann and Malkin, 1983a,b, 1984), in *Microcystis aeruginosa* one and the same enzyme brings about the dehydrogenation of both NADH and NADPH. It was suggested that the *Microcystis* enzyme differs from fd:NADP$^+$ oxidoreductase as it is unable to reduce NADP$^+$ at the expense of reduced ferredoxin (Viljoen et al., 1985). Soluble fd:NADP$^+$ oxidoreductase (EC 1.18.1.2), which is ubiquitous in plant-type photosynthetic cells, may display a variety of puzzling reactions including NADPH-specific diaphorase activity, NADPH:NAD$^+$ oxidoreductase (transhydrogenase) activity, and NADPH-dependent cytochrome *f* reductase activity (Rowell et al., 1981; Houchins and Hind, 1982; Serrano et al., 1984; Schrautemeier and Böhme, 1984).

To sum up, the following modes of electron transfer from reduced pyridine nucleotides to the electron transport chain of cyanobacteria emerge from the literature:

(a) Distinct membrane-bound dehydrogenases acting with NADH or NADPH (Matthijs et al., 1984a; Sandmann and Malkin, 1984a; Binder et al., 1984; Peschek, 1980a; Molitor and Peschek, 1986, see Table VII).

(b) A soluble dehydrogenase specific for NADH (Sandmann and Malkin, 1983b).

(c) A single, soluble dehydrogenase oxidizing both NADH and NADPH, yet preferring the latter over the former (Viljoen et al., 1985).

(d) In addition, NADPH dehydrogenation may be mediated by the soluble and ubiquitous ferredoxin:NADP$^+$ oxidoreductase ('FNR'), either directly reducing oxygen (Houchins and Hind, 1982), or feeding electrons first into the (respiratory) chain (Binder et al., 1984; Alpes et al., 1985). Rather unspecific electron transfer reactions catalyzed by this enzyme might obscure measurements of NADPH-oxidizing activities in isolated membranes which, though separated from the cytosol, may not be absolutely free of cytosolic proteins among them, e.g. the fd:NADP$^+$ oxidoreductase (Stürzl et al., 1984; Alpes et al., 1985), while extensive washing of the membrane may damage the membrane-bound NAD(P)H dehydrogenase (Stürzl et al., 1984).

2.2 Other membrane-bound dehydrogenases

An uptake hydrogenase found in cyanobacterial membranes, and capable of initiating an oxyhydrogen reaction similar to that demonstrated in non-photosynthetic H_2-oxidizing bacteria (Bowien and Schlegel, 1981) has been proposed to act as an electron donor to nitrogenase in heterocysts, recycling the H_2 constitutively formed during N_2 fixation (Bothe et al., 1977; Stewart, 1980; Houchins and Burris, 1981a,b; Lambert and Smith, 1981; Houchins, 1984). Yet, it appears to occur also in vegetative cells and in non-nitrogen-fixing species (Bothe et al., 1978; Eisbrenner et al., 1978; Peschek, 1979a,b,c; Howarth and Codd, 1985; see Peschek, 1984a). The question how many and what types of hydrogenases exist in cyanobacteria, and whether they occur soluble or membrane-bound, is yet unresolved.

Low rates of oxidation of malate (Biggins, 1969) and succinate (Biggins, 1969; Leach and Carr, 1969, 1970; Peschek, 1980a; Binder et al., 1984) by membrane preparations of cyanobacteria have also been reported. Electron transport to O_2 from all the substrates mentioned was shown to be coupled to oxidative phosphorylation in vitro, though yields and coupling efficiencies were invariably low (see Section 4).

2.3. Iron-sulfur proteins

The occurrence of iron-sulfur centers as components of photosynthetic electron transfer reactions in the membranes of cyanobacteria has been much investigated (see Chapters 1 and 2 of this volume). In marked contrast to photosynthesis, very little is known about iron-sulfur centers participating in respiratory electron transport of cyanobacteria. The Rieske iron-sulfur protein (Rieske et al., 1964) (with a redox potential of approx. +0.3 V) is a constituent of the cytochrome *b/c* complex of mitochondria (Engel et al., 1980), chloroplasts (Hurt and Hauska, 1981, 1982a,b), and photosynthetic bacteria (Gabellini et al., 1982). A cytochrome b_6/f complex composed of four polypeptides, among them apparently the Rieske Fe-S protein, was isolated from *A. variabilis* (Krinner et al., 1982). This complex was active in reconstituting electron transport from reduced plastoquinone to cytochrome *c*-553.

Solubilized and partly purified NAD(P)H dehydrogenase from *Synechocystis* 6714 (Sandmann and Malkin, 1983b) was found associated with only one redox active iron-sulfur center (Sandmann, 1985); by contrast, eukaryotic NADH dehydrogenases may contain up to seven Fe-S centers (Ohnishi, 1973; Ragan, 1976). Thermodynamic and kinetic properties of the cytochrome b_6/f complex, including the Rieske Fe-S protein, have been studied in whole cells of a thermophilic *Synechococcus* sp. (Nanba and Katoh, 1983, 1984, 1985).

Autoxidizable Fe-S centers similar to the Rieske type and involved in dark respiratory electron flow may be responsible for part of the cyanide-insensitive respiratory oxygen uptake frequently observed with many cyanobacteria. This would conform with the observation that oxidation of exogenous cyt c and TMPD in isolated membranes of *A. nidulans* is completely inhibited by very low concentrations (about 1 μM) of KCN (Molitor and Peschek, 1986; Trnka and Peschek, 1986) while the oxidation of NAD(P)H feeding electrons into the chain at a more negative redox potential still shows the usual 20% cyanide-insensitive component (Peschek, 1980a; Trnka et al., 1985). The latter is affected to some extent by strong metal-chelating agents such as hydroxamic acids and phenanthrolins, lipophilic bathophenanthrolin giving stronger inhibition than the more hydrophilic *o*-phenanthrolin. This might indicate that the redox-active non-heme iron is deeply buried in the lipid core of the membrane (Peschek, 1980a). Similar results are known for plant mitochondria, yet, no distinct cyanide-insensitive hemoprotein cytochrome oxidase has so far been identified in them (Solomos, 1977), nor could such an alternative terminal oxidase be clearly resolved in cyanobacterial membranes (Peschek, 1981a, b; Kienzl and Peschek, 1982; Peschek et al., 1981b, 1982a).

2.4. Quinones

A joint involvement of plastoquinone in both respiratory and photosynthetic electron transport in a thermophilic *Synechococcus* sp. is apparent from the results of Katoh and colleagues. They have demonstrated the reduction by respiratory electron donors in the dark of (soluble) cytochrome *c*-553, in whole cells, after pre-oxidation with far-red light flashes. This reduction was decreased in starved cells and stimulated by KCN (Hirano et al., 1980). Aoki and Katoh (1982) have followed the oxidation and reduction of plastoquinone by respiratory and photosynthetic electron transport, using the 'I-D dip', an early transient of fluorescence induction, as a means to monitor the redox changes of plastoquinone; they also estimated the plastoquinone pool sizes functioning in respiration and photosynthesis (Aoki and Katoh, 1983).

Experiments with membranes derived from *A. nidulans* that had been extracted with *n*-pentane followed by incorporation of various authentic quinones into the depleted membranes, showed that the ability of phylloquinone to restore dark respiratory electron transport, supported by various donor-acceptor couples, was slightly superior to that displayed by authentic plastoquinone-9 (Peschek, 1980b). Incorporation of ubiquinone-10, menaquinone-7, and tocopherylquinone was without effect. Similarly, when the light-dependent reduction of (exogenous) $NADP^+$ with H_2 was assayed in such membranes after reconstitution with authentic quinones only plastoquinone and phylloquinone showed reconstitutive activity (Table VI). These results, while demonstrating that

TABLE VI

Reconstitution of respiratory ($H_2 \rightarrow O_2$) and photosynthetic ($H_2 \rightarrow NADP^+$) electron transport with plastoquinone, phylloquinone, and ubiquinone in pentane-extracted membrane preparations from hydrogenase-induced *Anacystis nidulans*[a]

Donor-acceptor couple	Electron transport rate (neq.e^-/min per mg protein)				
	Native membranes	Extracted membranes	Extracted membranes reconstituted with		
			plastoquinone	phylloquinone	ubiquinone
Dark, $H_2 \rightarrow O_2$	110	4.2	35	75	15
Light, $H_2 \rightarrow NADP^+$	38	1.6	26	13	5

[a] Photoreduction of $NADP^+$ was measured spectrophotometrically (Peschek, 1979a) and oxygen uptake was followed polarographically (Peschek, 1979b) at 35°C. Extraction-reconstitution procedures were performed as described by Peschek (1980b); cf. Peschek and Kuntner (1987).

in *A. nidulans*, plastoquinone and phylloquinone can catalyze both respiratory and photosynthetic electron transport in isolated membranes, do not permit any quantitative conclusion as to the relative share of either quinone in the respective electron function in vivo. One could speculate that there might exist a 'differentiation' of electron flow at the quinone level with, e.g. plastoquinone exhibiting higher affinity for PSII and phylloquinone exhibiting higher affinity for the respiratory chain dehydrogenases. It is interesting to note in this context that an isolated and partially purified NAD(P)H dehydrogenase from *Synechocystis* 6714 showed a markedly higher affinity for naphthoquinones than for benzoquinones (Sandmann and Malkin, 1983b), and the same was true of a similar but soluble enzyme from *Microcystis aeruginosa* (Viljoen et al., 1985). The strikingly different maximum rates of respiratory and photosynthetic electron transport in cyanobacteria (Section 1.3.) may reflect, in part, the markedly different contents of plastoquinone and phylloquinone (between 25:1 and 15:1). A similar differentiation of respiratory and photosynthetic electron transport might occur at the level of the soluble *c*-type cytochromes that mediate electron transfer from the cytochrome b_6/f complex to P700 and/or cytochrome oxidase (Peschek and Schmetterer, 1982; Section 6).

One of the most straightforward demonstrations of a respiratory function of plastoquinone stems from dual-wavelength spectrophotometric measurements of the reduction of either cytochrome *f* or *b*-563 in membrane preparations from *Anacystis nidulans* (Peschek, 1983a) and *Nostoc* sp. strain MAC (Peschek and Schmetterer, 1982) that were physiologically oxidized by far-red light (predominantly absorbed in PSI), as well as by oxygen in the dark, and that were reduced both by PSII light and by respiratory electron donors such as NADH, NADPH, and H_2 in the dark. Both these two types of physiological re-reduction of the oxidized electron carriers, the photosynthetic and the respiratory one, were abolished after extraction of the membranes with *n*-pentane, but reappeared on reconstitution of the depleted membranes with authentic plastoquinone-9 (Section 6).

2.5. CYTOCHROMES

Cytochromes, like quinones and iron-sulfur proteins, are ubiquitous constituents of biological electron-transport chains in contemporary living cells. A striking exception may be the strictly anaerobic methane-producing bacteria in which neither quinones nor cytochromes could be so far detected (Zeikus, 1977; Thauer et al., 1977; Balch et al., 1979).

The complement of membrane-bound cytochromes in cyanobacteria is similar to that of chloroplasts. There is a *c*-type cytochrome analogous to cytochrome *f* of chloroplasts (Wood, 1977), and there is a *b*-type cytochrome (cytochrome *b*-563) analogous to cytochrome b_6 of chloroplasts. Both cytochromes are components of the cytochrome b_6/f complex (Hurt and Hauska, 1981, 1982a; Krinner et al., 1982) which, in addition, comprises a Rieske-type Fe-S protein (see Section 2.3.) and plastoquinone (Hurt and Hauska, 1982b). The function of another *b*-type cytochrome in cyanobacteria, cytochrome *b*-559, which can occur in both a low-potential (LP) and a high-potential (HP) form, is as unclear as it is in chloroplasts. Cytochrome *b*-559 is found associated with PSII preparations (Nakayama et al., 1979; Newman and Sherman, 1978; Koenig and Vernon, 1981), and is reported to be absent from heterocysts (Almon and Böhme, 1980). It has been speculated that most of the cytochrome $b\text{-}559_{LP}$ may result from native cytochrome $b\text{-}559_{HP}$ during rather harsh isolation procedures (for a discussion see Bendall, 1982).

Soluble *c*-type cytochromes are fairly abundant in cyanobacteria. Cytochrome *c*-553 (554) is functionally analogous to plastocyanin and, in some species (e.g. *Anabaena variabilis* and *Plectonema boryanum*) replaces the latter during growth on copper-deficient media (Sandmann and Böger,

1980). Other species, e.g. *Anacystis nidulans*, *Nostoc muscorum* (=*Anabaena* ATCC 29151), *Spirulina platensis*, and *Mastigocladus laminosus* appear to be constitutively incapable of synthesizing plastocyanin (see Sandmann and Böger, 1981); here, cytochrome *c*-553 is the physiological electron donor to P700. Plastocyanin is an acidic protein in chloroplasts (isoelectric point at pH 3.0; Spiegelmann et al., 1976) but is a basic protein in *Anabaena variabilis* (isoelectric point at pH 8.4; Davis et al., 1980). Similarly, cytochrome *c*-553 is a basic protein in most filamentous cyanobacteria (Davis et al., 1980); in unicellular species, however, it is an acidic protein (Holton and Myers, 1967a, b; Ho and Krogmann, 1984). The biochemistry of electron transfer from cytochrome *c*-553 (or plastocyanin) to P700 has been studied by Krogmann and colleagues in both chloroplasts (Davis et al., 1980) and cyanobacteria (Ho and Krogmann, 1984).

Few studies have been carried out on electron donation to (membrane-bound) cyanobacterial cytochrome oxidases, part of which may be situated in the thylakoid membrane, competing there with P700 for electrons from the dual functional photosynthetic/respiratory electron flow chain (Lockau, 1981; Peschek et al., 1981a; Kienzl and Peschek, 1982; Peschek and Schmetterer, 1982; Stürzl et al., 1982). With membranes from *Anabaena* ATCC 29413 cytochrome *c*-553 (from the same species) and horse heart cytochrome *c* effectively donate electrons to both P700 (in the light) and to the cytochrome oxidase (Lockau, 1981). By contrast, cytochrome *c*-553 from *Nostoc muscorum* (ATCC 29151) is a rather specific electron donor to both P700 and the cytochrome oxidase in membrane preparations of this organism and could not be replaced by cytochrome *c*-553 from *Spirulina platensis* or *Phormidium foveolarum*, and horse heart cytochrome *c* (Stürzl et al., 1982). Membranes from *Nostoc* sp. strain MAC accept horse heart cytochrome *c*, but not cytochrome *c*-553 (554) from *Anacystis nidulans* for the reconstitution of both photosynthetic $NADP^+$ reduction and respiratory oxygen uptake while in membranes from *A. nidulans* horse heart cytochrome *c* proved to be an excellent electron donor to the cytochrome oxidase but not to P700, with cytochrome *c*-554 (from *Anacystis*) showing the opposite behavior (Peschek and Schmetterer, 1982). The latter cytochrome was found to be a poor electron donor to mitochondrial cytochrome oxidase (Yamanaka et al., 1969). In general, the question of the native electron donor to cyanobacterial cytochrome oxidases must be regarded as unresolved at present.

A low-potential, auto-oxidizable, and CO-combining cytochrome *c*-549 was described in *A. nidulans* by Holton and Myers (1963, 1967a,b). Since then, similar soluble *c*-type cytochromes ('cytochromes *c*-550'; see Alam et al., 1984) have been found in many cyanobacteria and other organisms. No definite physiological function could as yet be assigned to cytochrome *c*-550. In vitro, cytochromes *c*-550 (viz., '*c*-548' of *Nostoc* sp. strain MAC, Pulich, 1977; and '*c*-549' of *A. nidulans*, P.F. Kienzl, G. Schmetterer and G.A. Peschek, unpublished) are rapidly reduced by NADPH (and, to a lesser extent, by NADH) in particle-free extracts of the respective species; however, in view of the low redox potential of cytochrome *c*-550 (E_0' around -0.25 V) a respiratory role for this cytochrome is unlikely. Cytochrome *c*-549 (of *A. nidulans*) is ineffective as an electron donor to the cytochrome oxidase in membrane preparations of the same species (Kienzl and Peschek, 1982) but catalyzes high rates of cyclic photophosphorylation, thereby replacing physiological ferredoxin under conditions of proper redox poising of the cyclic electron transport system (Kienzl and Peschek, 1983). These findings are suggestive of a cytoplasmic location of the cytochrome which would also conform to its possible role as a co-factor in nitrate reduction as suggested by Alam et al. (1984). However, Wood (1983) has recently rejected the idea of a cytoplasmic location of *c*-type cytochromes in general, his arguments being derived from evolutionary considerations.

It is clear from the foregoing discussion that the overwhelming attention so far has been paid to photosynthetic aspects of cyanobacterial cytochromes; comprehensive reviews of this topic are available (Cramer and Whitmarsh, 1977; Crofts and Wood, 1978; Böger and Böhme, 1982; Bendall, 1982; Ho and Krogmann, 1982, 1984; Peschek, 1984a). To the extent that the plastoquinone-cytochrome b_6/f electron transport sequence of cyanobacteria seems to be shared by both respiration and photosynthesis (see Section 6) cytochromes *f* and *b*-563 can be regarded as both respiratory and photosynthetic cytochromes (Peschek, 1984a). However, the only cytochrome with a straightforward respiratory (and only a respiratory) function, owing to its strongly positive redox potential, would be cytochrome *a*. Evidence for an *a*-type cytochrome as a component of the terminal oxidase in cyanobacteria will be discussed more fully in Section 2.6.

2.6. Cytochrome oxidase

The occurrence in cyanobacteria of a hemoprotein cytochrome oxidase, possibly similar to the mitochondrial type, was suggested already from early investigations on the inhibition of respiratory oxygen uptake by KCN in whole cells (Webster and Frenkel, 1953; Kratz and Myers, 1955). Results with other typical cytochrome oxidase inhibitors such as azide or carbon monoxide, tested on either intact cells or cell-free extracts gave rather variable and inconclusive results (Webster and Frenkel, 1953; Kratz and Myers, 1955; Smith et al., 1967; Biggins, 1969; Horton, 1968; Leach and Carr, 1970; Tang and Krogmann, 1972; Peschek, 1980a).

The chemical nature of the terminal oxidase in cyanobacteria was tentatively resolved only recently when extensive spectrophotometric characterization of the enzyme in isolated membranes from several axenic strains revealed it to be an aa_3-type cytochrome with features strikingly similar to the well-known mammalian-type cytochrome *c* oxidase (Peschek, 1981a,b; Kienzl and Peschek, 1982; Peschek et al., 1981a,b, 1982a,b,c; Peschek and Schmetterer, 1981; Peschek, 1984a). These properties include: reduced α- and γ-peaks at 604–605 and 445 nm, which are shifted, in the presence of CO, to 590 and 430 nm, respectively. Photo-action spectra for the relief of CO inhibition of cytochrome *c* oxidation and cytochrome *c*-supported oxygen uptake displayed peaks at 590 and 430 nm (Peschek, 1981a,b; Kienzl and Peschek, 1982; Peschek et al., 1982a). Photo-dissociation spectra showed that the CO complex of the membrane-bound terminal oxidase was decomposed only by light of wavelengths (e.g. 590 nm) absorbed by the complex (Peschek, 1981b; Peschek et al., 1982a). Rates of cytochrome *c* oxidation by the membrane-bound terminal oxidase from several cyanobacteria were compared using 17 different *c*-type cytochromes from both eukaryotic and prokaryotic sources (Peschek et al., 1981a; Kienzl and Peschek, 1982). Mitochondrial cytochrome c showed high affinity for the cyanobacterial cytochrome oxidases, while the plastidic cytochromes from *Euglena* and *Scenedesmus* were inactive.

Cytochrome oxidase activity in crude membrane preparations is inhibited at elevated ionic strength (Lockau, 1981; Kienzl and Peschek, 1982; Peschek et al., 1982a) and stimulated by (neutral) detergents such as Tween-20 and Tween-80 (Kienzl and Peschek, 1982; Peschek et al., 1982a; Peschek and Schmetterer, 1981). The use of ionic detergents such as sodium dodecylsulfate also leads to stimulation of the cytochrome *c*-oxidizing activity (Tang and Krogmann, 1972; Omata and Murata, 1985); this effect, however, was found to be due to non-enzymatic autoxidation of the cytochrome *c* catalyzed by heavy metal ions present as impurities in the SDS. SDS-mediated autoxidation of cytochrome *c* in the absence of membranes could be inhibited by mM concentrations of EDTA and KCN (!) both of which may scavenge part of the heavy metal ions (Molitor, and Peschek, 1986). The use of SDS for stimulating cytochrome oxidase activity should be discouraged in view of these findings.

Besides KCN, CO, azide, and sulfide, also EDTA and the Cu-chelating agent salicylaldoxim inhibited cytochrome *c* oxidation by membranes from cyanobacteria (Peschek et al., 1982a). The reaction was greatly diminished with membranes from *A. nidulans* that was grown in Cu-deficient media (Peschek et al., 1982a), which led to the suggestion that the enzyme might be a Cu-containing hemoprotein. The participation of tightly membrane-bound Cu in the oxidation of horse heart cytochrome *c* by crude membrane preparations from *Synechococcus* 6311 was indicated from EPR spectrometry (Fry et al., 1985). The characteristic temperature dependence and power saturation of the Cu^{2+} EPR signal excluded plastocyanin (which could not be detected in *Synechococcus* 6311; see Section 2.5.) or other 'dirty copper' as a possible source of the signal in the EDTA-washed membranes. Instead, its properties were similar to those known from other Cu-containing aa_3-type cytochrome oxidases (Fig. 1). From studies using EPR spectrometry and potentiometric titration,

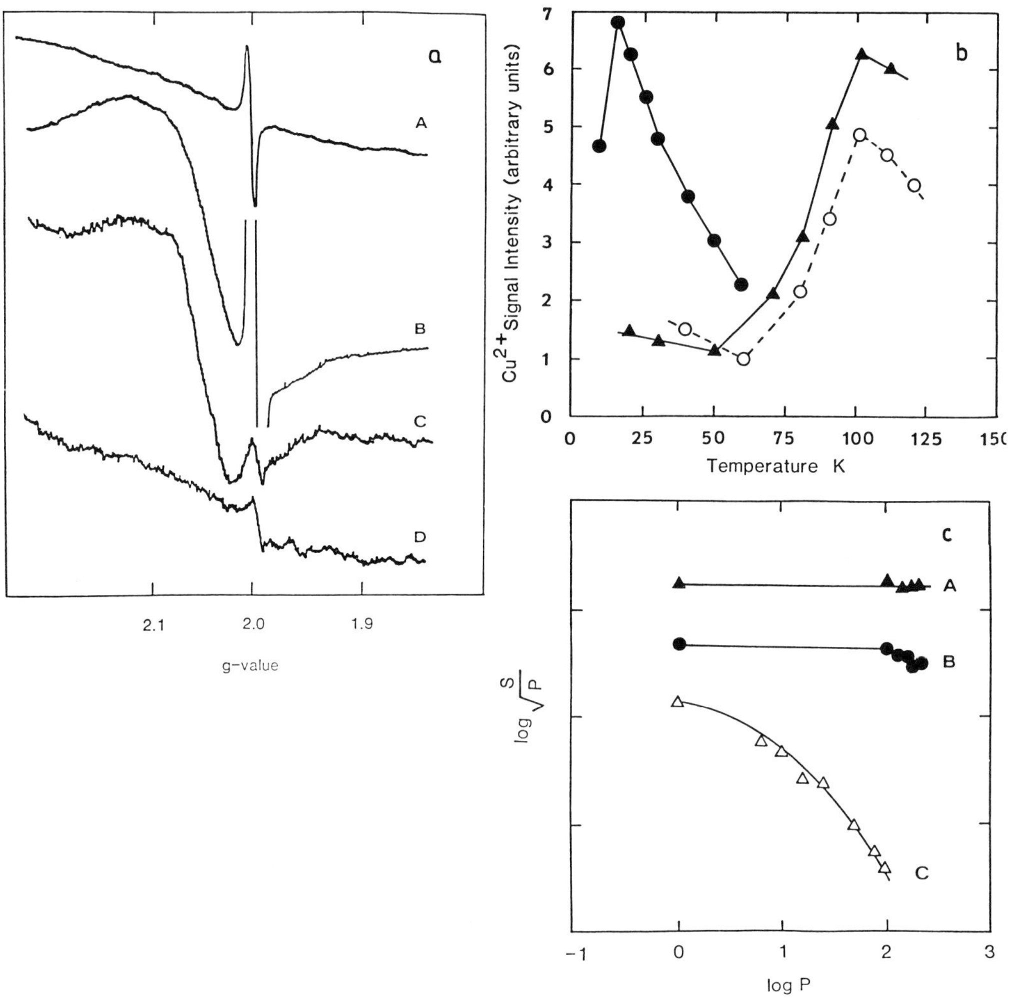

Fig. 1 (a, b, c). (a) Cu^{2+} EPR spectra of EDTA-washed membrane preparations of *Synechococcus* 6311 (A, B) and submitochondrial particles (C, D) in 10 mM Tes/10 mM Na_2EDTA buffer, pH 7.0, at 100 K in presence of 0.1% Triton X-100. The preparations were reduced with 10 mM ascorbate plus 25 μM TMPD in the presence of 5 mM KCN (A, D) and oxidized with 25 mM ferricyanide (B, C). (b) Temperature dependence of the Cu^{2+} EPR signals of *Synechococcus* 6311 membranes (▲ – ▲), submitochondrial particles (○—○), and *Gloeobacter violaceus* plastocyanin (●—●). (c) Microwave power saturation of the Cu^{2+} EPR signals of *Synechococcus* 6311 membranes (A) and submitochondrial particles (B) at 100 K, and of *Gloeobacter* plastocyanin (C) at 15 K. For details see Fry et al. (1985).

cytochrome aa_3 was suggested to occur in preparations of plasma membrane, but not of thylakoid membranes, of *Plectonema boryanum* (Matthijs, 1984; Matthijs et al., 1984b). The content of cytochrome aa_3 in membranes prepared from heterocysts of *Anabaena* ATCC 27893 was estimated spectrophotometrically to be roughly 6.4 nmol/μmol chl (Houchins and Hind, 1984) while in membranes from logarithmically growing cells of *A. nidulans* and other cyanobacteria similar estimates gave values between 0.05 and 0.5 nmol/μmol chl and between 0.1 and 1.6 nmol for cells harvested from late logarithmic cultures (Peschek et al., 1982a).

The subunit composition of bacterial aa_3-type cytochrome oxidases was invariably found to be much simpler than that of its mitochondrial counterpart, viz. 2–3 subunits as compared to at least 7 (Ludwig, 1980; Poole, 1983). Recently we isolated and purified plasma and thylakoid membrane fractions from *Anacystis nidulans*; using SDS polyacrylamide gel electrophoresis and Western blotting, we obtained two polypeptide bands from either fraction that showed cross reaction with specific antibodies against aa_3-type terminal oxidase from *Paracoccus denitrificans* (Ludwig and Schatz, 1980); antibody against the native enzyme as well as against subunits I and II, respectively, was used (Trnka et al., 1985; Trnka and Peschek, 1986; Molitor et al., 1987). The molecular masses of the two subunits of the *A. nidulans* aa_3-type cytochrome oxidase correspond to approximately 55 and 32 kD (Fig. 2). Table VII shows some properties of the separated membrane fractions and their cytochrome oxidase and reductase activities.

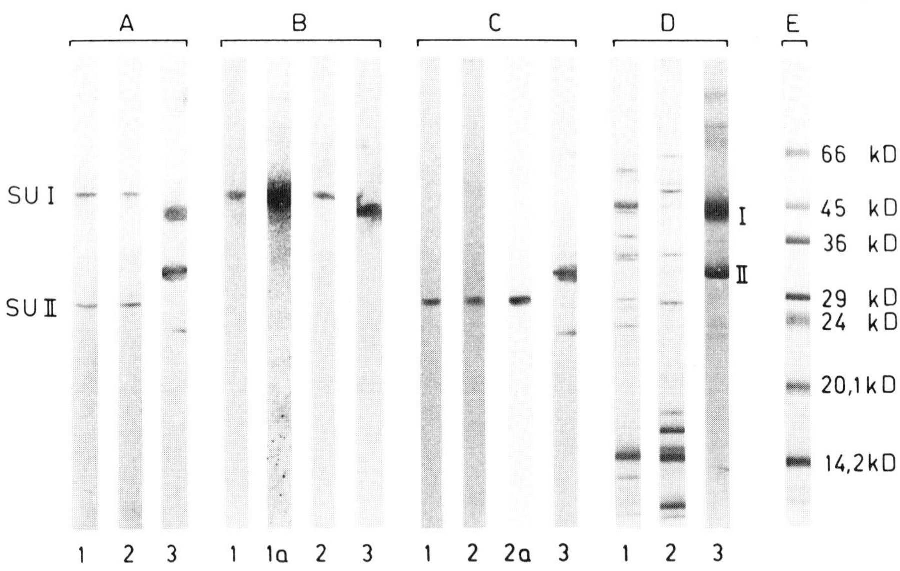

Fig. 2. (A–C) Immunological identification of aa_3-type cytochrome oxidase from *Anacystis nidulans* membranes and *Paracoccus denitrificans* after sodium dodecylsulfate polyacrylamide gel electrophoresis according to Laemmli (12.5% acrylamide) and Western blotting, using the cross reaction with specific antibodies against *P. denitrificans* cytochrome oxidase, and goat anti-rabbit-IgG-horse radish peroxidase (GAR-HRP) conjugate (lanes 1–3), ^{125}I-GAR-IgG (lane 1a) or ^{131}I-protein A (lane 2a). Antibodies against native *Paracoccus* oxidase (A), and against subunits I (B) and II (C) of the enzyme, were used. (D) Coomassie stain of membrane proteins of *A. nidulans* and of isolated *Paracoccus* oxidase. (E) Marker proteins. (A–C) (1, 1a) *Anacystis* plasma membranes (30 μg protein/lane); (2, 2a) *Anacystis* thylakoid membranes (75 μg protein/lane); (3) *Paracoccus* cytochrome oxidase (isolated enzyme, 1μg/lane). (D) *Anacystis* plasma (1) and thylakoid (2) membrane (30 μg protein/lane); (3) *Paracoccus* cytochrome oxidase (isolated enzyme, 20 μg protein). SU I, SU II, tentatively assigned to subunits I (55 kD) and II (32 kD) of the *Anacystis* cytochrome oxidase; I, II (lane D-3), subunits I and II of the *Paracoccus* oxidase.

TABLE VII

Oxidation and reduction of horse heart cytochrome *c* in membrane fractions separated from cell-free extracts of *Anacystis nidulans* by discontinuous sucrose density gradient centrifugation (131 500 × *g*, 16 h; the table gives lowest and highest values from up to 14 independent experimental runs (V. Molitor, M. Trnka and G.A. Peschek, unpublished; cf. Molitor and Peschek, 1986)

Fraction	% (w/w) sucrose	Chl/prot[a] (w/w)	nmol ferrocytochrome c^b oxidized/min per mg		nmol ferricytochrome c^c reduced/min per mg protein		% (w/w) protein recovered in the respective fraction
			protein	chlorophyll	NADH[d]	NADPH[d]	
(M)	(10)	0.1–0.2	20–120	100–1200	2.3–15.6	5.7–48.4	100
PhCy[e]	10/20	0.0	–	–	–	–	<0.05
(CM)[f]	20	0.0	–	–	–	–	0.05
CM	30	<0.05	6–15	120–300	5.1–27.4	2.5–2.8	1
CM + ICM	35	0.2	2–4	10–20	–	–	15
ICM	35/42	0.2–0.4	2–4	5–20	1.1–9.0	1.5–4.7	15–45
P	–	0.1–0.2	10–36	50–360	2.7	5.3	38–68

M, crude membranes; PhCy, phycocyanin; CM, plasma membrane; ICM, thylakoid membrane; P, pellet; [a] chlorophyll/protein ratios; [b] 15 μM; [c] 2 μM, rates determined by dual wavelength spectrophotometry; [d] 3.5 mM NAD(P)H; rates corrected for non-enzymatic cyt *c* reduction by NAD(P)H which always was less than 10% of membrane-catalyzed rates; [e] blue layer of phycocyanin, containing small CM fragments; [f] diffuse layer of small CM fragments. Cytochrome oxidase activity was inhibited >95% by 0.1 M NaCl, 1.2 μM KCN, 0.1 mM sulfide, 1.0 mM azide, 20 mM salicyl aldoxime, and CO (90%, v/v, in air), and stimulated up to 130% by 5% (w/v) Tween-80.

Carefully prepared, intact spheroplasts of *Anacystis nidulans* were found to oxidize exogenous *c*-type cytochromes (Peschek, 1983b, 1984b; Peschek et al., 1982b,c). The reaction showed the same inhibition profile as that catalyzed by crude membrane preparation, and was accompanied by stoichiometric proton extrusion (H^+/e^- near 1). KCN and CO equally inhibited H^+ translocation and electron transfer while DCCD inhibited the former more strongly than the latter. Similar findings have been reported for the H^+-translocating cytochrome oxidase in the inner mitochondrial membrane (Wikström and Krab, 1979; Casey et al., 1980; Prochaska et al., 1981; Azzi et al., 1984). From our findings we concluded that the plasma membrane of *A. nidulans* contains a proton-pumping cytochrome *c* oxidase chemically identical with that occurring in the thylakoid membrane (Peschek, 1984b; also see Section 7).

3. Proton translocation and proton electrochemical potential gradients

The protonmotive force (pmf) or proton electrochemical potential gradient ($\Delta\mu_{H^+}$), which is the ultimate driving force for ATPsynthesis according to the chemiosmotic theory (Mitchell, 1961, 1968; Nicholls, 1982), basically comprises both a proton concentration gradient (ΔpH) and an electrical potential gradient ($\Delta\psi$) across the membrane:

$$\Delta\mu_{H^+}(\text{mV}) = \Delta\psi - 60.\Delta\text{pH} \quad (\text{at } 35°\text{C}; \text{'}\Delta\text{'} = X_{in} - X_{out})$$

Depending on the experimental system, ΔpH and $\Delta\psi$ can contribute to varying degrees to the overall proton electrochemical potential. Either of the two components can be measured separately (see Rottenberg, 1979; Padan, 1981). Thus, while in thylakoids from chloroplasts and cyanobacteria ΔpH is the main component of the protonmotive force, it is the membrane potential in the internal membrane of intact mitochondria and in the intact bacterial cytoplasmic membrane.

ΔpH values between 2 and 3 units have been measured across the thylakoid membrane of intact *A. nidulans* (Falkner et al., 1976; Peschek et al., 1985a) and with vesicles derived from *Plectonema boryanum* (Padan and Schuldiner, 1978). Acidification has occurred inside the thylakoids upon energization of the cells (vesicles) either by light or by oxygen (in the dark), respectively. Similar values were obtained with *Anabaena variabilis* (Wax and Lockau, 1980), *Mastigocladus laminosus* (Bohler and Binder, 1980), *Coccochloris peniocystis* (Coleman and Colman, 1982), *Plectonema boryanum* (Masamoto and Nishimura, 1977; Barski et al., 1981), *Synechocystis* 6714 and *A. nidulans* (Ihlenfeldt and Gibson, 1977; Gibson, 1981).

Table VIII summarizes results on ΔpH across plasma and thylakoid membranes of cyanobacteria. Proper semantic discrimination between cytoplasmic and intrathylakoid pH would be desirable, but is rarely employed. Using suitable combinations of ΔpH- or $\Delta\psi$-indicating reagents, and applying appropriate formulae for calculation (see Heldt et al., 1973), it is possible to discriminate between ΔpH or $\Delta\psi$ across plasma and thylakoid membranes of intact cells. Such an approach was recently applied to dark aerobic and anaerobic cells of *A. nidulans* in order to shed light on

TABLE VIII

pH Gradients across cytoplasmic and thylakoid membranes in whole cells and vesicles of cyanobacteria.

pH in			Conditions	Species[a], methods[b], and references
suspension medium	cytoplasmic space	intrathylakoid space		
6.5–8.0	7.5	4.6	Light	*Anacystis nidulans*, whole cells; distribution of ^{14}C-labeled DMO and methylamine (Falkner et al., 1976)
	6.9	5.2	Dark	
6.8–8.0	7.6	–	Light	*Synechococcus* sp. and *Aphanocapsa* sp., whole cells; distribution of ^{14}C-labeled acetate (Gibson, 1981)
	7.1	–		
4.5	6.5	5.0	Dark, aerobic	*Anacystis nidulans*, whole cells; distribution of appropriate pairs of ^{14}C-labeled weak acids and bases (Peschek et al. 1985a); uptake of fluorescent amines (Hinterstoisser and Peschek, 1987)
	6.1	6.1	Dark, anaerobic	
5.3	6.8	5.05	Dark, aerobic	
	6.4	6.3	Dark, anaerobic	
7.5	7.1	5.15	Dark, aerobic	
	6.8	6.8	Dark, anaerobic	
9.0	7.25	5.25	Dark, aerobic	
	7.0	6.9	Dark, anaerobic	

TABLE VIII *(continued)*

pH in			Conditions	Species[a], methods[b], and references
suspension medium	cytoplasmic space	intrathylakoid space		
7–9.6	7	– (?)	Dark, aerobic	*Synechococcus leopoliensis* (= *Anacystis nidulans*), whole cells in presence of 5 mM external Na^+; distribution of ^{14}C-labeled methylamine (Miller et al., 1984)
4.7	4.7	–	Light	Thermophilic *Synechococcus* strain Y-7c-s, whole cells; ^{31}P NMR spectrometry (Kallas and Castenholz, 1982)
	4.7	–	Dark	
5.0	6.5	–	Light	
	6.0	–	Dark	
6.0	6.9	–	Light	
	6.6	–	Dark	
7.0	7.4	–	Light	
	7.2	–	Dark	
5.25	6.6	–	Light	*Coccochloris peniocystis*, whole cells; distribution of ^{14}C-labeled DMO and methylamine (Coleman and Colman, 1981)
7.0	7.6	–	Light	
8.0	7.9	–	Light	
10.0	7.9	–	Light	
6.4	–	4.5	Light	Thylakoid vesicles from *Plectonema boryanum;* distribution of 9-aminoacridine and ^{14}C-labeled methylamine (Padan and Schuldiner, 1978)
7.6	–	4.85	Light	
8.5	–	5.5	Light	
6.0	7.3	–	Light	*Anabaena variabilis*, whole cells; distribution of ^{14}C-labeled DMO (Reed et al., 1980)
7.0	7.7	–	Light	
8.0	7.9	–	Light	
6.0	8.0	–	–	*Escherichia coli*, whole cells; distribution of ^{14}C-labeled DMO and methylamine (Padan et al., 1976)
7.0	7.7	–	–	
8.0	– 7.7	–	–	
9.0	8.5	–	–	

[a] For comparison, *Escherichia coli*, a typically heterotrophic bacterium, is included; dash, not determined.
[b] DMO, 5,5-dimethyloxazolidine-2,4-dione; NMR, nuclear magnetic resonance. 'Dark' means 'Dark aerobic' unless otherwise specified.

the mode and extent of respiratory energization of either membrane (Peschek et al., 1985a; see Fig. 3). The protonmotive force across the plasma membrane was higher (more negative) in aerobic than in anaerobic conditions, increased with increasing respiration rates, and was virtually insensitive to DCCD which blocks the reversible ATPase thus leading to ATP-depleted cells. It was concluded that a major fraction of the trans-plasma membrane pmf can be sustained solely by (respiratory) electron transport in this membrane, and that in physiological conditions this gradient is used for the active extrusion of Na^+ from the cell rather than for ATP synthesis (Nitschmann and Peschek, 1982; Nitschmann et al., 1982; but see Paschinger, 1977). Preliminary characterization of the Na^+/H^+ antiporter activity in unicellular cyanobacteria points to the direct involvement of proton-translocating respiratory electron transport in the plasma membrane (Nitschmann et al., 1982; Blumwald et al., 1984; Fry et al., 1986). *Synechococcus leopoliensis* (=*A. nidulans*) was reported to maintain a neutral intracellular pH up to $pH_0 = 9.6$ only in the presence of (at least?) 5 mM external Na^+; the 'alkalotolerance' of the cells has been attributed partially to Na^+/H^+ antiporter activity (Miller et al., 1984). An Na^+/H^+ antiporter, however, would probably act in the opposite way, viz. depleting the cytosol of H^+ at the expense of Na^+ influx. Moreover, the use of CCCP-insensitive uptake of methylamine at pH_0 9.6 'across the plasma membrane' (and also across the thylakoid membrane; see Table VIII) as a measure of 'unspecific binding' of methylamine is open to criticism since CCCP ($pK_a = 6.0$) is unsuitable as an uncoupler at pH > 8–9.

Proton extrusion from energized cyanobacterial cells was observed already some time ago when intact *Anabaena variabilis* was illuminated, or oxygenated in the dark (Scholes et al., 1968). Barski et al. (1981), working with 9-aminoacridine as a fluorescent pH indicator, resolved a fast component of fluorescence quenching followed by a slower one of fluorescence enhancement in illuminated suspensions of *Plectonema boryanum*. The effect was attributed to more or less simultaneous, yet oppositely directed, H^+ movements across plasma and thylakoid membranes, respectively. Biphasic H^+ efflux from illuminated cyanobacteria comprised an initial, fast component possibly related to direct photoexcitation of the plasma membrane, followed by a slower component attributed to a unidirectional H^+-extruding ATP hydrolase which was inhibited by *ortho*-vanadate (Scherer and

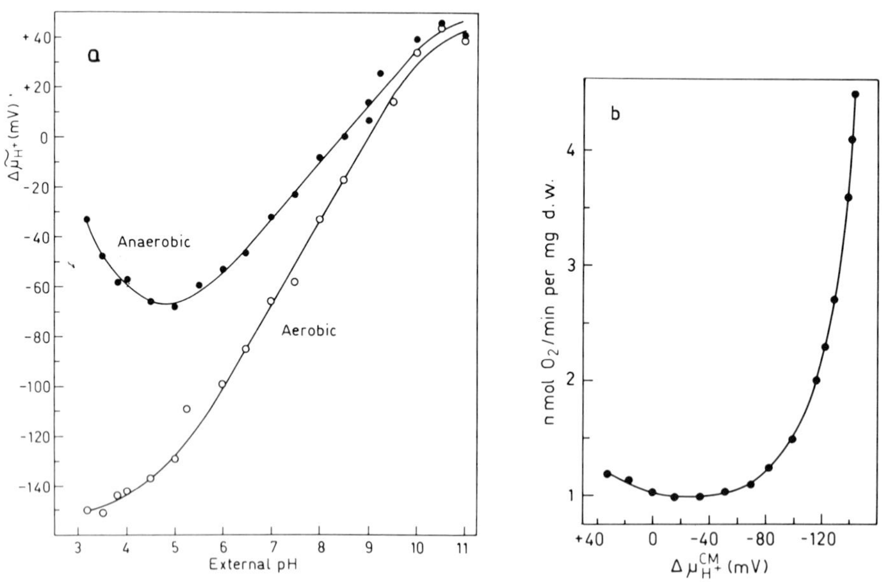

Fig. 3 (a, b). (a) Proton electrochemical potential gradients across the plasma membrane of dark aerobic (○) and anaerobic (●) cells of *Anacystis nidulans*. (b) Correlation between the protonmotive force across the plasma membrane and the rate of respiratory oxygen uptake observed at different external pH. (See Peschek et al., 1985a; also see Peschek et al., 1984)

Böger, 1984; Hinrichs et al., 1985). The same enzyme was claimed to be responsible for the likewise vanadate-sensitive H^+ efflux from oxygen-pulsed cyanobacteria in the dark (Scherer et al., 1984): there was a lag phase of 59 s between oxygenation and the onset of linear H^+ ejection, reflecting the time it took the cells to establish maximum ATP levels; from this the authors concluded that ATP as such, rather than directly respiratory electron transport in the plasma membrane was responsible for the H^+ extrusion observed.

The results obtained with oxygen-pulsed *Anacystis nidulans* and also, in part, with *Nostoc* sp. strain MAC and *Dermocarpa* ATCC 29372, point to major contributions from proton-translocating respiratory electron transport in the plasma membrane, together with a reversible H^+-ATPase (F_0F_1-ATPase) in the same membrane (Nitschmann and Peschek, 1984, 1985; see Figs. 4,5). This conclusion is derived from the following findings: (a) no major lag phase between oxygenation and the onset of linear H^+ extrusion; (b) non-linear correlation between H^+ ejection and intracellular ATP levels in the presence of increasing concentrations of electron transport inhibitors, such as CO and KCN; (c) less than 70% inhibition by saturating concentrations of DCCD (F_0 inhibitor) and NBD (F_1 inhibitor) which completely eliminate all oxidative phosphorylation; (d) distinctly different pH profiles of DCCD-sensitive and DCCD-insensitive H^+ extrusion; and (e) lack of inhibition by *ortho*-vanadate.

Reduction of the impermeable electron acceptor ferricyanide by intact *A. nidulans* in the dark was interpreted in terms of a respiratory chain in the plasma membrane; the reaction is enhanced by HOQNO which inhibits at the cyt *b*/*c* complex (Craig et al., 1984). It is faster in anaerobic than in aerobic conditions, and is stimulated by KCN in aerobic cells (Erber et al., 1986). The reaction was accompanied by proton ejection from the cells ($H^+/e^- = 0.7$ in anaerobic ferricyanide-reducing cells). Similar proton ejection was shown to accompany the oxidation of exogenous (and likewise impermeable) ferrocytochrome *c* by intact spheroplasts of *A. nidulans* (Peschek, 1983b, 1984b; Peschek et al., 1982b; see Section 2.6.). Comparison of the Arrhenius plots for the temperature dependence of ferricyanide reduction by intact spheroplasts with those for the oxidation of exogenous cytochrome *c* and other respiratory electron transport reactions showed that there is a discontinuity temperature (lipid phase transition temperature in the membranes) common to all these (respiratory) reactions which, therefore, appear to occur in the plasma membrane (Erber, Muchl and Peschek, unpublished; see Peschek et al., 1981d, 1982d). By contrast, from results with *Anabaena* ATCC 29413, *Mastigocladus laminosus*, *Phormidium foveolarum*, and *Synechocystis* 6714 it was concluded that the reduction of ferricyanide by these species does not need a respiratory chain in the plasma membrane (Scherer and Böger, 1985).

Light-induced H^+ efflux from *A. nidulans* was abolished by DCCD (Nitschmann and Peschek, 1984, 1985). Therefore, in *Anacystis* H^+ extrusion in the light was attributed entirely to a reversible H^+-ATPase in the plasma membrane which appears to be devoid of H^+-translocating photosynthetic electron transport. The absence of bulk amounts of photopigments from the plasma membrane of *A. nidulans* had previously been concluded from the finding that prolonged photooxidative conditions led to a reversible disintegration of the thylakoid membranes, but not of the plasma membrane (Peschek and Schmetterer, 1978; Schmetterer and Peschek, 1981; Schmetterer, 1982; Schmetterer et al., 1983). Also in isolated plasma membranes, separated from thylakoid membranes, no chlorophyll could be detected (Omata and Murata, 1984a; Trnka et al., 1985) and the same observation was made on isolated plasma membranes from *Anabaena* ATCC 29413 (Lockau and Pfeffer, 1982) and *Synechocystis* 6714 (Omata and Murata, 1984b).

The presence of a reversible H^+-ATPase in the plasma membrane of *A. nidulans* is apparent from the net ATP synthesis that was measured in dark anaerobic cells to which an artificial protonmotive force had been applied (Muchl and Peschek, 1983, 1984; Muchl et al., 1983). Fig. 6a shows

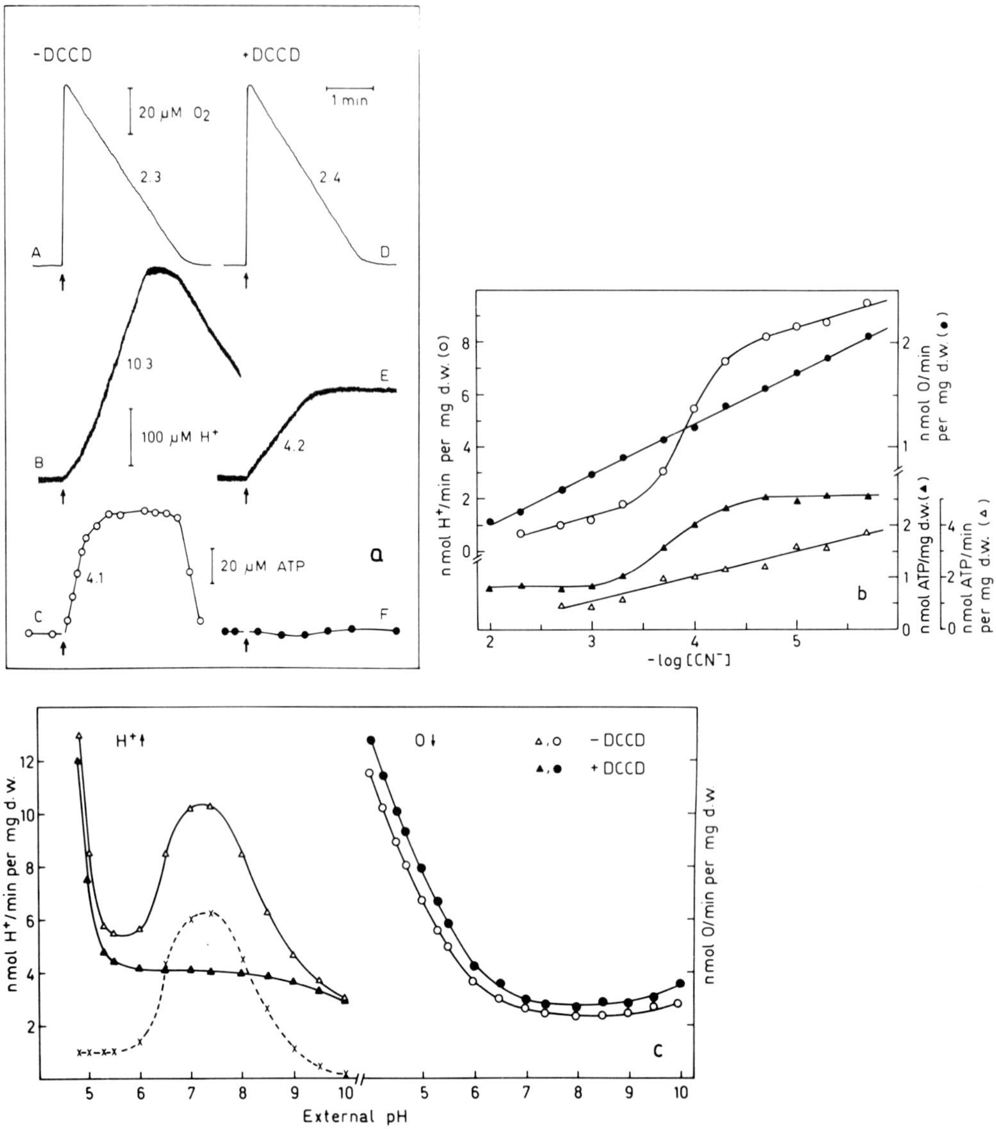

Fig. 4 (a, b, c). (a) Kinetics of oxygen uptake (A, D; downward deflection), proton extrusion (B, E; upward deflection), and intracellular ATP levels (C, F) in oxygen-pulsed suspensions of *Anacystis nidulans* in the absence (A–C) and in the presence (D–F) of a saturating DCCD concentration (20 nmol/mg dry weight) in the dark. Arrows indicate the admission of oxygen. (b) Respiratory oxygen uptake (●), H^+ ejection (○), and ATP synthesis (△) following oxygenation of dark anaerobic suspensions of *A. nidulans* in the presence of different concentrations of KCN; ▲ – ▲, steady state levels of ATP in the cells. (c) pH dependence of H^+ extrusion from, and oxygen uptake by, O_2-pulsed cells in the presence and in the absence of 20 nmol DCCD/mg dry weight. The broken line indicates the difference between H^+ efflux in the absence and in the presence of DCCD, thus reflecting ATPase-supported H^+ efflux per se. (See Nitschmann and Peschek, 1984, 1985; Peschek et al., 1985b)

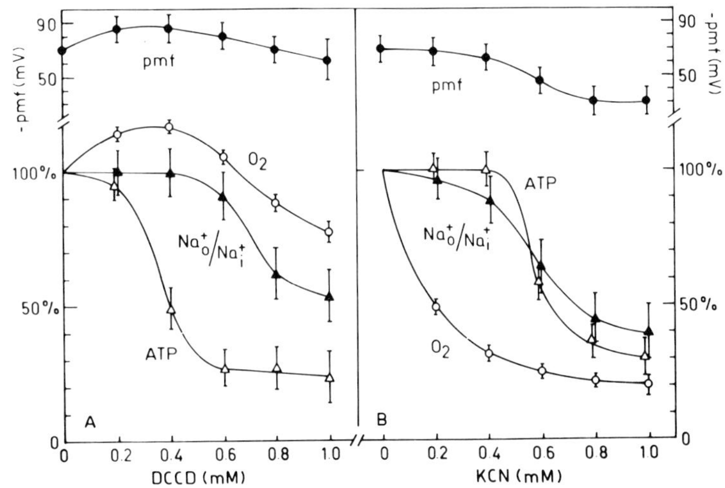

Fig. 5. Dependence of trans-plasma membrane protonmotive force (pmf; see Fig. 3), oxygen uptake, ATP content, and Na^+ exclusion factors (trans-plasma membrane concentration ratios of Na^+ at 150 mM external NaCl) on increasing concentrations of DCCD (A) and KCN (B) in dark aerobic suspensions of *Anacystis nidulans*. Aerobic control values (100%): O_2 uptake, 1.7 nmol O_2/min per mg dry weight; ATP content, 2.6 nmol/mg dry weight; Na^+_o/Na^+_i ratio, 5.9; corresponding anaerobic levels were 0.7 (ATP) and 2.1 (Na^+_o/Na^+_i). (Erber et al., 1986)

that initial rates of this 'Jagendorf-type' (Jagendorf and Uribe, 1966) ATP formation in *A. nidulans* were proportional to the protonmotive force applied regardless whether this consisted of a ΔpH (effected by acid jumps) or a $\Delta\psi$ (K^+ diffusion potential induced by valinomycin), or of a superposition of both. Fig. 6b gives a correlation of the magnitudes of the applied protonmotive force (from acid jumps) and the phosphorylation potentials in the cell. Clearly, the imposed proton electrochemical gradient must perform chemical work against the phosphorylation potential for ATP synthesis (Muchl and Peschek, 1983, 1984; Muchl et al., 1983). It is seen from Fig. 6b that, at a stationary intracellular phosphorylation potential corresponding to -400 mV (in the direction of ATP synthesis) net formation of ATP in the cells was obtained only when the applied protonmotive force was at least -200 to -250 mV. This suggests that the minimum proton stoichiometry of the plasma membrane-bound reversible proton ATPase in *A. nidulans* is around 2 (Peschek et al., 1986).

4. Oxidative phosphorylation

The efficiency of oxidative phosphorylation as expressed in terms of P/O ratio was estimated to be close to 3 for *Synechocystis* 6714 from a quantitative measurement of phosphorylated intermediates during glycogen formation from ^{14}C-glucose (Pelroy and Bassham, 1973). A value of 3 was found also for the obligately photoautotrophic *A. nidulans,* by comparing directly the initial rates of oxygen uptake and $\sim$P formation in Na^+-depleted oxygen pulsed cells ($\sim P = 2[ATP] + [ADP]$; see Nitschmann and Peschek, 1982, 1986; Nitschmann et al., 1982). Cells containing high levels of Na^+ showed greatly diminished P/O ratios (Nitschmann et al., 1982) probably because energy from respiration (or photosynthesis) was diverted to active sodium extrusion rather than being

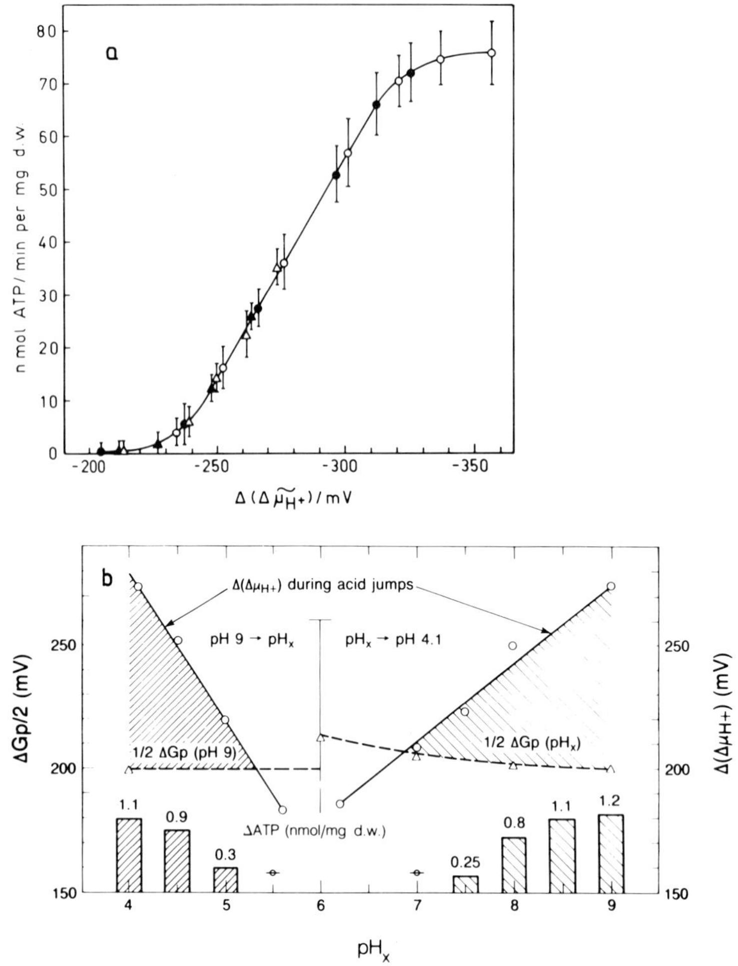

Fig. 6 (a, b). (a) Flow-force relationship between initial rates of pulse-induced ('Jagendorf-type') ATP synthesis and magnitudes of the artificial protonmotive forces imposed across the plasma membrane of dark anaerobic cells of *Anacystis nidulans*. Protonmotive forces were applied in the form of acid jumps (△, ▲), valinomycin pulse-induced K^+ diffusion potentials (○), or a combination of both (●). (b) Correlation of acid jump-induced ATP synthesis, size of the artificial protonmotive force applied ($\Delta(\Delta\mu_{H^+})$), and pre-existing steady-state phosphorylation potentials ΔG_P in dark anaerobic cells of *A. nidulans*. ΔG_P values were divided by 2 which may be the minimum requirement of H^+/ATP synthesized under these conditions; the assumption of $n=2$ would be in agreement with the data shown. ATP bars indicate the increase of ATP levels measured 5 s after the acid pulse. ΔG_P was calculated according to the equation $\Delta G_P = \Delta G^o + RT.\ln([ATP]/[ADP]\cdot[P_i])$, taken in the direction of ATP synthesis and converted into mV units by multiplying the thermodynamic values (in kJ) with 10.36 mV/kJ (see Nicholls, 1982). (Peschek et al., 1986; Muchl and Peschek, 1983, 1984)

used for ATP synthesis. Fig. 5 shows how intracellular ATP levels and respiratory electron transport as such may cooperate in maintaining low intracellular levels of Na^+ and high $\Delta\mu_{H^+}$(CM) in *A. nidulans*; a direct energization by respiratory electron transport in the plasma membrane, across which Na^+ has to be actively translocated, is evident.

Table IX summarizes rates of oxidative and, for comparison, photosynthetic phosphorylation in membrane preparations of various cyanobacteria. Rates and efficiencies of oxidative phosphorylation invariably are much lower than those of photophosphorylation. This is rather surprising in view of the fact that one and the same coupling factor ATPase is involved in both photosynthesis and respiration, and also in view of the partly identical electron transport pathways which generate protonmotive force, though at different rates, during respiratory and photosynthetic electron flow. Different rates of respiration and photosynthesis usually observed in cyanobacteria, are not uncommon also to eukaryotic algae (see Section 1.3.), and cannot explain the up to 100-fold differences in the rates and yields of oxidative and photosynthetic phosphorylation measured in membranes isolated from cyanobacteria. It may be noted, however, that in isolated heterocysts of *Nostoc muscorum* (*Anabaena* ATCC 29151) phosphorylation rates in the light were only 3- to 5-times higher than the rates of (endogenous) oxidative phosphorylation in the dark (Almon and Böhme, 1982; Ernst et al., 1983).

5. Coupling factor ATPases

Membrane-bound enzyme complexes that link transmembrane ΔpH to ATP synthesis (or hydrolysis) are known as reversible H^+-ATPases or F_0F_1-ATPases; they have been extensively studied in

TABLE IX

Oxidative and photosynthetic phosphorylation in membrane preparations of cyanobacteria[a]

Organism	Oxidative phosphorylation		References
	electron donor	rate	
Anabaena variabilis	NADPH	0.35–0.48	Leach and Carr, 1969, 1970
	NADH	0.13–0.29	
	Succinate	0.06–0.11	
	Malate	0.08–0.10	
	Isocitrate	0.05–0.09	
Anabaena variabilis	H_2	13.3	Peschek, 1982
	NADPH	10.5	
	NADH	5.1	
Anacystis nidulans	H_2	12.3	
	NADPH	8.3	
	NADH	4.2	
Nostoc sp.strain MAC	H_2	9.8	
	NADPH	6.8	
	NADH	3.0	
Mastigocladus laminosus	NADH	2.2–3.5	Frei et al., 1984
Plectonema boryanum	NADPH	1.5–11.5	Matthijs et al., 1984c

TABLE IX *(continued)*

Organism	Photophosphorylation		References
	electron acceptor or mediator[b]	rate	
Anacystis nidulans	PMS	228	Bothe, 1969
	Ferricyanide	45	
	Fd + NADP$^+$	53	
	PMS	160	Kienzl and Peschek, 1983
	Fd	54	
	Cyt *c*-549	38	
	None	5	
Phormidium luridum	Fd + NADP$^+$	208	Biggins, 1967
	Ferricyanide	302	
Anabaena variabilis	NADP$^+$	178	Lee et al., 1969
	PMS	833	
Spirulina maxima	PMS	1500	Lerma and Gomez-Lojero, 1982
Mastigocladus laminosus	MV	100–230	Bohler and Binder, 1980
	PMS	500–1000	

[a] Rates are in nmol ATP formed/min per mg protein.
[b] PMS, phenazine methosulfate; Fd, ferredoxin; MV, methylviologen.

chloroplasts, mitochondria, and bacteria (see McCarthy, 1978; Nelson, 1981a, b; Baird and Hammes, 1979). The possible function of a coupling factor in a cyanobacterium was first noted by Biggins (1967), who reported reconstitution of photophosphorylation in extensively washed particles from *Phormidium luridum* after addition of the supernatant to the membranes. The protein factor was purified some 150-fold and restored up to 50% of the original rate of cyclic photophosphorylation (Biggins, 1967; Binder et al., 1976). A reconstitutively active latent Ca^{2+}-ATPase (F_1?) was isolated and partially purified from *Spirulina platensis* (Owers-Narhi et al., 1979). Also the coupling factor ATPases of *Mastigocladus laminosus* (Binder and Bachofen, 1979; Binder et al., 1981a,b; Wolf et al., 1981; Bohler et al., 1982) and *Synechococcus lividus* (Lubberding et al., 1983; Van Walraven et al., 1983, 1984, 1985) were isolated, purified and extensively characterized. A Ca^{2+}-transporting ATPase activity distinct from the coupling factor was recently described in particle preparations of *Anabaena variabilis*; the activity appeared to be associated with the plasma membrane (Lockau and Pfeffer, 1982, 1983).

Besides electron transport phosphorylation, two further characteristics conferred to energy-transducing (respiratory) membranes by the reversible H^+-ATPase, namely reversed electron flow and respiratory control, have not yet received much attention with cyanobacteria: there are only two short reports demonstrating the presence of ATP-dependent reversed electron flow (Peschek, 1980b) and respiratory control (Peschek, 1976) in membranes from *A. nidulans* (see Peschek, 1984a).

6. Common links between respiration and photosynthesis

It has been shown in early studies that respiratory oxygen uptake by whole cells of cyanobacteria is inhibited by light (Brown, 1953; Brown and Webster, 1953). The inhibition, which was described in terms of a Kok effect (as reflected by a nonlinear plot of oxygen exchange vs. light intensity; see Kok, 1949), was attributed to a competition for a common pool of electrons between cytochrome oxidase and Photosystem I (Hoch et al., 1963; Jones and Myers, 1963). Light inhibition of respiration, which is characteristic also of other phototrophic prokaryotes, in particular of the facultatively chemoheterotrophic purple nonsulfur bacteria (see Keister, 1978), was investigated in detail with a variety of cyanobacteria by Scherer and co-workers (Scherer and Böger, 1982; Scherer et al., 1982), and their findings indicate that light inhibition of respiration is most pronounced in nitrogen-fixing species.

Cytochromes b_6 and *f* in membrane preparations of *A. nidulans* (Peschek and Schmetterer, 1982; Peschek, 1983a), *Nostoc* sp. strain MAC (Peschek and Schmetterer, 1982), *Synechocystis* 6714 (Sandmann and Malkin, 1984), and in isolated heterocysts of *Anabaena* ATCC 29413 (Böhme and Almon, 1983) could be physiologically oxidized either by illumination with PSI light or by oxygenation (in the dark). The oxidation, but not the reduction (either by PSII light or by respiratory electron donors), was abolished after extraction of plastoquinone from the membranes, but reappeared upon reconstitution of the depleted membranes with authentic plastoquinone (Peschek and Schmetterer, 1982; Fig. 7a). Therefore, a dual function of plastoquinone, similar to that of cytochromes b_6 and *f*, appears likely (Hirano et al., 1980; Aoki and Katoh, 1982, 1983; Eisbrenner and Bothe, 1979; Peschek, 1980b; see Section 2.3.). Joint participation of the cyt b_6/f complex in photosynthetic and respiratory electron flow of *Anabaena variabilis* was suggested by Lockau and co-workers (Krinner et al., 1982). The dual function of electron transport components is also supported by the findings of Manodori and Melis (1984) and Manodori et al. (1984) who studied oxidation-reduction kinetics of P700 in *Anacystis nidulans*. They showed rapid re-reduction of photooxidized P700 in the dark in whole cells (eventually by endogenous substrates), but not in washed membranes; when, however, NAD(P)H is added to the membranes the rapid dark reduction of $P700^+$ is restored (M. Huflejt, S. Spath, G.A. Peschek and A. Melis, unpublished).

Final electron donation to the terminal acceptor in the membranes, viz. P700 in the light or cytochrome oxidase in the dark, requires a soluble electron transport component. It is well known that this soluble (or easily solubilizable) electron carrier is cytochrome *c* in mitochondria and plastocyanin in (most) chloroplasts. In many cyanobacteria plastocyanin is absent and may be replaced by a soluble *c*-type cytochrome during growth in Cu-deficient media (see Section 2.5.). The cytochrome *c* mediating electron transfer from the cyt b_6/f complex to P700 and the cytochrome oxidase (Table X) may be a single cytochrome (cyt *c*-553) for both electron transfer processes in some species (Stürzl et al., 1982; Alpes et al., 1984; Sandmann and Malkin, 1984; Lockau, 1981; Peschek and Schmetterer, 1982) or two different cytochromes specific for either P700 or the cytochrome oxidase in others (Peschek and Schmetterer, 1982).

Cytochrome oxidase-mediated inhibition of photosynthetic electron transport by oxygen, and P700-mediated inhibition of respiratory electron transport by light can be directly demonstrated with membrane preparations of *A. nidulans* (Peschek and Schmetterer, 1982). Only electron donors not immediately reducing the terminal acceptor (P700 or cytochrome oxidase) were liable to competition (Fig. 7b). The results are consistent with plastoquinol:cyt b_6/f oxidoreductase being a common electron donor to both P700 (PSI) and cytochrome oxidase.

Two further possibilities for the interaction of photosynthesis and respiration in cyanobacteria should be mentioned. The first one is the allosteric regulation of key enzymes in either pathway

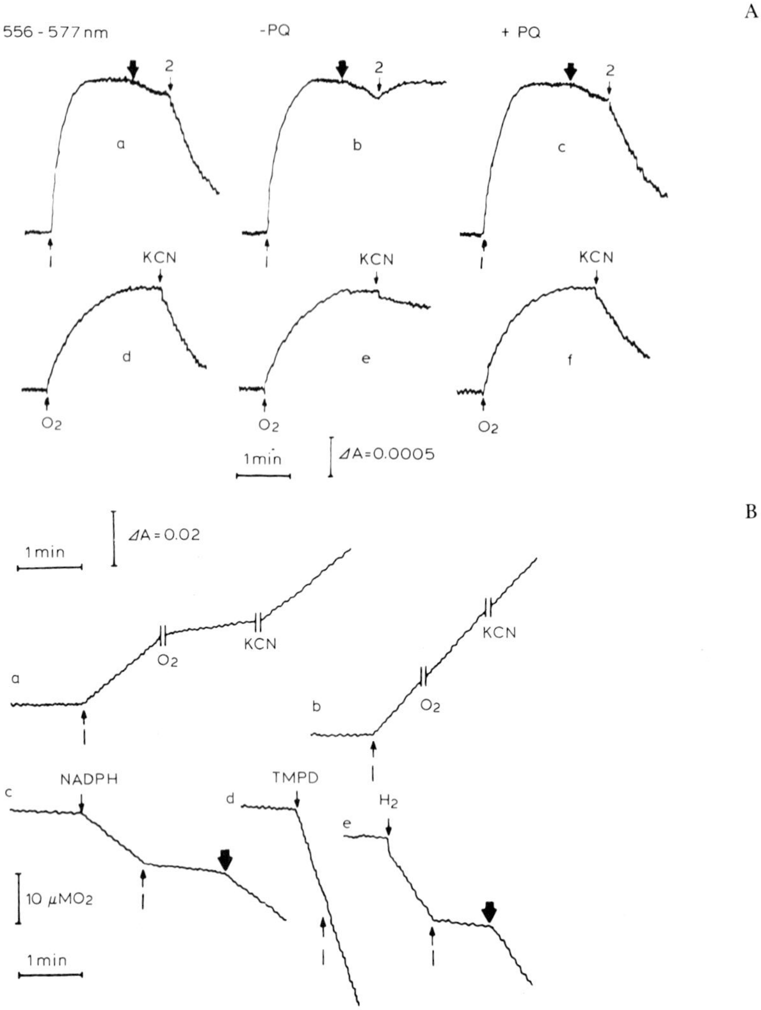

Fig. 7 (A, B). (A) Dual wavelength spectra of membranes prepared from *Nostoc* sp. strain MAC showing oxidation (upward deflection) and reduction (downward deflection) of cytochrome *f* under conditions of photosynthesis (a–c) and respiration (d–f). (a, d), Freshly prepared membranes; (b, e), membranes from which plastoquinone had been extracted with *n*-pentane; (c, f), depleted membranes after reconstitution with authentic plastoquinone. Arrow 1, 707 nm light on; arrow 2, 674 nm light on; bold arrow, light off; other additions as indicated. Reductants present were 3 mM diphenylcarbazide (a–c) and 3 mM NADPH or NADH (d–f), both of which gave essentially the same result. Similar results were obtained with membranes from *Anacystis nidulans* (Peschek, 1983a) and also when the spectra were recorded at 563–577 nm (cytochrome *b*-563). (B) Mutual inhibition of respiration and photosynthesis in membranes prepared from *A. nidulans*. (a, b), Anaerobic photoreduction of $NADP^+$ (3 mM) by 3 mM diaminobenzidine (a) or 8 mM ascorbate plus 0.5 mM DCPIP (b). (c–e) Dark reduction of oxygen supported by 3 mM NADPH (c), 8 mM ascorbate plus 0.5 mM TMPD (d) and H_2 (e). Arrow 1, 707 nm light on. Bold arrow, light off. Other additions as indicated. Similar results were obtained with membranes from *Nostoc* sp. strain MAC. For details see Peschek and Schmetterer, 1982; Peschek, 1983a.

TABLE X

Reconstitution of respiratory and photosynthetic electron transport reactions by exogenous *c*-type cytochromes in isolated membranes of cyanobacteria; rates expressed as nequiv. e^-/min per mg protein

Source of membranes	Added cytochrome (source)	Light	Dark	References
Anacystis nidulans	–	210[a]	48[b]	Peschek and Schmetterer, 1982
	c-550 (horse heart)	186[a]	104[b]	
	c-554 (*A. nidulans*)	384[a]	44[b]	
Nostoc sp. strain MAC	–	280[a]	132[b]	
	c-550 (horse heart)	354[a]	232[b]	
	c-554 (*A. nidulans*)	276[a]	124[b]	
Nostoc muscorum	–	98[c]	40[b]	Stürzl et al., 1982; Alpes et al., 1984
(=*Anabaena* ATCC 24151)	*c*-553 (*Nostoc*)	157[c]	57[b]	
	c-550 (horse heart)	80[c]	37[b]	
	c-553 (*Spirulina*)	87[c]	30[b]	
Anabaena ATCC 29413	–	Negligible		Lockau, 1981
	c-550 (horse heart)	2900[d]	30[d]	
	c-553 (*Anabaena*)	11700[d]	30[d]	
	Plastocyanin (*Anabaena*)	5800[d]	17[d]	

[a] Ascorbate/diaminodurol→$NADP^+$ (in the presence of added ferredoxin); [b] NADPH→O_2; [c] H_2O→$NADP^+$ (in the presence of added ferredoxin); [d] O_2 uptake supported by the reduced redox protein (5 μM); in the light, 0.1 mM methyl viologen was also present. Rates for *N. muscorum* were recalculated assuming a protein/chlorophyll ratio of 20.

by metabolites unique to the other pathway. Examples are the inhibition of ribulose-1,5-bisphosphate carboxylase/oxygenase by 6-phosphogluconate (Tabita and Colletti, 1979; Codd and Stewart, 1977), and the inhibition of glucose 6-phosphate dehydrogenase by NADPH and ATP (Grossman and McGowan, 1975; Schaeffer and Stanier, 1978; Cseke et al., 1981), and by elevated pH via the hysteretic properties of this enzyme (Schaeffer and Stanier, 1978).

The second possibility might be energy charge regulation conforming to the notion that, in an 'uncompartmentalized' prokaryotic cell, there is only a single pool of energy metabolites ATP, ADP and AMP. It is conceivable that, in this way, fast photophosphorylation overrides, and inhibits, slow oxidative phosphorylation which has to compete for the same substrate, namely ADP (Peschek, 1975). This type of regulation thus would involve the energy charge (E.C. = [ATP] + 0.5[ADP]/[ATP] + [ADP] + [AMP]) rather than simply ATP alone which does not decrease significantly under aerobic steady state conditions in the dark when compared to light conditions (Imafuku and Katoh, 1976; Bornefeld and Simonis, 1974; Ihlenfeld and Gibson, 1975; Pelroy and Bassham, 1972; however, see Lubberding and Schroten, 1984). Restoration of dark aerobic ATP levels during a few hours' period of dark anaerobic incubation was observed in *Anabaena cylindrica* and *Anabaenopsis circularis* (Bottomley and Stewart, 1976) and, in part, also in *Anacystis nidulans* (Nitschmann, 1982). The latter findings might point to some role of ATP production via fermentative pathways. Another indication of substrate-level phosphorylation in cyanobacteria is the presence of a Pasteur effect (Peschek, 1975, 1984a; Peavey et al., 1983). However, these mechanisms do not seem to play a decisive role in cyanobacteria which appear to rely mainly on electron transport regulation for adjusting the levels of their metabolic light and dark activities. (See Section 5.2. of Chapter 13 for other aspects of the interrelations between photosynthesis and respiration.)

7. Localization of respiratory electron transport

Both plasma and thylakoid membranes are potential candidates for housing (respiratory) electron transport. Since at least the bulk amount of chlorophyll is contained in the thylakoids, to which are also attached the phycobilisomes (except in *Gloeobacter violaceus* which does not form thylakoids under any growth conditions; Rippka et al., 1974), the thylakoid membrane is the site of photosynthetic electron transport. Since respiratory and photosynthetic electron flow share common components (Section 6), the thylakoid membrane also functions as a respiratory membrane (Peschek, 1984a). The idea that the thylakoid membranes are the exclusive site of both respiratory and photosynthetic electron transport in cyanobacteria is widespread in spite that there are only very few data in support of this assumption; e.g., Bisalputra et al. (1969) could not detect respiratory activity in the plasma membrane of *Nostoc sphaericum* by means of cytochemical techniques. There are however, more reports which point to electron transport in the plasma membrane, like those on DCCD-insensitive glucose uptake by *Plectonema boryanum* (Raboy and Padan, 1978), on potassium movements in *Anabaena variabilis* (Reed et al., 1981), and others (Scherer et al., 1981; Almon and Böhme, 1982). Fig. 8 compares electron-transporting and proton-translocating membrane functions in cyanobacteria (*A. nidulans*), chloroplasts, and mitochondria.

We have compiled substantial experimental evidence, along various independent lines, that the plasma membrane of *Anacystis nidulans* does contain a respiratory (but not a photosynthetic) electron transport system (see Peschek et al., 1984; Peschek, 1984a). These experiments have been performed with intact cells or spheroplasts, or with crude membrane preparations still composed of both plasma and thylakoid membranes. It is beyond the scope of this article to go into all the details but a brief summary of findings may be appropriate:

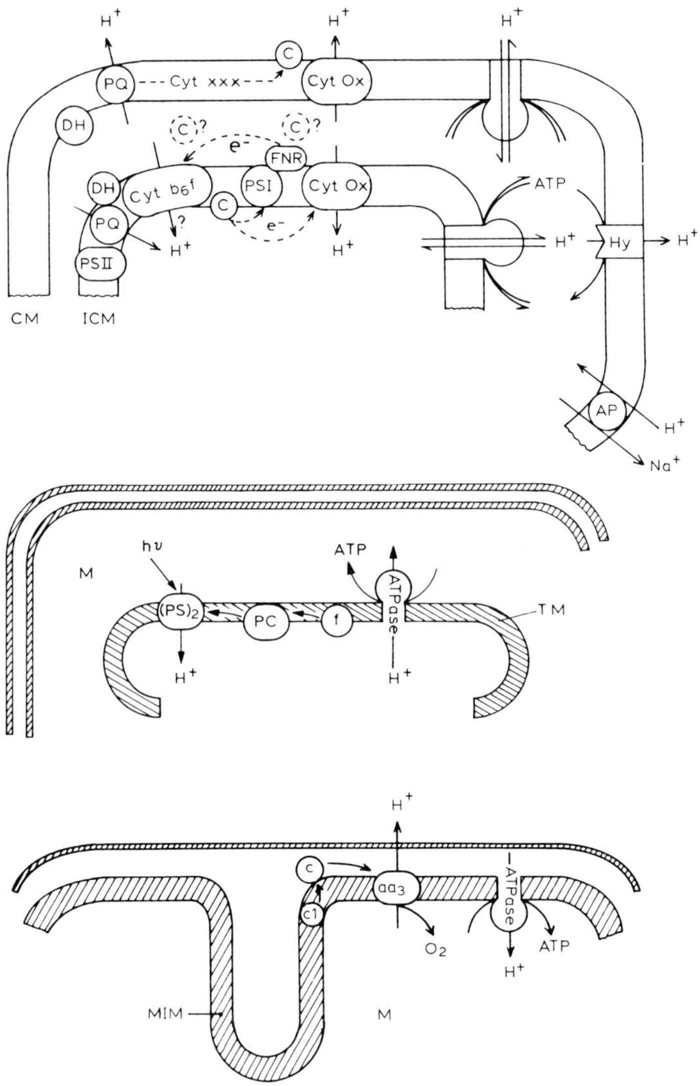

A

B

C

Fig. 8 (A, B, C). Polarity and sidedness of proton translocation, electron transport, and ATP synthesis/hydrolysis in the energy-transducing membranes of cyanobacteria (A), chloroplasts (B), and mitochondria (C). CM, plasma membrane; ICM (TM), thylakoid membrane; DH, dehydrogenases; PSI, PSII, Photosystem I and II; PQ, plastoquinone (in the schemes for mitochondria and chloroplasts the quinones are omitted); Cyt, cytochrome; CytOx (aa_3), cytochrome oxidase; c_1, cytochrome c_1; Cyt *xxx*, hitherto uncharacterized cytochromes in the plasma membrane of cyanobacteria; PC, plastocyanin (replaced by soluble *c*-type cytochromes in some algae and cyanobacteria); c, 'soluble' *c*-type cytochrome of intrathylakoid, periplasmic, and/or cytosolic (?) location (see Wood, 1983); FNR, ferredoxin: NADP$^+$-oxidoreductase; Hy, unidirectional ATP hydrolase; AP, antiporter; MIM, mitochondrial inner membrane; M, matrix or stroma space; broken lines indicate the path of electrons (e^-).

(a) Cytochemical detection of reduced tellurite and formazane, and of oxidized diaminobenzidine, associated with both plasma and thylakoid membranes of intact cells (Peschek et al., 1981c, 1982b).

(b) Distinct Arrhenius discontinuities for the temperature dependence of respiratory and photosynthetic electron transport corresponding to the distinct lipid phase transition temperatures in plasma and thylakoid membranes, respectively (Peschek et al., 1981d, 1982d; Ono and Murata, 1979; Verwer et al., 1978; Furtado et al., 1979; Wada et al., 1984).

(c) Efficiencies of oxidative phosphorylation as related to active Na^+ extrusion from the cells (Nitschmann et al., 1982; Nitschmann and Peschek, 1982).

(d) Proton-translocating oxidation of exogenous ferrocytochrome *c*, and reduction of FeCy, by intact spheroplasts (Peschek et al., 1982b,c; Peschek, 1983b, 1984b; Craig et al., 1984).

(e) Respiratory activity of thylakoid-disintegrated ('white') cells (Peschek and Schmetterer, 1978; Schmetterer and Peschek, 1981; Schmetterer et al., 1983).

(f) ATPase-independent respiratory energization of the plasma membrane effected by likewise ATPase-independent H^+ ejection from respiring cells (Nitschmann and Peschek, 1984, 1985; Peschek et al., 1985a,b; Erber et al., 1986).

We have succeeded recently with the isolation of plasma and thylakoid membranes from *Anacystis nidulans* (Trnka et al., 1985; Molitor and Peschek, 1986; Molitor et al., 1987) following up and improving the pioneering work of Murata and his colleagues (Murata et al., 1981; Omata and Murata, 1984a,b). Our cell membrane preparation, like that of Murata and his colleagues, was devoid of chlorophyll, but unlike that of Omata and Murata (1985) contained substantial cytochrome oxidase activity (Table VII, Section 2.6.). It is unclear at present how this discrepancy might be explained. However, using SDS-polyacrylamide gel electrophoresis and Western blotting, followed by cross reaction of the separated and transferred polypeptide bands with specific antibodies against *Paracoccus denitrificans* aa_3-type cytochrome oxidase (both subunit I and subunit II and native enzyme; see Ludwig and Schatz, 1980), we were able to show that two of the bands corresponded to subunits I and II of the aa_3-type oxidase. This result was obtained with both plasma and thylakoid membrane preparations (see Fig. 2).

Acknowledgements

Investigations in the author's laboratory were generously supported by grants from the Austrian Science Foundation. I thank all my co-workers for their knowledgeable and untiring cooperation, Miss A. Messner and Mr. O. Kuntner for invaluable technical assistance, and Mrs. Ch. Fisman and E. Berlakovich for patiently typing and re-typing the manuscript. Thanks are also due to Drs. L. Packer and S. Scherer for making available unpublished material.

References

Alam, J., Sprinkle, J., Hermodson, M.A. and Krogmann, D.W. (1984) Characterization of cytochrome *c*-550 from cyanobacteria. Biochim. Biophys. Acta 766, 317–321.

Almon, H. and Böhme, H. (1980) Composition and activity of the photosynthetic electron transport system of intact heterocysts isolated from the blue-green alga *Nostoc muscorum*. Biochim. Biophys. Acta 592, 113–120.

Almon, H. and Böhme, H. (1982) Photophosphorylation in isolated heterocysts from the blue-green alga *Nostoc muscorum*. Biochim. Biophys. Acta 679, 279–286.

Alpes, I., Stürzl, E., Scherer, S. and Böger, P. (1984) Interaction of photosynthetic and respiratory electron transport in blue-green algae: effect of a cytochrome *c*-553 specific antibody. Z. Naturforsch. 39c, 623–626.

Alpes, I., Schrautemeier, B., Scherer, S. and Böger, P. (1985) Different enzymes involved in NADH- and NADPH-dependent respiration in the cyanobacterium *Anabaena variabilis*. FEMS Microbiol. Lett. 26, 147–151.

Aoki, M. and Katoh, S. (1982) Oxidation and reduction of plastoquinone by photosynthetic and respiratory electron transport in a cyanobacterium *Synechococcus* sp. Biochim. Biophys. Acta 682, 307–314.

Aoki, M. and Katoh, S. (1983) Size of the plastoquinone pool functioning in photosynthetic and respiratory electron transport of *Synechococcus* sp. Plant Cell Physiol. 24, 1379–1386.

Azzi, A., Casey, R.P. and Nalecz, M.J. (1984) The effect of *N,N'*-dicyclohexylcarbodiimide on enzymes of bioenergetic relevance. Biochim. Biophys. Acta 768, 209–226.

Baird, B.A. and Hammes, C.G. (1979) Structure of oxidative and photo-phosphorylation coupling factor complexes. Biochim. Biophys. Acta 549, 31–54.

Balch, W.E., Fox, G.E., Magrum, L.J., Woese, C.R. and Wolfe, R.S. (1979) Methanogens: reevaluation of a unique biological group. Microbiol. Rev. 43, 260–296.

Barski, E.L., Gusev, M.V., Nikitina, K.A. and Samuilov, V.D. (1981) Light-induced proton-translocation through thylakoid and cytoplasmic membranes of *Plectonema boryanum*. Arch. Microbiol. 129, 105–108.

Bendall, D.S. (1982) Photosynthetic cytochromes of oxygenic organisms. Biochim. Biophys. Acta 683, 119–151.

Benedict, C.R. (1978) Nature of obligate photoautotrophy. Annu. Rev. Plant Physiol. 29, 67–93.

Biggins, J. (1967) Photosynthetic reactions by lysed protoplasts and particle preparations from the blue-green alga *Phormidium luridum*. Plant Physiol. 42, 1447–1456.

Biggins, J. (1969) Respiration in blue-green algae. J. Bacteriol. 99, 570–575.

Binder, A. (1982) Respiration by photosynthesis in energy-transducing membranes of cyanobacteria, J. Bioenerg. Biomembr. 14, 272–286.

Binder, A. and Bachofen, R. (1979) Isolation and characterization of a coupling factor I ATPase of the thermophilic blue-green alga (cyanobacterium) *Mastigocladus laminosus*. FEBS Lett. 104, 66–70.

Binder, A., Tel-Or, E. and Avron, M. (1976) Photosynthetic activities of membrane preparations of the blue-green alga *Phormidium luridum*. Eur. J. Biochem. 67, 187–196.

Binder, A., Bohler, M.C., Wolf, M. and Bachofen, R. (1981a) The proton ATPase of the thermophilic cyanobacterium *Mastigocladus laminosus*. In G. Akoyunoglou (ed.), Photosynthesis II: Electron Transport and Phosphorylation. Balaban International Science Services, Philadelphia, pp. 771–778.

Binder, A., Bohler, M.C., Wolf, M. and Muster, P. (1981b) Energy-transducing membranes in a thermophilic cyanobacterium. In B.R. Selman and S. Selman-Reimer (eds.), Energy Coupling in Photosynthesis, Elsevier/North-Holland, New York, pp. 313–321.

Binder, A., Hauser, R. and Krogmann, D. (1984) Respiration in energy-transducing membranes of the thermophilic cyanobacterium *Mastigocladus laminosus*. I. Relation of the respiratory and photosynthetic electron transport. Biochim. Biophys. Acta 765, 241–246.

Bisalputra, T., Brown, D.L. and Weier, T.E. (1969) Possible respiratory sites in a blue-green alga *Nostoc sphaericum* as demonstrated by potassium tellurite and tetranitro-blue tetrazolium reduction. J. Ultrastruct. Res. 27, 182–197.

Blumwald, E., Wolosin, J.M. and Packer, L. (1984) Na^+/H^+ exchange in the cyanobacterium *Synechococcus* 6311. Biochem. Biophys. Res. Commun. 122, 452–459.

Böger, P. and Böhme, H. (1982) Cytochromes of algae. In A. Mitsui and C.C. Black (eds.), Handbook of Biosolar Resources, Vol. I, Part 1, Basic Principles. CRC Press, Boca Raton, USA, pp. 147-152.

Bohler, M.C. and Binder, A. (1980) Photosynthetic activities of a membrane preparation of the thermophilic cyanobacterium *Mastigocladus laminosus*. Arch. Microbiol. 124, 155–160.

Bohler, M.C., Binder, A. and Bachofen, R. (1982) Reconstitution of photophosphorylation with the coupling factor AF_1 of a thermophilic cyanobacterium. FEMS Microbiol. Lett. 15, 115–118.

Böhme, H. and Almon, H. (1983) Reactions of hydrogen and oxygen with photosystem I of isolated heterocysts from *Anabaena variabilis* (ATCC 29413). Biochim. Biophys. Acta 722, 401–407.

Bornefeld, T. and Simonis, W. (1974) Effects of light, temperature, pH, and inhibitors on the ATP level of the blue-green alga *Anacystis nidulans*. Planta 115, 309–318.

Bothe, H. (1969) Ferredoxin als Kofaktor der cyclischen Photophosphorylierung in einem zellfreien System aus der Blaualge *Anacystis nidulans*. Z. Naturforsch. 246, 1574-1582.

Bothe, H., Tennigkeit, J. and Eisbrenner, G. (1977) The utilization of molecular hydrogen by the blue-green alga *Anabaena cylindrica*. Arch. Microbiol. 114, 43–49.

Bothe, H., Distler, E. and Eisbrenner, G. (1978) Hydrogen metabolism in blue-green algae. Biochimie 60, 277–289.

Bottomley, P.J. and Stewart, W.D.P. (1976) ATP pools and transients in the blue-green alga, *Anabaena cylindrica*. Arch. Microbiol. 108, 249–258.
Bottomley, P.J. and Van Baalen, C. (1978) Characteristics of heterotrophic growth in the blue-green alga *Nostoc* sp. strain MAC. J. Gen. Microbiol. 107, 309–318.
Bowien, B. and Schlegel, H.G. (1981) Physiology and biochemistry of aerobic hydrogen-oxidizing bacteria. Annu. Rev. Microbiol. 35, 405–452.
Brown, A.H. (1953) The effects of light on respiration using isotopically enriched oxygen. Am. J. Bot. 40, 719–729.
Brown, A.H. and Webster, G.C. (1953) The influence of light on the rate of respiration of the blue-green alga *Anabaena*. Am. J. Bot. 40, 753–758.
Carr, N.G. (1973) Metabolic control and autotrophic physiology. In N.G. Carr and B.A. Whitton (eds.), The Biology of Blue-Green Algae. Blackwell, Oxford, pp. 39–65.
Carr, N.G. and Whitton, B.A., Eds. (1973) The Biology of Blue-Green Algae. Blackwell, Oxford.
Carr, N.G. and Whitton, B.A., Eds. (1982) The Biology of Cyanobacteria. 2nd ed., Blackwell, Oxford.
Casey, R.P., Thelen, M. and Azzi, A. (1980) Dicyclohexylcarbodiimide binds specifically and covalently to cytochrome *c* oxidase while inhibiting its H^+-translocating activity, J. Biol. Chem. 255, 3994–4000.
Codd, G.A. and Stewart, W.D.P. (1977) D-Ribulose-1,5-diphosphate carboxylase from the blue-green alga *Aphanocapsa* 6308. Arch. Microbiol. 113, 105–110.
Coleman, J.R. and Colman, B. (1981) Inorganic carbon accumulation and photosynthesis in a blue-green alga as a function of external pH. Plant Physiol. 67, 917–921.
Conrad, R. and Schlegel, H.G. (1977) Different degradation pathways for glucose and fructose in *Rhodopseudomonas capsulata*. Arch. Microbiol. 112, 39–48.
Craig, T., Crane, F.L., Misra, P.C. and Barr, R. (1984) Trans-plasma membrane electron transport in *Anacystis nidulans*. Plant Sci. Lett. 35, 11–17.
Cramer, W.A. and Whitmarsh, J. (1977) Photosynthetic cytochromes. Annu. Rev. Plant Physiol. 28, 133–172.
Crofts, A.R. and Wood, P.M. (1978) Photosynthetic electron-transport of plants and bacteria and their role as proton pumps. Curr. Top. Bioenerg. 7, 175–244.
Croxdale, J.G. (1983) Quantitative measurements of phosphofructokinase in the shoot apical meristem, leaf primordia, and leaf tissues of *Dianthus chinensis* L. Plant Physiol. 73, 66–70.
Cséke, C., Balogh, A. and Farkas, G.L. (1981) Redox modulation of glucose 6-phosphate dehydrogenase in *Anacystis nidulans* and its 'uncoupling' by phage infection. FEBS Lett. 126, 85–88.
Davis, D.J., Krogmann, D.W. and San Pietro, A. (1980) Electron donation to photosystem I. Plant Physiol. 65, 697–702.
Doolittle, W.F. (1979) The cyanobacterial genome, its expression, and the control of that expression. Adv. Microbiol. Physiol. 20, 1–102.
Eisbrenner, G. and Bothe, H. (1979) Modes of electron transfer from molecular hydrogen in *Anabaena cylindrica*. Arch. Microbiol. 123, 37–45.
Eisbrenner, G., Distler, E., Floener, L. and Bothe, H. (1978) The occurrence of the hydrogenase in some blue-green algae. Arch. Microbiol. 118, 177–184.
Engel, W.D., Schägger, H. and von Jagow, G. (1980) Ubiquinol-cytochrome *c* reductase (EC 1.10.2.2): isolation in Triton X-100 by hydroxyapatite and gel chromatography. Structural and functional properties. Biochim. Biophys. Acta 592, 211–222.
Erber, W.W.A., Nitschmann, W.H., Muchl, R. and Peschek, G.A. (1986) Endogenous energy supply to the plasma membrane of dark aerobic cyanobacterium *Anacystis nidulans*. Arch. Biochem. Biophys. 247, 28–39.
Ernst, A., Böhme, H. and Böger, P. (1983) Phosphorylation and nitrogenase activity in isolated heterocysts from *Anabaena variabilis* (ATCC 29413). Biochim. Biophys. Acta 723, 83–90.
Falkner, G., Horner, F., Werdan, K. and Heldt, H.W. (1976) pH changes in the cytoplasm of the blue-green alga *Anacystis nidulans* caused by light-dependent proton flux into the thylakoid space. Plant Physiol. 58, 717–718.
Fewson, C.A., Al Hafidh, M. and Gibbs, M. (1962) Role of aldolase in photosynthesis I. Enzyme studies with photosynthetic organisms with special reference to blue-green algae. Plant Physiol. 37, 402–406.
Fogg, G.E., Stewart, W.D.P., Fay, P. and Walsby, A.E. (1973) The Blue-Green Algae. Academic Press, New York.
Frei, R., Binder, A. and Bachofen, R. (1984) Respiration in energy-transducing membranes of the thermophilic cyanobacterium *Mastigocladus laminosus*. II. Oxidative phosphorylation. Biochim. Biophys. Acta 765, 247–252.

Fry, I.V., Peschek, G.A., Huflejt, M. and Packer, L. (1985) EPR signals of redox active copper in EDTA washed membranes of the cyanobacterium *Synechococcus* 6311. Biochem. Biophys. Res. Commun. 129, 109–116.

Fry, I.V., Huflejt, M., Erber, W.W.A., Peschek, G.A. and Packer, L. (1986) The role of respiration in the adaptation of the freshwater cyanobacterium *Synechococcus* 6311 to a saline environment. Arch. Biochem. Biophys. 244, 686–691.

Furtado, D., Williams, W.P., Brain, A.P.R. and Quinn, P.J. (1979) Phase separations in membranes of *Anacystis nidulans* grown at different temperatures. Biochim. Biophys. Acta 555, 352–357.

Gabellini, N., Bowyer, J.R., Hurt, E., Melandri, B.A. and Hauska, G. (1982) Ubiquinol-cytochrome c_2 oxidoreductase from *Rhodopseudomonas sphaeroides* GA. Eur. J. Biochem. 126, 105–111.

Gaffron, H. (1939) Über Anomalien des Atmungsquotienten von Algen aus Zuckerkulturen. Biol. Zentr. 59, 288–302.

Genevois, L. (1927) Über Atmung und Gärung in grünen Pflanzen. Biochem. Z. 186, 461–473.

Gibson, J. (1981) Movement of acetate across the cytoplasmic membrane of the unicellular cyanobacteria *Synechococcus* and *Aphanocapsa*. Arch. Microbiol. 130, 175–179.

Grossman, A. and McGowan, R.E. (1975) Regulation of glucose 6-phosphate dehydrogenase in blue-green algae. Plant Physiol. 55, 658–662.

Haselkorn, R. (1978) Heterocysts. Annu. Rev. Plant Physiol. 29, 319–344.

Heldt, H.W., Werdan, K., Milovancev, M. and Geller, G. (1973) Alkalinization of the chloroplast stroma caused by light-dependent proton flux into the thylakoid space. Biochim. Biophys. Acta 314, 224–241.

Higgins, I.J., Best, D.J., Hammond, R.C. and Scott, D. (1981) Methane-oxidizing microorganisms. Microbiol. Rev. 45, 556–590.

Hinrichs, I., Scherer, S. and Böger, P. (1985) Two different mechanisms of light-induced proton efflux in the blue-green alga *Anabaena variabilis*. Physiol. Vég. 23, 717–724.

Hinterstoisser, B. and Peschek, G.A. (1987) Fluorimetric pH measurement in whole cells of dark aerobic and anaerobic cyanobacteria. FEBS Lett., in press.

Hirano, M., Satoh, K. and Katoh, S. (1980) Plastoquinone as a common link between photosynthesis and respiration in a blue-green alga. Photosynthesis Res. 1, 149–162.

Ho, K.K. and Krogmann, D.W. (1982) Photosynthesis. In N.G. Carr and B.A. Whitton (eds.), The Biology of Cyanobacteria, 2nd ed. Blackwell, Oxford, pp. 191–214.

Ho, K.K. and Krogmann, D.W. (1984) Electron donors to P700 in cyanobacteria and algae. An instance of unusual genetic variability. Biochim. Biophys. Acta 766, 310–316.

Hoch, G., Owens, O.H. and Kok, B. (1963) Photosynthesis and respiration. Arch. Biochem. Biophys. 101, 171–180.

Holm-Hansen, O. (1968) Ecology, physiology and biochemistry of blue-green algae. Annu. Rev. Microbiol. 21, 47–70.

Holton, R.W. and Myers, J. (1963) Cytochromes of a blue-green alga: Extraction of a *c*-type with a strongly negative redox potential. Science 142, 234–235.

Holton, R.W. and Myers, J. (1967a) Water-soluble cytochromes from a blue-green alga. I. Extraction, purification, and spectral properties of cytochromes *c* (549, 552 and 554, *Anacystis nidulans*). Biochim. Biophys. Acta 131, 362–374.

Holton, R.W. and Myers, J. (1967b) Water-soluble cytochromes from a blue-green alga. II. Physicochemical properties and quantitative relationships of cytochromes *c* (549, 552, and 554, *Anacystis nidulans*). Biochim. Biophys. Acta 131, 375–384.

Horton, A.A. (1968) NADH oxidase in blue-green algae. Biochem. Biophys. Res. Commun. 32, 839–845.

Houchins, J.P. (1984) The physiology and biochemistry of hydrogen metabolism in cyanobacteria. Biochim. Biophys. Acta 768, 227–255.

Houchins, J.P. and Burris, R.H. (1981a) Light and dark reactions of the uptake hydrogenase in *Anabaena* 7120. Plant Physiol. 68, 712–716.

Houchins, J.P. and Burris, R.H. (1981b) Physiological reactions of the reversible hydrogenase from *Anabaena* 7120. Plant Physiol. 68, 717–721.

Houchins, J.P. and Hind, G. (1982) Pyridine nucleotides and H_2 as electron donors to the respiratory and photosynthetic electron-transfer chains and to nitrogenase in *Anabaena* heterocysts. Biochim. Biophys. Acta 682, 86–96.

Houchins, J.P. and Hind, G. (1984) Concentration and function of membrane-bound cytochromes in cyanobacterial heterocysts. Plant Physiol. 76, 456–460.

Howarth, D.C. and Codd, G.A. (1985) The uptake and production of molecular hydrogen by unicellular cyanobacteria. J. Gen. Microbiol. 131, 1561–1569.

Hurt, E. and Hauska, G. (1981) A cytochrome *f*/b_6 complex of five polypeptides with plastoquinone-plastocyanin-oxidoreductase activity from spinach chloroplasts, Eur. J. Biochem. 117, 591–599.

Hurt, E. and Hauska, G. (1982a) Identification of the polypeptides in the cytochrome b_6/*f* complex from spinach chloroplasts with redox-center-carrying subunits. J. Bioenerg. Biomembr. 14, 405–424.

Hurt, E. and Hauska, G. (1982b) Involvement of plastoquinone bound within the isolated cytochrome b_6/*f* complex from chloroplasts in oxidant-induced reduction of cytochrome b_6. Biochim. Biophys. Acta 682, 466–473.

Ihlenfeldt, M.J.A. and Gibson, J. (1975) CO_2 fixation and its regulation in *Anacystis nidulans* (*Synechococcus*). Arch. Microbiol. 102, 13–21.

Ihlenfeldt, M.J.A. and Gibson, J. (1977) Acetate uptake by the unicellular cyanobacteria *Synechococcus* and *Aphanocapsa*. Arch. Microbiol. 113, 231–241.

Imafuku, H. and Katoh, T. (1976) Intracellular ATP level and light-induced inhibition of respiration in a blue-green alga, *Anabaena variabilis*. Plant Cell Physiol. 17, 515–524.

Ingram, L.O., Calder, J.A., Van Baalen, C., Plucker, F.E. and Parker, P.L. (1973) Role of reduced exogenous organic compounds in the physiology of the blue-green bacteria (algae): photoheterotrophic growth of a 'heterotrophic' blue-green bacterium. J. Bacteriol. 114, 695–700.

Jagendorf, A.T. and Uribe, E. (1966) ATP formation caused by acid-base transition of spinach chloroplasts. Proc. Natl. Acad. Sci. USA 55, 170–177.

Jones, L.W. and Myers, J. (1963) A common link between photosynthesis and respiration in a blue-green alga. Nature 199, 670–672.

Joset-Espardellier, F., Astier, C., Evans, E.H. and Carr, N.G. (1978) Cyanobacteria grown under photoautotrophic, photoheterotrophic, and heterotrophic regimes: sugar metabolism and carbon dioxide fixation. FEMS Microbiol. Letters 4, 261–264.

Kallas, T. and Castenholz, R.W. (1982) Internal pH and ATP-ADP pools in the cyanobacterium *Synechococcus* sp. during exposure to growth-limiting low pH. J. Bacteriol. 149, 229–236.

Keister, D.L. (1978) Respiration vs. photosynthesis. In R.K. Clayton and W.R. Sistrom (eds.), The Photosynthetic Bacteria. Plenum Press, New York, pp. 849–856.

Kelly, D.P. (1971) Autotrophy: Concepts of lithotrophic bacteria and their organic metabolism. Annu. Rev. Microbiol. 25, 177–210.

Kelly, G.J. and Latzko, E. (1977a) Chloroplast phosphofructokinase. I. Proof of phosphofructokinase activity in chloroplasts. Plant Physiol. 60, 290–294.

Kelly, G.J. and Latzko, E. (1977b) Chloroplast phosphofructokinase. II. Partial purification, kinetic and regulatory properties. Plant Physiol. 60, 295–299.

Kienzl, P.F. and Peschek, G.A. (1982) Oxidation of *c*-type cytochromes by the membrane-bound cytochrome oxidase (cytochrome aa_3) of cyanobacteria. Plant Physiol. 69, 580–584.

Kienzl, P.F. and Peschek, G.A. (1983) Cytochrome *c*-549 – an endogenous cofactor of cyclic photophosphorylation in the cyanobacterium *Anacystis nidulans*? FEBS Lett. 162, 76–80.

Koenig, F. and Vernon, L.P. (1981) Which polypeptides are characteristic of photosystem II? Analysis of active PSII particles from the blue-green alga *Anacystis nidulans*. Z. Naturforsch. 36c, 295–304.

Kok, B. (1949) On the interrelation of respiration and photosynthesis in green plants, Biochim. Biophys. Acta 3, 625–631.

Kratz, W.A. and Myers, J. (1955) Photosynthesis and respiration of three blue-green algae. Plant Physiol. 30, 275–280.

Krinner, M., Hauska, G., Hurt, E. and Lockau, W. (1982) A cytochrome *f*/b_6 complex with plastoquinol-cytochrome *c* oxidoreductase activity from *Anabaena variabilis*. Biochim. Biophys. Acta 681, 110–117.

Lambert, G.R. and Smith, G.D. (1981) The hydrogen metabolism of cyanobacteria (blue-green algae). Biol. Rev. 56, 589–660.

Lang, N.J. (1968) The fine structure of blue-green algae. Annu. Rev. Microbiol. 21, 15–46.

Leach, C.K. and Carr, N.G. (1969) Oxidative phosphorylation in an extract of *Anabaena variabilis*. Biochem. J. 112, 125–126.

Leach, C.K. and Carr, N.G. (1970) Electron transport and oxidative phosphorylation in the blue-green alga *Anabaena variabilis*. J. Gen. Microbiol. 64, 55–70.

Lee, S., Young, A.M. and Krogmann, D.W. (1969) Site-specific inactivation of photophosphorylation reactions of *Anabaena variabilis*. Biochim. Biophys. Acta 180, 130–136.

Lehmann, M. and Wöber, G. (1976) Accumulation and mobilization of glycogen in the blue-green bacterium *Anacystis nidulans*. Arch. Microbiol. 111, 93–97.

Lerma, C. and Gomez-Lojero, C. (1982) Photosynthetic phosphorylation by a membrane preparation of the cyanobacterium *Spirulina maxima*. Biochim. Biophys. Acta 680, 181–186.

Lockau, W. (1981) Evidence for a dual role of cytochrome *c*-553 and plastocyanin in photosynthesis and respiration of the cyanobacterium, *Anabaena variabilis*. Arch. Microbiol. 128, 336–340.

Lockau, W. and Pfeffer, S. (1982) A cyanobacterial ATPase distinct from the coupling factor of photophosphorylation. Z. Naturforsch. 37c, 658–664.

Lockau, W. and Pfeffer, S. (1983) ATP-dependent calcium transport in membrane vesicles of the cyanobacterium *Anabaena variabilis*. Biochim. Biophys. Acta 733, 124–132.

Lubberding, H.J. and Schroten, W. (1984) The ATP level in the thermophilic cyanobacterium *Synechococcus* 6716 during light-dark transition and in the presence of some specific inhibitors. FEMS Microbiol. Lett. 22, 93–96.

Lubberding, H.J., Zimmer, G., Van Walraven, H.S., Schrickx, J. and Kraayenhof, R. (1983) Isolation, purification and characterization of the ATPase complex from the thermophilic cyanobacterium *Synechococcus* 6716. Eur. J. Biochem. 137, 95–99.

Ludwig, B. (1980) Heme aa_3-type cytochrome oxidase from bacteria. Biochim. Biophys. Acta 594, 177–189.

Ludwig, B. and Schatz, G. (1980) A two-subunit cytochrome *c* oxidase (cytochrome aa_3) from *Paracoccus denitrificans*. Proc. Natl. Acad. Sci. USA, 77, 196–200.

Manodori, A., Alhadeff, M., Glazer, A.N. and Melis, A. (1984) Photochemical apparatus organization in *Anacystis nidulans* (Cyanophyceae). Effect of CO_2 concentration during cell growth. Plant Physiol. 74, 67–71.

Masamoto, K. and Nishimura, M. (1977) Estimation of the internal pH in cells of blue-green algae in the dark and under illumination, J. Biochem. 82, 483–487.

Matin, A. (1978) Organic nutrition of chemolithotrophic bacteria. Annu. Rev. Microbiol. 32, 433–468.

Matthijs, H.C.P. (1984) Energy metabolism in the cyanobacterium *Plectonema boryanum*. Ph.D. Thesis, Vrije Universiteit, Amsterdam.

Matthijs, H.C.P., Luderus, E.M.E., Löffler, H.J.M., Scholts, M.J.C. and Kraayenhof, R. (1984a) Energy metabolism in the cyanobacterium *Plectonema boryanum*. Participation of the thylakoid photosynthetic electron transfer chain in the dark respiration of NADPH and NADH. Biochim. Biophys. Acta 766, 29–37.

Matthijs, H.C.P., van Hoek, A.N., Löffler, H.J.M. and Kraayenhof, R. (1984b) Presence of cytochromes in the thylakoid and cell membranes of the cyanobacterium *Plectonema boryanum* cultured in normal and copper depleted growth medium. In C. Sybesma (ed.), Advances in Photosynthesis Research, Vol. II. Martinus Nijhoff/Dr. W. Junk Publ., The Hague, The Netherlands, pp. 643–646.

Matthijs, H.C.P., Luderus, E.M.E., Scholts, M.J.C. and Kraayenhof, R. (1984c) Energy metabolism in the cyanobacterium *Plectonema boryanum*. Oxidative phosphorylation and respiratory pathways. Biochim. Biophys. Acta 766, 38–44.

McCarty, R.E. (1978) The ATPase complex of chloroplasts and chromatophores. Curr. Top. Bioenerg. 7, 245–278.

Meyerhof, O. (1925) Über den Einfluß des Sauerstoffs auf die alkoholische Gärung der Hefe, Biochem. Z. 162, 43–86.

Miller, A.G. Turpin, D.H. and Canvin, D.T. (1984) Na^+ requirement for growth, photosynthesis, and pH regulation in the alkalotolerant cyanobacterium *Synechococcus leopoliensis*. J. Bacteriol. 159, 100–106.

Mitchell, P. (1961) Coupling of phosphorylation to electron and hydrogen transfer by a chemiosmotic type of mechanism. Nature 191, 144–148.

Mitchell, P. (1968) Chemiosmotic Coupling and Energy Transduction. Glynn Research, Bodmin, England.

Molitor, V. (1984) Radiorespirometrische Versuche zur Feststellung des Abbauweges positionsmarkierter ^{14}C-Glucose und ^{14}C-Fructose bei *Anacystis nidulans*, *Anabaena variabilis* und *Nostoc* MAC. M. Sc. Thesis, University of Vienna, Austria.

Molitor, V. and Peschek, G.A. (1986) Respiratory electron transport in plasma and thylakoid membrane preparations from the cyanobacterium *Anacystis nidulans*. FEBS Lett. 195, 145–150.

Molitor, V., Trnka, M. and Peschek, G.A. (1987) Isolated und purified plasma and thylakoid membranes of the cyanobacterium *Anacystis nidulans* contain immunologically cross-reactive aa_3-type cytochrome oxidase. Curr. Microbiol., in press.

Muchl, R. and Peschek, G.A. (1983) pH-jump-induced phosphorylation of ADP in the cyanobacterium *Anacystis nidulans*. FEBS Lett 164, 116–120.

Muchl, R. and Peschek, G.A. (1984) Valinomycin pulse-induced phosphorylation of ADP in dark anaerobic cells of the cyanobacterium *Anacystis nidulans*. Curr. Microbiol. 11, 179–182.

Muchl, R., Schmetterer, G., Nitschmann, W.H. and Peschek, G.A. (1983) Energy transduction by the cytoplasmic membrane of intact cells of the cyanobacterium *Anacystis nidulans*. Naturwissenschaften 70, 615–616.

Murata, N., Sato, N., Omata, T. and Kuwabara, T. (1981) Separation and characterization of thylakoid and cell envelope of the blue-green alga (cyanobacterium) *Anacystis nidulans*. Pl. Cell Physiol. 22, 855–866.

Nakayama, K., Yamaoka, T. and Katoh, S. (1979) Chromatographic separation of photosystem I and photosystem II from the thylakoid membranes isolated from a thermophilic blue-green alga. Plant Cell Physiol. 20, 1565–1576.

Nanba, M. and Katoh, S. (1983) Reaction kinetics of P-700, cytochrome *c*-553 and cytochrome *f* in the cyanobacterium, *Synechococcus* sp. Biochim. Biophys. Acta 725, 272–279.

Nanba, M. and Katoh, S. (1984) Effects of dibromothymoquinone on oxidation-reduction reactions and the midpoint potential of the Rieske iron-sulfur center in photosynthetic electron transport of *Synechococcus* sp. Biochim. Biophys. Acta 767, 396–403.

Nanba, M. and Katoh, S. (1985) Electron transport from cytochrome b_6-f complexes to Photosystem I reaction center complexes in *Synechococcus* sp. Is cytochrome c-553 a mobile electron carrier? Biochim. Biophys. Acta 808, 39–45.

Nelson, N. (1981a) Proton ATPase of chloroplasts. Curr. Top. Bioenerg. 11, 1–33.

Nelson, N. (1981b) Structure and function of CF_0. In B.R. Selman and S. Selman-Reimer (eds.), Energy Coupling in Photosynthesis, Elsevier/North-Holland, New York, pp. 261–268.

Neuer, G. and Bothe, H. (1982) The pyruvate: ferredoxin oxidoreductase in heterocysts of the cyanobacterium *Anabaena cylindrica*. Biochim. Biophys. Acta 716, 358–365.

Newman, P.J. and Sherman, L.A. (1978) Isolation and characterization of photosystem I and II membrane particles from the blue-green alga *Synechococcus cedrorum*. Biochim. Biophys. Acta 503, 343–361.

Nicholls, D. (1982) Bioenergetics: An Introduction to Chemiosmotic Theory. Academic Press, New York.

Nitschmann, W.H. (1982) Oxidative Phosphorylierung bei Blaualgen (Cyanobakterien), Doctoral thesis, University of Vienna, Austria.

Nitschmann, W.H. and Peschek, G.A. (1982) Oxidative phosphorylation in intact cyanobacteria. FEBS Lett. 139, 77–80.

Nitschmann, W.H. and Peschek, G.A. (1984) Vanadate and dicyclohexylcarbodiimide insensitive proton extrusion from oxygen pulsed cells of the cyanobacterium *Anacystis nidulans*. Biochem. Biophys. Res. Commun. 123, 358–364.

Nitschmann, W.H. and Peschek, G.A. (1985) Modes of proton translocation across the cell membrane of respiring cyanobacteria. Arch. Microbiol. 141, 330–336.

Nitschmann, W.H. and Peschek, G.A. (1986) Oxidative phosphorylation and energy buffering in cyanobacteria. J. Bacteriol. 168, 1205–1211.

Nitschmann, W.H., Schmetterer, G., Muchl, R. and Peschek, G.A. (1982) Active sodium extrusion reduces net efficiencies of oxidative phosphorylation in the strictly photoautotrophic cyanobacterium *Anacystis nidulans*. Biochim. Biophys. Acta 682, 293–296.

Ohnishi, T. (1973) Mechanism of electron transport and energy conservation in the site I region of the respiratory chain. Biochim. Biophys. Acta, 301, 105–128.

Omata, T. and Murata, N. (1984a) Cytochromes and prenylquinones in preparations of cytoplasmic and thylakoid membranes from the cyanobacterium (blue-green alga) *Anacystis nidulans*. Biochim. Biophys. Acta 766, 395–402.

Omata, T. and Murata, N. (1984b) Isolation and characterization of three types of membranes from the cyanobacterium (blue-green alga) *Synechocystis* PCC 6714. Arch. Microbiol. 139, 113–116.

Omata, T. and Murata, N. (1985) Electron transport reactions in cytoplasmic and thylakoid membranes prepared from the cyanobacteria (blue-green algae) *Anacystis nidulans* and *Synechococcus* PCC 6714. Biochim. Biophys. Acta 810, 354–361.

Ono, T.A. and Murata, N. (1979) Temperature-dependence of the photosynthetic activities in the thylakoid membranes from the blue-green alga *Anacystis nidulans*. Biochim. Biophys. Acta 502, 477–485.

Owers-Narhi, L., Robinson, S.J., Selvius-DeRoo, C. and Yocum, C.F. (1979) Reconstitution of cyanobacterial photophosphorylation by a latent Ca^{2+}-ATPase. Biochem. Biophys. Res. Commun. 90, 1025–1031.

Padan, E. (1979) Facultative anoxygenic photosynthesis in cyanobacteria. Annu. Rev. Plant Physiol. 30, 27–40.

Padan, E. (1981) pH homeostasis in bacteria. Biochim. Biophys. Acta 650, 151–166.

Padan, E. and Schuldiner, S. (1978) Energy transduction in the photosynthetic membranes of the cyanobacterium (blue-green alga) *Plectonema boryanum*. J. Biol. Chem. 253, 3281–3286.

Padan, E., Zilberstein, D. and Rottenberg, H. (1976) The proton electrochemical gradient in *Escherichia coli* cells. Eur. J. Biochem. 63, 533–541.

Paschinger, H. (1977) DCCD-induced sodium uptake by *Anacystis nidulans*. Arch. Microbiol. 113, 285–291.

Pearce, J. and Carr, N.G. (1967) The metabolism of acetate by the blue-green algae *Anabaena variabilis* and *Anacystis nidulans*. J. Gen. Microbiol. 49, 301–313.

Pearce, J. and Carr, N.G. (1969) The incorporation and metabolism of glucose by *Anabaena variabilis*. J. Gen. Microbiol. 54, 451–462.

Pearce, J., Leach, C.K. and Carr, N.G. (1969) The incomplete citric acid cycle in the blue-green alga *Anabaena variabilis*. J. Gen. Microbiol. 49, 301–313.

Peavey, D.G., Gfeller, R.P. and Gibbs, M. (1983) Presence of a Pasteur effect in algae. Physiol. Veg. 21, 337–342.

Pelroy, R.A. and Bassham, J.A. (1972) Photosynthetic and dark carbon metabolism in unicellular blue-green algae. Arch. Mikrobiol. 86, 25–38.

Pelroy, R.A. and Bassham, J.A. (1973) Efficiency of energy conversion by aerobic glucose metabolism in *Aphanocapsa* 6714. J. Bacteriol. 115, 937–942.

Pelroy, R.A., Rippka, R. and Stanier, R.Y. (1972) Metabolism of glucose by unicellular blue-green algae. Arch. Mikrobiol. 87, 303–322.

Peschek, G.A. (1975) Light inhibition of respiration and fermentation and its reversal by inhibitors and uncouplers of photophosphorylation in whole cells of a prokaryotic alga, *Anacystis nidulans*. In M. Avron (ed.), Proceedings of the 3rd International Congress on Photosynthesis. Vol. II, Elsevier, Amsterdam, pp. 921–928.

Peschek, G.A. (1976) Respiratory activities of whole cells and cell-free extracts of *Anacystis nidulans*. In G.A. Codd and W.D.P. Stewart (eds.), Abstracts of the 2nd International Symposium on Photosynthetic Prokaryotes. FEMS, Dundee, Scotland, pp. 209–212.

Peschek, G.A. (1979a) Anaerobic hydrogenase activity in *Anacystis nidulans:* H_2-dependent photoreduction and related reactions. Biochim. Biophys. Acta 548, 187–202.

Peschek, G.A. (1979b) Aerobic hydrogenase activity in *Anacystis nidulans:* the oxyhydrogen reaction. Biochim. Biophys. Acta 548, 203–215.

Peschek, G.A. (1979c) Evidence for two functionally distinct hydrogenases in *Anacystis nidulans*. Arch. Microbiol. 123, 81–92.

Peschek, G.A. (1979d) The role of the Calvin cycle for anoxygenic CO_2 photo-assimilation in *Anacystis nidulans*. FEBS Lett. 106, 34–38.

Peschek, G.A. (1980a) Electron transport reactions in respiratory particles of hydrogenase-induced *Anacystis nidulans*. Arch. Microbiol. 125, 123–131.

Peschek, G.A. (1980b) Restoration of respiratory electron transport reactions in quinone-depleted particle preparations of *Anacystis nidulans*. Biochem. J. 186, 515–523.

Peschek, G.A. (1981a) Spectral properties of a cyanobacterial cytochrome oxidase: Evidence for cytochrome aa_3. Biochem Biophys. Res. Commun. 98, 72–79.

Peschek, G.A. (1981b) Occurrence of cytochrome aa_3 in *Anacystis nidulans*. Biochim. Biophys. Acta 635, 470–475.

Peschek, G.A. (1982) The cytochrome oxidase-hydrogenase relationship in cyanobacteria. Naturwissenschaften 69, 599.

Peschek, G.A. (1983a) The cytochrome *f/b* electron-transport complex: a common link between photosynthesis and respiration in the cyanobacterium *Anacystis nidulans*. Biochem. J. 210, 269–272.

Peschek, G.A. (1983b) Proton pump coupled to cytochrome *c* oxidase in the cyanobacterium *Anacystis nidulans*. J. Bacteriol. 153, 539–542.

Peschek, G.A. (1984a) Structure and function of respiratory membranes in cyanobacteria (blue-green algae). Subcell. Biochem. 10, 85–191.

Peschek, G.A. (1984b) Characterization of the proton-translocating cytochrome *c* oxidase activity in the plasma membrane of intact *Anacystis nidulans* spheroplasts, Plant Physiol. 75, 968–973.

Peschek, G.A. and Broda, E. (1973) Utilization of fructose by a unicellular blue-green alga, *Anacystis nidulans*. Naturwissenschaften 60, 479–480.

Peschek, G.A. and Schmetterer, G. (1978) Reversible photooxidative loss of pigments and of intracytoplasmic membranes in the blue-green alga *Anacystis nidulans*. FEMS Microbiol. Lett. 3, 295–297.

Peschek, G.A. and Schmetterer, G. (1981) Effects of lipids on the membrane-bound cytochrome oxidase of cyanobacteria. Naturwissenschaften 68, 575.

Peschek, G.A. and Schmetterer, G. (1982) Evidence for plastoquinol-cytochrome *f/b*-563 reductase as a common electron donor to P700 and cytochrome oxidase in cyanobacteria. Biochem. Biophys. Res. Commun. 108, 1188–1195.

Peschek, G.A. and Sleytr, U.B. (1983) Thylakoid morphology of the cyanobacteria *Anabaena variabilis* and *Nostoc* MAC grown under light and dark conditions, J. Ultrastruct. Res. 82, 233–239.

Peschek, G.A. and Kuntner, O. (1987) Differential effects of plastoquinone and phylloquinone in photosynthetic and respiratory electron transfer reactions of *n*-pentane-extracted membrane preparations from cyanobacteria. In A. Favre (ed.), From Photophysics to Photobiology. Elsevier Science Publ., Amsterdam, in press.

Peschek, G.A., Kienzl, P.F. and Schmetterer, G. (1981a) Reactivity of membraneous cytochrome oxidase (cytochrome aa_3) from *Anacystis nidulans* towards *c*-type cytochromes. FEBS Lett. 131, 11–16.

Peschek, G.A., Schmetterer, G., Lockau, W. and Sleytr, U.B. (1981b) Photosynthesis and respiration in blue-green algae: electron-transporting functions of thylakoids and plasmamembrane, and occurrence of cytochrome aa_3 in *Anacystis nidulans*. In G. Akoyunoglou (ed.), Photosynthesis, V: Chloroplast Development. Balaban, Philadelphia pp. 707–719.

Peschek, G.A., Schmetterer, G. and Sleytr, U.B. (1981c) Possible respiratory sites in the plasma membrane of *Anacystis nidulans*: ultracytochemical evidence. FEMS Microbiol. Lett. 11, 121–124.

Peschek, G.A., Muchl, R. and Schmetterer, G. (1981d) Temperature dependence of horse heart cytochrome *c* oxidation by membrane preparations of *Anacystis nidulans*: characterization of two distinct Arrhenius discontinuities, Curr. Microbiol. 6, 233–237.

Peschek, G.A., Schmetterer, G., Lauritsch, G., Nitschmann, W.H., Kienzl, P.F. and Muchl, R. (1982a) Do cyanobacteria contain 'mammalian-type' cytochrome oxidase? Arch. Microbiol. 131, 261–265.

Peschek, G.A., Schmetterer, G., Lauritsch, G., Muchl, R., Kienzl, P.F. and Nitschmann, W.H. (1982b) Proton-pumping cytochrome oxidase in the cytoplasmic membrane of *Anacystis nidulans*. In G.C. Papageorgiou and L. Packer (eds.), Photosynthetic Procaryotes: Cell Differentiation and Function. Elsevier/North-Holland, New York, pp. 147–162.

Peschek, G.A., Schmetterer, G. and Wagesreiter, H. (1982c) Oxidation of exogenous c-type cytochromes by intact spheroplasts of *Anacystis nidulans*. Arch. Microbiol. 133, 222–224.

Peschek, G.A., Muchl, R., Kienzl, P.F. and Schmetterer, G. (1982d) Characteristic temperature dependence of respiratory and photosynthetic electron transport activities in membrane preparations from *Anacystis nidulans*. Biochim. Biophys. Acta 697, 35–43.

Peschek, G.A., Schmetterer, G., Muchl, R., Nitschmann, W.H. and Riedler, M. (1984) Topography and function of cytochrome oxidase in a cyanobacterium. In C. Sybesma (ed.), Advances in Photosynthesis Research, Vol. III. Martinus Nijhoff/Dr. W. Junk Publ., The Hague, The Netherlands, pp. 335–342.

Peschek, G.A., Czerny, T., Schmetterer, G. and Nitschmann, W.H. (1985a) Transmembrane proton electrochemical gradients in dark aerobic and anaerobic cells of the cyanobacterium (blue-green alga) *Anacystis nidulans*. Plant Physiol. 79, 278–284.

Peschek, G.A., Nitschmann, W.H. and Springer-Lederer, H. (1985b) Respiration-linked proton extrusion from the unicellular cyanobacterium *Anacystis nidulans*. Naturwissenschaften 72, 273–274.

Peschek, G.A., Hinterstoisser, B., Riedler, M., Muchl, R. and Nitschmann, W.H. (1986) Exogenous energy supply to the plasma membrane of dark anaerobic cyanobacterium *Anacystis nidulans*. Arch. Biochem. Biophys., 247, 40–48.

Poole, R.K. (1983) Bacterial cytochrome oxidases. A structurally and functionally diverse group of electron-transfer proteins. Biochim. Biophys. Acta 726, 205-243.

Prochaska, L.J., Bisson, R., Capaldi, R.A., Steffens, G.C.M. and Buse, G. (1981) Inhibition of cytochrome *c* oxidase function by dicyclohexylcarbodiimide. Biochim. Biophys. Acta 637, 360–373.

Pulich, W.Jr. (1977) Cytochrome c_{548} in *Nostoc* sp. (Cyanophyceae): an electron acceptor from reduced NADP in the dark. J. Phycol. 13, 40–45.

Raboy, B. and Padan, E. (1978) Active transport of glucose and α-methyl-glucoside in the cyanobacterium *Plectonema boryanum*. J. Biol. Chem. 253, 3287–3291.

Raboy, B., Padan, E. and Shilo, M. (1976) Heterotrophic capacities of *Plectonema boryanum*. Arch. Microbiol. 110, 77–85.

Ragan, C.I. (1976) NADH-ubiquinone oxidoreductase. Biochim. Biophys. Acta 456, 249–290.

Reed, R.H., Rowell, P. and Stewart, W.D.P. (1980) Components of the proton electrochemical potential gradient in *Anabaena variabilis*. Biochem. Soc. Trans. 8, 707–708.

Reed, R.H., Rowell, P. and Stewart, W.D.P. (1981) Characterization of the transport of potassium ions in the cyanobacterium *Anabaena variabilis* Kütz. Eur. J. Biochem. 116, 323–330.

Rieske, J.S., MacLennan, D.H. and Coleman, R. (1964) Isolation and properties of an iron-protein from the (reduced coenzyme Q)-cytochrome *c* reductase complex of the respiratory chain. Biochem. Biophys. Res. Commun. 15, 338–344.

Rippka, R. (1972) Photoheterotrophy and chemoheterotrophy among unicellular blue-green algae. Arch. Mikrobiol. 87, 93–98.

Rippka, R., Waterbury, J. and Cohen-Bazire, G. (1974) A cyanobacterium which lacks thylakoids, Arch. Mikrobiol. 100, 419–436.

Rippka, R., Deruelles, J., Waterbury, J.B., Herdman, M. and Stanier, R.Y. (1979) Generic assignments, strain histories and properties of pure cultures of cyanobacteria. J. Gen. Microbiol. 31, 225–274.

Rittenberg, S.C. (1969) The roles of exogenous matter in the physiology of chemolithotrophic bacteria. Adv. Microbiol. Physiol. 3, 159–196.

Rittenberg, S.C. (1972) The obligate autotroph – the demise of a concept. Antonie van Leeuwenhoek J. Microbiol. Serol. 38, 457–478.

Rottenberg, H. (1979) The measurement of membrane potential and pH in cells, organelles and vesicles, Methods Enzymol. 55, 547–569.

Rowell, P., Diez, J., Apte, S.K. and Stewart, W.D.P. (1981) Molecular heterogeneity of ferredoxin:$NADP^+$ oxidoreductase from the cyanobacterium *Anabaena cylindrica*. Biochim. Biophys. Acta 657, 507–516.

Sandmann, G. (1985) Consequences of iron deficiency on photosynthetic and respiratory electron transport in blue-green algae. Photosyn. Res. 6, 261–271.

Sandmann, G. and Böger, P. (1980) Copper-induced exchange of plastocyanin and cytochrome *c*-553 in cultures of *Anabaena variabilis* and *Plectonema boryanum*. Plant Sci. Lett. 17, 417–424.

Sandmann, G. and Böger, P. (1981) Plastocyanin and cytochrome *c*-553, two different electron donors to photosystem I in algae. In G. Akoyunoglou (ed.), Photosynthesis, II. Electron Transport and Photophosphorylation. Balaban International Science Services, Philadelphia, pp. 623–632.

Sandmann, G. and Malkin, R. (1983a) NADH and NADPH as electron donors to respiratory and photosynthetic electron transport in the blue-green alga, *Aphanocapsa*. Biochim. Biophys. Acta 725, 221–224.

Sandmann, G. and Malkin, R. (1983b) NADH and NADPH dehydrogenases from the blue-green alga, *Aphanocapsa*. Arch. Microbiol. 136, 49–53.

Sandmann, G. and Malkin, R. (1984) Light inhibition of respiration is due to a dual function of the cytochrome b_6-*f* complex and the plastocyanin/cytochrome *c*-553 pool in *Aphanocapsa*. Arch. Biochem. Biophys. 234, 105–111.

Schaeffer, F. and Stanier, R.Y. (1978) Glucose-6-phosphate dehydrogenase of *Anabaena* sp. Arch. Microbiol. 116, 9–19.

Scherer, S. and Böger, P. (1982) Respiration of blue-green algae in the light. Arch. Microbiol. 132, 329–332.

Scherer, S. and Böger, P. (1984) Vanadate-sensitive proton efflux by filamentous cyanobacteria. FEMS Microbiol. Letters 22, 215–218.

Scherer, S. and Böger, P. (1985) Fructose-stimulated ferricyanide reduction by intact blue-green algae. Z. Naturforsch. 40c, 138–141.

Scherer, S., Stürzl, E. and Böger, P. (1981) Arrhenius plots indicate localization of photosynthetic and respiratory electron transport in different membrane regions of *Anabaena*. Z. Naturforsch. 36c, 1036–1040.

Scherer, S., Stürzl, E. and Böger, P. (1982) Interaction of respiratory and photosynthetic electron transport in *Anabaena variabilis* Kütz, Arch. Microbiol. 132, 333–337.

Scherer, S., Stürzl, E. and Böger, P. (1984) Oxygen-dependent proton efflux in cyanobacteria (blue-green algae). J. Bacteriol. 158, 609–614.

Schlegel, H.G. (1981) Allgemeine Mikrobiologie, 5. Auflage, Georg Thieme Verlag, Stuttgart-New York.

Schmetterer, G. (1982) Formation of hydrocarbons by photobleaching cyanobacterium, *Anacystis nidulans*. Z. Naturforsch. 37c, 205–209.

Schmetterer, G. and Peschek, G.A. (1981) Treatments effecting reversible photo-bleaching and thylakoid degradation in the blue-green alga *Anacystis nidulans*. Biochem. Physiol. Pflanz. 176, 90–100.

Schmetterer, G., Peschek, G.A. and Sleytr, U.B. (1983) Thylakoid degradation during photooxidative bleaching of the cyanobacterium *Anacystis nidulans*. Protoplasma 115, 202–207.

Scholes, P., Mitchell, P. and Moyle, J. (1969) The polarity of proton translocation in some photosynthetic micro-organisms. Eur. J. Biochem. 8, 450–454.

Schrautemeier, B. and Böhme, H. (1984) Different functions assigned to NAD(H) and NADP(H) in light-dependent nitrogen fixation by heterocysts of *Anabaena variabilis*. FEMS Microbiol. Letters 25, 215–218.

Serrano, A., Rivas, J. and Losada, M. (1984) Studies on the one- and two-electron FAD-mediated reactions by ferredoxin:$NADP^+$ oxidoreductase from *Anabaena*. FEBS Lett. 170, 85–88.

Smith, A.J. (1982) Modes of cyanobacterial carbon metabolism. In N.G. Carr and B.A. Whitton (eds.), The Biology of Cyanobacteria, 2nd. ed. Blackwell, Oxford, pp. 47–85.

Smith, A.J. and Hoare, D.S. (1977) Specialist phototrophs, litotrophs, and methylotrophs: a unity among a diversity of procaryotes? Bacteriol. Rev. 41, 419–448.

Smith, A.J., London, J. and Stanier, R.Y. (1967) Biochemical basis of obligate autotrophy in blue-green algae and thiobacilli. J. Bacteriol. 94, 972–983.

Solomos, T. (1977) Cyanide-resistant respiration in higher plants. Annu. Rev. Plant Physiol. 28, 279–297.

Spiegelmann, M.H., Rasched, I.R., Kunert, K., Kroneck, P. and Böger, P. (1976) Plastocyanin: possible significance of quaternary structure. Eur. J. Biochem. 64, 131–140.

Stanier, R.Y. (1973) Autotrophy and heterotrophy in unicellular blue-green algae. In N.G. Carr and B.A. Whitton (eds.), The Biology of Blue-Green Algae, Blackwell, Oxford, pp. 501–518.

Stanier, R.Y. and Cohen-Bazire, G. (1977) Phototrophic prokaryotes: the cyanobacteria. Annu. Rev. Microbiol. 31, 225–274.

Stewart, W.D.P. (1980) Some aspects of structure and function in nitrogen-fixing cyanobacteria. Annu. Rev. Microbiol. 34, 497–536.

Stürzl, E., Scherer, S. and Böger, P. (1982) Reconstitution of electron transport by cytochrome *c*-553 in a cell-free system of *Nostoc muscorum*. Photosynthesis Res. 3, 191–201.

Stürzl, E., Scherer, S. and Böger, P. (1984) Interaction of respiratory and photosynthetic electron transport, and evidence for membrane-bound pyridine-nucleotide dehydrogenases in *Anabaena variabilis*. Physiol. Plant 60, 479–483.

Tabita, F.R. and Colletti, C. (1979) Carbon dioxide assimilation in cyanobacteria: regulation of ribulose-1,5-bisphosphate carboxylase. J. Bacteriol. 140, 452–458.

Tang, F.L.M. and Krogmann, D.W. (1972) Cytochrome oxidase activity in cell-free preparations from blue-green algae. Plant Physiol. 49, 264–266.

Thauer, R.K., Jungermann, K. and Decker, K. (1977) Energy conservation in chemotrophic anaerobic bacteria. Bacteriol. Rev. 41, 100–180.

Trnka, M. and Peschek, G.A. (1986) Immunological identification of aa_3-type cytochrome oxidase in membrane preparations of the cyanobacterium *Anacystis nidulans*. Biochem. Biophys. Res. Commun., 136, 235–241.

Trnka, M., Molitor, V., Kuntner, O. and Peschek, G.A. (1985) Immunological identification of aa_3-type cytochrome oxidase in isolated and purified plasma and thylakoid membranes from *Anacystis nidulans*. In H. Zuber (ed.), Abstracts of the Fifth International Symposium on Photosynthetic Prokaryotes. Grindelwald, Switzerland, p. 246.

Van Liere, L., Mur, L.R., Gibson, C.E. and Herdman, M. (1979) Growth and physiology of *Oscillatoria agardhii* Gomont cultivated in continuous culture with a light-dark cycle. Arch. Microbiol. 123, 315–318.

Van Walraven, H.S., Lubberding, H.J., Marvin, H.J.P. and Kraayenhof, R. (1983) Characterization of reconstituted ATPase complex proteoliposomes prepared from the thermophilic cyanobacterium *Synechococcus* 6716. Eur. J. Biochem. 137, 101–106.

Van Walraven, H.S., Marvin, H.J.P., Koppenaal, E. and Kraayenhof, R. (1984) Proton movements and electric potential generation in reconstituted ATPase proteoliposomes from the thermophilic cyanobacterium *Synechococcus* 6716. Eur. J. Biochem. 144, 555–561.

Van Walraven, H.S., Hagendorn, M.J.M., Krab, K., Haak, N.P. and Kraayenhof, R. (1985) ATP synthesis by ATPase proteoliposomes from the thermophilic cyanobacterium *Synechococcus* 6716 by ionophore-induced electric potentials and proton gradients. Biochim. Biophys. Acta 809, 236–244.

Verwer, W., Ververgaert, P.H.J.T., Leunissen-Bijvelt, J. and Verkleij, A.J. (1978) Particle aggregation in photosynthetic membranes of the blue-green alga *Anacystis nidulans*. Biochim. Biophys. Acta 504, 231–234.

Viljoen, C.C., Cloete, F. and Scott, W.E. (1985) Isolation and characterization of an NAD(P)H dehydrogenase from the cyanobacterium, *Microcystis aeruginosa*. Biochim. Biophys. Acta 827, 247–259.

Wada, H., Hirasawa, R., Omata, T. and Murata, N. (1984) The lipid phase of thylakoid and cytoplasmic membranes from the blue-green algae (cyanobacteria), *Anacystis nidulans* and *Anabaena variabilis*. Pl. Cell Physiol. 25, 907–911.

Wax, E. and Lockau, W. (1980) Stoichiometric photophosphorylation in thylakoids from the blue-green alga, *Anabaena variabilis*. Z. Naturforsch. 35c, 98–105.

Webster, G.C. and Frenkel, A.W. (1953) Some respiratory characteristics of the blue-green alga, *Anabaena*. Plant Physiol. 28, 63–69.

Whittenbury, R. and Kelly, D.P. (1977) Autotrophy: a conceptual phoenix. Symp. Soc. Gen. Microbiol. 27, 121–149.

Wikström, M. and Krab, K. (1979) Proton-pumping cytochrome *c* oxidase. Biochim. Biophys. Acta 549, 177–222.

Wolf, M., Binder, A. and Bachofen, R. (1981) Influence of growth temperature on structure, thermostability and kinetic properties of the ATPase coupling factor AF_1 of a thermophilic blue-green alga (cyanobacterium). Eur. J. Biochem. 118, 423–427.

Wolk, C.P. (1973) Physiology and cytological chemistry of blue-green algae. Bacteriol. Rev. 37, 32–101.
Wolk, C.P. and Shaffer, P.W. (1976) Heterotrophic micro- and macrocultures of a nitrogen-fixing cyanobacterium. Arch. Microbiol. 110, 145–147.
Wood, P.M. (1977) The role of c-type cytochromes in algal photosynthesis. Eur. J. Biochem. 72, 605–612.
Wood, P.M. (1983) Why do *c*-type cytochromes exist? FEBS Lett. 164, 223–226.
Yamanaka, T., Takenami, S. and Okunuki, R. (1969) Reactivity of *Anacystis nidulans* cytochrome *c*-554 with two cytochrome oxidases. Biochim. Biophys. Acta 180, 193–195.
Zeikus, J.G. (1977) The biology of methanogenic bacteria. Microbiol. Rev. 41, 514–541.

P. Fay and C. Van Baalen (Eds.) The Cyanobacteria

6

Assimilation of inorganic nitrogen

MIGUEL G. GUERRERO and CATALINA LARA

Departamento de Bioquímica, Facultad de Biología y C.S.I.C. Apdo. 1095, 41080-Sevilla, Spain

1. Introduction

Nitrogen accounts for as much as 8–10% of the total dry weight of exponentially growing cyanobacterial cells. Availability of nitrogen has been identified in many instances as the factor limiting productivity in natural habitats. The most abundant forms of nitrogen present, in both terrestrial and aquatic habitats, are the inorganic species dinitrogen (N_2), nitrate (NO_3^-) and ammonium (NH_4^+), in decreasing order of availability. In general, cyanobacteria can use nitrate or ammonium as the sole nitrogen source for growth, and many have the additional ability to fix N_2. Few data are available about the relative contribution of cyanobacteria to global nitrogen assimilation, though it might be very significant, especially in the oceans and littoral regions.

Under autotrophic conditions, the utilization of any form of inorganic nitrogen by cyanobacteria is strictly dependent on the availability of light and CO_2. The light dependence corresponds to the requirement for reductant and ATP in the assimilatory processes involved. The CO_2 requirement results from a set of interactions between carbon and nitrogen metabolism, some of which are regulatory in nature and determine that the assimilation of C and N are balanced. A preference in the utilization of inorganic nitrogen forms is observed when two or more are simultaneously available. This is a reflection of the effective control to which inorganic nitrogen assimilation, a major component of biosynthetic metabolism in cyanobacteria, is subject.

In this article we attempt to summarize and discuss some aspects of the biochemistry and physiology of inorganic nitrogen assimilation by cyanobacteria. Space limitations preclude a comprehensive coverage of all aspects of the processes under consideration. Consequently, this chapter will probably reflect our own bias towards nitrate metabolism. Those particularly interested in N_2 fixation are referred to Chapter 7 of this volume and to several recent reviews on the topic (Stewart, 1980a; Wolk, 1980, 1982; Bothe, 1982; Houchins, 1984).

2. Utilization of nitrate, dinitrogen and ammonium

For the present purpose, inorganic nitrogen utilization will be considered to comprise the set of events that allow the conversion of extracellular nitrogen into the α-amino group of an amino acid. In the utilization of any form of inorganic nitrogen (Fig. 1), a series of basic steps can be distinguished. First, the nitrogenous compound in the outer medium should enter the cell. Second, as only ammonium is the nitrogen form directly incorporated into amino acids, dinitrogen and nitrate must be reduced to ammonium. Finally, ammonium is incorporated as an α-amino group to carbon skeletons. These basic steps will be sequentially discussed in this section. Unless explicitly indicated, both ammonium ion and ammonia would collectively be referred to as ammonium throughout the text.

2.1. Entrance of inorganic nitrogen compounds

2.1.1. Nitrate uptake

Since cyanobacteria are devoid of organelles in which ions could be stored, the in vivo experimental separation of the entrance step from the subsequent metabolism of the ions is very difficult. In the particular case of nitrate, attempts to measure its intracellular accumulation have failed. Even the use of extremely sensitive techniques, involving the exposure of cells to ^{13}N-nitrate and high pressure liquid chromatographic analysis of the acid cell lysate after centrifugation through silicone oil (Thayer and Huffaker, 1982), have proven unsuccessful when applied to *Anabaena* species (Meeks et al., 1983). Most of the studies on the entrance of nitrate have, therefore, been performed by following the uptake of the ion from the external medium by intact cells.

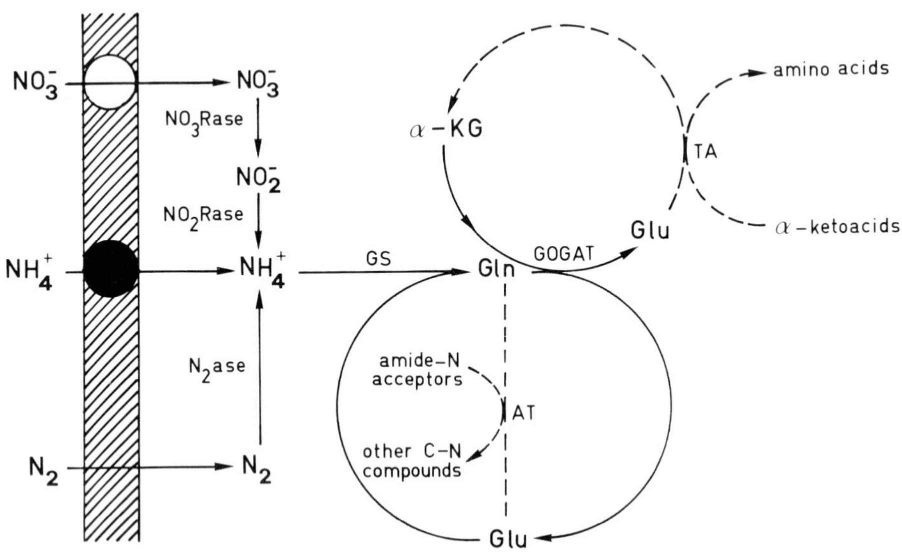

Fig. 1. Pathways of nitrate, ammonium and N_2 utilization in cyanobacteria. Dotted lines indicate reactions catalyzed by transaminases (TA) and amide transferases (AT), complementary to the glutamine synthetase (GS)-glutamate synthase (GOGAT) pathway. The hatched area on the left represents the cell membrane. Glu, glutamate; Gln, glutamine; α-KG, α-ketoglutarate; N_2ase, nitrogenase; NO_3Rase, ferredoxin-nitrate reductase; NO_2Rase, ferredoxin-nitrite reductase. See text for details.

Nitrate is a relatively voluminous anion and its passing across the polarized plasma-membrane by simple diffusion does not seem feasible on thermodynamic grounds, especially at the low concentrations (usually well below 1 mM) at which nitrate is found in natural environments. In fact, nitrate uptake by various cyanobacterial strains exhibits saturation kinetics with K_S values below 30 μM (Zevenboom et al., 1980; Flores et al., 1983a; Meeks et al., 1983; Tischner and Schmidt, 1984). The participation of a nitrate transport system in nitrate uptake by cyanobacteria appears, therefore, a plausible hypothesis, which is currently being investigated. Furthermore, some aspects of the regulation of nitrate utilization by ammonium and CO_2 (see below) are best explained assuming not only the existence of such a transport system, but also its active participation in controlling the rate of nitrate uptake. In this context, it should be stressed that nitrate transport into the cell is subject to interferences which might arise from the operation of control mechanisms, and special precaution must therefore be taken when selecting the experimental conditions to be used in studies of nitrate transport.

Indications of the nature of nitrate transport systems have been obtained for *Anacystis nidulans* cells (Flores et al., 1983a) treated with L-methionine-D,L-sulfoximine (MSX), an inhibitor of ammonium assimilation that selectively inactivates glutamine synthetase and releases nitrate uptake from the CO_2 requirement (Section 4 in this chapter). MSX-treated cells are able to take up nitrate and reduce it to ammonium, which is stoichiometrically released to the outer medium (Flores et al., 1980, 1983a; Ramos et al., 1982). This represents a useful system for studying the uptake and reduction of nitrate separately from the energy-requiring assimilation of the resulting ammonium. Both in normal and MSX-treated cells, nitrate uptake is sensitive to the uncoupler carbonyl cyanide *p*-trifluoromethoxyphenylhydrazone (FCCP) and to the ATPase inhibitor dicyclohexylcarbodiimide (DCCD) at concentrations that do not significantly affect the activities of either the nitrate-reducing enzymes or the photosynthetic electron flow. The involvement in nitrate uptake of an active nitrate transport system responsible for the sensitivity of the process to FCCP and DCCD has thus been inferred (Flores et al., 1983a). Sensitivity of nitrate uptake to uncouplers and ATPase inhibitors has also been reported for intact untreated cells of *Anabaena cylindrica* (Ohmori et al., 1977) and *Nostoc muscorum* (Rai et al., 1981), although in these cases the observed inhibition of nitrate uptake could also be ascribed to other side effects, such as inhibition of CO_2 fixation. It is worth mentioning that secondary active transport systems are thought to operate in nitrate uptake by the duckweed *Lemna* (Ullrich, 1983) and the acidophilic eukaryotic alga *Cyanidium caldarium* (Fuggi, 1985).

The uptake of nitrite by cyanobacteria shows similarity to that of nitrate with regard to the requirement for both light and CO_2, as well as in relation to the inhibition by ammonium (Flores, 1982). Nevertheless, a passive component corresponding to the diffusion of nitrous acid ($pK_a = 3.4$) also contributes to nitrite uptake, as indicated by the pH-dependent sensitivity of the process to uncouplers and ATPase inhibitors (Flores, 1982; Flores et al., 1983a). The situation seems to be similar in eukaryotic algae (Syrett, 1981; Ullrich, 1983). The possible involvement of a common active system in the transport of both nitrate and nitrite should be considered.

2.1.2. Diffusion of N_2

Molecular nitrogen, like other gases, can diffuse across biological membranes, and this seems to be the way by which it enters the cyanobacterial cell. It has been estimated that even in heterocysts, whose thick envelope may slow down the diffusion of gases, gas permeability is high enough to maintain maximum rate of N_2 fixation (Walsby, 1985). Actually, a method has been developed to measure cyanobacterial N_2 fixation by determining N_2 uptake using mass spectrometry (Jensen and Cox, 1985).

2.1.3. Ammonium uptake

Biological membranes are usually impermeable to the ammonium ion, but permeable to the unprotonated base ammonia ($pK_a = 9.3$). If ammonia were the species taken up by cyanobacteria, the uptake rate should be pH-dependent, and following its protonation within the cell, ammonium would accumulate, provided that internal pH is lower than the pH of the outer medium. Independence of the ammonium uptake rate on the external pH, in the range between 6.2 and 8.7, has been reported for *Anacystis* R2. At pH 7.2–7.6, these cells are able to accumulate ammonium at concentrations up to 1000-fold higher than those in the outer medium (Boussiba et al., 1984a). These results indicate the existence of an ammonium transport system.

As in many other microorganisms (see Kleiner, 1985), ammonium effectively inhibits methylammonium uptake in cyanobacteria, supporting the idea of ammonium being the natural substrate of the methylammonium uptake system. The characterization of methylammonium uptake could therefore provide a useful approach to the understanding of the ammonium transport system. The kinetics of methylammonium uptake in both *Anacystis* R2 (Boussiba et al., 1984b) and *Anabaena variabilis* (Rai et al., 1984) exhibit a biphasic pattern, with a rapid high-affinity phase, related to the entry of methylammonium, and a slower phase, associated with its metabolism via glutamine synthetase. Interestingly, only the slower phase is abolished by treatment of cells with the glutamine synthetase inhibitor, MSX, which affects only slightly the fast initial phase. Methylammonium accumulation is an energy-requiring process, sensitive to uncouplers and ATPase inhibitors. Inhibition by K^+ and triphenylmethylphosphonium indicates that the accumulation is dependent on the membrane potential and apparently independent of the pH gradient across the plasma-membrane (Boussiba et al., 1984b; Rai et al., 1984). It should be mentioned that prolonging the incubation of cells with MSX eventually results in complete suppression of methylammonium accumulation (Turpin et al., 1984; Boussiba and Gibson, 1985a), and this has been taken as evidence of a direct inhibitory action of MSX on the methylammonium (ammonium) transport system (Turpin et al., 1984). Nevertheless, the simplest explanation for the long-term effect of MSX on methylammonium uptake is that, under conditions of sustained glutamine synthetase inactivation, intracellular ammonium generated by protein degradation (Boussiba et al., 1984c) is the actual inhibitor of methylammonium transport (Boussiba and Gibson, 1985a).

Some differences appear to exist, however, between the uptake of ammonium and that of methylammonium. In *Anacystis* R2, the latter is repressed by growth on ammonium, requires Na^+ and is inhibited by K^+, whereas ammonium uptake is not affected by any of these factors (Boussiba et al., 1984b). These differences leave open the possibility that more than one transport system may contribute to ammonium uptake by the cyanobacterial cell. The ammonium-repressible system would transport either methylammonium or ammonium, whereas the other, highly specific for ammonium, would be constitutive (Boussiba et al., 1984b; Boussiba and Gibson, 1985b).

2.2 Assimilatory nitrate reduction

Prior to being transformed into organic nitrogen, intracellular nitrate must be converted into ammonium. This involves an 8-electron reduction of the nitrogen atom, which changes its formal charge of $+5$, as it is in nitrate, to -3, the oxidation level of nitrogen in ammonium and in organic nitrogenous compounds. Nitrate reduction in cyanobacteria occurs immediately after the ion has entered the cell, in such a way that the intracellular nitrate level in metabolically active cells remains very low (see Section 2.1.1). The assimilatory reduction of nitrate to ammonium in cyanobacteria, as in other organisms able to assimilate nitrate (Guerrero et al., 1981; Flores et al., 1983b), occurs in two sequential steps: (a) the 2-electron reduction of nitrate to nitrite, catalyzed by nitrate reductase; and (b) the 6-electron reduction of nitrite to ammonium, catalyzed by nitrite reductase (Fig. 1).

2.2.1. Reduction of nitrate to nitrite

The reaction catalyzed by cyanobacterial nitrate reductase can be written as:

$$NO_3^- + 2\ Fd_{red} + 2\ H^+ \rightarrow NO_2^- + 2\ Fd_{ox} + H_2O$$

This equation reflects present evidence for reduced ferredoxin being the physiological electron donor for nitrate reduction in a variety of cyanobacteria (Hattori and Myers, 1967; Manzano et al., 1976; Ortega et al., 1976; Candau, 1979, Flores et al., 1983b; Mikami and Ida, 1984), and the experimentally determined stoichiometry between ferredoxin oxidized and nitrate reduced (Candau, 1979; Mikami and Ida, 1984). The ferredoxin dependence of cyanobacterial nitrate reductase is in agreement with the situation proposed to occur in other prokaryotes, and is clearly distinct from that established for nitrate reductases from eukaryotic organisms in which NAD(P)H rather than ferredoxin represents the effective physiological electron donor for the reaction they catalyze (Vennesland and Guerrero, 1979; Guerrero et al., 1981). In this context, it is interesting to note that reports concerning the presence of NADH- and NADPH-dependent nitrate reductase activity in cyanobacterial extracts (Hattori and Myers, 1967; Candau, 1979) can be explained in terms of the additional presence of ferredoxin and ferredoxin-$NADP^+$ oxidoreductase or NADH-diaphorase in such preparations. Candau (1979) has actually shown that the ability of *Anacystis* extracts to reduce nitrate with NAD(P)H is lost following ion-exchange chromatography, with nitrate reductase activity now becoming absolutely dependent upon reduced ferredoxin.

The efficiency of ferredoxin to act as reductant for nitrate reductase in cyanobacteria is fairly well documented for those cases in which this redox protein is reduced with illuminated thylakoid preparations (Hattori and Myers, 1967; Hattori and Uesugi, 1968a; Candau et al., 1976; Manzano et al., 1976; Ortega et al., 1976), with NADPH plus ferredoxin-$NADP^+$ oxidoreductase (Hattori and Myers, 1967; Manzano et al., 1976; Ortega et al., 1976; Candau, 1979), or with illuminated 5-deazariboflavin in the presence of a suitable electron donor (Candau et al., 1980). Flavodoxin, a flavoprotein which acts as the physiological substitute for ferredoxin in a variety of metabolic reactions under conditions of iron starvation, effectively replaces ferredoxin in donating electrons to nitrate reductase of *Anacystis nidulans* when reduced photosynthetically (Candau, 1979; Peschek, 1979a).

The apparent inability of ferredoxin to act as an electron donor for the cyanobacterial enzyme when reduced with excess dithionite (Hattori and Myers, 1967; Hattori, 1970; Manzano et al., 1976, 1978; Candau, 1979; Ida and Mikami, 1983; Mikami and Ida, 1984), or, in some cases, even when enzymatically reduced with NADPH (Hattori, 1970), led to the proposal of the involvement of a labile electron carrier in the transfer of electrons between reduced ferredoxin and nitrate reductase (Hattori and Myers, 1967; Hattori, 1970). Nevertheless, chemically reduced ferredoxin has been shown to behave as efficient donor for the nitrate reductase of *Anacystis* and *Plectonema* (Candau, 1979; Mikami and Ida, 1984) under anaerobic conditions, provided that dithionite is present only in limiting amounts. Apparently, in the presence of an excess of this strong chemical reductant, ferredoxin and cyanobacterial nitrate reductase interact rather strongly forming a stable complex which is ineffective for nitrate reduction (Manzano et al., 1978; Candau, 1979). This behavior has facilitated purification of *Anacystis* nitrate reductase by redox-affinity chromatography on a ferredoxin-Sepharose gel (Manzano et al., 1978; Candau, 1979). It has also been suggested that treatment of ferredoxin-nitrate reductase with excess dithionite results in the conversion of the native enzyme into an inactive form (Mikami and Ida, 1984), but sufficient experimental evidence supporting this proposal is so far lacking.

Ferredoxin-nitrate reductase (ferredoxin:nitrate oxidoreductase) is an original nitrate-reducing enzyme that still awaits recognition by the Enzyme Commission (Nomenclature Committee of IUB, 1984). The enzyme from cyanobacteria has been purified to homogeneity and characterized as a molybdo-iron protein consisting of a single polypeptide chain of about 75000 daltons for *Anacystis nidulans* (Candau, 1979; Guerrero et al., 1981) and about 85000 daltons for *Plectonema boryanum* (Mikami and Ida, 1984). Nitrate reductase from *Plectonema* contains one atom of molybdenum, four non-heme iron and four acid-labile sulfur atoms per enzyme molecule (Mikami and Ida, 1984). While no definite assignment for the type of iron-sulfur cluster has been made, an arrangement of two [2Fe-2S] clusters appears plausible on the basis of circular dichroism and magnetic circular dichroism spectra (Mikami and Ida, 1984). Ferredoxin-nitrate reductase is thus clearly distinct from the eukaryotic pyridine nucleotide-dependent nitrate reductase (NADH: and NAD(P)H:nitrate oxidoreductase, EC 1.6.6.1 and 1.6.6.2), a multimeric enzyme composed of at least two identical subunits of about 100000 daltons, containing FAD and a heme group of the *b*-557 type, in addition to molybdenum (Guerrero et al., 1981; Kleinhofs et al., 1985).

Ferredoxin-nitrate reductase from several cyanobacteria appears to be tightly attached to thylakoid membranes, prolonged sonication or treatment with detergents being required to release the enzyme in a soluble form (Hattori and Myers, 1967; Manzano et al., 1976; Ortega et al., 1976; Ida and Mikami, 1983; Mikami and Ida, 1984). In heterocystous N_2-fixing cyanobacteria, nitrate reductase appears to be confined to the vegetative cells, with no or negligible activity present in extracts of purified heterocyst preparations (Ramos and Guerrero, 1983).

2.2.2. *Reduction of nitrite to ammonium*

In cyanobacteria, as in photosynthetic eukaryotes, nitrite reductase depends on reduced ferredoxin (or flavodoxin) as electron donor (Hattori and Uesugi, 1968b; Manzano et al., 1976; Manzano, 1977; Peschek, 1979a; Vega et al., 1980; Guerrero et al., 1981; Méndez and Vega, 1981; Méndez et al., 1981), and catalyzes the following reaction:

$$NO_2^- + 6\ Fd_{red} + 8\ H^+ \rightarrow NH_4^+ + 6\ Fd_{ox} + 2\ H_2O$$

Ferredoxin-nitrite reductase (ferredoxin:nitrite oxidoreductase, EC 1.7.7.1) from the unicellular *Anacystis nidulans* and from the filamentous N_2-fixing *Anabaena* 7119 and *Anabaena cylindrica* (Hattori and Uesugi, 1968b; Manzano et al., 1976; Manzano, 1977; Méndez and Vega, 1981; Méndez et al., 1981) has been described as an iron protein consisting of a single polypeptide chain of 52000–68000 daltons. Mainly on the basis of spectrophotometric studies of partially purified preparations, it has been suggested that *Anabaena* nitrite reductase is a heme-protein (Méndez and Vega, 1981). It remains to be established whether siroheme and/or iron-sulfur cluster(s) are the actual prosthetic groups of cyanobacterial nitrite reductase (Méndez and Vega, 1981), as is the case for the enzyme in photosynthetic eukaryotes (Vega et al., 1980; Guerrero et al., 1981).

Ferredoxin-nitrite reductase is tightly attached to thylakoid membranes in *Anacystis* (Candau et al., 1976; Manzano et al., 1976; Manzano, 1977), but only loosely bound to membranes in *Anabaena* and *Nostoc* (Hattori and Myers, 1966; Hattori and Uesugi, 1968b; Ortega et al., 1976; Méndez and Vega, 1981).

2.3. N_2 FIXATION

The ability to fix N_2 is not a general feature among cyanobacteria. Heterocystous and a few non-heterocystous species appear able to assimilate N_2 under aerobic conditions. Certain other strains fix N_2 under anaerobic or microaerobic conditions only (Stewart, 1978, 1980a; Gallon, 1980;

Bothe, 1982). The basis of these differences, intimately related to the O_2-sensitivity of nitrogenase, will not be discussed here, as it is treated in detail in Section 6 of Chapter 7 of this volume. Therefore, just an overview of general aspects of cyanobacterial N_2 fixation will be presented in this section.

Intracellular N_2 must be reduced to ammonium as a previous step to its conversion into organic nitrogen (Fig. 1). In this process, six electrons and a high but still uncertain number of energy-rich phosphate-bonds are needed for the reductive splitting of the N_2 molecule into two molecules of ammonium. The reaction is catalyzed by the enzyme-complex nitrogenase which, as a general feature of this catalyst (Lowe et al., 1985), reduces simultaneously N_2 and protons, according to the equation:

$$N_2 + 10\ H^+ + 8\ Fd_{red} \xrightarrow{>12\ ATP} 2\ NH_4^+ + H_2 + 8\ Fd_{ox}$$

Present evidence shows that reduced ferredoxin is the physiological electron donor to nitrogenase (Smith et al., 1971; Houchins and Hind, 1982; Schrautemeier et al., 1984, 1985; Schrautemeier and Böhme, 1985). Under conditions of iron-deficiency, flavodoxin could substitute for ferredoxin in its role as electron donor to nitrogenase (Bothe, 1982; Houchins, 1984). The major source of ATP under phototrophic conditions is photophosphorylation, whereas ATP for dark N_2 fixation is provided by respiration (Bothe, 1982; Houchins, 1984; Ohmori, 1984).

Nitrogenase (EC 1.18.6.1) from cyanobacteria is analogous to the enzyme complex from other N_2-fixing organisms (Bothe, 1982; Houchins, 1984; Lowe et al., 1985). It consists of two redox proteins, a tetrameric Fe-Mo-containing protein and a smaller dimeric Fe-containing protein. The Fe-Mo component isolated from *Anabaena cylindrica* has an estimated molecular weight of 223000, being composed of subunits (α_2,β_2) of 55000 and 53000 daltons and containing about 2 Mo, 20 non-heme iron and 20 acid-labile sulfur atoms per molecule of component. The Fe-component has a molecular weight of about 60000 and is composed of two apparently identical subunits of 30000 daltons (Hallenbeck et al., 1979). The flow of electrons through nitrogenase is from ferredoxin to the Fe-protein, and then to the Fe-Mo protein by a mechanism coupled to ATP hydrolysis. The Fe-Mo component contains the binding site for N_2 and protons, being therefore the protein that *sensu stricto* catalyzes N_2 reduction (Lowe et al., 1985).

In heterocystous cyanobacteria, nitrogenase is confined to the heterocysts (Wolk, 1982). Immunochemical studies with *Anabaena cylindrica* have shown that, even under strict anaerobic conditions, the Fe-Mo component of nitrogenase is located exclusively in the heterocysts, indicating that even under environmental conditions that enable nitrogenase to function, the vegetative cells do not express the *nif* genes (Murry et al., 1984; see also Golden et al., 1985).

2.4. Ammonium assimilation

Ammonium taken up from the outer medium or internally generated by either nitrate reduction or dinitrogen fixation is assimilated through its combination with carbon skeletons. In parallel to this primary assimilation, the ammonium resulting from the photorespiratory conversion of glycine to serine (Codd and Stewart, 1973; Bergman, 1984) or from protein degradation followed by deamination and deamidation of amino acids (Boussiba et al., 1984c), is also reassimilated.

2.4.1. Pathways

Available evidence indicates that for a wide range of cyanobacteria, the glutamine synthetase(GS)-glutamate synthase (GOGAT) enzyme system is the quantitatively most important pathway for the assimilation of ammonium, either exogenously supplied or internally generated. In this route (see Fig. 1), ammonium-N is initially incorporated into the amide position of glutamine, from where it is transferred to the α-position of α-ketoglutarate. One ATP molecule and two reducing equivalents are required for the net formation of one glutamate unit from ammonium and α-ketoglutarate (Stewart, 1980a; Wolk, 1980; Meeks, 1981; Lea et al., 1982; Flores et al., 1983b). Evidence supporting this includes: (a) ^{13}N (Thomas et al., 1975, 1977; Wolk et al., 1976; Meeks et al., 1977, 1978, 1985; Meeks, 1981) and ^{15}N (Stewart et al., 1975) tracer studies; (b) use of glutamate analogs, inhibitors of either enzyme of the pathway, such as MSX, azaserine, 5-hydroxylysine, 6-diazo-5-oxo-norleucine, and phosphinotricine (Stewart and Rowell, 1975; Wolk et al., 1976; Meeks et al., 1977, 1978; Thomas et al., 1977; Flores et al., 1980, 1983b; Ramos et al., 1982; Bergman, 1984; Boussiba et al., 1984c; Lea et al., 1984); and (c) detection of the enzymes and evaluation of their cellular activity level (Haystead et al., 1973; Stewart et al., 1975; Stewart, 1980b).

The net glutamate unit resulting from ammonium assimilation through the concerted participation of GS and GOGAT can donate its amino group to a variety of carbon skeletons in reactions catalyzed by transaminases. Some of the glutamine synthesized by the action of GS does not transfer its amide group to α-ketoglutarate, but to various substrates of glutamine-amide transferases other than GOGAT (Fig. 1).

In addition to the GS-GOGAT system, other enzymes like glutamate dehydrogenase and alanine dehydrogenase may be active in ammonium assimilation in cyanobacteria (Meeks et al., 1977; Stewart, 1980b; Lea et al., 1982; Meeks, 1981; Flores et al., 1983b). Glutamate dehydrogenase activity was detectable at low levels in some cyanobacteria but not in most strains assayed (Stewart, 1980b). There is almost general agreement that this enzyme plays a negligible role (if any) in primary ammonium assimilation in most cyanobacteria. Though glutamate dehydrogenase has been reported as the main ammonium assimilating pathway in *Anacystis nidulans* strain UTEX 625 when grown on nitrate (Meeks et al., 1977; Meeks, 1981) GS-GOGAT appears to be the main route in nitrate-grown *Anacystis nidulans* strain 1402-1 (Flores et al., 1983b). Attempts to detect glutamate dehydrogenase activity in some strains of *Anacystis* (Hoare et al., 1967; Flores et al., 1983b) have been unsuccessful. The involvement of alanine dehydrogenase in the primary assimilation of ammonium by cyanobacteria remains a matter of discussion (Stewart, 1980b). Results from ^{13}N tracer studies (Meeks et al., 1977; Thomas et al., 1977; Meeks, 1981) point to its active participation in *Cylindrospermun licheniforme*, *Anabaena cylindrica*, *Plectonema boryanum* and *A. nidulans* strain UTEX 625, but not in *Anabaena variabilis*. Carbamyl phosphate synthetase does not seem to significantly contribute to ammonium assimilation by cyanobacteria (Meeks et al., 1977).

2.4.2. *Enzymes*

Glutamine synthetase (L-glutamate:ammonia ligase (ADP forming), EC 6.3.1.2) catalyzes the following reaction:

$$\text{L-glutamate} + NH_4^+ + \text{ATP} \xrightarrow{Mg^{2+}} \text{L-glutamine} + \text{ADP} + P_i$$

The enzyme from both filamentous N_2-fixing (Stacey et al., 1977; Sampaio et al., 1979; Orr et al., 1981) and unicellular non N_2-fixing (Florencio and Ramos, 1985) cyanobacteria has been purified to homogeneity and characterized. Cyanobacterial glutamine synthetase is structurally similar to the enzyme isolated from chemotrophic bacteria, has a molecular weight in the order of 600000 and is composed of 12 identical subunits of about 50000 daltons, arranged in two superimposed hexagonal rings (Stacey et al., 1977; Sampaio et al., 1979; Haselkorn et al., 1980; Orr et al., 1981).

In heterocystous N_2-fixing cyanobacteria, glutamine synthetase is found in both vegetative cells and heterocysts (Stewart et al., 1985), although the amount of enzyme per cell seems to be about two-fold higher in the heterocysts (Dharmawardene et al., 1973; Haselkorn et al., 1980). In *Anacystis nidulans*, glutamine synthetase appears to be a membrane-bound enzyme (Florencio and Ramos, 1985).

Glutamate synthase (L-glutamine:2-oxoglutarate amino-transferase (GOGAT), or L-glutamine: ferredoxin oxidoreductase (transaminating), EC 1.4.7.1) catalyzes the transfer of the amide group of glutamine to α-ketoglutarate, yielding glutamate, in a reaction which in cyanobacteria requires reduced ferredoxin as the electron donor (Lea and Miflin, 1975):

$$\alpha\text{-ketoglutarate} + \text{L-glutamine} + 2\ Fd_{red} + 2\ H^+ \rightarrow 2\ \text{L-glutamate} + 2\ Fd_{ox}$$

In higher plants and algae, two different glutamate synthases, one ferredoxin-dependent and another NADH-dependent, are present, whereas only a NAD(P)H-glutamate synthase seems to exist in chemotrophic bacteria (Lea et al., 1982). Cyanobacterial GOGAT may therefore be more closely related to the Fd-dependent enzyme of photosynthetic eukaryotes than to GOGAT from other prokaryotes. Since, however, no data are available regarding the properties of GOGAT isolated from cyanobacteria, direct comparisons cannot be made at present.

Isolated heterocysts have been shown to metabolize $[^{13}N]N_2$ and $^{13}NH_4^+$ essentially only to glutamine, and first localization studies indicated the apparent confinement of GOGAT activity in the vegetative cells of heterocystous cyanobacteria (Thomas et al., 1977). This would imply that glutamine formed in heterocysts from N_2-derived ammonium must move into vegetative cells in order to undergo amide transfer, and that most of the glutamate from which glutamine is synthesized is ultimately derived from vegetative cells (Thomas et al., 1977; Meeks, 1981; Wolk, 1980, 1982). There are some recent reports, however, suggesting that the enzyme can also be present in heterocysts (Gupta and Carr, 1981; Häger et al., 1983).

Glutamate dehydrogenase (L-glutamate:NAD(P)$^+$ oxidoreductase (deaminating), EC 1.4.1.3) catalyzes the reductive amination of α-ketoglutarate to glutamate with ammonium as the substrate and NAD(P)H as the electron donor:

$$\alpha\text{-ketoglutarate} + NH_4^+ + NAD(P)H + H^+ \rightarrow \text{L-glutamate} + NAD(P)^+ + H_2O$$

The equilibrium constant for this reaction is close to one, the reaction in the deaminating direction being thus equally possible. Low but consistent NADPH-dependent glutamate dehydrogenase activity has been found in *Anabaena variabilis* (Stewart et al., 1975), but to the best of our knowledge no report concerning the characterization of homogenous glutamate dehydrogenase from cyanobacteria is available.

Alanine dehydrogenase (L-alanine:NAD$^+$ oxidoreductase (deaminating), EC 1.4.1.1) catalyzes an easily reversible reaction, analogous to that catalyzed by glutamate dehydrogenase, but with pyruvate as the ammonium receptor:

$$\text{pyruvate} + NH_4^+ + NADH + H^+ \rightarrow \text{L-alanine} + NAD^+ + H_2O$$

Its presence in cyanobacteria is well documented (Stewart, 1980b; Lea et al., 1982; Flores et al., 1983b). The enzyme from *Anabaena cylindrica* has been characterized as an oligomer with a molecular weight of 270 000 daltons composed of six identical 43 000 dalton subunits. The enzyme uses NAD(H) but not NADP(H) in both the aminating and deaminating directions. While oxaloacetate can partially substitute for pyruvate, L-aspartate is not a substrate for the deaminating activity (Rowell and Stewart, 1976).

2.5. Photosynthetic nature of inorganic nitrogen assimilation

The terms photosynthesis and CO_2 fixation are commonly used interchangeably. Nevertheless, ATP and reduced ferredoxin generated in photosynthetic reactions are utilized for a number of assimilatory processes other than CO_2 fixation. This applies to many of the reactions involved in the assimilation of inorganic nitrogen by cyanobacteria in the light. A direct utilization of photosynthetically generated reducing power appears evident for both nitrate reduction and ammonium assimilation, which involve reductive reactions catalyzed by the ferredoxin-dependent enzymes nitrate reductase, nitrite reductase and GOGAT. The ATP required in the uptake of these nitrogen sources and for the GS reaction can be provided by photophosphorylation (Flores et al., 1983b). This appears also to be the ultimate source of the large amounts of ATP required in N_2-fixation, while supply of the immediate reductant for nitrogenase appears to be only indirectly linked to water photolysis. This is particularly apparent in the case of heterocystous cyanobacteria, in which N_2-fixation in the heterocysts depends upon photosynthate provided by the vegetative cells (for further information see Section 3 of Chapter 7).

Although nitrogen assimilation may appear quantitatively of minor significance in comparison to CO_2 fixation, this is in fact not the case. While the assimilation of a carbon dioxide unit to the level of carbohydrate requires four electrons, the generation of an ammonium unit from nitrate needs eight electrons (less if derived from N_2 fixation, but the ATP requirement is higher; Section 2.3), and its further conversion into amino-N requires two more electrons. Since the C/N ratio in cyanobacteria is around five, on an average and net basis, the proportion of electrons coming from water photolysis and channeled to the assimilation of nitrogen is as much as one-half of that required for the assimilation of carbon (Flores et al., 1983b). It follows that inorganic nitrogen assimilation actually represents the second major process in the photosynthetic metabolism of cyanobacteria.

The photosynthetic nature of nitrate reduction merits special attention, since this process has traditionally been considered as indirectly dependent upon light, with recent products of CO_2 fixation acting as obligate intermediates between photosynthetically-generated reductant and the reduction of nitrate (see Fogg et al., 1973; Losada et al., 1981; Syrett, 1981). This proposal would be hardly compatible with the ferredoxin-dependent character of nitrate and nitrite reductases. Present evidence shows, moreover, that nitrate reduction in cyanobacteria is a genuine photosynthetic process, which directly uses photosynthetically generated reductant (Candau et al., 1976; Flores et al., 1983a, b; see also Section 5.3.1. of Chapter 13). The coupling of nitrate reduction with the photolysis of water was conclusively shown in thylakoid-containing fractions of *Anabaena* 7119 exhibiting nitrate reductase activity (Ortega et al., 1976) and of *Anacystis nidulans* with associated nitrate and nitrite reductases (Candau et al., 1976). More recently, intact *Anacystis nidulans* cells treated with MSX have been used for the *in vivo* analysis of the photosynthetic character of nitrate reduction, without the interferences that might arise from the subsequent assimilation of the resulting ammonium (Flores et al., 1983a, b). The results indicate a coupling between water photolysis and nitrate reduction, without the need for carbon metabolites derived from CO_2 fixation. Stoichiometry values of two mol oxygen evolved per mol nitrate reduced to ammonium (1.5 mol oxygen per mol nitrite) have been experimentally determined for MSX-treated *Anacystis*, in agreement with the expected values if the required reducing equivalents were provided via water photolysis, according to the equation (Flores et al., 1983a, b):

$$NO_3^- + 4\,H_2O + 2\,H^+ \rightarrow NH_4^+ + 3\,H_2O + 2\,O_2$$

Further support for the existence of a close relationship between the photochemical reactions of photosynthesis and nitrate reduction in cyanobacteria is the nitrate- and nitrite-induced quenching of chlorophyll fluorescence demonstrated with cells of *Anabaena* 7119 and *Nostoc* 6719 (Serrano et al., 1981, 1982). Photoreduction of nitrate to nitrite by a mutant strain of *Agmenellum quadruplicatum*, defective in nitrite reduction, was shown to proceed unimpaired for several hours in the presence of DCMU, suggesting independence of nitrate reduction from PS II activity (Stevens and Van Baalen, 1973). The cells used in these studies were extremely N-starved and rich in carbohydrates, which would easily account for the existence of a pool of reductant able to provide electrons to PSI, thus allowing photoreduction of nitrate by means of anoxygenic photosynthesis. Indeed, light-dependent reduction of nitrite in anoxygenic photosynthesis with H_2 as the reductant, has been achieved with whole cells of *Anabaena cylindrica* (Hattori, 1963) and *Anacystis nidulans* (Peschek, 1979b). Particulate cell-free fractions from these cyanobacterial strains can also photoreduce nitrate and/or nitrite when supplied with a suitable electron donor to PSI, such as ascorbate-reduced dichlorophenolindophenol (Hattori and Uesugi, 1968a; Guerrero et al., 1974; Manzano et al., 1976; Ortega et al., 1976) or H_2 (Peschek, 1979a, b).

3. Interactions in the utilization of various nitrogen forms

Usually, and provided that light and CO_2 are not limiting, pH is kept constant and the nitrogen source is present in adequate amounts, the rate of nitrogen assimilation, and also the growth rate, for a given cyanobacterium is much the same regardless of the nature of the nitrogen form supplied. This applies not only to nitrate and ammonium, but also to N_2 in the case of N_2-fixing strains (see e.g. Ohmori and Hattori, 1972; Fogg et al., 1973; Bottomley et al., 1979). Moreover, if more than one nitrogen source is available, the total rate of nitrogen assimilation would not increase. Although concurrent assimilation of two nitrogen sources can take place, at least transitorily, the general rule is that ammonium is utilized in preference to nitrate, the latter being in turn preferred to N_2. This behavior is in agreement with the logics of cell economy, since the energy requirement for the assimilation of these nitrogenous substrates increases in the order ammonium, nitrate, N_2, and it points to the existence of regulatory interactions between the assimilatory processes involved (nitrogen control).

3.1. Interactions in nitrate utilization

When a cyanobacterium is supplied with both nitrate and ammonium, nitrate is not utilized until ammonium has disappeared. This preferential assimilation of ammonium over nitrate is related to the regulation of nitrate assimilation, which is exerted at various levels. Thus, the uptake and reduction systems involved in nitrate utilization by cyanobacteria are fully expressed when nitrate represents the sole source of nitrogen, but their activities are suppressed, or at least decreased, when ammonium is present, either alone or in combination with nitrate. This behavior is common to both non-N_2 fixers and N_2-fixing strains. In the absence of combined nitrogen, non-N_2-fixing cyanobacteria exhibit significant levels of the functions needed for nitrate utilization, as though they were ready to use any nitrate which might reach the cell. The N_2-fixing strains, however, do not exhibit an active nitrate-utilization capacity when grown in the absence of combined nitrogen, but only when nitrate is available. There exist, therefore, different strategies according to the capability of the organism to use N_2 as a nitrogen source (Herrero et al., 1981, 1985; Flores et al., 1983b).

3.1.1. Control of nitrate uptake

When cell suspensions of cyanobacteria actively utilizing nitrate are supplemented with ammonium at concentrations as low as 10–50 μM, the ability of cells to take up nitrate is promptly and severely hampered. The inhibition caused by ammonium is reversible, the uptake of nitrate being resumed at the original rate as soon as the ammonium added is exhausted (Ohmori et al., 1977; Flores et al., 1980, 1983b). As nitrate reductase is not affected in the short-term, the ammonium effect has been ascribed to operate at the level of the nitrate transport system (Ohmori et al., 1977; Flores et al., 1980, 1983b), as in eukaryotic microalgae, which respond to ammonium in a similar fashion (Guerrero et al., 1981; Syrett, 1981; Ullrich, 1983).

The ability of ammonium to inhibit nitrate uptake requires its prior assimilation into organic nitrogen. This is substantiated by experiments with *Anacystis nidulans* using several glutamate analogs which inhibit either GS (MSX or 5-hydroxylysine) or GOGAT and other glutamine-amide transferases (azaserine) (Flores et al., 1980, 1983b). All these compounds, by preventing ammonium assimilation, obviate the ammonium inhibition of nitrate uptake (Table I). Although under normal conditions the effect of ammonium on nitrate uptake is vigorous and immediate, this is not always so, and it appears that the degree of the ammonium inhibition somehow correlates with the rate of its assimilation by the cells. Results in Table I also show that in the absence of added ammonium, some of the glutamate analogs effect an increase in the rate of nitrate uptake above that measured in the control untreated cells (see also Flores et al., 1980). This suggests that even in the absence of exogenous ammonium the rate of nitrate uptake is regulated by the ammonium produced during nitrate reduction (Flores et al., 1980, 1983b). Thus, the effect of ammonium addition upon nitrate uptake may be seen as an exaggeration of a feed-back control continuously modulating the nitrate uptake rate.

Some observations indicate that the activity and the development of the nitrate transport system is effected by the nitrogen source in N_2-fixing heterocystous cyanobacteria. N_2-grown *Anabaena var-*

TABLE I

Effect of different glutamate analogs on the uptake of nitrate and ammonium and on the ammonium-inhibition of nitrate uptake in *Anacystis nidulans* 1402-1

Compounds	Nitrate uptake (% of control)		Ammonium uptake (% of control)
	minus NH_4^+	*plus* NH_4^+	
None	100	0	100
MSX, 1 mM	102	101	2
5-Hydroxylysine, 2.5 mM	121	32	65
5-Hydroxylysine, 10 mM	153	121	3
Azaserine, 1 mM	146	145	5

Nitrate and ammonium uptake were determined as described in Romero et al. (1985a). Initial concentrations of KNO_3 and NH_4Cl were 0.25 mM and 0.50 mM, respectively. $NaHCO_3$ was present at 10 mM concentration. Prior to starting the experiments by addition of the corresponding nitrogenous substrate(s), the cells were preincubated for 15 min in the light in the absence or presence of the indicated glutamate analog. Rates of ammonium or nitrate uptake in control untreated cells (100%) were 33 ± 2 μmol·mg^{-1} (chl)·h^{-1}. MSX, L-methionine D,L-sulfoximine. We are grateful to José M. Romero for these data.

iabilis and *Nostoc muscorum*, though exhibiting moderate levels of nitrate reductase activity, cannot utilize nitrate immediately following their transfer to nitrate medium, but may acquire this ability after some time in a process that is not dependent on de novo protein synthesis (Bagchi and Singh, 1984; Avissar, 1985). It would appear that the nitrate transport system is latent in N_2-grown cells and becomes activated by nitrate. Ammonium-grown *Nostoc* cells also develop the capacity to take up nitrate some time after their transfer to a nitrate medium, but in this case the process is dependent on protein synthesis, indicating that the nitrate transport system is repressed by ammonium and/or induced by nitrate (Bagchi and Singh, 1984).

3.1.2. Control of the amount of active enzyme

Ammonium also exhibits a negative effect on the cellular activity levels of both nitrate reductase and nitrite reductase, acting as a nutritional repressor of the synthesis of these enzymes (Herrero et al., 1981, 1985; Herrero and Guerrero, 1986; Flores et al., 1983b). Ammonium-promoted repression of nitrate reductase, apparently at the transcriptional level, has been shown to operate in the unicellular non-N_2 fixing strains *Agmenellum quadruplicatum*, *Anacystis nidulans* and *Synechocystis* 6714 (Stevens and Van Baalen, 1974; Herrero et al., 1981, 1985; Flores et al., 1983b). Nitrate reductase synthesis in N_2-fixing cyanobacteria is also negatively affected by ammonium. In addition, and contrary to the situation with the non-N_2 fixers, nitrate acts as an inducer of nitrate reductase (Hattori, 1962; Ohmori and Hattori, 1970; Herrero et al., 1981, 1985; Flores et al., 1983b; Ramos and Guerrero, 1983). This requirement for nitrate precludes a definite conclusion about the actual role played by ammonium in the N_2-fixing strains, i.e. whether it is promoting repression of nitrate reductase synthesis as in the non-N_2 fixer strains, or causing inhibition of the entry of the inducer nitrate into the cells (Herrero et al., 1981, 1985; Flores et al., 1983b). There are differences among cyanobacterial strains with regard to the responsiveness of the nitrate assimilation system to the negative effects of ammonium, being very strong in *Anacystis nidulans*, *Anabaena* 33047 (*Anabaena* CA), *Anabaena variabilis* and *Anabaena* 7120, but weaker in *Anabaena* 7119, *Nostoc* 6719 and *Anabaena cylindrica* (Herrero et al., 1985). As in the case of inhibition of nitrate uptake, the repression of nitrate reductase synthesis by ammonium requires ammonium assimilation. This conclusion is based on the observation that MSX allows synthesis of nitrate reductase in the presence of ammonium (Herrero et al., 1981, 1985; Ramos et al., 1982; Flores et al., 1983b; Ramos and Guerrero, 1983).

Although in non-N_2 fixers nitrate is not required as inducer of nitrate reductase synthesis, it protects preformed enzyme from inactivation. This would explain the higher nitrate reductase levels found in cells incubated with nitrate than in the absence of any nitrogen source (Herrero et al., 1981, 1984, 1985; Flores et al., 1983b).

Available data concerning regulation of nitrite reductase in cyanobacteria are rather scanty. In *Anacystis nidulans* and *Anabaena* 7119, the nature of the nitrogen source effects nitrite reductase levels in a similar way as those of nitrate reductase (Méndez et al., 1981; Herrero and Guerrero, 1986). Repression of nitrite reductase synthesis by ammonium in *Anacystis* is also prevented by MSX, and the positive action of nitrate on the synthesis of nitrite reductase does not require its prior conversion into nitrite (Herrero and Guerrero, 1986). Nitrite reductase from *Anabaena cylindrica* appears to be induced however by nitrite but not by nitrate (Ohmori and Hattori, 1970).

3.2. Interactions in N_2 fixation

The process of N_2 fixation in cyanobacteria is also subject to regulation by the nature of the available nitrogen source. This nitrogen control has an additional feature in heterocystous strains, which differentiate a fraction of their cells into heterocysts in response to N-starvation. The subject of heterocyst differentiation will not be considered here in detail, and the reader is referred to comprehensive reviews on the topic (Fay, 1973; Haselkorn, 1978; Wolk, 1982) and to Chapter 7 of this volume. The aspect of nitrogen control addressed in this section deals with the effect of different forms of inorganic nitrogen on the nitrogenase level, and seems to be common to the generality of the N_2-fixing cyanobacteria, although most of the information available corresponds to heterocystous strains.

It is well known that N_2 is not required as an inducer for the development of N_2-fixing capacity in cyanobacteria, and that exogenous combined nitrogen, either nitrate or ammonium, inhibits N_2 fixation. This inhibition is generally thought to result from effects exerted at the level of protein synthesis, involving repression of nitrogenase or, at least, of some proteins of the N_2 fixation system (Fleming and Haselkorn, 1973; Ohmori and Hattori, 1974; Scherer et al., 1980; Bothe, 1982; Ramos et al., 1985; Stewart et al., 1985). Some authors, however, interpret these effects in terms of inhibition of nitrogenase activity (Murry and Benemann, 1979; Yoch and Gotto, 1982; Murry et al., 1983).

As is the general case for enzymes, the actual nitrogenase level in an N_2-fixing cyanobacterial cell should be viewed as resulting from the balance between the rate of nitrogenase synthesis and that of its inactivation. Under a given O_2 tension, e.g. that of air, the nitrogenase level stabilizes at different values depending on the absence or presence of N_2, nitrate or ammonium, indicating different rates of nitrogenase synthesis under the various conditions. Nitrogenase activity is particularly high in the absence of any nitrogen source (e.g. under argon), lower under N_2-fixing conditions, and still lower or totally absent in cells grown in the presence of combined nitrogen (Ohmori and Hattori, 1972, 1974). An implication of these results is that nitrogen control is operative even in the absence of exogenously added combined nitrogen, i.e. when N_2 represents the sole nitrogen source (Ramos et al., 1985). In general ammonium strongly represses nitrogenase synthesis (and heterocyst formation), repression by nitrate being less marked and variable in extent and depending on the strain (Ohmori and Hattori, 1972; Meeks et al., 1983). An exception to this general pattern is that of a marine strain of *Anabaena* (CA = ATCC 33047), in which nitrate but not ammonium completely prevents N_2 fixation and heterocyst formation (Bottomley et al., 1979; see Section 5 of Chapter 7 for details).

The negative effect of ammonium on nitrogenase synthesis seems to be dependent on its assimilation. Stewart and Rowell (1975) first demonstrated for *Anabaena cylindrica* that inactivation of glutamine synthetase by MSX allows full expression of nitrogenase activity (and heterocyst development) in cyanobacteria regardless of the presence of ammonium. Since then, this has been verified with various strains by using not only MSX (Meeks et al., 1977, 1978; Ownby, 1977; Bottomley et al., 1979; Singh et al., 1983a, b; Ramos et al., 1985), but other GS inhibitors, as 5-hydroxylysine (Ladha et al., 1978) or phosphinotricine (Lea et al., 1984). Although some workers have favored the idea of glutamine synthetase having a direct role in nitrogenase regulation (Ownby, 1977; Singh et al., 1983a), this contention has been superseded by that of products of ammonium assimilation being the actual effectors. Some authors, however, still claim a key role for ammonium per se as the inhibitor of nitrogenase and heterocyst formation (Singh et al., 1983b; Turpin et al., 1984), but their arguments have been disputed (Stewart et al., 1985). Inhibitors of ammonium assimilation also abolish nitrogenase repression induced by nitrate, indicating that the nitrate effect is indirect and due to the ammonium generated in nitrate reduction. *Anabaena* 33047 cells treated with MSX and suspended in a nitrate medium take up and reduce nitrate and N_2 simultaneously at high rates, without any apparent competition for assimilatory power, releasing large amounts of ammonium to the medium (Ramos and Guerrero, 1983).

A word of caution should be added about the influence of the metabolic state (e.g. C/N ratio) of the cells on their response to combined nitrogen. Thus, in N-starved filaments in *Anabaena*, the inhibitory effects of ammonium on nitrogenase synthesis can be absent or delayed until the cells recover from nitrogen starvation (Ramos et al., 1985). Nitrogenase development in *Anabaena* cells maintained in the absence of N_2 (argon:CO_2 atmosphere) for 14–16 h is not sensitive to ammonium inhibition (Murry and Benemann, 1979; Murry et al., 1983; Ramos et al., 1985). Cells from different stages of growth (Murry and Benemann, 1979; Murry et al., 1983) or exposed to different light intensities (Ohmori and Hattori, 1974) also show a differential behavior with regard to the responsiveness of nitrogenase to ammonium. A common factor effected by these conditions could be the availability of carbon skeletons. A high supply of carbohydrates could counteract the negative effect of ammonium derivatives on nitrogenase synthesis (Ramos et al., 1985), in a similar way as proposed for the regulation of nitrate uptake (see Section 4).

3.3. Regulatory signals in nitrogen control

At this stage we would like to raise the point that the different expressions of the nitrogen control considered above may share a common factor. This proposal may or may not prove true eventually, but at least no conclusive evidence to the contrary appears to be available. The common factor is the nature of the signal, whose intensity is a precise reflection of the nitrogen status of the cell and that induces different responses in the various systems related to the utilization of oxidized forms of inorganic nitrogen. In support of the involvement of a common factor in the regulation of these systems, is the fact that inhibition of ammonium assimilation in all cases results in the release from the negative effects of ammonium. In an elegant piece of work, Meeks et al. (1983) have shown for two strains of *Anabaena* clear differences in their responsiveness to nitrogen control, with nitrogen metabolism being subject to a tighter regulation by the nitrogen source in *Anabaena* 7120 than in *Anabaena cylindrica*. In the latter strain ammonium is less active in preventing nitrate utilization and, simultaneously, nitrate behaves as a poorer repressor of nitrogenase synthesis and heterocyst differentiation than in strain 7120. These and other observations bespeak a link among the systems involved with regard to the operative mechanisms of control. An additional assumption should be that the various systems exhibit different thresholds for the intensity of the signal, since certain situations exist in which the function of some of them is expressed whereas that of others is impeded. This is the case with N_2 fixers grown on nitrate, in which all systems involved in nitrate utilization are synthesized and fully operative, whereas heterocyst differentiation and nitrogenase activity are partially or totally affected.

Glutamine appears to be a good candidate for the regulatory signal, especially when it is considered that treatment with MSX or other GS inhibitors, which reverse the negative effects of ammonium, leads to decreased glutamine levels (Rowell et al., 1977; Ladha et al., 1978; Flores et al., 1980, 1983b). This has in fact been proposed in several instances, either with or without the reservation that glutamine derivatives rather than glutamine itself might be the actual effectors, or that other metabolites are needed in addition to glutamine (Rowell et al., 1977; Thomas et al., 1977; Ladha et al., 1978; Flores et al., 1980, 1983b; Haselkorn et al., 1980; Stewart, 1980a; Wolk, 1982; Ramos and Guerrero, 1983; Ramos et al., 1985). Results of experiments in which the effect of added amino acids on nitrogenase activity, heterocyst differentiation and nitrate uptake has been tested using both, normal untreated and MSX-treated cells, also point to an active role of glutamine and/or glutamine derivatives (Rowell et al., 1977; Haselkorn et al., 1980; Romero et al., 1985a).

Some nitrogenous metabolites generated in reactions which use glutamine as the amide-N donor may be involved in the ammonium effect, instead of or in addition to glutamine. Asparagine might well represent such a metabolite, as it can be synthesized in an amide-transferase reaction with glutamine as a substrate, behaving otherwise as an effective inhibitor of nitrate utilization in MSX-treated *Anacystis nidulans* cells. In this system, glycine also inhibits nitrate uptake (Romero et al., 1985a). Alternative candidates, in synthesis of which glutamine is involved, include tryptophan or intermediates of its biosynthesis (Bottomley et al., 1980; Wolk, 1982) and carbamyl phosphate (Lawrie, 1979; Gallon, 1980). None of these compounds has unequivocally been identified, however, with the key regulatory signal in nitrogen control, whose nature remains elusive.

4. Interactions between carbon and nitrogen assimilation

In the utilization of any form of inorganic nitrogen, an obligate requirement for carbon skeletons, mostly α-ketoacids, arises at the level of ammonium assimilation in the GS-GOGAT pathway and related reactions (see Fig. 1). This requirement manifests in the dependence of ammonium utilization upon CO_2 fixation or on the mobilization of accumulated carbohydrates. Sustained ammonium uptake in nitrate-grown *Anacystis* is dependent upon the availability of CO_2 (Boussiba et al., 1984a), and cannot proceed in cells in which CO_2 fixation has been specifically inhibited by treatment with D,L-glyceraldehyde (Lara and Romero, 1986).

A different aspect of carbon dependence is associated with N_2 fixation, especially when the process is carried out by heterocystous cyanobacteria. Heterocysts appear totally dependent on the vegetative cells for the provision of organic carbon compounds. While the exact nature of the imported molecules has not yet been unambiguously identified (Haury and Spiller, 1981; Jüttner, 1983), their oxidation in the heterocyst is the major source of reducing equivalents for nitrogenase under autotrophic conditions (Stewart, 1980a; Bothe, 1982; Stewart et al., 1985). In addition to the metabolic requirement for α-ketoglutarate to incorporate the resulting ammonium, N_2 fixation in the heterocyst is dependent upon CO_2 fixation in the vegetative cell as a means of indirectly receiving electrons from water photooxidation. An analogous situation is also apparent for dark N_2 fixation, either in heterocystous species (Fay, 1976; Ohmori, 1984) or in *Gloeothece* (Mullineaux et al., 1980), in which the assimilatory power stored in carbohydrates synthesized during the light period is the source of energy for the process. In all cases, organic carbon compounds are obligate intermediates between photosynthetically generated assimilatory power and N_2 fixation (see Sections 3 and 5 of Chapter 7 for more details).

In contrast, the utilization of nitrate in cyanobacteria does not depend on carbohydrates as a source of reductant (see Section 2.5.), although it exhibits stringent dependence upon CO_2 fixation, much stronger than in case of ammonium utilization (Lara and Romero, 1986). The correlation between rates of nitrate utilization and CO_2 fixation in *Anacystis nidulans* becomes evident from a series of observations. The dependence of nitrate uptake upon CO_2 availability exhibits saturation kinetics, with the CO_2 concentration allowing half-maximum rate of nitrate uptake (3.7 μM) being very close to the K_m for CO_2 of CO_2-dependent oxygen evolution (4 μM) (Flores et al., 1983c). The progressive inhibition of CO_2 fixation by increasing concentrations of D,L-glyceraldehyde results in a gradual decrease in the rate of nitrate utilization (Lara et al., 1984; Romero et al., 1985b). The light-saturation profile of nitrate utilization matches that of CO_2 fixation, with half-saturation photon fluency values for both processes of about 100 μeinstein·m^{-2}·s^{-1}. The correlation between CO_2 fixation and nitrate utilization is far tighter than that exhibited either by uncontrolled nitrate uptake and reduction in MSX-treated cells or by ammonium utilization. This indicates that the coupling between CO_2 fixation and nitrate utilization does not arise from the supply of photosynthetic assimilatory power or of carbon skeletons to incorporate ammonium, but that it may rather reflect a regulatory link between CO_2 fixation and nitrate uptake (Lara and Romero, 1986).

As is the case for eukaryotic microalgae (see Syrett, 1981), the CO_2-dependence of nitrate uptake in cyanobacteria is intimately linked to the existing C/N ratio of the cells. In this respect, N-starved cells of a mutant strain of *Agmenellum quadruplicatum*, unable to assimilate nitrite and exhibiting a high C/N ratio, were shown to take up nitrate and excrete nitrite transiently in the absence of CO_2 (Stevens and Van Baalen, 1973), presumably as long as the high carbohydrate reserve was being mobilized. Similar results have been obtained with N-starved cells of *Anacystis nidulans*, which were able to utilize nitrate under a CO_2-free atmosphere (Flores et al., 1983c). Thus, carbohydrate reserves accumulated in N-starved cells can apparently substitute for fresh products of CO_2 fixation in stimulating nitrate uptake.

Indications of the nature of the carbon dependence of nitrate uptake have been obtained in studies of the effects of inhibitors of ammonium assimilation on the CO_2-requirement of nitrate utilization in *Anacystis nidulans*. It has been shown that MSX-treated cells take up nitrate and reduce it to ammonium in the complete absence of CO_2 (Flores et al., 1983a,c). Treatment with MSX or azaserine also relieves the inhibition of nitrate uptake caused by the CO_2-fixation inhibitor D,L-glyceraldehyde (Lara et al., 1984; Romero et al., 1985b). Thus, prevention of ammonium assimilation not only results in the abolishment of the ammonium-induced inhibition of nitrate uptake (see Table I and Section 3.1.1), but also in the release of the process from carbon dependence. This is illustrated in Table II. At light intensities saturating for CO_2 fixation, maximal rates of nitrate uptake were measured in both, untreated and MSX-treated *Anacystis nidulans* cells. Limiting light intensity to a value that only allows CO_2 fixation to proceed at about 60% of its maximal capacity, the rate of nitrate uptake declines to the same extent in normal untreated cells. In cells treated with MSX nitrate utilization at limiting light intensity proceeds at maximal rates while CO_2 fixation remains limited by light (Lara and Romero, 1986). The results suggest a common link between the mechanism of the CO_2-dependence and that of the ammonium inhibition of nitrate uptake. In the presence of MSX, the ability to utilize nitrate by cells previously exposed to added ammonium was regained immediately after CO_2 addition, whereas in the absence of CO_2, restoration of the nitrate utilization capacity took about 40–50 min (Flores et al., 1983c). This shows that CO_2 acts as an antagonist of ammonium in its effect on nitrate uptake.

The results have been interpreted within the framework of a model for the regulation of nitrate uptake (Fig. 2) involving the products of both ammonium and CO_2 assimilation (Flores et al.,

TABLE II

Release by MSX of the coupling between CO_2 fixation and nitrate uptake in *Anacystis nidulans* 1402-1

Photon flux density ($\mu E \cdot m^{-2} \cdot s^{-1}$)	Treatment	CO_2 fixation	NO_3^- uptake
		($\mu mol \cdot mg^{-1}(chl) \cdot h^{-1}$)	
235	none	422	63
92	none	242	37
235	MSX	397	58
92	MSX	232	59

CO_2 fixation and nitrate uptake were determined as in Romero et al. (1985b). Initial concentrations of $[^{14}C]NaHCO_3$ and KNO_3 were 10 and 0.25 mM, respectively. Treatment with MSX was performed by preincubating the cells with the inhibitor (1 mM) in the light for 15 min. MSX, L-methionine-D,L-sulfoximine. We are grateful to José M. Romero for these data.

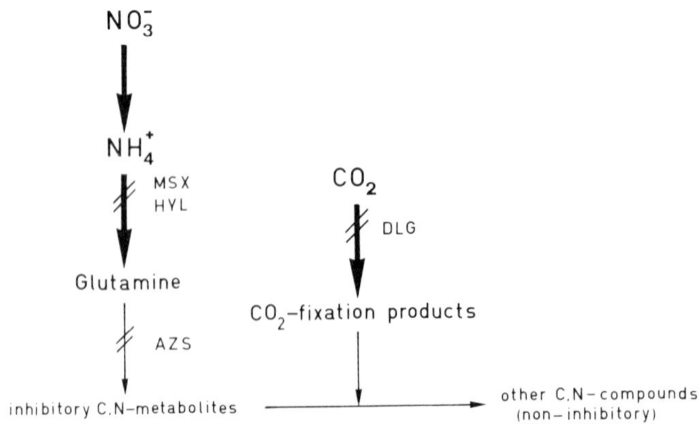

Fig. 2. A working hypothesis on the generation and removal of negative effectors involved in the feed-back regulation of nitrate utilization. AZS, azaserine; DLG, D,L-glyceraldehyde; HYL, 5-hydroxylysine; MSX, L-methionine D,L-sulfoximine. See text for details. (After Flores et al., 1983c)

1983c; Romero et al., 1985b). According to this scheme, nitrate uptake is regulated by a negative feed-back control exerted by certain organic nitrogenous compounds produced during nitrate assimilation via ammonium. The intracellular concentration of these inhibitory metabolites is determined partly by the supply of ammonium from which they are generated and partly by the supply of certain CO_2-fixation products, which in the course of normal metabolism would combine with the inhibitory ammonium derivatives, thus removing them while generating other non-inhibitory C,N-metabolites. By either withdrawing CO_2 or impeding its assimilation, accumulation of the inhibitory metabolites is favored and inhibition of nitrate uptake occurs. On the other hand, by inhibiting ammonium assimilation and the generation of negative effectors, nitrate uptake is relieved from the requirement for CO_2 fixation products. Such a regulatory system would allow a fine modulation of nitrate utilization exerted not only by accumulation of some of its own products, but also through the availability of carbon metabolites, exerting a close control of nitrate utilization by carbon assimilation.

A similar interaction between photosynthetic carbon and nitrogen metabolism appears also to be applicable to eukaryotic microalgae (Syrett, 1981), and may perhaps represent a general mechanism for green cells. Recent results from our laboratory (Lara, Romero and Guerrero, unpublished) provide strong support to the idea of nitrate transport being the target of the nitrogen-carbon control.

Acknowledgements

We are very grateful to Prof. Manuel Losada for continued advice and support. His critical comments on the manuscript and those of Enrique Flores, Antonia Herrero and José M. Romero have been very helpful. The patient and skillful assistance of Antonia Friend and Pepa Pérez de León is much appreciated. Research cited from the authors' laboratory has been supported by grants from Comisión Asesora de Investigación and Fundación Ramón Areces.

References

Avissar, Y.J. (1985) Induction of nitrate assimilation in the cyanobacterium *Anabaena variabilis*. Physiol. Plant. 63, 105–108.

Bagchi, S.N. and Singh, H.N. (1984) Genetic control of nitrate reduction in the cyanobacterium *Nostoc muscorum*. Mol. Gen. Genet. 193, 82–84.

Bergman, B. (1984) Photorespiratory ammonium release by the cyanobacterium *Anabaena cylindrica* in the presence of methionine sulfoximine. Arch. Microbiol. 137, 21–25.

Bothe, H. (1982) Nitrogen fixation. In N.G. Carr and B.A. Whitton (eds.), The Biology of Cyanobacteria, Blackwell, Oxford, pp. 87–104.

Bottomley, P.J., Grillo, J.F., Van Baalen, C. and Tabita, F.R. (1979) Synthesis of nitrogenase and heterocysts by *Anabaena* sp. CA in the presence of high levels of ammonia. J. Bacteriol. 140, 938–943.

Bottomley, P.J., Van Baalen, C. and Tabita, P.R. (1980) Heterocyst differentiation and tryptophan metabolism in the cyanobacterium *Anabaena* sp. CA. Arch. Biochem. Biophys. 203, 204–213.

Boussiba, S. and Gibson, J. (1985a) The role of glutamine synthetase activity in ammonium and methylammonium transport in *Anacystis nidulans* R-2. FEBS Lett. 180, 13–16.

Boussiba, S. and Gibson, J. (1985b) The regulation of methylammonium/ammonium transport in *Anacystis nidulans* R-2. In Abstracts of the Vth International Symposium on Photosynthetic Prokaryotes, Grindelwald, Switzerland, p. 27.

Boussiba, S., Resch, C.M. and Gibson, J. (1984a) Ammonia uptake and retention in some cyanobacteria. Arch. Microbiol. 138, 287–292.

Boussiba, S., Dilling, W. and Gibson, J. (1984b) Methylammonium transport in *Anacystis nidulans* R-2. J. Bacteriol. 160, 204–210.

Boussiba, S., Liu, X.-Q. and Gibson, J. (1984c) Endogenous ammonia production by *Anacystis nidulans* R-2 induced by methionine sulfoximine. Arch. Microbiol. 138, 217–219.

Candau, P. (1979) Purificación y propiedades de la ferredoxina-nitrato reductasa de la cianobacteria *Anacystis nidulans*. Ph.D. Thesis, University of Sevilla, Sevilla.

Candau, P., Manzano, C. and Losada, M. (1976) Bioconversion of light energy into chemical energy through reduction with water of nitrate to ammonia. Nature 262, 715–717.

Candau, P., Manzano, C., Guerrero, M.G. and Losada, M. (1980) Ferredoxin-dependent enzymatic reduction of nitrate with a deazaflavin photosystem. Photobiochem. Photobiophys. 1, 167–174.

Codd, G.A. and Stewart, W.D.P. (1973) Pathways of glycollate metabolism in the blue-green alga *Anabaena cylindrica*. Arch. Microbiol. 119, 1–5.

Dharmawardene, M.W.N., Haystead, A. and Stewart, W.D.P. (1973) Glutamine synthetase of the nitrogen-fixing alga *Anabaena cylindrica*. Arch. Mikrobiol. 90, 281–295.

Fay, P. (1973) The heterocyst. In N.G. Carr and B.A. Whitton (eds.), The Biology of Blue-Green Algae, Blackwell, Oxford, pp. 238–259.

Fay, P. (1976) Factors influencing dark nitrogen fixation in a blue-green alga. Appl. Environ. Microbiol. 31, 376–379.

Fleming, H. and Haselkorn, R. (1973) Differentiation in *Nostoc muscorum*: nitrogenase is synthesized in heterocysts. Proc. Natl. Acad. Sci. USA 70, 2727–2731.

Florencio, F.J. and Ramos, J.L. (1985) Purification and characterization of glutamine synthetase from the unicellular cyanobacterium *Anacystis nidulans*. Biochim. Biophys. Acta 838, 39–48.

Flores, E. (1982) Utilización de nitrato por la cianobacteria *Anacystis nidulans*. Mecanismo y regulación. Ph.D. Thesis, University of Sevilla, Sevilla.

Flores, E., Guerrero, M.G. and Losada, M. (1980) Short-term ammonium inhibition of nitrate utilization by *Anacystis nidulans* and other cyanobacteria. Arch. Microbiol. 128, 137–144.

Flores, E., Guerrero, M.G. and Losada, M. (1983a) Photosynthetic nature of nitrate uptake and reduction in the cyanobacterium *Anacystis nidulans*. Biochim. Biophys. Acta 722, 408–416.

Flores, E., Ramos, J.L., Herrero, A. and Guerrero, M.G. (1983b) Nitrate assimilation by cyanobacteria. In G.C. Papageorgiou and L. Packer (eds.), Photosynthetic Prokaryotes: Cell Differentiation and Function, Elsevier/North-Holland, Amsterdam, pp. 363–387.

Flores, E., Romero, J.M., Guerrero, M.G. and Losada, M. (1983c) Regulatory interaction of photosynthetic nitrate utilization and carbon dioxide fixation in the cyanobacterium *Anacystis nidulans*. Biochim. Biophys. Acta 725, 529–532.

Fogg, G.E., Stewart, W.D.P., Fay, P. and Walsby, A.E. (1973) The Blue-green Algae, Academic Press, London-New York, pp. 190–213.

Fuggi, A. (1985) Mechanism of proton-linked nitrate uptake in *Cyanidium caldarium*, an acidophilic non-vacuolated alga. Biochim. Biophys. Acta 815, 392–398.

Gallon, J.R. (1980) Nitrogen fixation by photoautotrophs. In W.D.P. Stewart and J.R. Gallon (eds.), Nitrogen Fixation. Academic Press, London, pp. 197–238.

Golden, J.W., Robinson, S.J. and Haselkorn, R. (1985) Rearrangement of nitrogen fixation genes during heterocyst differentiation in the cyanobacterium *Anabaena*. Nature 314, 419–423.

Guerrero, M.G., Manzano, C. and Losada, M. (1974) Nitrite photoreduction by a cell-free preparation of *Anacystis nidulans*. Plant Sci. Lett. 3, 273–278.

Guerrero, M.G., Vega, J.M. and Losada, M. (1981) The assimilatory nitrate-reducing system and its regulation. Ann. Rev. Plant Physiol. 32, 169–204.

Gupta, M. and Carr, N.G. (1981) Detection of glutamate synthase in heterocysts of *Anabaena* sp. strain 7120. J. Bacteriol. 148, 980–982.

Häger, K.-P., Danneberg, G. and Bothe, H. (1983) The glutamate synthase in heterocysts of *Nostoc muscorum*. FEMS Microbiol. Lett. 17, 179–183.

Hallenbeck, P.C., Kostel, P.J. and Benemann, J.R. (1979) Purification and properties of nitrogenase from the cyanobacterium *Anabaena cylindrica*. Eur. J. Biochem. 98, 275–284.

Haselkorn, R. (1978) Heterocysts. Ann. Rev. Plant. Physiol. 29, 311–344.

Haselkorn, R., Mazur, B., Orr, J., Rice, D., Wood, N. and Rippka, R. (1980) Heterocyst differentiation and nitrogen fixation in cyanobacteria (blue-green algae). In W.E. Newton and W.H. Orme-Johnson (eds.), Nitrogen Fixation, Vol. II, University Park Press, Baltimore, pp. 259–278.

Hattori, A. (1962) Adaptive formation of nitrate reducing system in *Anabaena cylindrica*. Plant Cell Physiol. 3, 371–377.

Hattori, A. (1963) Effect of hydrogen on nitrite reduction by *Anabaena cylindrica*. In Japanese Society of Plant Physiologists (ed.), Studies on Microalgae and Photosynthetic Bacteria. The University of Tokyo Press, Tokyo, pp. 485–492.

Hattori, A. (1970) Solubilization of nitrate reductase from the blue-green alga *Anabaena cylindrica*. Plant Cell Physiol. 11, 975–978.

Hattori, A. and Myers, J. (1966) Reduction of nitrate and nitrite by subcellular preparations of *Anabaena cylindrica*. I. Reduction of nitrite to ammonia. Plant Physiol. 41, 1031–1036.

Hattori, A. and Myers, J. (1967) Reduction of nitrate and nitrite by subcellular preparations of *Anabaena cylindrica*. II. Reduction of nitrate to nitrite. Plant Cell Physiol. 8, 327–337.

Hattori, A. and Uesugi, I. (1968a) Ferredoxin-dependent photoreduction of nitrate and nitrite by subcellular preparations of *Anabaena cylindrica*. In H. Shibata, A. Takamiya, A.T. Jagendorf and R.C. Fuller (eds.), Comparative Biochemistry and Biophysics of Photosynthesis. University of Tokyo Press, Tokyo, pp. 201–205.

Hattori, A. and Uesugi, I. (1968b) Purification and properties of nitrite reductase from the blue-green alga *Anabaena cylindrica*. Plant Cell Physiol. 9, 689–699.

Haury, J.F. and Spiller, H. (1981) Fructose uptake and influence on growth of and nitrogen fixation by *Anabaena variabilis*. J. Bacteriol. 147, 227–235.

Haystead, A., Dharmawardene, M.W.N. and Stewart, W.D.P. (1973) Ammonia assimilation in a nitrogen-fixing blue-green alga. Plant Sci. Lett. 1, 439–445.

Herrero, A. and Guerrero, M.G. (1986) Regulation of nitrite reductase in the cyanobacterium *Anacystis nidulans*. J. Gen. Microbiol. 132, 2463–2468.

Herrero, A., Flores, E. and Guerrero, M.G. (1981) Regulation of nitrate reductase levels in the cyanobacteria *Anacystis nidulans*, *Anabaena* sp. strain 7119, and *Nostoc* sp. strain 6719. J. Bacteriol. 145, 175–180.

Herrero, A., Flores, E. and Guerrero, M.G. (1984) Regulation of the nitrate reductase level in *Anacystis nidulans:* Activity decay under nitrogen stress. Arch. Biochem. Biophys. 234, 454–459.

Herrero, A., Flores, E. and Guerrero, M.G. (1985) Regulation of nitrate reductase cellular levels in the cyanobacteria *Anabaena variabilis* and *Synechocystis* sp. FEMS Microbiol. Lett. 26, 21–25.

Hoare, D.S., Hoare, S.L. and Moore, R.B. (1967) The photoassimilation of organic compounds by autotrophic blue-green algae. J. Gen. Microbiol. 73, 317–326.

Houchins, J.P. (1984) The physiology and biochemistry of hydrogen metabolism in cyanobacteria. Biochim. Biophys. Acta 768, 227–255.

Houchins, J.P. and Hind, G. (1982) Pyridine nucleotides and H_2 as electron donors to the respiratory and photosynthetic electron-transfer chains and to nitrogenase in *Anabaena* heterocysts. Biochim. Biophys. Acta 682, 86–96.

Ida, S. and Mikami, B. (1983) Purification and characterization of assimilatory nitrate reductase from the cyanobacterium *Plectonema boryanum*. Plant Cell Physiol. 24, 649–658.

Jensen, B.B. and Cox, R.P. (1985) Cyanobacterial dinitrogen uptake measurements using membrane-inlet mass spectrometry. In R.K. Poole and C.S. Dow (eds.), Microbial Gas Metabolism. Academic Press, London, pp. 279–285.

Jüttner, J. (1983) ^{14}C-labeled metabolites in heterocysts and vegetative cells of *Anabaena cylindrica* filaments and their presumptive function as transport vehicles of organic carbon and nitrogen. J. Bacteriol. 155, 628–633.

Kleiner, D. (1985) Bacterial ammonium transport. FEMS Microbiol. Rev. 32, 87–100.

Kleinhofs, A., Warner, R.L. and Narayanan, K.R. (1985) Current progress towards an understanding of the genetics and molecular biology of nitrate reductase in higher plants. Oxford Surveys Plant Mol. Cell Biol. 2, 91–121.

Ladha, J.K., Rowell, P. and Stewart, W.D.P. (1978) Effects of 5-hydroxylysine on acetylene reduction and NH_4^+-assimilation in the cyanobacterium *Anabaena cylindrica*. Biochem. Biophys. Res. Commun. 83, 688–696.

Lara, C. and Romero, J.M. (1986) Distinctive light and CO_2-fixation requirements of nitrate and ammonium utilization by the cyanobacterium *Anacystis nidulans*. Plant Physiol. 81, 686–688.

Lara, C., Romero, J.M., Flores, E., Guerrero, M.G. and Losada, M. (1984) Regulation of nitrate utilization by CO_2 fixation products in the cyanobacterium *Anacystis nidulans*. In C. Sybesma (ed.), Advances in Photosynthesis Research, Vol. II. Martinus Nijhoff/Dr. Junk, The Hague, pp. 715–718.

Lawrie, A. (1979) Effect of carbamoyl phosphate on nitrogenase in *Anabaena cylindrica* Lemm. J. Bacteriol. 139, 115–119.

Lea, P.J. and Miflin, B.J. (1975) Glutamate synthase in blue-green algae. Biochem. Soc. Trans. 3, 381–384.

Lea, P.J., Mills, W.R., Wallsgrove, R.M. and Miflin, B.J. (1982) Assimilation of nitrogen and synthesis of amino acids in chloroplasts and cyanobacteria (blue-green algae). In J.A. Schiff (ed.), On the Origins of Chloroplast. Elsevier/North-Holland, Inc., New York, pp. 149–178.

Lea, P.J., Joy, K.W., Ramos, J.L. and Guerrero, M.G. (1984) The action of 2-amino-4-(methylphosphinyl)-butanoic acid (phosphinothricin) and its 2-oxo-derivative on the metabolism of cyanobacteria and higher plants. Phytochemistry 23, 1–6.

Losada, M., Guerrero, M.G. and Vega, J.M. (1981) The assimilatory reduction of nitrate. In H. Bothe and A. Trebst (eds.), Biology of Inorganic Nitrogen and Sulfur. Springer, Berlin, pp. 30–63.

Lowe, D.J., Thorneley, R.N.F. and Smith, B.E. (1985) Nitrogenase. In P.M. Harrison (ed.), Metalloproteins, Part 1. Verlag Chemie, Weinheim, pp. 207–249.

Manzano, C. (1977) La reducción fotosintética del nitrato en el alga verdeazulada *Anacystis nidulans*. Ph.D. Thesis, University of Sevilla, Sevilla.

Manzano, C., Candau, P., Gómez-Moreno, C., Relimpio, A.M. and Losada, M. (1976) Ferredoxin-dependent photosynthetic reduction of nitrate and nitrite by particles of *Anacystis nidulans*. Mol. Cell. Biochem. 10, 161–169.

Manzano, C., Candau, P. and Guerrero, M.G. (1978) Affinity chromatography of *Anacystis nidulans* ferredoxin-nitrate reductase and NADP reductase on reduced ferredoxin-sepharose. Anal. Biochem. 90, 408–412.

Meeks, J.C. (1981) Radiotracer studies of nitrogen metabolism in cyanobacteria. In J.W. Root and K.A. Krohn (Eds.), Advances in Chemistry, Series 197. American Chemical Society, Washington, pp. 269–294.

Meeks, J.C., Wolk, C.P., Thomas, J., Lockau, W. and Shaffer, P.W. (1977) The pathways of assimilation of $^{13}NH_4^+$ by the cyanobacterium, *Anabaena cylindrica*. J. Biol. Chem. 252, 7894–7900.

Meeks, J.C., Wolk, C.P., Lockau, W., Schilling, N., Shaffer, P.W. and Chien, W.-S. (1978) Pathways of assimilation of ^{13}N-N_2 and $^{13}NH_4^+$ by cyanobacteria with and without heterocysts. J. Bacteriol. 134, 125–130.

Meeks, J.C., Wycoff, K.L., Chapman, J.S. and Enderlin, C.S. (1983) Regulation of expression of nitrate and dinitrogen assimilation by *Anabaena* species. App. Environ. Microbiol. 45, 1351–1359.

Meeks, J.C., Steinberg, N., Joseph, C.M., Enderlin, C.S., Jorgensen, P.A. and Peters, G.A. (1985) Assimilation of exogenous and dinitrogen-derived $^{13}NH_4^+$ by *Anabaena azollae* separated from *Azolla caroliniana* willd. Arch. Microbiol. 142, 229–233.

Méndez, J.M. and Vega, J.M. (1981) Purification and molecular properties of nitrite reductase from *Anabaena* sp. 7119. Physiol. Plant 52, 7–14.

Méndez, J.M., Herrero, A. and Vega, J.M. (1981) Characterization and catalytic properties of nitrite reductase from *Anabaena* sp. 7119, Z. Pflanzenphysiol. 103, 305–315.

Mikami, B. and Ida, S. (1984) Purification and properties of ferredoxin-nitrate reductase from the cyanobacterium *Plectonema boryanum*. Biochim. Biophys. Acta 791, 294–304.

Mullineaux, P.M., Gallon, J.R. and Chaplin, A.E. (1980) Effects of a light to dark transition on carbon reserves, nitrogen fixation and ATP concentrations in cultures of *Gloeocapsa* (*Gloeothece*) sp. 1430/3. J. Gen. Microbiol. 102, 227–232.

Murry, M.A. and Benemann, J.R. (1979) Nitrogenase regulation in *Anabaena cylindrica*. Plant Cell Physiol. 20, 1391–1401.

Murry, M.A., Jensen, D.B. and Benemann, J.R. (1983) Role of ammonia in regulation of nitrogenase synthesis and activity in *Anabaena cylindrica*. Biochim. Biophys. Acta 756, 13–19.

Murry, M.A., Hallenbeck, P.C. and Benemann, J.R. (1984) Immunochemical evidence that nitrogenase is restricted to the heterocysts in *Anabaena cylindrica*. Arch. Microbiol. 137, 194–199.

Nomenclature Committee of the International Union of Biochemistry (1984) Enzyme Nomenclature. Academic Press Inc., Orlando.

Ohmori, M. (1984) Effects of preillumination on dark nitrogen fixation and respiration by *Anabaena cylindrica*. Plant Cell Physiol. 25, 125–130.

Ohmori, K. and Hattori, A. (1970) Induction of nitrate and nitrite reductases in *Anabaena cylindrica*. Plant Cell Physiol. 11, 873–878.

Ohmori, M. and Hattori, A. (1972) Effect of nitrate on nitrogen-fixation by the blue-green alga *Anabaena cylindrica*. Plant Cell Physiol. 13, 589–599.

Ohmori, M. and Hattori, A. (1974) Effect of ammonia on nitrogen-fixation by the blue-green alga *Anabaena cylindrica*. Plant Cell Physiol. 15, 131–142.

Ohmori, M., Ohmori, K. and Strotmann, H. (1977) Inhibition of nitrate uptake by ammonia in a blue-green alga, *Anabaena cylindrica*. Arch. Microbiol. 114, 225–229.

Orr, J., Keefer, L.M., Keim, P., Nguyen, T.D., Wellems, T., Heinrikson, L. and Haselkorn, R. (1981) Purification, physical characterization and NH_2-terminal sequence of glutamine synthetase from the cyanobacterium *Anabaena* 7120. J. Biol. Chem. 256, 13091–13098.

Ortega, T., Castillo, F. and Cárdenas, J. (1976) Photolysis of water coupled to nitrate reduction by *Nostoc muscorum* subcellular particles. Biochem. Biophys. Res. Commun. 71, 885–891.

Ownby, J.D. (1977) Effects of amino acids on methionine-sulfoximine-induced heterocyst formation in *Anabaena*. Planta 136, 277–279.

Peschek, G.A. (1979a) Nitrate and nitrite reductase and hydrogenase in *Anacystis nidulans* grown in Fe- and Mo-deficient media. FEMS Microbiol. Lett. 6, 371–374.

Peschek, G.A. (1979b) Anaerobic hydrogenase activity in *Anacystis nidulans*. H_2-dependent photoreduction and related reactions. Biochim. Biophys. Acta 584, 187–202.

Rai, A.K., Kashyap, A.K. and Gupta, S.L. (1981) ATP-dependent uptake of nitrate in *Nostoc muscorum* and inhibition by ammonium ions. Biochim. Biophys. Acta 674, 78–86.

Rai, A.N., Rowell, P. and Stewart, W.D.P. (1984) Evidence for an ammonium transport system in free-living and symbiotic cyanobacteria. Arch. Microbiol. 137, 241–246.

Ramos, J.L. and Guerrero, M.G. (1983) Involvement of ammonium metabolism in the nitrate inhibition of nitrogen fixation in *Anabaena* sp. strain ATCC 33047. Arch. Microbiol. 136, 81–83.

Ramos, J.L., Guerrero, M.G. and Losada, M. (1982) Photoproduction of ammonia from nitrate by *Anacystis nidulans* cells. Biochim. Biophys. Acta 679, 323–330.

Ramos, J.L., Madueño, F. and Guerrero, M.G. (1985) Regulation of nitrogenase levels in *Anabaena* sp. ATCC 33047 and other filamentous cyanobacteria. Arch. Microbiol. 141, 105–111.

Romero, J.M., Flores, E. and Guerrero, M.G. (1985a) Inhibition of nitrate utilization by amino acids in intact *Anacystis nidulans* cells. Arch. Microbiol. 142, 1–5.

Romero, J.M., Lara, C. and Guerrero, M.G. (1985b) Dependence of nitrate utilization upon active CO_2 fixation in *Anacystis nidulans*: a regulatory aspect of the interactions between photosynthetic carbon and nitrogen metabolism. Arch. Biochem. Biophys. 237, 396–401.

Rowell, P. and Stewart, W.D.P. (1976) Alanine dehydrogenase of the N_2-fixing blue-green alga, *Anabaena cylindrica*. Arch. Microbiol. 107, 115–124.

Rowell, P., Enticott, S. and Stewart, W.D.P. (1977) Glutamine synthetase and nitrogenase activity in the blue-green alga *Anabaena cylindrica*. New Phytol. 79, 41–54.

Sampaio, M.J.A.M., Rowell, P. and Stewart, W.D.P. (1979) Purification and some properties of glutamine synthetase from the nitrogen-fixing cyanobacteria *Anabaena cylindrica* and a *Nostoc* sp. J. Gen. Microbiol. 111, 181–191.

Scherer, S., Kerfin, W. and Böger, P. (1980) Increase of nitrogenase activity in the blue-green alga *Nostoc muscorum* (cyanobacterium). J. Bacteriol. 144, 1017–1023.

Schrautemeier, B. and Böhme, H. (1985) A distinct ferredoxin for nitrogen fixation isolated from heterocysts of the cyanobacterium *Anabaena variabilis*. FEBS Lett. 184, 304–308.

Schrautemeier, B., Böhme, H. and Böger, P. (1984) In vitro studies on pathways and regulation of electron transport to nitrogenase with a cell free extract from heterocysts of *Anabaena variabilis*. Arch. Microbiol. 137, 14–20.

Schrautemeier, B., Böhme, H. and Böger, P. (1985) Reconstitution of a light-dependent nitrogen-fixing and transhydrogenase system with heterocysts thylakoids. Biochim. Biophys. Acta 807, 147–154.

Serrano, A., Rivas, J. and Losada, M. (1981) Nitrate and nitrite as 'in vivo' quenchers of chlorophyll fluorescence in blue-green algae. Photosynthesis Res. 2, 175–184.

Serrano, A., Rivas, J. and Losada, M. (1982) Changes in fluorescence spectra by nitrate and nitrite in blue-green algae. Photobiochem. Photobiophys. 4, 257–264.

Singh, H.N., Rai, U.N., Rao, V.V. and Bagchi, S.N. (1983a) Evidence for ammonia as an inhibitor of heterocyst and nitrogenase formation in the cyanobacterium *Anabaena cycadeae*. Biochem. Biophys. Res. Commun. 111, 180–187.

Singh, H.N., Singh, R.K. and Sharma, R. (1983b) An L-methionine-D,L-sulfoximine-resistant mutant of the cyanobacterium *Nostoc muscorum* showing inhibitor-resistant γ-glutamil-transferase, defective glutamine synthetase and producing extracellular ammonia during N_2 fixation. FEBS Lett. 154, 10–14.

Smith, R.V., Noy, R.J. and Evans, M.C.W. (1971) Physiological donor systems to the nitrogenase of the blue-green alga *Anabaena cylindrica*. Biochim. Biophys. Acta 253, 104–109.

Stacey, G., Tabita, F.R. and Van Baalen, C. (1977) Nitrogen and ammonia assimilation in the cyanobacteria: purification of glutamine synthetase from *Anabaena* sp. strain CA. J. Bacteriol. 132, 596–603.

Stevens, S.E. and Van Baalen, C. (1973) Characteristics of nitrate reduction in a mutant of the blue-green alga *Agmenellum quadruplicatum*. Plant Physiol. 51, 350–356.

Stevens, S.E. and Van Baalen, C. (1974) Control of nitrate reductase in a blue-green alga. The effects of inhibitors, blue light, and ammonia. Arch. Biochem. Biophys. 161, 146–152.

Stewart, W.D.P. (1978) Nitrogen-fixing cyanobacteria and their associations with eukaryotic plants. Endeavour, New Series, 2, 170–179.

Stewart, W.D.P. (1980a) Some aspects of structure and function in N_2-fixing cyanobacteria. Ann. Rev. Microbiol. 34, 497–536.

Stewart, W.P.D. (1980b) Transport and utilization of nitrogen sources by algae. In J.W. Payne (ed.) Microorganisms and Nitrogen Sources. John Wiley, New York, pp. 577–607.

Stewart, W.D.P. and Rowell, P. (1975) Effects of L-methionine-DL-sulphoximine on the assimilation of newly fixed NH_3, acetylene reduction and heterocyst production in *Anabaena cylindrica*. Biochem. Biophys. Res. Commun. 65, 846–856.

Stewart, W.D.P., Haystead, A. and Dharmawardene, M.W.N. (1975) Nitrogen assimilation and metabolism in blue-green algae. In W.D.P. Stewart (ed.), Nitrogen Fixation by Free-Living Microorganisms. Cambridge University Press, Cambridge, pp. 129–158.

Stewart, W.D.P., Rowell, P., Cossak, J.D. and Kerby, N.W. (1985) Physiological studies on N_2-fixing cyanobacteria. In P.W. Ludden and J. Burris (eds.), Nitrogen Fixation and CO_2 Metabolism. Elsevier Science Publishing, Amsterdam, pp. 269–279.

Syrett, P.J. (1981) Nitrogen metabolism of microalgae. Can. Bull. Fish. Aquat. Sci. 210, 182–210.

Thayer, J.R. and Huffaker, R.C. (1982) Kinetic evaluation, using ^{13}N, reveals two assimilatory nitrate transport systems in *Klebsiella pneumoniae*. J. Bacteriol. 149, 198–202.

Thomas, J., Wolk C.P. and Shaffer, P.W. (1975) The initial organic products of fixation of ^{13}N-labeled nitrogen gas by the blue-green alga *Anabaena cylindrica*. Biochem. Biophys. Res. Commun. 67, 501–507.

Thomas, J., Meeks, J.C., Wolk, C.P., Shaffer, P.W., Austin, S.M. and Chien, W.-S. (1977) Formation of glutamine from [^{13}N]ammonia,[^{13}N]dinitrogen, and [^{14}C]glutamate by heterocysts isolated from *Anabaena cylindrica*. J. Bacteriol. 129, 1545–1555.

Tischner, R. and Schmidt, A. (1984) Light mediated regulation of nitrate assimilation in *Synechococcus leopoliensis*. Arch. Microbiol. 137, 151–154.

Turpin, D.M., Edie, S.A. and Canvin, D.T. (1984) In vivo nitrogenase regulation by ammonium and methylamine and the effect of MSX on ammonium transport in *Anabaena flos-aquae*. Plant Physiol. 74, 701–704.

Ullrich, W.R. (1983) Uptake and reduction of nitrate: algae and fungi. In A. Läuchli and R.L. Bieleski (eds.), Inorganic Plant Nutrition, Encyclopedia of Plant Physiology, New Series, Vol. 15A. Springer, Berlin, pp. 376–397.

Vega, J.M., Cárdenas, J. and Losada, M. (1980) Ferredoxin-nitrite reductase. In A. San Pietro (ed.), Methods in Enzymology, Vol. 69, Part C. Academic Press, New York, pp. 255–270.

Vennesland, B. and Guerrero, M.G. (1979) Reduction of nitrate and nitrite. In M. Gibbs and L. Latzko (eds.), Photosynthesis II, Encyclopedia of Plant Physiology, New Series, Vol. 6. Springer, Berlin, pp. 425–444.

Walsby, A.E. (1985) Permeability of heterocysts to the gases N_2 and O_2. In Abstracts of the Vth International Symposium on Photosynthetic Prokaryotes, Grindelwald, Switzerland, p. 27.
Wolk, C.P. (1980) Heterocysts, ^{13}N, and N_2-fixing plants. In W.E. Newton and W.H. Orme-Johnson (eds.), Nitrogen Fixation, Vol. II. University Park Press, Baltimore, pp. 279–292.
Wolk, C.P. (1982) Heterocysts. In N.G. Carr and B.A. Whitton (eds.), The Biology of Cyanobacteria. Blackwell Sci. Publ., Oxford, pp. 359–386.
Wolk, C.P., Thomas, J., Shaffer, P.W., Austin, S.M. and Galonsky, A. (1976) Pathway of nitrogen metabolism after fixation of ^{13}N-labeled nitrogen gas by the cyanobacterium, *Anabaena cylindrica*. J. Biol. Chem. 251, 5027–5034.
Yoch, D. and Gotto, J.W. (1982) Effect of light intensity and inhibitors of nitrogen assimilation on NH_4^+ inhibition of nitrogenase activity in *Rhodospirillum rubrum* and *Anabaena* sp. J. Bacteriol. 151, 800–806.
Zevenboom, W., De Groot, G.J. and Mur, L.R. (1980) Effects of light on nitrate-limited *Oscillatoria agardhii* in chemostat cultures. Arch. Microbiol. 125, 59–65.

P. Fay and C. Van Baalen (Eds.) The Cyanobacteria

7

Nitrogen fixation

CHASE VAN BAALEN

The University of Texas Marine Science Institute, Port Aransas, Texas 78373, USA

1. Introduction

The blue-green algae, now also known as the cyanobacteria, include several cell types that carry out the reduction of the dinitrogen molecule to ammonia. When combined nitrogen is absent, members of the *Nostoc-Anabaena* group differentiate specialized cells, termed heterocysts, for aerobic nitrogen fixation. To confound the issue, there are also several bonafide non-heterocystous forms, *Gleocapsa* (Wyatt and Silvey, 1969; Rippka et al., 1971) and *Aphanothece* (Singh, 1973), that carry out an acceptable aerobic nitrogen fixation albeit at much slower growth rates than heterocystous forms. Further details of aerobic and of microaerophilic nitrogen fixation, as well as other facets of nitrogen fixation in cyanobacteria, have been admirably reviewed (e.g. Stewart, 1980; Bothe, 1982), and this chapter will strive not to be iterative. Three major unresolved problems will be discussed: reductant and ATP supply to nitrogenase, oxygen protection of nitrogenase, and features of the regulation of nitrogenase. An attempt will be made to integrate work with whole cells with selected observations from isolated heterocysts and enzyme studies. In addition, since nitrogen fixation in the cyanobacteria is primarily a light-dependent reaction, the increasing interest in the role of light and of the photosynthetic pigments will be examined.

2. Overall requirements

A general equation for nitrogen fixation can be written:

$$N_2 + 4\ NADPH + 6\ H^+ + 12\ ATP \rightarrow 2\ NH_4^+ + H_2 + 12\ ADP + 4\ NADP^+$$

(Jensen, 1983) with ferredoxin the more likely in vivo electron donor (Stewart, 1980). There are several interesting aspects of this conceptualization of nitrogen fixation. The formation of hydrogen is believed to be an obligatory part of the mechanism, with hydrogen evolution occurring even under 50 atmospheres of nitrogen (Simpson and Burris, 1984). Secondly, the cost of nitrogen fixation is very similar to the cost of forming NH_4^+ via the reduction of NO_3^- (Postgate and Cannon,

1981); the latter process requiring in whole cells about one-fifth of the electron flow needed for the fixation of carbon dioxide (Guerrero et al., 1981; see Section 2.5, of Chapter 6). With the reasonable assumption that nitrogen assimilation limits growth, these efficiency considerations can be examined in terms of autotrophic growth rates. For six rapidly growing *Anabaena* strains, the average of the generation times under optimum growth conditions were 5.2 h on N_2, 4.9 h on $NaNO_3$, and 3.8 h on NH_4Cl (Stacey et al., 1977; Gotto et al., 1979a). In *Anabaena variabilis* grown on fructose in the dark, Jensen (1983) found that the ratio between cell carbon and fructose carbon was 0.38 in diazotrophically-grown and nitrate-grown cells and 0.52 in ammonia-grown cells. Both these cases fit the above predictions.

3. Substrate effects on nitrogenase in whole cells

Although the cyanobacteria are largely photoautotrophic and any nitrogen fixation is, therefore, necessarily light dependent, there are several heterotrophs that grow on N_2 in darkness. In early studies sugars were found to support dark growth but at very slow growth rates. In *Chlorogloea fritschii* sucrose followed by glucose was the best substrate supporting heterotrophic growth on N_2 with a generation time estimated as 3 weeks (Fay, 1965). In *Anabaenopsis circularis* glucose, fructose or sucrose supported growth on N_2 in the dark (Watanabe and Yamamoto, 1967). The growth rate estimated from the yield of cells on glucose was 9 days, without allowance for any initial lag or the growth allowed by combined nitrogen contamination of the medium. Later work in the area is more convincing. Wolk and Shaffer (1976) grew *Anabaena variabilis* to high cell density on fructose + N_2 in the dark at a generation time of 36 h. Jensen (1983), using the same strain of *A. variabilis*, observed a generation time on fructose + N_2 of 25 h at 35°C or 40% of the maximum autotrophic rate (conditions unspecified). The case is then made that in one or more cyanobacteria, respiratory electron transport can substitute for photoreactions in driving nitrogenase, as it can in the reduction of nitrate to ammonia during heterotrophic growth of *Nostoc* strain MAC (Bottomley and Van Baalen, 1978). However, in the light both these pathways are ferredoxin-linked (Stewart, 1980; Guerrero et al., 1981; see Section 2.2. of Chapter 6), and it remains to be seen if this is so in darkness. Several ferredoxins have been isolated from *Nostoc* strain MAC (Hutson and Rogers, 1975) and a distinct ferredoxin has been found in heterocysts (Schrautemeier and Böhme, 1985; see Section 1.4. of Chapter 2). Glyoxylate was found to stimulate nitrogenase activity and photosynthesis, but not growth in *Anabaena cylindrica, Anabaena variabilis* and *Nostoc muscorum*, with the maximum effect seen after 15–20 h incubation in the light (Bergman, 1980). The reason for the long expression time for the glyoxylate effect was unknown. It was postulated that glyoxylate may exert some effect on photorespiration, freeing up more reductant for nitrogenase. Erythrose stimulates acetylene reduction in whole filaments of *Anabaena* sp. strain 7120, but only under conditions of limited reductant flow, i.e. in the presence of 3′-(3,4-dichlorophenyl)-1,1′-dimethyl urea (DCMU) or in darkness (Privalle, 1984). Sucrose, fructose or glucose were without effect. Erythrose also stimulated light-driven acetylene reduction as measured over 20 min in heterocysts isolated from the same organism (Privalle and Burris, 1984). Sucrose, D-glucose, and D-glucose-6-phosphate plus $NADP^+$ were 40% as active as erythrose. In addition, it was observed that H_2-supported acetylene reduction in these isolated heterocysts was inhibited in the presence of optimal erythrose concentration (5 mM). In contrast, erythrose does not inhibit the H_2-supported acetylene reduction in isolated heterocysts from *Anabaena* sp. strain CA (Kumar and Van Baalen, unpublished observations). Heterocysts from strain CA are some four to five times more active than heterocysts isolated from *Anabaena* sp. strain 7120, and this may be part of the explanation for the difference in the results. Taken as a whole, these results with substrates shed little light on the nature of the reductant supply in heterocysts. A continuous supply of a small molecule(s) transferred from a vegetative cell to a heterocyst is most unlikely. Heterocysts can be readily labeled by $^{14}CO_2$ added to a culture growing on N_2, and the incorporated radioactivity is not lost after repeated washings, which indicates the insoluble nature of the labeled product(s). In addition, heterocysts isolated from strain CA after incubation of the whole filaments in argon had a very high endogenous activity (Kumar et al., 1983). Such heterocysts form upward of 10 μmol of ethylene per mg dry weight of heterocysts without benefit of any external supply of reductant or an ATP-generating system. It remains to be seen if the high endogenous activity of heterocysts from strain CA is all accounted for by stored material that can be converted to reductant.

The fructose and erythrose observations may indicate an intermediary carbon pathway akin to the anaerobic breakdown of glucose in *Bifidobacterium bifidum* (Poupard et al., 1973) operating in the heterocyst. However, the fate of any acetate or lactate formed in the heterocyst is difficult to assess.

In short time (20 s) labeling experiments with $^{14}CO_2$ of intact filaments of *Anabaena cylindrica*, heterocysts had a higher percentage of label in alanine, glutamine, glutamate and, to a lesser extent, in glucose-6-phosphate (Juttner, 1983). These results were interpreted as supportive of the oft-quoted transport of compounds from vegetative cells to heterocysts (Wolk, 1968). In this work labeled ribulose-1,5-bis-phosphate was also found in heterocysts at a two-fold higher level than in the whole filaments. Unless this is some type of artifact, it is very difficult to explain a need for this substrate since ribulose-1,5-bis-phosphate carboxylase/oxygenase activity is generally thought to be absent from the heterocysts (see Section 4.4. of Chapter 4).

4. Thioredoxin and glutathione regulation

The thioredoxin system is a protein-mediated mechanism that links light to the control of enzyme activity in oxygenic photosynthesis. The light signal via chlorophyllous membranes, ferredoxin and ferredoxin-thioredoxin reductase brings about changes in the sulfhydryl status of target enzymes. Cell-free preparations of *Nostoc muscorum* strain 7119 have been shown to contain the two different thioredoxins, f and m, and ferredoxin-thioredoxin reductase. Each of these components is interchangeable with their spinach counterparts in activation of thioredoxin-modulated enzymes (Schmidt, 1980; Yee et al., 1981; Crawford et al., 1984). Isocitrate dehydrogenase isolated and purified 400-fold from *Anacystis nidulans* (Tx20) had properties characteristic of hysteretic enzymes. It responded slowly to rapid changes in ligand concentration, $NADP^+$, isocitrate, Mg^{2+} ions, or rapid changes in temperature (Friga and Farkas, 1981). No evidence was found for redox control. In later work on isocitrate dehydrogenase in crude extracts from *Nostoc muscorum* and *Anabaena cylindrica*, the activity level was higher in the heterocyst than in the vegetative cell debris (Papen et al., 1983). The enzyme could be negatively regulated by reduction charge and was scarcely affected by oxoglutarate. Thioredoxin preparations from spinach chloroplasts or from *A. cylindrica* activate isocitrate dehydrogenase from either heterocysts or vegetative cells of *A. cylindrica*. The position was taken in this paper that isocitrate dehydrogenase, despite its higher activity in heterocysts, was not supplying reductant to nitrogenase, but rather was related to heterocyst differentiation, or possibly to supplying oxoglutaric for glutamate synthesis. Glucose-6-phosphate dehydrogenase (G6PDH) isolated from vegetative cells and heterocysts of *Anabaena* sp. strain PCC 7120 was more active in its oxidized than in reduced forms (Udvardy et al., 1984a). A protein from *Anabaena* with thioredoxin activity when reduced by dithiothreitol, inactivated the G6PDH. Curiously, deactivated G6PDH was not oxidatively reactivated by oxidized thioredoxin or O_2, but was reactivated by oxidized glutathione or H_2O_2. If G6PDH is active in supplying reductant in the heterocyst, these data imply that there must be a mechanism to maintain it in an oxidized form. The heterocyst is generally viewed as an anaerobic to microaerophilic system which implies that any oxidative activation may be restricted. The H_2O_2 activation of G6PDH may point to a photoreduction of O_2 in heterocysts forming H_2O_2 as in Photosystem I (Furbank, 1984). Oxidative photoreactions in heterocysts should receive more attention. Specific antisera raised against purified thioredoxins from *Anabaena cylindrica* and *Anabaena variabilis* were used to determine the localization of thioredoxin in whole filaments (Cossar et al., 1985). Heterocysts from both organisms showed substantially less labelling than vegetative cells. Whether a diminished level of thioredoxin in the heterocyst in some manner augments G6PDH activity in the heterocyst remains to be seen. Fe^{3+}EDTA chelates, and the less avid Fe^{3+} complexes of naturally occurring compounds (ATP, ADP, GTP, oxalate, pyridoxal 5′-phosphate, P_i, PP_i) were found to mediate the oxidative modulation of several cyanobacterial and chloroplast enzymes (Udvardy et al., 1984b). Glucose-6-phosphate dehydrogenase, previously deactivated by treatment with dithiothreitol plus thioredoxin, was reactivated in the presence of Fe^{3+} chelates under aerobic conditions in the light. Aerobic photodegradation of Fe^{3+} EDTA is well known (Lockhart and Blakeley, 1975), and one or more of the products formed could be activating G6PDH. It is not known if the presence of Fe^{3+} EDTA affects the activity of heterocysts.

Glutathione has been found in cyanobacteria and it may be the dominant thiol (Fahey et al., 1978; Giddings et al., 1981). Glutathione functions, inter alia, in the protection of cells against free radicals and of reactive oxygen intermediates, and in the reduction of sulfide linkages in proteins (Meister, 1983). Glutathione reductase activity with $NADPH^+$ has been noted in extracts of vegetative cells and heterocysts of *Nostoc muscorum*, and it may alleviate the toxicity of oxygen reduction products formed via superoxide dismutase or in a photoreaction (Karni et al., 1984). Glutathione reductase has been purified to electrophoretic homogeneity from *Anabaena* sp. strain 7119 (Serrano et al., 1984). The purified enzyme was characterized as a dimeric flavin adenine dinucleotide protein with an absolute specificity for $NADPH^+$. Glyceraldehyde-3-phosphate dehydrogenase from *Anacystis nidulans* was activated up to five-fold by reduced glutathione in a physiological concentration range, 0.1 to 2 mM (Udvardy et al., 1982). Reduced glutathione may be a highly suitable compound to protect enzymes against oxidative stress and the system glutathione-glutathione reductase can provide an important mechanism to control the intracellular disulfide-sulfhydryl couple.

5. Regulation of nitrogenase in cells

Anabaena strains CA (ATCC 33047) and 1F are very similar organisms with respect to their general growth requirement and their growth rates (Stacey et al., 1977; Gotto et al., 1979a) but are very different in the response of their nitrogenase in vivo to NH_4^+ (see Section 3 of Chapter 6). Strain 1F is of the classical type. In the presence of NH_4Cl, nitrogenase and heterocysts are lost. Strain CA, on the other hand, expresses heterocysts and nitrogenase (acetylene reduction), even when grown in 20 mM NH_4Cl (Bottomley et al., 1979). The generation times, nitrogenase activity, heterocyst and proheterocyst frequency were very similar whether cells in the presence of 20 mM NH_4Cl were grown on an atmostphere of argon-O_2-CO_2 (78:21:1%, by vol.) – wherein NH_4^+ was the nitrogen source – or on nitrogen-O_2-CO_2 (78:21:1%, by vol.) wherein both NH_4^+ and N_2 were available. The presence of nitrate is required for complete suppression of nitrogenase and heterocyst expression in strain CA. The level of NH_4Cl in which strain CA will grow, 20 mM (1 g/l), is unprecedented among the cultures of cyanobacteria used as experimental tools where toxicity is usually evident at 2 mM (Van Baalen, unpublished observations). The NH_4^+ metabolism of strain CA, independent of nitrogenase effects, needs close examination. In another light-driven nitrogen fixation, that of purple non-sulfur bacteria, there is a rapid and complete inactivation of nitrogenase activity by NH_4^+ through covalent modification by adenosine diphosphoribosylation of the Fe protein (Pope et al., 1985). In a comparative study of the short-time effects of NH_4^+ at high and low light on nitrogenase activity (acetylene reduction) in *Rhodospirillum rubrum* and *Anabaena* sp. strain 1F (Yoch and Gotto, 1982), inhibition was especially pronounced in *R. rubrum* at low light intensities (5 $\mu E \cdot m^{-2} \cdot s^{-1}$), and not readily reversed by subsequent exposure of the cells at the high light (140 $\mu E \cdot m^{-2} \cdot s^{-1}$). In *Anabaena* sp. strain 1F (a typical NH_4^+ sensitive organism) the small inhibition observed at low light after 1 h was rapidly reversed by shifting cells to high light. It was concluded that the covalent modification mechanism of nitrogenase during inhibition by NH_4^+ in *R. rubrum* was not operating in *Anabaena* sp. strain 1F. The NH_4^+ repression of nitrogenase activity in strain 1F occurs over a period of hours, as it does in *Anabaena cylindrica* (Ohmori and Hattori, 1974) and other organisms, reflecting suppression of synthesis of new nitrogenase while existing enzyme is slowly diluted out. Although the rapid inactivation of nitrogenase by covalent modification that is triggered by NH_4^+ in *R. rubrum* is apparently not present in cyanobacteria, an analogous type of regulation may exist, but triggered by a different agent, e.g. oxygen. The inactivation of nitrogenase under 100% O_2 in strains CA and IF was rapid, 5–10 min, and appeared not to involve irreversible denaturation and de novo synthesis of nitrogenase itself (Pienkos et al., 1983). Rather, there was recovery back to the original activity within several hours.

The strong regulation of nitrogenase by nitrate in *Anabaena* sp. strain CA has been examined using several approaches. Several chlorate-resistant mutants were isolated that did not grow on nitrate when cultured under a nitrogen-free atmosphere (79% Ar, 20% O_2, 1% CO_2) and had no nitrate reductase activity. When grown on N_2 these mutants produced heterocysts and reduced acetylene even in the presence of 10 mM nitrate (Tabita et al., 1981). The simple conclusion that regulation of heterocysts and nitrogenase synthesis in strain CA is mediated by a reduction product of nitrate or by some property of nitrate reductase itself is difficult to understand since these nitrate reductase-less mutants were fully regulated during growth on NH_4NO_3 (Tabita et al., 1981). Subsequent work using the classical inhibitor of glutamine synthetase, L-methionine-D,L-sulfoximine (MSX), showed that MSX alleviated the repressive effect of nitrate on nitrogenase synthesis which suggests that the nitrate effect is indirect through metabolites generated from the ammonium via nitrate reduction (Ramos and Guerrero, 1983).

There is no doubt that *Anabaena* sp. strain CA shows unique properties in regulation of nitrogenase. Even at 20 mM NH_4Cl it continues to form heterocysts and to express 25% of control level of nitrogenase, but in the presence of KNO_3, heterocysts and nitrogenase are not formed (Bottomley et al., 1979). It is possible that these observations in strain CA can be reconciled through a broad specificity monoxygenase such as methane monoxygenase acting as an ammonia monooxygenase (Woodland and Dalton, 1984). An oxidation product of ammonia or a partial reduction product of nitrate would then become the true regulatory species which could act directly on nitrogenase. Oxygen is becoming increasingly attractive as a cellular regulatory agent and the interesting suggestion that glutamine synthetase requires oxygenation via a p-450 mediated oxygenase prior to proteolysis (Levine et al., 1981; Levine, 1983) could have direct bearing on the regulation of nitrogenase in cyanobacteria.

Nitrogenase activity (acetylene reduction) in photoautotrophically grown cultures rapidly diminishes when cells are transferred to darkness (Fay, 1976). An argument was made that continued nitrogen fixation in the dark required build-up of photosynthate accumulated during a light period which was then used via respiration to supply ATP, but only for a limited time period (Ohmori, 1984) (see Section 4 of Chapter 6). In a more extensive study using *Anabaena* ATCC 33047 (same as *Anabaena* sp. strain CA), nitrogenase activity fell to nearly zero 2 h after transfer to darkness (Ramos et al., 1985). Recovery of full activity upon transfer to light was linear and took approximately 1 h. The light recovery was inhibited by chloramphenicol or rifampin, indicating that de novo protein synthesis was required. Analogous results were also obtained with *Anabaena* PCC 7119, *Anabaena* 29143 and *Nostoc* PCC 6719. The function in nitrogen fixation of this protein that is inactivated in the dark is unknown. However, its existence does provide strong evidence for a special role of light in regulating nitrogenase in cyanobacteria.

6. The effects of oxygen

Nitrogen fixation is an anaerobic process, with both components of nitrogenase being rapidly and irreversibly destroyed by oxygen (Robson and Postgate, 1980). The cyanobacteria are the only organisms that have solved the special problem of protecting nitrogenase during simultaneous photosynthetic O_2 evolution. Because of this situation the cyanobacteria have probably evolved the most efficient nitrogenase oxygen protection systems. Indeed, in the classic genus *Azotobacter* growth under nitrogen-fixing conditions and nitrogenase activity in vivo are inhibited by excess oxygen (Dalton and Postgate, 1969).

Most cyanobacteria that grow diazotrophically elaborate the specialized cell known as the heterocyst. Heterocysts can be isolated that have a high endogenous nitrogenase (acetylene reduction) activity (Kumar et al., 1983). These heterocysts require no external source of reductant or ATP and, as noted before, can accumulate up to 10 μmol C_2H_4 per mg dry weight in 3 h. They provide a clear affirmation that nitrogenase, the oxygen protection mechanism, the necessary photosynthetic pigments and light reactions, and whatever provides electrons, all reside in the heterocyst. Most of the data discussed in this section is from experimental work with heterocystous cyanobacteria, but it should be kept in mind that there are several examples of non-heterocystous cyanobacteria that grow diazotrophically. Wyatt and Silvey (1969) reported that a strain of *Gleocapsa* grew aerobically in the light without a combined nitrogen source. These early observations have been amply confirmed (Rippka et al., 1971; Hamadi and Gallon, 1981). Although not yet in pure culture, the genus *Trichodesmium (oscillatoria)* provides further evidence that appreciable aerobic nitrogen fixation can occur in non-heterocystous cyanobacteria (see Section 3.2. of Chapter 15). Aerobic activities as high as 3% of the maximum rate of *Anabaena* sp. strain CA (30 to 40 nmol ethylene formed per mg dry weight per min) were noted in field-collected fresh bundles of *Trichodesmium* sp. (Saino and Hattori, 1982). The color and cell morphology of this material was excellent with no microscopic indication of any decay along the filament. It should be noted that 50% of this activity remained even when the bundles were purposely disrupted into filaments and exposed to air level oxygen. Evidently the oxygen protective mechanism(s) in *Gloeocapsa* and *Trichodesmium* are well advanced and merit study.

There are presently two experimental approaches to observing the effects of oxygen on nitrogenase activity in whole cells. In the first, cells growing diazotrophically at air oxygen levels are purposely shifted to higher or lower oxygen levels (Mackey and Smith, 1983). Cells of *Anabaena cylindrica* ATCC 29414 grown in air, then sparged with 100% N_2 and returned to air, became more sensitive to oxygen, suggesting that under lowered oxygen cells partially lost the ability to protect nitrogenase from oxygen. This same type of observation was made when cells of *Anabaena* sp. strain

CA growing diazotrophically on 1% CO_2–99% N_2 were shifted to 1% CO_2-in-air, wherein there was 30-min lag in re-establishing the maximum rate of acetylene reduction (Gotto and Van Baalen, unpublished observations). If chloramphenicol was added just before the shift to 1% CO_2-in-air then recovery was blocked. The other side of the coin, recovery of nitrogenase activity after exposure of cells to 99 to 100% O_2 is also useful, although untoward effects not related to nitrogenase are likely. When *Anabaena* sp. strains CA or 1F growing on N_2 were shifted from 1% CO_2-in-air to 1% CO_2–99% O_2, nitrogenase activity declined rapidly to about 5% of the control and then recovered to 70 to 80% of the control after 3 h (Pienkos et al., 1983). The recovery phase was again blocked by chloramphenicol. In either transition, from 1% CO_2–99% N_2 then to 1% CO_2-in-air or 1% CO_2-in-air to 1% CO_2–99% O_2, there is then a requirement for protein synthesis to protect against oxygen toxicity. These experiments provide strong evidence that a protein associated with nitrogenase rapidly responds to external O_2 tension. Whether this is the elusive 'oxygen-protection protein' which imparts conformation protection (Robson and Postgate, 1980) or whether the protein is of some regulatory nature remains to be seen. In any event, the way seems now open to finally learning about the nitrogenase oxygen protection mechanism(s) in the cyanobacteria.

A second approach to the problem of oxygen protection of nitrogenase which complements the studies in whole cells and which in the long run may prove even more valuable, is the isolation of mutants with increased or even decreased sensitivity to the oxygen concentration external to the cells. Haury and Wolk (1978) observed three classes of *N*-methyl-*N'*nitro-*N*-nitrosoguanidine-(MNNG)-induced mutants of *Anabaena variabilis* (ATCC 29413) that were phenotypically oxygen-sensitive. Mutants deficient in heterocyst envelope glycolipids had little or no nitrogenase activity when assayed under aerobic conditions. Apparent revertants capable of aerobic growth on N_2 had increased levels of glycolipids. Another phenotype expressed high nitrogenase activity only at low oxygen tensions. Three other mutants showed the very interesting property of high nitrogenase, but only in the presence of dithionite and, may be, impaired in electron transport. Murry (1983), using several of the mutants isolated by Haury and Wolk, examined the question of whether mutants unable to fix nitrogen under aerobic conditions still synthesized nitrogenase. Under aerobic conditions, although nitrogenase activity could not be detected, an immunological assay showed that Fe-Mo protein was present at levels similar to those found in the wild-type and in the mutant strains induced microaerophilically. The results indicated that O_2 does not repress Fe-Mo protein synthesis, at least in these mutants. Mutants altered in their ability to grow under nitrogen-fixing conditions have also been isolated from *Anabaena* sp. strain CA after treatment with MNNG (Grillo et al., 1979; Gotto et al., 1979b). The most striking characteristic of these Nif-impaired mutants was the varied sensitivity of their nitrogen-fixing activities to O_2 in comparison with the parent strain, which is scarcely affected. Five of the mutants isolated seemed to represent a gradation of sensitivity to O_2, as suggested by their different responses to atmospheric O_2 in terms of growth and nitrogenase activity. One mutant, GM9, displayed the interesting characteristic of complete inhibition of nitrogenase activity upon transfer to air, and immediate recovery of acetylene reduction after a short period of gassing with Ar. Apparently in this mutant nitrogenase is switched on and off depending on the external O_2 tension. This may be related to some feature of electron transport. However, it could also be a subtle kind of regulation protecting nitrogenase from short-time high oxygen effects, as might occur on a microscale in the environment.

One gains the impression, even from the limited study of these few mutants from two organisms, that there are at least several different phenomena involved in O_2 protection of nitrogenase. The importance of this area cannot be over-emphasized. Oxygen protection mechanisms for nitrogenase most likely evolved along with the oxygen side of photosynthesis in some ancestral cyanobacteria.

At the molecular level they may share some common features, especially some related proteins. The oxygen side of photosynthesis must include some types of protection against the potentially lethal effects of reactive species of oxygen and some of these mechanisms may be similar to or the same needed to protect nitrogenase.

7. The photosynthetic pigments

There has been a long-standing controversy on the presence and possible function of the accessory pigments, the phycobiliproteins, in heterocysts. In early studies Fay (1969) found phycocyanin, some 73% reduced in amount, in heterocysts of *Anabaena cylindrica* as compared to vegetative cells and he supposed that phycocyanin was probably absent from fully developed heterocysts. Thomas (1970) used microspectrophotometric technique to compare the pigments in old and young vegetative cells and heterocysts. He concluded that all heterocysts, irrespective of type, contain very little *c*-phycocyanin and no *allo*-phycocyanin. In still later work it was claimed that isolated heterocysts lack phycobiliproteins and phycobilisomes altogether (Alberte et al., 1980). In contrast to these negative reports, it now seems that phycobiliproteins are present in heterocysts of certain species, and, in addition, that they can participate in energy transfer to chlorophyll *a*. Although not directly isolated and characterized, the occurrence of phycobiliproteins in heterocysts of the endophytic *Anabaena azollae* was inferred from fluorescence microscopy of intact filaments (Tyagi et al., 1981). In addition, in this work it was found that quanta absorbed by phycobiliproteins were as effective in driving acetylene reduction as quanta absorbed by chlorophyll *a*. Peterson et al. (1981) also noted energy transfer from phycobiliproteins to Photosystem I (chlorophyll *a*) in vegetative cells and heterocysts of *Anabaena variabilis* (see Section 5.1. of Chapter 1). Yamanaka and Glazer (1983), in a detailed study of isolated heterocysts of *Anabaena* PCC 7119, found by emission fluorescence spectrophotometry that heterocysts exhibited spectra characteristic of both phycocyanin and chlorophyll *a*. Excitation at 580 nm, a wavelength strongly absorbed by phycocyanin, resulted in a maximum emission at 649 nm and a broad, but smaller peak near 720 nm. On the basis of these types of results and sodium dodecyl sulfate (SDS) gel electrophoresis of phycobiliprotein particles they concluded that the phycobiliproteins in heterocysts are present as multimeric complexes, attached to thylakoid membranes but lacking *allo*-phycocyanin, and are capable of transferring absorbed light energy between individual biliproteins. They also concluded that phycobiliproteins of heterocysts were not synthesized de novo during heterocyst differentiation, implying that they are carried over from the vegetative cell. In isolated, highly active, heterocysts from *Anabaena* sp. strain CA, however, both *c*-phycocyanin and *allo*-phycocyanin were isolated and purified on DEAE-cellulose albeit in reduced amount (50% and 20%, respectively) than in vegetative cells (Kumar et al., 1982) (see also Section 5.3.2. of Chapter 13). There also seems to be some discrepancy in the chlorophyll *a* levels reported in heterocysts as compared to whole filaments of various *Anabaena* species. Both Fay (1969) and Alberte et al. (1980) suggested that heterocysts had about one-half the chlorophyll *a* content of vegetative cells, while isolated heterocysts of strain CA contained a similar level of chlorophyll *a* as in vegetative cells (Kumar et al., 1982), and later work indicates that the chlorophyll *a* content of these isolated heterocysts may even be 10–20% higher than that of vegetative cells (Smith and Van Baalen, unpublished observations). Overall, the carotenoids in the heterocysts seem little diminished from the levels in vegetative cells (Fay, 1969; Thomas, 1970).

8. The role of light

The question of what light does in the heterocyst continues to be elusive. The older work emphasized the specialization of the heterocysts in having Photosystem I only (Donze et al., 1972; Lex and Stewart, 1973) and therefore no oxygen evolution. It is certainly true that no one has recorded net oxygen evolution in isolated heterocysts, which would imply the absence of a functional Photosystem II. However, if the heterocyst relies upon oxidation of reduced carbon transferred and stored from the vegetative cells, the oxidative pathway used and the oxygen source remain unknown. Action spectra do little to resolve the conundrum. Fay (1970) proposed that the action spectra of acetylene reduction corresponded to the maximum light absorption by chlorophyll *a*, and he suggested that Photosystem I was primarily involved in nitrogen fixation by blue-green algae. However, as pointed out by Tyagi et al. (1981), the 625 nm light absorbed by phycobiliproteins appears at least 70% as effective as light absorbed by chlorophyll *a*. The presence of phycobiliproteins in heterocysts and their possible involvement in the action spectrum of acetylene reduction strongly suggest that the light reactions in the heterocyst are not as simple as is commonly supposed. It is, furthermore, disquieting that isolated heterocysts can endogenously give such high yields of ethylene per mg of heterocysts (Kumar et al., 1983). In order to accomplish this the heterocyst would need to oxidize all its carbohydrate. One speculation that seems not examined is that the heterocyst operates something like the violaxanthin cycle (Sapozhnikov, 1973) and can thereby reversibly bind and release any oxygen formed in the light in a tightly controlled manner.

References

Alberte, R.S., Tel-Or, E., Packer, L. and Thornber, J.P. (1980) Functional organization of the photosynthetic apparatus in heterocysts of nitrogen fixing cyanobacteria. Nature (London) 284, 481–483.

Bergman, B. (1980) Stimulation of nitrogenase activity and photosynthesis in some cyanobacteria by glyoxylate. Physiol. Plant 49, 398–404.

Bothe, H. (1982) Nitrogen fixation. In N.G. Carr and B.A. Whitton (eds.), The Biology of Cyanobacteria. The University of California Press, Berkeley-Los Angeles, pp. 87–104.

Bottomley, P.J. and Van Baalen, C. (1978) Characteristics of heterotrophic growth in the blue-green alga, *Anabaena* sp. strain MAC. J. Gen. Microbiol. 107, 309–318.

Bottomley, P.J., Grillo, J.F., Van Baalen, C. and Tabita, F.R. (1979) Synthesis of nitrogenase and heterocysts by *Anabaena* sp. CA in presence of high levels of ammonia. J. Bacteriol. 140, 938–943.

Cossar, J.D., Darling, A.J., Ip, S.M., Rowell, P. and Stewart, W.P.D. (1985) Immunocytochemical localization of thioredoxins in the cyanobacteria *Anabaena cylindrica* and *Anabaena variabilis*. J. Gen. Microbiol. 131, 3029–3035.

Crawford, N.A., Sutton, C.W., Yee, B.C., Johnson, T.C., Carlson, D.C. and Buchanan, B.B. (1984) Contrasting modes of photosynthetic enzyme regulation in oxygenic and anoxygenic prokaryotes. Arch. Microbiol. 139, 124–129.

Dalton, H. and Postgate, J.R. (1969) Effect of oxygen on growth of *Azotobacter chroococcum* in batch and continuous cultures. J. Gen. Microbiol. 54, 463–473.

Donze, M., Haveman, J. and Schiereck, P. (1972) Absence of Photosystem II in heterocysts of the blue-green alga *Anabaena*. Biochim. Biophys. Acta 256, 157–161.

Fahey, R.C., Brown, W.C., Davis, W.B. and Worsham, M.B. (1978) Occurrence of glutathione in bacteria. J. Bacteriol. 133, 1126–1129.

Fay, P. (1965) Heterotrophy and nitrogen fixation in *Chlorogloea fritschii*. J. Gen. Microbiol. 39, 11–20.

Fay, P. (1969) Cell differentiation and pigment composition in *Anabaena cylindrica*. Arch. Mikrobiol. 67, 62–70.

Fay, P. (1970) Photostimulation of nitrogen fixation in *Anabaena cylindrica*. Biochim. Biophys. Acta 216, 353–356.

Fay, P. (1976) Factors influencing dark nitrogen fixation in a blue-green alga. Appl. Environ. Microbiol. 31, 376–379.
Friga, G.M. and Farkas, G.L. (1981) Isolation and properties of an isocitrate dehydrogenase. Arch. Microbiol. 129, 331–334.
Furbank, R.T. (1984) Photoreduction of oxygen in higher plants: mechanisms and physiological functions. What's New Plant Physiol. 15, 33–36.
Giddings, T.H., Wolk, C.P. and Shomer, I.A. (1981) Metabolism of sulfur compounds by whole filaments and heterocysts of *Anabaena variabilis*. J. Bacteriol. 146, 1067–1074.
Gotto, J.W., Tabita, F.R. and Van Baalen, C. (1979a) Isolation and characterization of rapidly-growing, marine, nitrogen-fixing strains of blue-green algae. Arch. Microbiol. 121, 155–159.
Gotto, J.W., Tabita, F.R. and Van Baalen, C. (1979b) Mutants of *Anabaena* strain CA altered in their ability to grow under nitrogen fixing conditions. J. Bacteriol. 140, 327–332.
Grillo, J.F., Bottomley, P.J., Van Baalen, C. and Tabita, F.R. (1979) A mutant of *Anabaena* sp. CA with oxygen sensitive nitrogenase activity. Biochem. Biophys. Res. Commun. 89, 685–693.
Guerrero, M.G., Vega, J.M. and Losada, M. (1981) The assimilatory nitrate-reducing system and its regulation. Ann. Rev. Plant Physiol. 32, 169–204.
Hamadi, A.F. and Gallon, J.R. (1981) Calcium ions, oxygen and acetylene reduction (nitrogen fixation) in the unicellular cyanobacterium *Gloeocapsa* sp. 1430/3. J. Gen. Microbiol. 125, 391–398.
Haury, J.F. and Wolk, C.P. (1978) Classes of mutant with oxygen sensitive nitrogenase activity. J. Bacteriol. 136, 688–692.
Hutson, K.G. and Rogers, L.J. (1975) Two plant-type ferredoxins in light-grown and dark-grown cells of the blue-green bacterium *Nostoc* strain MAC. Biochem. Soc. Trans. 3, 377–379.
Jensen, B.J. (1983) Energy requirement for diazotrophic growth of the cyanobacterium *Anabaena variabilis* determined from growth yields in the dark. J. Gen. Microbiol. 129, 2633–2640.
Juttner, F. (1983) ^{14}C-labeled metabolites in heterocysts and vegetative cells of *Anabaena cylindrica* filaments and their presumptive function as transport vehicles of organic carbon and nitrogen. J. Bacteriol. 155, 628–633.
Karni, L., Moss, S.J. and Tel-Or, E. (1984) Glutathione reductase activity in heterocysts and vegetative cells of the cyanobacterium *Nostoc muscorum*. Arch. Microbiol. 140, 215–217.
Kumar, A., Tabita, F.R. and Van Baalen, C. (1982) Isolation and characterization of heterocysts from *Anabaena* sp. strain CA. Arch. Microbiol. 133, 103–109.
Kumar, A., Tabita, F.R. and Van Baalen, C. (1983) High endogenous nitrogenase activity in isolated heterocysts of *Anabaena* sp. strain CA after nitrogen starvation. J. Bacteriol. 155, 493–497.
Levine, R.L. (1983) Oxidative modification of glutamine synthetase. I. Inactivation is due to loss of one histidine residue. J. Biol. Chem. 258, 11823–11827.
Levine, R.L., Oliver, C.N., Fulks, R.M. and Stadtman, E.R. (1981) Turnover of bacterial glutamine synthetase: oxidative inactivation precedes proteolysis. Proc. Natl. Acad. Sci. USA 78, 2120–2124.
Lex, M. and Stewart, W.P.D. (1973) Algal nitrogenase, reductant pools and photosystem I activity. Biochim. Biophys. Acta 292, 436–443.
Lockhart, H.B. and Blakeley, R.V. (1975) Aerobic photodegradation of Fe (III)-(ethylenedinitrilo) tetraacetate (ferric EDTA), Environ. Sci. Technol. 9, 1036–1038.
Mackey, E.J. and Smith, G.D. (1983) Adaptation of the cyanobacterium *Anabaena cylindrica* to high oxygen tensions. FEBS Letts. 156, 108–112.
Meister, A. (1983) Selective modification of glutathione metabolism. Science 220, 470–478.
Murry, M.A. (1983) Evidence for the lack of oxygen repression of Fe-Mo protein in mutants of *Anabaena variabilis*. Curr. Microbiol. 8, 159–163.
Ohmori, M. (1984) Effects of preillumination on dark nitrogen fixation and respiration by *Anabaena cylindrica*. Plant Cell Physiol. 25, 125–130.
Ohmori, M. and Hattori, A. (1974) Effect of ammonia on nitrogen fixation by the blue-green alga *Anabaena cylindrica*. Plant Cell Physiol. 15, 131–142.
Papen, H., Neuer, G., Refaian, M. and Bothe, H. (1983) The isocitrate dehydrogenase from cyanobacteria. Arch. Microbiol. 134, 73–79.
Peterson, R.B., Dolan, E., Calvert, H.E. and Ke, B. (1981) Energy transfer from phycobiliproteins to photosystem I in vegetative cells and heterocysts of *Anabaena variabilis*. Biochim. Biophys. Acta 634, 237–248.
Pienkos, P.T., Bodmer, S. and Tabita, F.R. (1983) Oxygen inactivation and recovery of nitrogenase activity in cyanobacteria. J. Bacteriol. 153, 182–190.
Pope, M.R., Murrell, S.A. and Ludden, P.W. (1985) Covalent modification of the iron protein of nitrogen-

ase from *Rhodospirillum rubrum* by adenosine diphosphoribosylation of a specific arginine residue. Proc. Natl. Acad. Sci. USA 82, 3173–3177.

Postgate, J.R. and Cannon, F.C. (1981) The molecular and genetic manipulation of nitrogen fixation. Phil. Trans. Roy. Soc. Lond. B 292, 589–599.

Poupard, J.A., Husain, I. and Norris, R.F. (1973) Biology of the *Bifidobacteria*. Bacteriol. Rev. 37, 136–165.

Privalle, L.S. (1984) Effects of D-erythrose on nitrogenase activity in whole filaments of *Anabaena* sp. strain 7120. J. Bacteriol. 160, 794–796.

Privalle, L.S. and Burris, R.H. (1984) D-Erythrose supports nitrogenase activity in isolated heterocysts of *Anabaena* sp. strain 7120. J. Bacteriol. 157, 350–356.

Ramos, J.L. and Guerrero, M.G. (1983) Involvement of ammonium metabolism in the nitrate inhibition of nitrogen fixation in *Anabaena* sp. strain ATCC 33047. Arch. Microbiol. 136, 81–83.

Ramos, J.L., Madueno, F. and Guerrero, M.G. (1985) Regulation of nitrogenase levels in *Anabaena* sp. ATCC 33047 and other filamentous cyanobacteria. Arch. Microbiol. 141, 105–111.

Rippka, R., Neilson, A., Kunisawa, R. and Cohen-Bazire, G. (1971) Nitrogen fixation by unicellular blue green algae. Arch. Mikrobiol. 76, 341–348.

Robson, R.L. and Postgate, J.R. (1980) Oxygen and hydrogen in biological nitrogen fixation. Ann. Rev. Microbiol. 34, 183–307.

Saino, T. and Hattori, A. (1982) Aerobic nitrogen fixation by the marine non-heterocystous cyanobacterium *Trichodesmium (oscillatoria)* spp.: its protective mechanism against oxygen, Marine Biol. 70, 251–254.

Sapozhnikov, D.I. (1973) Investigation of the violaxanthin cycle. Pure Appl. Chem. 35, 47–61.

Schmidt, A. (1980) Isolation of two thioredoxins from the cyanobacterium *Synechococcus* 6301. Arch Microbiol. 127, 259–265.

Schrautemeier, B. and Böhme, H. (1985) A distinct ferredoxin for nitrogen fixation isolated from heterocysts of the cyanobacterium *Anabaena variabilis*. FEBS Letts. 184, 304–308.

Serrano, A., Rivas, J. and Losada, M. (1984) Purification and properties of glutathione reductase from the cyanobacterium *Anabaena* sp. strain 7119. J. Bacteriol. 158, 317–324.

Simpson, F.B. and Burris, R.H. (1984) A nitrogen pressure of 50 atmospheres does not prevent evolution of hydrogen by nitrogenase. Science 224, 1095–1097.

Singh, P.K. (1973) Nitrogen fixation by the unicellular blue-green alga *Aphanothece*. Arch. Mikrobiol. 92, 59–62.

Stacey, G., Van Baalen, C. and Tabita, F.R. (1977) Isolation and characterization of a marine *Anabaena* sp. capable of rapid growth on molecular nitrogen. Arch. Microbiol. 114, 197–201.

Stewart, W.P. (1980) Some aspects of structure and function in N_2-fixing cyanobacteria. Ann. Rev. Microbiol. 34, 497–536.

Tabita, F.R., Ramsden, D.K. and Van Baalen, C. (1981) Nitrate control of heterocyst and nitrogenase synthesis in *Anabaena* sp. CA. In A.H. Gibson and W.E. Newton (eds.), Current Perspectives in Nitrogen Fixation. Elsevier/North-Holland, Amsterdam, p. 452.

Thomas, J. (1970) Absence of pigments of photosystem II of photosynthesis in heterocysts of a blue-green alga. Nature (London) 228, 181–183.

Tyagi, V.V.S., Ray, T.B., Mayne, B.C. and Peters, G.A. (1981) The *Azolla-Anabaena azollae* relationship. XI. Phycobiliproteins in the action spectrum for nitrogenase catalyzed acetylene reduction. Plant Physiol. 68, 1479–1484.

Udvardy, J., Balogh, A. and Farkas, G.L. (1982) Modulation of glyceraldehyde-3-phosphate dehydrogenase in *Anacystis nidulans*. Arch. Microbiol. 133, 2–5.

Udvardy, J., Borbely, G., Juhasz, A. and Farkas, G.L. (1984a) Thioredoxins and the redox modulation of glucose-6-phosphate dehydrogenase in *Anabaena* strain PCC 7120 vegetative cells and heterocysts. J. Bacteriol. 157, 681–683.

Udvardy, J., Borbely, G., Juhasz, A. and Farkas, G.L. (1984b) Fe^{3+} chelates mediate the oxidative modulation of cyanobacterial and chloroplast enzymes, FEBS Letts. 172, 11–16.

Watanabe, A. and Yamamoto, Y. (1967) Heterotrophic nitrogen fixation by the blue-green alga, *Anabaenopsis circularis*. Nature 214, 738.

Wolk, C.P. (1968) Movement of carbon from vegetative cells to heterocysts in *Anabaena cylindrica*. J. Bacteriol. 96, 2138–2143.

Wolk, C.P. and Shaffer, P.W. (1976) Heterotrophic micro- and macrocultures of a nitrogen-fixing cyanobacterium. Arch. Microbiol. 110, 145–147.

Woodland, M.P. and Dalton, H. (1984) Purification and characterization of component A of the methane monooxygenase from *Methylococcus capsulatus* (Bath). J. Biol. Chem. 259, 53–59.

Wyatt, J.J. and Silvey, J.H.G. (1969) Nitrogen fixation by *Gloeocapsa*. Science 165, 908–909.
Yamanka, G. and Glazer, A.N. (1983) Phycobiliproteins in *Anabaena* 7119 heterocysts. In G.P. Papageorgiou and L. Packer (eds.), Photosynthetic Prokaroytes: Cell Differentiation and Function. Elsevier, Amsterdam.
Yee, B.C., de la Torre, A., Crawford, N.A., Lara, C., Carlson, D.E. and Buchanan, B.B. (1981) The ferredoxin/thioredoxin system of enzyme regulation in a cyanobacterium. Arch. Microbiol. 130, 14–18.
Yoch, D.C. and Gotto, J.W. (1982) Effect of light intensity and inhibitors of nitrogen assimilation on NH_4^+ inhibition of nitrogenase activity in *Rhodospirillum rubrum* and *Anabaena* sp. J. Bacteriol. 151, 800–806.

P. Fay and C. Van Baalen (Eds.) The Cyanobacteria

8

Inclusion bodies in the cyanobacteria: cyanophycin, polyphosphate, polyhedral bodies

ROBERT D. SIMON

Department of Biology, State University of New York at Geneseo, Geneseo, NY 14454, USA

1. Introduction

The long-term availability of nutrients including carbon, nitrogen, and phosphorus is essential for the growth and survival of the cyanobacteria. Therefore these organisms have evolved a variety of mechanisms to accumulate, store, and regulate such materials. The major storage form of these nutrients in the cyanobacteria, as in many other prokaryotes, is as insoluble polymers localized in characteristic 'inclusion bodies'. This chapter will deal with the nature of several of these inclusion bodies, and the role they play in the metabolism of the cyanobacteria.

Several reviews of inclusion bodies in prokaryotes (Codd and Marsden, 1984; Dawes and Senior, 1973; Harold, 1966; Shively, 1974; Shively et al., 1973a) and cyanobacteria (Allen, 1984; Fogg et al., 1973; Fuhs, 1973; Jensen, 1984a,b, 1985) have been published, as well as individual reviews on specific inclusion types (see below). As a basis for considering these bodies in the cyanobacteria, it should be noted that inclusion bodies in prokaryotes share several common characteristics:

(a) Each has a relatively homogeneous composition and represents the deposition of a limited number of biologically formed polymers.

(b) The inclusion is not the site of a functional system equivalent to the organelles found in the eukaryotes, but usually consists of a material lacking enzymatic function. However, the packaging of a limited number of active or potentially active enzymes may occur (e.g., polyhedral bodies).

(c) Inclusion bodies are generally not bounded by cell membranes. In the few cases where there are limiting structures (e.g., polyhedral body, gas vacuoles), they consist of an unusual protein.

(d) Inclusion bodies are most often thought of as serving a 'reserve' function, being limited to those containing materials which can be used by the cell, and are distinguished from the waste deposits which are made by some bacteria (Foster, 1947). Inclusion biopolymers are often produced in either times of abundance or of unbalanced growth which results in the excess of a particular nutrient. They are used in times of need. There are bodies which seem to be primarily carbon reserves (glycogen granules, poly-β-OH butyrate), nitrogen reserves (cyanophycin granules, phycobilisomes (Boussiba and Richmond, 1980)), and phosphate reserves (polyphosphate). Occasionally the bodies play other roles. For example, variations in glycogen and gas vacuoles can alter the buoyancy properties of the cell (Krompkamp and Mur, 1984; Thomas and Walsby, 1985), and are perhaps involved in the physiological ecology of the organism. Such a 'functional' role for inclusions is less frequent, although there are a number of inclusion bodies which have limited distribution or are rarely seen, and which are as yet uncharacterized (Jensen, 1984b).

(e) Storage of material as an inclusion is beneficial to the prokaryote because the stockpiled material is effectively out of solution: the polymer is less osmotically active than the same amount of the free monomer; no energy is required to maintain a high internal concentration; there are no large pools of an unwanted compound to effect metabolic regulation; and, the cell can control the use of the reserve by regulating the enzymes which metabolize the polymer.

(f) The formation of inclusion biopolymers may be the central focus of metabolism at certain times in the life cycle, both in the rate and amount of material being processed (Clark and Jensen, 1969; Fogg, 1951).

(g) The bodies can serve as 'dynamic reservoirs' (Allen, 1984) in nutrient assimilation. With material being continually stored and reused, they can play an important role in buffering the cell against environmental variation (Morita, 1982). For example, the cyanobacteria are primarily obligate autotrophs (Rippka et al., 1979) and successful adaptation to the daily light/dark cycles requires energy storage in the form of glycogen reserves.

The primary objective of this review is to examine the relationship between the structure and function of the material present in several cyanobacterial inclusion bodies. This is now possible, because the most commonly observed inclusion bodies have been isolated and chemically characterized. This review will focus on three inclusions: cyanophycin or structured granules; polyphosphate bodies; and carboxysomes or polyhedral bodies. This will not be a comprehensive discussion since the literature is far too extensive, and several general reviews already exist. However, the present author hopes to provide some new ideas about and enthusiasm for these cyanobacterial cell inclusions.

2. Cyanophycin

Cyanophycin granules (Cyanoficina – Borzi (Borzi, 1886)) are highly refractile granules in the cells of the cyanobacteria (Fig. 1). They vary in size and shape but have been reported in almost all species: unicellular and filamentous, nitrogen-fixing and non-nitrogen-fixing. The structures have been described by light microscopists since the late 1800's and are characterized by their solubility in dilute acid and their cytochemical staining by agents specific for arginine (Borzi, 1886; Fogg, 1951; Hegler, 1901). Electron micrographs show the absence of a membrane around the inclusion but suggest that there is some structure to the internal contents of the inclusion, although the amount of substructure seen is a direct function of the fixation and staining procedure (Fig. 2) (Brown and Bisalputra, 1969; Lang and Fisher, 1969). The name 'structured granule' has become the accepted name for the inclusion seen in the electron microscope (Lang, 1968). Lang et al. (1972) showed that isolated cyanophycin granules were equivalent to the structured granules seen in the electron microscope.

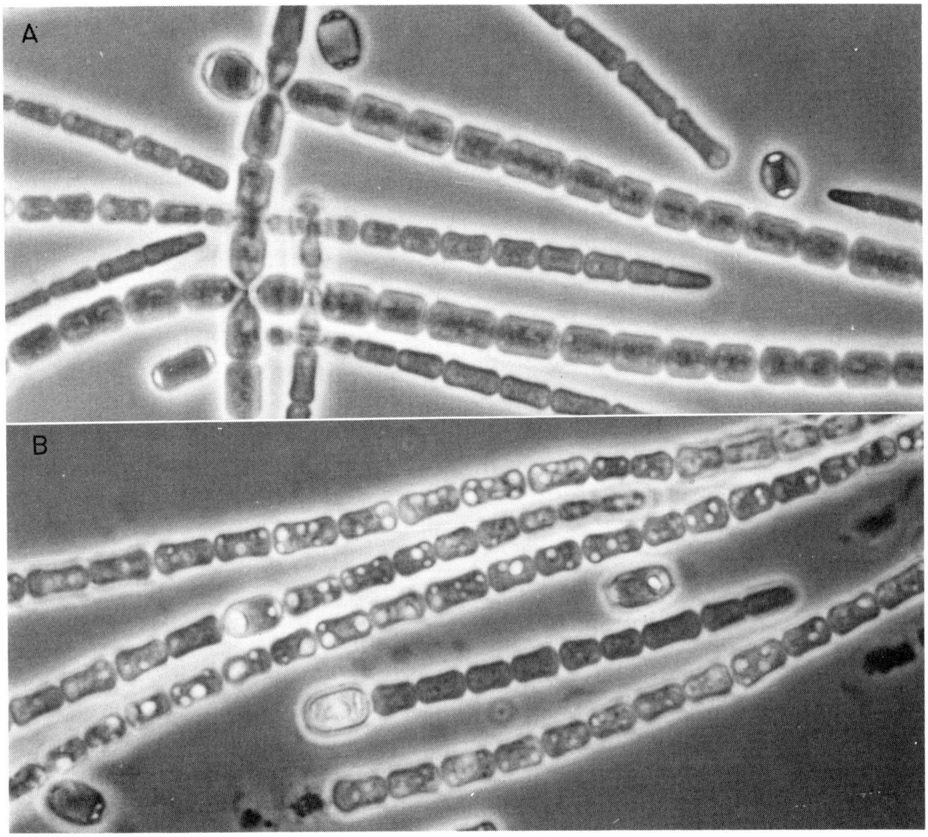

Fig. 1. Phase-contrast light micrographs of *Anabaena cylindrica* grown on mineral medium containing 1.0 g/l $NaNO_3$. Panel A is from cells taken during exponential growth. Cells in panel B are from a stationary phase culture, and show large numbers of cyanophycin granules.

2.1. Isolation and characterization

Simon isolated intact cyanophycin granules from the filamentous cyanobacterium *Anabaena cylindrica* by differential centrifugation, and showed that they consisted of a high molecular mass polypeptide (25 000–100 000 daltons) containing L-aspartic acid and L-arginine in a 1:1 molar ratio (Simon, 1971b). Subsequently, the cyanophycin polypeptide has been isolated from cyanobacterial cell lysates by taking advantage of its unique solubility properties (Simon, 1973b).

2.1.1. *The primary structure*

It was not possible to determine the primary structure of cyanophycin granule polypeptide using traditional sequencing techniques. The polypeptide is resistant to tryptic hydrolysis under all conditions examined, and is unaffected by pronase, pepsin, chymotrypsin, and α- or β-carboxypeptidase (Simon and Weathers, 1976). The polypeptide could also not be degraded from the N-terminus using the Edman procedure (Laursen, 1972). However, studies on the linkages present in the polypeptide provide an explanation for these observations. Several lines of evidence suggest that

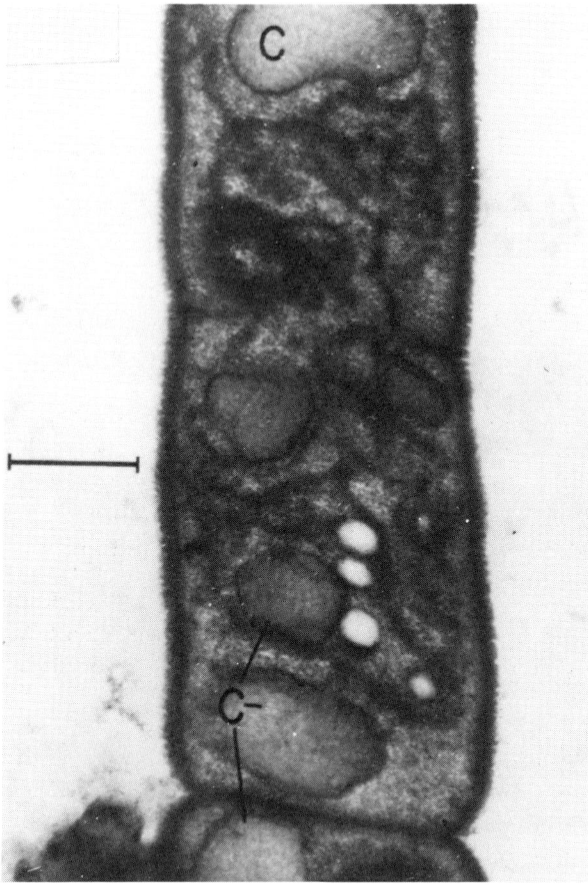

Fig. 2. Thin section electron micrograph of *Plectonema boryanum* grown in phosphate-deficient mineral medium. Scale bar = 500 nm; C, cyanophycin granules. (From Lawry and Simon, 1982)

the polypeptide is a highly branched structure consisting of a polyaspartic acid core to which arginyl residues are attached at each potentially free carboxyl group of the polyaspartic acid (Fig. 3):

(a) Arginine is preferentially released following the hydrolysis of the polypeptide for short times in dilute acid (0.03 N acetic acid, 105° C, 30 min) (Simon and Weathers, 1976). Aspartic acid forms extremely acid-labile peptide bonds, and is often the first amino acid to be released intact upon acid hydrolysis of a polypeptide (Blackburn, 1970). Preferential release of arginine indicates an unusual primary structure.

(b) Linkage analysis was carried out by reducing the free carboxyl groups in the polypeptide and characterizing the products. The polypeptide was reduced with sodium borohydride, hydrolyzed in 6 N HCl at 105° C for 20 h, and the products analyzed. Aspartic acid was unaffected by reduction, indicating that both the aspartic acid carboxyl groups must be involved in covalent linkages. However, the arginine reduction product argininol was present indicating that the alpha carboxyl group of arginine is not involved in covalent linkage. Thus, the linear model could not

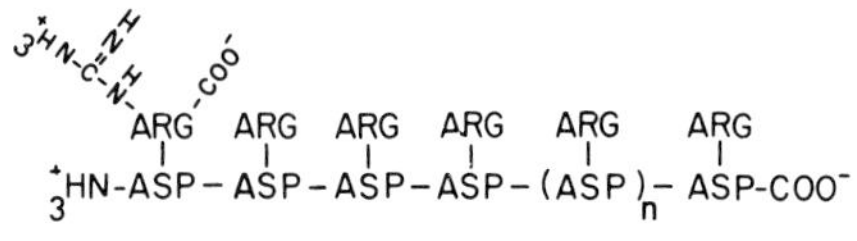

Fig. 3. Proposed model for the structure of the primary constituent of cyanophycin granules, multi-L-arginyl-poly(L-aspartic acid). ARG, arginine; ASP, aspartic acid; $_3{}^+$HN-ASP and ASP-COO$^-$, amino and carboxyl terminal aspartic acids, respectively.

be correct and a branched model for the primary structure was proposed (Simon and Weathers, 1976). Since aspartic acid and arginine are in a 1:1 molar ratio and since no aspartic acid appears to be reduced, every free carboxyl group from the polyaspartic acid core must be arginylated and each arginine branch must be no more than one residue long.

(c) Stark (1972) developed a chemical method for cleaving the carboxyl terminal amino acids from a polypeptide. According to the proposed model (Fig. 3), every arginine in the polypeptide is equivalent to a C-terminal amino acid. Treatment of the isolated cyanophycin granule polypeptide by the Stark procedure resulted in the removal of approximately 75% of the arginine from the polypeptide. Thus, all or almost all of the arginyl residues are at carboxyl-terminal positions.

The branched structure of the cyanophycin granule polypeptide is similar in form to synthetically produced multichain polyamino acids. Using the nomenclature for describing multichain polyamino acids, it has been proposed that the cyanophycin granule polypeptide be called multi-L-arginyl-poly(L-aspartic acid) (Arg-poly(Asp)) (Simon and Weathers, 1976).

2.1.2. Higher order structure

The molecular mass of the polypeptide isolated from cyanophycin granules is of a heterogeneous distribution between 25 000 and 100 000 daltons (Simon, 1971b). This conclusion is based upon the results from the electrophoresis of Arg-poly(Asp) in sodium dodecyl sulfate (SDS) containing polyacrylamide gels and the chromatographic separation of Arg-poly(Asp) on Sephadex G-200. Detailed analysis of SDS gels shows that polypeptides under 25 000 daltons make up less than 5% of the isolated material and that the distribution of weight sizes is not unimodal. There appears to be three distinct classes of Arg-poly(Asp) with peaks centered at approximately 39 000 daltons, 69 000 daltons, and 100 000 daltons. The significance of this distribution of material is not known, although an enzyme which elongates pre-existing chains of Arg-poly(Asp) (see below) incorporates amino acids into all three weight classes (Simon, 1976). In addition, various molecular weight classes separated on Sephadex all have a 1:1 molar ratio of arginine to aspartic acid indicating that the different molecular weight sizes do not represent radically different polypeptide chains (Simon, 1971a).

Isoelectric focusing of Arg-poly(Asp) in 6.0 M urea separates two forms: a major form containing more than 97% of the material at pH 4.75 and a minor band at pH 6.10 (Simon and Weathers, 1976). An enzyme which elongates Arg-poly(Asp) incorporates radioactive arginine or aspartic acid into both fractions. Since the enzyme will not incorporate arginine or aspartic acid into other polypeptides which contain amino- or carboxyl-terminal arginine or aspartic acid, the minor band is probably not a previously unseparated contaminant. This observation is difficult to reconcile with the proposed primary structure. The major charged species in the polypeptide are the free carboxyl and guanido groups of arginine which have pK's of approximately 3.0–4.0 and 13.0, respectively (Tanford, 1962). Thus, the isoelectric point of Arg-poly(Asp) should be somewhere in the range of pH 7–8. The shifted isoelectric point may result from the occasional removal of an arginine branch residue during the biosynthesis, storage, isolation, or manipulation of Arg-poly(Asp). If a single molecule was to lose just a few such branches it would still have an arginine/aspartic acid ratio similar to 1.0, but its excess aspartic carboxyl groups would cause it to focus in the acid range. The minor band at 6.10 may represent that fraction in which all arginine branches are intact.

Arg-poly(Asp) from *Anabaena cylindrica* has been examined by circular dichroism (CD) and laser Raman spectroscopy (Simon et al., 1980). The polypeptide exhibits an ordered structure which is environmentally sensitive. It shows a defined secondary structure in acidic solution but not in alkaline medium. CD spectra suggest that β-sheets exist in the acid soluble form, and by inference, in the insoluble cellular material. The CD spectrum not only supports the basic branch model for Arg-poly(Asp), but is in agreement with the suggestion that all arginine residues are linked via the β-carboxylate groups of aspartic acid, i.e. the polyaspartic acid core contains only α-linked peptide bonds. Raman spectra show that the structure of the insoluble (storage?) form of the polypeptide is similar but not identical to that of the acid-soluble form.

2.2. The enzymology of multi-l-arginyl-poly(l-aspartic acid)

The metabolic pathways involved in the synthesis and degradation of Arg-poly(Asp) have been examined. Since the addition of the protein synthesis inhibitor chloramphenicol to an exponentially growing culture of the cyanobacterium *Anabaena cylindrica* resulted in the accumulation of Arg-poly(Asp) (Simon, 1973a), it was concluded that synthesis was via a non-ribosomal route. Characterization of an enzyme which would elongate pre-existing chains of Arg-poly(Asp) (Simon, 1976) confirmed this notion. In addition, although Arg-poly(Asp) is resistant to a wide variety of commercially available proteases (Simon and Weathers, 1976), enzymes have been identified which will hydrolyze it (Allen et al., 1984; Gupta and Carr, 1982; Simon, 1971a).

2.2.1. *Arg-poly(Asp) synthesis*

An ATP-dependent enzyme will incorporate aspartic acid or arginine into pre-existing chains of Arg-poly(Asp) (Simon, 1976). The enzyme was purified 92-fold using ultracentrifugation, ammonium sulfate precipitation and chromatography on DEAE cellulose. The ability to incorporate aspartic acid and arginine purified together suggesting that one enzyme was responsible for both activities. The reaction required ATP, KCl, MgCl, a sulfhydral reagent, and Arg-poly(Asp). No peptide or polypeptide could replace the requirement for Arg-poly(Asp) and ATP was the only nucleotide which would support the reaction. The high molecular weight product was identified as Arg-poly(Asp) based upon its solubility, resistance to tryptic hydrolysis, isoelectric point, and size distribution. Although the product had a high molecular mass (20 000–100 000 daltons), the amount of arginine or aspartic acid incorporated was not enough to produce such large polypeptides. Therefore, the activity measured must have been an elongation reaction. Whether different reaction conditions or an entirely different enzyme is responsible for the 'de novo' synthesis of Arg-poly(Asp) remains to be determined.

Kinetic experiments (Simon, unpublished) on the enzyme catalyzing the synthetic reaction showed that the K_m for arginine was 1.0×10^{-6} M while the K_m for aspartic acid is significantly higher, 2.0×10^{-4} M. The enzymatic incorporation was unaffected by the addition of either RNAse or DNAse or by the addition of protein synthesis inhibitors. Since the nucleases would degrade ^{3}H-labeled RNA and DNA added to reaction mixtures, the nucleases were active under the conditions of the assay. These indicated that amino-acylated transfer RNAs were not involved in the reaction mechanism as they are in the synthesis of peptidoglycan (Plapp and Strominger, 1969). The enzymatic incorporation of ^{3}H-arginine is dependent upon the presence of cold aspartic acid but not vice versa, suggesting that the mechanism of chain elongation consists of the addition of an aspartic acid residue to the core polyaspartic acid and then the addition of an arginyl residue to the newly added aspartate. Since the enzyme would not add arginine to polyaspartic acid or to any other peptide or protein containing aspartic acid, the sequence of addition must be quite specific.

2.2.2. Arg-poly(Asp) degradation

Arg-poly(Asp) is not digested by commercially available proteases or by arginase (EC 3.3.3.1), an enzyme which will cleave urea from carboxyl-terminal arginine (Greenberg, 1960), even though all the arginine residues in the polypeptide are effectively carboxyl-terminals (Fig. 3). However, the cyanobacteria do degrade Arg-poly(Asp) under a variety of conditions and enzyme activities degrading the polypeptide have been reported. When a stationary phase culture of *Anabaena cylindrica* with cells containing stored Arg-poly(Asp) was diluted to reinitiate growth, a cyanophycin-degrading enzyme activity could be shown (Simon, 1971a). The release of 10% trichloroacetic acid (TCA) soluble arginine from Arg-poly(Asp) has been used as an assay for the activity. The enzyme is found in the soluble fraction following high speed centrifugation, and could be further purified by ammonium sulfate precipitation. No activity was seen below pH 6.0 and there was a broad maximum of activity between pH 8.8 and 10.0. Only free arginine and aspartic acid were found as degradation products, and even at short times no peptides were found in the TCA soluble extracts. Gupta and Carr (1982) also demonstrated the presence of a 'cyanophycinase' from a nitrogen-fixing *Anabaena*. The enzyme activity was higher in heterocysts than in extracts from the vegetative cells of the organism. It was suggested that the enzyme was an exopeptidase and that the final product of degradation was the dipeptide Arg-Asp. The enzyme was inhibited by free arginine and aspartic acid, and had a pH optimum of 8.5.

Extracts from the unicellular cyanobacterium *Aphanocapsa* 6308 also showed a cyanophycin protease (Allen et al., 1982). Using a sensitive assay for measuring the release of radioactive amino acids from ^{14}C-labeled cyanophycin it was possible to show the presence of an enzyme which would release free arginine and aspartic acid. The enzyme was found in cells under conditions where the reserves were being degraded. The pH optimum of the enzyme was 8.0. All information about Arg-poly(Asp) degradation points to an alkaline protease with unusual specificity. The question of regulation and end product formation will not be settled until the protease has been purified free of contaminants.

2.3. Metabolism of the cyanophycin granule

An assay based upon the physical and chemical properties of cyanophycin granule has been developed to measure the cellular content of Arg-poly(Asp) (Simon, 1973b). The assay is relatively insensitive since it requires a chemical measurement, and experiments are often done with relatively dilute suspensions of cyanobacteria. The development of a more sensitive assay would be useful. Since Arg-poly(Asp) is antigenic (Simon, unpublished), the development of immunoassays should be possible. The chemical assay, however, has been used to examine the control mechanisms regulating the cellular content of Arg-poly(Asp) in several organisms.

It is generally considered that the main function of Arg-poly(Asp) is nitrogen storage. Nitrogen is often the limiting nutrient in the aquatic environment (Howarth and Cole, 1985), and many cyanobacteria spend an extraordinary amount of energy to fix nitrogen (Bothe, 1982; Orme-Johnson, 1985). If one were designing a nitrogen reserve material it would be difficult to come up with a polymer as good as cyanophycin. Five molecules of nitrogen are stored for every two amino acids and it is insoluble at the pH and ionic strength inside the cell. Polyarginine would store more nitrogen per weight of reserve, but this basic polypeptide would be soluble and could have devastating effects on the nucleic acid metabolism of the cell.

The metabolism of Arg-poly(Asp) reflects this proposed role as a nitrogen reserve. Arg-poly-(Asp) concentration is low in exponentially growing cells of *Anabaena cylindrica*, but accumulates in cells of stationary phase cultures and may comprise as much as 10–12% of the dry weight (dw) of stationary phase cells (Simon, 1973b). The material also accumulates in cells following the addition of a variety of nitrogen-containing compounds to the medium, including: ammonium chloride, urea, glycine, aspartic acid, and arginine (Lawry and Simon, 1982). Species differ with respect to whether cyanophycin reserves are produced in response to exogenous nitrogen sources. *Anabaena cylindrica* produces reserves given aspartic acid but not arginine (Lawry and Simon, 1982), while *Aphanocapsa* 6308 produces more reserves when given exogenous arginine (Allen et al., 1980). This observation may result from differences in amino acid transport in the two strains.

Stored reserves can be degraded. Arg-poly(Asp) content decreases faster than can be accounted for by dilution alone following the resumption of growth after stationary phase cultures are diluted or when protein synthesis inhibitors are removed (Fig. 4) (Simon, 1973b). Hegler reported the gradual disappearance of cyanophycin granules in dark (Hegler, 1901). Structured granules in akinetes disappear when the akinetes germinate (Miller and Lang, 1968) (see also Section 2.5.3. of Chapter 9).

While the amount of reserve present in a cell must be tuned to the metabolic flow of carbon and nitrogen, the controlling mechanisms must certainly depend upon other parameters as well. In *Aphanocapsa* 6308, Arg-poly(Asp) is degraded when growth resumes after the addition of inorganic nitrogen to nitrogen starved cultures (nitrogen repletion) (Allen, 1984), a time at which one would think that nitrogen reserves could be spared. The strategy which cells use for regulating the storage of reserves are not well worked out. Perhaps, organisms may store materials in anticipation of future starvation periods (Cohen and Parnas, 1976).

Unbalanced growth in the cyanobacteria, as for example after starvation for a required nutrient, often leads to the extensive deposition of cyanophycin granules and/or other inclusion materials. Studies on the effects of sulfur starvation and readdition upon the content of Arg-poly(Asp) in *Anabaena cylindrica* (Lawry and Simon, 1982) show that sulfur starvation inhibits protein synthesis but not the production of Arg-poly(Asp). The reintroduction of SO_4 to sulfur starved cells, while temporarily inhibiting cell growth, causes a significant breakdown of cell material and a rapid buildup of reserve. After growth resumes however, the elevated levels of reserve (up to 18% dw) are rapidly degraded to amounts normally found in cells (0.77% dw).

Some information is available, however, on the control mechanisms operating at the cellular level. Experiments on the effect of protein synthesis inhibitors in vivo suggest that the enzyme(s) responsible for the synthesis of Arg-poly(Asp) are present in the organism, regardless of the reserve content (Simon, 1973a). The addition of chloramphenicol to exponentially growing cells results in the inhibition of protein synthesis and in the simultaneous production of Arg-poly(Asp). Thus, either the enzyme(s) to make the reserve are themselves synthesized on special chloramphenicol insensitive ribosomes, or the enzyme is always present but not active. Such a situation is reasonable in that an enzyme which is expected to form an amino acid containing reserve in the absence of protein synthesis should not itself depend upon protein synthesis. Recent evidence supports this hypothesis. The presence of at least one of the synthetic enzymes has been shown to be independent of the cellular reserve content. The activity of the enzyme which elongates Arg-poly(Asp) has been measured as a function of culture growth in *Anabaena variabilis*. The reserve content is low during exponential growth, yet the synthetic activity is higher in exponentially growing cells than in cells from older cultures which are in the process of producing Arg-poly(Asp) (Fig. 5). The results are also in agreement with the more general observations that many enzymes of amino acid and car-

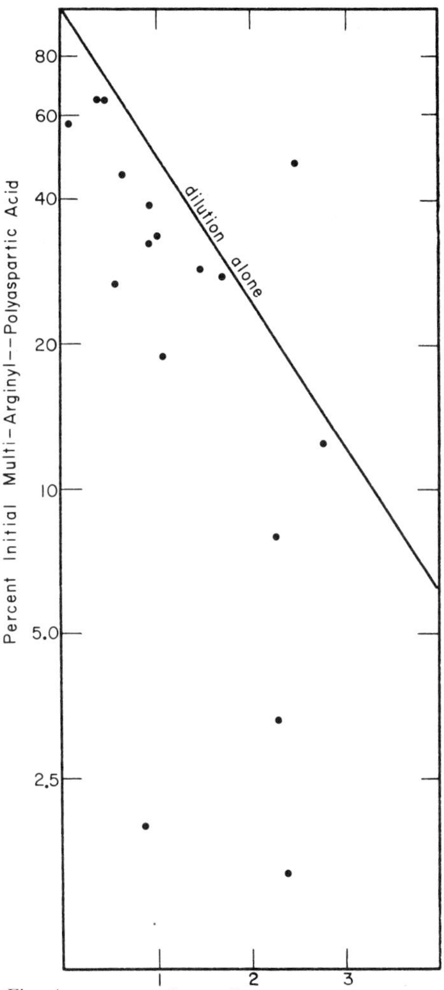

Fig. 4.

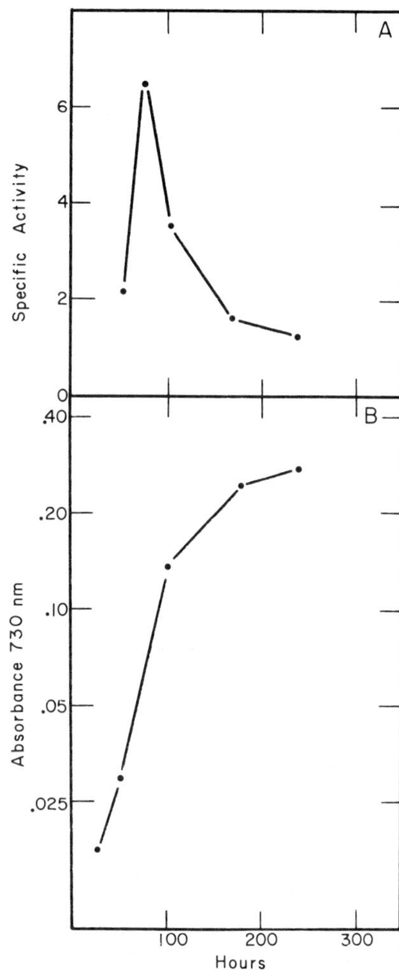

Fig. 5

Fig. 4. Loss of Arg-poly(Asp) reserves following the removal of chloramphenicol from a culture of *Anabaena cylindrica*. Reserve is given as a percentage of that present at the time of drug removal, and the time is given in generations. One generation is equivalent to the time necessary to double cell mass. The line labeled 'dilution alone' is the result expected if synthesis terminates immediately following drug removal, and there is no reserve degradation.

Fig. 5. The activity of the Arg-poly(Asp) elongating enzyme as a function of culture density. Panel A shows the specific activity in nm incorporated $h^{-1} \cdot mg^{-1}$ protein for the enzyme which elongates chains of Arg-poly(Asp) as a function of the age of the culture. In panel B, the growth of the culture is measured by plotting its absorbance at 730 nm.

bon metabolism in the cyanobacteria are not regulated by transcriptional controls (Carr, 1973). These findings also suggest that the control of synthesis does not involve a radical activation – inactivation of at least the synthetic enzyme since activity is measurable in vitro after the addition of substrates.

The simplest mechanism regulating reserves would be a kinetic one: i.e. reserve does not accumulate until the concentrations of amino acids or the ATP is above a certain level. When growth becomes limited, such as in stationary phase cultures or following the addition of protein synthesis inhibitors, perhaps intracellular pools enlarge and Arg-poly(Asp) is produced. In support of this hypothesis, it has been shown that under nitrogen-fixing conditions *Anabaena cylindrica* contains high levels of glutamic acid, glutamine, and aspartic acid, but that arginine is present only in trace amounts (Dharmawardene et al., 1972). Following the addition of protein synthesis inhibitors to exponentially growing cultures (which induces Arg-poly(Asp)), the total amino acid pool rises 5.2-fold in 6 h (Simon, unpublished).

2.4. Distribution of cyanophycin and taxonomy

The identification of the cyanophycin granule is dependent either upon the isolation of Arg-poly-(Asp) using its unique solubility properties, or by the examination of electron micrographs for bodies with a characteristic morphology (Fig. 2). The analysis of electron micrographs makes possible broad surveys because one can identify small amounts of cyanophycin in a cell. In the search for cyanophycin, it is necessary to examine organisms from all culture stages and after growth under conditions which are known to induce the storage of Arg-poly(Asp): addition of an exogenous nitrogen source, the inhibition of protein synthesis with chloramphenicol, or, unbalanced growth brought on nutrient limitation.

Cyanophycin is the only inclusion whose distribution appears limited to the cyanobacteria. No ultrastructural evidence exists for the occurrence of similar bodies in other prokaryotes, including the filamentous bacteria such as *Beggiatoa* (Lawry and Simon, 1982), which were at one time considered as forms of colorless cyanobacteria. It has not been possible to isolate material with the characteristics of Arg-poly(Asp) from other prokaryotes, or from the eukaryotic green algae (*Chlamydomonas, Scenedesmus*), *Euglena*, or red algae (*Porphyridium*) (Simon, unpublished). Of course it is not possible to prove that something does not exist, but if such polymers are found in other organisms, they are rare indeed.

Cyanophycin is found in all members of the cyanobacteria with the exception of strains of the unicellular group *Synechococcus*. It has not been possible to find cyanophycin in *S. cedorum* (UTEX 1191) or *S. leopoliensis* (UTEX 625, *Anacystis nidulans*), even after extensive attempts to induce the formation of the reserve (Fig. 6) (Lawry and Simon, 1982). Whether this is a characteristic of all the members of the group *Synechococcus* as defined by Rippka et al. (1979) remains to be seen.

Cyanophycin granules have been found in all the filamentous cyanobacteria, and in many they are often found along the transverse cell wall. The presence and location of this inclusion has been used in the taxonomic description of a number of species (Stam, 1980). The inclusion may have no direct value in taxonomy with the exception perhaps of defining the group *Synechococcus*, since the amount of reserve per cell can be greatly affected by environmental conditions. However, the localization along the transverse cell wall may be significant. It is interesting to note that when strains with such a localization pattern are grown under conditions which greatly increase the reserve levels, the relationship with the transverse cell wall remains (Lawry and Simon, 1982). The mechanism of inclusion localization within a cell is not known, but since cells can generate electrical fields, perhaps inclusion electrophoresis is involved (DeLoof, 1985).

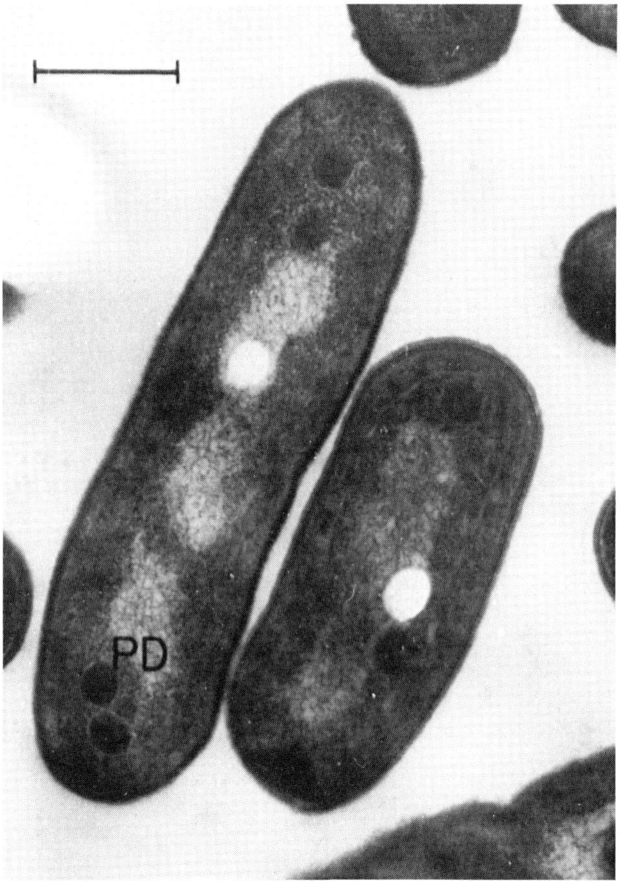

Fig. 6. Thin section electron micrograph of *Anacystis nidulans* (*Synechococcus leopoliensis*) grown in nitrate-containing mineral medium. Scale = 500 nm; PD, polyhedral bodies. (From Lawry and Simon, 1982)

2.5. Metabolism of Arg-poly(Asp) during cell differentiation

The gram-negative vegetative cells of certain filamentous cyanobacteria can, under the proper circumstances, differentiate into alternative forms; heterocysts, akinetes and hair cells (Adams and Carr, 1981). While there has been only a limited study of the role of the storage of reserve materials during cell differentiation, Arg-poly(Asp) contents do change during the formation of both heterocysts and akinetes.

Heterocysts are specialized to carry out aerobic nitrogen fixation, and their formation represents an event of terminal differentiation. The initial work of Lang (1965) showed that significant ultrastructural changes took place during heterocyst formation including the loss of most inclusion bodies, such as cyanophycin granules, dramatic changes in internal membrane configuration, the deposition of additional wall layers (envelope), and the formation of elaborate pore plugs at the connections with adjacent undifferentiated vegetative cells. Heterocysts are formed in response to nitrogen limitation. This results in a dramatic number of alterations in cellular physiology including the induction of protease activity, and the breakdown of phycobiliproteins. It is not surprising that changes also occur in the metabolism of cyanophycin, a primary nitrogen storage polymer. The large refractile polar nodule, a characteristic heterocyst structure (Fogg, 1951) at the channel connecting the heterocyst with adjacent vegetative cells, may consist of Arg-poly(Asp) based upon its characteristic appearance in both the light and the electron microscopes (Fogg, 1951; Lang et al., 1972). What role the pore plug may play in cyanobacterial intercellular traffic or heterocyst O_2 protection mechanisms is not known.

Electron micrographs show that akinetes appear to have the same inclusions as vegetative cells, and that there are no major internal cellular modifications except a dramatic increase in the number of structured granules (Clark and Jensen, 1969; Miller and Lang, 1968; see Section 2.3.5. of Chapter 9). Akinetes of *Anabaena cylindrica* contain 7.0% of their dry weight as cyanophycin. Isoelectric focusing and determination of the range of molecular weights by SDS electrophoresis show that the reserve from akinetes and vegetative cells is identical (Simon, unpublished). Cyanophycin granules disappear when the akinetes germinate, and may be used to drive the early stages of akinete outgrowth.

Filaments of *A. cylindrica* growing on N_2 as a nitrogen source show no measurable reserve during exponential growth and only low levels at later stages. When cultures are supplied with $NaNO_3$, sporulation is inhibited and cyanophycin granules accumulate in stationary phase cells. Such stationary phase cells contain up to 7.5% of the dry weight as reserves. Vegetative cells containing high levels of cyanophycin, therefore, have a larger N/C ratio than akinetes, but they lack the structures found in akinetes (envelope?) necessary for survival in unfavorable environments.

3. Polyphosphate

Polyphosphate bodies showing characteristic cytological properties have been seen for 100 years. The reaction of polyphosphate with basic dyes produces a shift in the absorption spectrum, hence the term metachromatic. Because of the resulting 'reddish' staining properties with aged toluidine blue and methylene blue, polyphosphate bodies have been called 'volutin granules' or 'metachromatic bodies' (Harold, 1966). Electron micrographs show no limiting structure around the body (Fig. 7). Polyphosphate has been seen or measured in many bacteria, including the cyanobacteria, as well as lower eukaryotes such as fungi, yeasts, and algae (*Chlamydomonas*, *Scenedesmus*, *Euglena*). It is one of the few prokaryotic inclusion materials found in higher plants and animals.

Several reviews of polyphosphate metabolism have appeared (Dawes and Senior, 1973; Harold, 1966; Kulaev, 1975), and it is still instructive to examine the classic work of Harold (Harold, 1963, 1966; Harold and Sylvan, 1963). There is a great deal of information from prokaryotes about the biochemistry, metabolism, and genetics of polyphosphate, although most comes from studies on heterotrophic bacteria. Reviews relating to the cyanobacteria have also been published (Allen, 1984; Fogg et al., 1973; Fuhs, 1973; Jensen, 1984a,b, 1985). Little is known about the enzymology in higher plants and animals, which is unfortunate because of the evolutionary relationship between the cyanobacteria and eukaryotic plant cell chloroplasts.

3.1. Chemistry and analysis

Polyphosphate is a polymer of *ortho*phosphate with phosphoanhydride linkages equivalent to those in ATP. Polyphosphate exists in several forms, but the polyphosphate in inclusion bodies consists primarily of high molecular weight linear polyphosphate species. The number of phosphate residues varies from a few to thousands.

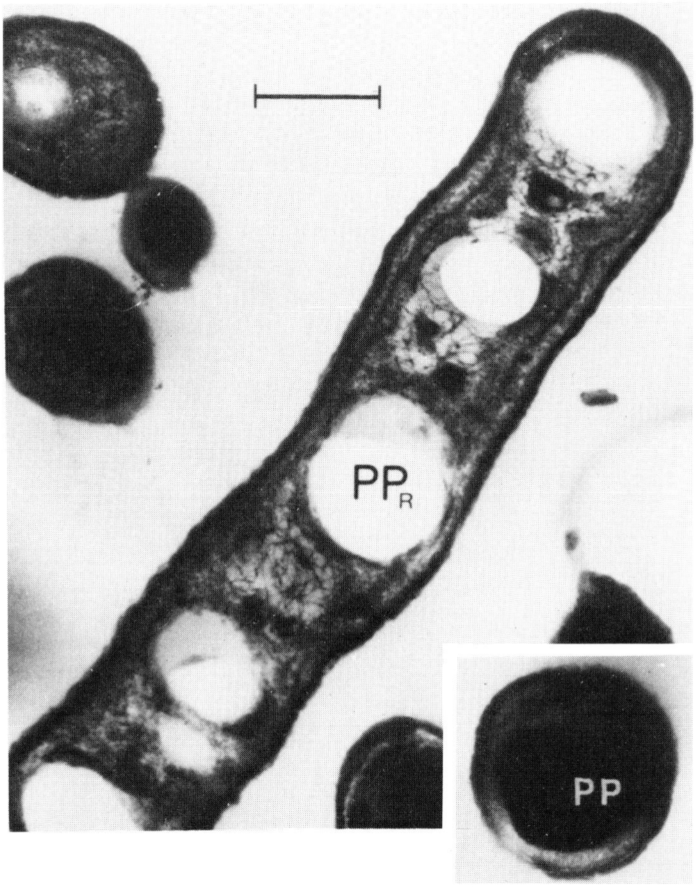

Fig. 7. Thin section electron micrograph of *Anacystis nidulans* (*Synechococcus leopoliensis*) grown in sulfur-deficient mineral medium. Inset shows a transverse section with an intact polyphosphate body. Scale = 500 nm; PP, polyphosphate body; PP$_R$ polyphosphate body remnant. (From Lawry and Simon, 1982)

Polyphosphate is a polyanion, and the number of negative charges must be balanced either by organic or inorganic cellular cations. The concentration of balancing cationic species can be substantial. In addition, because of the high binding constants between polyphosphate and certain cations, insoluble precipitates can form. Ba^{3+} salts have been used to precipitate polyphosphate, and it may be that polyphosphate bodies act as traps for specific metal ions. Whatever their other roles, the ion-exchange properties of polyphosphate must be considered.

Analysis of polyphosphate levels can be carried out in several ways. Both light and electron microscopy (EM) offer a means for quantitative analysis. Polyphosphate stained with lead nitrate and ammonium sulfide results in the deposition of insoluble PbS at the site of the inclusion bodies. It is then possible to count the number of bodies to estimate the relative ability of strains to sequester phosphate. This has been done for the analysis of environmental samples of *Chamaesiphon incrustans* and *Fischerella musicola* as well as diatoms, green and red algae (Stevenson and Stoermer, 1982). The behavior of polyphosphate in the electron beam of an EM is characteristic (Fig. 7), and such microscopy provides a convienient way to screen strains for the presence of the polyphosphate as well as to determine the effects of cell treatments upon reserve levels. Cell morphometry, though time consuming, can provide quantitative measures of reserve levels. Cytochemistry also provides a way to correlate the presence of specific enzyme activities with particular inclusion bodies. The ability to carry out X-ray elemental analysis on sections prepared for electron microscopy has not only confirmed that the bodies identified as polyphosphate contain high concentrations of phosphate (Kessel, 1977; Webster et al., 1984), but offers a convenient way to monitor the types and amounts of inorganic cations which bind to polyphosphate in the cell (Kachlin et al., 1984; Sicko-Goad et al., 1975).

Solubility properties and sensitivity to acid hydrolysis form the basis of the chemical analysis of polyphosphate. Polyphosphates are insoluble in alkaline hypochlorite (Poindexter and Eley, 1983), but depending upon chain length, may or may not be soluble in alkali and/or acid. The material is hydrolyzed by treatment with 1.0 M HCl for 15 min at 100° C. More recently, analysis of ^{31}P nuclear magnetic resonance (NMR) signals have been used to estimate relative cellular polyphosphate levels. Polyphosphate in *Synechococcus* strain Y-30-s shows a broad resonance peak with chemical shift (ppm) +22.5 – +28.8, which apparently represents a heterogeneous array of polyphosphate compounds (Kallas and Castenholz, 1982; Kallas and Dahlquist, 1981). It is possible to measure such resonance peaks to follow metabolic changes in polyphosphate accumulation (Rao et al., 1985; Suresch et al., 1985).

Most polyphosphates are isolated from cells by taking advantage of the differential solubility properties, much as one would use to isolate DNA or RNA. This author is unaware of attempts to gently isolate intact polyphosphate bodies from the cyanobacteria, and until this is done it is not possible to rule out the presence of other specific components being associated with polyphosphate bodies.

3.2. Metabolism of polyphosphate

3.2.1. Biosynthesis

The only enzyme shown to be involved in the synthesis of polyphosphate is polyphosphate kinase ($ATP + (P_i)_n \rightarrow ADP + (P_i)_{n+1}$). This enzyme was first detected in yeast and has been purified from *E. coli* (Dawes and Senior, 1973; Kulaev, 1975). Using mutants of *Aerobacter aerogenes*, it is possible to show that this is the only synthetic pathway in this heterotrophic bacterium (Harold, 1963; Harold and Sylvan, 1963). The enzyme adds terminal phosphates, requires Mg^{2+}, and is inhibited by fluoride and ADP. Low ATP/ADP ratios cause the enzyme to run in reverse.

In the cyanobacteria, the enzyme shows slightly different activity. Polyphosphate kinase has been measured in *Anacystis nidulans* (Vaillancourt et al., 1978) using γ-labeled ^{32}P-ATP incorporation into acid-insoluble material. The detergent cetyltrimethylammonium bromide (CTAB) was added at 0.01% to the reaction mixtures to permeabilize whole cells so that the measurements could be carried out. CTAB at 1.0% killed all activity. Mutant analysis suggests that this is the primary pathway of polyphosphate synthesis.

In *Synechococcus* (Grillo and Gibson, 1979) polyphosphate synthetase required Mg^{2+} and the optimal ratio of Mg^{2+}/ATP was 1:1. The enzyme was soluble and was unstable after cell breakage. A polyphosphate primer was not required for the reaction, and since ADP inhibited the reaction it was concluded that the reaction was reversible. The solubility, hydrolytic products, and the metachromatic nature of the reaction product confirmed that it was polyphosphate.

3.2.2. *Degradation*

In contrast to the single means of synthesizing polyphosphate, a diverse set of pathways have been found for the degradation of polyphosphate. In some organisms free phosphate is released, while in others the phosphate is transferred to an organic acceptor. The enzymes identified include polyphosphate kinase, polyphosphate-AMP-phosphotransferase, polyphosphate-glucokinase, polyphosphate-fructokinase, and several polyphosphatases (Dawes and Senior, 1973; Harold, 1966; Kulaev, 1975).

In the cyanobacteria there is, as yet, no evidence for the presence of enzymes which use polyphosphate to phosphorylate sugars. Polyphosphatases are measured by assaying for inorganic phosphate release from polyphosphate. No activity of polyphosphatase was measured in *Anacystis nidulans* (Vaillancourt et al., 1978), although it may be that this is because the cells analyzed were not breaking down the reserve. In *Synechococcus* a stable soluble enzyme could be measured (Grillo and Gibson, 1979). Mg^{2+} was required for the reaction, and because the rate of hydrolysis was linear until 70–80% of the polyphosphate was digested, it has been suggested that the enzyme was an exophosphatase. This is supported by an analysis of the hydrolysis products of the reaction. Inorganic phosphate does not inhibit the reaction, and this suggests that the enzyme is not regulated by the size of the free phosphate pool.

Acid phosphatase (EC 3.1.3.2, *ortho*phosphate monoester phosphorylase) may also play a role in polyphosphate degradation. The enzyme was measured in *Nostoc* at pH 6.5 using *para*nitrophenyl phosphate (PNP) as substrate. Localization studies at the EM level showed that the enzyme was present around the perimeter of the polyphosphate bodies with some localized in the cytoplasm. With exogenously added PNP, activity was localized at the plasmalemma. Dark treatment of *Nostoc* resulted in a breakdown of polyphosphate, and after 72 h in the dark, activity was localized throughout the polyphosphate body matrix.

3.3. Regulation of polyphosphate levels

Polyphosphate concentration varies markedly depending upon the culture conditions. In phosphate-starved cells it may be undetectable, whereas under other conditions it may constitute more than 50% of the total cell phosphorus. Studies on several members of the cyanobacteria show that the general framework of phosphate metabolism is the same as that in heterotrophic prokaryotes (Harold, 1966).

Polyphosphate is produced during balanced exponential growth. In studies on *Synechococcus* (Grillo and Gibson, 1979), polyphosphate levels were highest in nitrate limited chemostat cultures at the slowest growth rates where the reserve was 20% of cell phosphorus, and the material was still synthesized even when growth was phosphate limited. Indeed, at all rates of phosphate-limited growth, polyphosphate was about 10% of the total cellular phosphorus. These results suggest either that polyphosphate plays a role in the cell other than simply that of reserve material, or that cellular metabolism anticipates future needs and stores reserves even when the nutrient being stored is the one which is limiting growth (Cohen and Parnas, 1976).

There are two patterns of excessive polyphosphate accumulation: when growth is unbalanced and when inorganic phosphate is added back to phosphate-starved cells. Unbalanced growth can occur in a variety of ways. Polyphosphate accumulates during the transition of a batch-grown culture into stationary phase or when growth is limited either by starvation or by extreme environmental conditions. Sulfur starvation can cause large amounts of polyphosphate to accumulate in members of the cyanobacteria (Lawry and Jensen, 1979; Lawry and Simon, 1982). However, not all cells respond to starvation conditions in the same manner. For example, *Gloeocapsa alpicola* and

Anabaena variabilis produce massive amounts of cyanophycin when starved for phosphorus or sulfur, but little polyphosphate when sulfur starved. Conversely, sulfur starvation can result in large amounts of polyphosphate in strains of *Synechococcus* which are unable to make cyanophycin. There are a number of cyanobacteria such as *Plectonema boryanum* which respond to sulfur starvation with massive cyanophycin and polyphosphate deposition (Lawry and Simon, 1982). Perhaps the variation in response to starvation is linked to the ability of cells to transport or synthesize the necessary reserve monomers.

If nucleic acid synthesis is inhibited by the addition of rifampicin, polyphosphate is formed (Rodriguez-Lopez et al., 1971). Presumably intracellular phosphate is channeled to either nucleic acid synthesis or polyphosphate, and when access to one cell sink is blocked, flow increases into the other major sink. In heterotrophic bacteria, inhibition of growth by nutrient limitation using mutants requiring uracil or an amino acid results in polyphosphate accumulation. Such mutants are available in the cyanobacteria (Ladha and Kumar, 1978), but no such physiological/genetic studies have been carried out. When nucleic acid synthesis is restarted, polyphosphate is broken down. It can be shown in *Synechococcus* that there is a transfer of phosphate from polyphosphate to nucleic acid (Grillo and Gibson, 1979).

Polyphosphate accumulates rapidly when cells which have been grown under phosphate-starved condition are given phosphate (Harold, 1966). This phenomenon is known as 'polyphosphate overplus', and probably results from alterations in polyphosphate kinase levels. Studies with *Synechococcus* (Grillo and Gibson, 1979) show that starvation increases both phosphate transport rates and the levels of polyphosphate kinase as much as 30–50-fold, while there is no change in the level of polyphosphatases. Addition of phosphate to starved cultures results in the rapid accumulation of polyphosphate and the fall in the level of kinase. Thus, regulation of reserve is at the level of the availability of monomers (phosphate) and synthesis. Alkaline phosphatase levels also rise in cyanobacteria under conditions of phosphate starvation as the cells attempt to scavenge cellular phosphorus for required synthesis (Doonan and Jensen, 1980; Fitzgerald and Nelson, 1966; Healey, 1982).

Changes in polyphosphate during cell differentiation in the filamentous cyanobacteria are not as dramatic as changes in cyanophycin levels. Polyphosphate bodies disappear during heterocyst formation (Lang, 1965), although it is not possible to exclude the presence of small amounts of polyphosphate, especially that which may be associated with membrane fractions. Membrane-bound polyphosphate has been demonstrated (Suresch et al., 1985), and marked changes in intracellular membrane configuration take place during heterocyst differentiation (Lang and Fay, 1971). Polyphosphate is present in akinetes at levels similar to that found in vegetative cells (Clark and Jensen, 1969; Miller and Lang, 1968).

3.4. Function of polyphosphate

There is no naturally occurring member of the cyanobacteria which has been shown to be unable to make polyphosphate. However, polyphosphate mutants have been isolated in *Anacystis nidulans* following treatment with mutagens and penicillin enrichment for strains unable to grow on phosphate-free media (Vaillancourt et al., 1978). This phenotype had a reversion rate of 1.1×10^{-9}–1.4×10^{-10}. All the mutants examined except one lacked polyphosphate when studied in the electron microscope, and several strains had altered polyphosphate kinase activity. All mutants were able to take up phosphate in short-term experiments at the same rate, although several were unable to take up phosphate rapidly after the later stages of sulfur starvation (unbalanced growth), and all but one appeared unable to synthesize polyphosphate. Since the growth rates were similar for mutant and wild type, polyphosphate is not required for normal growth.

The role of polyphosphate in an organism may vary depending upon environmental conditions and cell history. It is certainly possible that polyphosphate has one or more functions at any given time.

3.4.1. Phosphate reserve

Polyphosphate is a source of inorganic phosphorus for the synthesis of nucleic acids and lipids, and its role as a phosphate storage depot is probably one of the main functions served by this material. The regulation of its level and the enzyme activities responsible for its synthesis and utilization are consistent with this proposal. The advantages of storing material as polymers were considered above, and it is clear that it would be best to avoid the high internal concentrations of inorganic phosphate which would undoubtedly affect the enzymes regulating cellular energy charge. The utilization of polyphosphate may occur under conditions of rapid growth or phosphate starvation. Because polyphosphate is stored even where phosphate is limiting for growth (Grillo and Gibson, 1979), there must be a complex strategy involving storage. There have been suggestions that the pathway of phosphorus assimilation requires obligatory passage through a polyphosphate pool (Niemeyer and Richter, 1969), but evidence for this remains scanty.

3.4.2. Energy storage

At the molecular level, polyphosphate is a microbial phosphogen. Phosphate bond energy is stored, and in theory could be utilized for the synthesis of ATP, either by the reversal of polyphosphate kinase or by the formation of organic phosphoryl compounds. Several lines of evidence, however, argue against this as the major role for polyphosphate.

There is no evidence for enzymes in cyanobacteria which could utilize polyphosphate bond energy other than polyphosphate kinase, and in the latter case there is no evidence that the enzyme actually runs backwards under normal cellular conditions. What is known about energy charge in the cyanobacteria would make the reverse process highly unlikely.

In *Nostoc*, polyphosphate does not completely break down if the energy supply is limited or blocked. Incubation of the cells with CO_2 in the dark does result in some breakdown as well as the elaboration of alkaline phosphatase, but a significant fraction remains intact (DuBois et al., 1984). Of course, it may be that an organism anticipates the need for an energy source in order to be able to adapt when conditions again become favorable for growth (Cohen and Parnas, 1976).

It is possible to do calculations on how much energy could be available from polyphosphate reserves. A useful measure would be to compare the potential stored in relation to the basal metabolic rate. The rate of respiration given as ATP requirement per min, has been calculated for *Oscillatoria princeps* (Halfen and Castenholz, 1971). If polyphosphate is 2% of cellular dry weight, and there is 100% efficiency in conversion of the bond energy, reserves could supply an amount of ATP less than that produced by 0.9 h of basal respiration. This is not to suggest that polyphosphate bond energies may have no metabolic role. Indeed, the same calculations done above for true muscle phosphogens would show that they could not provide nearly enough energy for long-term movements. But they are most useful over short-time periods with high energy requirement. Perhaps polyphosphate plays the same role as an energy buffer, by regulating the energy charge of the cell over short periods of time.

3.4.3. Detoxification mechanism

The ability of polyphosphate to form insoluble salts with various heavy metal ions provides a potential means for removing these ions from the cell cytoplasm. Analysis of the composition of polyphosphate bodies with electron microprobe X-ray analysis indicates that salts of polyphosphate do exist in situ (Jensen et al., 1982; Rachlin et al., 1984) and, for example calcium, is a significant elemental component of polyphosphate in *Plectonema boryanum* (Sicko-Goad et al., 1975). There has, therefore, been some interest in the formation of polyphosphate metal salts, not only because it would provide an effective passive detoxification mechanism for inhibitory levels of heavy metals, but also because it might yield a potential mechanism for the bioconcentration of commercially desirable metals.

Studies have been carried out to determine whether the exposure to low levels of heavy metals results in the production of polyphosphate (Sicko-Goad, 1982). This work shows that significant polyphosphate production is not a general response to heavy metal exposure. Concentrations of 10 ppb Pb and 5 ppb Cu had no effect on the level of polyphosphate determined by morphometric analysis in *Anacystis cyanea* after 24 h (Sicko-Goad, 1982). Exposure to Cd^+ caused an increase in the volume, but not the number of polyphosphate bodies in *Plectonema boryanum* (Rachlin et al., 1982). Perhaps this is explained by the fact that Cd^+ inhibits the uptake of phosphate in *Anacystis nidulans* (Singh and Yanava, 1984). A cadmium detoxification mechanism using polyphosphate seems less likely. Of course higher concentrations of inhibitory metals which do effect growth or phosphate uptake may well result in the elaboration of inclusion bodies as a general response. This area deserves much greater attention.

3.4.4. Counter-ion storage

Fungal polyphosphate is primarily localized in vacuoles (Miller, 1984; Tijssen and Steveninck, 1985). In *Neurospora crassa* not only is 78% of the polyphosphate in vacuoles, but over 97% of soluble free basic amino acids (arginine, lysine, and ornithine) are found in the same location (Cramer et al., 1980). Since synthetic polyphosphate can concentrate arginine from dilute solution (Cramer et al., 1980), it has been suggested that polyphosphate production may be the driving mechanism for the accumulation of basic amino acids. Although more recent data indicate that there may not be an obligatory coupling between polyphosphate and basic amino acids (Cramer and Davis, 1984), the fact remains that the storage of polyphosphate will also require the storage of counter-ions. Basic amino acids or polyamines are logical organic counter-ions, and could provide the cell with a mechanism to store nitrogen in a form which is effectively out of solution – i.e., like that in inclusion bodies. This may be of particular importance in strains of *Synechococcus* which are unable to synthesize cyanophycin. X-ray microprobe analysis (Jensen et al., 1982; Rachlin et al., 1984) can indicate the elemental composition of polyphosphate salts in vivo. However, the analysis of potential organic counter-ions will have to await the gentle isolation of intact polyphosphate bodies.

4. Polyhedral bodies

Polyhedral bodies have been seen as an ultrastructural feature of the cyanobacteria since the earliest investigations of this group with the electron microscope (Lefort, 1960; Pankratz and Bowen, 1963). Jensen and Bowen (1961) described them in *Nostoc pruniforme* and termed them polyhedral bodies because of their characteristic shape. They appear as polyhedral objects with a diameter ranging from 0.1 μm to 0.7 μm (Fig. 6), and are surrounded by a membranous layer, that in thin section differs significantly from typical membrane protein-lipid bilayer. Polyhedral bodies do not appear to contain lipid (Biedermann and Westphal, 1979). They are most often located in or associated with the nuclear region of the cyanobacterial cell (nucleoplasm) (Edwards et al., 1968; Pankratz and Bowen, 1963), although the reason for this relationship remains unexplained.

Polyhedral bodies are not limited to the cyanobacteria, but structures similar in size, shape, and location have been found in a variety of facultative chemolithotrophic bacteria including members of the groups: *Thiobacillus*, *Nitrobacter*, and *Nitrosomonas* (Codd and Marsden, 1984). Although a rarity in the photosynthetic bacteria, the purple non-sulphur bacterium *Rhodomicrobium vannielii* also appears to contain polyhedral bodies, and they have been found in the thermophilic hydrogen bacteria *Pseudomonas thermophilia*. Their isolation from *Thiobacillus neopolitanus* by Shively et al. (1973a) resulted in the demonstration that they consisted in large part of the enzyme ribulose-1,5-bisphosphate carboxylase-oxygenase (EC 4.1.1.39, RuBPC/O or Rubisco), the initial enzyme of carbon dioxide incorporation in autotrophic organisms and the enzyme responsible for the first step in the process of photorespiration (Lorimer, 1981; Miziorko and Lorimer, 1983). The presence of this enzyme suggested a potential role of these bodies in carbon fixation, and thus the designation of these bodies as 'carboxysomes' (see Chapter 4 for details).

There are several general reviews of work in this area (Codd and Marsden, 1984; Shively, 1974) as well as reviews focusing on the inclusion in the cyanobacteria (Allen, 1984; Codd et al., 1984; Fogg et al., 1973; Fuhs, 1973; Jensen, 1984a). The discussion below will focus primarily on this inclusion body in the cyanobacteria, although there is no evidence that the polyhedral bodies in the cyanobacteria differ in any way from those in the other prokaryotes.

4.1. Distribution

Most species of cyanobacteria examined in the electron microscope appear to contain polyhedral bodies. Polyhedral bodies are found in the chlorophyll-*b* containing *Prochloron* (Whatley, 1977), an unusual prokaryote that may be a member of the cyanobacteria. Structures which could be similar to polyhedral bodies are found in the blue-green derived 'cyanelles' (Lambert et al., 1985) of *Glaucocystis* (Hall and Claus, 1967) and *Cyanophora* (Hall and Claus, 1963), both eukaryotes (see Section 3.3. of Chapter 11). There are no inclusions similar to the polyhedral body in the chloroplasts or plastids of other eukaryotes. However, the pyrenoid, a structure found in the chloroplasts of many members of the algae (Griffiths, 1970), appears to consist primarily of Rubisco (Holdsworth, 1971; Kerby and Evans, 1981). Perhaps the need for storing this enzyme is a general phenomenon?

While polyhedral bodies may be universal in the cyanobacteria (Allen, 1984; Codd and Marsden, 1984), their distribution and presence within a species may vary. The vegetative cells of heterocyst-containing filamentous nitrogen-fixing cyanobacteria have polyhedral bodies (Stewart and Codd, 1975). However, the heterocysts of these species lack these inclusions, and the disappearance of polyhedral bodies during heterocyst differentiation has been previously noted (Lang, 1965; Lang and Fay, 1971; Roussard-Jacquemin, 1983; but see Section 2.6. of Chapter 21). The loss of polyhedral bodies is accompanied by the disappearance of Rubisco activity (Stewart and Codd, 1975; Winkenbach and Wolk, 1973). Heterocysts are not capable of photosynthetic carbon fixation (Fay and Walsby, 1966; Wolk, 1968) and its associated oxygen production, and therefore the loss of polyhedral bodies is not unexpected. Several non-heterocystous nitrogen-fixing strains are also capable of aerobic nitrogen fixation. In these strains there may also be a spatial separation of nitrogen fixation and carbon fixation, and perhaps polyhedral bodies are absent from those regions unable to fix CO_2 (Bryceson and Fay, 1981; Carpenter and Price, 1976).

There can be significant variations in both the amount of Rubisco and the relative rate of carbon fixation in vitro. These variations are paralleled by changes in the numbers of polyhedral bodies in the cyanobacterial cells. Environmental variations which can affect the rate of photosynthesis, such as changes in carbon dioxide concentration, do not significantly change the relative amount of polyhedral bodies in *Synechococcus* (Miller and Holt, 1977), although there were responses to light intensity (Dohler, 1976) which may have an effect on the number of these inclusion bodies present. In iron-starved cells of *Anacystis nidulans* R2, polyhedral bodies increase in CO_2-starved cells following the addition of iron, but there is no alteration in the numbers of polyhedral bodies in cells grown on air (Sherman and Sherman, 1983).

4.2. Isolation and characterization

Polyhedral bodies have been isolated from a limited number of cyanobacterial species including: *Anabaena cylindrica* (Codd and Stewart, 1976), *Chlorogloeopsis fritschii* (Lanaras and Codd, 1981a), and *Coccochloris peniocystis* (Coleman et al., 1982). Polyhedral bodies from *Chlorogloeopsis* have greater stability in media of low ionic strength, and the isolation of this inclusion has been accomplished by the lysis of cells in the presence of low molarity Tris buffer (Codd et al., 1984). Mg^{2+} is required for the stability of the particulate enzyme from *Coccochloris*, and even under the best of conditions isolated polyhedral bodies released soluble Rubisco at a slow rate ($t_{1/2}$=2–3 h) (Coleman et al., 1982). Purification of polyhedral bodies is accomplished by differential centrifugation followed by density gradient centrifugation in sucrose or Percoll plus sucrose.

Electron micrographs of isolated polyhedral bodies show that they retain the characteristic polyhedral structure, and therefore must have a rigid coating (Shively et al., 1973a,b). In addition, under high magnification an internal particle substructure can be seen that resembles isolated Rubisco complex (Lanaras and Codd, 1981b; Shively et al., 1973b).

The purified polyhedral bodies are surrounded by a thin membrane (3.5 nm in *Thiobacillus*; Shively et al., 1973). SDS gel electrophoresis of the isolated polyhedral bodies show an average of from 7 to 13 protein bands, depending on the species being examined (Codd et al., 1984; Codd and Marsden, 1984). The two major bands in each gel which usually account for more than 50% of the total protein represent the L and S subunits of Rubisco. In *Thiobacillus neapolitanus* two peptides of 10 000 daltons and 15 000 daltons appear to be associated with the membrane (Cannon and Shively, 1983). It is not known which of the minor proteins demonstrated in the cyanobacterial polyhedral bodies (Codd et al., 1984) are associated with the membrane, and indeed the identity and location of peptides other than the L and S subunits of Rubisco are not known (Lanaras et al., 1985; Marsden et al., 1984).

A small circular double stranded DNA has been seen in the polyhedral bodies from *Nitrobacter agilis*, and there are effects of plasmid curing agents on the numbers of polyhedral bodies in this organism (Westphal et al., 1979). There are a few other inconclusive reports which suggest nucleic acids in polyhedral bodies (Cannon et al., 1981), but no indication of such in polyhedral bodies from the cyanobacteria. The role of the DNA remains enigmatic. The nucleic acid appeared to be inside the polyhedral bodies since the treatment of isolated inclusions with DNAse did not remove the DNA. The regular association of the polyhedral bodies with the nuclear region suggests that perhaps nucleic acids play a unique role in polyhedral body metabolism.

4.3. Function of the polyhedral body

Because of their composition, the major questions regarding the function of polyhedral bodies are concerned with whether their constituent enzyme(s) are active.

4.3.1. Carbon fixation

Polyhedral bodies might be the equivalent of a subcellular organelle, playing a role in the process of autotrophic carbon metabolism. Beudeker has reported that polyhedral bodies in *Thiobacillus neapolitanus* contain Rubisco and the additional enzymes in the Calvin-Benson cycle which are necessary for carbon fixation (Beudeker and Kuenen, 1981). However, other work on *Thiobacillus* does not support the idea that there are other photosynthetic enzymes localized in the polyhedral bodies (Cannon and Shively, 1983), and attempts to show the presence of such additional enzymes in the cyanobacterial polyhedral bodies have been unsuccessful. For example, phosphoribulokinase, the only enzyme known to be involved in the synthesis of ribulose bisphosphate, has not been shown in isolated polyhedral bodies (Marsden et al., 1984). Immunogold electron microscopy confirms this and shows that the enzyme is localized in the cytoplasm (Hawthornthwaite et al., 1985).

If the entire process of photosynthesis is not in the polyhedral body, the question remains open as to whether the Rubisco in this inclusion is used during carbon fixation. Because the enzyme isolated from polyhedral bodies is active in vitro without special treatment, this is a distinct possibility. The amount of soluble Rubisco is highest during exponential growth when photosynthetic rates are highest, and Rubisco in non-growing stationary phase cells is primarily in the particulate form (Lanaras and Codd, 1982). This suggests that the soluble fraction is used in photosynthesis and that the particulate form may be involved in storage. However, it is still possible that the smaller amount of enzyme in the particulate fraction is active. Technical difficulties make it hard to evaluate this. The rates of carbon fixation in the organism can be compared with the in vitro enzyme activities; however, carbon fixation in vivo is a function of many factors, including both the carboxylase and oxygenase activities of Rubisco, pH and ionic concentrations, etc., so it is difficult to relate these parameters to the actual conditions in different cyanobacterial cell compartments.

The polyhedral bodies have the potential for involvement in photorespiration, and the enzymes normally associated with the metabolism of glycollate which are present in the cyanobacteria (Codd et al., 1980; see Section 4.3. of Chapter 4). However there is no evidence of their localization in the polyhedral bodies. Measurement of the kinetic properties of carbon fixation at different oxygen concentrations does indicate that Rubisco in intact polyhedral bodies is less susceptible to oxygen inhibition than the soluble enzyme (Coleman et al., 1982). This suggests that the polyhedral body membrane keeps the oxygen concentration inside the inclusion less than that outside. Perhaps this argues for an active role of polyhedral body Rubisco only under certain (high O_2) conditions?

It is unlikely that the polyhedral body could play a role in the process of carbon dioxide accumulation. The cyanobacteria have an active bicarbonate transport system (Miller and Colman, 1980); see Section 3 of Chapter 4) which is not localized in the polyhedral body. There is a polypeptide in the polyhedral bodies from *Chlorogloeopsis* which has a molecular weight similar to that of the enzyme carbonic anhydrase (CA). However, recent studies show that while most of the CA activity is particulate, it is not localized in the polyhedral body (Lanaras et al., 1985).

4.3.2. Storage

Rubisco stored in polyhedral bodies could serve as a potential nitrogen source. Polyhedral bodies are broken down during nitrogen starvation in *Anabaena cylindrica* (De Vasconcelos and Fay, 1974) and *Spirulina platensis* (Eykelenburg, 1980), but not in *Chlorogloeopsis fritschii* (Peat and Whitton, 1967) or *Agmenellum quadruplicatum* (Stevens et al., 1981a). The reason for these differences to the same stimulus is not clear. Starvation for sulfur results in the loss of polyhedral bodies in *Plectonema boryanum* and *Oscillatoria prolifera*, but they are still present in *Anacystis nidulans* and may even be present in larger numbers in *Gloeocapsa alpicola* (Lawry and Simon, 1982). It is interesting that in the latter case, sulfur starvation also results in the massive accumulation of the nitrogen reserve cyanophycin, but does not cause a decrease in the number of polyhedral bodies. There is no direct evidence to suggest that the disappearance of polyhedral bodies results from protein catabolism.

Polyhedral bodies are occasionally found in cells which are not capable of carbon fixation, and perhaps in these cases the bodies represent a storage form of the enzyme. In akinetes, the enzyme could be available for immedate use when germination took place. Polyhedral bodies are also found in the symbiotic cyanobacteria of cycad roots, even though the latter cannot fix carbon dioxide because they are below the soil surface (Lindblad et al., 1985). The endophytic *Nostoc* isolated from the coralloid roots of the cycad *Macrozamia lucida* can grow photo-autotrophically or heterotrophically in the dark (Hoare et al., 1971). In both cases the ultrastructural appearance of the cyanobacterium is the same, and polyhedral bodies are present in similar numbers.

5. Conclusion

The cyanobacteria may well be among the most diverse group of prokaryotes, yet at the ultrastructural level they are similar to other prokaryotes in containing characteristic subcellular inclusions. The presence or absence of these inclusions may vary from species to species, and from one environmental condition to another, but each type is similar in morphology and biochemistry.

While much is known about the major inclusion bodies that have been isolated and characterized, this review will conclude by reminding the reader that there are good reasons for the continued study of inclusion bodies:

(a) These materials can be important in the metabolism of the cell during all stages of a growth cycle, including times of cell differentiation. The ability to successfully occupy specific niches and to survive environmental constraints may well rest on the ability to store and use inclusion materials.

(b) Studies with cyanobacteria may serve as models for understanding inclusion formation in other prokaryotes, and may also provide useful information about the relationships and evolution of the prokaryotes. It may be easier to experimentally manipulate the cyanobacteria than other prokaryotes since they grow on dilute media, and simple environmental shifts can cause dramatic alterations in inclusion metabolism.

(c) Cyanobacterial inclusion metabolism is interesting biochemistry. It deals with biological products which are essentially out of solution, i.e. with the enzymology of reactions at solid surfaces.

(d) Work on inclusion materials has a potential role in biotechnology (Sinskey, 1983). The cyanobacteria make many very specific polymers. They can synthesize materials with unusual monomers or linkages, and with unique proteinaceous membranes. Such materials may be difficult if not impossible to make by the traditional routes of organic chemistry. Cells, but not chemists, can also make supramolecular structures – ordered aggregates of polymers with useful properties. Indeed, one of the objects of biotechnology and genetic engineering is the production of massive amounts of specific polymers. In the best engineered strains, overproduction of proteins or peptides can result in the formation of inclusion bodies containing these products. Yet this process is a natural occurrence in the formation of cyanobacterial inclusion bodies. Perhaps as information is obtained about inclusion metabolism and genetics, it will be possible to redirect the cellular biochemistry so as to produce desired products.

Acknowledgements

The author would like to thank numerous colleagues, both for providing reprints/preprints, and for the opportunity to talk about this interesting area of cyanobacteriology. My thanks to Maxine Lapiana for help with the manuscript and to Dr. Nelson Lawry for Figures 2,6, and 7 taken from N.H. Lawry and R.D. Simon, Journal of Phycology 18: 391–399, 1982.

References

Adams, D.G. and Carr, N.G. (1981) The developmental biology of heterocyst and akinete formation in cyanobacteria. Crit. Rev. Microbiol. 9, 45–100.

Allen, M.M. (1984) Cyanobacterial cell inclusions. Ann. Rev. Microbiol. 38, 1–25.

Allen, M.M., Hutchison, F. and Weathers, P.J. (1980) Cyanophycin granule polypeptide formation and degradation in the cyanobacterium *Aphanocapsa* 6308. J. Bacteriol. 141, 687–693.

Allen, M.M., Zimmerman, W., Morris, R. and Slonicki, R. (1982) Cyanophycin granule polypeptide protease in *Aphanocapsa* 6308. Northeast Section, ASPP Cornell University, P-02.

Allen, M.M., Morris, R. and Zimmerman, W. (1984) Cyanophycin granule polypeptide protease in a unicellular cyanobacterium. Arch. Microbiol. 138, 119–123.

Beudeker, R.F. and Kuenen, J.G. (1981) Carboxysomes: 'Calvinosomes'?. FEBS Lett. 131, 269–274.

Biedermann, M. and Westphal, K. (1979) Chemical composition and stability of Nb1-particles from *Nitrobacter agilis*. Arch. Microbiol. 121, 187–192.

Blackburn, S. (1970) Protein Sequence Determination. Marcel Dekker, New York, pp. 55–61.

Borzi, A. (1886) Le comunicazioni intracellulari delle Nostochinee. Malpighia 1, 28–74.

Bothe, H. (1982) Nitrogen fixation. In N.G. Carr and B.A. Whitton (eds.), Biology of the Cyanobacteria. Blackwell, Oxford, pp. 87–104.

Boussiba, S. and Richmond, A.E. (1980) C-phycocyanin as a storage protein in the blue-green alga *Spirulina platensis*. Arch. Microbiol. 125, 143–147.

Brown, D.L. and Bisalputra, T. (1969) Fine structure of the blue-green alga *Nostoc sphaericum:* the structured granule. Phycologia 8, 119–126.

Bryceson, I. and Fay, P. (1981) Nitrogen fixation in *Oscillatoria erythraea* in relation to bundle formation and trichrome differentiation. Marine Biol. 61, 159–166.

Cannon, G.C. and Shively, J.M. (1983) Characterization of a homogeneous preparation of carboxysomes from *Thiobacillus neopolitanus*. Arch. Microbiol. 134, 52–59.

Cannon, G.C., Toth, M.N. and Shively, J.M. (1981) Structural studies of carboxysomes from *Thiobacillus neapolitanus*. Abstr. Ann. Meeting Am. Soc. Microbiol. 93.

Carpenter, E.J. and Price, C.C. (1976) Marine *Oscillatoria* (*Trichodesmium*): explanation for aerobic nitrogen fixation without heterocysts. Science 191, 1278–1280.

Carr, N.G. (1973) Metabolic control and autotrophic physiology. In N.G. Carr, B.A. Whitton (eds.), The Biology of the Blue-green Algae. University of California Press, Berkeley, pp. 39–63.

Clark, R.L. and Jensen, T.E. (1969) Ultrastructure of akinete development in a blue-green alga. *Cylindrospermum* sp. Cytologia 34, 439–448.

Codd, G.A. and Marsden, W.J.M. (1984) The carboxysomes (polyhedral bodies) of autotrophic prokaryotes. Biol. Rev. 59, 389–422.

Codd, G.A. and Stewart, W.D.P. (1976) Polyhedral bodies and ribulose 1,5-diphosphate carboxylase in the blue-green alga *Anabaena cylindrica*. Planta 130, 323–326.

Codd, G.A., Okabe, K. and Stewart, W.D.P. (1980) Cellular compartmentation of photosynthetic and photorespiratory enzymes in the heterocystous cyanobacterium *Anabaena cylindrica*. Arch. Microbiol. 124, 149–154.

Codd, G.A., Lanaras, T. and Leadbeater, L. (1984) Intracellular compartmentation of ribulose bisphosphate carboxylase in cyanobacteria and other autotrophic prokaryotes. In A. Barth, R. Haschen, F.H. Laibach and H. Possin (eds.), Molecular and Cellular Regulation of Enzyme Activity, Part III. Martin-Luther-Universität, Halle (Saale), pp. 273–314.

Cohen, D. and Parnas, H. (1976) An optimal policy for the metabolism of storage materials in unicellular algae. J. Theor. Biol. 56, 1–18.

Coleman, J.R., Seemann, R. and Berry, J.A. (1982) RuBP carboxylase in carboxysomes of blue-green algae. Carnegie Inst. Wash. Year Book 81, 83–87.

Cramer, C.L. and Davis, R.H. (1984) Polyphosphate-cation interaction in the amino acid containing vacuole of *Neurospora crassa*. J. Biol. Chem. 259, 5152–5157.

Cramer, C.L., Vaughn, L.E. and Davis, R.H. (1980) Basic amino acids and inorganic polyphosphates in *Neurospora crassa*: independent regulation of vacuolar pools. J. Bacteriol. 142, 845–952.

Dawes, E.A. and Senior, P.J. (1973) The role and regulation of energy reserve polymers in microorganisms. Adv. Microbiol. Physiol. 10, 135–266.

DeLoof, A. (1985) The cell as a miniature electrophoresis chamber. Comp. Biochem. Physiol. 80A, 453–459.

De Vasconcelos, L. and Fay, P. (1974) Nitrogen metabolism and ultrastructure in *Anabaena cylindrica*. I. The effect of nitrogen starvation. Arch. Microbiol. 96, 271–279.

Dharmawardene, M.W.N., Stewart, W.D.P. and Stanley, S.D. (1972) Nitrogenase activity, amino acid pool patterns and amination in blue-green algae. Planta 108, 133–145.

Dohler, G. (1976) Photosynthetische Carboxylierungsreaktionen verschieden pigmentierter *Anacystis* Zellen. Planta 131, 129–133.

Doonan, B.B. and Jensen, T.E. (1980) Ultrastructural localization of alkaline phosphatase in the cyanobacteria *Coccochloris peniocystis* and *Anabaena cylindrica*. Protoplasma 102, 189–198.

DuBois, J.D., Roberts, K.R. and Kapustka, L.A. (1984) Polyphosphate body and acid phosphatase localization in *Nostoc* sp. Can. J. Microbiol. 30, 8–15.

Edwards, M.R., Berns, D.S., Ghiorse, W.C. and Holt, S.C. (1968) Ultrastructure of the thermophilic blue-green alga *Synechococcus lividus* Copeland. J. Phycol. 4, 283–298.

Eykelenburg, V.A.N. (1980) Ecophysiological studies on *Spirulina platensis* effect of temperature, light intensity and nitrate concentration on growth and ultrastructure. Antonie Leeuwenhoek J. Microbiol. 46, 113–128.

Fay, P. and Walsby, A.E. (1966) Metabolic activities of isolated heterocysts of the blue-green alga *Anabaena cylindrica*. Nature (Lond.) 209, 94–95.

Fitzgerald, G.P. and Nelson, T.C. (1966) Extractive and enzymatic analysis for limiting or surplus phosphorus in algae. J. Phycol. 2, 32–37.

Fogg, G.E. (1951) Growth and heterocyst production in *Anabaena cylindrica* Lemm. III. The cytology of heterocysts. Ann. Bot. 15, 23–35.

Fogg, G.E., Stewart, W.D.P., Fay, P. and Walsby, A.E. (1973) The Blue-green Algae. Academic Press, New York, pp. 35–77.

Foster, J.W. (1947) Some introspections on mold metabolism. Bacteriol. Rev. 11, 167–188.

Fuhs, F.W. (1973) Cytochemical examination. In N.G. Carr and B.A. Whitton (eds.), The Biology of Blue-green Algae. University of California Press, Berkeley, pp. 117–143.

Greenberg, D.M. (1960) Arginase. In P.D. Boyer, H. Lardy and K. Myback (eds.), The Enzymes. Academic Press, New York, pp. 257–267.

Griffiths, D.J. (1970) The pyrenoid. Bot. Rev. 36, 29–58.

Grillo, J.F. and Gibson, J. (1979) Regulation of phosphate accumulation in the unicellular cyanobacterium *Synechococcus*. J. Bacteriol. 140, 508–517.

Gupta, M. and Carr, N.G. (1982) Enzyme activities related to cyanophycin metabolism in heterocysts and vegetative cells of *Anabaena* spp. J. Gen. Microbiol. 125, 17–23.

Halfen, L.N. and Castenholz, R.W. (1971) Energy expenditure for gliding motility in a blue-green alga. J. Phycol. 7, 258–260.

Hall, W.T. and Claus, G. (1963) Ultrastructural studies on the blue-green algal symbiont in *Cyanophora paradoxa* Korschikoff. J. Cell Biol. 19, 551–563.

Hall, W.T. and Claus, G. (1967) Ultrastructural studies on the cyanelles of *Glaucocystis nostochinearum* Itzigsohn. J. Phycol. 3, 37–51.

Harold, F.M. (1963) Accumulation of inorganic polyphosphate in *Aerobacter aerogenes*: I. Relationship to growth and nucleic acid synthesis. J. Bacteriol. 86, 216–221.

Harold, F.M. (1966) Inorganic polyphosphates in biology: structure, metabolism, and function. Bacteriol. Rev. 30, 772–794.

Harold, F.M. and Sylvan, S. (1963) Accumulation of inorganic polyphosphate in *Aerobacter aerogenes*: II. Environmental control and the role of sulfur compounds. J. Bacteriol. 86, 222–231.

Hawthornthwaite, A.M., Lanaras, T. and Codd, G.A. (1985) Immunoelectronmicroscopic localization of Calvin cycle enzymes in *Chlorogloeopsis fritschii*. J. Gen. Microbiol. 131, 2497–2500.

Healey, F.P. (1982) Phosphate. In N.G. Carr and B.A. Whitton (eds.), Biology of the Cyanobacteria. Blackwell Publ., Oxford, pp. 105–124.

Hegler, R. (1901) Untersuchungen über die Organisation der Phycochromaceenzellen. Jahrb. Wissensch. Botanik. 36, 229–354.

Hoare, D.S., Ingram, L.O., Thurston, E.L. and Walkup, R. (1971) Dark heterotrophic growth of an endophytic blue-green alga. Arch. Microbiol. 78, 310–321.

Holdsworth, R.H. (1971) The isolation and partial characterization of pyrenoid protein of *Eremosphaera viridis*. J. Cell Biol. 51, 499–513.

Howarth, R.W. and Cole, J.J. (1985) Molybdenum availability, nitrogen limitation, and phytoplankton growth in natural waters. Science 229, 653–655.

Jensen, T.E. (1984a) Cell inclusions in the cyanobacteria. In N.N. Scott (ed.), Ann. NY Acad. Sci. NY Acad. Sci., New York, pp. 279–282.

Jensen, T.E. (1984b) Cyanobacterial inclusions of irregular occurrence: systematic and evolutionary implications. Cytobios 39, 35–62.

Jensen, T.E. (1985) Cell inclusions in the cyanobacteria. Arch. Hydrobiol. Suppl. 71, 33–73.

Jensen, T.E. and Bowen, C.C. (1961) Organization of the centroplasm in *Nostoc pruniforme*. Iowa Acad. Sci. 68, 86–89.

Jensen, T.E., Baxter, M., Rachlin, W. and Jani, V. (1982) Uptake of heavy metals by *Plectonema boryanum* (Cyanophyceae) into cellular components, especially polyphosphate bodies: an X-ray energy dispersive study. Environ. Pollution (Ser. A) Ecol. Biol. 27, 119–127.

Kallas, T. and Castenholz, R.W. (1982) Internal pH and ATP-ADP pools in the cyanobacterium *Synechococcus* sp. during exposure to growth-inhibiting low pH. J. Bacteriol. 149, 229–236.

Kallas, T. and Dahlquist, F.W. (1981) Phosphorous-31 nuclear magnetic resonance analysis of internal pH during photosynthesis in the cyanobacterium *Synechococcus*. Biochemistry 20, 5900–5907.

Kerby, N.W. and Evans, L.V. (1981) Pyrenoid protein from the brown alga *Pilayella littoralis*. Planta 151, 469–475.

Kessel, M. (1977) Identification of a phosphorous-containing storage granule in the cyanobacterium *Plectonema boryanum* by electron microscope X-ray microanalysis. J. Bacteriol. 129, 1502–1505.

Krompkamp, J.C. and Mur, L.R. (1984) Buoyant density changes in the cyanobacterium *Microcystis aeruginosa* due to changes in the cellular carbohydrate content. FEMS Microbiol. Lett. 25, 105–110.

Kulaev, I.S. (1975) Biochemistry of inorganic polyphosphates. Rev. Physiol. Biochem. Pharmacol. 73, 131–158.

Ladha, J.K. and Kumar, H.D. (1978) Genetics of blue-green algae. Biol. Rev. 53, 355–386.

Lambert, D.H., Bryant, D.A., Stirewalt, V.L., Dubbs, J.M., Stevens, S.E. and Porter, R.D. (1985) Gene map for the *Cyanophora paradoxa* genome. J. Bacteriol. 164, 659–664.

Lanaras, T. and Codd, G.A. (1981a) Structural and immunoelectrophoretic comparison of soluble and particulate ribulose bisphosphate carboxylases from the cyanobacterium *Chlorogloeopsis fritschii*. Arch. Microbiol. 130, 213–217.

Lanaras, T. and Codd, G.A. (1981b) Ribulose 1,5-bisphosphate carboxylase and polyhedral bodies of *Chlorogloeopsis fritschii*. Planta 153, 279–285.

Lanaras, T. and Codd, G.A. (1982) Variations in ribulose 1,5-bisphosphate carboxylase protein levels, activities and subcellular distribution during photoautotrophic batch culture of *Chlorogloeopsis fritschii*. Planta 154, 284–288.

Lanaras, T., Hawthornthwaite, A.M. and Codd, G.A. (1985) Localization of carbonic anhydrase in the cyanobacterium *Chlorogloeopsis fritschii*. FEMS Microbiol. Lett. 26, 285–288.

Lang, N.J. (1965) Electron microscopic study of heterocyst development in *Anabaena azollae*. J. Phycol. 1, 127–134.

Lang, N.J. (1968) The fine structure of blue-green algae. Ann. Rev. Microbiol. 22, 15–46.

Lang, N.J. and Fay, P. (1971) The heterocysts of the blue-green algae. II. Details of ultrastructure. Proc. Roy. Soc. Ser. B 178, 193–203.

Lang, N.J. and Fisher, K.A. (1969) Variation in the fixation image of 'Structured Granules' in *Anabaena*. Arch. Microbiol. 67, 173–181.

Lang, N.J., Simon, R.D. and Wolk, C.P. (1972) Correspondence of cyanophycin granules with structured granules in *Anabaena cylindrica*. Arch. Microbiol. 83, 313–320.

Laursen, R.A. (1972) Automatic solid phase Edman degradation. In C.H.W. Hirs and S.N. Timasheft (eds.), Methods in Enzymology. Academic Press, New York, pp. 344–359.

Lawry, N.H. and Jensen, T.E. (1979) Deposition of condensed phosphate as an effect of varying sulfur deficiency in the cyanobacterium *Synechococcus* sp. (*Anacystis nidulans*). Arch. Microbiol. 120, 1–7.

Lawry, N.H. and Simon, R.D. (1982) The normal and induced occurrence of cyanophycin inclusion bodies in several blue-green algae. J. Phycol. 18, 391–399.

Lefort, M. (1960) Structure inframicroscopique du chromatoplasma de quelques cyanophycées. Compt. Rend. Acad. Sci. 250, 1525–1527.

Lindblad, P., Bergman, B., Hofsten, A.V., Hallbom, L. and Nylund, J.E. (1985) The cyanobacterium-*Zamia* symbiosis: an ultrastructural study. New Phytol. 101, 707–716.

Marsden, W.J.N., Lanaras, T. and Codd, G.A. (1984) Subcellular segregation of phosphoribulokinase and ribulose-1,5-bisphosphate carboxylase/oxygenase in the cyanobacterium *Chlorogloeopsis fritschii*. J. Gen. Microbiol. 130, 2089–2093.

Miller, A.G. and Colman, B. (1980) Active transport and accumulation of bicarbonate by a unicellular cyanobacterium. J. Bacteriol. 143, 1253–1259.

Miller, J.J. (1984) In vitro experiments concerning the state of polyphosphate in the yeast vacuole. Can. J. Microbiol. 30, 236–246.

Miller, L.S. and Holt, S.C. (1977) Effects of carbon dioxide on pigment and membrane content in *Synechococcus lividus*. Arch. Microbiol. 115, 185–198.

Miller, M.M. and Lang, N.J. (1968) The fine structure of akinete formation and germination in *Cylindrospermum*. Arch. Microbiol. 60, 303–313.

Morita, R.Y. (1982) Starvation-survival of heterotrophs in the marine environment. In K.C. Marshall (ed.), Advances in Microbial Ecology. Plenum Press, New York, pp. 171–198.

Niemeyer, R. and Richter, G. (1969) Schnellmarkierte Polyphosphate und Metaphosphate bei der Blaualge *Anacystis nidulans*. Arch. Microbiol. 69, 54–59.

Orme-Johnson, W.H. (1985) Molecular basis of nitrogen fixation. Ann. Rev. Biophys. Biophys. 14, 419–459.

Pankratz, H.S. and Bowen, C.C. (1963) Cytology of the blue-green algae I. The cells of *Symploca muscorum*. Am. J. Bot. 50, 387–399.

Peat, A. and Whitton, B.A. (1967) Environmental effects on the structure of the blue-green alga *Chlorogloea fritschii*. Arch. Microbiol. 57, 155–180.

Plapp, R. and Strominger, J.L. (1969) Biosynthesis of the peptidoglycan of bacterial cell walls. XVIII. Biosynthesis of peptidoglycan and of interpeptide bridges in *Lactobacillus viridescens*. J. Biol. Chem. 245, 3667–3674.

Poindexter, J.S. and Eley, L.F. (1983) Combined procedure for assays of poly-β-hydroxybutyric acid and inorganic polyphosphate. J. Microbiol. Methods 1, 1–17.

Rachlin, J.W., Jensen, T.E., Baxter, M. and Jani, V. (1982) Utilization of morphometric analysis in evaluating response of *Plectonema boryanum* (Cyanophyceae) to exposures to eight heavy metals. Arch. Environ. Contam. Toxicol. 11, 323–333.

Rachlin, J.W., Jensen, T.E. and Warkentine, B. (1984) The toxicological response of the alga *Anabaena flos-aquae* (Cyanophyceae) to cadmium. Arch. Environ. Contam. Toxicol. 13, 143–152.

Rao, N.N., Roberts, M.F. and Torriani, A. (1985) Amount and chain length of polyphosphates in *Escherichia coli* depend on cell growth conditions. J. Bacteriol. 162, 242–247.

Rippka, R., Deruelles, J., Waterbury, J.B., Herdman, M. and Stanier, R.Y. (1979) Generic assignments, strain histories and properties of pure cultures of cyanobacteria. J. Gen. Microbiol. 111, 1–61.

Rodriguez-Lopez, M., Munoz-Calvo, M.L. and Gomez-Acebo, J. (1971) The effect of rifamycins in the ultrastructure of *Anacystis montana*. J. Ultrastruct. Res. 36, 595–602.

Roussard-Jacquemin, M. (1983) Etude ultrastructurale de la différentiation des hétérocystes chez la cyanobacterie *Anabaena cylindrica* Lemm. Can. J. Microbiol. 29, 1564–1575.

Sherman, D.N. and Sherman, L.A. (1983) Effect of iron deficiency and iron restoration on ultrastructure of *Anacystis nidulans*. J. Bacteriol. 156, 393–401.

Shively, J.M. (1974) Inclusion bodies in procaryotes. Ann. Rev. Microbiol. 28, 167–187.

Shively, J.M., Ball, F., Brown, D.H. and Saunders, R.E. (1973a) Functional organelles in procaryotes: polyhedral inclusions (Carboxysomes) of *Thiobacillus neapolitanus*. Science 182, 584–586.

Shively, J.M., Ball, F.L. and Kline, B.W. (1973b) Electron microscopy of the carboxysomes (polyhedral bodies) of *Thiobacillus neapolitanus*. J. Bacteriol. 116, 1405–1411.

Sicko-Goad, L. (1982) A morphometric analysis of algal response to low dose, short term heavy metal exposure. Protoplasma 110, 75–86.

Sicko-Goad, L., Crang, R.E. and Jensen, T.E. (1975) Phosphate metabolism in blue-green algae. IV. In situ analysis of polyphosphate bodies by X-ray energy dispersive analysis. Cytobiologie 11, 430–437.

Simon, R.D. (1971a) Isolation, characterization, and physiology of the cyanophycin granules from the blue-green alga *Anabaena cylindrica* Lemm. Thesis, Michigan State University, East Lansing, pp. 1–114.

Simon, R.D. (1971b) Cyanophycin granules from the blue-green alga *Anabaena cylindrica*: a reserve material consisting of copolymers of aspartic acid and arginine. Proc. Natl. Acad. Sci. USA 68, 265–267.

Simon, R.D. (1973a) The effect of chloramphenicol on the production of cyanophycin granule polypeptide in the blue-green alga *Anabaena cylindrica*. Arch. Microbiol. 92, 115–122.

Simon, R.D. (1973b) Measurement of the cyanophycin granule polypeptide contained in the blue-green alga *Anabaena cylindrica*. J. Bacteriol. 114, 1213–1216.

Simon, R.D. (1976) The biosynthesis of multi-l-arginyl-poly(l-aspartic acid) in the filamentous cyanobacterium *Anabaena cylindrica*. Biochim. Biophys. Acta 422, 407–418.

Simon, R.D. and Weathers, P. (1976) Determination of the structure of the novel polypeptide containing aspartic acid and arginine which is found in cyanobacteria. Biochim. Biophys. Acta 420, 165–176.

Simon, R.D., Lawry, N.H. and McLendon, G.L. (1980) Structural characterization of the cyanophycin granule polypeptide of *Anabaena cylindrica* by circular dichroism and Raman spectroscopy. Biochim. Biophys. Acta 626, 277–281.

Singh, S.P. and Yanava, Y. (1984) Cadmium induced inhibition of ammonium and phosphate uptake in *Anacystis nidulans*: interactions with other cations. J. Gen. Appl. Microbiol. 30, 79–86.

Sinskey, A.J. (1983) Organic chemicals from biomass: an overview. In D.L. Wise (ed.), Organic Chemicals from Biomass. Benjamin/Cummings, Menlo Park, pp. 1–67.

Stam, W.T. (1980) Relationships between a number of filamentous blue-green algal strains (Cyanophyceae) revealed by DNA-DNA hybridization. Arch. Hydrobiol. 56, 351–374.

Stark, G.R. (1972) Sequential degradation of peptides and proteins from their COOH termini with ammonium thiocyanate and acetic anhydride. In C.H.W. Hirs and S.N. Timasheft (eds.), Methods in Enzymology. Academic Press, New York, pp. 369–384.

Stevens, S.E., Balkwill, D.L. and Paone, D.A.M. (1981a) The effects of nitrogen limitation on the ultrastructure of the cyanobacterium *Agmenellum quadruplicatum*. Arch. Microbiol. 130, 204–212.

Stevens, S.E., Paone, D.A.M. and Balkwill, D.L. (1981b) Accumulation of cyanophycin granules as a result of phosphate limitation in *Agmenellum quadruplicatum*. Plant Physiol. 67, 716–719.

Stevenson, R.J. and Stoermer, E.E. (1982) Luxury consumption of phosphorus by benthic algae. Bioscience 32, 682–683.

Stewart, W.D.P., and Codd, G.A. (1975) Polyhedral bodies (carboxysomes) of nitrogen-fixing blue-green algae. Br. J. Phycol. 10, 273–278.

Suresch, N., Warburg, R., Timmermen, M., Wells, J., Coccia, M., Roberts, M.F. and Halvorson, H.O. (1985) New strategies for the isolation of microorganisms responsible for phosphate accumulation. Water Sci. Technol. 17, 99–112.

Tanford, C. (1962) The interpretation of hydrogen ion titration curves of proteins. Adv. Protein Chem. 17, 69–165.

Thomas, R.H. and Walsby, A.E. (1985) Buoyancy regulation in a strain of *Microcystis*. J. Gen. Microbiol. 131, 799–810.

Tijssen, J.P.F. and Van Steveninch, J. (1985) Cytochemical staining of a yeast polyphosphate fraction, localized outside the plasma membrane. Protoplasma 125, 124–128.

Vaillancourt, S., Beauchemin-Newhouse, N. and Cedergren, R.J. (1978) Polyphosphate deficient *Anacystis nidulans*. Can. J. Microbiol. 24, 112–116.

Webster, J.D., Fay, D.D., Costa, J.L., Jones, P.M. and Hugh, R. (1984) Elemental composition of bacterial metachromatic inclusions determined by electron microprobe X-ray analysis. J. Bacteriol. 158, 441–446.

Westphal, K., Bock, E., Cannon, G. and Shively, J.M. (1979) Deoxyribonucleic acid in *Nitrobacter* carboxysomes. J. Bacteriol. 140, 285–288.

Whatley, J.M. (1977) The fine structure of *Prochloron*. New Phytol. 79, 309–313.

Winkenbach, F. and Wolk, C.P. (1973) Activities of enzymes of the oxidative and the reductive pentose phosphate pathways in heterocysts of a blue-green alga. Plant Physiol. 52, 480–483.

Wolk, C.P. (1968) Movement of carbon from vegetative cells to heterocysts in *Anabaena cylindrica*. J. Bacteriol. 96, 2138–2143.

P. Fay and C. Van Baalen (Eds.) The Cyanobacteria

9

Akinetes: structure and function

MICHAEL HERDMAN

Unité de Physiologie Microbienne, Institut Pasteur, 28 rue du Docteur Roux, 75724 Paris Cedex 15, France

1. Introduction: akinetes 1856–1956

Cyanobacterial akinetes are considered to be resting cells, normally resistant to low temperatures and to desiccation (as first shown by Janczewski, 1874) but rarely surviving exposure to high temperature (see Section 2.2. of this chapter). They were observed to remain viable for long periods under such adverse conditions (Bornet and Flahault, 1886; Borzi, 1878; Bristol, 1919; Lipman, 1941). Akinetes are often called 'spores', particularly in the early literature, but there now seems to be good agreement that the word akinete (Greek. *Akinetos*: motionless), first used by Wille (1887), be used to differentiate these unique structures from the variety of spores, cysts, endospores and exospores observed in other microorganisms and from the 'endospores' (baeocytes: Waterbury and Stanier, 1978) of pleurocapsalean cyanobacteria.

Cyanobacteria are easily detected in the natural environment, because they are pigmented and many form macrocolonies; in addition, many possess large cells, easily seen in the light microscope. It is, therefore, not surprising that cyanobacterial akinetes were first described (Carter, 1856; Bary, 1863; Rabenhorst, 1865) long before the bacterial endospore was discovered by Cohn (1877) and Koch (1877). Resting cells of other prokaryotes were described even later: the microcyst of *Myxococcus* (Thaxter, 1892) and the cyst of *Azotobacter* (Beijerinck, 1901). The same characteristics permitted the study of the structure of akinetes, which were shown to be enlarged, possess a thick wall, to contain cyanophycin granules as reserve materials and to lack photosynthetic pigments (Geitler, 1925). The thickened wall was apparently composed of two layers (Gomont, 1888; Fritsch, 1905) and was surrounded by a thick outer envelope which often appeared to be composed of several layers, the outer layer being 'traversed by numerous radiating lines' (Fritsch, 1918); this feature is seen clearly in the light microscope (Fig. 1).

Akinetes were found to be restricted to certain genera of filamentous heterocystous cyanobacteria (Brand, 1903). They were observed to arise either adjacent to heterocysts in e.g. *Gloeotrichia, Cylindrospermum* and *Anabaena* or, as particularly emphasized by Geitler (1925), midway between two heterocysts (in, e.g., *Nostoc* spp.) where a new heterocyst would be expected to appear; in the latter case, a long chain of akinetes was often seen in this intercalary position. The apparent spatial relationship between heterocysts and akinetes led to the suggestion (see Fritsch, 1945) "that the former might act as store-houses for reserve food (Hieronymus, 1892; Hegler, 1901; Brand, 1903; Fritsch, 1904a) but there is reason to believe that this phenomenon has some deeper significance". Akinetes were thought to develop under conditions of nitrogen starvation (Glade, 1914; Harder, 1917c); the significance of the spatial patterns of the two types of differentiated cell had to await the discovery (Fay et al., 1968) of the function of the heterocyst.

On the return of conditions favourable for growth, akinetes were seen to germinate, either immediately (Fritsch, 1904b; Spratt, 1911; Harder, 1917a; Bharadwaja, 1933) or only after a resting period (Rose, 1934). Harder (1917a,b) determined that akinetes of most species require light for germination, the rate of germination being proportional to light intensity; in some strains, however, germination occurred in the dark in the presence of 'cane sugar'. Morphological changes during germination were well described by Fritsch (1945): cell division sometimes occurred before the young germling was liberated from the akinete envelope; often, however, a small pore was observed in the envelope before the akinete contents were liberated as a single cell or as a dividing filament.

2. Recent advances: akinetes 1956–1986

The invention of the electron microscope did little to extend the careful descriptions of the microscopists of the previous century. In contrast, many physiological studies of akinetes, often giving conflicting results, have been described. Such studies were facilitated by the availability of suitable axenic strains of cyanobacteria to which, for reasons which escape comprehension, microbiologists of the earlier period did not have access.

2.1. Structure of akinetes

The structure of cyanobacterial akinetes, well reviewed previously (Nichols and Carr, 1978; Nichols and Adams, 1982), will not be described in detail here. In summary, akinetes can be conveniently divided into three classes (see Fig. 1): the very large elongated akinetes which arise immediately adjacent to the heterocysts of e.g., *Anabaena*, *Gloeotrichia* and *Cylindrospermum* spp.; the smaller spherical ones which first appear in the centre of the interheterocyst interval in e.g., *Nostoc* spp.; and the akinetes of a number of planktonic species which nearly always differentiate from the third vegetative cell away from the heterocyst (see Section 2.6.1.). Exceptions to this rule are probably *Fischerella* and *Chlorogloeopsis*, morphologically complex cyanobacteria (see e.g., Rippka et al., 1979; Rippka and Herdman, 1985), whose akinete spatial patterns have not been thoroughly studied.

Akinetes of all types generally possess a thickened cell wall and a multilayered extracellular envelope (Wildon and Mercer, 1963; Leak and Wilson, 1965; Miller and Lang, 1968; Clark and Jensen, 1969; Jensen and Clark, 1969; Sutherland et al., 1979; Braune, 1980; Grilli Caiola and De Vecchi, 1980). Exceptions to this rule are the akinetes of *Aphanizomenon flos-aquae*, where the envelope is replaced by a thin sheath (Wildman et al., 1975) and those of *Anabaena variabilis*, which lack the thickened cell wall (Braune, 1980). The structure of the cell wall and envelope of akinetes of *Nostoc* PCC 7524 is shown in Fig. 2A.

Akinetes accumulate cyanophycin, the reserve polymer of arginine plus aspartate (see Section 2 of Chapter 8), which is easily seen in the light microscope (Miller and Lang, 1968; Clark and Jensen, 1969; Jensen and Clark, 1969; Sutherland et al., 1979). In all other respects, akinetes are similar to vegetative cells in structure except that, in organisms which produce gas vacuoles, these structures disappear during akinete differentiation (see Rother and Fay, 1977).

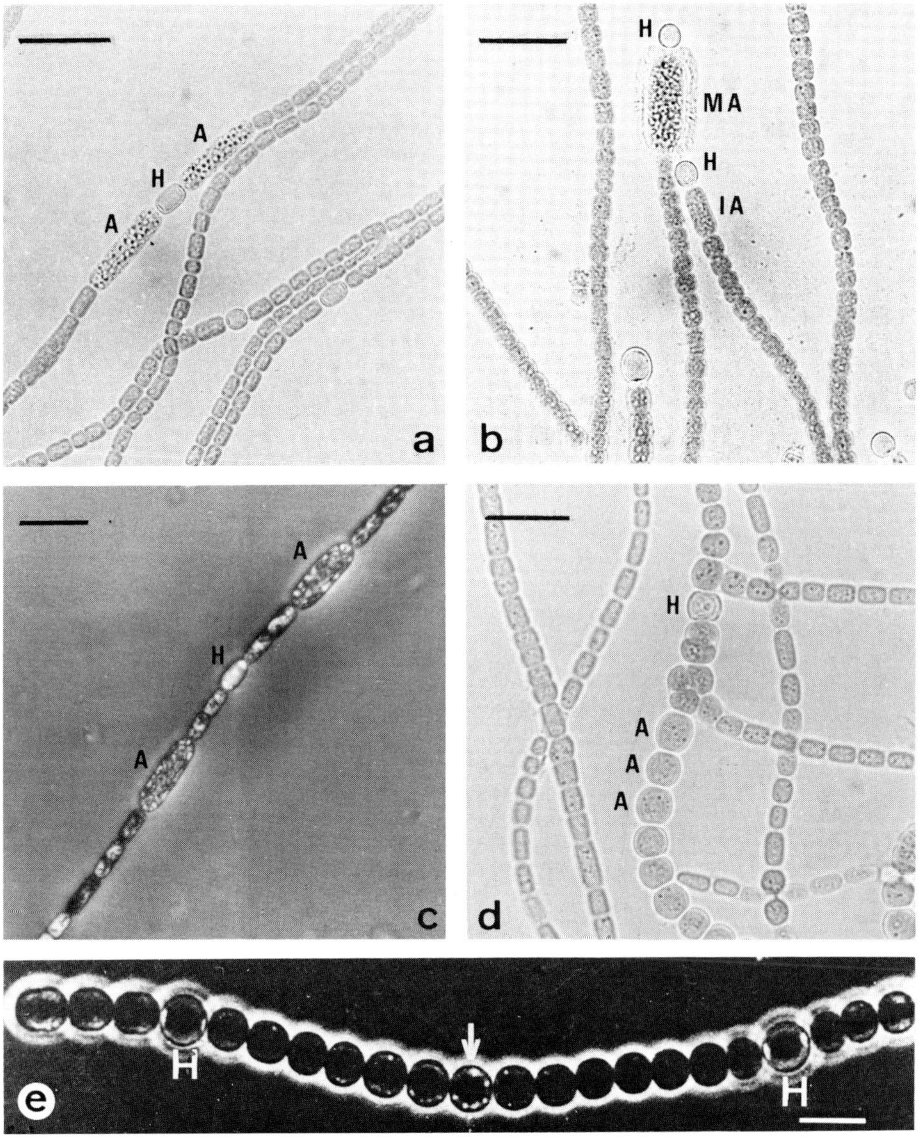

Fig. 1. The spatial pattern of akinetes (A) relative to heterocysts (H) in (a) *Anabaena cylindrica* PCC 7122, (b) *Cylindrospermum* PCC 7417, (c) *Anabaena circinalis*, (d) *Fischerella* PCC 7414 and (e) *Nostoc* PCC 7524. Note the mature (MA) and immature (IA) akinetes in *Cylindrospermum*. The refractile granules in (e) (arrow) are cyanophycin granules; they appear first in the cell midway between two heterocysts and mark the first stage of akinete differentiation. Note also the chain of akinetes in *Fischerella* and the lateral branches composed of vegetative cells. (a), (b) and (d) bright-field (Courtesy of R. Rippka); (c) phase contrast (Courtesy of P. Fay); (e) phase contrast (from Sutherland et al., 1979, with permission). Bar markers represent 20 μm in (a)–(d), 5 μm in (e).

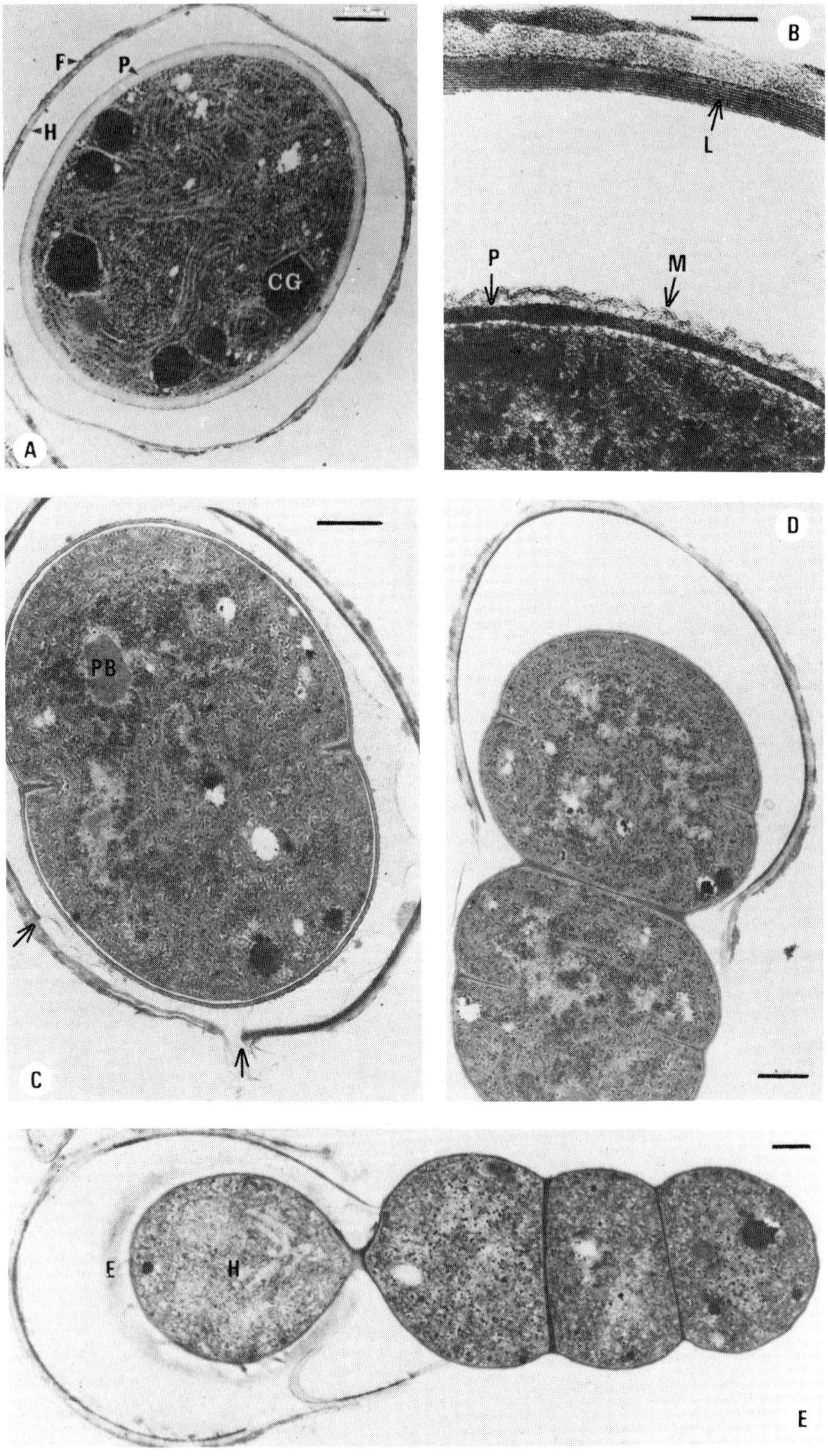

Fig. 2.

2.2. Function of akinetes

Akinetes of *Anabaena cylindrica* were shown to be resistant to desiccation, surviving at least 5 years in the dark in the dry state, in contrast to vegetative cells which survived no longer than 2 weeks (Yamamoto, 1975). Glade (1914) had shown that akinetes of *Cylindrospermum* survived exposure to elevated temperature for short periods, and Wieringa (1968) used this property to purify akinete-forming cyanobacteria. However, akinetes are unable to survive prolonged exposure to high temperature; those of *A. cylindrica* died at temperatures above 55°C (Yamamoto, 1976) and the viability of akinetes of *Nostoc* PCC 7524 was reduced to 5% following incubation for 10 min at 45°C, only 4°C higher than the maximum temperature for growth of vegetative cells (Sutherland et al., 1979). In this respect, therefore, akinetes resemble the cysts of *Azotobacter* and differ from the heat-resistant bacterial endospore. In contrast, 95% of the akinetes of *Nostoc* PCC 7524 survived storage in the dark at 4°C for 15 months, while over 90% of the vegetative cells became inviable within 7 days at this temperature (Sutherland et al., 1979); approximately 45% of the akinetes survived freezing at −20°C and 22% survived two cycles of freezing and thawing (see also Chauvat and Joset-Espardellier, 1981). Like akinetes of *A. cylindrica* (Yamamoto, 1975), those of *Nostoc* PCC 7524 were resistant to desiccation, 45 to 80% surviving after 6 months (Sutherland et al., 1979).

These properties suggest that akinetes serve a perennating role, enabling the organism to survive cold winters or dry summers. It has been shown that akinetes, but not vegetative cells, survive in surface mud sediments during winter and germinate in spring (Rose, 1934; Roelofs and Oglesby, 1970; Wildman et al., 1975). However, Reynolds (1972, 1975) suggested that the number of akinetes is so low that their germination contributes little to the spring population; a small stock of vegetative cells survived in the water column, sufficient to account for the spring growth. In addition, as pointed out by Rother and Fay (1977), cyanobacteria which are incapable of forming akinetes (e.g., *Microcystis*) nevertheless survive the winter in the sediments of deep lakes, giving rise to planktonic populations in spring. After an extensive study of two lakes, Rother and Fay (1977) demonstrated that in one, overwintering akinetes contributed significantly to the development of the population in spring; in the second lake the population survived the winter in the vegetative state. The value of akinetes as overwintering structures would therefore appear to be questionable; Rother and Fay (1977) concluded that since akinetes of planktonic strains, formed in dense surface blooms in the summer, germinate soon after maturation, their function is to permit survival during periods of high light intensity, when vegetative cells are subject to photo-oxidative death. The process of akinete formation removes the organism from the plankton, because the disappearance

←

Fig. 2. Germination of akinetes of *Nostoc* PCC 7524. (A) Mature akinete showing the thickened peptidoglycan layer (P) of the cell wall and the extracellular envelope composed of homogeneous (H) and fibrous (F) layers. CG, cyanophycin granule. Bar = 500 nm. (B) An akinete several hours after the initiation of germination: the peptidoglycan layer (P) of the cell wall is thinner than that of the mature akinete but has not yet completely reverted to the thickness of that of the vegetative cell; it is surrounded by an outer membrane (M) and a new laminated layer (L) has appeared on the inside of the envelope. Bar: 100 nm. (C) An early stage of formation of the septum during the first cell division. Note the wide break (indicated by arrows) in the laminated layer of the envelope; the gap is still covered by the outer layers. Polyhedral bodies (PB) are visible within the cell and few cyanophycin granules remain. Bar = 500 nm. (D) Division of the two daughter cells is initiated synchronously; the germling is still partly inside the envelope. Bar = 500 nm. (E) A germinating akinete at the 4-celled stage. A heterocyst (H) has begun to differentiate from the terminal cell of the young filament, still trapped inside the akinete envelope; the heterocyst envelope (E) is not yet completely formed. Bar = 500 nm. (Fig. 2A reproduced from Sutherland et al., 1979; Figs. 2B–2E from Sutherland et al., 1985b, with permission) of gas vacuoles and accumulation of reserve materials lead to a considerable increase in density of the cell, which therefore sinks.

Akinetes of *Anabaena circinalis* were much more resistant to digestive enzymes than were vegetative cells; 10–25 % survived undamaged after passage through the gut of a natural predator, *Daphnia* (Fay, 1985). The akinetes were also more resistant to toxic chemicals and survived prolonged incubation in the dark under N_2 at 8°C, conditions simulating those in lake sediment.

In conclusion, akinetes do permit survival of the organism under adverse conditions, although they are not always necessary since many strains (including those capable of producing akinetes) can survive in the vegetative state. They certainly ensure the long-term survival of the organism: Livingstone and Jaworski (1980) isolated viable akinetes from sediments as old as 64 years.

2.3. Macromolecular composition of akinetes

2.3.1. *Macromolecular changes during akinete differentiation*

Akinetes of *Nostoc* PCC 7524 (Sutherland et al., 1979) developed after the end of exponential growth (see Section 2.6.2.). During this period heterocyst differentiation ceased and their frequency decreased from 9 % to 4.5 % as vegetative cell numbers doubled. Chlorophyll, total protein, phycocyanin and total nitrogen contents of the culture increased with culture density until the onset of akinete differentiation; their synthesis ceased during the phase of akinete maturation. Dry weight and total carbon content continued to increase, however, and cyanophycin accumulated at a rapid rate. This unbalanced macromolecular synthesis was the result of differences in the rates of fixation of CO_2 and N_2: although the former decreased by 40 % prior to the onset of akinete development, remaining constant during maturation, the rate of N_2 fixation decreased dramatically. The accumulation of glycogen and cyanophycin, and the increase in dry weight and total carbon in the absence of protein synthesis, were the most striking physiological changes during akinete differentiation. Increase in dry weight of the culture during akinete development was also observed in *A. cylindrica* (Fay, 1969a; Simon, 1977a) without protein synthesis (Simon, 1977a).

Fay et al. (1984) also observed a decrease, of approximately 50 %, of the heterocyst frequency during akinete differentiation in cultures of *A. circinalis* in post-exponential phase; again, the rate of N_2 fixation declined before that of CO_2 fixation and the C:N ratio increased from its level during mid-exponential growth to a maximum at the end of akinete differentiation. The same general trends are, therefore, apparent during akinete development in *Nostoc* PCC 7524 and *A. circinalis*, and may be common to all cyanobacteria. The changes in macromolecular composition which result are described below.

2.3.2. *The akinete envelopes*

C.P. Wolk and co-workers (Dunn and Wolk, 1970; Cardemil and Wolk, 1976, 1979) elucidated the chemical composition of the akinete wall of *A. cylindrica*. It contained (as a percentage of dry weight) 41 % carbohydrate and 24 % amino compounds, in comparison to 18 % and 65 %, respectively, in the vegetative cell wall. The glucose-rich carbohydrate (glucose:mannose:galactose:xylose, 75:20:3:4) was similar to that of the heterocysts but very different from that of the vegetative cells (35:50:5:8, respectively). The apparent close relationship between akinetes and heterocysts was strengthened by the discovery (Cardemil and Wolk, 1979) that both contain identical repeated sequences of sugars in their polysaccharide: these sequences are composed of glucose and mannose, linked by $\beta(1\rightarrow3)$ glycosidic bonds; side branches contain galactose, glucose, mannose and xylose.

2.3.3. *Pigments*

The precise pigment content of akinetes appears to be a controversial subject. Fay (1969a) demonstrated that akinetes of *A. cylindrica* lacked phycocyanin, contained an increased amount of xanthophylls and that chlorophyll was mostly replaced by pheophytin. However, Wolk and Simon (1969) reported that the pigment content of akinetes of the same organism was similar to that of the vegetative cells. Akinetes of *Anabaena doliolum* apparently lost both chlorophyll and phycocyanin following incubation in the dark for 3 to 4 days (Singh and Sunita, 1974). It would seem that these different results stem from the fact that as akinetes age they progressively degrade phycocyanin (Sutherland et al., 1985a) and, possibly, chlorophyll.

Newly-formed akinetes of *Nostoc* PCC 7524 contained the same amount of phycocyanin per cell as did the vegetative cells but the chlorophyll content (per cell) increased 2-fold; however, since the dry weight of the akinete was double that of the vegetative cell, chlorophyll remained constant, while phycocyanin decreased by 50 %, as a proportion of the dry weight (see Table I). Fay et al. (1984) showed a dramatic decrease in the phycocyanin content (expressed as percentage of dry weight) in cultures of *A. circinalis* immediately after the initiation of akinete differentiation, while the chlorophyll content decreased less markedly; values of the cellular contents of these pigments were not provided.

Björn et al. (1983) provided evidence for the presence of phycochromes *b* and *d*, and of allophycocyanin-B, in akinetes of *A. variabilis*; in vivo fluorescence measurements suggested that akinetes lack a functional Photosystem II although the reaction centre chlorophyll was present and connected to the phycobilisomes.

2.3.4. *Glycogen, lipids and fatty acids*

Yamamoto (1972) demonstrated that the content of lipids and fatty acids of akinetes of *A. cylindrica* was reduced to 85 % and 70 %, respectively, in comparison to the vegetative cells, and that the fatty acids were about 30 % more saturated. Glycogen accumulated in akinetes of *Anabaena* sp. (Sarma and Kanta, 1979) and *Nostoc* PCC 7524 (Sutherland et al., 1979) (see Table I), increasing 4- and 16-fold, respectively, above the levels in the vegetative cells. However, such increases in glycogen content also occur in cyanobacterial vegetative cells in stationary phase, and are therefore not unique to akinetes (see Sutherland et al., 1985a).

2.3.5. *Cyanophycin and free amino acids*

Although it is generally agreed that akinetes contain large amounts of cyanophycin, most observations are based on microscopic studies and it seems that only one quantitative analysis of cyanophycin content during growth and subsequent akinete differentiation has been performed (Sutherland et al., 1979): levels of cyanophycin were low during exponential growth of *Nostoc* PCC 7524, increasing concomitantly with the number of akinetes during post-exponential phase; akinetes contained 8 times more of this reserve material than did vegetative cells (Table I). However, cyanophycin content also increases in the vegetative cells of cyanobacteria, commonly accumulating in the range 8–16 % of dry weight after the end of exponential growth (Simon, 1973; Allen et al., 1980; see Section 2 of Chapter 8). These values are higher than that (5.1 %, see Table I) observed in the akinetes of *Nostoc* PCC 7524; cyanophycin accumulation is therefore not a fundamental process associated with akinete differentiation but rather, as is the case for glycogen, should be considered to be a normal physiological response of the cell to the environmental change.

The pools of free amino acids in akinetes of *A. cylindrica* were different to those of vegetative cells (Yamamoto, 1976). Levels of aspartate and proline increased, while threonine, glutamate, glycine, alanine, valine, leucine, isoleucine, tyrosine and phenylalanine decreased significantly. Lysine, histidine, arginine, cysteine and methionine were either not detected, or occurred only at very low levels, in both akinetes and vegetative cells. The significance of the changes in pools of free amino acids is not clear, but should be considered in relation to the decrease in the rate of N_2 fixation in the culture at the time of akinete differentiation (see Section 2.3.1.).

2.3.6. *Nucleic acids and protein*

The reported values of the content of DNA and RNA of akinetes vary widely. Ueda and Sawada (1972) suggested that akinetes of *Cylindrospermum* sp. contained 30 times more DNA than did vegetative cells; DNA content was measured in vivo by a fluorescence method, and absolute values were not given. Simon (1980) described the difficulties in applying such techniques to the measurement of cyanobacterial DNA content.

The first accurate measurements of the levels of DNA and RNA in akinetes were provided by Simon (1977a). Akinetes of *A. cylindrica* contained on average 57 genome equivalents (2.3×10^{-13} g) of DNA; the vegetative cells from which they developed (in late exponential growth) contained 25 genome equivalents (1×10^{-13} g) whereas cells growing exponentially contained a similar amount (2.2×10^{-13} g) of DNA as the akinetes. These values for genome content require slight modification, because they were based on a genome size of 2.47×10^9 daltons (determined by Herdman and Carr (1974) on a different strain of *A. cylindrica*); the correct genome size of the strain (*A. cylindrica* PCC 7122, UTEX 629) examined by Simon (1977a) is 3.17×10^9 daltons (Herdman et al., 1979) and therefore the akinetes contain 44 genome equivalents of DNA, compared to 20 in vegetative cells in the late exponential phase of growth. In contrast, the RNA content of the akinetes (1.3×10^{-11} g) increased markedly in comparison to vegetative cells growing exponentially (3×10^{-12} g) or in the late exponential phase of growth (1.3×10^{-12} g). These cell types contained, respectively, 4.9×10^{-11}, 1.7×10^{-11} and 1.1×10^{-11} g of protein (Simon, 1977a).

Since the macromolecular composition of vegetative cells of cyanobacteria changes dramatically with growth rate (see e.g. Mann and Carr, 1974), cells in the post-exponential phase of growth should contain less DNA and RNA than those growing exponentially. It follows that the macromolecular composition of akinetes, which develop after the end of exponential growth, can only be validly compared to that of the vegetative cells from which they differentiated. According to the data of Simon (1977a) therefore, akinetes of *A. cylindrica* contain 10 times more RNA, 5 times more protein and 2.3 times more DNA, than their parental cells. Such increases are not surprising, since the volume of the akinete in this organism is 10-fold larger than that of the vegetative cells (Fay, 1969a,b). Unlike *A. cylindrica*, *A. variabilis* UTEX 1444 produces akinetes which are only approximately twice the volume of the vegetative cells (Simon, 1980). The DNA content of the akinetes (5×10^{-14} g), however, was again similar to that of exponentially-growing vegetative cells; the DNA content of cells in late exponential phase was, unfortunately, not determined. Assuming a genome size of 3.5×10^9 daltons, the akinetes of this organism contain 8–9 genome equivalents of DNA.

The akinetes of *Nostoc* PCC 7524 contained the same amount of DNA (10 genome equivalents) as the vegetative cells from which they differentiated, and their content of RNA did not change significantly (Sutherland et al., 1979). The 2-fold increase in dry weight of these akinetes was not accompanied by an increase in protein (Table I), and could not be entirely accounted for by the marked increases in cellular contents of cyanophycin, glycogen and chlorophyll.

It is unfortunate that the complete macromolecular composition of akinetes is available for only one cyanobacterial strain, *Nostoc* PCC 7524 (Sutherland et al., 1979; see Table I), most authors having confined their analyses to one or two macromolecules.

TABLE I

Cellular composition of akinetes and vegetative cells of *Nostoc* PCC 7524

Component	Mean cellular content (pg)			
	akinete		vegetative cell[a]	
Chlorophyll	0.31	(2.1)	0.15	(2.1)
Phycocyanin	0.82	(5.5)	0.88	(12.4)
Total protein[b]	4.47	(30.0)	4.76	(67.0)
Cyanophycin	0.81	(5.4)	0.10	(1.4)
Glycogen	2.30	(15.4)	0.14	(2.0)
DNA	0.09	(0.63)	0.10	(1.4)
RNA	2.0	(13.4)	1.9	(26.8)
Total nitrogen	0.88	(5.9)	0.75	(10.6)
Total carbon	6.3	(42.2)	3.0	(42.3)
Total dry weight	14.9		7.1	

[a]Values refer to exponentially growing cells.
[b]Values for total protein exclude cyanophycin, which is not measured by the Lowry method (Simon, 1977a).
Numbers in parentheses indicate the content as a percentage of the dry weight.
(Reproduced from Sutherland et al., 1979)

2.4. Metabolism of akinetes

The studies of Fay (1969a,b) on purified akinetes of *A. cylindrica* showed that they fixed CO_2 at a reduced rate (15% of that observed in intact filaments), consistent with their reduced content of photosynthetic pigments. In contrast, the akinetes exhibited elevated rates of respiration (see also Yamamoto, 1976) but were incapable of fixing nitrogen (Fay, 1969b). Similar results for the fixation of CO_2 and N_2 were obtained by Sutherland et al. (1985a) with akinetes of *Nostoc* PCC 7524; since these two strains represent two different classes of akinete-forming cyanobacteria (with respect to the position of the akinetes relative to the heterocysts), it would appear that these characteristics are typical of all cyanobacterial akinetes. However, Chauvat et al. (1982) reported that aged akinetes, after being maintained in culture for long periods of time, also lost most of their respiratory capacity. Further studies on the metabolism of akinetes of different ages would therefore appear to be necessary.

On transfer to normal growth conditions, akinetes of *A. doliolum* were capable of immediate synthesis of phycocyanin and chlorophyll (Singh and Sunita, 1974), suggesting that the resting akinete may contain the enzymes necessary for such synthesis; however, it cannot be excluded that de novo synthesis of these enzymes occurs immediately on transfer of the akinetes to germination conditions. Similarly, akinetes of *Nostoc* PCC 7524 were able to synthesize protein, phycocyanin, glycogen and RNA, fix CO_2 (Sutherland et al., 1985a) and respire (Chauvat et al., 1982), immediately following transfer to fresh medium in the light. Synthesis of DNA and of chlorophyll, however, commenced after a lag of 80 min and 20 h, respectively, after the induction of germination.

2.5. Akinete germination

2.5.1. Energetic aspects: photosynthesis and respiration

Cyanobacterial akinetes, having survived conditions which are not favourable for the vegetative cells, germinate to produce a new filament. The major environmental change required for germination appears to be an increase in light intensity. Akinetes of *Anabaenopsis arnoldii* (Reddy et al., 1975), *Anabaena cylindrica* (Yamamoto, 1976) and *Anabaena variabilis* (Braune, 1979) were able to germinate efficiently only in light of wavelengths which supported photosynthesis, and Yamamoto (1976) demonstrated the requirement for Photosystem II activity by showing that germination did not occur in the presence of 3′-(3,4-dichlorophenyl)-1′,1′-dimethylurea (DCMU), a specific inhibitor of this photosystem. The rate of germination of akinetes of *A. cylindrica* depended on light intensity (Yamamoto, 1976). Braune (1979) showed that the optimum intensity of white light required for germination was between 5 and 20 Wm^{-2}, higher intensities being inhibitory, and that akinetes were able to germinate, although only slowly, at low light intensities which supported only slow growth of the vegetative cells of *A. variabilis*. The detailed studies of Braune (1979), using monochromatic light, demonstrated that the action spectrum for germination was similar to that of photosynthesis: light of 625 nm (absorbed by phycocyanin) and 670 nm (absorbed by chlorophyll) was most efficient in inducing germination.

Braune (1979) showed that DCMU had relatively little effect on the release of akinetes of *A. variabilis* from their envelopes (contrary to the results of Yamamoto, 1976), although it reduced CO_2 fixation by 95%; however, the young germlings subsequently died either immediately or after the first cell division. These results, therefore, suggest that Photosystem II activity is not always necessary for germination, which can also occur at the expense of cyclic photophosphorylation around Photosystem I, using the endogenous glycogen reserves (see Section 2.3.3.) as carbon source in place of CO_2. The possibility that cyclic photophosphorylation and/or respiration can support akinete germination was examined in *Nostoc* PCC 7524 (Chauvat et al., 1982; see Table II). Rapid and complete germination was only observed when both Photosystems I and II were operating, all of the akinetes having germinated within 22 h. Germination did not occur in the dark in the presence of sucrose, a carbon source which permitted heterotrophic growth of this strain (see Rippka et al., 1979). Under anaerobic conditions in the light in the presence of DCMU (conditions under which only cyclic photophosphorylation can operate), very few (4%) of the akinetes were able to germinate. The addition of air to such cultures, permitting respiration to occur, increased the efficiency to 20%; the further addition of sucrose allowed all of the akinetes to germinate by 192 h, although sucrose had no effect under anaerobic conditions (Table II). These results indicate that when Photosystem II activity is inhibited, germination requires both respiration and cyclic photophosphorylation; respiration alone (in the dark) does not permit germination while cyclic photophosphorylation alone is very inefficient. Since few akinetes germinated aerobically in the absence of sucrose when Photosystem II was inhibited it is clear that the endogenous respiratory substrate (glycogen) is insufficient, in the majority of akinetes, to support germination.

Chauvat et al. (1982) further showed that, although the akinetes were unable to germinate in the dark, preincubation in the dark permitted a small increase in respiratory and photosynthetic activities and decreased the lag period before the first cell division following transfer to the light. This implies that the early stages of germination can be initiated, but not completed, in the dark. Similarly, when akinetes were illuminated for various periods and then transferred to the dark, the degree of germination in the dark was proportional to the length of the light period. In summary, the above results indicate that light is absolutely necessary for germination to occur but that respiration can play an important role.

TABLE II

The influence of culture conditions on germination of akinetes of *Nostoc* PCC 7524

Culture conditions	Akinetes (%) germinated after			
	22 h	46 h	120 h	192 h
Light				
N_2/CO_2	100			
air/CO_2	100			
DCMU/N_2/CO_2	0	4.3	4.0	4.0
DCMU/N_2/CO_2/sucrose	0	4.0	4.5	4.1
DCMU/air/CO_2	0	20.7	21.3	21.0
DCMU/air/CO_2/surcrose	8.6	28.5	76.0	100
Dark				
air/CO_2	0	0	1.3	1.1
air/CO_2/sucrose	0	0	1.3	1.0

(Reproduced from Chauvat et al., 1982)

In contrast, Kaushik and Kumar (1970) claimed that germination of akinetes of *A. doliolum* required 'non-photosynthetic light' (blue or green). The results provided by these authors, however, did not support this conclusion: germination occurred most rapidly in white and red light, and was delayed in yellow, blue and green light; the filters employed to provide the latter three spectral regions transmitted a broad range of wavelengths, sufficient to support slow growth of the vegetative cells in the absence of an exogenous carbon source. The 'non-photosynthetic light' therefore did support photosynthesis, although at a low rate. Another report, implying that a phytochrome is involved in germination, must be criticized. Reddy et al. (1975) demonstrated relatively efficient germination of akinetes in red light (approximately 50% of that of the controls in white light, measured after 144 h); far red light did not promote germination. Exposure to red light for 24 h, followed by far red light for 120 h, was claimed to 'inhibit' germination (less than 1% of the akinetes germinated under these conditions), whereas the opposite shift 'stimulated' germination (although the results presented showed clearly that fewer akinetes germinated after this shift from far red light than in cultures which had been exposed continuously to red light). The authors interpreted these results as indicating the involvement of a phytochrome in the process of germination, by analogy with the phenomenon of photoreversal found in higher plants. The correct interpretation of the results, however, is simple: short exposure to red light is not sufficient to permit germination in the far red light supplied subsequently; conversely, exposure to red light for 120 h after 24 h of far red illumination still permits germination.

2.5.2. Other factors implicated in akinete germination

The pH optimum (7–8) and optimum temperature (27°C) for germination of akinetes of *A. cylindrica* were similar to those for the growth of vegetative cells (Yamamoto, 1976). Among a wide range of organic compounds examined, only acetate stimulated the germination of akinetes of this organism; the role of acetate was not clear, since the rates of respiration and photosynthesis were unchanged (Yamamoto, 1976). Heat shock, which breaks the dormancy of bacterial and fungal spores, did not stimulate the germination of *A. cylindrica* akinetes (Yamamoto, 1976). In contrast, akinetes of *Anabaena* strain 1616 were reported to require a short period of desiccation before germinating (Stulp and Stam, 1982).

2.5.3. Morphological changes during germination

In *Cylindrospermum* sp. (Miller and Lang, 1968) and *Nostoc* sp. (Grilli Caiola and De Vecchi, 1980) the young germling is liberated from the surrounding envelope by one of three, apparently random, mechanisms: (1) a single pore develops at one end of the envelope, through which the germling emerges after one or two cell divisions; (2) the germling may be released through the pore before undergoing cell division; (3) rarely, the entire envelope is dissolved. Miller and Lang (1968) suggested that these patterns may reflect different stages of akinete maturity, although I interpret them as meaning that cell division and rupture of the envelope are not well coordinated. Akinetes of numerous *Anabaena* spp. (Stulp and Stam, 1982) and of *Nostoc* PCC 7524 (Sutherland et al., 1985b) mostly germinated via a terminal pore in the envelope, with cell division often occurring before liberation of the germling. However, Braune (1980) showed that the akinete of *A. variabilis* was normally ejected from the envelope before dividing; ejection appeared to result from the deposition of an expanding electron dense layer inside the akinete envelope.

Cyanophycin granules were not present in the germinating akinetes of *Cylindrospermum* (Miller and Lang, 1968), but were observed in e.g., *A. variabilis* (Braune, 1980) and *Nostoc* PCC 7524 (Sutherland et al., 1985b). This difference is probably related to the age of the akinetes, which degrade cyanophycin if they are incubated in the light in dense culture (where they are unable to germinate) following maturation (Sutherland et al., 1985a). Cyanophycin granules of *A. variabilis* were 'corroded' at the periphery (Braune, 1980) and those of *Aphanizomenon flos-aquae* stained less intensively than in mature akinetes (Wildman et al., 1975), during germination, suggesting that this reserve material is mobilised during the process of germination (see Section 2.5.4.). Germinating akinetes accumulated glycogen and lipid deposits (Miller and Lang, 1968); the physiological basis for this is described in Section 2.5.4.

The true sequence of morphological changes which occur during germination can only be unequivocally established if akinetes can be induced to germinate synchronously. Stulp and Stam (1982) observed synchronized germination of akinetes of *Anabaena* strain 1617; cell division started 25–40 h after transfer to fresh medium and 90 % of the akinetes had germinated by 43 h. After 94 h, one terminal cell in each young filament had differentiated into a heterocyst; heterocysts subsequently developed in intercalary positions. Sutherland et al. (1985b) reported synchronous germination of akinetes of *Nostoc* PCC 7524, following dilution into fresh medium in the light. In the absence of a source of combined nitrogen, the first synchronized cell division occurred 12 h after dilution; one of the two daughter cells divided at 16 h while the other differentiated into a heterocyst (observed in a mature form at 19 h), producing a germling comprised of three cells with the heterocyst in a terminal position (Fig. 2 C–E). Heterocyst differentiation was well synchronized, having occurred in 50 % of the young filaments at 19 h and in all of the filaments by 24 h. Subsequently, the two vegetative cells divided synchronously, and the other terminal cell of the filament differentiated into a heterocyst between 30 and 40 h.

The factors which control the spatial pattern of heterocyst development in the young filament are unknown. As described above, the heterocysts invariably arise from the terminal cells of the germling of *Nostoc* PCC 7524, and only after further division of the vegetative cells do they develop in an intercalary position (Sutherland et al., 1985b). However, in *Anabaena* CA, each germling develops a heterocyst only when it is composed of about seven cells (see Nichols and Adams, 1982); the heterocysts differentiate at apparently random positions within the filament and never at the end. Such major differences deserve further study.

Studies with the electron microscope (Sutherland et al., 1985b) showed that changes were occurring within 2 h of the induction of germination: there were fewer cyanophycin granules, the peptidoglycan layer of the cell wall was thinner and, concomitantly, an electron-dense layer appeared on the inside of the envelope. This layer was composed of alternating light and dark lamellae with a combined width of 4 nm (Fig. 2B). The envelope ruptured (Fig. 2C) during germination; following the second (synchronized) cell division (Fig. 2D) heterocysts often differentiated before the germling was released from the akinete envelope (Fig. 2E). Braune (1980) had also described the development of a laminated layer in the envelope during germination of akinetes of *A. variabilis*, although the distance between the electron-dense layers was 5 nm. The 4 nm spacing between laminae in *Nostoc* PCC 7524 is identical to that observed in the heterocyst wall of *A. cylindrica* by Winkenback et al. (1972). Furthermore, this layer was absent in the akinetes of a mutant of *Nostoc* PCC 7524 incapable of forming heterocysts, suggesting that it is common to both akinetes and heterocysts (Sutherland et al., 1985b) (see also Section 2.3.2.).

2.5.4. Biochemical changes during akinete germination

Akinetes of *A. doliolum*, after storage in the dark, germinated relatively synchronously in the light, with 90 % reaching the first cell division after 72–80 h (Singh and Sunita, 1974). Synthesis of RNA commenced between 4 and 8 h after the initiation of germination, followed by synthesis of chlorophyll (commencing at 8 h), phycocyanin (12 h) and DNA (24 h). DNA synthesis appeared to be quite well synchronized, increasing 4-fold (suggesting two consecutive rounds of replication) during a period of about 10 h; there followed a lag of approximately 12 h before DNA synthesis recommenced.

Chauvat et al. (1982) showed that akinetes of *Nostoc* PCC 7524 underwent the first cell division about 10 h after transfer to the light, before significant synthesis of photosynthetic pigments had occurred. During this period, both photosynthetic and respiratory capacities increased markedly; the increase in respiration did not require de novo synthesis of RNA or protein, whereas that of photosynthesis was inhibited by rifampicin and chloramphenicol. It may be suggested that the increase in respiratory activity (measured as O_2 consumption in the dark in samples of akinetes which had been illuminated for various periods up to 10 h) resulted simply from the large accumulation of glycogen in the light (see below); this respiratory substrate becomes less limiting as germination proceeds and, therefore, future studies on the role of respiration during akinete germination should concentrate on the activities of the enzymes involved in respiration rather than measuring the rate of O_2 consumption per se.

Sutherland et al. (1985a) studied the sequence of macromolecular synthesis during synchronous germination of akinetes of *Nostoc* PCC 7524. In the absence of an exogenous supply of combined nitrogen, the first and second cell divisions occurred at 12 and 16 h, respectively, after the initiation of germination (see Section 2.5.3.). Growth during this period was unbalanced: the 60 % increase in dry weight was almost entirely accounted for by the increase in glycogen content of the cells, which resulted from the high rate of CO_2 fixation in the absence of N_2 fixation. Although N_2 fixation did not commence until 19 h, following heterocyst differentiation, protein synthesis started immediately after induction of germination and continued for 11 h. The nitrogen reserves, cyanophycin and phycocyanin, did not serve as nitrogen sources for protein synthesis, even though the akinetes contained increased levels of cyanophycin; degradation of the latter commenced only after 6 h and was associated with the synthesis of phycocyanin. Since the total nitrogen content of the culture remained constant until the onset of N_2 fixation, although protein was synthesized, it was proposed that the akinetes contained an additional nitrogen reserve. This may be the thick peptidoglycan layer of the cell wall (Fig. 2A) which was broken down during germination to liberate a lipopolysaccharide-like product: the laminated layer, which accumulated inside the akinete envelope (Fig. 2B). The presence of an additional nitrogen reserve was confirmed in studies of aged akinetes, which had been maintained in normal culture conditions for 7 days following their differentiation. Such akinetes contained decreased levels of cyanophycin and phycocyanin, but were still able to germinate normally in the absence of an exogenous source of combined nitrogen; although degradation of cyanophycin did not occur, synthesis of protein commenced after 8 h, long before the induction of N_2 fixation at 19 h.

Synthesis of RNA commenced immediately after the induction of germination, whereas that of DNA was initiated only after a lag of 80 min. (Sutherland et al., 1985a). In the presence of phenethylalcohol, an inhibitor of DNA synthesis, akinetes were able to germinate, and produced filaments of 10 vegetative cells if NO_3^- was provided as nitrogen source (J. Reaston and M. Herdman, unpublished); heterocysts did not develop. This extensive growth in the absence of DNA replication was possible because the akinetes contained 10 genome equivalents of DNA (Sutherland et al., 1979); cell division apparently continued until each cell contained only one genome. DNA synthesis did not occur in the presence of rifampicin, indicating a requirement for de novo synthesis of RNA for the priming of DNA synthesis (J. Reaston and M. Herdman, unpublished).

The time course of morphological and biochemical changes which occur during the germination of akinetes of *Nostoc* PCC 7524 in the absence of an exogenous source of combined nitrogen is summarised in Fig. 3.

In the presence of NO_3^-, the pattern of germination of akinetes of *Nostoc* PCC 7524 was similar to that observed under N_2, except that synthesis of phycocyanin commenced immediately and that of chlorophyll after a lag of only 10 h (Sutherland et al., 1985a). Growth of the young filaments was more rapid under these conditions and the ready availability of nitrogen resulted in a rapid decline in the glycogen content of the culture after the first cell division, in contrast to akinetes germinating with N_2 as sole nitrogen source.

Since akinetes of *Nostoc* PCC 7524 were capable of immediate fixation of CO_2 on transfer to the light, and accumulated large quantities of glycogen during the early stages of germination, their large glycogen reserves (see Table I) appeared to be unnecessary for germination under the conditions employed (Sutherland et al., 1985a), although they may be important for germination under low light intensities and for meeting the maintenance energy requirement of the akinete in the dark. The glycogen reserve is not a special feature of the akinete; it accumulates during akinete differentiation at the end of exponential growth, just as it would in vegetative cells under the same conditions. However, the accumulated glycogen was partly degraded in the later stages of germination, when CO_2 fixation decreased before the onset of N_2 fixation; again, degradation of glycogen is not unique to the process of akinete germination since its use as an endogenous carbon source in cyanobacterial vegetative cells is well documented (Lehmann and Wöber, 1976; Rippka and Stanier, 1978; Van Liere et al., 1979). Similarly, the degradation of cyanophycin in the akinetes and (just before the differentiation of the first heterocysts) the breakdown of phycocyanin in the young filaments (Sutherland et al., 1985a) are not unique to akinetes and their process of germination. Except for the structure of the cell wall and the presence of the extracellular envelope (see Section 2.1.), features which probably contribute to their resistance to cold and to desiccation, akinetes show no unique features and closely resemble their parental vegetative cells in physiological properties (Sutherland et al., 1985a).

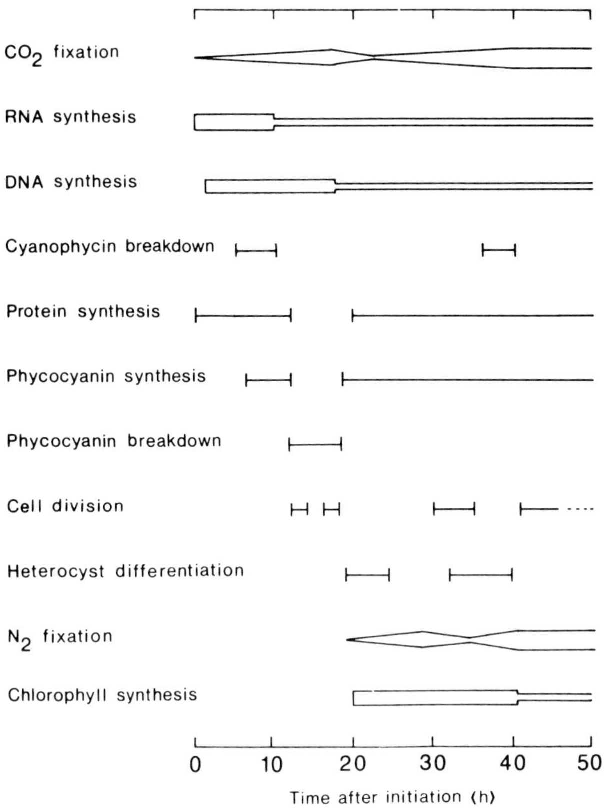

Fig. 3. Summary scheme of the morphological and biochemical changes during germination of akinetes of *Nostoc* PCC 7524, with N_2 as sole nitrogen source. The relative rates of fixation of CO_2 and N_2, and of macromolecular synthesis, are indicated by the thickness of the bars. (From Sutherland et al., 1985a, with permission)

2.6. The control of akinete differentiation

2.6.1. The spatial pattern

As described in Section 2.1., akinetes develop (in different species) either (a) adjacent to heterocysts, (b) midway between two heterocysts or (c) at a position three cells away from the heterocyst. Many authors have stressed the fixed spatial relationship between akinetes and heterocysts, implicating the heterocyst in the control of the akinete spatial pattern.

Experimental studies of the control of pattern formation have been confined to three organisms: *A. cylindrica* (type a) and *Anabaena* CA and *Nostoc* PCC 7524 (both type b). Cultures of *A. cylindrica* grown in the presence of NH_4Cl do not produce mature heterocysts until the end of exponential growth when heterocysts and proheterocysts develop at low frequency; in such filaments, akinetes differentiate not only in their normal positions next to the heterocysts but also in isolated positions more than 20 cells from the nearest heterocyst (Nichols et al., 1980), implying that heterocysts are not necessary for akinete development. However, Wolk (1966) had clearly shown that when heterocysts were detached from the filament, the vegetative cells which had been adjacent to the heterocyst could no longer differentiate into an akinete. Although it cannot be excluded that heterocyst detachment caused damage to the vegetative cells, these results seem to indicate a dependency. Wolk (1966), however, also observed a low proportion of akinetes developing at long distances from the heterocysts. These apparent contradictions could be explained if it is assumed that the latter akinetes develop adjacent to a proheterocyst which is still too young to be recognised in the light microscope, and whose further development is repressed by light limitation in the dense culture; this may well be the case, since the akinetes developed about 20 cells from a heterocyst (Nichols et al., 1980) where the next proheterocyst would be expected to develop. This implies that the role of the heterocyst in akinete differentiation can be exerted at a very early stage of heterocyst development.

Wolk (1965) debated the possible mechanisms by which heterocysts of *A. cylindrica* could be involved in the control of akinete differentiation and concluded that the heterocysts produce a compound which abates a sporulation-inhibiting influence which exists in the vegetative cells. Subsequently, Hirosawa and Wolk (1979a,b) identified a compound, present in culture supernatants of *Cylindrospermum licheniforme*, which stimulated akinete differentiation when added to young cultures. It is not known whether this compound is produced in the heterocysts or in the vegetative cells. In the former case, the compound could diffuse along the filament, forming a concentration gradient so that the vegetative cells nearest to the heterocyst would be the first to differentiate into akinetes; subsequently the second and third vegetative cell would differentiate, producing a short chain of akinetes showing a gradient of maturity as had previously been observed (see Wolk, 1965) and later quantified (Simon, 1977b). However, when added to the culture medium, the compound should be taken up by all the vegetative cells, not only by those adjacent to the heterocysts; Hirosawa and Wolk (1979a) suggested that it enters the filament only at the connections between heterocysts and vegetative cells. It was also shown that H_2 stimulated akinete formation by a factor of 2.5; the location of the uptake hydrogenase mainly in the heterocyst is in favour of the hypothesis that a compound is reduced by the hydrogenase in the heterocyst, and then moves into the vegetative cells and stimulates akinete formation, although it cannot be excluded that H_2 increases the inactivation of an inhibitory substance (Hirosawa and Wolk, 1979a). The effect of H_2 was not mediated by electron donation to nitrogenase, since akinete frequency increased even in the absence of N_2.

Anabaena CA is certainly capable of forming akinetes when heterocyst differentiation is repressed (by nitrate), and indeed most of the vegetative cells are converted to akinetes under these conditions (see Nichols and Adams, 1982). A statistical study in *Nostoc* PCC 7524 (Sutherland et al., 1979) showed that under N_2-fixing conditions, 42% of the first-formed akinetes developed exactly at the centre of the interheterocyst interval, 45% within two cells from the centre and the rest within three cells from the centre, the mean interval between heterocysts being 23 cells. In filaments grown in NH_4Cl, which did not contain heterocysts, akinetes arose at increased frequency and in random positions. Thus, in both *Anabaena* CA and *Nostoc* PCC 7524, heterocysts are not required for akinete differentiation although their presence does impose a fixed pattern. In contrast, Fay et al. (1984) showed that akinetes of *A. circinalis* did not appear in the presence of nitrate, urea or casamino acids, all of which inhibited heterocyst differentiation.

Further analysis (Sutherland et al., 1979) showed that the interheterocyst intervals in *Nostoc* PCC 7524 varied from less than 18 to more than 30 cells. In the shortest intervals, akinetes did not develop. In intervals of mean length 18 cells, akinetes invariably developed at a single site equidistant from the heterocysts, whereas in longer intervals akinetes arose at two sites, 9 or 10 vegetative cells away from the nearest heterocyst. The statement (see above) that akinetes arise midway between two heterocysts is therefore not strictly true; they arise at a *fixed distance* (9–10 cells) from the heterocyst and, in most intervals, appear to be at the centre only because the majority

(61%) of the intervals are about 18 cells long (Sutherland et al., 1979). This situation is therefore similar to that in the planktonic strains described by Fay et al. (1984), except that the mean akinete-heterocyst interval in these organisms is reduced to two cells. According to present information, therefore, cyanobacteria can be re-classified into two major groups with respect to the position of their akinetes: (a) those where the akinetes first arise from the cell immediately adjacent to the heterocyst and (b) those where the akinete first arises at a fixed distance (3 or 9 cells) from the heterocyst.

The control mechanism which is involved in the initial pattern of akinete differentiation rapidly breaks down, however, since in *Nostoc* PCC 7524 akinetes appear successively on either side of the first, giving a chain of akinetes which exhibit a gradient of maturity; also, akinetes eventually develop in the shortest interheterocyst intervals which initially did not contain them (Sutherland et al., 1979). These gradients of maturity have their counterpart in *A. cylindrica*, studied in detail by Simon (1977b); in both organisms the gradient resulted from different lags in the time of initiation of differentiation in the vegetative cells.

Since heterocysts fix atmospheric N_2 and export the product to the vegetative cells (Thomas et al., 1977), one would predict the existence of gradients of nitrogenous compounds along the interheterocyst interval, the lowest levels occurring in the centre where the next heterocyst is due to develop. Such gradients are not involved in the control of the pattern of akinete formation in *Nostoc* PCC 7524: cultures grown under argon:CO_2 (with nitrate as nitrogen source) still produce akinetes in their normal positions relative to the heterocyst, in the absence of N_2 fixation and under conditions where all vegetative cells have access to the exogenous nitrate (Sutherland et al., 1979). In addition, akinetes of *A. cylindrica* can develop next to proheterocysts and mature heterocysts in the presence of NH_4Cl, which represses N_2 fixation (Nichols et al., 1980).

Akinetes of *Nostoc* PCC 7524 developed at the end of exponential growth (see below), at a position where the next heterocyst would be expected to appear if growth had continued, suggesting that the spatial pattern of differentiation of the two cell types is under the control of the same mechanism. That this statement may be true was shown by the effect of 7-azatryptophan, a tryptophan analogue which induces a major change (increase in heterocyst frequency and the induction of multiple, adjacent, heterocysts) in the heterocyst pattern (see Mitchison and Wilcox, 1973). This compound had an identical effect on the patterns of heterocyst and akinete differentiation in *Nostoc* PCC 7524 (Sutherland et al., 1979). If a common control mechanism exists for the spacing of heterocysts and akinetes in *Nostoc* PCC 7524, this would appear not to be true for *A. cylindrica*, where the akinetes normally appear next to the heterocysts. The arginine analogue canavanine disturbs the control of akinete formation in this organism, since in its presence not only did the akinete frequency increase 5-fold but akinetes also developed in random positions between heterocysts. Another analogue of arginine, cyanoalanine, and the serine analogue aminobutyric acid, also resulted in an increased frequency of akinetes but caused no change in their pattern (Nichols et al., 1980). Studies with analogues of amino acids therefore show great promise in the elucidation of the metabolic control of akinete spatial patterns.

2.6.2. Environmental factors affecting akinete differentiation

(a) Nutrient supply. Variations in the supply of an extremely wide range of nutrients have been reported to induce or to inhibit akinete differentiation. The same compound often appears to have opposite effects in different organisms, although misinterpretation of the experimental results appears to be a common failure and in many cases it is not clear if the cultures employed were axenic. Axenic cultures are of obvious importance in work of this type because contaminants may grow

at the expense of the added nutrient, utilising other components of the culture medium which repress akinete differentiation; once utilised, repression would be relieved and akinete differentiation would be attributed (falsely) to the added nutrient.

Early reports (see Wolk, 1965, for a review) implicated glutamate, NaCl, sulphate or nitrate, and the absence of combined nitrogen, in the induction of akinete differentiation. None of these conditions promoted akinete differentiation in *A. cylindrica* (Wolk, 1965) but acetate sometimes stimulated and akinete formation significantly increased in the presence of calcium glucuronate. Wolk (1965) further showed that phosphate starvation induced, and phosphate inhibited, akinete differentiation in *A. cylindrica*; phosphate deprivation also resulted in increased akinete frequency in *Gloeotrichia ghosei* (Sinclair and Whitton, 1977). In contrast, Rother and Fay (1977, 1979) showed that akinete formation in a number of planktonic cyanobacteria occurred, both in the natural environment and in pure culture, in the presence of high concentrations of phosphate and was not stimulated by phosphate deficiency. Fay et al. (1984) analysed the effects of the major inorganic nutrients and showed that phosphate stimulated akinete formation in *A. circinalis*; akinetes of *Nostoc* PCC 7524 were never produced in the absence of phosphate (Sutherland et al., 1979). Nitrate markedly inhibited akinete development in *A. circinalis* (Fay et al., 1984), similar to the findings that nitrate, nitrite and ammonia inhibited akinete differentiation in *A. doliolum* (Singh and Srivastava, 1968; Tyagi, 1974) and in a strain named '*A. circinalis*' (Tyagi, 1978) which does not at all resemble the *A. circinalis* described by Fay et al. (1984). In *Nostoc* PCC 7524, however, concentrations of NH_4^+ sufficient to inhibit heterocyst differentiation had no inhibitory effect on akinete formation (Sutherland et al., 1979).

Studies with metal ions suggest a role of Fe in akinete development. In *A. cylindrica*, akinete frequency increased under conditions of Fe deficiency (Sinclair and Whitton, 1977), and Fe completely suppressed akinete formation in *A. circinalis* (Fay et al., 1984). However, Fe deficiency resulted in a decreased akinete frequency in *Gloeotrichia ghosei* (Sinclair and Whitton, 1977). Other metal ions had relatively minor effects in the above organisms and in *Aphanizomenon flos-aquae* neither addition nor deficiency of metal ions had any effect on akinete frequency (Rother and Fay, 1979).

In *C. licheniforme*, akinete development is stimulated by a compound which accumulated in the culture medium (see Section 2.6.1.); similarly, culture filtrates of *A. circinalis* increased akinete frequency by up to 50% in a concentration-dependent manner (Fay et al., 1984). The possibility that the filtrates contained a compound that complexes metal ions that are inhibitory to akinete differentiation, or transports essential ions into the cell, deserves further study. In contrast, culture filtrates of *Aphanizomenon flos-aquae* (Rother and Fay, 1979), *Nostoc* PCC 7524 (Sutherland et al., 1979) or *A. cylindrica* (Nichols and Adams, 1982) did not have a stimulatory effect on akinete differentiation when added to young cultures. The importance of intracellular metabolism in the control of akinete germination is beginning to emerge from studies with a variety of potential growth substrates. Akinete differentiation in *A. doliolum* (Tyagi, 1974) and '*A. circinalis*' (Tyagi, 1978) was stimulated by glucose. The stimulatory effect of H_2 on akinete differentiation (Hirosawa and Wolk, 1979a) is described in Section 2.6.1. The same authors showed that tryptophan, aspartate, phenylalanine, proline and isoleucine all increased akinete formation and that cyclic nucleotides had little or no effect. The importance of amino acid metabolism is perhaps indicated by the action of amino acid analogues in *A. cylindrica* (Nichols et al., 1980; see Section 2.6.1.). In *A. circinalis*, akinete frequencies increased in the presence of mixtures of organic acids and alcohols, and following the addition of fumarate, citrate or acetate; these latter compounds were metabolised by the cells since the growth rate of the culture increased in their presence (Fay et al., 1984). Ethanol also markedly increased akinete frequency in this organism; akinetes appeared in random positions within the filament but never matured (Fay et al., 1984). In contrast, malate, oxaloacetate, urea and casamino acids were inhibitory to akinete development.

There are, therefore, a large number of reports concerning the effect of nutrients on akinete formation, many of them conflicting. Obviously, different organisms may respond to different stimuli. However, in many cases the experiments appear to have been carelessly performed; many authors studied the effects of a specific compound by adding it to cultures at a certain stage of growth and then examining the culture after a fixed time interval. There is now general agreement that akinetes are formed after the end of exponential growth (see below); if the compound decreases the growth rate, akinetes might still be formed, but later than in the control cultures. Examination of the cultures only at the time of formation of akinetes in the control culture would not permit their detection in the treated culture and the compound would be listed as being inhibitory. The reverse is true for compounds which enhance the growth rate. A good example of these problems was given by Sutherland et al. (1979): the addition of sucrose (a utilisable carbon source) to cultures of *Nostoc* PCC 7524 prolonged exponential growth and the cultures reached a higher maximum density; akinete differentiation therefore occurred later in time than in the controls, when the culture had reached the end of exponential growth (Fig. 4), but their frequency was unchanged. Sucrose cannot, therefore, be considered to be inhibitory to akinete formation (although Nichols and Adams (1982) considered it to be inhibitory); it merely delays akinete differentiation until the cells reach the same physiological state (resulting from energy limitation, see below) which induces this process in the control cultures. Clearly, therefore, the term 'inhibition' should be used with care.

(b) Light. Akinetes were first described 130 years ago; it is an astonishing fact that it took about 120 years to discover the one factor which would be expected to play a crucial role in their development: light. Fay (1969a) showed that akinetes of *A. cylindrica* developed at the end of exponential growth of the vegetative cells; these findings were confirmed by Simon (1977b) and Nichols et al. (1980) with the same organism. In the natural environment, differentiation in planktonic strains

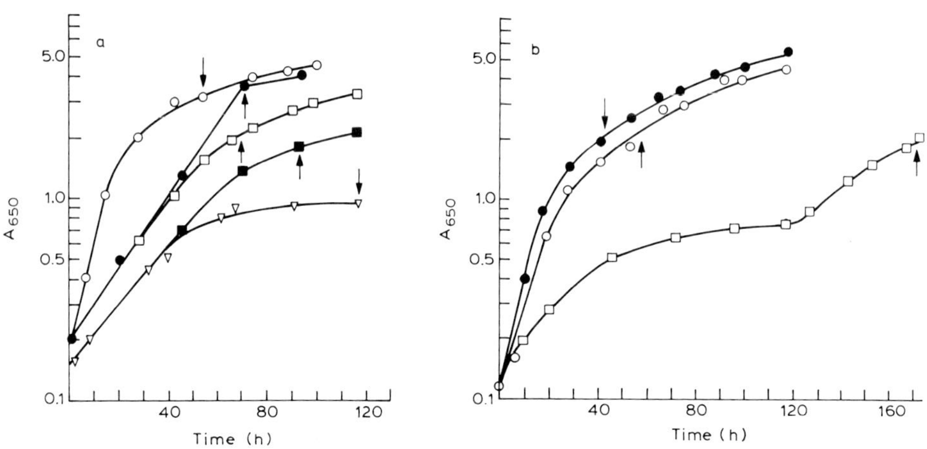

Fig. 4. (a) Growth and the time of akinete differentiation (indicated by arrows) in N_2-grown cultures of *Nostoc* PCC 7524 at light intensities of 10 000 lx (○), 8 000 lx (□) and 5 000 lx (▽), and at 8 000 lx in the presence of sucrose (●) or glucose (■). (b) As in (a), but at 8 000 lx in the presence of NH_4^+ (○) or glutamine (●), or in the absence of phosphate (□); phosphate was re-added to the latter culture at 110 h. (From Sutherland et al., 1979, with permission)

occurred in dense surface blooms (Rother and Fay, 1977) where many of the cells are subjected to light limitation, as are the cells at the end of exponential growth in laboratory culture.

The role of light- (and therefore energy-) limitation in the differentiation of akinetes was conclusively demonstrated by Sutherland et al., 1979. Under N_2-fixing conditions, at a light intensity of 8 000 lx, *Nostoc* PCC 7524 exhibited a short (30 h) period of exponential growth (doubling time 8.5 h) followed by a period of 100 h of non-exponential growth. Akinete differentiation was initiated at 70 h and their frequency increased to 60 % of the total cell population when growth ceased at 130 h. At 8 000 lx, akinete differentiation commenced at an OD_{650} of 2.0 (Fig. 4). At lower light intensities, differentiation was initiated at lower culture densities (OD_{650} 0.94 at 5 000 lx). When the light intensity was increased to 10 000 lx, exponential growth was prolonged and akinete differentiation was delayed until the culture density reached 3.2. Akinete differentiation was therefore not induced by limitation of inorganic nutrients at the normal culture density (2.0) at which it occurred since, if it were, differentiation would have occurred at the same culture density even at higher light intensity. The effect of increased light intensity was mimicked by the addition of sucrose, a utilisable carbon source for photoheterotrophic growth of this strain, but not by glucose or acetate (which were not utilisable). Energy limitation was therefore the only major factor involved in the induction of akinete differentiation.

Increased light intensity, and the addition of sources of combined nitrogen (Fig. 4) caused akinetes to appear sooner (at 53 and 40–60 h, respectively) than in control cultures grown under N_2 at 8 000 lx. In the absence of data on growth rate and culture density, these results would falsely be interpreted as a stimulation of akinete formation. However, the only effect of these conditions was to increase the rate of exponential growth; in all cases akinetes developed in the post-exponential phase of growth when light became limiting, and at the same OD (in the presence and absence of sources of combined nitrogen) if light intensity was held constant.

Nichols et al. (1980) found a similar relationship between the light intensity supplied and the culture density at which akinete differentiation was initiated (again at the end of exponential growth), in *A. cylindrica*, and showed this relationship to be linear. Similarly, Fay et al. (1984) demonstrated the induction of akinete differentiation in *A. circinalis* to be a result of light (energy) limitation: these workers grew cells at a fixed light intensity and then concentrated the samples by varying degrees to decrease light penetration into the cultures; akinete differentiation began first in the most concentrated cultures followed progressively by the less dense cultures. In every case, differentiation occurred at a time when the culture density had reached an OD of 0.25–0.275.

The recent results of three different groups with three very different organisms (with respect to their spatial pattern of akinete formation) therefore lead to the same conclusion: light energy limitation is the most important environmental trigger for akinete differentiation. However, in *C. licheniforme* (Glade, 1914) and *A. cylindrica* (Wolk, 1965), phosphate limitation is also important in the control process. *A. circinalis* liberates compounds which appear to overcome the photocontrol mechanism (Fay et al., 1984), reminiscent of the stimulatory factor which overcomes phosphate inhibition of akinete formation in *C. licheniforme* (Hirosawa and Wolk, 1979a); the possibility that these compounds inhibit photosynthetic electron transport, thereby causing energy limitation, has not yet been examined.

Acknowledgements

I thank Rosmarie Rippka and Peter Fay for providing original photographs of the strains shown in Figure 1, and Germaine Cohen-Bazire for critical comments on the manuscript. This chapter is dedicated to a close friend and colleague, Chase Van Baalen.

References

Allen, M.M., Hutchinson, F. and Weathers, P.J. (1980) Cyanophycin granule polypeptide formation and degradation in the cyanobacterium *Aphanocapsa* 6308. J. Bacteriol. 141, 687–693.

Beijerinck, M.W. (1901) Ueber oligonitrophile Mikroben. Centralblatt für Bakteriologie, Parasitenkunde und Infektionskrankheiten, Abt. 27, 561–582.

Bharadwaja, Y. (1933) Contributions to our knowledge of the myxophyceae of India. Ann. Bot. 47, 117–143.

Björn, G.S., Braune, W. and Björn, L.O. (1983) Photochromic pigments in akinetes and pigment characteristics of akinetes in comparison to vegetative cells of *Anabaena variabilis*. Physiol. Plant 59, 493–500.

Bornet, E. and Flahault, C. (1886) Révision des Nostocacées hétérocystées. I. Ann. Sci. Nat. Bot. VII 3, 323–381.

Borzi, A. (1878) Note alla morfologia e biologia delle alghe Ficocromacee. Nuov. Giorn. Bot. Ital. 10, 236–288.

Brand, F. (1903) Morphologisch-physiologische Betrachtungen über Cyanophyceen. Beih. Bot. Centralbl. 15, 31–64.

Braune, W. (1979) C-phycocyanin – the main photoreceptor in the light-dependent germination process of *Anabaena* akinetes. Arch. Microbiol. 122, 289–295.

Braune, W. (1980) Structural aspects of akinete germination in the cyanobacterium *Anabaena variabilis*. Arch. Microbiol. 126, 257–261.

Bristol, B.M. (1919) On the retention of vitality of algae from old stored soils. New Phytol. 18, 92–107.

Cardemil, L. and Wolk, C.P. (1976) The polysaccharides from heterocyst and spore envelopes of a blue-green alga. Methylation analysis and structure of the backbones. J. Biol. Chem. 251, 2967–2975.

Cardemil, L. and Wolk, C.P. (1979) The polysaccharides from heterocyst and spore envelopes of a blue-green alga. Structure of the basic repeating unit. J. Biol. Chem. 254, 736–741.

Carter, H.J. (1856) Notes on the freshwater infusoria of the island of Bombay. No. 1. Organisation. Ann. Mag. Nat. Hist. 18 (2nd ser.), 115–132, 221–249.

Chauvat, F. and Joset-Espardellier, F. (1981) A freezing method for the purification and storage of akinete suspensions of the cyanobacterium *Nostoc* PCC 7524. FEMS Microbiol. Lett. 10, 319–321.

Chauvat, F., Corre, B., Herdman, M. and Joset-Espardellier, F. (1982) Energetic and metabolic requirements for the germination of akinetes of the cyanobacterium *Nostoc* PCC 7524. Arch. Microbiol. 133, 44–49.

Clark, R.L. and Jensen, T.E. (1969) Ultrastructure of akinete development in a blue-green alga, *Cylindrospermum* sp. Cytologica 34, 439–448.

Cohn, F. (1877) Untersuchungen über Bakterien. IV. Beiträge zur Biologie der Bacillen. Beitr. Biol. Pflanz. 2, 249–276.

De Bary, A. (1863) Beitrag zur Kenntnis der Nostocaceen insbesondere der *Rivularien*. Flora, Jena 35, 553–560.

Dunn, J.H. and Wolk, C.P. (1970) Composition of the cellular envelopes of *Anabaena cylindrica*. J. Bacteriol. 103, 153–158.

Fay, P. (1969a) Cell differentiation and pigment composition in *Anabaena cylindrica*. Arch. Mikrobiol. 67, 62–70.

Fay, P. (1969b) Metabolic activities of isolated spores of *Anabaena cylindrica*. J. Exp. Bot. 20, 100–109.

Fay, P. (1985) Akinete formation and function in the natural environment. Proceedings of the Vth Internat. Symposium on Photosynthetic Prokaryotes, Institut für Molekularbiologie und Biophysik, Zürich, p. 133.

Fay, P., Stewart, W.D.P., Walsby, A.E. and Fogg, G.E. (1968) Is the heterocyst the site of nitrogen fixation in blue-green algae? Nature (Lond.) 220, 810–812.

Fay, P., Lynn, J.A. and Majer, S.C. (1984) Akinete development in the planktonic blue-green alga *Anabaena circinalis*. Br. Phycol. J. 19, 163–173.

Fisher, R.W. and Wolk, C.P. (1976) Substance stimulating the differentiation of spores of the blue-green alga *Cylindrospermum licheniforme*. Nature (Lond.) 259, 394–395.

Fritsch, F.E. (1904a) Some points in the structure of *Anabaena*. New Phytol. 3, 85–96.

Fritsch, F.E. (1904b) Some points in the reproduction of *Anabaena*. New Phytol. 3, 216–228.

Fritsch, F.E. (1905) Structure of the investment and spore-development in some Cyanophyceae. Beih. Bot. Centralbl. I, 18, 194–214.

Fritsch, F.E. (1918) Contributions to our knowledge of the freshwater algae of Africa. II. Ann. S. Afr. Mus. 9, 483–611.

Fritsch, F.E. (1945) Structure and Reproduction of the Algae, Vol. II. Cambridge University Press, p. 799.
Geitler, L. (1925) Synoptische Darstellung der Cyanophyceen in morphologischer und systematischer Hinsicht. Beih. Bot. Centralbl. II, 41, 163–294.
Glade, R. (1914) Zur Kenntnis der Gattung *Cylindrospermum*. Beitr. Biol. Pflanz. 12, 295–346.
Gomont, M. (1888) Recherches sur les enveloppes cellulaires des Nostocacées filamenteuses. Bull. Soc. Bot. France 35, 204–236.
Grilli Caiola, M. and De Vecchi, L. (1980) Akinete ultrastructure of *Nostoc* species isolated from cycad coralloid roots. Can. J. Bot. 58, 2513–2519.
Harder, R. (1917a) Ueber die Beziehung der Keimung von Cyanophyceensporen zum Licht. Ber. Deutsch. Bot. Ges. 35, 58–64.
Harder, R. (1917b) Ueber die Beziehung des Lichtes zum Keimung von Cyanophyceensporen. Jahrb. Wissensch. Bot. 58, 237–294.
Harder, R. (1917c) Ernährungsphysiologische Untersuchungen an Cyanophyceen, hauptsächlich dem endophytischen *Nostoc punctiforme*. Zeitschr. Bot. 9, 145–242.
Hegler, R. (1901) Untersuchungen über die Organisation der Phycochromaceenzellen. Jahrb. Wissensch. Bot. 36, 229–354.
Herdman, M. and Carr, N.G. (1974) Estimation of the genome size of blue-green algae from DNA renaturation rates. Arch. Mikrobiol. 99, 251–254.
Herdman, M., Janvier, M., Rippka, R. and Stanier, R.Y. (1979) Genome size of cyanobacteria. J. Gen. Microbiol. 111, 73–85.
Hieronymus, G. (1892) Beiträge zur Morphologie und Biologie der Algen. Beitr. Biol. Pflanz. 5, 461–495.
Hirosawa, T. and Wolk, C.P. (1979a) Factors controlling the formation of akinetes adjacent to heterocysts in the cyanobacterium *Cylindrospermum licheniforme* Kütz. J. Gen. Microbiol. 114, 423–432.
Hirosawa, T. and Wolk, C.P. (1979b) Isolation and characterization of a substance which stimulates the formation of akinetes in the cyanobacterium *Cylindrospermum licheniforme* Kütz. J. Gen. Microbiol. 114, 433–441.
Janczewski, E. (1874) Observations sur la reproduction de quelques Nostocacées. Ann. Sci. Nat. Bot. Ser. 5 19, 119–130.
Jensen, T.E. and Clark, R.L. (1969) Cell wall and coat of the developing akinete of a *Cylindrospermum* species. J. Bacteriol. 97, 1494–1495.
Kaushik, M. and Kumar, H.D. (1970) The effect of light on growth and development of two nitrogen-fixing blue-green algae. Arch. Mikrobiol. 74, 52–57.
Koch, R. (1877) Untersuchungen über Bakterien. V. Die Aetiologie der Milzbrand-Krankheit, begründet auf der Entwicklungsgeschichte des *Bacillus anthracis*. Beitr. Biol. Pflanz. 2, 277–310.
Leak, L.V. and Wilson, G.B. (1965) Electron microscopic observations on a blue-green alga, *Anabaena* sp. Can. J. Genet. Cytol. 7, 237–249.
Lehmann, M. and Wöber, G. (1976) Accumulation, mobilization and turnover of glycogen in the blue-green bacterium *Anacystis nidulans*. Arch. Microbiol. 111, 93–97.
Lipman, C.B. (1941) The successful revival of *Nostoc commune* from a herbarium specimen eighty-seven years old. Bull. Torrey Bot. Club 68, 664–666.
Livingstone, D. and Jaworski, G.H.M. (1980) The viability of akinetes of blue-green algae recovered from the sediments of Rostherne Mere. Br. Phycol. J. 15, 357–364.
Mann, N. and Carr, N.G. (1974) Control of macromolecular composition and cell division in the blue-green alga, *Anacystis nidulans*. J. Gen. Microbiol. 83, 399–405.
Miller, M.M. and Lang, N.J. (1968) The fine structure of akinete formation and germination in *Cylindrospermum*. Arch. Mikrobiol. 60, 303–313.
Mitchison, G.J. and Wilcox, M. (1973) Alteration in heterocyst pattern of *Anabaena* produced by 7-azatryptophan. Nature, New Biol. 246, 229–233.
Nichols, J.M. and Adams, D.G. (1982) Akinetes. In N.G. Carr and B.A. Whitton (eds.), The Biology of Cyanobacteria. Blackwell Scientific Publ., Oxford, pp. 387–412.
Nichols, J.M. and Carr, N.G. (1978) Akinetes of cyanobacteria. In C.H. Chambliss and J.C. Vary (eds.), Spores, VII. American Society for Microbiology, Washington D.C., pp. 335–343.
Nichols, J.M., Adams, D.G. and Carr, N.G. (1980) Effect of canavanine and other amino acid analogues on akinete formation in the cyanobacterium *Anabaena cylindrica*. Arch. Microbiol. 127, 67–75.
Rabenhorst, L. (1865) Flora Europaea Algarum, Vol. 2. 319 pp., Leipzig.
Reddy, P.M., Rao, P.S.N. and Talpasayi, E.R.S. (1975) Effect of red and far red illuminations on the germination of spores of two blue-green algae. Curr. Sci. 44, 678–679.

Reynolds, C.S. (1972) Growth, gas vacuolation and buoyancy in a natural population of a planktonic blue-green alga. Freshwat. Biol. 2, 87–106.

Reynolds, C.S. (1975) Interrelations of photosynthetic behaviour and buoyancy regulation in a natural population of a blue-green alga. Freshwat. Biol. 5, 323–338.

Rippka, R. and Herdman, M. (1985) Division patterns and cellular differentiation in cyanobacteria. Ann. Microbiol. (Inst. Pasteur) 136A, 33–39.

Rippka, R. and Stanier, R.Y. (1978) The effects of anaerobiosis on nitrogenase synthesis and heterocyst development by nostocacean cyanobacteria. J. Gen. Microbiol. 105, 83–94.

Rippka, R., Deruelles, J., Waterbury, J.B., Herdman, M. and Stanier, R.Y. (1979) Generic assignments, strain histories and properties of pure cultures of cyanobacteria. J. Gen. Microbiol. 111, 1–61.

Roelofs, T.D. and Oglesby, R.T. (1970) Ecological observations on the planktonic cyanophyte *Gloeotrichia echinulata*. Limnol. Oceanogr. 15, 224–229.

Rose, E.T. (1934) Notes on the life-history of *Aphanizomenon flos-aquae*. Univ. Iowa Stud. Nat. Hist. 16, 129–140.

Rother, J.A. and Fay, P. (1977) Sporulation and the development of blue-green algae in two Salopian meres. Proc. Roy. Soc. Ser. B. 196, 317–332.

Rother, J.A. and Fay, P. (1979) Blue-green algal growth and sporulation in response to simulated surface bloom conditions. Br. Phycol. J. 14, 59–68.

Sarma, T.A. and Kanta, S. (1979) Biochemical studies on sporulation in blue-green algae. Zeitsch. Allg. Microbiol. 19, 571–575.

Simon, R.D. (1973) Measurement of the cyanophycin granule polypeptide content in the blue-green alga *Anabaena cylindrica*. J. Bacteriol. 114, 1213–1216.

Simon, R.D. (1977a) Macromolecular composition of spores from the cyanobacterium *Anabaena cylindrica*. J. Bacteriol. 129, 1154–1155.

Simon, R.D. (1977b) Sporulation in the filamentous cyanobacterium *Anabaena cylindrica*. The course of spore formation. Arch. Microbiol. 111, 283–288.

Simon, R.D. (1980) DNA content of heterocysts and spores of the filamentous cyanobacterium *Anabaena variabilis*. FEMS Microbiol. Lett. 8, 241–245.

Sinclair, C. and Whitton, B.A. (1977) Influence of nutrient deficiency on hair formation in the Rivulariaceae. Br. Phycol. J. 12, 297–313.

Singh, H.N. and Srivastava, B.S. (1968) Studies on morphogenesis in a blue-green alga. I. Effect of inorganic nitrogen sources on developmental morphology of *Anabaena doliolum*. Can. J. Microbiol. 14, 1341–1346.

Singh, H.N. and Sunita, K.M. (1974) A biochemical study of spore germination in the blue-green alga, *Anabaena doliolum*. J. Exp. Bot. 25, 837–845.

Spratt, E.R. (1911) Some observations on the life-history of *Anabaena cycadeae*. Ann. Bot. 25, 369–380.

Stulp, B.K. and Stam, W.T. (1982) General morphology and akinete germination of a number of *Anabaena* strains (Cyanophyceae) in culture. Arch. Hydrobiol. Suppl. 63,1 (Algological Studies 30), 35–52.

Sutherland, J.M., Herdman, M. and Stewart, W.D.P. (1979) Akinetes of the cyanobacterium *Nostoc* PCC 7524: macromolecular composition, structure and control of differentiation. J. Gen. Microbiol. 115, 273–287.

Sutherland, J.M., Reaston, J., Stewart, W.D.P. and Herdman, M. (1985a) Akinetes of the cyanobacterium *Nostoc* PCC 7524: macromolecular and biochemical changes during synchronous germination. J. Gen. Microbiol. 131, 2855–2863.

Sutherland, J.M., Stewart, W.D.P. and Herdman, M. (1985b) Akinetes of the cyanobacterium *Nostoc* PCC 7524: morphological changes during synchronous germination. Arch. Microbiol. 142, 269–274.

Thaxter, R. (1892) On the Myxobacteriaceae: a new order of Schizomycetes. Bot. Gazette 17, 389–406.

Thomas, J., Meeks, J.C., Wolk, C.P., Shaffer, P.W., Austin, S.M. and Chien, W.-S. (1977) Formation of glutamine from [^{13}N]ammonia, [^{13}N]dinitrogen and [^{14}C]glutamate by heterocysts isolated from *Anabaena cylindrica*. J. Bacteriol. 129, 1545–1555.

Tyagi, V.V.S. (1974) Some observations on the pattern of sporulation in a blue-green alga, *Anabaena doliolum*. Ann. Bot. 38, 1107–1111.

Tyagi, V.V.S. (1978) The pattern of sporulation in *Anabaena circinalis* and comments on the role of heterocysts in sporulation in blue-green algae. Zeitsch. Allgem. Mikrobiol. 18, 281–286.

Ueda, K. and Sawada, M. (1972) DNA content of akinete cells in *Cylindrospermum*. Cytologia 37, 519–523.

Van Liere, L., Mur, L.R., Gibson, C.E. and Herdman, M. (1979) Growth and physiology of *Oscillatoria agardhii* Gomont cultivated in continuous culture with a light-dark cycle. Arch. Microbiol. 123, 315–318.

Waterbury, J.B. and Stanier, R.Y. (1978) Patterns of growth and development in pleurocapsalean cyanobacteria. Microbiol. Rev. 42, 2–44.

Wieringa, K.T. (1968) A new method for obtaining bacteria-free cultures of blue-green algae. Antonie van Leeuwenhoek 34, 54–56.

Wildman, R.B., Loescher, J.H. and Winger, C.L. (1975) Development and germination of akinetes of *Aphanizomenon flos-aquae*. J. Phycol. 11, 96–104.

Wildon, D.C. and Mercer, F.V. (1963) The ultrastructure of the heterocyst and akinete of the blue-green algae. Arch. Mikrobiol. 47, 19–31.

Wille, N. (1887) Ueber Akineten und Aplanosporen. Jahrb. Wissensch. Bot. 18, 492–514.

Winkenbach, F., Wolk, C.P. and Jost, M. (1972) Lipids of membranes and of the cell envelope in heterocysts of a blue-green alga. Planta (Berl.) 107, 69–80.

Wolk, C.P. (1965) Control of sporulation in a blue-green alga. Dev. Biol. 12, 15–35.

Wolk, C.P. (1966) Evidence of a role of heterocysts in the sporulation of a blue-green alga. Am. J. Bot. 53, 260–262.

Wolk, C.P. and Simon, R.D. (1969) Pigments and lipids of heterocysts. Planta (Berl.) 86, 92–97.

Yamamoto, Y. (1972) The fatty acid composition of akinetes, heterocysts and vegetative cells of *Anabaena cylindrica*. Pl. Cell Physiol. (Tokyo) 13, 913–915.

Yamamoto, Y. (1975) Effect of desiccation on the germination of akinetes of *Anabaena cylindrica*. Pl. Cell Physiol. (Tokyo) 16, 749–752.

Yamamoto, Y. (1976) Effect of some physical and chemical factors on the germination of akinetes of *Anabaena cylindrica*. J. Gen. Appl. Microbiol. 22, 311–323.

P. Fay and C. Van Baalen (Eds.) The Cyanobacteria

10

Advances in cyanobacterial molecular genetics

N. TANDEAU DE MARSAC and J. HOUMARD

Institut Pasteur, Unité de Physiologie Microbienne, Département de Biochimie et Génétique Moléculaire, 28 rue du Docteur Roux, 75724 Paris Cedex 15, France

1. Introduction

Until recently, the development of cyanobacterial genetics was limited by our inability to transfer genetic material, both within a given species and between different cyanobacterial species. About 300 cyanobacterial strains have been isolated as axenic cultures but less than 10, all unicellular species, have been shown to be transformable by DNA. Moreover, while specific cyanobacteriophages have been described, gene transduction has not yet been demonstrated.

Molecular genetics offers new possibilities for studying various properties of cyanobacteria, such as oxygenic photosynthesis, aerobic and anaerobic nitrogen fixation, light-regulated gene expression and cell differentiation, properties for which biochemical and physiological data are already available. An increasing number of laboratories are now studying this widespread and very large group of microorganisms using such genetic approaches. Recent reports that conjugation can occur in *Anabaena/Nostoc* strains pave the way for the transfer of genetic material among filamentous cyanobacteria and strengthen the work already initiated in order to understand, at the molecular level, the important processes mentioned above.

In this chapter, we will review current knowledge on cyanobacterial plasmids and restriction enzymes, such data being of special interest for both molecular genetics and the development of DNA transfer systems, and we will describe which cyanobacterial genes have been cloned and characterized. Finally, we will present some of the features which have emerged from both sequence data and gene expression experiments in homologous and/or heterologous hosts, as well as the methodological approaches used.

Due to the state of cyanobacterial taxonomy, the literature is somewhat confused by different

---

Abbreviations: Ap, ampicillin; bp, base pair; CCC, covalently closed circular; CCCP, carbonyl cyanide-*m*-chlorophenyl hydrazone; Cm, chloramphenicol; DCMU, 3-(3,4-dichlorophenyl)-1,1-dimethylurea; Em, erythromycin; Kb, kilobase; kD, kilodalton; Km, kanamycin; MeA, methyladenine; MeC, methylcytosine; Nm, neomycin; Sm, streptomycin.

names being appended by different authors to the same strain. In order to allow easier reading of this chapter, whenever possible we have referred to specific strains by their number in the Pasteur Culture Collection (PCC) and the designation given to them by Guglielmi and Cohen- Bazire (1984), Rippka et al. (1979), Rippka and Cohen-Bazire (1983) and R. Rippka (personal communication).

2. Occurrence, distribution and characterization of cyanobacterial plasmids and sequence-specific endonucleases

2.1. Plasmids

Since the discovery of small extrachromosomal covalently closed circular (CCC) DNA in a *Synechococcus* species by Asato and Ginoza (1973), much effort has been put into the characterization of plasmids in cyanobacteria. Despite the technical difficulties often encountered in plasmid isolation, more than 50 unicellular and filamentous strains belonging to the major taxonomic subgroups (Guglielmi and Cohen-Bazire, 1984; Herdman, 1982; Rippka et al., 1979; Rippka and Cohen-Bazire, 1983) have now been screened. About 60% of the strains examined contain one or several (up to eight) plasmid types whose sizes, as determined from gel electrophoresis and, in some cases, from electron microscopic analyses, range from 1.3 Kb to about 130 Kb (Table I). Two strains, *Synechocystis* PCC 6711 and *Nostoc* PCC 8009, initially found not to contain plasmid DNA (Lau et al., 1980a) have been reexamined both in our laboratory (unpublished data) and by Lambert and Carr (1982) using slightly different extraction techniques and shown to harbor several plasmid species (Table I).

It is worth mentioning here that at least four closely related *Synechococcus* strains PCC 6311, 6908, 7942 and 7943 (Table I) of different geographical origin, but similar morphology and chromosomal GC % content (Rippka et al., 1979; Rippka and Cohen-Bazire, 1983; Wilmotte and Stam, 1984) possess two plasmids of identical size and restriction endonuclease digestion pattern (Van den Hondel et al., 1979; Lau and Doolittle, 1979; Van den Hondel and Van Arkel, 1980). However the results concerning *Synechococcus* PCC 6301 are contradictory. Van den Hondel et al. (1979) found that its plasmids were identical to those of the strains mentioned above, while Shinozaki et al. (1982) observed slight differences in the restriction map of the small plasmid, pBA1. A more precise comparison of the restriction maps of these plasmids is needed to determine the extent of these differences and to draw a clear-cut conclusion. More interestingly, the same two plasmids were found in *Cyanobium* PCC 6707 which shows a 13% higher chromosomal GC % content (Table I) (Van den Hondel et al., 1979). Taken together with the widespread geographical occurrence of these strains this indicates possible interspecific, or even intergeneric, plasmid transfer in nature.

Whether or not all the plasmids found in the cyanobacterial strains screened correspond to different plasmid species cannot be ascertained. More extensive studies may demonstrate that some of them represent multimeric forms of a single plasmid species. For example, *Nostoc* PCC 7524 contains three plasmids pDU1, pDU2 and pDU3 of differing sizes (6.1, 11.8 and 37.3 Kb, respectively), but pDU2 is a dimeric form of pDU1, the restriction endonuclease digestion patterns of these two plasmids being identical (Reaston et al., 1980).

The abundance of the different plasmid types varies with the metabolic state of the cells. The copy number of small plasmids has been found to be lower in exponentially growing cells than in cells in stationary growth phase (Lambert and Carr, 1982; Lambert et al., 1984; Roberts and

TABLE I

Cyanobacterial plasmids

Organism[a]	Number in culture collection[b]				DNA base ratio mol % GC[c]	Number of plasmids	Size of plasmids[d] (Kb)	Reference
	PCC	ATCC	CCAP	UTEX				
SECTION I:								
GENUS *Synechococcus*								
Synechococcus sp.	6301	27144	1405/1	625,1550,0100	55.1	2	7.0, 55	Simon, 1978
(*Anacystis nidulans*,						2	7.9 (pUH1), 48.8 (pUH2)	Van den Hondel et al., 1979
Anacystis nidulans TX20,						2	8.5, 45.5	Lambert and Carr, 1982
Synechococcus leopoliensis)						2	8.0, 48.7	Engwall and Gendel, 1985
						2	7.9[e], 45.5	Lau and Doolittle, 1979
								Lau et al., 1980a
(*Anacystis nidulans*)				2196		2	8.0, 48.7	Engwall and Gendel, 1985
Synechococcus sp.	6311	27145		1549	54.8	2	8.0 (pDF3), 50	Friedberg and Seijffers, 1979, 1983
(*Anacystis* sp.)						2	7.9, 49.3	Van den Hondel and Van Arkel, 1980
Synechococcus sp.	6908	27146		1191	56.0	2	7.9 (pUH5), 48.4 (pUH6)	Van den Hondel et al., 1979
(*Synechococcus cedrorum*)						2	7.6, 45.2	Lau and Doolittle, 1979
Synechococcus sp.	7942				55	2	7.9 (pUH24)[f], 49.3 (pUH25)	Van den Hondel et al., 1980
(*Anacystis nidulans* R-2)						2	8.0, 48.7	Engwall and Gendel, 1985
						2	8.0 (pANS), 48.7 (pANL)	Gendel et al.,1983a
								Laudenbach et al., 1983
Synechococcus sp.	7943					2	7.9, 49.3	Van den Hondel and Van Arkel, 1980
(*Anacystis nidulans* 602)								
Synechococcus sp.	7002	27264			49.1	6	4.5 (pAQEI), 9.7, 15.4, 30,	Robert and Koths, 1976
(*Agmenellum quadruplicatum* PR-6)							36.9, 112	Buzby et al., 1983
						5–7	5.2, 9.3, 15.8, 38.8, ≥ 74.6	Lau et al., 1980a
Synechococcus sp.	73109	29404			49.0	4–5	8.8[c], 15.8, ≥ 74.6	Lau et al., 1980a
(*Agmenellum quadruplicatum* BG-1)								
Synechococcus sp.	6312	27167			50.2	1	23.7 (pUH9)	Van den Hondel et al., 1979
(*Anacystis* sp.)								
Synechococcus sp.	7335	29403			47.4	1	32	Our laboratory (unpublished data)

TABLE I (*continued*)

Organism[a]	Number in culture collection[b]				DNA base ratio mol % GC[c]	Number of plasmids	Size of plasmids[d] (Kb)	Reference
	PCC	ATCC	CCAP	UTEX				
Synechococcus sp.	7336					2	5.4, 32	Our laboratory (unpublished data)
Synechococcus sp. (*Coccochloris elabens* 17-AR)	7003	27265			49.3	5	2.8, 4.3, 5.2, 49.3, ≥ 74.6	Lau et al., 1980a
Synechococcus sp. (*Microcystis* sp.)	7425	29141			48.6	2	8.1 (pUH7), 35.8 (pUH8)	Van den Hondel et al., 1979
GENUS *Cyanobium*								
Cyanobium sp. (*Synechococcus* sp.)	6707				67.7	2	7.9 (pUH3), 49.3 (pUH4)	Van den Hondel et al., 1979
GENUS *Synechocystis*								
Synechocystis sp. (*Aphanocapsa* HA)	6701	27170			35.8	8	2-40	Lau et al., 1980a
						8	5.0 (pSCY1), 7.5 (pSCY2), 13.5 (pSCY3), 15 (pSCY4) + 4 other bands	Anderson and Eiserling (personal communication)
						6	4.1, 6.4, 8.4, 14.9, 18.1, 22.7	Our laboratory (unpublished data)
Synechocystis sp. (*Aphanocapsa* sp.)	6711	27175			37	3	2.1, 16.8, 27	Our laboratory (unpublished data)
Synechocystis sp. (*Aphanocapsa* sp.)	6714	27178			47.9	1	44.8	Lau et al., 1980a
Synechocystis sp. (*Aphanocapsa* sp.)	6803	27184			47.5	2	2.3 (pUG1), 5.2 (pUG2)	Chauvat et al., 1985
						4	2.5, 5.2[e], 50, 100	Our laboratory (unpublished data)
						3	2.1 (pSS2), 4.5, 90	Shestakov et al., 1985
Synechocystis sp. (*Aphanocapsa* sp.)	6808	27170			36	3	3.8, 22, 26	Our laboratory (unpublished data)
Synechocystis sp. (*Microcystis aeruginosa*)	7005	27153	1450/1		45.4	1	3.0 (pUH10)	Van den Hondel et al., 1979

TABLE I (*continued*)

Organism[a]	Number in culture collection[b]				DNA base ratio mol % GC[c]	Number of plasmids	Size of plasmids[d] (Kb)	Reference
	PCC	ATCC	CCAP	UTEX				
SECTION III:								
GENUS *Oscillatoria*								
Oscillatoria limnetica						1	7.6	Friedberg and Seijffers, 1979
GENUS *Pseudanabaena*								
Pseudanabaena sp. (*Synechococcus* sp.)	6802	27183			45.9	2	2.5, 3.7	Our laboratory (unpublished data)
Pseudanabaena sp. (L.P.P.)	7409	29541			43.6	4	2.7, 4.2, 6.6, 10	Our laboratory (unpublished data)
L.P.P. GROUP								
(*Plectonema boryanum*)	6306	27894	1463/1	581,1542	45.8; 48.8	1	14	Simon, 1978
						3	1.3, 14.9, ~17.9	Potts, 1984
(*Phormidium luridum* var. *olivacea*	7602		1462/2	426	46	2	<3.1, 13.4	Simon, 1978
						3	1.3, 14.9, ~17.9	Potts, 1984
(*Plectonema notatum*, *Phormidium*, *Lyngbya*, *Plectonema boryanum*)	73110	29407	482,485,487 488,594,596 597,790		45.4; 49.9	3	1.3, 14.9, ~17.9	Potts, 1984
(*Plectonema* FS180)						3	1.3 (pMP1), 14.9 (pMP2), ~17.9 (pMP3)	Potts, 1984
(*Phormidium faveolarum*)				427	45.7	3	1.3, 14.9, ~17.9	Potts, 1984
(*Plectonema*)	6402	27902		1541	46.3; 47.4	3	4.5, 14.9, >45.8	Potts, 1984
(*Plectonema calothricoides*)				598	45.4	2	14.9, ~17.9	Potts, 1984
SECTION IV:								
GENUS *Anabaena*								
Anabaena sp. (*Anabaena variabilis*, *Cylindrospermum* sp.)	7118	27892			41–43	2	5.4 (pGL1), 20.9	Lambert and Carr, 1982; Lambert et al., 1984

TABLE I *(continued)*

Organism[a]	Number in culture collection[b]				DNA base ratio mol % GC[c]	Number of plasmids	Size of plasmids[d] (Kb)	Reference
	PCC	ATCC	CCAP	UTEX				
Anabaena sp. (*Nostoc muscorum* ISU)	7120	27893			42.5	4	4.9, 49.3, 63, 110	Simon, 1978
Anabaena sp. (*Anabaena doliolum*)	7936					3	8.5, 38.8, 54	Simon, 1978
Anabaena sp. (*Anabaena variabilis* UW, *Anabaena flos-aquae*)	7937	29413	1403/13a	1444		2	37.3, 47.8	Simon, 1978
Anabaena sp. (*Anabaena cylindrica*, *Anabaena inequalis*)	7938	29414	1446/1c	381		5	20.9, 35.8, 60, 79, 110	Simon, 1978
GENUS *Nostoc*								
Nostoc sp.	6705	29131			43.4	3	3.9 (pGL2), 20.9, 60	Lambert and Carr, 1982 Lambert et al., 1984
Nostoc sp.	7524	29411			39	3	6.1 (pDU1), 11.8 (pDU2), 37.3 (pDU3)	Reaston et al., 1980, 1982b
Nostoc sp. (*Nostoc* sp. MAC)	8009				n.d.	3	8.4 (pDC1), 30, 60	Lambert and Carr, 1982, 1983 Lambert et al., 1984
						4	7.9, 30, 38, 40	Our laboratory (unpublished data)
GENUS *Calothrix*								
Calothrix sp. (*Fremyella diplosiphon*)	7601			481	42.7	4	35.8, 47.8, 60, 72	Simon, 1978
						n.d.	15.5 (pFdB), 15.9 (pFdB′), 18.5 (pFdA) + others	Bogorad et al., 1983
						4	9.1, 15, 19.1, 90,(A)[g]	Our laboratory (unpublished data)
						4	11.9, 15, 16.7, 19.7 (B)[g]	Our laboratory (unpublished data)

TABLE I *(continued)*

[a] Strains maintained in the Pasteur Culture Collection are designated according to Guglielmi and Cohen-Bazire (1984), Rippka et al. (1979), Rippka and Cohen-Bazire (1983), and R. Rippka, personal communication. Strain designations previously published, if different, are mentioned in parentheses. Other strains are designated by the name given in the corresponding cited reference.

[b] PCC, Pasteur Culture Collection, Paris, France.

ATCC, American Type Culture Collection.

CCAP, Culture Collection of Algae and Protozoa, Cambridge, England.

UTEX, Culture Collection of Algae at the University of Texas, USA (Starr, 1978).

[c] From Lachance (1981), Rippka et al. (1979), Stam (1980), Stanier et al. (1971) and Wilmotte and Stam (1984).

[d] Published plasmid designations are indicated in parentheses. Plasmid sizes were estimated either by gel electrophoresis and/or from contour length electron micrographs. Values published in megadaltons were converted into kilobases by dividing by the factor 0.67.

[e] Plasmids which can be lost spontaneously in clonally derived strains.

[f] The complete nucleotide sequence of pUH24 is determined (Weisbeek et al., 1985).

[g] For explanations concerning the patterns A and B see Section 2.1.

n.d., not determined.

Strains that have been found not to contain plasmids are: *Cylindrospermum* sp. (UTEX 942) and *Cylindrospermum licheniforme* (UTEX 1828), *Nodularia* PCC 7103, *Tolypothrix* PCC 7101, *Fischerella* PCC 73103 and 7115 (Simon, 1978), *Synechococcus* PCC 6715 (Van den Hondel et al., 1979); *Synechococcus* PCC 6307 and 7502, *Gloeobacter* PCC 7421, *Gloeothece* PCC 6909, *Chroococcidiopsis* PCC 6712, *Microcystis aeruginosa* NRC-1, *Gloeocapsa alpicola* W, *Pseudanabaena* PCC 6901, *Chlorogloea fritschii*, *Anabaena flos-aquae* (CCAP 1403/13D; 1403/13F; 1403/13G), and some natural isolates (4 strains) (Lau et al., 1980a); *Gloeocapsa alpicola* (A. Smith), *Anabaena variabilis* (CCAP 1403/12), *Anabaena flos-aquae* (UTEX 1555), *Nostoc* PCC 73102 and 7422 (Lambert and Carr, 1982); *Chroococcus* S24 and N41(Potts et al., 1983).

Koths, 1976; our own observations) but nothing is known about the mechanism of cyanobacterial plasmid replication nor about the molecular control of copy number. Experiments aimed at elucidating these points would be very valuable.

For a given cyanobacterial species, plasmid types and sizes determined in different laboratories are generally in good agreement. There is, however, one exception, *Calothrix* PCC 7601 (*Fremyella diplosiphon*), for which striking differences in plasmid patterns have been obtained. This filamentous strain possesses most of the interesting and attractive physiological properties of cyanobacteria, such as a capacity to grow heterotrophically, to fix molecular nitrogen and the ability to undergo complementary chromatic adaptation and photomorphogenesis (Tandeau de Marsac, 1983). Simon (1978) first reported that *Calothrix* PCC 7601 contained four CCC DNAs having sizes between 35.8 Kb and 72 Kb. Later on, the plasmid content of the cells has been re-examined by both Bogorad's group and our own laboratory. Both the sizes and the number of plasmids found differed from those first determined by Simon. According to Bogorad et al. (1983), electron microscopic analysis revealed a population of molecules with sizes ranging from 5.4 Kb to 78 Kb. The relative abundance of some of the plasmids was shown to depend on the detergent used during extraction. When Brij was replaced by Sarkosyl the plasmid pFdA (18.5 Kb) remained the most abundant species, but about three times more DNA of two smaller plasmids pFdB (15.5 Kb) and pFdB′ (15.9 Kb) were obtained in the preparation. Based on hybridization experiments, plasmids pFdB and B′ were shown to be closely related to pFdA, although some DNA sequences were absent and others rearranged. In our laboratory, analysis of plasmids from *Calothrix* PCC 7601 also gave rise to unexpected results. When the same extraction procedure was used on eight subcultures of cells grown in white light, two distinct patterns could be observed. Patterns 'A' (seven subcultures) and 'B' (one subculture) differed in four plasmid bands (Table I). These variations in plasmid patterns could not be related to any obvious phenotypic changes. However, these results along with our previous observations (Tandeau de Marsac, 1983) that spontaneous pigment mutants arise with an unexpectedly high frequency (10^{-3} to 10^{-4}) in *Calothrix* PCC 7601 strongly suggest the existence of mobile genetic elements in this strain (transposons, insertion elements or Mu-type phages). In other cyanobacteria, sequence homologies have also been found between two distinct plasmids within a single species, as well as between plasmids of differing sizes and restriction endonuclease digestion patterns which had been isolated from different strains (Lau et al., 1980a). So far, the first evidence for the existence of genomic rearrangements in cyanobacteria are those occurring in the filamentous nitrogen-fixing strain *Anabaena* PCC 7120 during the differentiation of vegetative cells into heterocysts (J.W. Golden et al., 1985a,b). However, the presence of mobile genetic elements could well be as widespread in cyanobacterial genomes as it is in other prokaryotic or eukaryotic organisms.

Many functions, such as resistance to antibiotics (Flores and Wolk, 1985b; Van den Hondel et al., 1979), to high salt concentrations or to heavy metals (Van den Hondel et al., 1979), gas vacuolation (Walsby, 1977), toxin production (Hauman, 1981; Van den Hondel et al., 1979) and restriction/modification enzyme synthesis (Whitehead and Brown, 1985a,b), have been assumed to be plasmid-borne, but no definitive proof has been established so far. Examining a number of filamentous strains, Simon (1978) could not correlate the lack of CCC DNA in certain species to metabolic defects, structural modifications and/or differentiation capabilities. Similarly, it has been observed that spontaneous loss of small plasmids from different unicellular strains, such as *Synechococcus* PCC 6301 and 73109 (Lau and Doolittle, 1979; Lau et al., 1980a), and *Synechocystis* PCC 6803 (our laboratory, unpublished data) could occur without causing any obvious phenotypic change. However, Lau et al. (1980b) were able to obtain in vivo transcription products of the 7.9 Kb plasmid from *Synechococcus* PCC 6301, and, based on DNA-RNA hybridizations Bogorad et al. (1983) reported that plasmid pFdA from *Calothrix* PCC 7601 contained some so-called 'verdogenes' having green-light induced expression. Neither the gene products nor the precise control of verdogenes expression have however been elucidated. Chauvat et al. (1983) mentioned that a pUH24-less derivative of *Synechococcus* PCC 7942 has a 30 times lower transformation capacity than the wild-type strain. These authors postulated that a pUH24 encoded function could favor DNA uptake into the cells of this strain, an assumption that remains to be proven.

In conclusion, although plasmids are widespread in cyanobacteria and have been well characterized in several strains, in molecular terms, they are all cryptic in the sense that their coding functions and thus their physiological role(s) remain unknown. Indeed, the discovery of plasmid-borne genetic markers would be of great help in the development of molecular genetic studies, in particular for the construction of cloning vehicles and in the understanding of genetic transfer mechanism(s) (see Section 4 of this chapter). Last but not least is the information that it would provide about the adaptative evolution of this large and diverse group of photosynthetic prokaryotes.

2.2. Sequence-specific endonucleases

Whilst no systematic search has been carried out so far for type I and III endonucleases, during the last 3 years an increasing number of cyanobacterial strains have been analyzed for their type II sequence specific endonuclease content. About 50 strains belonging to nine different genera have been shown to contain between one and five type II sequence specific endonucleases, otherwise commonly referred to as restriction enzymes (Table II). Among the restriction endonucleases characterized for their recognition sequences, 17 new enzymes have been found, some of which have already proved particularly useful for recombinant DNA studies. As in other bacterial groups, isoschizomers (enzymes recognizing the same nucleotide sequences) have been found to occur even in unrelated strains, indicating that they cannot be used as taxonomic markers (De Waard and Duyvesteyn, 1980). Since some isoschizomers, such as the *Ava* I or *Ava* II-type enzymes, are widely distributed among different genera, it is tempting to assume that these enzymes have a common evolutionary origin and together with their modification methylase counterparts could have been acquired by the transfer of plasmid- or cyanophage-encoded genes. So far, however, no correlation has been established between the plasmid content of cyanobacterial strains, their phage sensitivity, and their restriction endonuclease content. More information is needed to test this hypothesis.

Chromosomal DNA from different filamentous and unicellular cyanobacteria was shown to be surprisingly resistant to cleavage by a number of restriction endonucleases, far more than expected from their type II restriction enzyme content (Herrero et al., 1984; Lambert and Carr, 1984; Van den Hondel et al., 1983). In filamentous strains as diverse as *Nostoc*, *Gloeotrichia* and *Plectonema* which contain either different types of restriction endonuclease or none (for example in *Nostoc* PCC 73102), a very similar pattern of extensive DNA modification has been observed (Lambert and Carr, 1984). In contrast, in five unicellular strains these authors observed a great variation in the susceptibility of the genomes to cleavage by a given set of restriction endonucleases. Use of isoschizomers has clearly established that resistance of the DNA to cleavage could either be the result of a host-controlled restriction/modification system as exists in other bacterial groups, or due to the presence of a *dam* (DNA adenine methylase) enzyme of the *Escherichia coli* type (Geier and Modrich, 1979). Examples of possible restriction-modification systems operating in vivo were described for *Anabaena* PCC 7937 (Currier and Wolk, 1979), and for *Synechococcus* PCC 6301 (Szekeres, 1981; Szekeres et al., 1983). Cyanophage N-1 grown in *Anabaena* PCC 7120 formed plaques on *Anabaena* PCC 7937 only at a very low efficiency, while surviving progeny of cyanophage N-1 grown in *Ana-*

TABLE II

Sequence-specific endonucleases from cyanobacteria

Organism[a]	Number in culture collection[b]				Endonuclease	Isoschizomer of:	Recognition sequence[c]	Reference
	PCC	ATCC	CCAP	UTEX				
SECTION I:								
GENUS *Synechococcus*								
Synechococcus sp. (*Agmenellum quadruplicatum* PR-6)	7002	27264			*Ava* I	*Ava* I	C↓PyCGPuG	Lau and Doolittle, 1980
Synechococcus sp. (*Coccochloris elabens* 17a)	7003	27265			*Cel* I	*Bam* HI	GGATCC	F. Calleja and A. De Waard (personal communication)
					Cel II	*Esp* I	GCTNAGC	
Synechococcus sp. (*Anacystis nidulans* R-2)	7942				*Ani* I		n.d.	Gallagher and Burke, 1985a
GENUS *Cyanobacterium*								
Cyanobacterium stanieri (*Synechococcus cedrorum*)	7202	29140	1479/2a and 2b		*Sce* I	*Fnu* DII	CGCG	F. Calleja and A. De Waard (personal communication)
GENUS *Cyanothece*								
Cyanothece halophytica (*Synechococcus* sp. *Aphanothece halophytica*)	7418	29534			*Aha* I	*Cau* II	$CC^{C}_{G}GC$	Whitehead and Brown, 1985b
					Aha II	*Acy* I	GPu↓CGPyC	
					Aha III	*Dra* I	TTT↓AAA	Whitehead and Brown, 1982
GENUS *Gloeothece*								
Gloeothece sp. (*Gloeocapsa* sp.)	6909	27152	1430/3	795	*Gsp* I	*Pvu* II	CAGCTG	F. Calleja and A. De Waard (personal communication)
GENUS *Synechocystis*								
Synechocystis sp. (*Aphanocapsa* HA)	6701	27170			*Sec* I		C↓CNNGG	Calleja et al., 1985
					Sec II	*Msp* I	CCGG	
					Sec III	*Mst* II	CCTNAGG	
Synechocystis sp. (*Aphanocapsa* sp.)	6711	27175			*Sci* I	*BstE* II	GGTNACC	F. Calleja and A. De Waard (personal communication)
					Sci II	*Pvu* II	CAGCTG	

TABLE II *(continued)*

Organism[a]	Number in culture collection[b]				Endonuclease	Isoschizomer of:	Recognition sequence[c]	Reference
	PCC	ATCC	CCAP	UTEX				
Synechocystis sp.	6906	27266			*Esp* I		GC↓TNAGC	Calleja et al., 1984
(*Eucapsis* sp.)					*Esp* II		n.d.	
SECTION III:								
GENUS *Pseudanabaena*								
Pseudanabaena sp.	6901	27263			*Psp* I	*Asu* I	GGNCC	B. Mulligan and M. Szekeres
(*Synechococcus* sp.)								(unpublished data)
Pseudanabaena sp.	7409	29541			*Pse* I	*Asu* I	GGNCC	A. De Waard
(L.P.P.)								(personal communication)
L.P.P. Group								
Microcoleus sp.				2220	*Mst* I		TGCGCA	Gingeras et al., 1978
					Mst II	*Sau* I	CC↓TNAGG	I. Schildkraut
								(unpublished data)
SECTION IV:								
GENUS *Anabaena*								
Anabaena sp.	6309	29211	1403/4b	377	*Asu* I		G↓GNCC	Hughes et al., 1980
(*Anabaena subcylindrica*)					*Asu* II		TT↓CGAA	De Waard and Duyvesteyn,
					Asu III		GPuCGPyC	1980
Anabaena sp.	6411	27898		1597	*Asp* (6411) I	*Ava* I	CPyCGPuG	Duyvesteyn et al., 1983
(*Nostoc* sp.)					*Asp* (6411) II	*Ava* II	GG$_{T}^{A}$CC	
Anabaena sp., strain W	7108	29208			*Ast* W I	*Acy* I	GPuCGPyC	De Waard and Duyvesteyn, 1980
								Wolk and Kraus, 1982
Anabaena sp.	7118	27892			*Ava* I		C↓PyCGPuG	Hughes and Murray, 1980
(*Anabaena variabilis*,					*Ava* II		G↓G$_{T}^{A}$CC	Hughes and Murray, 1980
Cylindrospermum sp.)								Sutcliffe and Church, 1978
					Ava III		ATGCAT	Roizes et al., 1979

TABLE II *(continued)*

Organism[a]	Number in culture collection[b]				Endonuclease	Isoschizomer of:	Recognition sequence[c]	Reference
	PCC	ATCC	CCAP	UTEX				
Anabaena sp. (*Nostoc muscorum* UW)	7119	29151			*Asp* (7119) I	*Ava* I	CPyCGPuG	Duyvesteyn et al., 1983
					Asp (7119) II	*Ava* II	GG$^{A}_{T}$CC	
Anabaena sp. (*Nostoc muscorum* ISU)	7120	27893			*Asp* (7120) I	*Ava* I	CPyCGPuG	Duyvesteyn et al., 1983
					Asp (7120) II	*Ava* II	GG$^{A}_{T}$CC	
Anabaena sp. (*Anabaena cylindrica*)	7122	27899	1403/2a	629	*Acy* I		GPu↓CGPyC	De Waard et al., 1978
					Acy II		n.d.	
Anabaena sp. (*Anabaena variabilis* UW, *Anabaena flos-aquae*)	7937	29413	1403/13a	1444	*Avr* I	*Ava* I	C↓PyCGPuG	Duyvesteyn et al., 1983
					Avr II		CCTAGG	Wolk and Kraus, 1982
								E.C. Rosenvold (unpublished data, cited in: Szybalski and Szybalski, 1979)
Anabaena catenula			1403/1		*Aca* I	*Asu* II	TTCGAA	F. Calleja and A. De Waard (personal communication)
					Aca II	*Bam* HI	GGATCC	
					Aca III	*Mst* I	TGCGCA	
					Aca IV	*Hae* III	GGCC	
Anabaena 'oscillarioides'			1403/9		*Aoc* I	*Mst* II	CC↓TNAGG	F. Calleja and A. De Waard (personal communication)
					Aoc II	*Sdu* I	G$\begin{smallmatrix}G\\A\\T\end{smallmatrix}GC\begin{smallmatrix}C\\T\\A\end{smallmatrix}$C	
Anabaena oscillarioides			1403/11		*Aos* I	*Mst* I	TGC↓GCA	De Waard et al., 1979
					Aos II	*Acy* I	GPu↓CGPyC	
					Aos III	*Sac* II	CCGCGG	A. De Waard (personal communication)
Anabaena sp., strain H			1403/12		*Asp* H I	*Asu* II	TTCGAA	A. De Waard (personal communication)
					Asp H II	*Mst* I	TGCGCA	
Anabaena flos-aquae			1403/13f		*Afl* I	*Ava* II	G↓G$^{A}_{T}$CC	Whitehead and Brown, 1985a
					Afl II		C↓TTAAG	
					Afl III		A↓CPuPyGT	

TABLE II *(continued)*

Organism[a]	Number in culture collection[b] PCC	ATCC	CCAP	UTEX	Endonuclease	Isoschizomer of:	Recognition sequence[c]	Reference
Anabaena inequalis			1446/1a		*Ain* I *Ain* II	*Pst* I *Bam* HI	CTGCAG GGATCC	F. Calleja and A. De Waard (personal communication)
Anabaena sp., strain TAi					*Asp* T Ai I *Asp* T Ai II *Asp* T Ai III	*Pst* I *Bam* HI *Hae* III	CTGCAG GGATCC GGCC	F. Calleja and A. De Waard (personal communication)
GENUS *Nostoc*								
Nostoc sp. (*Anabaena* sp.)	6302	27897		1551	*Nsp* (6302) I	*Mst* I	TGCGCA	F. Calleja and A. De Waard (personal communication) Wolk and Kraus, 1982
Nostoc sp. (*Anabaena spiroides*)	6310	27896		1552	*Nsp* (6310) I *Nsp* (6310) II	*Ava* I	CPyCGPuG n.d.	Wolk and Kraus, 1982 E. Flores and C.P. Wolk (personal communication)
Nostoc sp., strain B	6705	29131			*Nsp* B I *Nsp* B II	*Asu* II	TTCGAA $C^{A}_{C}G{\downarrow}C^{T}_{G}G$	Duyvesteyn et al., 1983
Nostoc sp. (*Nostoc muscorum* UC142)	6719	29105			*Nsp* (6719) I *Nsp* (6719) II	*Ava* I *Ava* II	CPyCGPuG $GG^{A}_{T}CC$	Duyvesteyn et al., 1983
Nostoc sp. (*Anabaenopsis circularis*)	6720	27895	1402/1		*Acr* I *Acr* II	*Ava* I *Bst* EII	CPyCGPuG GGTNACC	F. Calleja and A. De Waard (personal communication)
Nostoc sp., strain H	7413	29106			*Nsp* H I *Nsp* H II	*Nsp* C I *Ava* II	PuCATG↓Py $GG^{A}_{T}CC$	Duyvesteyn et al., 1983
Nostoc sp., strain C	7524	29411			*Nsp* C I *Nsp* C II *Nsp* C III *Nsp* C IV *Nsp* C V	*Nsp* H I *Sdu* I *Ava* I *Asu* I *Asu* II	PuCATG↓Py $GA^{G}_{T}GCT^{C}_{A}{\downarrow}C$ CPyCGPuG GGNCC TTCGAA	Reaston et al.,1982a
Nostoc sp. (*Nostoc* sp. MAC)	8009				*Nsp* MAC I	*Bgl* II	A↓GATCT	Lau et al., 1985

TABLE II *(continued)*

Organism[a]	Number in culture collection[b] PCC	ATCC	CCAP	UTEX	Endonuclease	Isoschizomer of:	Recognition sequence[c]	Reference
Nostoc linckia					*Nli* I	*Ava* I	CPyCGPuG	Duyvesteyn et al., 1983
					Nli II	*Ava* II	$GG^{A}_{T}CC$	
Nostoc muscorum, strain M-131-G					*Nmu* I	*Ava* I	CPyCGPuG	Duyvesteyn et al., 1983
					Nmu II	*Ava* II	$GG^{A}_{T}CC$	
Nostoc sp., strain 19-6C-C					*Nsp* 19-6C-C I	*Ava* II	$GG^{A}_{T}CC$	C. Karreman and A. De Waard (personal communication) Wolk and Kraus, 1982
Nostoc sp., strain 23-9B					*Nsp* 23-9B I	*Ava* II	$GG^{A}_{T}CC$	F. Calleja and A. De Waard (personal communication) Wolk and Kraus, 1982
Nostoc sp., strain 78-12B					*Nsp* 78-12B I	*Asu* I	GGNCC	F. Calleja and A. De Waard (personal communication) Wolk and Kraus, 1982
Nostoc sp., strain UM-3					*Nsp* UM-3 I	*Mst* I	TGCGCA	C. Karreman and A. De Waard (personal communication) Wolk and Kraus, 1982
					Nsp UM-3 II	*Asu* I	GGNCC	
					Nsp UM-3 III		n.d.	
					Nsp UM-3 IV		n.d.	
GENUS *Cylindrospermum*								
Cylindrospermum 'licheniforme'		29412		1828	*Clc* I	*Pst* I	CTGCAG	F. Calleja and A. De Waard (personal communication)
					Clc II	*Mst* I	TGCGCA	
Cylindrospermum licheniforme				2014	*Cli* I	*Ava* II	$GG^{A}_{T}CC$	C. Karreman and A. De Waard (personal communication)
					Cli II	*Mst* I	TGCGCA	Wolk and Kraus, 1982
					Cli III		n.d.	

TABLE II *(continued)*

Organism[a]	Number in culture collection[b]				Endonuclease	Isoschizomer of:	Recognition sequence[c]	Reference
	PCC	ATCC	CCAP	UTEX				
GENUS *Calothrix*								
Calothrix sp. (*Tolypothrix tenuis*)	7101	27914	1482/3a and b		*Tte* I	*Hae* III	GGCC	B. Siegelman (unpublished data)
Calothrix sp. (*Fremyella diplosiphon*)	7601		1429/1	481	*Fdi* I *Fdi* II	*Ava* II *Mst* I	G↓G$^{A}_{T}$CC TGC↓GCA	Van den Hondel et al., 1983
Calothrix scopulorum			1410/5		*Csc* I	*Sac* II	CCGCGG	Duyvesteyn et al., 1981
SECTION V:								
GENUS *Fischerella*								
Fischerella sp. (*Fischerella musicola*)	73103	29114	1427/1	1301	*Fsp* I *Fsp* II	*Mst* I *Asu* II	TGCGCA TTCGAA	M. Szekeres (N.E. Biolabs, 1985)
Fischerella sp. (*Mastigocladus laminosus*)	7414	29161	1447/1		*Mla* I	*Asu* II	TT↓CGAA	Duyvesteyn and De Waard, 1980

[a] See note [a] of Table I for designation of strains.

[b] See note [b] of Table I for abbreviations.

[c] Arrows indicate the cleavage site in the recognition sequence.

n.d. not determined.

baena PCC 7937 had a higher relative efficiency of plaque formation on this same strain. This suggests that DNA from cyanophage N-1 which escaped restriction by endonucleases present in *Anabaena* PCC 7937 could have been modified by the host cells. On the other hand, Szekeres (1981) performed in vitro experiments with *Synechococcus* PCC 6301 showing that an endonuclease produced by cells after infection by AS-1 cyanophage could degrade *Synechococcus* PCC 6301 DNA, the infecting cyanophage DNA being resistant to cleavage. Moreover, the cyanophage DNA appeared unprotected after cloning in a plasmid vector and propagation in an *E. coli* strain. These results were taken as evidence that a restriction/modification like system was induced in *Synechococcus* PCC 6301 by AS-1 cyanophage, in order to eliminate selectively the host DNA. The modification process involved remains unknown. Very recently, the modification enzyme M.*Nsp*MACI was purified from *Nostoc* PCC 8009, a strain which contains the restriction enzyme *Nsp*MACI, an isoschizomer of *Bgl*II (Reaston and Carr, 1985). The precise modification site of M.*Nsp*MACI has not yet been established. Interestingly however, preliminary observations showed that *Bgl*II cleaves M.*Nsp*MACI modified DNA, while *Nsp*MACI failed to cleave the same DNA. This suggests that the modification sites of M.*Bgl*II and M.*Nsp*MACI are different (Reaston and Carr, 1985). According to Weisbeek et al. (1985), a modification enzyme would also exist in *Synechococcus* PCC 7942, which would methylate the central adenine residue in GATC sequences. However, the recognition sequence of the restriction enzyme (*Ani*I) discovered in this strain has not been determined nor has the corresponding modification enzyme been purified (Gallagher and Burke, 1985a).

Even if, as appears most likely, modification processes by means of methylation exist in cyanobacteria, they are not sufficient to explain the lack of cleavage by a number of restriction endonucleases. Possible explanations have been proposed by Lambert and Carr (1984), such as the presence of inhibitors and/or DNA binding proteins in the DNA preparations, the presence of unusual bases in the DNA (other than MeC and MeA) and the absence of specific recognition sequences in the genome of certain cyanobacteria. A paucity of sites for some restriction endonucleases has also been suggested by Herrero et al. (1984) to occur in the DNA of a derivative of *Anabaena* PCC 7937. From the collective results obtained so far, it is obviously impossible to draw any clear-cut conclusions about the mechanism(s) which lead(s) to the peculiar resistance to restriction cleavage of cyanobacterial DNA. Moreover, it remains to be determined whether the high degree of methylation found in DNA from some cyanobacteria, and especially from filamentous strains, plays a role in differentiation processes, similarly to what has been suggested to occur in eukaryotic cells during the developmental regulation of gene expression (Chisholm, 1982).

3. Construction of cloning vehicles

3.1. Cyanobacterial plasmid vectors

The first attempt to use cyanobacterial plasmids for the construction of cloning vectors was described by Van den Hondel et al. (1980) using the highly transformable and fast-growing cyanobacterium *Synechococcus* PCC 7942. As mentioned previously, this strain harbors only two plasmids, pUH25 (49.3 Kb) and pUH24 (7.9 Kb) (Table I). However, in the absence of any known coding function, no selective marker was available. The fact that *E. coli* plasmids could be transferred but not maintained into *Synechococcus* PCC 7942 cells* was used to introduce in vivo, into pUH24, the

*In a recent report, Daniell et al. (1985) claim to have obtained transformation and stable maintenance of the *E. coli* plasmids pBR322 and pHUB4 in *Synechococcus* PCC 7942 cells. It is thus necessary to reconsider the widespread idea concerning the inability of the *E. coli* plasmids to replicate in cyanobacterial cells.

transposon Tn*901* (Ap^R) initially carried on the *E. coli* plasmid, pRI46. The recombinant plasmids thus obtained (pCH1 to pCH5) were able to express ampicillin resistance in the transformants. This provided the first cyanobacterial plasmid bearing a selectable marker. Since then, a number of other vectors have been constructed in vitro (Table III).

Although both pCH1 and the other recombinant plasmids have proved very useful in obtaining Met^- and Nar^- mutants by transposition mutagenesis and, subsequently, in cloning the corresponding wild-type genes from *Synechococcus* PCC 7942 (Kuhlemeier et al., 1984a,b; Tandeau de Marsac et al., 1982), their use has been limited by their inability to replicate in *E. coli*. Indeed, because recombinant DNA technology has been so largely developed using *E. coli*, it has become of great advantage to have shuttle vectors able to replicate both in *E. coli* and the cyanobacterium under study.

3.2. Shuttle cloning vectors

All the shuttle cloning vectors constructed so far are presented in Table IV. Kuhlemeier et al. (1981) constructed, by in vitro recombination, two hybrid plasmids, pUC104 and pUC105, consisting of the *E. coli* vector pACYC184 and a cyanobacterial plasmid, pUC1, derived from pUH24. These two plasmids differ only by the location and orientation of pACYC184 in the recombinant molecules. They transform both *E. coli* and *Synechococcus* PCC 7942 with high frequency. Plasmid pUC104 is stably maintained in *Synechococcus* PCC 7942, irrespective of the selection marker used; in contrast, pUC105 is only stable when chloramphenicol is used as a selective agent, giving under ampicillin selection a deletion-derivative plasmid identical to pUC1. Plasmid pUC104 was further converted into a cosmid (pPUC29) by insertion of a *Bgl*II fragment carrying the cohesive ends of bacteriophage lambda (Tandeau de Marsac et al., 1982). This cosmid allows the cloning of large DNA fragments, ranging from about 25 to 40 Kb; it can be easily transferred to *E. coli* by infection after in vitro DNA packaging, as well as to *Synechococcus* PCC 7942 by transformation. Further modification of pUC105 made by Dzelzkalns et al. (1984) led to the construction of a derivative

TABLE III

Characteristics of the plasmid vectors from *Synechococcus* PCC 7942 (= *Anacystis nidulans* R-2)[a]

Designation	DNA origin	Size (Kb)	Selectable markers	Unique restriction sites[b]
pCHI-pCH5	pUH24 + Tn*901*	12.4	Ap^R	*Xho*I
pUCl	pCHI derivative	8.2	Ap^R	*Xho*I, *Bam*HI
pUC12[c]	pUCI + *Hpa*I-*Pvu*II fragment from pR1477S (= RSF1010)	9.8	Ap^R, Sm^R	*Xho*I, *Bam*HI
pUC13[c]	pUC12 derivative	6.9	Sm^R	*Xho*I, *Bgl*I, *Hind*III
pUC14[c]	pUC12 derivative	7.8	Sm^R	*Xho*I, *Bgl*II

[a] Data taken from Van den Hondel et al. (1980) and Kuhlemeier et al. (1983).

[b] None of the unique restriction sites are located within antibiotic resistance markers.

[c] Previously designated pUC2, pUC3 or pUC4, respectively.

TABLE IV

Cyanobacterial cloning vectors

Organism[a]	Cyanobacterial plasmid origin		*E. coli* plasmid origin		Shuttle vector or hybrid plasmid		Selectable markers	Unique restriction sites	Reference
	designation	size (Kb)	designation	size (Kb)	designation	size (Kb)			
Synechococcus PCC 6301 (*Anacystis nidulans*)	pBA1	7.8	pBR322	4.36	pBAS18[b]	12.1	Ap^R	*Eco*RI, *Xho*I	Shinozaki et al., 1982
Synechococcus PCC 7942 (*Anacystis nidulans* R-2)	pUC1	8.2	pACYC184	4.0	pUC104	12.2	Ap^R, Cm^R	*Eco*RI*, *Xho*I, *Sal*I	Kuhlemeier et al., 1981
					pUC105	12.2	Ap^R, Cm^R	*Eco*RI*, *Xho*I, *Sal*I, *Bgl*II, *Bam*HI	Kuhlemeier et al., 1981
					pPUC29 (pUC104 + a 1.7 Kb *Bgl*II fragment from pHC79 housing the *cos* sequence)	14.2	Ap^R, Cm^R	*Eco*RI*, *Xho*I, *Sal*I, *Xba*I, *Bst*EII	Tandeau de Marsac et al., 1982; Kuhlemeier et al., 1984b
	pUC13	6.9	pACYC184	4.0	pUC303	11.0	Cm^R, Sm^R	*Eco*RI*, *Xho*I, *Sal*I	Kuhlemeier et al., 1983
	pUH24	8.0	pBR322::Cm		pLS103	10.1	Ap^R	*Hind*III, *Sal*I	Sherman and Van de Putte, 1982
	pUH24	8.0	pBR328	4.9	pSG111	12.9	Ap^R, Cm^R	*Eco*RI*, *Xho*I, *Sal*I, *Sph*I	Golden and Sherman, 1983
	pANS (=pUH24)	8.0	pBR325	5.7	pECANI	13.5	Ap^R, Cm^R	*Eco*RI*, *Sal*I	Gendel et al., 1983a
	pANS	8.0	pDPL13	2.5	pPLANB1 and 2[c]	10.2	Ap^R	*Xba*I, *Sma*I, *Nar*I, *Stu*I, *Eco*RI, *Sac*I, *Sal*I	Gendel et al., 1983a,b
					pCB4 (pPLANB2 deleted of a 4 Kb *Xho*I fragment)	6.7	Ap^R	*Bam*HI, *Sma*I, *Xba*I, *Nar*I, *Xho*I	Gendel et al., 1983a
	pANS (*Bam*HI-*Xho*I fragment)	4.65	pDPL13	2.5	pXB7	7.2	Ap^R	*Bam*HI, *Eco*RI, *Mst*I, *Nae*I, *Nar*I, *Nru*I, *Sac*I, *Sma*I, *Stu*I, *Xba*I	Lau and Straus, 1985
	pANS (*Bam*HI-*Xho*I fragment)	4.65	pUC8	2.65	pECAN8	7.3	Ap^R	*Eco*RI, *Sma*I, *Bam*HI, *Sal*I	Lau and Straus, 1985
	pANS (*Bam*HI-*Xho*I fragment)	4.65	pDPLK	3.15	pKBX	7.8	Km^R	*Bam*HI, *Eco*RI, *Mst*I, *Nae*I, *Nar*I, *Nru*I, *Sac*I, *Sma*I, *Stu*I, *Xba*I, *Ava*I	Lau and Straus, 1985

TABLE IV *(continued)*

Organism[a]	Cyanobacterial plasmid origin		*E. coli* plasmid origin		Shuttle vector or hybrid plasmid		Selectable markers	Unique restriction sites	Reference
	designation	size (Kb)	designation	size (Kb)	designation	size (Kb)			
Synechococcus PCC 6311 (*Anacystis nidulans*)	pDF3	8.0	pBR325	5.7	pDF30	14.0	Ap^R, Cm^R	*Xho*I, *Sal*I, *Eco*RI*	Friedberg and Seijffers, 1983
Synechococcus PCC 7002 (*Agmenellum quadruplicatum* PR-6)	pAQ1	4.6	pBR322	4.36	pAQE1 and 2[c]	8.9	Ap^R	*Pvu*II, *Ava*I, *Sal*I, *Bam*HI, *Eco*RI	Buzby et al., 1983
					pAQE5 (*Ava*I site deletion of pAQE2)	8.3	Ap^R	*Pvu*II, *Sal*I, *Bam*HI, *Eco*RI	Buzby et al., 1983
					pAQE12 (pAQE2 deleted of 3 Kb)	5.9	Ap^R	*Hind*III, *Eco*RI, *Hinc*II*	R. Porter (personal communication)
					pAQE15 (pAQE12 derivative containing the polylinker MCS7 from pUC7)	5.9	Ap^R	*Hind*III, *Eco*RI, *Bam*HI, *Sal*I, *Acc*I	R. Porter (personal communication)
					pAQE17 (pAQE15 derivative containing the Km^R/Nm^R gene from pRZ102)	7.3	Ap^R Km^R/Nm^R	*Hind*III, *Eco*RI, *Bam*HI, *Sal*I, *Acc*I, *Ava*I, *Sma*I, *Bgl*II	R. Porter (personal communication)
					pAQE17L (pAQE17 derivative containing the *lac* operon from pBRP2)	18.3	Ap^R Km^R/Nm^R Lac^+	*Hind*III, *Bam*HI	R. Porter (personal communication)
					pAQE17DL (pAQE17L deleted of 3 Kb)	15.3	Ap^R Km^R/Nm^R Lac^+	*Hind*III, *Bgl*II, *Sal*I	R. Porter (personal communication)
	pAQ1	4.6	pBR325	5.7	pAQE9 and 10[c]	10.3	Ap^R, Cm^R	*Ava*I, *Sal*I, *Bam*HI	Buzby et al., 1983
					pAQE11 (*Ava*I site deletion of pAQE10)	9.4	Ap^R, Cm^R	*Sal*I, *Bam*HI, *Pvu*II*	Buzby et al., 1983

TABLE IV *(continued)*

Organism[a]	Cyanobacterial plasmid origin		*E. coli* plasmid origin		Shuttle vector or hybrid plasmid		Selectable markers	Unique restriction sites	Reference
	designation	size (Kb)	designation	size (Kb)	designation	size (Kb)			
Synechocystis PCC 6803	pUG1 (*Hpa*I fragment)	1.7	pACYC177	3.7	pUF3[d]	5.7	Km^R	*Xho*I*, *Cla*I*, *Sma*I*, *Bam*HI, *Nco*I, *Xba*I, *Pst*I, *Mst*I, *Bgl*I, *Acc*I	Chauvat et al., 1985 Chauvat (personal communication)
					pUF311 (result of *in vivo* recombination between pUF3 and pUG1)	8.0	Km^R	*Xho*I*, *Cla*I*, *Sma*I*, *Bam*HI, *Pst*I, *Mst*I, *Bgl*I, *Acc*I	Chauvat et al., 1985 Chauvat (personal communication)
					pFCLV7 (pUF311 derivative containing Cm^R gene from pACYC184)	9.35	Km^R, Cm^R	*Xho*I*, *Cla*I*, *Sma*I*, *Eco*RI*, *Pvu*II*, *Bam*HI, *Pst*I, *Mst*I, *Bgl*I, *Acc*I	Chauvat et al., 1985 Chauvat (personal communication)
	pUG1 (*Hpa*I fragment)	1.7	pACYC177	3.7	pUF12	5.75	Km^R	*Xho*I*, *Cla*I*, *Sma*I*, *Bam*HI, *Nco*I, *Xba*I, *Pst*I, *Mst*I, *Bgl*I, *Acc*I	Chauvat et al., 1985 Chauvat (personal communication)
	pSS2 (*Sau*3A fragment)	1.9	pACYC184	4.0	pSE7	5.9	Cm^R	not mentioned	Shestakov et al., 1985
					pSE76 (result of *in vivo* recombination between pSE7 and pSS2)	8.2	Cm^R	*Sal*I*	Shestakov et al., 1985
Plectonema PCC 6306	pGL3	1.4	pBR328	4.9	pGL5[b]	6.3	Ap^R, Cm^R	*Pst*I*, *Pvu*II*, *Eco*RI*, *Hpa*I, *Hind*III, *Bam*HI, *Sph*I, *Sal*I, *Ava*I	Lambert et al., 1984
Nostoc PCC 6705	pGL2	3.9	pBR328	4.9	pGL4[b]	8.8	Ap^R, Cm^R	*Pst*I, *Pvu*II, *Eco*RI, *Hpa*I, *Xba*I, *Sph*I, *Sal*I, *Ava*I, *Bam*HI	Lambert et al., 1984

TABLE IV *(continued)*

Organism[a]	Cyanobacterial plasmid origin		*E. coli* plasmid origin		Shuttle vector or hybrid plasmid		Selectable markers	Unique restriction sites	Reference
	designation	size (Kb)	designation	size (Kb)	designation	size (Kb)			
Nostoc PCC 7524	pDU1	6.28	pBR322	4.36	pVW1[b]	10.6	Ap^R	not mentioned	Wolk et al., 1984
					pVW1C (pVW1 derivative containing the Cm^R gene from pBR328)[b]	11.6	Ap^R, Cm^R	not mentioned	Wolk et al., 1984
					pRL1 (pVW1C deleted of *Ava*II *Ava*II, *Avr*II sites)	9.3	Cm^R	*Bal*I*, *Nco*I*, *Nae*I, *Nde*I, *Cla*I, *Bcl*I, *Bgl*I	Wolk et al., 1984
					pRL5 (pRL1 derivative containing the Sm^R gene from R300B)	11.1	Cm^R, Sm^R	*Ava*I*, *Sst*I*, *Nde*I, *Cla*I, *Bcl*I	Wolk et al., 1984
					pRL6 (pRL1 derivative containing the Km^R/Nm^R gene from ColEI::Tn*5*)	11.3	Cm^R Km^R/Nm^R	*Ava*II*, *Sph*I*, *Bgl*II* *Nde*I, *Cla*I	Wolk et al., 1984
					pRL8 (pRL1 derivative containing the Cm^R gene from pE194)	11.5	Cm^R, Em^R	*Nco*I*, *Bal*I*, *Nae*I, *Ava*II*, *Sst*I*, *Cla*I, *Bgl*I	Wolk et al., 1984

[a] Strains are designated according to Rippka et al. (1979) and R. Rippka, personal communication. Published strain designations, if different, are mentioned in parentheses.
[b] Hybrid plasmids not shown to behave as stable shuttle vectors.
[c] Plasmids with the insert in different orientations.
[d] pUF3 needs to recombine with the 2.27 Kb endogenous plasmid, pUG1, to be stably maintained in *Synechocystis* PCC 6803.
*Unique restriction sites located within antibiotic resistance markers.

which encodes chloramphenicol resistance by means of a promoterless bacterial *cat* (chloramphenicol acetyl transferase) gene fused to a chloroplast promoter.

In order to improve gene cloning systems in *Synechococcus* PCC 7942, Kuhlemeier et al. (1983) constructed other hybrid plasmids, notably pUC303, a shuttle vector carrying two antibiotic markers Sm^R Cm^R, which were able to transform both organisms with high frequency and contained a convenient single *Eco*RI site in the *cam* gene for the selection of recombinant molecules by means of insertional inactivation. However, in *Synechococcus* PCC 7942, pUC303 frequently loses its non selected marker, presumably because of recombinational events occurring between the cyanobacterial part of the vector and the resident small plasmid pUH24. To overcome this problem, a *Synechococcus* PCC 7942 derivative strain cured of pUH24 was isolated (strain Spc). The plasmid pUC303 transformed the strain Spc at high frequency ($\sim 10^6$ colonies per μg of DNA) and was stably maintained in the recipient cells. With the *Synechococcus* PCC 7942 strain Spc, the plasmid pUC303 fulfills the major requirements for a host vector system suitable for gene cloning.

A number of other shuttle vectors have been developed by different groups for the same strain, *Synechococcus* PCC 7942. Sherman's group produced hybrid vectors basically similar to those of Kuhlemeier et al. (1983) and Tandeau de Marsac et al. (1982), but in which the *E. coli* plasmid pACYC184 was replaced by pBR322 derivatives (Golden and Sherman, 1983; Sherman and Van de Putte, 1982). Three shuttle vectors were constructed by Gendel et al. (1983a) and Lau and Straus (1985) which combined by in vivo manipulation the small *Synechococcus* PCC 7942 plasmid, called by these authors pANS (Table I), with a 2.5 Kb pBR322 derivative, pDPL13. Interestingly, these hybrid Ap^R vectors contain a multisite polylinker. A fourth shuttle vector was formed by addition of the origin of replication of pANS to a newly constructed *E. coli* recombinant plasmid pDPLK. The resulting vectors pPLANB2 and pCB4 (Gendel et al., 1983a), as well as pXB7 and pKBX (Lau and Straus, 1985) are of reasonably small size ($\leqslant 10$ Kb) with several unique restriction sites suitable for insertion of a variety of foreign DNA fragments, but none of the cloning sites are located within a selectable marker. One of the major advantages of the shuttle vector pECAN8, derived from in vitro recombination between pANS and pUC8 (Lau and Straus, 1985), is that this small vector efficiently transforms *Synechococcus* PCC 7942 cells and contains four unique cloning sites within the *lacZ* gene, allowing rapid screening in *E. coli* of insert-carrying vectors by means of selection for β-galactosidase expression. From the endogenous plasmid of *Synechoccoccus* PCC 6311, called pDF3, Friedberg and Seijffers (1983) generated a chimeric vector, pDF30, similar to pECAN1 constructed by Gendel et al. (1983a). Restriction mapping indicated that plasmid pDF3 is indeed identical to pANS from *Synechococcus* PCC 7942. In an attempt to construct an expression vector, Friedberg and Seijffers (1985) have linked an operator-promoter region and the temperature-sensitive repressor gene *cI*857 of bacteriophage lambda to a *Synechococcus* PCC 7942 replicon. When cloned in such a plasmid the *cat* gene is expressed from the lambda promoter and regulated by the thermosensitive repressor. This is the first example of a cloning system allowing gene-controlled expression in a cyanobacterial host.

Cloning vectors have been developed for *Synechococcus* PCC 7002. This strain is easily transformable by homologous DNA and presents the advantage of being able to grow heterotrophically, a property which might be helpful for the study of genes involved in photosynthesis. A series of biphasic plasmids has been constructed by joining the smallest endogenous plasmid pAQ1 to either pBR322 or pBR325, followed by several in vitro manipulations to shorten the hybrid plasmids and to add both more useful antibiotic markers and several unique restriction cloning sites (Buzby et al., 1983). Interestingly, it was noticed that the transformation efficiency was considerably increased (100- to 1000-fold) by the use either of multimeric forms of the biphasic plasmids or of plasmid DNA extracted from *Synechococcus* PCC 7002 transformants, or of DNA in which the *Ava*I recognition site was deleted. This clearly indicated that the *Aqu*I restriction endonuclease (isoschizomer of *Ava*I) present in *Synechococcus* PCC 7002 strongly affects plasmid transformation. In an attempt to circumvent this problem, ectopic mutagenesis was performed and mutants with enhanced transformation efficiency by *Aqu*I-sensitive plasmids were selected. Mutations in the modification enzyme system rather than in its restriction counterpart had most likely occurred (R.D. Porter; personal communication). The same research group has pushed their vector construction a step further by introducing the *lac* operon into one of their shuttle vectors. The resulting pAQE17DL vector expressed the β-galactosidase gene even in the cyanobacterial recipient cells; this opens the possibility of expressing other foreign gene products under the control of the *lac* promoter in this photosynthetic prokaryote. Moreover, Porter's group has recently constructed plasmid vectors bearing a gene fusion between the promoter for the phycocyanin apoprotein subunit genes (*cpc*) from *Synechococcus* PCC 7002 and a promoterless *lacZ* gene from *E. coli*. This construction permitted to test in vivo the gene expression, from a cyanobacterial promoter, by means of β-galactosidase assays and showed that, in this cyanobacterium, the *cpc* promoter was several times more efficient than the *lacZ* promoter (Porter et al., 1985).

Two different groups have chosen the smallest plasmid of *Synechocystis* PCC 6803, termed either pUG1 or pSS2 (Table I), to establish a gene-cloning system for this cyanobacterium which like *Synechococcus* PCC 7002 can grow heterotrophically. Chauvat et al. (1985) have constructed the shuttle plasmids, pUF12 and pUF311, which consist of a part of the endogenous plasmid pUG1 ligated to the *E. coli* plasmid pACYC177. When compared to pUF12, the plasmid pUF311 shows a better expression of the Km resistance gene and is stably maintained in the *Synechocystis* PCC 6803 cells even without antibiotic selection. Consequently, pUF311 was used for the further construction of the shuttle vector pFCLV7 which contains, in addition, the Cm resistance gene from the *E. coli* plasmid pACYC184. This vector possesses convenient single cloning sites located either in the Km or the Cm resistance genes. However, the transformation efficiency of *Synechocystis* PCC 6803 cells by pFCLV7 is low (3–$6 \cdot 10^{-7}$ on a per cell basis). Unexpectedly, this efficiency is enhanced about 100-fold when the vector pFCLV7 is transferred into *Synechocystis* PCC 6803 cells already containing the hybrid plasmid pUF311 as a resident plasmid. Shestakov et al. (1985) constructed the recombinant plasmid pSE7 (Cm^R) using the *E. coli* plasmid pACYC184 and a part of the *Synechocystis* PCC 6803 small plasmid, pSS2. From the cyanobacterial Cm^R transformants, plasmid pSE76 was obtained which resulted from an in vivo recombination between pSE7 and the resident plasmid pSS2. The vector pSE76 can replicate autonomously in both *E. coli* and *Synechocystis* PCC 6803, and still carries the Cm resistance marker.

Wolk et al. (1984) reported the construction of a family of shuttle vectors able to replicate in *E. coli*, as well as in different filamentous cyanobacteria of *Anabaena*/*Nostoc* species, via conjugative transfer. This genetic exchange system was developed on the basis that RP4, a bacterial conjugative plasmid, could promote the transfer of pBR322 derivatives if, in addition, transacting factors were provided by *E. coli* plasmids, such as pGJ28 or pDS4101. For this purpose, hybrid vectors were constructed in vitro. These vectors contained the replication origin of pBR322 and its *bom* (basis for mobility) region left intact, plus the plasmid pDU1 from *Nostoc* PCC 7524 (a replicon functional in *Anabaena*) and one or two different antibiotic selectable markers of various bacterial origins. Selective removal of the recognition sites corresponding to restriction endonucleases (*Ava*I and *Ava*II) present in the cyanobacterial host cells was also carried out to minimize possible degradation of the incoming DNA. These shuttle plasmid vectors (pRL family) can be transferred very efficiently by conjugation from *E. coli* to several strains of *Anabaena*, as well as to other filamentous cyanobacterial strains (Flores and Wolk, 1985a), and are stably maintained in the recipient cells. Interestingly, Wolk et al. (1985) have recently constructed pRL derivatives which contain promoterless DNA fragments encoding *E. coli* β-galactosidase, β-lactamase (from pBR322) or luciferase (from *Vibrio fischerii*). All three genes are well expressed in *Anabaena* and therefore offer potential means to monitor transcription in cyanobacterial cells either using chromogenic substrates or bioluminescence of the luciferase.

Since the pRL vectors cannot replicate in the unicellular cyanobacterium *Synechococcus* PCC 7942, Wolk et al. (1984, 1985) developed a mobilizable derivative of the shuttle vector pSG111 (Golden and Sherman, 1983), which contains a *Synechococcus* PCC 7942 replicon. This vector (pSG111M) can be transferred to *Synechococcus* PCC 7942 by conjugation and replicate in this host. Thanks to Cobley (1985), a similar conjugal vector was recently created for the chromatically adapting strain *Calothrix* PCC 7601.

Hybrid plasmids have been constructed by fusion of *E. coli* replicons and endogenous plasmids from *Synechococcus* PCC 6301 (Shinozaki et al., 1982), *Nostoc* PCC 6705 and *Plectonema* PCC 6306 (Lambert et al., 1984). The physical maps of the recombinant plasmids have been established, but their ability to behave as shuttle vectors has not yet been tested.

4. Gene transfer

Reviews on genetic transfer systems in cyanobacteria have been published recently (Herdman, 1982; Kück and Mulligan, 1984; Porter, 1985). We will recall here the major conclusions which can be drawn from the collective results obtained so far, and emphasize the aspects which appear to be the most important for gene cloning. Two major genetic transfer systems have been developed for cyanobacteria: transformation by purified chromosomal or plasmid DNA, described for a few numbers of unicellular strains (for reviews see Herdman, 1982; Porter, 1985) and, more recently, conjugation for some filamentous strains (Flores and Wolk, 1985a; Wolk et al., 1984), as well as for the unicellular strain *Synechococcus* PCC 6301 (Reichelt and Delaney, personal communication).

4.1. Transformation

On the basis of the data so far obtained, transformation would appear to be a 'natural' process which requires no particular chemical treatment. There is one exception, however, for *Synechococcus* PCC 6308 (Devilly and Houghton, 1977) whose cells, surrounded by a slime layer, could be transformed only after treatment with $CaCl_2$ in a manner analogous to that used for *E. coli*. Optimal conditions were investigated to improve transformation efficiencies of *Synechococcus* PCC 7942 (Chauvat et al., 1983; Golden and Sherman, 1984) using chromosomal or plasmid DNAs. The best transformation efficiencies obtained range from 10^6 to 10^7 transformants per μg of DNA.

Whether there exists a competence state for cyanobacterial cells remains unclear. According to Chauvat et al. (1983), maximal efficiency was obtained for the transformation of *Synechococcus* PCC 7942 cells with hybrid plasmids pUC12 and pUC13, at the transition between the first and the second exponential growth phases. This is consistent with previous observations on transformation of *Synechococcus* PCC 6301 (Orkwiszewski and Kaney, 1974) and *Synechocystis* PCC 6714 (Astier and Espardellier, 1976) with chromosomal DNA. In contrast to observations by Chauvat et al. (1983), Golden and Sherman (1984), as well as Williams and Szalay (1983), did not detect any stringent period of competence during the various growth phases of the same cyanobacterium. Similarly, Stevens and Porter (1980) obtained good transformation efficiency with chromosomal DNA throughout the exponential growth phase of *Synechococcus* PCC 7002 without distinct peaks of competence being observed. Since Golden and Sherman (1984) noticed some variations in transformability between different cellular clones of *Synechococcus* PCC 7942, the different results obtained with this organism could possibly be due to the origin of the isolates and/or to the culture conditions used for its maintenance in different laboratories.

While necessary during the preculture of the *Synechococcus* PCC 7942 cells, light does not seem to be required during the incubation period in the presence of DNA, which, depending on the authors, varied from 15 min to 6 h (Chauvat et al., 1983; Golden and Sherman, 1984; Williams and Szalay, 1983). According to Golden and Sherman (1984), a higher transformation frequency was in fact obtained when the incubation was performed in the dark or in the presence of photosynthesis inhibitors (DCMU or CCCP). Most interestingly, these same authors showed that iron starvation during cell growth prior to transformation significantly enhanced the efficiency of DNA uptake. Since alteration occurs in the photosynthetic apparatus under such growth conditions, this observation could be applied to study the expression of cloned genes coding for membrane proteins of the photosynthetic apparatus which are not rapidly turned over. DNA concentrations of hybrid plasmids generally used varied from 35 ng to 1 μg per 10^8–10^9 cells and, under these conditions, transformation efficiency seemed to depend more on the nature of the selective marker used than on plasmid size (Golden and Sherman, 1984). This is probably related to the level of expression of the antibiotic resistance marker in the recipient cells. Streptomycin and chloramphenicol appear to be better selective agents than ampicillin which confers very low resistance level in *Synechococcus* PCC 7942 (0.5 to 2 μg/ml versus at least 50 μg/ml in *E. coli*) (Kuhlemeier et al., 1981 and 1983).

Although it had been shown that expression of selectable markers occurred in liquid medium when cells were transformed by chromosomal DNA (Astier and Espardellier, 1976; Herdman, 1973), better results with hybrid plasmids were obtained when expression of the transferred genes and selection of transformants were performed on solid medium with progressive exposure of the cells to increasing concentrations of the selective agents. Different procedures have been devised for this purpose, such as the addition of antibiotics either underneath the agar (Chauvat et al., 1983; Van den Hondel et al., 1980), or within a soft agar overlayer (R.D. Porter, personal communication), or growth of cells on a membrane filter followed by their transfer onto an antibiotic containing agar plate (Gendel et al., 1983a; Williams and Szalay, 1983).

As in other bacteria such as *Streptococcus pneumoniae* (Saunders and Guild, 1981a,b) and *Bacillus subtilis* (Canosi et al., 1978; Mottes et al., 1979), oligomerization is possibly an important step in the mechanism of DNA uptake during transformation. Indeed, Buzby et al. (1983) showed that multimeric forms of the hybrid plasmids of the pAQE series were more efficient at transforming *Synechococcus* PCC 7002 cells than the corresponding monomeric forms. Clearly, a more detailed study is needed to establish the transformation mechanism in cyanobacterial cells.

Detailed studies performed by Williams and Szalay (1983) have established the conditions under which stable integration of foreign DNA into the chromosome of *Synechococcus* PCC 7942 could be achieved by transformation, using donor molecules devoid of a replication origin which would be functional in the recipient cells. This provides a powerful means of modifying precisely and extensively the cyanobacterial genome in order to study, in this organism, various processes such as for example, oxygenic photosynthesis. Different hybrid circular molecules were constructed by combining a *Synechococcus* PCC 7942 chromosomal DNA fragment with the vector pBR322. A chloramphenicol resistance gene from *E. coli* was inserted within the cyanobacterial DNA or linked to it as a flanking sequence. Their results clearly showed that in order to obtain efficient and stable integration of these chimeric DNA molecules into *Synechococcus* PCC 7942, the best configuration for the incoming DNA was one in which the foreign DNA was inserted within the cyanobacterial DNA. A second requirement was the use of the genetic marker carried by the inserted DNA, as unique selective agent. Under such conditions, integration occurred via a non-reciprocal exchange between plasmid and chromosome, resulting in the replacement of chromosomal DNA by the homologous sequence containing the foreign insert and the concomitant loss of the vector part of the donor molecule. Subsequently Kolowsky et al. (1984) used similar donor molecules that contained different lengths of flanking or inserted cyanobacterial DNA. They showed that the probability of DNA integration from the chimeric molecule decreased by half for each 2 to 3 Kb increase in length of the foreign DNA insert. Nonetheless, pieces of foreign DNA as large as 20 Kb could be stably maintained in the chromosome of *Synechococcus* PCC 7942 when it had been integrated by the replacement mechanism. According to Williams and Szalay (1983), similar transformation efficiencies were obtained when the chimeric donor molecules were linearized, provided that the foreign DNA was linked at both ends to cyanobacterial DNA. Consequently, these authors assumed that some similarities might exist between the transformation process in *Synechococcus* PCC 7942 and in the Gram-positive *Bacillus subtilis*, since in the latter, donor DNA molecules are linearized concomitant with uptake.

4.2. Conjugation

Only one example of transfer of broad host-range conjugative plasmids into a cyanobacterium has been described so far. Reichelt and Delaney (personal communication) have demonstrated that plasmids from the P1 incompatibility group (RP1, RP4, R68 and R68.45) can be transferred, although at a very low frequency (10^{-6} to 10^{-10}), from *E. coli* to the unicellular strain *Synechococcus* 6301. By restriction and hybridization analysis of plasmids isolated from transconjugants, these authors have shown that, in most of them, free R68.45 plasmid was present together with a cointegrate of this R-plasmid and the 48.8 Kb endogenous plasmid pUH2; in some others, the R-plasmid was found to be integrated in the chromosomal DNA. In both cases the integration mechanism was mediated via transposable elements from the R-plasmid.

An important breakthrough was made by Wolk and his co-workers who have set up a conjugative transfer system for cyanobacteria that was first developed for *Anabaena* strains (Wolk et al., 1984). In their system, pRL vectors (see Table IV) were first used to transform the *E. coli* strain HB101 containing a helper plasmid (either pGJ28 or pDS4101). Then triparental matings were performed with (a) *E. coli* HB101 containing pRL vectors and the helper plasmid, (b) *E. coli* J53 harboring the RP4 plasmid, and (c) *Anabaena* strains. Cyanobacterial exconjugants were selected on solid medium containing antibiotic. These cells were freed of *E. coli* contamination by streaking on cyanobacterial mineral medium containing the selective agent. Plasmids pRL were recovered from exconjugants without any alteration in their restriction patterns and could be re-transferred into *E. coli* cells. The success of this system was dependent on two major factors: the absence (or elimination), in the plasmid vector, of the recognition sites corresponding to the restriction endonucleases synthesized by the cyanobacterial strain and the ability of the cyanobacterial replicon to function in the recipient cells. Three points should be emphasized in this study. Firstly, this is the only gene transfer system currently available for filamentous cyanobacteria; secondly, this conjugative system devised for gene cloning purposes via pRL shuttle vectors allows gene transfer at a very high frequency ($\sim 10^{-3}$); thirdly, its use has been recently extended to other filamentous strains belonging to the *Nostoc* genus (Flores and Wolk, 1985a) and could be applied to other cyanobacterial strains provided that the cyanobacterial replicon of the conjugative plasmid is functional in the cyanobacterial strain under study. Indeed, such conjugal vectors have now been developed for *Synechococcus* PCC 7942 (Wolk et al., 1985) and *Calothrix* PCC 7601 (Cobley, 1985). Obviously, this conjugative system is very promising for the development of molecular genetic studies in cyanobacteria and will be particularly helpful for the study of processes such as cell differentiation, nitrogen fixation and complementary chromatic adaptation which are encountered in filamentous and unicellular strains for which no other gene transfer system is yet available.

5. Molecular cloning of cyanobacterial genes

Gene cloning by selection for phenotypic complementation of a mutant is, in principle, very simple and straightforward. However, since the transfer of genetic material into cyanobacteria still poses a major problem, this kind of strategy has been scarcely employed. Cyanobacterial genes have been cloned mainly by using heterologous DNA probes to screen genomic libraries, and more recently by using synthetic oligonucleotides, the sequence of the latter being derived from the amino acid sequence of a part of the corresponding protein. Approximately seventy cyanobacterial genes have been cloned so far. They are listed in Table V according to the organism from which they originate and will be discussed below in relation to the various methods developed for their isolation.

5.1. Direct cloning using mutants

With the construction of specific cloning vehicles and the simultaneous obtainment of stable mutants by means of transposons (W.E. Borrias, personal communication), the direct cloning of genes by phenotypic complementation becomes feasible for *Synechococcus* PCC 7942. Indeed, four genes have been directly isolated from genomic libraries constructed in shuttle vectors: a *met* gene (Tandeau de Marsac et al., 1982) and three genes involved in nitrate assimilation, *narA, narB* and *narC* (Kuhlemeier et al., 1984a,b). The *met* and *narB* genes were cloned using Tn*901*-induced mutants. The ampicillin resistance encoded by the transposon was used to select the DNA fragment corresponding to the inactivated gene which in turn allowed the cloning of the wild-type gene (Tandeau de Marsac et al., 1982; Kuhlemeier et al., 1984a). The two other *nar* genes were picked up by screening for direct complementation of *Synechococcus* PCC 7942 *nar* mutants (Kuhlemeier et al., 1984a,b). All these cloning experiments were performed in recombination-proficient hosts. Importantly, Kuhlemeier et al. (1984a) noticed that for stable maintenance of a cloned gene in a plasmid vector, double selection for both the vector and the gene to be cloned was needed. Integration of the cloned DNA into the chromosome via an allele replacement mechanism has indeed been observed in most cases (Williams and Szalay, 1983).

Since cyanobacteria are Gram-negative prokaryotes, direct cloning by phenotypic complementation of well-characterized *E. coli* mutants offers, a priori, an attractive alternative strategy. This strategy implies that the protein encoded by the cloned gene is synthesized in the heterologous host and is able to fulfill the function(s) otherwise undertaken by the wild-type protein. Very recently, Kodaki et al. (1985) succeeded in cloning, from a *Synechococcus* strain, a gene which encodes phosphoenol pyruvate carboxylase (*ppc*). This enzyme supplies oxaloacetate to the tricarboxylic acid cycle but, in addition, is involved in the initial reaction of photosynthetic CO_2 assimilation at least in C_4 plants. Chromosomal *Sau*3A DNA fragments of *Synechococcus* ligated into the *Bam*HI site of pBR322 were used to transform an *E. coli ppc* mutant. Five transformants out of 12 000 Ap^R colonies showed a wild-type phenotype. All five hybrid plasmids had in common an identical 5Kb DNA fragment which was always inserted with the same orientation. This last observation suggests that the *Synechococcus ppc* gene was expressed under the control of a pBR322 promoter in these transformants. Porter et al. (1985) also succeeded in cloning *argE* and *leuB* genes from *Synechococcus*

TABLE V

Cloned cyanobacterial genes

Organism	Cloned gene	Gene product	Isolation procedure	Sequence	Reference
Synechococcus PCC 6301 (*Anacystis nidulans*)	*rbcL*	ribulose bisphosphate carboxylase large subunit	hybridization with spinach DNA probe	yes	Reichelt and Delaney, 1983 Shinozaki et al., 1983
	rbcS	ribulose bisphosphate carboxylase small subunit	hybridization with pea cDNA probe	yes	Shinozaki and Sugiura, 1983
	rrnA	5S, 16S, 23SrRNAs,$tRNA^{Ile}$ and $tRNA^{Ala}$	rRNA-DNA hybridization	yes	Tomioka et al., 1981 Tomioka and Sugiura, 1983 Kumano et al., 1983 Tomioka and Sugiura, 1984
	rrnB	ibid.	rRNA-DNA hybridization	yes	Williamson and Doolittle, 1983 Douglas and Doolittle, 1984a,b
	apcA,apcB	allophycocyanin α and β subunits	hybridization with cyanelle DNA probe	yes	Houmard et al., 1985
	apcC	7.8kD linker polypeptide located in the core of the phycobilisome	found downstream of *apcB* by sequencing	yes	ibid.
	ppc	phosphoenolpyruvate carboxylase	complementation of *E. coli* *ppc* mutation	yes	Kodaki et al., 1985 Katagiri et al., 1985
	cpcB	phycocyanin β subunit	hybridization with synthetic specific oligonucleotides	yes	Lind et al., 1985 Kalla et al., 1985
Synechococcus PCC 7942 (*Anacystis nidulans* R-2)	*met*	methionine biosynthesis	use of Tn*901*-induced mutation	no	Tandeau de Marsac et al., 1982
	narB	involved in nitrate reduction	use of Tn*901*-induced mutation	no	Kuhlemeier et al., 1984a
	narA	ibid.	complementation of *Synechococcus* *nar* mutation	no	Kuhlemeier et al., 1984a
	narC	ibid.	ibid.	no	Kuhlemeier et al., 1984b
	psbAI, AII, AIII	32kD thylakoid membrane protein	hybridization with spinach chloroplast DNA probe	yes	S.S. Golden et al., 1985
Synechococcus PCC7002 (*Agmenellum quadruplicatum* PR6)	*apcA, apcB*	allophycocyanin α and β subunits	hybridization with cyanelle DNA probe	yes	D.A. Bryant (personal communication)
	cpcA, cpcB	phycocyanin α and β subunits	hybridization with synthetic specific oligonucleotides	yes	de Lorimier et al., 1984 Pilot and Fox, 1984

TABLE V *(continued)*

Organism	Cloned gene	Gene product	Isolation procedure	Sequence	Reference
	cpcC, *cpcD*	33kD and 9kD linker polypeptides located in the rods of phycobilisomes	found downstream of *cpcA* by sequencing	yes	Bryant et al., 1985b
	argE	acetylornithinase	complementation of *E. coli argE* mutation	no	Porter et al., 1985
	leuB	isopropylmalate dehydrogenase	complementation of *E. coli leuB* mutation	no	ibid.
	psaA, *psaB*	presumed chlorophyll a-binding apoproteins of the PSI/P700 complex	hybridization with maize DNA probes	no	D.A. Bryant (personal communication)
	recA	DNA repair enzyme	hybridization with *E. coli* DNA probe	no	ibid.
	rbcL, *rbcS*	ribulose bisphosphate carboxylase large and small subunits	hybridizaton with cyanelle DNA probe	no	S.E. Stevens, Jr. (personal communication)
Synechocystis PCC 6803 (*Aphanocapsa* sp.)	*psbB*	47kD chlorophyll binding protein	ibid	yes	Vermaas et al., 1985
	psbC	43kD chlorophyll binding protein	hybridization with plastid DNA probe	?	Williams, 1985
	psbD	D2 protein	ibid	?	ibid.
	psbE	cytochrome b_{559}	hybridization with spinach chloroplast DNA probe	yes	Pakrasi et al., 1985
Spirulina platensis C1 (G. Florenzano)	*rbcL*, *rbcS*	ribulose bisphosphate carboxylase large and small subunits	hybridization with *Chlamydomonas* DNA probe	no	Tiboni et al., 1984
	tuf, *fus*	EF-Tu, EF-G elongation factors	hybridization with *E. coli* DNA probes	no	Tiboni and Ciferri, 1985
Pseudanabaena PCC 7409 (*L.P.P.*)	*apcA*, *apcB*	allophycocyanin α and β subunits	hybridization with cyanelle DNA probe	no	D.A. Bryant (personal communication)
	cpcAI, *cpcBI*	phycocyanin (PC-1) α and β subunits present in both green and red light	hybridization with *Synechococcus* PCC 7002 DNA probe	no	Bryant et al., 1985b
	cpcAII,*cpcBII*	phycocyanin (PC-2) α and β subunits only present in red light	ibid.	no	ibid.
	cpeA, *cpeB*	phycoerythrin α and β subunits	ibid.	partly	ibid.

TABLE V *(continued)*

Organism	Cloned gene	Gene product	Isolation procedure	Sequence	Reference
Anabaena PCC 7120 (*Nostoc muscorum* ISU)	*nif H*	nitrogenase reductase	hybridization with *K. pneumoniae* DNA probe	yes	Rice et al., 1982 Mevarech et al., 1980
	*nif H**	ibid.	ibid.	yes	Robinson et Haselkorn, 1985
	nif D	nitrogenase α subunit	ibid.	yes	Rice et al., 1982 Lammers and Haselkorn, 1983
	nif K	nitrogenase β subunit	ibid.	yes	Rice et al., 1982 Mazur and Chui, 1982
	nif S	involved in the maturation of the nitrogenase reductase	ibid.	yes	Rice et al., 1982 Haselkorn, 1985
	glnA	glutamine synthetase	hybridization with *E. coli* DNA probe	yes	Fisher et al., 1981 Tumer et al., 1983
	rbcL	ribulose bisphosphate carboxylase large subunit	hybridization with corn and *Chlamydomonas* DNA probe	yes	Curtis and Haselkorn, 1983
	rbcS	ribulose bisphosphate carboxylase small subunit	hybridization with pea cDNA probe	yes	Nierzwicki-Bauer et al., 1984
	psbAI, AII	32kD thylakoid membrane protein	hybridization with spinach chloroplast DNA probe	yes	Curtis and Haselkorn, 1984
	als	acetolactate synthase	hybridization with yeast DNA probe	no	B.J. Mazur (personal communication)
	xisA	excisase	located between *nif D* and *nif K* by expression in *E. coli*	yes	J.W.Golden et al., 1985b
Calothrix PCC 7601 (*Fremyella diplosiphon*)	*psbA* (*ps2BI, II*)	32kD thylakoid membrane protein	hybridization with maize plastid DNA probe	yes	Mulligan et al., 1984
	gvpA, gvpB	gas vesicle proteins	hybridization with specific synthetic oligonucleotide	yes	Tandeau de Marsac et al., 1985a,b
	apcA, apcB	allophycocyanin α and β subunits	hybridization with cyanelle DNA probe	partly	Grossman et al., 1985
	cpcAI, cpcBI	phycocyanin (PC-1) α and β subunits present in both green and red light	ibid.	partly	Conley et al., 1985
	cpcAII, cpcBII	phycocyanin (PC-2) α and β subunits only present in red light	ibid.	partly	ibid.

TABLE V *(continued)*

Organism	Cloned gene	Gene product	Isolation procedure	Sequence	Reference
	cpcCII	PC-2 associated linker polypeptide only present in red light	hybridization with specific synthetic oligonucleotide	partly	Lomax et al., 1985
	thrB	homoserine kinase	complementation of *E. coli* *thrB* mutation	yes	Our laboratory (unpublished data)

PCC 7002 by their ability to complement corresponding *E. coli* auxotrophic mutants. However, they were unable to complement other *E. coli* mutations, such as *argB6*, *metBI* and *trpA* for example, as well as *metB*, *trpC*, *aroB*, *leuB*, *hisH* and *tyrB B. subtilis* mutations. Similarly, we have isolated a recombinant plasmid which carries the *Calothrix* PCC 7601 *thrB* gene since it is able to restore prototrophy in an *E. coli thrB* mutant (our laboratory, unpublished data).

5.2. Use of heterologous hybridization

Cyanobacteria are prokaryotes which, like higher plants and algae, evolve oxygen during photosynthesis. Consequently, heterologous DNA probes of both prokaryotic and eukaryotic origin have been used to clone cyanobacterial genes.

Although being oxygen-evolving photosynthetic organisms, many cyanobacteria are able to fix nitrogen (see Chapter 7). Biochemical analyses have shown, however, that the enzymes of their nitrogenase complexes are not less oxygen-sensitive than those of the other diazotrophs. In filamentous cyanobacteria heterocysts differentiate along the filaments of vegetative cells when cultures are deprived of combined nitrogen under aerobic conditions. Heterocysts provide an anaerobic environment and are the exclusive site of nitrogen fixation in aerobic cultures. Such a differentiation process requires a large number of biochemical and morphological changes related to both nitrogen metabolism and photosynthetic functions. By cloning representative genes, the mechanisms involved in the sequential repression and activation of various sets of genes in the filamentous heterocystous *Anabaena* PCC 7120 can be elucidated (Haselkorn et al., 1985). These genes were selected from a genomic library, using prokaryotic DNA probes for the genes involved in nitrogen metabolism (*nifH*, *nifD*, *nifK*, *nifS* and *glnA*), and eukaryotic ones for the genes related to photosynthesis (*rbcL*, *rbcS*, *psbAI* and *psbAII*). In addition, the gene-encoding acetolactate synthase (ALS) which, like the 32 kD thylakoid membrane protein (*psbA* gene), is a target for some herbicides, was recently identified in an *Anabaena* PCC 7120 genomic library using a yeast *als* probe (B. Mazur, personal communication).

Among the 17 genes which constitute the *nif* cluster of *Klebsiella pneumoniae*, only four (*nifH*, *D*, *K*, and *S*) have been found to share, with the *Anabaena nif* genes, sequence homologies sufficient to allow their isolation by heterologous hybridization. The *nifH*, *D* and *K* genes encode the structural components of the nitrogenase complex while the *nifS* gene is one of the genes required in the maturation of the nitrogenase reductase. None of the other genes, which in *K. pneumoniae* are involved either in protein maturation or specific electron transfer and regulation of the expression of the *nif* operons, gave hybridization with *Anabaena* PCC 7120 DNA. A peculiar rearrangement of *nif* genes occurs in *Anabaena* PCC 7120 heterocysts (Golden et al., 1985a). The 11 kb DNA fragment which exists between *nifD* and *nifK* in the chromosome of vegetative cells is excised during heterocyst differentiation leading to the formation of a *nifHDK* operon. The *xisA* gene, encoding the excisase involved in this rearrangement, has been located on the 11 kb DNA fragment adjacent to *nifK*.

In higher plants, the large subunit of ribulose 1,5-bisphosphate (RuBP) carboxylase (L) is coded for and synthesized in the chloroplast, whereas the small subunit (S) is encoded by nuclear genes and transported across the chloroplast membrane. Consequently, chloroplast DNA probes were used to clone the *rbcL* gene while cDNA probes were used to clone the *rbcS* gene (see Table V for references). In addition to those of *Anabaena* PCC 7120, the *rbcL* and *rbcS* genes have been isolated from a filamentous non-heterocystous strain, *Spirulina platensis*, and two unicellular ones, *Synechococcus* PCC 6301 and PCC 7002 (see Sections 4.1. and 4.2. of Chapter 4). At the protein level, the *Anabaena* PCC 7120 small subunit of RuBP carboxylase (*rbcS*) is 61% homologous to that of *Syne-*

chococcus PCC 6301 and about 40% homologous to the eukaryotic small-subunit proteins. In contrast, the *Anabaena* PCC 7120 large-subunit protein (*rbcL*) is about 80% homologous to both *Synechococcus* PCC 6301 and the eukaryotic large-subunit proteins. Thus, the latter, which are encoded by chloroplast genes, are more homologous to their cyanobacterial counterparts than the small subunits, encoded by nuclear genes. Mulligan et al. (1984) have reported the cloning and sequencing of a gene for the B protein of Photosystem II (*ps2B-1* = *psbA* gene) from a filamentous heterocystous strain, *Calothrix* PCC 7601. This gene is about 80% homologous to the corresponding spinach chloroplast gene, as are the *Anabaena* PCC 7120 genes (*psbAI* and *II*). In *Synechococcus* PCC 7942, three copies of the *psbA* gene have been identified and cloned by S.S. Golden et al. (1985) (see Section 2 of Chapter 1). Finally, plastid DNA probes have also allowed the isolation of genes the products of which are part of the Photosystem I and II reaction centers: *psaA* and *psaB* from *Synechococcus* PCC 7002 (D.A. Bryant, personal communication), *psbC*, *psbD* and *psbE* from *Synechocystis* PCC 6803 (Williams, 1985; Pakrasi et al., 1985).

Other cyanobacterial genes involved in photosynthesis have been cloned using heterologous DNA probes, initially obtained from *Cyanophora paradoxa*. This cryptomonad harbors organelles, termed cyanelles, which contain a photosynthetic apparatus closely resembling that of both cyanobacteria and red algae (see Chapter 11). In particular, they possess, as major light-harvesting pigments, phycobiliproteins (mainly phycocyanin and allophycocyanin) (see Chapter 3). These proteins are encoded by the cyanelle genome which is at least 20 times smaller than cyanobacterial genomes. The cyanelle *apcA* and *apcB* genes, encoding the α and β allophycocyanin subunits respectively, have recently been cloned (Bryant et al., 1985a). They have been used to select the corresponding genes from two unicellular strains, *Synechococcus* PCC 6301 and PCC 7002, and from a filamentous one, *Pseudanabaena* PCC 7409, which is a chromatic adapter. In addition, by DNA sequencing, the *apcC* gene from *Synechococcus* PCC 6301 has been found downstream of the *apcB* gene (Houmard et al., 1985). This gene encodes a 7.8 kD linker polypeptide which is associated with the allophycocyanin subunits (*apcA* and *apcB* genes) in the core of the phycobilisomes (Glazer, 1985). In this strain, these three genes are thus located in the following order, *apcA*, *apcB* and *apcC*. Grossman et al. (1985) have also identified the *apcA*, *apcB*, *cpcAI*, *cpcBI*, *cpcAII* and *cpcBII* genes of the chromatic adapter, *Calothrix* PCC 7601, by using the *cpc* genes from cyanelles as probes. Similarly, a *psbB* gene (encoding a 47 kD chlorophyll binding protein) has been cloned from *Synechocystis* PCC 6803 using the corresponding cyanelle gene (Vermaas et al., 1985). Finally, as will be discussed below, De Lorimier et al. (1984) have been able to isolate the *Synechococcus* PCC 7002 *cpcA* and *cpcB* genes. These genes were subsequently used to pick up the *cpcAI*, *cpcBI*, *cpcAII*, *cpcBII*, *cpeA* and *cpeB* genes, namely those encoding the α and β subunits of both types of phycocyanin (PC-1 and PC-2) and of phycoerythrin, from *Pseudanabaena* PCC 7409 (Bryant et al., 1985b).

5.3. Screening with synthetic oligonucleotides

The recent development of oligodeoxyribonucleotide synthesis offers new possibilities for the direct cloning of genes whose products are at least partially sequenced (Itakura et al., 1984). Amino acid sequence data allow corresponding oligonucleotide probes to be synthesized and used to screen genomic libraries. Since the genetic code is degenerated, either multiple short oligomers (14- to 17-mers) corresponding to all possible base combinations are synthesized or a single longer oligonucleotide (25 bases or more) is constructed on the basis of the sequence expected to be the most probable from codon usage considerations.

Genes for the α and β subunits of phycocyanin (PC), a major phycobiliprotein found in all cyanobacteria, have been recently isolated from *Synechococcus* PCC 7002 using synthetic oligonucleotides

(De Lorimier et al., 1984; Pilot and Fox, 1984). De Lorimier et al. (1984) synthesized 8 of the 16 possible tetradecamers corresponding to PC α apoprotein (amino acid residues 107–111). Since PC ceases to be synthesized in cyanobacteria during nitrogen starvation but reappears after nutrient replenishment, they assumed that PC mRNA would be proportionally enriched in N_2 repleted cells. Out of the eight unique tetradecameric probes separately hybridized to such total RNA, one gave a stronger signal than the others and was assumed to be the best candidate for a PCα specific probe. The screening of a genomic library constructed in pBR325, with this selected tetradecamer, allowed the authors to isolate recombinant plasmids which carried the *cpcA* gene (phycocyanin α subunit). Maxicell experiments and sequencing data have shown that the cloned DNA fragment directs the synthesis of both PCα and PCβ apoproteins in *E. coli* (Bryant et al., 1985c). Adjacent on the chromosome, these authors have found the *cpcB* gene (phycocyanin β subunit) upstream of *cpcA*, and downstream from it, the *cpcC* and *cpcD* genes which encode respectively the 33 kD and 9 kD linker polypeptides associated with the phycocyanin in the rods of phycobilisomes. At the same time, Pilot and Fox (1984) used two mixtures of oligonucleotides, each containing 12 tetradecamers to screen a genomic library, constructed in pUC9, of the same *Synechococcus* PCC 7002. The sequence of the first probe was derived from the sequence of the five NH_2-terminal amino acids of the PC β subunit, and that of the second from the sequence of residues 87–91 of the same polypeptide. The use of two sets of oligonucleotides corresponding to distant regions of the same protein allows the elimination of false-positives which result from non-specific hybridization. These authors were able to clone a DNA fragment bearing the *cpcB* gene and found downstream from it, the *cpcA* gene. Similarly, using two sets of oligonucleotides the *cpcB* gene of *Synechococcus* PCC 6301 has been cloned (Lind et al., 1985). On the other hand, the *cpcII* gene, which encodes a linker polypeptide associated with the rods of the phycobilisomes in red light grown *Calothrix* PCC 7601 cells, has been identified by combining immunological and oligonucleotide screening (Lomax et al., 1985).

Many different aquatic prokaryotes contain gas-vacuoles which are gas-filled structures providing the cells with buoyancy and allowing their positioning at the most favorable depth for growth (see Chapter 14). Gas-vacuolate prokaryotes are most commonly planktonic and synthesize gas-vacuoles constitutively (Walsby, 1981). However, there exist nonplanktonic filamentous cyanobacteria in which gas vacuole formation occurs at a specific stage in their developmental cell cycle, and is restricted to specialized filaments called hormogonia which are involved in the dispersal of the species in their natural habitat. Changes in environmental factors, such as light intensity and/or wavelength, have been shown to induce formation of gas-vacuolate hormogonia (Walsby, 1981; Tandeau de Marsac, 1983; Rippka and Herdman, 1985; G. Guglielmi, personal communication). A gas-vacuole protein (GVP) was partially sequenced from different cyanobacteria and found to be highly homologous (Walker et al., 1984). Based on the partial sequence from *Anabaena flos-aquae* GVP, a unique oligonucleotide (29-mer) was synthesized which corresponded to the most probable sequence as deduced from the codon usage found for the *Anabaena* PCC 7120 and *Calothrix* PCC 7601 genes already sequenced (Table V). A lambda genomic library of the filamentous *Calothrix* PCC 7601 was screened with this 29-mer probe and recombinant phages selected (Tandeau de Marsac et al., 1985a). The cloned *gvpA* gene encodes a protein of 70 amino acids which is highly homologous to the *Anabaena flos-aquae* GVP. 25 out of the 29 nucleotides of the probe perfectly matched the corresponding *gvpA* base sequence. Codon usage was thus as expected for most of the amino acids chosen to construct the synthetic oligonucleotide. Interestingly, a second gene, *gvpB*, was found 104 base pairs (bp) downstream from the *gvpA* gene (Tandeau de Marsac et al., 1985b). At the amino acid level, the two proteins are 100% homologous but nucleotide substitutions occur mostly at the third position for 18 out of the 71 codons. This suggests that the second gene may have appeared by gene duplication followed by subsequent silent mutations or could have been acquired from an other gas-vacuolate organism by intergeneric transfer (see Section 2.1 of Chapter 14).

5.4. Cloning using RNA/DNA hybridization

The endosymbiotic theory implying that eukaryotic chloroplasts were derived from an ancestral photosynthetic prokaryote is becoming widely accepted (Gray and Doolittle, 1982). Evidence supporting this idea came in part from comparison of chloroplasts and *E. coli* rRNAs. However, the comparison between cyanobacterial and plastid RNAs is obviously more pertinent. The rRNA genes (*rrn*) of *Synechococcus* PCC 6301 were cloned by homologous RNA-DNA hybridization. Two gene clusters have been found, *rrnA* and *rrnB*, which share the same organization (Tomioka et al., 1981). The *rrnA* cluster encodes the 16S RNA, $tRNA^{Ile}$ and $tRNA^{Ala}$, the 23S and 5S RNAs, the sequences of which have been fully determined (Tomioka et al., 1981; Kumano et al., 1983; Tomioka and Sugiura, 1983, 1984). Independently, Doolittle and co-workers have sequenced the *rrnB* cluster (Williamson and Doolittle, 1983; Douglas and Doolittle, 1984a,b). The two *rrn* operons are highly conserved; the sequences of the two tRNAs present in the 16S-23S spacer regions are identical and only eight positions differ between the 23S sequences which are 2876 nucleotides long. Interestingly, both the tRNA and rRNA primary and secondary structures of *Synechococcus* PCC 6301 resemble more closely those of chloroplast RNAs than those of their *E. coli* counterparts (Douglas and Doolittle, 1984a; Tomioka and Sugiura, 1983, 1984). These comparisons reinforce the endosymbiont hypothesis and, in addition, show that, except for the last one, the 15 nucleotides present at the 3′-end of the 16S rRNA are the same for *E. coli, Synechococcus* PCC 6301 and tobacco chloroplasts. The transcripts in these three organisms are thus more likely to share the same ribosome binding sites.

6. Regulation of gene expression

Gene cloning and sequencing are the first steps necessary for molecular studies of both genetics and physiology. Until now, about 70 different cyanobacterial genes have been isolated, either from unicellular or filamentous cyanobacteria and, in some instances, the same genes have been cloned from both types (Table V). About half of these genes have been sequenced and, in a few cases, the transcription initiation sites have been mapped and the regulation of gene expression examined in either cyanobacteria or *E. coli*.

Before discussing the regulation of gene expression, it is worth noting that the expression of a given gene into a heterologous host implies that: (a) the gene must have a promoter region which shares structural similarities with that of the genes of the recipient cells (the promoter which is functional in the cyanobacterium, a vector promoter or any sequence susceptible to be recognized by the host transcriptional machinery); (b) the transcript should carry upstream from the translation initiation site a sequence able to bind host ribosomes; and (c) both, the organism from which the gene originates and the host, must have compatible genetic codes and codon usages.

6.1. Features emerging from sequence data

Sequence data are not sufficient to define the portions of the 5′-flanking sequences which regulate the expression of a given gene. This holds particularly true for cyanobacterial genes since very few genes have been sequenced and regulatory regions may be poorly conserved among different species. In addition, since at least some cyanobacteria can undergo cell differentiation and exhibit light-regulated gene expression, particular nucleotide sequences probably exist which govern the expression of the genes involved in these processes. These considerations underline the necessity to pinpoint transcriptional starts for cloned cyanobacterial genes, and the need for a precise quantitation of the variations in mRNA levels under various physiological conditions.

The *Anabaena* PCC 7120 *nifH* promoter was the first cyanobacterial promoter to be well characterized. Both S1-mapping and primer extension studies agreed with a transcript starting 128 bases upstream from the *nifH* gene initiator codon (Haselkorn et al., 1985). Similar studies have been carried out for the *Anabaena* PCC 7120 *nif K* gene. Results show that sequences upstream from the start sites for transcription are conserved in the -10 and -35 regions, defining a *nif* gene promoter for that strain (see Table VI). The -35 and -10 regions thus defined correspond in *E. coli* to very poor promoter sequences (Haselkorn et al., 1985). This observation is perfectly consistent with the requirement for specific transcriptional activators already mentioned for the expression of *nif* genes in other diazotrophs (Beynon et al., 1983). The regulatory regions of *Anabaena* PCC 7120 *glnA* gene, another gene involved in nitrogen metabolism, were also defined. Interestingly, two promoters were found in front of the gene (Tumer et al., 1983). One transcript starts at position -93 (RNA_I) while a second starts at position -155 (RNA_{II}). Under nitrogen-fixing conditions, only RNA_I is present while in ammonia-grown cells both RNA_I and RNA_{II} are found, the major transcript being RNA_{II}. The -10 and -35 regions corresponding to RNA_I are very different from the

TABLE VI

Promoter sequences of cyanobacterial genes[a]

Organism	Gene	'-35 region'	'-10 region'	Reference
Synechococcus PCC 6301 (*Anacystis nidulans*)	*rbcL-rbcS*	TTCAAT -38 to -36	TAGGTT -13 to -11	Shinozaki and Sugiura, 1985
Synechococcus PCC 7002 (*Agmenellum quadruplicatum* PR 6)	*cpcB-cpcA*	TGTTTA -36	TATAAT -13	Pilot and Fox, 1984
		ATGACA -35	TAATGT -12	Bryant et al., 1985c
Anabaena PCC 7120	*nifH-nifD*	CATAAC -41	TCTACT -14	Haselkorn et al., 1985
	glnA '*nif*-like'	CAAAAC -35	TCTACG -13	Haselkorn et al., 1985
	glnA '*E. coli*-like'	TTGTGC -37	TAATAT -13	Haselkorn et al., 1985
	rbcL-rbcS	GAATAA -35	TATATT -12	Haselkorn et al., 1985
	psbAI	n.d.	TAAATT -10	Curtis and Haselkorn, 1984
Escherichia coli	consensus sequence	TTGACA -36 to -34	TATAAT -14 to -11	Rosenberg and Court, 1979

[a] This compilation is restricted to genes for which transcripts have been mapped (S-1 mapping, primer extension, or both). n.d., not determined.

E. coli consensus sequences but resemble those of the *Anabaena nif* promoter, (see Table VI). In contrast, the -10 and -35 regions upstream from the start of RNA_{II} are similar to *E. coli* promoters and indeed allow transcription of the *Anabaena glnA* gene in *E. coli* (Tumer et al., 1983).

The transcription of the genes encoding the two subunits of ribulose 1,5-bisphosphate carboxylase (*rbcL* and *rbcS*; Section 5.2.) has been studied in both the filamentous heterocystous *Anabaena* PCC 7120 and the unicellular *Synechococcus* PCC 6301 (Nierzwicki-Bauer et al., 1984; Haselkorn et al., 1985; Shinozaki and Sugiura, 1985) (see Section 4.1. of Chapter 4). In both organisms co-transcription of the genes was observed, the *rbcS* gene being located downstream of the *rbcL* coding region. In *Synechococcus* PCC 6301, the 5′-end of the mRNA was mapped about 160 bp upstream from the initiator codon of the *rbcL* gene, the 3′-end being about 230 bp downstream from the UAA termination codon of the *rbcS* gene. As shown in Table VI no significant sequence homologies are found between the promoter regions of *Anabaena* PCC 7120 and *Synechococcus* PCC 6301, while sequences in the coding region of the *rbcL* gene are 77% homologous at the nucleotide level and 84% at the amino acid level (Curtis and Haselkorn, 1983). In *Anabaena*, the start point for transcription is at position -504 with respect to the start of the *rbcL* coding region. The corresponding -10 sequence is a good -10 sequence for *E. coli* or *B. subtilis* but that centered around -35 would not be active in *E. coli* unless activated by an auxiliary protein. In vitro experiments showed that *E. coli* RNA polymerase does not transcribe these genes while the *Anabaena* enzyme transcribes them efficiently from the identified promoter (Haselkorn et al., 1985). It thus seems likely that the *rbcL* gene promoter can be transcribed by the vegetative RNA polymerase without specific auxiliary activator proteins. The RNA corresponding to the *rbcL* gene disappears from *Anabaena* cells induced for nitrogenase because of inhibition of transcription and/or rapid degradation of the RNA (Haselkorn et al., 1983). On the other hand, *Synechococcus* PCC 6301 does not fix nitrogen nor differentiates heterocysts. So, the large size difference observed for the untranslated leader RNA of *rbc* genes between *Synechococcus* and *Anabaena* may reflect the additional gene expression regulation found in *Anabaena*.

As mentioned above (see Section 5.2.) two *psbA* genes encoding a 32 kD thylakoid membrane protein from *Anabaena* PCC 7120 have been cloned and sequenced (Curtis and Haselkorn, 1984). Genes *psbAI* and *psbAII* differ in their nucleotide sequence by only 10% within the open reading frame but the 5′-flanking regions are totally dissimilar with the exception of residues -2 to -14 which are identical. Only one transcript was detected and it appears that it corresponds to the *psbAI* gene starting at residue -67 relative to the coding sequence. An mRNA of about 1.3 Kb was found for the *Calothrix* PCC 7601 *ps2B-1* gene (=*psbA*), a size similar to that of the *Anabaena psbAI* transcript, but the transcription start site has not been mapped (Mulligan et al., 1984).

Unexpectedly, the transcript of the gene encoding the β subunit of *Synechococcus* PCC 7002 phycocyanin starts about 350 bp upstream from the coding region (Pilot and Fox, 1984; Bryant et al., 1985c) while in most prokaryotes, the 5′ untranslated region of mRNA is generally no longer than 200 bp (Kozak, 1983). In fact, an open reading frame 114 bp long starts about 100 bp downstream from the transcription initiation site and ends 112 upstream from the AUG of the *cpcB* gene. It is currently unknown whether the 114 bp open reading frame is translated into a polypeptide.

As seen from Table VI, only few cyanobacterial promoter sequences have been published so far and they do not share any real common features. The *nifH-nifD* and *nif*-like *glnA* promoters are rather similar but very different from the *E. coli* consensus promoter. In particular, the sequences in the -35 region seem quite unrelated to that assigned to the *E. coli* consensus promoter. In the -10 region, the sequence TTCTA resembles more that of the *K. pneumoniae nifA*-activated *nif* promoters (TTGCA) (Beynon et al., 1983) than the classical TATAAT Pribnow box. The other cyanobacterial genes appear to have a -10 region which resemble each other and the putative

RNA polymerase binding consensus sequence (TATAAT) but the -35 regions clearly differ. However, one should note that these cyanobacterial genes, especially those from *Anabaena*, are subject to particular expression regulation in the heterocyst. Examination of promoter regions of genes involved in functions common to all prokaryotes is needed to define homologies which might exist between regulatory regions of cyanobacterial genes and those of the other prokaryotic organisms. Genes of the amino acid biosynthesis pathway, as well as the recently identified *recA*, *tuf* and *fus* genes involved in DNA metabolism (see Table V), are already accessible for such a study.

Once the transcript is made, ribosomes must bind to it at a proper location in order to start translation into a polypeptide chain. The step-by-step assembly of initiation complexes has not yet been studied in cyanobacteria nor have the protein and cofactor needs been defined. The only available information dealing with translation initiation is derived from sequence data. Shine and Dalgarno (1974) have postulated that, in *E. coli*, base pairing might occur between a purine-rich sequence upstream from the AUG initiator codon of an mRNA and a complementary sequence near the 3′-end of the ribosomal 16S RNA. Various experimental data now support the role of mRNA-rRNA base pairing in the selection of initiation sites by *E. coli* ribosomes, as well as by Bacilli and by chloroplast ribosomes (Kozak, 1983). It thus seems reasonable to start with the working hypothesis that a Shine-Dalgarno mechanism also exists in cyanobacteria. To date, the only cyanobacterial 16S rRNA sequenced is the one from *Synechococcus* PCC 6301 (Tomioka and Sugiura, 1983). Its 3′-end is identical to that of the *E. coli* 16S rRNA for the 14 nucleotides preceding the last base where a U is present instead of the A found in *E. coli*. In particular, the sequence CCUC-CU, most often involved for prokaryotes in the Shine-Dalgarno interaction, is perfectly conserved. Unfortunately, most of the cyanobacterial genes which have been sequenced (see Table V) originate from strains for which the sequences of the 3′-end of the 16S rRNA have not been determined yet. Sequences such as GGAG, AGGA or even GAGA, about 10 nucleotides upstream from the AUG initiator codon, have been reported as potential Shine-Dalgarno sequences for the genes that were sequenced from the various cyanobacteria (for references, see Table V). For three of these genes, namely *Anabaena* PCC 7120 *rbcS* and *psbA* genes from both *Anabaena* PCC 7120 and *Calothrix* PCC 7601, no typical prokaryotic ribosome-binding site sequences have yet been identified. However, it should be noted that: (a) a similar absence of Shine-Dalgarno sequence was reported for the *psbA* genes isolated from both *Nicotiana* and spinach chloroplasts (Zurawski et al., 1982) and (b) 13 out of the 14 nucleotides upstream from the postulated AUG initiator codon are perfectly conserved within the nucleotide sequences of both *Anabaena* PCC 7120 and *Calothrix* PCC 7601 *psbA* genes. The absence of a Shine-Dalgarno sequence has also been reported for the mRNA encoding the bacteriorhodopsin from the archaebacterium *Halobacterium halobium*; this mRNA starts only three nucleotides upstream from the initiator codon (Dunn et al., 1981). On the other hand, there are evidences that support the hypothesis that eukaryotic ribosomes initiate via a scanning mechanism (Kozak, 1983). The question then arises: "Does a second mechanism for translation initiation exist in prokaryotes?"

To date, all cyanobacteria seem to use, for translating nucleotide sequences into polypeptide chains, the genetic code common to both the other prokaryotes and the eukaryotic chloroplasts. Since, within a given organism, all the genes are translated by the same machinery it is tempting to correlate cellular tRNA levels and codon frequencies, but a comparison of codon usage, gene by gene, demonstrates that large variations can occur and that summing the data for several genes can lead to erroneous conclusions (Wain-Hobson et al., 1981). Nevertheless, since the introduction of synthetic oligonucleotides as an efficient tool for gene cloning (see Section 5.3.), it has become important to know which codons are more often used, especially when long but unique sequences are to be synthesized. Table VII presents a compilation, organism by organism, of the data so far

TABLE VII

Codon usage in cyanobacterial genes[a]

Organisms (GC% content)		*Synechococcus* PCC 6301 (55.1%)						*Synechococcus* PCC 7002 (49.1%)		*Anabaena* PCC 7120 (42.5%)								*Calothrix* PCC 7601 (42.7%)		
Amino acid	Gene codon	*rbcL*	*rbcS*	*ppc*	*apcA*	*apcB*	*apcC*	*cpcA*	*cpcB*	*nifH*	*nifD*	*nifK*	*rbcL*	*rbcS*	*glnA*	*psbAI*	*psbAII*	*ps2BI*	*gvpA*	*gvB*
Arg	CGA	0	0	16	0	0	1	0	1	0	1	0	0	1	0	0	0	0	0	0
	CGC	11	3	45	3	6	2	2	4	5	5	9	9	0	3	4	5	5	0	0
	CGG	3	0	15	2	0	2	0	1	0	1	3	1	0	3	1	0	2	1	1
	CGU	13	5	6	7	4	1	5	6	11	14	5	18	5	8	5	4	7	2	2
	AGA	0	0	1	0	0	1	0	0	1	2	3	2	1	1	2	2	0	0	0
	AGG	0	0	1	0	0	0	0	0	1	1	0	0	0	0	0	0	0	0	0
Leu	CUA	0	0	8	3	0	0	1	0	4	4	4	0	0	6	3	7	3	2	2
	CUC	10	3	40	2	6	2	9	10	2	8	7	10	3	6	1	0	0	0	0
	CUG	22	3	61	6	7	3	0	0	2	3	6	23	3	7	3	2	3	0	0
	CUU	1	0	8	2	0	0	4	2	2	6	0	1	0	2	0	0	0	1	1
	UUA	0	0	1	2	0	0	0	0	7	7	4	4	0	8	5	7	9	2	2
	UUG	10	1	24	0	4	2	2	0	7	5	16	15	1	10	18	14	17	2	2
Ser	UCA	0	0	6	0	2	0	0	0	1	2	2	1	0	2	1	3	1	2	0
	UCC	5	0	7	2	0	0	0	3	5	4	9	4	1	5	11	11	8	1	4
	UCG	10	2	19	5	5	0	0	0	0	0	0	0	0	0	0	0	0	0	0
	UCU	2	0	6	0	1	0	2	6	4	17	11	9	4	14	8	3	12	3	2
	AGC	2	5	19	2	4	2	5	3	2	3	9	0	3	5	4	5	4	1	1
	AGU	0	0	9	1	3	0	0	0	0	2	1	0	0	2	1	0	1	0	0
Thr	ACA	0	0	5	0	0	0	0	0	1	3	8	10	2	13	3	3	3	1	1
	ACC	24	1	26	7	9	4	6	7	14	10	20	18	4	8	12	14	13	0	0
	ACG	4	1	24	1	0	1	1	0	0	0	0	2	0	2	0	0	0	0	0
	ACU	1	2	9	1	1	2	8	4	1	4	8	4	0	3	1	0	2	2	2

TABLE VII (*continued*)

Organisms (GC% content)		*Synechococcus* PCC 6301 (55.1%)						*Synechococcus* PCC 7002 (49.1%)		*Anabaena* PCC 7120 (42.5%)								*Calothrix* PCC 7601 (42.7%)		
Amino acid	Gene codon	*rbcL*	*rbcS*	*ppc*	*apcA*	*apcB*	*apcC*	*cpcA*	*cpcB*	*nifH*	*nifD*	*nifK*	*rbcL*	*rbcS*	*glnA*	*psbAI*	*psbAII*	*ps2BI*	*gvpA*	*gvpB*
Pro	CCA	1	0	7	0	0	1	0	0	3	3	7	4	2	13	2	5	1	0	0
	CCC	7	5	19	2	2	2	3	4	2	7	3	5	3	6	4	5	5	0	0
	CCG	9	1	24	1	1	1	0	0	0	0	0	0	0	0	0	0	0	0	0
	CCU	5	2	7	2	0	0	2	0	4	12	16	11	3	12	10	6	10	1	1
Ala	GCA	7	1	6	2	5	0	5	3	8	3	4	13	1	10	11	18	13	3	4
	GCC	8	0	23	4	5	0	2	3	2	5	0	4	1	7	1	3	1	2	0
	GCG	10	0	20	2	1	1	2	4	1	0	4	6	0	2	4	1	2	0	0
	GCU	18	2	17	9	10	2	14	17	16	21	25	25	1	17	23	17	22	6	7
Gly	GGA	0	0	6	1	0	1	0	1	0	2	1	3	1	3	3	5	1	0	0
	GGC	17	3	27	9	6	2	4	1	3	6	7	6	1	9	5	7	4	0	0
	GGG	4	0	7	0	0	0	0	0	0	4	0	2	0	0	0	0	0	0	0
	GGU	23	1	11	6	6	1	9	10	26	29	32	31	2	24	25	21	28	3	3
Val	GUA	0	0	4	2	0	0	1	6	11	16	11	15	2	6	13	14	14	4	3
	GUC	13	2	21	8	8	0	1	0	0	3	1	3	1	3	1	0	3	1	0
	GUG	10	2	26	2	3	0	0	0	0	4	2	1	0	1	2	2	3	0	0
	GUU	4	1	10	3	0	2	5	10	6	13	15	11	6	11	8	6	5	6	8
Ile	AUA	0	0	0	0	0	0	0	0	1	0	0	0	0	1	0	1	0	0	0
	AUC	21	7	23	7	11	5	5	5	15	17	21	19	4	18	26	26	19	4	4
	AUU	1	0	28	5	1	0	1	3	6	14	9	6	3	13	5	5	8	3	3
Lys	AAA	17	2	10	3	4	1	4	4	6	16	13	18	3	23	1	2	0	3	1
	AAG	9	3	16	3	2	4	3	2	8	23	19	7	3	8	0	0	1	0	2
Asn	AAC	13	2	19	2	4	1	8	4	14	15	22	17	1	20	21	23	20	0	0
	AAU	1	1	12	1	1	2	1	3	2	4	2	0	1	1	1	1	0	1	1

TABLE VII (*continued*)

Organisms (GC% content)		*Synechococcus* PCC 6301 (55.1%)						*Synechococcus* PCC 7002 (49.1%)		*Anabaena* PCC 7120 (42.5%)								*Calothrix* PCC 7601 (42.7%)		
Amino acid	Gene codon	*rbcL*	*rbcS*	*ppc*	*apcA*	*apcB*	*apcC*	*cpcA*	*cpcB*	*nifH*	*nifD*	*nifK*	*rbcL*	*rbcS*	*glnA*	*psbAI*	*psbAII*	*ps2BI*	*gvpA*	*gvpB*
Gln	CAA	8	5	29	3	5	2	4	3	9	9	10	13	7	11	9	10	9	0	1
	CAG	4	2	50	1	0	3	2	3	1	3	4	2	2	4	2	1	2	1	0
His	CAC	15	1	15	0	0	0	1	0	7	12	10	12	3	7	10	10	10	0	0
	CAU	1	1	9	0	0	0	1	0	0	0	2	1	0	3	0	1	0	0	0
Glu	GAA	24	3	39	10	4	2	5	8	22	28	32	23	7	24	14	14	12	6	6
	GAG	5	8	52	4	3	1	1	2	3	8	4	8	2	8	3	4	4	0	0
Asp	GAC	25	2	18	9	6	0	6	5	16	15	13	16	2	11	4	5	2	1	2
	GAU	6	3	27	1	3	1	4	5	1	14	17	12	0	17	3	2	6	2	1
Tyr	UAC	12	6	23	8	10	1	9	4	7	17	18	15	6	13	13	12	14	0	1
	UAU	2	0	13	0	2	0	2	1	2	4	3	4	2	9	1	0	0	2	1
Cys	UGC	5	4	6	1	1	1	1	2	4	5	5	3	2	2	5	4	5	0	0
	UGU	2	0	4	0	0	0	1	1	2	3	1	4	0	4	0	0	0	0	0
Phe	UUC	19	7	25	2	1	2	5	4	6	15	19	12	2	14	23	26	21	0	0
	UUU	5	1	18	1	0	2	0	2	0	6	9	8	3	11	3	2	5	0	0
Met	AUG	12	4	16	3	4	2	3	6	15	14	18	8	1	12	11	11	10	1	1
Trp	UGG	9	2	13	0	0	1	1	0	0	8	8	9	2	8	10	10	10	1	1
Stop	UAA	1	1	0	1	1	0	1	1	0	1	0	1	1	1	1	1	1	1	1
	UAG	0	0	0	0	0	1	0	0	1	0	1	0	0	0	0	0	0	0	0
	UGA	0	0	1	0	0	0	0	0	0	0	0	1	0	0	0	0	0	0	0

[a] References are given in Table V.

obtained for cyanobacterial genes. From this table, it appears that: (a) the two unicellular *Synechococcus* strains share a rather similar codon usage, as well as the two filamentous strains (*Anabaena* and *Calothrix*); (b) the correlation between codon usage and GC% content does not seem very meaningful, although the two *Synechococcus* strains which have higher GC% contents use more frequently codons ending with a G or a C than the two filamentous strains; (c) qualitatively, the preferred synonymous codons of cyanobacterial genes generally match those of the *E. coli* genes but the preferred proline (CCG) and leucine (CUG) codons of *E. coli* are infrequently used in *Synechococcus* PCC 7002, *Anabaena* PCC 7120 and *Calothrix* PCC 7601. Obviously, more genes have to be sequenced in order to obtain significant statistics and to draw more general conclusions.

6.2. Expression of cyanobacterial genes in *E. coli*

Studies of gene expression in *E. coli* and/or in vitro transcription with *E. coli* RNA polymerase have been carried out for a few cyanobacterial genes. In agreement with the predictions one can make from the sequence data presented in Table VI, the only genes shown to be expressed are *Anabaena* PCC 7120 *glnA* (Tumer et al., 1983), *Synechococcus* PCC 7002 *cpcA,B* (Bryant et al., 1985c) and *Synechococcus* PCC 6301 *rbcL,S* genes (Shinozaki and Sugiura, 1985). For these genes, the -10 and -35 regions fit approximately *E. coli* consensus sequences. In *E. coli*, a *glnA* transcript was found which originated only from the '*E. coli*-like' promoter but its level did not respond to the regulatory mechanisms operating in *Anabaena* cells (Tumer et al., 1983). Maxicell experiments showed that the *Synechococcus cpcA,B* genes were transcribed in *E. coli* from a promoter located on the cyanobacterial DNA fragment (Bryant et al., 1985c). Similarly, in vitro transcription studies with *E. coli* and *Synechococcus* PCC 6301 RNA polymerases revealed that transcripts for the *rbcL,S* genes start at the same initiation site for both enzymes, i.e. the in vivo start, thus suggesting that the *Synechococcus* promoter should be functional in *E. coli* (Shinozaki and Sugiura, 1985).

Despite the lack of heterologous hybridization, Kodaki et al. (1985) succeeded in cloning the *Synechococcus ppc* gene, which encodes phosphoenol pyruvate carboxylase, using phenotypic complementation of an *E. coli ppc* mutant. Interestingly, in all the selected recombinants the cyanobacterial DNA fragment was inserted into pBR322 in the same orientation. This observation suggests that the expression of the gene in *E. coli* might be dependent on a vector promoter. Few additional genes have more recently been cloned by complementation of *E. coli* mutants (Porter et al., 1985; our laboratory, unpublished data), while other attempts have failed. No sequence data of the promoters nor mapping of the transcripts has been published so that the relationship between promoter sequences and gene expression in *E. coli*, as well as the rationale which governs the expression of a given gene, remain to be established. Since cyanobacteria seem to share with *E. coli* the same genetic code, the reasons which sustain the lack of complementation might well be: (a) the absence of DNA sequences susceptible to be recognized by the host RNA polymerase, or/and of ribosome binding sites; and (b) a difference in the metabolic pathway under study or in the regulation of the activity of the gene product by specific effectors.

6.3. Regulation of gene expression in cyanobacteria

At least three levels of regulation can be distinguished in gene expression: transcriptional, translational and post-translational. Biochemical experiments have shown that the induction of specific proteolytic enzymes can occur in response to specific physiological conditions. For example, it has been known for some time that nitrogen starvation induces a specific proteolytic degradation of phycobiliproteins and/or cyanophycin (Wood and Haselkorn, 1980; Gupta and Carr, 1981). Protein processing also exists in cyanobacteria such as that required to anchor prosthetic groups to polypeptides so as to obtain functional phycobiliproteins, nitrate reductase, or nitrogenase, for example. However, the mechanisms of these regulations have not been elucidated in detail, and a lack of genetic tools has for a long time prevented extensive studies of the regulation of gene expression in cyanobacteria. To our knowledge, no data have been reported yet concerning translational regulation. At present, most of the information about the regulation of gene expression in cyanobacteria concerns transcriptional regulation.

Processes such as attenuation have not been demonstrated in cyanobacteria but the occurrence of polycistronic messages has been observed. The genes encoding the two polypeptides of phycocyanin (*Synechococcus* PCC 7002 *cpcB* and *cpcA*) as well as those encoding the large and small subunits of ribulose bisphosphate carboxylase (*Synechococcus* PCC 6301 and *Anabaena* PCC 7120 *rbcL* and *rbcS* genes) are cotranscribed (Pilot and Fox, 1984; Shinozaki and Sugiura, 1985; Nierzwicki-Bauer et al., 1984). The *nifH* and *nifD* genes are also cotranscribed, together with *nifK* following a very peculiar developmentally regulated DNA excision, occurring under nitrogen-fixing conditions in *Anabaena* PCC 7120 cells (J.W. Golden et al., 1985a). Coordinated transcription, which is a well-exemplified mechanism of regulation for genes whose products have to meet a precise stoichiometry, thus occurs in cyanobacteria as well as in other prokaryotes. The data obtained by Haselkorn and co-workers have led these authors to postulate that the *Anabaena* PCC 7120 RNA polymerase is modified, during heterocyst differentiation, in order to recognize new promoter sequences (Haselkorn et al., 1985). Transcription corresponding to *nif* genes can be detected, in *Anabaena*, only under nitrogen-fixing conditions, while that of the *psbA* gene is roughly as abundant in cells grown on ammonia or starved for nitrogen. The expression of the *nif* genes is thus regulated at the transcriptional level, as for example in *Klebsiella pneumoniae* (Beynon et al., 1983). However, the transcription of the *rbcL-rbcS* genes, in contrast to what has been found in some plant species, is not light-regulated, at least in *Anabaena* PCC 7120 cells (Nierzwicki-Bauer et al., 1984). Depending on the physiological conditions, it was noticed that two kinds of *glnA* transcripts can be detected in *Anabaena* PCC 7120 cells (Tumer et al., 1983). These observations can be correlated with the promoter sequence features of these genes (Table VI) and implies a differential mode of action for the RNA polymerase. Transcriptional control of gene expression was also postulated to occur in the chromatic adaptation phenomenon (Tandeau de Marsac, 1983; see Section 4.2.1. of Chapter 13). The recent cloning of genes encoding phycobiliprotein polypeptides from strains such as *Pseudanabaena* PCC 7409 and *Calothrix* PCC 7601 (Table V), which are chromatic adapters, allows the study of this phenomenon at a molecular level. Probing Northern blots with cloned phycocyanin genes shows that specific mRNAs can be detected in total RNA obtained from red light grown *Calothrix* PCC 7601, while they are absent in RNA from cells grown under green light (Conley et al., 1985).

Gene families have been demonstrated for *nifH* in *Anabaena* PCC 7120 and for *psbA* genes in all cyanobacteria studied so far (Haselkorn et al., 1983; Curtis and Haselkorn, 1984; Mulligan et al., 1984; S.S. Golden et al., 1985). However, the rationale of these reiterations is unknown and the data concerning the expression of the *nifH* genes differ from that found for the *psbA* genes. Under appropriate conditions both *nifH* genes were shown to be transcribed, but we do not know whether the additional copy of the *nifH* gene is translated (Robinson and Haselkorn, 1985). Curtis and Haselkorn (1984) found at least four *psbA* genes in *Anabaena* PCC 7120 but only the transcript corresponding to the *psbAI* gene was detected when specific internal probes were used. In contrast, in *Synechococcus* PCC 7942, the three *psbA* copies which have been cloned seem to be expressed (S.S. Golden et al., 1985).

It is noteworthy that expression of heterologous genes has been observed in cyanobacteria. The gene encoding ampicillin resistance carried by Tn*901* is expressed in *Synechococcus* PCC 7942 although to a far lesser extent than in *E. coli* and has served to develop cloning vehicles for this organism (Van den Hondel et al., 1980). Few other cloning vectors have recently been constructed which make use of the *lacZ* or lambda *cI857* expression in cyanobacteria (see Section 3.2.). In addition, Elanskaya et al. (1985) and Gallagher and Burke (1985b) were able to demonstrate the expression of genes originating from Gram-positive bacteria into *Synechococcus* PCC 7942. Both antibiotic resistance encoded by the staphylococcal plasmid pUB110 and the secretion, into the periplasm of the cyanobacterium, of α-amylase encoded by a DNA fragment originating from *Bacillus amyloliquefaciens* have been noted. Another important result was the demonstration by Dzelzkalns et al. (1984) that the expression of the chloramphenicol acetyl transferase (*CAT*) gene can be driven, in *Synechococcus* PCC 6301, by a chloroplast promoter. The promoter of the chloroplast *psbA* gene, believed to be a strong promoter in that organelle, allows accurate transcription initiation in the cyanobacterium of the promoterless bacterial *cat* gene. This result once again underlines the possible prokaryotic (cyanobacterial?) origin of chloroplast genes and paves the way for testing the ability of chloroplast gene products to function in the cyanobacterial photosynthetic apparatus or vice versa, when and if DNA transfer into chloroplast becomes feasible.

7. Concluding remarks

The important and peculiar physiological properties of cyanobacteria render them attractive for both studies aimed at extending our basic knowledge and developing approaches to genetic engineering. In this respect, gene cloning, transfer of these genes back into the organism from which they originate or into a specially designed organism, and studies on the control of their expression represent the major immediate research goals. In the last few years, insights have been gained in each of these areas but a great deal of research remains to be done.

Initially, vectors designed for *E. coli* were used to clone cyanobacterial genes but were found not to allow the construction of merodiploids since they were not capable of autonomous replication in cyanobacteria. As mentioned in this chapter, nearly all cyanobacteria possess endogenous plasmids suitable for the construction of cloning vehicles but, since no function encoded by cyanobacterial plasmids has been identified so far, specific genetic markers have had to be introduced into these molecules before they could become useful tools. Some antibiotic resistance genes from both Gram-negative and Gram-positive bacteria can be expressed in cyanobacteria and used as genetic markers. The use of some specific and readily expressed cyanobacterial markers, however, may well be more suitable, and this line of research should be encouraged. For a few strains, various cloning vectors, including shuttle vectors (*E. coli*-cyanobacterium), are now available but information dealing with plasmid replication and control of plasmid copy number in cyanobacteria is still insufficient.

Classical molecular genetic techniques have been used to clone cyanobacterial genes but the data so far obtained indicate that there is probably no ubiquitous method. Heterologous hybridization has most often been used for gene cloning whether or not the gene could then be expressed in *E. coli*. Such an approach was successfully used with both prokaryotic and eukaryotic probes but has failed, for example, for the cloning of the *ppc* gene which has been isolated by direct phenotypic complementation of a heterologous host. Since biochemical rather than genetic studies of cyanobacteria have predominated over many years, the use of synthetic oligonucleotides, which have already been successfully used, seems the most promising approach to the cloning of genes for which amino acid sequence data of their products are available. Despite its power and its availability for use in some cyanobacteria, the use of transposon-induced mutants for the cloning of the corresponding gene is impaired by the occasional high frequency of spontaneous mutation, probably due to the presence of mobile genetic elements in some cyanobacterial strains. The same considerations hold true for the direct cloning by phenotypic complementation of mutants, and obviously more studies have to be undertaken along these lines.

In order to study the regulation of the expression of a given gene, it is necessary to transfer defined DNA fragments which contain this gene into the organism from which it originates. This very important step was long restricted to the few cyanobacteria which could be transformed by DNA, and is still limited to unicellular strains. The development of a conjugation system available for filamentous strains is very helpful since some of these cyanobacteria, such as some *Anabaena* species for example, have an important ecological role. Once the problem of transfer of genetic material is overcome, the availability of more sophisticated tools, such as recombination-deficient host cells, strains cured of some of their endogenous plasmids, various vectors allowing either fusion to promoterless genes or repression and derepression of the expression of cloned genes under specific conditions, will be of constantly increasing value.

Finally, it is worth noting that: (a) the close relationship found between cyanobacterial RNA genes and their chloroplast counterparts, together with some observations derived from both sequence data and gene expression studies, reinforce the endosymbiont hypothesis concerning the origin of eukaryotic chloroplasts; (b) the Ti plasmid of *Agrobacterium tumefaciens* already allows the insertion of heterologous DNA into plant nuclear genome, as well as into the chloroplast genome. Exchange, via DNA manipulation, of important physiological functions between cyanobacteria and chloroplasts, or vice versa, may thus no longer be considered unrealistic.

Acknowledgements

We are deeply grateful to Pr. A. De Waard for the compilation of the cyanobacterial restriction enzymes presented in Table II. Our thanks also go to our colleagues who provided us with information in advance of publication. We are very much indebted to Drs. G. Cohen-Bazire, I. Saint-Girons, and C. Parsot for their helpful suggestions and to Dr. P. Avner and Dr. J. de Gunzburg for critical reading of the manuscript.

References

Asato, Y. and Ginoza, H.S. (1973) Separation of small circular DNA molecules from the blue-green alga *Anacystis nidulans*. Nature 244, 132–133.

Astier, C. and Espardellier, F. (1976) Mise en évidence d'un système de transfert génétique chez une cyanophycée du genre *Aphanocapsa*. C.R. Acad. Sci. Paris 282, 795–797.

Beynon, J., Cannon, M., Buchanan-Wollaston, V. and Cannon, F. (1983) The *nif* promoters of *Klebsiella pneumoniae* have a characteristic primary structure. Cell 34, 665–671.

Bogorad, L., Gendel, S.M., Haury, J.H. and Koller, K.P. (1983) Photomorphogenesis and complementary chromatic adaptation in *Fremyella diplosiphon*. In G.C. Papageorgiou and L. Packer (eds.), Photosynthetic Prokaryotes: Cell Differentiation and Function. Elsevier, New York, pp. 119–126.

Bryant, D.A., De Lorimier, R., Lambert, D.H., Dubbs, J.M., Stirewalt, V.L., Stevens, S.E., Jr., Porter, R.D., Tam, J. and Jay, E. (1985a) Molecular cloning and nucleotide sequence of the α and β subunits of allophycocyanin from the cyanelle genome of *Cyanophora paradoxa*. Proc. Natl. Acad. Sci. USA 82, 3242–3246.

Bryant, D.A., De Lorimier, R., Guglielmi, G., Stirewalt, V.L., Dubbs, J.M., Illman, B., Porter, R.D. and Stevens, S.E., Jr. (1985b) Genes for phycobilisome components in *Synechococcus* 7002, *Pseudanabaena* 7409, and *Mastigocladus laminosus*. In Vth International Symposium on Photosynthetic Prokaryotes. Grindelwald, Switzerland. Abstract p. 103.

Bryant, D.A., Dubbs, J.M., Fields, P.I., Porter, R.D. and De Lorimier, R. (1985c) Expression of phycobiliprotein genes in *Escherichia coli*. FEMS Microbiol. Lett. 29, 343–349.
Buzby, J.S., Porter, R.D. and Stevens, S.E., Jr. (1983) Plasmid transformation in *Agmenellum quadruplicatum* PR-6: construction of biphasic plasmids and characterization of their transformation properties. J. Bacteriol. 154, 1446–1450.
Calleja, F., Dekker, B.M.M., Coursin, T. and De Waard, A. (1984) A new sequence specific endonuclease *Esp*I, of cyanobacterial origin. FEBS Lett. 178, 69–72.
Calleja, F., Tandeau de Marsac, N., Coursin, T., Van Ormondt, H. and De Waard, A. (1985) A new endonuclease recognizing the deoxynucleotide sequence C^CNNGG from the cyanobacterium *Synechocystis* 6701. Nucl. Acids Res. 13, 6745–6751.
Canosi, U., Morelli, G. and Trautner, T.A. (1978) The relationship between molecular structure and transformation efficiency of some *Staphylococcus aureus* plasmids isolated from *Bacillus subtilis*. Mol. Gen. Genet. 166, 259–267.
Chauvat, F., Astier, C., Vedel, F. and Joset-Espardellier, F. (1983) Transformation in the cyanobacterium *Synechococcus* R-2: improvement of efficiency; role of the pUH24 plasmid. Mol. Gen. Genet. 191, 39–45.
Chauvat, F., De Vries, L., Van der Ende, A. and Van Arkel, G. (1985) A host vector system for gene cloning in the cyanobacterium *Synechocystis* 6803. In Vth International Symposium on Photosynthetic Prokaryotes, Grindelwald, Switzerland. Abstract p. 327.
Chisholm, R.L. (1982) Methylation and developmental regulation of gene expression. Trends Biochem. Sci. 7, 421.
Cobley, J. (1985) Chromatic adaptation in *Fremyella diplosiphon*. I. Mutants hypersensitive to green light. II. Construction of mobilizable vectors lacking sites for *Fdi*I and II. In Vth International Symposium on Photosynthetic Prokaryotes, Grindelwald, Switzerland. Abstract p. 105.
Conley, P.B., Lemaux, P.G. and Grossman, A.R. (1985) Cyanobacterial light-harvesting complex subunits are encoded in two red light-induced transcripts. Science 230, 550–553.
Currier, T.C. and Wolk, C.P. (1979) Characteristics of *Anabaena variabilis* influencing plaque formation by cyanophage N-1. J. Bacteriol. 139, 88–92.
Curtis, S.E. and Haselkorn, R. (1983) Isolation and sequence of the gene for the large subunit of ribulose-1,5-biphosphate carboxylase from the cyanobacterium *Anabaena* 7120. Proc. Natl. Acad. Sci. USA 80, 1835–1839.
Curtis, S.E. and Haselkorn, R. (1984) Isolation, sequence and expression of two members of the 32 kd thylakoid membrane protein gene family from the cyanobacterium *Anabaena* 7120. Plant Mol. Biol. 3, 249–258.
Daniell, H., Sarojini, G. and McFadden, B.A. (1985) Characterization of transformation of cyanobacteria by the plasmids pBR322, pHUB4, and pCH1. In: Vth International Symposium on Photosynthetic Prokaryotes, Grindelwald, Switzerland. Abstract p. 358.
De Lorimier, R., Bryant, D.A., Porter, R.D., Liu, W.Y., Jay, E. and Stevens, S.E. Jr. (1984) Genes for the α and β subunits of phycocyanin. Proc. Natl. Acad. Sci. USA 81, 7946–7950.
Devilly, C.I. and Houghton, J.A. (1977) A study of genetic transformation in *Gloeocapsa alpicola*. J. Gen. Microbiol. 98, 277–280.
De Waard, A. and Duyvesteyn, M. (1980) Are sequence-specific deoxyribonucleases of value as taxonomic markers of cyanobacterial species? Arch. Microbiol. 128, 242–247.
De Waard, A., Korsuize, J., Van Beveren, C.P. and Maat, J. (1978) A new sequence-specific endonuclease from *Anabaena cylindrica*. FEBS Lett. 96, 106–110.
De Waard, A., Van Beveren, C.P., Duyvesteyn, M. and Van Ormondt, H. (1979) Two sequence-specific endonucleases from *Anabaena oscillarioides*. FEBS Lett. 101, 71–76.
Douglas, S.E. and Doolittle, W.F. (1984a) Complete nucleotide sequence of the 23S rRNA gene of the cyanobacterium, *Anacystis nidulans*. Nucl. Acids Res. 12, 3373–3386.
Douglas, S.E. and Doolittle, W.F. (1984b) Nucleotide sequence of the 5S rRNA gene and flanking regions in the cyanobacterium *Anacystis nidulans*. FEBS Lett. 166, 307–310.
Dunn, R., McCoy, J., Simsek, M., Majumdar, A., Chang, S.M., Rajbhandary, U.L. and Khorana, H.G. (1981) The bacteriorhodopsin gene. Proc. Natl. Acad. Sci. USA 78, 6744–6748.
Duyvesteyn, M.G.C. and De Waard, A. (1980) A new sequence-specific endonuclease from a thermophilic cyanobacterium, *Mastigocladus laminosus*. FEBS Lett. 111, 423–426.
Duyvesteyn, M.G.C., Korsuize, J. and De Waard, A. (1981) Isolation and characterization of a sequence-specific deoxyriboendonuclease from *Calothrix scopulorum*. Plant Mol. Biol. 1, 75–79.
Duyvesteyn, M.G.C., Korsuize, J., De Waard, A., Vonshak, A. and Wolk, C.P. (1983) Sequence-specific endonucleases in strains of *Anabaena* and *Nostoc*. Arch. Microbiol. 134, 276–281.

Dzelzkalns, V.A., Owens, G.C. and Bogorad, L. (1984) Chloroplast promoter driven expression of the chloramphenicol acetyltransferase gene in a cyanobacterium. Nucl. Acids Res. 23, 8917–8925.

Elanskaya, I., Morzunova, I. and Shestakov, S. (1985) Expression of Gram-positive bacteria genes in cyanobacterium *Anacystis nidulans* R2. In Vth International Symposium on Photosynthetic Prokaryotes, Grindelwald, Switzerland. Abstract p. 372.

Engwall, K.S. and Gendel, S.M. (1985) A rapid method for identifying plasmids in cyanobacteria. FEMS Microbiol. Lett. 26, 337–340.

Fisher, R., Tuli, R. and Haselkorn, R. (1981) A cloned cyanobacterial gene for glutamine synthetase functions in *Escherichia coli*, but the enzyme is not adenylated. Proc. Natl. Acad. Sci. USA 78, 3393–3397.

Flores, E. and Wolk, C.P. (1985a) Identification of facultatively heterotrophic, N_2-fixing cyanobacteria able to receive plasmid vectors from *Escherichia coli* by conjugation. J. Bacteriol. 162, 1339–1341.

Flores, E. and Wolk, C.P. (1985b) Production of bacteriocins by filamentous, nitrogen-fixing cyanobacteria. In Vth International Symposium on Photosynthetic Prokaryotes, Grindelwald, Switzerland. Abstract p. 210.

Friedberg, D. and Seijffers, J. (1979) Plasmids in two cyanobacterial strains. FEBS Lett. 107, 165–168.

Friedberg, D. and Seijffers, J. (1983) A new hybrid plasmid capable of transforming *Escherichia coli* and *Anacystis nidulans*. Gene 22, 267–275.

Friedberg, D. and Seijffers, J. (1985) Regulated gene expression in *Anacystis nidulans* R2 by lambda regulatory signals. In Vth International Symposium on Photosynthetic Prokaryotes, Grindelwald, Switzerland. Abstract p. 330.

Gallagher, M.L. and Burke, W.F., Jr. (1985a) Sequence-specific endonuclease from the transformable cyanobacterium *Anacystis nidulans* R-2. FEMS Microbiol. Lett. 26, 317–321.

Gallagher, M.L. and Burke, W.F., Jr (1985b) Expression by *Anacystis nidulans* R-2 of an antibiotic resistance gene originating from a Gram-positive organism. Abstract H155, American Society for Microbiology 85th ann. meeting.

Geier, G.E. and Modrich, P. (1979) Recognition sequence of the *dam* methylase of *Escherichia coli* K12 and mode of cleavage of *Dpn*I endonuclease. J. Biol. Chem., 254, 1408–1413.

Gendel, S., Straus, N., Pulleyblank, D. and Williams, J. (1983a) Shuttle cloning vectors for the cyanobacterium *Anacystis nidulans*. J. Bacteriol. 156, 148–154.

Gendel, S., Straus, N., Pulleyblank, D. and Williams, J. (1983b) A novel shuttle cloning vector for the cyanobacterium *Anacystis nidulans*. FEMS Microbiol. Lett. 19, 291–294.

Gingeras, T.R., Milazzo, J.P. and Roberts, R.J. (1978) A computer assisted method for the determination of restriction enzyme recognition sites. Nucl. Acids Res. 5, 4105–4127.

Glazer, A.N. (1985) Light harvesting by phycobilisomes. Ann. Rev. Biophys. Biophys. Chem. 14, 47–77.

Golden, J.W., Robinson, S.J. and Haselkorn, R. (1985a) Rearrangement of nitrogen fixation genes during heterocyst differentiation in the cyanobacterium *Anabaena*. Nature 314, 419–423.

Golden, J.W., Lammers, P.J., Mulligan, M.E. and Haselkorn, R. (1985b) Rearrangement of *Anabaena* nitrogen fixation genes. In Vth International Symposium on Photosynthetic Prokaryotes, Grindelwald, Switzerland. Abstract p. 121.

Golden, S.S. and Sherman, L.A. (1983) A hybrid plasmid is a stable cloning vector for the cyanobacterium *Anacystis nidulans* R-2. J. Bacteriol. 155, 966–972.

Golden, S.S. and Sherman, L.A. (1984) Optimal conditions for genetic transformation of the cyanobacterium *Anacystis nidulans* R-2. J. Bacteriol. 158, 36–42.

Golden, S.S., Brusslan, J.A. and Haselkorn, R. (1985) Herbicide resistance and the expression of *psbA* genes in *Anacystis nidulans* R2. In Vth International Symposium on Photosynthetic Prokaryotes, Grindelwald, Switzerland. Abstract p. 110.

Gray, M.W. and Doolittle, W.F. (1982) Has the endosymbiont hypothesis been proven? Microbiol. Rev. 46, 1–42.

Grossman, A., Conley, P.B., Lemaux, P.G. and Lomax, T. (1985) Arrangement and transcription of genes encoding phycobilisome polypeptides of *Fremyella diplosiphon*. In Vth International Symposium on Photosynthetic Prokaryotes, Grindelwald, Switzerland. Abstract p. 104.

Guglielmi, G. and Cohen-Bazire, G. (1984) Etude taxonomique d'un genre de cyanobactérie Oscillatoriacée: le genre *Pseudanabaena* Lauterborn. I. Etude ultrastructurale. Protistologica XX, 377–391.

Gupta, M. and Carr, N.G. (1981) Enzyme activites related to cyanophycin metabolism in heterocysts and vegetative cells of *Anabaena* sp. J. Gen. Microbiol. 125, 17–23.

Haselkorn, R. (1985) Molecular genetics of cyanobacteria. In Vth International Symposium on Photosynthetic Prokaryotes, Grindelwald, Switzerland. Abstract p. 96.

Haselkorn, R., Rice, D., Curtis, S.E. and Robinson, S.J. (1983) Organization and transcription of genes important in *Anabaena* heterocyst differentiation. Ann. Microbiol. (Inst. Pasteur) 134B, 181–193.
Haselkorn, R., Curtis, S.E., Golden, J.W., Nierzwicki-Bauer, S.A., Robinson, S.J. and Tumer, N.E. (1985) Organization and transcription of *Anabaena* genes regulated during heterocyst differentiation. In L. Van Vlotten Doting, G.S.T. Groot and T.C. Hall (eds.), Molecular Form and Function of the Plant Genome. Plenum-Press, New York-London, in press.
Hauman, J.H. (1981) Is a plasmid(s) involved in the toxicity of *Microcystis aeruginosa?* In W.W. Carmichael (ed.), The Water Environment. Algal Toxins and Health, Vol. 20. Plenum Press, New York-London, pp. 97–102.
Herdman, M. (1973) Mutations arising during transformation in the blue-green alga *Anacystis nidulans*. Mol. Gen. Genet. 120, 369–378.
Herdman, M. (1982) Evolution and genetic properties of cyanobacterial genomes. In N.G. Carr and B.A. Whitton (eds.), The Biology of Cyanobacteria, Botanical Monographs, Vol. 19. Blackwell Scientific Publications, Oxford, pp. 263–305.
Herrero, A., Elhai, J., Hohn, B. and Wolk, C.P. (1984) Infrequent cleavage of cloned *Anabaena variabilis* DNA by restriction endonucleases from *A. variabilis*. J. Bacteriol. 160, 781–784.
Houmard, J., Mazel, D., Bryant, D.A. and Tandeau de Marsac, N. (1985) Molecular cloning and sequencing of the α and β allophycocyanin subunits from *Synechococcus* PCC 6301. In Vth International Symposium on Photosynthetic Prokaryotes, Grindelwald, Switzerland. Abstract p. 339.
Hughes, S.G. and Murray, K.H. (1980) The nucleotide sequences recognized by endonucleases *Ava*I and *Ava*II from *Anabaena variabilis*. Biochem. J. 185, 65–75.
Hughes, S.G., Bruce, T. and Murray, K.H. (1980) The isolation and characterization of a sequence-specific endonuclease from *Anabaena subcylindrica*. Biochem. J. 185, 59–63.
Itakura, K., Rossi, J.J. and Wallace, R.B. (1984) Synthesis and use of synthetic oligonucleotides. Ann. Rev. Biochem. 53, 323–356.
Kalla, S.R., Lind, L., Lönneborg, A., Lidholm, J., Oquist, G. and Gustafsson, P. (1985) Structural and transcriptional organisation of phycocyanin subunit encoding genes from the cyanobacterium *Anacystis nidulans*. In Vth International Symposium on Photosynthetic Prokaryotes, Grindelwald, Switzerland. Abstract p. 341.
Katagiri, F., Kodaki, T., Fujita, N., Izui, K. and Katsuki, H. (1985) Nucleotide sequence of the phosphoenolpyruvate carboxylase gene of the cyanobacterium *Anacystis nidulans*. Gene 38, 265–269.
Kodaki, T., Katagiri, F., Asano, M., Izui, K. and Katsuki, H. (1985) Cloning of phosphoenolpyruvate carboxylase gene from a cyanobacterium, *Anacystis nidulans*, in *Escherichia coli*. J. Biochem. 97, 533–539.
Kolowsky, K.S., Williams, J.G.K. and Szalay, A.A. (1984) Length of foreign DNA in chimeric plasmids determines the efficiency of its integration into the chromosome of the cyanobacterium *Synechococcus* R-2. Gene 27, 289–299.
Kozak, M. (1983) Comparison of initiation of protein synthesis in procaryotes, eucaryotes and organelles. Microbiol. Rev. 47, 1–45.
Kück, U. and Mulligan, B.J. (1984) Recombination. DNA transformation in cyanobacteria, algae and fungi: trends and perspectives. Progress in Botany, Vol. 46. Springer-Verlag, Berlin-Heidelberg, pp. 192–207.
Kuhlemeier, C.J., Borrias, W.E., Van den Hondel, C.A.M.J.J. and Van Arkel, G.A. (1981) Vectors for cloning in cyanobacteria: construction and characterization of two recombinant plasmids capable of transformation to *Escherichia coli* K12 and *Anacystis nidulans* R-2. Mol. Gen. Genet. 184, 249–254.
Kuhlemeier, C.J., Thomas, A.A.M., Van der Ende, A., Van Leen, R.W., Borrias, W.E., Van den Hondel, C.A.M.J.J. and Van Arkel, G.A. (1983) A host-vector system for gene cloning in the cyanobacterium *Anacystis nidulans* R-2. Plasmid 10, 156–163.
Kuhlemeier, C.J., Logtenberg, T., Stoorvogel, W., Van Heugten, H.A.A., Borrias, W.E. and Van Arkel, G.A. (1984a) Cloning of nitrate reductase genes from the cyanobacterium *Anacystis nidulans*. J. Bacteriol. 159, 36–41.
Kuhlemeier, C.J., Teeuwsen, V.J.P., Janssen, M.J.T. and Van Arkel, G.A. (1984b) Cloning of a third nitrate reductase gene from the cyanobacterium *Anacystis nidulans* R-2 using a shuttle cosmid library. Gene 31, 109–116.
Kumano, M., Tomioka, N. and Sugiura, M. (1983) The complete nucleotide sequence of 23S rRNA gene from a blue-green alga, *Anacystis nidulans*. Gene 24, 219–225.
Lachance, M.A. (1981) Genetic relatedness of heterocystous cyanobacteria by deoxyribonucleic acid-deoxyribonucleic acid reassociation. Int. J. Syst. Bacteriol. 31, 139–147.
Lambert, G.R. and Carr, N.G. (1982) Rapid small-scale plasmid isolation by several methods from filamentous cyanobacteria. Arch. Microbiol. 133, 122–125.

Lambert, G.R. and Carr, N.G. (1983) A restriction map of plasmid pDC1 from the filamentous cyanobacterium *Nostoc* sp. MAC PCC 8009. Plasmid 10, 196–198.

Lambert, G.R. and Carr, N.G. (1984) Resistance of DNA from filamentous and unicellular cyanobacteria to restriction endonuclease cleavage. Biochim. Biophys. Acta 781, 45–55.

Lambert, G.R., Scott, J.G. and Carr, N.G. (1984) Characterization and cloning of extrachromosomal DNA from filamentous cyanobacteria. FEMS Microbiol. Lett. 21, 225–231.

Lammers, P.J. and Haselkorn, R. (1983) Sequence of the *nifD* gene coding for the α subunit of dinitrogenase from the cyanobacterium *Anabaena*. Proc. Natl. Acad. Sci. USA 80, 4723–4727.

Lau, R.H. and Doolittle, W.F. (1979) Covalently closed circular DNAs in closely related unicellular cyanobacteria. J. Bacteriol. 137, 648–652.

Lau, R.H. and Doolittle, W.F. (1980) *Aqu*I: a more easily purified isoschizomer of *Ava*I. FEBS Lett. 121, 200–202.

Lau, R.H. and Straus, N.A. (1985) Versatile shuttle cloning vectors for the unicellular cyanobacterium *Anacystis nidulans* R-2. FEMS Microbiol. Lett. 27, 253–256.

Lau, R.H., Sapienza, C. and Doolittle, W.F. (1980a) Cyanobacterial plasmids: their widespread occurrence, and the existence of regions of homology between plasmids in the same and different species. Mol. Gen. Genet. 178, 203–211.

Lau, R.H., Sapienza, C. and Doolittle, W.F. (1980b) Plasmids in cyanobacteria. In C. Stuttard and K.R. Rozee (eds.), Plasmids: Environmental effects and maintenance mechanisms. Academic Press, New York, pp. 263–273.

Lau, R.H., Visentin, L.P., Martin, S.M., Hofman, J.D. and Doolittle, W.F. (1985) Site specific restriction endonuclease from the filamentous cyanobacterium *Nostoc* sp. MAC PCC 8009. FEBS Lett. 179, 129–132.

Laudenbach, D.E., Straus, N.A., Gendel, S. and Williams, J.P. (1983) The large endogenous plasmid of *Anacystis nidulans*: mapping, cloning and localization of the origin of replication. Mol. Gen. Genet. 192, 402–407.

Lind, L.K., Kalla, S.R., Lönneborg, A., Oquist, G. and Gustafsson, P. (1985) Cloning of the β-phycocyanin gene from *Anacystis nidulans*. FEBS Lett. 188, 27–32.

Lomax, T.L., Conley, P.B. and Grossman, A.R. (1985) Isolation and localization of phycobilisome linker polypeptide genes from *Fremyella diplosiphon*. In Vth International Symposium on Photosynthetic Prokaryotes, Grindelwald, Switzerland. Abstract p. 338.

Mazur, B.J. and Chui, C.F. (1982) Sequence of the gene coding for the β-subunit of dinitrogenase from the blue-green alga *Anabaena*. Proc. Natl. Acad. Sci. USA 79, 6782–6786.

Mevarech, M., Rice, D. and Haselkorn, R. (1980) Nucleotide sequence of a cyanobacterial *nifH* gene coding for nitrogenase reductase. Proc. Natl. Acad. Sci. USA 77, 6476–6480.

Mottes, M., Grandi, G., Sgaramella, V., Canosi, U., Morelli, G. and Trautner, T.A. (1979) Different specific activities of the monomeric and oligomeric forms of plasmid DNA in transformation of *B. subtilis* and *E. coli*. Mol. Gen. Genet. 174, 281–286.

Mulligan, B., Schultes, N., Chen, L. and Bogorad, L. (1984) Nucleotide sequence of a multiple-copy gene for the B protein of photosystem II of a cyanobacterium. Proc. Natl. Acad. Sci. USA 81, 2693–2697.

Nierzwicki-Bauer, S.A., Curtis, S.E. and Haselkorn, R. (1984) Cotranscription of genes encoding the small and large subunits of ribulose-1,5-bisphosphate carboxylase in the cyanobacterium *Anabaena* 7120. Proc. Natl. Acad. Sci. USA 81, 5961–5965.

Orkwiszewski, K.G. and Kaney, A.R. (1974) Genetic transformation of the blue-green bacterium, *Anacystis nidulans*. Arch. Microbiol. 98, 31–37.

Pakrasi, H.B., Williams, J.G.K. and Arntzen, C.J. (1985) Genetic analysis of cytochrome B559 in *Synechocystis* 6803. In Vth International Symposium on Photosynthetic Prokaryotes, Grindelwald, Switzerland. Abstract p. 362.

Pilot, T.J. and Fox, J.L. (1984) Cloning and sequencing of the genes encoding the α and β subunits of C-phycocyanin from the cyanobacterium *Agmenellum quadruplicatum*. Proc. Natl. Acad. Sci. USA 81, 6983–6987.

Porter, R.D. (1985) Transformation in cyanobacteria. Crit. Rev. Microbiol. 13, 111–132.

Porter, R.D., Buzby, J.S., Pilon, A., Fields, P.I., Dubbs, J. and Stevens, S.E., Jr. (1985) Gene cloning and merodiploid analysis in *Agmenellum quadruplicatum*, PR-6. In Vth International Symposium on Photosynthetic Prokaryotes, Grindelwald, Switzerland. Abstract p. 331.

Potts, M. (1984) Distribution of plasmids in cyanobacteria of the LPP group. FEMS Microbiol. Lett. 24, 351–354.

Potts, M., Ocampo-Friedmann, R., Bowman, M.A. and Tözun, B. (1983) *Chroococcus* S24 and *Chroococcus* N41 morphological, biochemical and genetic characterization and effects of water stress on ultrastructure. Arch. Microbiol. 135, 81–90.

Reaston, J. and Carr, N.G. (1985) Purification of the modification enzyme M. *Nsp* MACI from the filamentous cyanobacterium *Nostoc* PCC 8009. In Vth International Symposium on Photosynthetic Prokaryotes, Grindelwald, Switzerland. Abstract p. 332.

Reaston, J.R., Van den Hondel, C.A.M.J.J., Van der Ende, A., Van Arkel, G.A., Stewart, W.D.P. and Herdman, M. (1980) Comparison of plasmids from the cyanobacterium *Nostoc* PCC 7524 with two mutant strains unable to form heterocysts. FEMS Microbiol. Lett. 9, 185–188.

Reaston, J., Duyvesteyn, M.G.C. and De Waard, A. (1982a) *Nostoc* PCC 7524, a cyanobacterium which contains five sequence-specific deoxyribonucleases. Gene 20, 103–110.

Reaston, J., Van den Hondel, C.A.M.J.J., Van Arkel, G.A. and Stewart, W.D.P. (1982b) A physical map of plasmid pDU1 from the cyanobacterium *Nostoc* PCC 7524. Plasmid 7, 101–104.

Reichelt, B.Y. and Delaney, S.F. (1983) The nucleotide sequence for the large subunit of ribulose 1,5-bisphosphate carboxylase from a unicellular cyanobacterium, *Synechococcus* PCC 6301. DNA 2, 121–129.

Rice, D., Mazur, B.J. and Haselkorn, R. (1982) Isolation and physical mapping of nitrogen fixation genes from the cyanobacterium *Anabaena* 7120. J. Biol. Chem. 257, 13157–13163.

Rippka, R. and Cohen-Bazire, G. (1983) The cyanobacteriales: a legitimate order based on the type strain *Cyanobacterium stanieri*? Ann. Microbiol. (Inst. Pasteur) 134B, 21–36.

Rippka, R. and Herdman, M. (1985) Division patterns and cellular differentiation in cyanobacteria. Ann. Microbiol. (Inst. Pasteur) 136A, 33–39.

Rippka, R., Deruelles, J., Waterbury, J.B., Herdman, M. and Stanier, R.Y. (1979) Generic assignments, strain histories and properties of pure cultures of cyanobacteria. J. Gen. Microbiol. 111, 1–61.

Roberts, T.M. and Koths, K.E. (1976) The blue-green alga *Agmenellum quadruplicatum* contains covalently closed DNA circles. Cell 9, 551–557.

Robinson, S.J. and Haselkorn, R. (1985) Second copy of the gene for nitrogenase reductase in *Anabaena*. In Vth International Symposium on Photosynthetic Prokaryotes, Grindelwald, Switzerland. Abstract p. 346.

Roizes, G., Nardeux, P.C. and Monier, R. (1979) A new specific endonuclease from *Anabaena variabilis*. FEBS Lett. 104, 39–44.

Rosenberg, M. and Court, D. (1979) Regulatory sequences involved in the promotion and termination of RNA transcription. Ann. Rev. Genet. 13, 319–353.

Saunders, C.W. and Guild, W.R. (1981a) Pathway of plasmid transformation in *Pneumococcus*: open circular and linear molecules are active. J. Bacteriol. 146, 517–526.

Saunders, C.W. and Guild, W.R. (1981b) Monomer plasmid DNA transforms *Streptococcus pneumoniae*. Mol. Gen. Genet. 181, 57–62.

Sherman, L.A. and Van de Putte, P. (1982) Construction of a hybrid plasmid capable of replication in the bacterium *Escherichia coli* and the cyanobacterium *Anacystis nidulans*. J. Bacteriol. 150, 410–413.

Shestakov, S., Elanskaya, I. and Bibikova, M. (1985) Vectors for gene cloning in *Synechocystis* sp. 6803. In Vth International Symposium on Photosynthetic Prokaryotes, Grindelwald, Switzerland. Abstract p. 109.

Shine, J. and Dalgarno, L. (1974) The 3′-terminal sequence of *Escherichia coli* 16S ribosomal RNA: complementary to nonsense triplets and ribosome binding sites. Proc. Natl. Acad. Sci. USA 71, 1342–1346.

Shinozaki, K. and Sugiura, M. (1983) The gene for the small subunit of ribulose-1,5-bisphosphate carboxylase/oxygenase is located close to the gene for the large subunit in the cyanobacterium *Anacystis nidulans* 6301. Nucl. Acids Res. 11, 6957–6964.

Shinozaki, K. and Sugiura, M. (1985) Genes for the large and small subunits of ribulose-1,5-bisphosphate carboxylase/oxygenase constitute a single operon. Mol. Gen. Genet. 200, 27–32.

Shinozaki, K., Tomioka, N., Yamada, C. and Sugiura, M. (1982) Cloning and characterization of a plasmid DNA from *Anacystis nidulans* 6301. Gene 19, 221–224.

Shinozaki, K., Yamada, C., Takahata, N. and Sugiura, M. (1983) Molecular cloning and sequence analysis of the cyanobacterial gene for the large subunit of ribulose-1,5-bisphosphate carboxylase/oxygenase. Proc. Natl. Acad. Sci. USA 80, 4050–4054.

Simon, R.D. (1978) Survey of extrachromosomal DNA found in the filamentous cyanobacteria. J. Bacteriol. 136, 414–418.

Stam, W.T. (1980) Relationships between a number of filamentous blue-green algal strains (Cyanophyceae) revealed by DNA-DNA hybridization. Arch. Hydrobiol./Suppl. 56, Algological Studies 25, 351–374.

Stanier, R.Y., Kunisawa, R., Mandel, M. and Cohen-Bazire, G. (1971) Purification and properties of unicellular blue-green algae (order Chroococcales). Bacteriol. Rev. 35, 171–205.

Starr, R. (1978) The culture collection of algae at the University of Texas at Austin. J. Phycol. 14, Suppl., 47–100.

Stevens, S.E., Jr. and Porter, R.D. (1980) Transformation in *Agmenellum quadruplicatum*. Proc. Natl. Acad. Sci. USA 77, 6052–6056.

Sutcliffe, J.G. and Church, G.M. (1978) The cleavage site of the restriction endonuclease *Ava*II. Nucl. Acids Res. 5, 2312–2319.
Szekeres, M. (1981) Phage-induced development of a site-specific endonuclease in *Anacystis nidulans*, a cyanobacterium. Virology 111, 1–10.
Szekeres, M., Szmidt, A.E. and Török, I. (1983) Evidence for a restriction/modification-like system in *Anacystis nidulans* infected by cyanophage AS-1. Eur. J. Biochem. 131, 137–141.
Szybalski, E.H. and Szybalski, W. (1979) A comprehensive molecular map of bacteriophage lambda. Gene 7, 217–270.
Tandeau de Marsac, N. (1983) Phycobilisomes and complementary chromatic adaptation. Bull. Inst. Pasteur, 81, 201–254.
Tandeau de Marsac, N., Borrias, W.E., Kuhlemeier, C.J., Castets, A.M., Van Arkel, G.A. and Van den Hondel, C.A.M.J.J. (1982) A new approach for molecular cloning in cyanobacteria: cloning of an *Anacystis nidulans met* gene using a Tn*901*-induced mutant. Gene 20, 111–119.
Tandeau de Marsac, N., Mazel, D., Bryant, D.A. and Houmard, J. (1985a) Molecular cloning and nucleotide sequence of a developmentally regulated gene from the cyanobacterium *Calothrix* PCC 7601: a gas vesicle protein gene. Nucl. Acids Res. 13, 7223–7236.
Tandeau de Marsac, N., Mazel, D., Bryant, D.A. and Houmard, J. (1985b) Molecular studies of developmentally regulated genes from the cyanobacterium *Calothrix* PCC 7601: the gas vacuole protein genes. In Vth International Symposium on Photosynthetic Prokaryotes, Grindelwald, Switzerland. Abstract p. 32.
Tiboni, O. and Ciferri, O. (1985) Cloning of the genes involved in protein synthesis and CO_2 fixation from the cyanobacterium *S. platensis*. In Vth International Symposium on Photosynthetic Prokaryotes, Grindelwald, Switzerland. Abstract p. 361.
Tiboni, O., Di Pasquale, G. and Cifferi, O. (1984) Cloning and expression of the genes for ribulose-1,5-bisphosphate carboxylase from *Spirulina platensis*. Biochim. Biophys. Acta 783, 258–264.
Tomioka, N. and Sugiura, M. (1983) The complete nucleotide sequence of a 16S ribosomal RNA gene from a blue-green alga, *Anacystis nidulans*. Mol. Gen. Genet. 191, 46–50.
Tomioka, N. and Sugiura, M. (1984) Nucleotide sequence of the 16S–23S spacer region in the *rrn*A operon from a blue-green alga, *Anacystis nidulans*. Mol. Gen. Genet. 193, 427–430.
Tomioka, N., Shinozaki, K. and Sugiura, M. (1981) Molecular cloning and characterization of ribosomal RNA genes from a blue-green alga, *Anacystis nidulans*. Mol. Gen. Genet. 184, 359–363.
Tumer, N.E., Robinson, S.J. and Haselkorn, R. (1983) Different promoters for the *Anabaena* glutamine synthetase gene during growth using molecular or fixed nitrogen. Nature 306, 337–342.
Van den Hondel, C.A.M.J.J. and Van Arkel, G.A. (1980) Development of a cloning system in cyanobacteria. Antonie van Leeuwenhoek 46, 228–229.
Van den Hondel, C.A.M.J.J., Keegstra, W., Borrias, W.E. and Van Arkel, G.A. (1979) Homology of plasmids in strains of unicellular cyanobacteria. Plasmid 2, 323–333.
Van den Hondel, C.A.M.J.J., Verbeek, S., Van der Ende, A., Weisbeek, P.J., Borrias, W.E. and Van Arkel, G.A. (1980) Introduction of transposon Tn*901* into a plasmid of *Anacystis nidulans*: preparation for cloning in cyanobacteria. Proc. Natl. Acad. Sci. USA 77, 1570–1574.
Van den Hondel, C.A.M.J.J., Van Leen, R.W., Van Arkel, G.A., Duyvesteyn, M. and De Waard, A. (1983) Sequence-specific nucleases from the cyanobacterium *Fremyella diplosiphon*, and a peculiar resistance of its chromosomal DNA towards cleavage by other restriction enzymes. FEMS Microbiol. Lett. 16, 7–12.
Vermaas, W.F.J., Williams, J.G.K. and Arntzen, C.J. (1985) Genetic manipulation of the 47-kDA chlorophyll-binding photosystem II protein (CP47) in the cyanobacterium *Synechocystis* 6803. In Vth International Symposium on Photosynthetic Prokaryotes, Grindelwald, Switzerland. Abstract p. 112.
Wain-Hobson, S., Nussinov, R., Brown, R.J. and Sussman, J.L. (1981) Preferential codon usage in genes. Gene 13, 355–364.
Walker, J.E., Hayes, P.K. and Walsby, A.E. (1984) Homology of gas vesicle proteins in cyanobacteria and halobacteria. J. Gen. Microbiol. 130, 2709–2715.
Walsby, A.E. (1977) Absence of gas vesicle protein in a mutant of *Anabaena flos-aquae*. Arch. Microbiol. 114, 167–170.
Walsby, A.E. (1981) Cyanobacteria: planktonic gas-vacuolate forms. In M. Starr, H. Stolp, H. Truper, A. Balow and H.G. Schlegel (eds.), The Prokaryotes. Springer-Verlag, New York, pp. 224–235.
Weisbeek, P.J., Teerstra, R., Van Dijk, M., Bloemheuvel, G., De Boer, D., Van der Plas, J., Borrias, W.E. and Van Arkel, G.A. (1985) Modification and sequence analysis of the small plasmid pUH24 of *Anacystis nidulans* R2. In Vth International Symposium on Photosynthetic Prokaryotes, Grindelwald, Switzerland. Abstract p. 328.
Whitehead, P.R. and Brown, N.L. (1982) *Aha*III: a restriction endonuclease with a recognition sequence containing only A:T base pairs. FEBS Lett. 143, 296–300.

Whitehead, P.R. and Brown, N.L. (1985a) Three restriction endonucleases from *Anabaena flos-aquae*. J. Gen. Microbiol. 131, 951–958.

Whitehead, P.R. and Brown, N.L. (1985b) A simple and rapid method for screening bacteria for type II restriction endonucleases: enzymes in *Aphanothece halophytica*. Arch. Microbiol. 141, 70–74.

Williams, J.G.K. (1985) Directed mutagenesis of the PSII genes *psbC* and *psbD* in the cyanobacterium *Synechocystis* 6803. In Vth International Symposium on Photosynthetic Prokaryotes, Grindelwald, Switzerland. Abstract p. 111.

Williams, J.G.K. and Szalay, A.A. (1983) Stable integration of foreign DNA into the chromosome of the cyanobacterium *Synechococcus* R-2. Gene 24, 37–51.

Williamson, S.E. and Doolittle, W.F. (1983) Genes for $tRNA^{Ile}$ and $tRNA^{Ala}$ in the spacer between the 16S and 23S rRNA genes of a blue-green alga: strong homology to chloroplast tRNA genes and tRNA genes of the *E. coli rrn*D gene cluster. Nucl. Acids Res. 11, 225–235.

Wilmotte, A.M.R. and Stam, W.T. (1984) Genetic relationships among cyanobacterial strains originally designated as '*Anacystis nidulans*' and some other *Synechococcus* strains. J. Gen. Microbiol. 130, 2737–2740.

Wolk, C.P. and Kraus, J. (1982) Two approaches to obtaining low, extracellular deoxyribonuclease activity in cultures of heterocyst-forming cyanobacteria. Arch. Microbiol. 131, 302–307.

Wolk, C.P., Vonshak, A., Kehoe, P. and Elhai, J. (1984) Construction of shuttle vectors capable of conjugative transfer from *Escherichia coli* to nitrogen-fixing filamentous cyanobacteria. Proc. Natl. Acad. Sci. USA 81, 1561–1565.

Wolk, C.P., Flores, E., Schmetterer, G., Herrero, A. and Elhai, J. (1985) Development of the genetics of nitrogen-fixing cyanobacteria. In Vth International Symposium on Photosynthetic Prokaryotes, Grindelwald, Switzerland. Abstract p. 122.

Wood, N.B. and Haselkorn, R. (1980) Control of phycobiliprotein proteolysis and heterocyst differentiation in *Anabaena*. J. Bacteriol. 141, 1375–1385.

Zurawski, G., Bohnert, H.J., Whitfeld, P.R. and Bottomley, W. (1982) Nucleotide sequence of the gene for the M_r 32 000 thylakoid membrane protein from *Spinacia oleracea* and *Nicotiana debneyi* predicts a totally conserved primary translation product of M_r 38 950. Proc. Natl. Acad. Sci. USA 79, 7699–7703.

Note added in proof: Relevant articles published since this chapter was submitted:

Anderson, L.K. and Eiserling, F.A. (1985) Plasmids of the cyanobacterium *Synechocystis* 6701. FEMS Microbiol. Lett. 29, 193–195.

Buzby, J.S., Porter, R.D. and Stevens, S.E., Jr. (1985) Expression of the *E. coli lacZ* gene on a plasmid vector in a cyanobacterium. Science 230, 805–807.

Chauvat, F., De Vries, L., Van der Ende, A. and Van Arkel, G.A. (1986) A host-vector system for gene cloning in the cyanobacterium *Synechocystis* PCC 6803. Mol. Gen. Genet. 204, 185–191.

Golden, S.S., Brusslan, J. and Haselkorn, R. (1986) Expression of a family of *psbA* genes encoding a photosystem II polypeptide in the cyanobacterium *Anacystis nidulans* R2. EMBO J. 5, 2789–2798.

Houmard, J., Mazel, D., Moguet, C., Bryant, D.A. and Tandeau de Marsac, N. (1986) Organization and nucleotide sequence of genes encoding core components of the phycobilisomes from *Synechococus* 6301. Mol. Gen. Genet. 205, 404–410.

Karreman, C., Tandeau de Marsac, N. and De Waard, A. (1986) Isolation of a deoxycytidylate methyl transferase capable of protecting DNA uniquely against cleavage by endonuclease R.*Aqu* I (isochizomer of *Ava* I. Nucl. Acids Res. 14, 5199–5205.

Mazel, D., Guglielmi, G., Houmard, J., Sidler, W., Bryant, D.A. and Tandeau de Marsac, N. (1986) Green light induces transcription of the phycoerythrin operon in the cyanobacterium *Colothrix* 7601. Nucl. Acids Res. 14, 8279–8290.

Porter, R.D., Buzby, J.S., Pilon, A., Fields, P.I., Dubbs, J.M. and Stevens, S.E., Jr. (1986) Genes from the cyanobacterium *Agmenellum quadruplicatum* isolated by complementation: characterization and production of merodiploids. Gene 41, 249–260.

Rebière, M.-C., Castets, A.-M., Houmard J. and Tandeau de Marsac, N. (1986) Plasmid distribution among unicellular and filamentous cyanobacteria: occurrence of large and mega-plasmids. FEMS Microbiol. Lett. 37, 269–275.

Schmetterer, G., Wolk, C.P. and Elhai, J. (1986) Expression of luciferases from *Vibrio harveyi* and *Vibrio fisheri* in filamentous cyanobacteria. J. Bacteriol. 167, 411–414.

P. Fay and C. Van Baalen (Eds.) The Cyanobacteria

11

Cyanelles: organization and molecular biology

CATHERINE C. WASMANN[a], WOLFGANG LOEFFELHARDT[b] and HANS J. BOHNERT[a,c]

Departments of [a]Biochemistry, [c]Molecular and Cellular Biology, and Plant Sciences, University of Arizona, BioSciences West, Tucson, AZ, 85721, USA and [b]Institut für Allgemeine Biochemie und Ludwig-Boltzmann-Forschungsstelle für Biochemie, Universität Wien, A 1090 Wien, Austria

1. Introduction

Symbiotic relationships between cyanobacteria and eukaryotic host organisms were called syncyanoses (Pascher, 1914, 1929) and the individual association has been termed the cyanome. The symbiotic character is easily recognized in those cases in which the cyanobacteria are located intercellularly, and both partners are able to grow separately. A number of organisms have been described in which the association between the partners is obligate suggesting that a high degree of integration has been established. The few species of this enigmatic group which have been investigated in some detail are *Paulinella chromatophora, Gloeochaete wittrockiana, Glaucocystis nostochinearum, Glaucosphaera vacuolata*, and *Cyanophora paradoxa*. Their photosynthetic organelles, termed 'cyanelles' by Pascher (Pascher, 1929), were originally viewed as 'blue-green' endosymbionts derived from free-living cyanobacteria which had either invaded an apoplastidic organism or persisted within the host cell after non-destructive phagocytosis. Pascher (1914) also coined the words 'endocyanome' for the whole organism and 'endocyanosis' for this form of endosymbiosis. The endosymbiont hypothesis, advanced by Mereschkowsky (1905) to explain the establishment of organelles, helped to foster the idea that these endocyanomes represent an association of two species. For example, *Cyanophora paradoxa* (Korschikoff, 1924) has been described as an association between a biflagellated protist and a cyanobacterium, named *Cyanocyta korschikoffiana* (Hall and Claus, 1963). One goal of this review is to show that the view that these organisms are endosymbiotic associations must certainly be abandoned for *Cyanophora* and, most likely, should be abandoned for the other cyanelle-containing species as well.

If endocyanomes are not an association of two organisms, which potentially can live independently, then how do we include them in a classification system? There is evidence (discussed below) that the host cells belong to different classes supporting the assumption that endocyanomes were established several times. In contrast the cyanelles in all these organisms appear to be similar enough to warrant their classification as a single class. One possibility is that endocyanomes are relics of endosymbiotic events. Multiple events would explain the different host organisms. Alternatively, it may be argued that cyanelles are an evolutionarily distinct class of plastids, analogous to chloroplasts or rhodoplasts, and as such might be called cyanoplasts. Proponents of these views accept to varying degrees the endosymbiont hypothesis for the establishment of cell organelles. In light of this hypothesis, it has been suggested that cyanelles can serve as a model for the acquisition of plastids.

In recent years the endosymbiont hypothesis has gained credibility (Whatley and Whatley, 1981; Margulis, 1981), as comparative data on protein sequences, oligonucleotide catalogs, and nucleotide sequences have become available (Schwartz and Dayhoff, 1981; Fox et al., 1980; Gray and Doolittle, 1982; Koessel et al., 1983). Chloroplast and cyanobacterial genes have been shown to be highly homologous. In addition, chloroplasts and cyanobacteria show striking similarities both in the structures of the photosynthetic reaction centers, and in the primary biochemistry of photosynthesis (see Chapter 1 of this volume). In a puristic view, these apparent coincidences might be imposed by functional constraints and may thus represent analogies rather than true homologies. Recent reports show that the gene order within operons is conserved. In some cases the conserved gene order has been found in eubacteria and in cyanobacteria and in the cyanelle as well as in chloroplast chromosomes (see below). Such similarities in gene order are at least consistent with, if not proof of, evolutionary relatedness.

One of the questions which remains is whether host/cyanelle associations are ancient remnants of a process which led to plastids, or whether they are representatives of an ongoing process. Whereas it may be impossible to verify one or the other experimentally, it may be possible to determine whether or not the various organisms originated from different endosymbiotic events. By comparing structural and functional characteristics of genes and operons one may be able to trace the lineage from the cyanobacteria to the various plastid types. One major field of research will be to study the mechanisms which coordinate gene expression between host cell and cyanelle. These mechanisms will have to be compared with the mechanisms controlling the expression of genes in the chloroplast and nucleus of higher plants.

In this review we have compiled data on the structural organization of the different cyanelles and on their biochemical characteristics. Several other reviews have been devoted to this topic (Trench, 1979; Trench, 1982a,b; Reisser, 1984). It appears justified to add another survey and summary which includes molecular biological data on the cyanelle from *Cyanophora paradoxa*, the only cyanelle that has been studied in this respect. Over the last 3 years many data concerning genome size, gene organization and gene content of this cyanelle have been reported. Other reviews which describe our knowledge about the structure of chloroplast chromosomes and genes are available (Ellis, 1981; Bohnert et al., 1982a; Whitfeld and Bottomley, 1983; Stutz, 1984; Crouse et al., 1984, 1985; Palmer, 1985).

2. Taxonomic position and host organization

The discovery of endocyanomes and the riddle which they present has led to several contradictory suggestions about how they might be classified. It is hoped that the integration of ultrastructural, biochemical, and molecular data will provide a comprehensive view. This stage has, however, not yet been reached. In the endocyanomes *Cyanophora*, *Glaucocystis*, *Glaucosphaera*, *Gloeochaete*, and *Paulinella*, the host organisms are true eukaryotes (Skuja, 1954; Kies, 1979, 1980) containing a nucleus, mitochondria, vacuoles, dictyosomes, and starch granules. Despite the wealth of structural data, the taxonomic affinities of these host organisms have remained unclear. The confusion concerning their position is exemplified by *Cyanophora paradoxa*. Attempts to classify the flagellated host of this organism have been largely unsatisfactory. One widely held view was that the organism is related to cryptomonad algae. This concept was refuted by Mignot et al. (1969) who, after an electron microscopic study, concluded that the host was not related to cryptomonads but rather to primitive dinoflagellates. In an attempt to clarify the taxonomic position of *Cyanophora*, Pickett-Heaps (1972) studied cell division in the flagellate. He found that the host possesses a typical mitotic spindle which is distinct from the unusual mitotic spindle of dinoflagellates. Both Mignot et al. (1969) and Pickett-Heaps (1972) noted that the basal body of the flagellum of *Cyanophora* was unlike that of any alga which had been examined previously. The unique structure of the flagellar roots of *Cyanophora* was also recognized by Kies (1976). Skuja (1954) had suggested previously that a new class, the Glaucophyceae, be established to include several endocyanomes. This view was revived by Kies (1976, 1980, 1984), who compared the ultrastructure of *Cyanophora paradoxa*, *Gloeochaete wittrockiana*, *Glaucocystis nostochinearum*, and *Glaucosphaera vacuolata*. In light of their ultrastructural studies Kies and his colleagues (Kies and Kremer, 1979) considered Skuja's proposal justified and suggested that *Cyanophora paradoxa* be included in the Glaucophyceae. Two different views have been presented with regard to the relationship of the Glaucophyceae and other algal classes. One proposal placed them close to the Chlorophyceae (Moestrup, 1978). The second view which emphasizes the particular endosymbiotic event, suggested that the Glaucophyceae are related to the Rhodophyta (Cavalier-Smith, 1981). However, the Rhodophyta do not possess any flagellated stages. In addition, the pattern of primary assimilates of photosynthetic carbon dioxide fixation in the endocyanomes studied differs from that of the red algae (see below).

Another endocyanome which is taxonomically unique is *Paulinella chromatophora*. Because of its ultrastructural characteristics Kies (1980, 1984) classified the host as a thecamoeba. Kies (1980) stated that of all cyanelles those in *Paulinella* most closely resemble free-living cyanobacteria. However, Kies also cautioned that cyanelles, because of their uniformity, should not be used for classification of endocyanomes.

Another organism of controversial classification is *Cyanidium caldarium*. Some authors have considered this thermoacidophilic alga to be an endocyanome (Raven, 1970; Kremer et al., 1978; Kremer, 1983). Others have viewed it as a primitive rhodophyte (Seckbach, 1972). *Cyanidium* possesses a single plastid with cyanobacterial pigment composition and thylakoid structure, and a single mitochondrion. No flagellated forms are known. In a recent review this organism and the other members of the (sub)class cyanidia (three species) are seen as a phylogenetic 'bridge' (Seckbach et al., 1983) linking prokaryotic cyanobacteria and the primitive red algae. Based on a comparative study of the photoassimilatory products of *C. caldarium*, *Cyanophora paradoxa*, the cyanobacterium *Anacystis nidulans*, and the unicellular red-alga *Rhodella violacea*, Kremer (1983) supported the view that *Cyanidium caldarium* is an endocyanome.

3. Ultrastructural organization of cyanelles

The ultrastructure of cyanelles has been studied by several groups (Hall and Claus, 1963, 1967; Schnepf et al., 1966; Schnepf and Brown, 1971; Robinson and Preston, 1971; Kies, 1979, 1980; Trench et al., 1978; Giddings et al., 1983; Scott et al., 1984). The essential features of cyanelle ultrastructure are compiled in Table I.

TABLE I
Characteristics of endocyanomes

Species	Proposed taxonomic position of host[a]	Cyanelle		Cyanelle characteristics			
		number	size[b]	independent growth	cell wall	central body	pigments
Cyanophora paradoxa	dinoflagel. cryptomonad protomastign.	1-4(8)	2–4	no	yes	yes	chlorophyll *a*, zeaxanthin C-phycocyanin, β-carotene allophycocyanin β-cryptoxanthin
Glaucocystis nostochinearum	Chlorophyta	up to 100	3–5	no	yes	yes	as *C. paradoxa*
Gloeochaete wittrockiana	Chlorophyta	many	2–3	no	yes	yes	as *C. paradoxa*
Glaucosphaera vacuolata	Chlorophyta	many	1–2	no	no(?)	no	as *C. paradoxa* + R-phycocyanin
Paulinella chromatophora	Rhizopoda (thecamoebae)	2	up to 20	not known	yes	several small	not known
Cyanidium caldarium[c]	Rhodophyta	1	2	not known	no	no	chlorophyll *a* + *d*, R/C-phycocyanin, R/B-phycoerythrin β-carotene, xanthophylls
Geosiphon pyriforme[c]	phycomycetes true symbiosis (*Nostoc sp.*)	many	–	yes	yes	yes	C-phycocyanin, β-carotene chlorophyll *a*, echinenone myxoxanthophyll

Several other species have been described which have been studied only marginally. The list includes *Cardiomonas printzii*, *Cryptella cyanophora*, *Cryptoeonas gemma*, *Cyanophora tetracyanea*, *Gymnodinium lantzschii*, *Massartia crassifilum*, *M. hiemalis*, *M. edax*, *M. montana*, *Peliaiana cyanea*, *Sphaeraspis pascheri*, *Strobobilomonas cyanea*. All these species are viewed as the result of an endosymbiotic event between a flagellate host and a cyanobacterium. Other species where the host is believed to be a eukaryotic alga are *Chalarodora azurea*, *Cyanoptyche gloeocystis*, and *C. dispersal*. These species are described by Geitler (1959). The classification of endocyanomes as members of a separate class Glaucophyceae has been proposed (Skuja, 1954; Kies, 1980).

[a]Different authors have placed *Cyanophora paradoxa* into different phyla. [b]Size in micrometer. [c]Included for comparative purposes.

3.1. Cell wall

All cyanelles, except perhaps those of *Glaucosphaera*, appear to be surrounded by a rudimentary cell wall. The presence of this wall is most easily demonstrated in *Cyanophora paradoxa*. In this organism the host cell, but not the cyanelle, may be ruptured by mild osmotic shock. The osmotic stability of the cyanelle is destroyed by treatment with lysozyme (Schenk, 1970). The lysozyme sensitivity of this wall was explained by the observation that the cyanelles of *Cyanophora* and *Glaucocystis* contain peptidoglycan (Aitken and Stanier, 1979; Scott et al., 1984). The thickness of the wall varies in the different endocyanomes, being in the order of 7–10 nm (Hall and Claus, 1963; Giddings et al., 1983). When compared with the cell wall of cyanobacteria the rudimentary wall of the cyanelles appears to consist only of the inner peptidoglycan layer. The lipopolysaccharide layer found in cyanobacteria is absent (Drews and Weckesser, 1982). Based on freeze-fracture data, Giddings et al. (1983) suggested that some lipopolysaccharide may be present in the cyanelle wall of *C. paradoxa*.

3.2. Thylakoid structure

The pigment composition (Chapman, 1966; reviewed by Trench, 1982a,b), and the structure of the thylakoids are reminiscent of those of cyanobacteria (Gantt, 1981; Glazer, 1984). Thylakoid arrangement is peripheral and concentric. The phycobilisomes are attached to the outer surface of the thylakoids. A chromatoplasm is clearly distinguishable from the thylakoid-free centroplasm. Lipid droplets, sometimes in regular arrays, ribosomes and polyphosphate granules are distributed throughout the chromatoplasm.

3.3. Centroplasm

Cyanelles, except those of *G. vacuolata*, possess a dense central body, which often appears as a crystalline structure (Schnepf and Brown, 1971; Trench et al., 1978; Trench, 1982a). This structure is surrounded by a membrane in *G. wittrockiana*, but not in *C. paradoxa*. The size of the crystalline body varies in different cyanelles. In *G. nostochinearum* it occupies a peripheral position. The cytoplasm surrounding the crystalline part of the centroplasm seems denser than in the other regions. The cyanelle DNA appears to be imbedded in this denser cytoplasmic layer. The DNA in the cyanelles of *Cyanophora paradoxa* has been visualized by DNA-specific cytochemical fluorescence, DAPI staining (James and Jope, 1978). Light microscopic observations showed that the fluorescence signal was located primarily in the center of the cyanelle (Trench, 1982a; Bohnert and Loeffelhardt, 1984). Electron microscopy of DNA preparations from gently lysed cyanelles showed that the DNA is associated with the central crystalline body (Bohnert and Loeffelhardt, 1984).

Although the central body has not been characterized biochemically, several ideas have been advanced regarding its nature. The prevailing view is that it contains ribulose 1,5-bisphosphate carboxylase (Trench, 1982a,b; Kies, 1980). Kies suggested that it is a polyhedral body or carboxysome, a structure similar to that found in most cyanobacteria (Codd and Stewart, 1976, 1977; see also Section 4 of Chapter 8). Other suggestions regarding this structure are that it is a pyrenoid (Schnepf and Brown, 1971) or a nucleoid (Margulis, 1981).

3.4. Division

Typically several cyanelles are found in each cell. The number of cyanelles is assumed to be controlled by the nucleus and to be coupled to mitosis of the host cell. *Cyanophora* commonly contains either two or four cyanelles. The cells may contain as many as eight cyanelles. During division of the host cell the cyanelles are divided equally between the daughter cells. After division of the host cell each cyanelle divides once. *Paulinella* has two cyanelles, each daughter cell receives one cyanelle during division of the host. In *Glaucocystis* and *Gloeochaete* as many as 100 cyanelles may be present in each cell. In these organisms degeneration (digestion) of cyanelles (Kies, 1976) or extrusion of groups of cyanelles has been observed. Loss of cyanelles due to non-synchronized growth or unequal distribution to the daughter cells has been observed in *Peliaina* (Pascher, 1929).

Cyanelles divide by constriction, a mechanism similar to that found in coccoid cyanobacteria (Hall and Claus, 1963; Pickett-Heaps, 1972). The plane of division is always perpendicular to the longer axis of the cyanelle. Division starts with an increased deposition of peptidoglycan localized at the future plane of division. As division proceeds, the ring of peptidoglycan constricts, interrupting the thylakoids and bisecting the central body. The daughter cyanelles separate, provoking division of the vesicle surrounding the original organelle (Kies, 1976, 1984). The nature of this vesicle, which is missing in *G. vacuolata* (Kies, 1976), is not known. Kies (1984) suggests that the vesicular membrane is provided by the host cell.

4. Biochemistry and metabolism of cyanelles

Cyanelle biochemistry has been reviewed recently by Trench (1982a,b). In this review, we will restrict ourselves to a short summary and description of the characters which distinguish cyanelles from both higher plant chloroplasts and the free-living cyanobacteria. Special emphasis is placed on newer results, derived mainly from studies of *Cyanophora paradoxa*.

4.1. Photosynthesis, pigments and proteins of thylakoids

For biochemical studies *Cyanophora paradoxa* may be grown axenically in synthetic media. Using light-synchronized cultures Marten et al. (1983) studied development of the photosynthetic membranes. Under these conditions, thylakoid development appeared to be synchronized. During the dark period pigment synthesis ceased and stacks of thylakoids accumulated at defined regions. Although these membranes appeared to be associated with biliproteins, their structure differed from that of functional thylakoids. The stacks of membranes disappeared concomitantly with the light-induced synthesis of pigments. Marten et al. (1983) suggested that the membranous material is a storage form which can be converted into functional thylakoids during the light period.

The pigments of cyanelles (Table I) are chlorophyll *a*, β-carotene, zeaxanthin, β-cryptoxanthin as well as the phycobiliproteins C-phycocyanin and allophycocyanin. A unique R-phycocyanin was found in *Glaucosphaera* (Gantt et al., 1979). Two carotenoids, echinenone and myxoxanthophyll, found in many cyanobacteria are absent in cyanelles. Chlorophyll *c*, which is present in cryptomonad algae and dinoflagellates, is not present in cyanelles. The most obvious differences between cyanelles and chloroplasts is the lack of chlorophyll *b* in the cyanelles and the presence of phycocyanins as the major accessory pigment for Photosystem II (Chapman, 1966; Trench and Ronzio, 1978; Trench et al., 1978; Schmidt et al., 1979).

The biliproteins C-phycocyanin and allophycocyanin have been purified from the cyanelles of *Cyanophora* (Trench and Ronzio, 1978). The C-phycocyanin was found to consist of two subunits with apparent molecular masses of 13 200 and 14 500 daltons. The allophycocyanin was found to be composed of two subunits of similar molecular mass, 12 600 daltons. After cross-linking with dimethylsuberimidate (DMSI) the biliprotein dimers had apparent molecular masses of 28 000 daltons (allophycocyanin) and 31 000 daltons (C-phycocyanin). These values are similar to those reported for cyanobacteria (Bogorad, 1975; Gantt, 1981; Yamanaka et al., 1978).

4.2. Metabolic functions of cyanelles

Schenk (1977) reported that the rate of oxygen evolution by *Cyanophora* is comparable to that of chloroplasts. Carbon dioxide fixation in the dark was negligible. Isolated cyanelles sustained CO_2-dependent oxygen evolution for a short time only and only at a fractional amount of the fixation of intact cells (Trench et al., 1978; Floener and Bothe, 1982). Rates of carbon dioxide fixation are of the same order of magnitude as has been reported for free-living cyanobacteria.

Analysis of the primary products of carbon dioxide fixation showed that all cyanelles produced predominantly glucose, organic phosphates (intermediates of the Krebs-cycle), amino acids, and lipids. As much as 30% of the total radioactivity assimilated from $^{14}CO_2$ may be present in the lipid fraction (Kies and Kremer, 1979). Overall, the pattern of products labeled by $NaH^{14}CO_3$ is consistent with the operation of the reductive pentose phosphate cycle (Kremer et al., 1979). Unlike cyanobacteria, cyanelles labeled for short-time periods do not synthesize free disaccharides or oligosaccharides. Oligosaccharide synthesis appears to be a function of the host cell (Kremer et al., 1979). In contrast to the results of Kremer et al. (1979), Trench et al. (1978) observed that in addition to glucose, a disaccharide was among the primary products of CO_2 fixation. Without identifying the compound, these authors suggested that it was sucrose. Radiolabeling for longer periods resulted in appreciable amounts of activity in the insoluble fraction. Hydrolysis of this material yielded free glucose and amino acids. Kies and Kremer (1979) suggest that these labeled compounds are deposited in a polyglucan-type cell wall.

The enzyme catalysing the initial step in the Calvin cycle, ribulose 1,5-bisphosphate carboxylase/oxygenase, has been purified from whole cells and from isolated cyanelles (Codd and Stewart, 1977; Reisfeld and Edelman, 1976; Burnap and Trench, 1984). The holoenzyme, like that of cyanobacteria and plants, consists of eight large and eight small subunits (see Section 4 of Chapter 4). Based on the amino acid sequence deduced from the nucleotide sequence, the molecular weights of the two subunits are 52 800 and 12 400 (Starnes et al., 1985; Wasmann, 1985; see below). The molecular weights of the subunits as determined by sodium dodecylsulfate (SDS)-gel electrophoresis have varied but are consistent with those obtained from the nucleotide sequence.

Host/cyanelle interaction has been studied using inhibitors which differentially affect protein synthesis occurring on cytoplasmic and cyanelle (as well as chloroplast) ribosomes. This approach has the advantage that duplicate samples of cells may be treated with different inhibitors during the same experiment. Information can thereby be gained regarding the sites of synthesis of proteins both in the cytoplasm of the host and in the cyanelle. There are, however, several disadvantages associated with this approach. A major problem is uncertainty regarding the specificity of the inhibitors used. The two drugs, chloramphenicol and cycloheximide, have been reported to perturb cellular processes other than protein synthesis. For example, photophosphorylation is inhibited by chloramphenicol in isolated spinach chloroplasts (Ellis, 1977; Crouse et al., 1984). The difficulty of determining the concentration of the drug at the target site and the mutual dependence of organelle and cytosolic metabolic function also complicate the interpretation of the results of inhibitor studies. Despite these shortcomings, several groups have used inhibitors of protein synthesis to study host/cyanelle interaction in *C. paradoxa* (Trench and Siebens, 1978; Siebens and Trench, 1978; Marten et al., 1981; Neu-Mueller and Schenk, 1983; Marten et al., 1983). Both chloramphenicol and cycloheximide inhibited synthesis of chlorophyll *a*, C-phycocyanin and allophycocyanin. Oxygen evolution was not affected when intact cells were exposed to either inhibitor for short periods of time. However, when cells were exposed to either chloramphenicol or cycloheximide for longer periods of time, it was found that chloramphenicol, but not cycloheximide, inhibited oxygen evolution (Brandt and Marten, 1983; Marten et al., 1983). Trench and Siebens (1978) reported that chloramphenicol, but not cycloheximide, inhibited cyanelle rRNA synthesis. Cytoplasmic rRNA synthesis was inhibited by cycloheximide, but not by chloramphenicol. Conclusions about the degree of integration that exists between the cyanelles and the host cell were based on such studies though one may hesitate, without questioning the data, to assign to such experiments particular significance because of the difficulties (discussed above) of interpreting the results of inhibitor studies.

4.3. Nitrogen metabolism

In higher plants, the assimilatory nitrate reductase is a cytosolic enzyme; nitrite reductase is located in the chloroplast. In cyanobacteria both enzymes are bound to the thylakoid membrane (see Section 2.2. of Chapter 6). Nitrite and nitrate reduction have been studied in *Cyanophora paradoxa*. Bothe and his colleagues (Bothe and Floener, 1978; Floener et al., 1982) have reported that cyanelles, which have been carefully purified from cytosolic contaminants, lacked the enzyme nitrate reductase. One NADH-dependent isoform of the enzyme was found in whole cell extracts (Floener et al., 1982; Boettcher et al., 1982). The location of nitrite reductase in *C. paradoxa* is controversial. Floener et al. (1982) found that nitrite reductase was present exclusively in the cyanelle fraction and that it is a ferredoxin-dependent thylakoid-bound enzyme. The earlier observation of Bothe and Floener (1978) that nitrite reductase activity is stimulated by light appears to support their contention that the enzyme is located in the cyanelle. On the basis of their results, Bothe and coworkers proposed that the pathway of assimilatory nitrate reduction in *Cyanophora* is similar to that of photosynthetic eukaryotes and unlike the pathway in cyanobacteria. In contrast to the results of Floener et al. (1982), Boettcher et al. (1982) found a significant portion of the nitrite reductase activity in the cytoplasm of the host.

Ammonia assimilation in *Cyanophora* proceeds via the glutamine synthetase/glutamate synthase pathway (Floener and Bothe, 1982). Glutamate synthase activity is present both in the cyanelle and in the host cytoplasm. In addition, Floener and Bothe (1982) found two glutamine synthetases which differed in their specific activities. The enzyme having the higher specific activity is localized in the cytoplasm. These characteristics are consistent with the view that cyanelles and chloroplasts are closely related biochemically. If cyanelles are endosymbiotic cyanobacteria the possibility exists that they might be capable of nitrogen fixation. Experiments by Bothe and Floener (1978) suggest, however, that *C. paradoxa* is unable to synthesize nitrogenase, not even in a 'microaerobic' environment, though anaerobically grown organisms do synthesize hydrogenase.

With respect to other biosynthetic pathways which occur in plant chloroplasts, information on their counterparts in *C. paradoxa* is scanty. A 3′-phosphoadenosine-5′-phosphosulfate (PAPS) sulfotransferase was partially purified from cyanelles and characterized (Schmidt and Christen, 1979). In the absence of information about other forms of sulfotransferases in the whole cell it is impossible to judge what role the cyanelles have in assimilatory sulfate reduction.

5. Molecular biology of cyanelles

5.1. Genome size and complexity

The single member of the Glaucophyceae for which data are available is *Cyanophora paradoxa*. Herdman and Stanier (1977) first isolated DNA from the cyanelles and showed that the kinetic complexity, about 175 Kbp (kilo-basepairs), of the cyanelle DNA was only 5 to 10% of that of free-living cyanobacteria (Herdman et al., 1979a,b; Mucke et al., 1980; Jaynes et al., 1981). The small size of the cyanelle genome may explain why attempts to grow isolated cyanelles from *Cyanophora paradoxa* failed (Trench et al., 1978; Klein et al., 1981a,b). Determinations based on restriction endonuclease digestions indicate that the sizes of the cyanelle DNA from two strains of *Cyanophora* are approximately 127 Kbp ('Pringsheim' strain; LB555 UTEX) and approximately 139 Kbp ('Kies' strain, number 1555), respectively (Bohnert et al., 1983; Loeffelhardt et al., 1983). Comparable values are given by others (Wasmann, 1985 — 128 Kbp; Lambert et al., 1985 — 133 Kbp). Electron microscopic data and physical mapping of the DNA showed that the cyanelle genome is a circular molecule of the size predicted by restriction endonuclease digestion (Bohnert et al., 1982b, 1985; Wasmann, 1985; Lambert et al., 1985). In *Cyanophora* the cyanelle DNA is present in approximately 40 copies per cyanelle (Mucke et al., 1980). The assignment of DNA buoyant density classes to the DNAs from different cellular compartments of *Cyanophora* has been problematic. Values of either 1.692 g/cm^3 (between 1.690 and 1.695 g/cm^3) or 1.715 g/cm^3 have been reported for the cyanelle DNA (Herdman and Stanier, 1977; Edelman et al., 1967). It now seems clear that the cyanelle DNA has a buoyant density of 1.692 g/cm^3 and that the host mitochondrial DNA has a buoyant density of 1.715 g/cm^3. Restriction endonuclease digestion of the DNA from isolated host mitochondria produces a distinct set of DNA fragments with a combined complexity of approximately 250 Kbp (Bohnert et al., 1982b; Loeffelhardt W., unpublished).

Estimates of the size of the cyanelle DNA based on renaturation kinetics disagree with estimates based on electron microscopy or on digestion with restriction endonucleases. This disparity may be due to the presence of closed circular DNA molecules of approximately 50 Kbp which frequently contaminate the preparations of cyanelle DNA. This contaminant may be removed by DNAase treatment which does not degrade the DNA in the cyanelles. The buoyant density of this contaminating DNA is 1.688 g/cm^3 (Bohnert et al., 1982b). It was suggested that this DNA represented a nuclear or a mitochondrial plasmid or that it is derived from a contaminating organism. It is unlikely that this DNA is of mitochondrial origin, because the DNA isolated from mitochondria both lacks this component and has a different buoyant density.

5.2. Chromosome structure

Restriction maps of the cyanelle DNA have been presented by several authors (Wasmann, 1985; Lambert et al., 1985; Bohnert and Loeffelhardt, 1982; Bohnert et al., 1985). In agreement with the electron microscopic images showing a circular molecule, the physical map of restriction endonuclease cleavage sites is also circular. Like many algal and higher plant chromosomes, the cyanelle chromosome is characterized by two segments of a repeat unit arranged in inverted orientation (Whitfeld and Bottomley, 1983; Palmer, 1985). The size of this repeat unit is 11 ± 0.5 Kbp. The two copies of the repeat unit are separated by regions which contain single copy DNA. Fig. 1a shows the structure of this chromosome. For comparison a map of the spinach chloroplast DNA is shown in Fig. 1b.

The structure of the spinach chloroplast DNA is the most common type of chloroplast chromo-

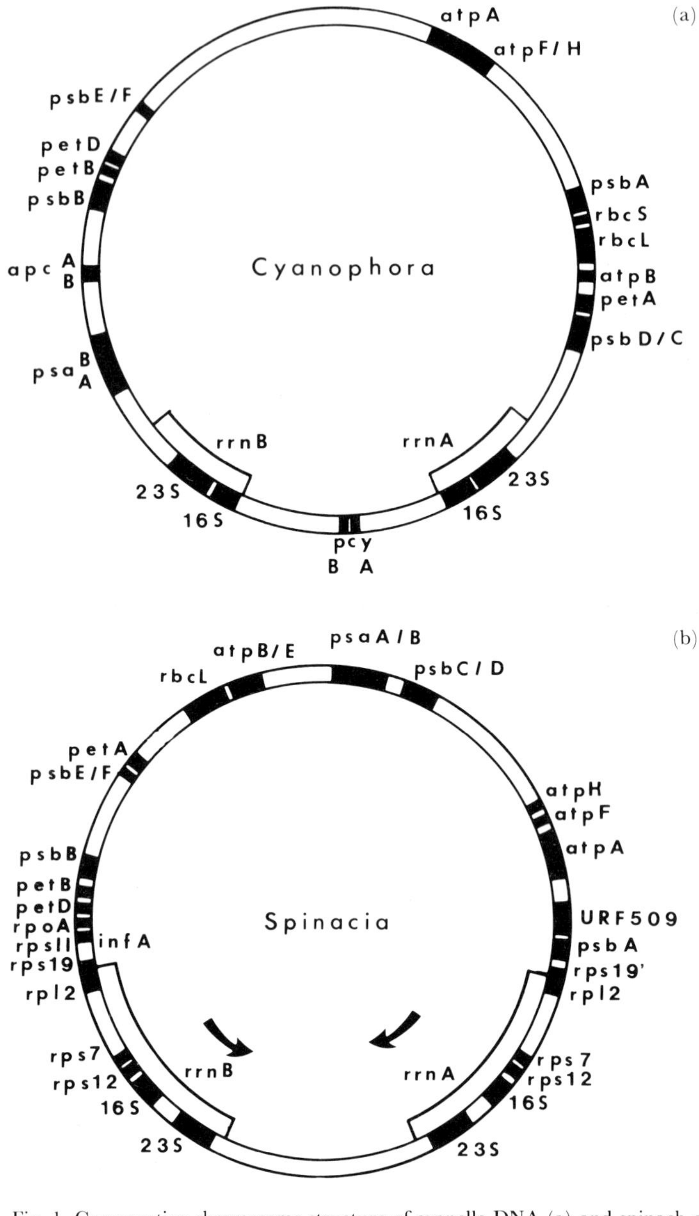

Fig. 1. Comparative chromosome structure of cyanelle DNA (a) and spinach chloroplast DNA (b). The symbols of genes (see Table II) are included at the position mapped or sequenced.

some organization. It is found in most higher plants and several green algae, e.g., *Chlamydomonas* and *Chlorella* (see Bohnert et al., 1982a; Palmer, 1985, for reviews; Lemieux et al., 1985; Yamada, 1983). At present only three different types of chloroplast chromosomes are known: the spinach-type (Fig. 1b); the *Euglena*-type, which is distinguished by the presence of a variable number of tandem repeats (Hallick et al., 1985; Koller and Delius, 1982; Ravel-Chapuis et al., 1984); and the pea-type in which one copy of the inverted repeat unit has been deleted (Koller and Delius, 1980).

The surprising conservation of the chromosome structure of higher plant chloroplast DNAs led Bedbrook and Kolodner (1979) to propose that the large inverted repeats act to stabilize this structure. Recombination between the repeats, which is presumed to occur in chloroplast DNAs with inverted repeats, has been shown to occur in *Cyanophora paradoxa* (Bohnert and Loeffelhardt, 1982), in common bean (Palmer, 1983; Mubumbila et al., 1983), and in *Chlamydomonas reinhardtii* (Aldrich et al., 1985). The chloroplast or cyanelle DNA of these organisms exists as a mixed population of two orientation isomers. The presence of more than one isomeric form of organelle DNA may indicate the presence of a recombination system which recognizes sequences within the inverted repeat units as targets. A similar 'flip-flop' recombination mechanism has been found in yeasts that contain the 2-micron plasmid (Broach et al., 1982) and in a mitochondrial DNA (Hudspeth et al., 1983).

The data on the chromosome structure of cyanelle DNA reviewed here were obtained with the so-called 'Pringsheim' strain (LB555 UTEX, Culture Collection University of Texas), which is the strain used by most investigators. Differences exist between this strain and another strain of *Cyanophora paradoxa* isolated by Kies (1979). The 'Kies'-strain, number 1555, is distinguished by a larger, 139 Kbp, cyanelle DNA. Whereas the pattern of restriction endonuclease cleavage sites of the single copy regions of strain 1555 is different from that of strain LB555 UTEX, the cleavage sites in the inverted repeat units appear to be conserved (Loeffelhardt et al., 1983; Loeffelhardt et al., unpublished).

5.3. Gene localization

At present no information is available on the chromosome or gene structure of the nuclear DNA of the host cells. Recently, some data have been provided concerning genes located on the cyanelle DNA from *Cyanophora paradoxa*. The rationale for assigning genes to restriction fragments of cyanelle DNA is based on the demonstration that genes from chloroplasts share considerable homology with genes from cyanobacteria and eubacteria (Loeffelhardt et al., 1985; see also Section 5.4. of Chapter 10). Thus, DNA/DNA hybridization can be used to find sequence homologs. Homologies (see below) also have been reported for rRNA and tRNA genes (Gray and Doolittle, 1982; Bohnert et al., 1982a; Koessel et al., 1983). The genes located in this way are listed in Table II. The nomenclature of the genes used here follows that used for chloroplast genes (Hallick and Bottomley, 1983). Hybridization probes were obtained from cloned chloroplast or cyanobacterial DNA fragments. The probes used include probes for genes of Photosystems I and II, and the cytochrome b_6/f complex, components of the chloroplast ATP-synthase complex, and chloroplast ribosomal protein genes (Lambert et al., 1985; Bohnert et al., 1985; Wasmann, 1985).

The *rbcL* gene from *C. paradoxa*, coding for the large subunit of ribulose 1,5-bisphosphate carboxylase/oxygenase (Rubisco), has been localized on the cyanelle DNA and the gene has been sequenced (Wasmann, 1985). A partial sequence has been reported by Lambert et al. (1985). The localization of the gene *rbcS*, for the small subunit of Rubisco is somewhat ambiguous. Based on hybridizations with a pea cDNA clone of *rbcS* Heinhorst and Shively (1983) and Bohnert et al. (1983) reported that it is located on the cyanelle chromosome approximately 3 to 7 Kbp from the *rbcL* gene. More recently, DNA sequencing (Lambert et al., 1985; Wasmann, 1985) has shown that an *rbcS* gene is located immediately adjacent to the 3′-end of *rbcL* (cf. Section 4.1. of Chapter 4). The nature of the additional hybridization signal obtained by using *rbcS* probes from higher plants is unclear. Although the possibility exists that a second *rbcS* gene is present, this case exemplifies

TABLE II

Genes located on the cyanelle DNA from *Cyanophora paradoxa* and genes located on chloroplast DNAs

Gene symbol	Description	Cyanelle location	Cyanelle sequence	Remarks/references
Components of Photosystem I				Chloroplast genes sequenced
psaA	apoprotein P700a	+	–	three genes identified in chloroplasts; psaA/B
psaB	apoprotein P700b	+	–	presumably in one operon; (Alt et al., 1984)
Components of Photosystem II				Chloroplast and cyanobacterial genes sequenced
psbA	Q_B, herbicide binding protein, D-1 protein	+	–	located and sequenced in several chloroplast DNAs and cyanobacteria Zurawski et al., 1982b; Erickson et al., 1984; Curtis and Haselkorn, 1984
psbB	apoprotein P680, 51 kD-protein	+	–	Morris and Herrmann, 1984
psbC	apoprotein, 44 kD-protein	+	–	Holschuh et al., 1984; Alt et al., 1984
psbD	D-2 protein	+	–	Holschuh et al., 1984; Alt et al., 1984
psbE	apoprotein cytochrome *b*-559	+	–	Westhoff et al., 1985b
ATP-synthase complex				Chloroplast and cyanobacterial genes sequenced
atpA	α subunit, CF_1	+	–	Deno et al., 1983
atpB	β subunit, CF_1	+	partial	Zurawski et al., 1982a; Wasmann, 1985
atpE	ε subunit, CF_1	+	–	Zurawski et al., 1982a
atpF	subunit I, CF_1	+	–	Alt et al., 1983b
atpH	subunit III, CF_0 or DCCD-binding protein	+	–	Alt et al., 1983a
Cytochrome-b_6/f complex				Chloroplast genes sequenced
petA	apoprotein cytochrome *f*	+	–	Alt and Herrmann, 1984; Wiley et al., 1984
petB	apoprotein cytochrome b_6	+	–	Heinemeyer et al., 1984
petD	b_6/f complex subunit IV	+	–	Heinemeyer et al., 1984
Ribulose 1,5-bisphosphate carboxylase/oxygenase				Chloroplast and cyanobacterial genes sequenced
rbcL	large subunit	+	+	chloroplast gene in higher plants and Chlorophyceae, Zurawski et al., 1981; Starnes et al., 1985; Wasmann, 1985
rbcS	small subunit	+	+	nuclear gene in higher plants and *Chlamydomonas*, Starnes et al., 1985; Wasmann, 1985
Ribosomal proteins/translation machinery				Chloroplast genes sequenced
rps4	ribosomal small subunits, S4	–	–	Subramanian et al., 1983
rps7	ribosomal small subunits, S7	–	–	Montandon and Stutz, 1984

TABLE II *(continued)*

Gene symbol	Description	Cyanelle location	Cyanelle sequence	Remarks/references
rps11	ribosomal small subunits, S11	–	–	Sijben-Mueller et al., 1986
rps12	,, ,, ,, S12	–	–	Montandon and Stutz, 1984
rps19	,, ,, ,, S19	+	–	Sugita and Sugiura, 1983; Zurawski et al., 1984
rpl2	ribosomal large subunit, L2	+	–	Zurawski et al., 1984
infA	initiation factor IF-1	–	–	Sijben-Mueller et al., 1986
rpoA	RNA-polymerase, α subunit	–	–	Sijben-Mueller et al., 1986
Phycobiliprotein apoproteins				Cyanobacterial and cyanelle genes sequenced
pcyA	α subunit, C-phycocyanin	+	+	Lemaux and Grossman, 1985
pcyB	β subunit, C-phycocyanin	+	+	Lemaux and Grossman, 1984, 1985; De Lorimer et al., 1984
apcA	α subunit, allophycocyanin	+	+	Lemaux and Grossman, 1985; Bryant et al., 1985
apcB	β subunit, allophycocyanin	+	+	Lemaux and Grossman, 1985; Bryant et al., 1985

Genes of cyanelle DNA and the origin and isolation of probes used for heterologous hybridizations are described by Lemaux and Grossman (1984, 1985), Starnes et al. (1985), Lambert et al. (1985), Bohnert et al. (1985), Wasmann (1985). For the nomenclature of genes see Hallick and Bottomley (1983).

the hazards encountered when using heterologous probes. It is advisable to employ other means of identification (see below).

Kuntz et al. (1984) have separated approximately 40 distinct tRNAs from the cyanelles of *C. paradoxa* by two-dimensional gel electrophoresis. The tRNA from the individual spots was identified by charging the tRNA with radiolabeled amino acids in vitro. The radioactively labeled tRNAs were then hybridized against cyanelle DNA fragments. In total 29 isoacceptor tRNA species were identified and at least 19 tRNA genes were localized on the chromosome. Most of the tRNAs hybridized with chloroplast DNA fragments at positions where homologous chloroplast tRNA genes had been located (Kuntz et al., 1984; Crouse et al., 1985). These results suggest that both the number and sequence of the cyanelle tRNA genes are very similar to those of chloroplasts. For example, trnA and trnI genes of the cyanelles are found in the transcribed spacer region which separates the 16S and 23S rRNA genes (see below). These tRNA genes are found at analogous locations in chloroplasts, cyanobacteria, and in four of seven rRNA operons in *E. coli* (see Bohnert et al., 1982a; Palmer, 1985, for reviews).

The rRNA genes which have been located on the cyanelle DNA of *C. paradoxa* are the 16S rRNA of the small subunit and 23S and 5S rRNAs of the large subunit of the ribosome (Siebens and Trench, 1978; Loeffelhardt et al., 1980, 1983). As noted by Siebens and Trench (1978), these rRNAs migrate indistinguishably from their counterparts in *E. coli* and in higher plant chloroplasts. Another characteristic which the cyanelle 23S rRNA molecule shares with chloroplast 23S rRNA (Wollgiehn and Parthier, 1980) is the presence of preformed nicks which render it prone to degradation into specific subfragments. The genes for the 16S and 23S rRNAs have been localized within the inverted repeat segment of the cyanelle DNA (Loeffelhardt et al., 1980, 1983) by hybridization (with identical results) with rRNAs purified from spinach chloroplast, *E. coli*, or cyanelles. The 5′-end of the 16S rRNA gene is approximately 500 bp from the border of the repeat units that are adjacent to the small single-copy DNA region (Fig. 1a). The 16S rRNA gene is separated from the 23S rRNA gene by a spacer DNA segment of 300–600 bp. A trnA (isoacceptor tRNA for alanine) and trnI (isoleucine) gene have been located in this spacer (Kuntz et al., 1984). Preliminary hybridization results (unpublished) suggest that a sequence with homology to the chloroplast 5S rRNA is located not more than 1 Kbp downstream from the 23S rRNA gene. Preliminary evidence also suggests that the rRNA genes of the cyanelle constitute an operon, as do the rRNA genes of *E. coli*, several cyanobacteria, and chloroplasts (Crouse et al., 1984; Whitfeld and Bottomley, 1983).

5.4. Gene structure

Cyanelle gene structure has been studied recently by several authors (Lemaux and Grossman, 1984, 1985; Lambert et al., 1985; Starnes et al., 1985; Wasmann, 1985). Wasmann (1985) sequenced the *rbcL* and *rbcS* genes from *C. paradoxa*. Upon sequencing the *rbcL* gene, a *rbcS* gene was found downstream of the *rbcL* coding sequence. It was identified by its homology with the sequences of the higher plant and cyanobacterial *rbcS* genes. Compared with the amino acid sequence of the *Cyanophora* enzyme, the sequences of the enzymes from the cyanobacteria *Anabaena* and *Anacystis* are 57% and 50% homologous, respectively. A part of the nucleotide sequence of *rbcL* and *rbcS* genes was also reported by Starnes et al. (1985). These authors demonstrated that the genes for the two subunits of ribulose 1,5-bisphosphate carboxylase/oxygenase are cotranscribed. This gene arrangement is the same as in two cyanobacterial species, *Anacystis nidulans* and *Anabaena* 7120 (Curtis and Haselkorn, 1983; Nierzwicki-Bauer et al., 1984; Shinozaki and Sugiura, 1983, 1985; Reichelt and Delany, 1983) (see also Section 4.1. of Chapter 4 and Sections 5.2. and 6.1. of Chapter 10). In addition, Wasmann (1985) found that the gene for *atpB* (β-subunit of the ATP-synthase CF_1 complex) is located close to the *rbcL* gene and is transcribed from the opposite DNA strand. Although the sequence of this gene is not yet completed, the sequence showed a high degree of homology with chloroplast *atpB* genes (Zurawski et al., 1982a). The location and divergent transcription of the *rbcL* and *atpB* genes in the cyanelle DNA is similar to the location and direction of transcription of these genes on chloroplast DNAs. In cyanobacteria the genes *atpB* and *rbcL* are not close to each other (Curtis and Haselkorn, 1983).

Reading frames have been found for the α and β subunit proteins of C-phycocyanin and for α and β subunits of allophycocyanin apoproteins (Lemaux and Grossman, 1984). Antibodies against a mixture of biliproteins from *Cyanidium caldarium* were used to screen a library of cyanelle DNA for specific cloned fragments which expressed the biliproteins. The cloned fragments were subsequently sequenced. Two genes for β subunits of C-phycocyanin and two genes for subunits of allophycocyanin proteins were found on the cyanelle DNA from *C. paradoxa*. By using oligonucleotide probes for conserved regions of phycobiliprotein genes Bryant et al. (1985) were able to confirm the results of Lemaux and Grossman (1984).

In those cases where the cyanelle gene structure is known, sequence homology with cyanobacterial and chloroplast genes suggest that they are closely related. It might be argued that such homologies are indicative of common function rather than of common ancestry. Perhaps a stronger argument for evolutionary relationship comes from the results that show the chromosomal arrangement of genes. In both the cyanelle DNA and higher plant chloroplast DNA the *atpB/E* and *rbcL* genes are close to each other and are transcribed from opposite strands of the DNA. Comparison of cyanobacterial and cyanelle *rbcL/S* operons shows that they have a very similar structure. Hybridization results (Lambert et al., 1985; Bohnert et al., 1985) suggest that a number of other genes namely, the *psbB* and *psbD* genes, the *atpA* and *atpF/G/H* genes, and the *psaA* and *psaB* genes, which are close together on the chloroplast chromosome are located similarly on the cyanelle DNA (Fig. 1a, b). Final identification of the hybridization signals obtained with the cyanelle DNA is, however, still necessary.

Cyanelle transcription has not yet been studied in depth. Siebens and Trench (1978) characterized the cyanelle ribosomal ribonucleic acids and monitored the effects that inhibitors of transcription and translation have on rRNA synthesis. They found that cyanelle 23S and 16S rRNAs migrated indistinguishably from *E. coli* rRNAs. Chloroplast rRNAs (Crouse et al., 1984) behaved similarly. The inhibitors rifampicin and chloramphenicol interfered with the synthesis of cyanelle rRNA. Synthesis of host rRNA (25S and 18S rRNA) was inhibited by cycloheximide. Recently, Starnes et al. (1985) have found that the region of the cyanelle chromosome which contains both the *rbcL* and *rbcS* genes (Wasmann, 1985) is synthesized as a single primary transcript. Similar results have been reported for transcription of these genes in cyanobacteria (Nierzwicki-Bauer et al., 1984; Shinozaki and Sugiura, 1983, 1985). Multigene transcripts have also been proposed for the *pcyA/B* and the *apcA/B* genes (Lemaux and Grossman, 1985; Bryant et al., 1985).

6. Conclusions

For several decades cyanelles and their 'host' organisms presented an enigma to taxonomists. Viewed as curiosities they have received little attention. In part this neglect may be attributed to the difficulties encountered in cultivating many of the endocyanomes. To date, only five species have been studied, and only one, *Cyanophora paradoxa*, has been studied extensively.

The 'endosymbiont hypothesis' of eukaryotic cell evolution (Frederick, 1981; Margulis, 1981; Gray and Doolittle, 1982; Whatley and Whatley, 1979, 1981) proposes that eukaryotic cells are chimaeras, their various organelles derived from once free-living prokaryotes that were engulfed by heterotrophic cells. Arguments have also been advanced for an autogenous origin of organelles (Uzzell and Spolsky, 1974; Bogorad, 1975; Cavalier-Smith, 1975; Woese, 1977). This hypothesis proposes that compartmentation of a single photosynthetic cell gave rise to the organellar genomes, which diverged intracellularly during or after the formation of the first protoeukaryote. Without elaborating on the dialectics of the reasoning we refer the reader to the discussion of the problem by Gray and Doolittle (1982). Their review deals with the relative significance of 'proofs' brought forward in support of the endosymbiont hypothesis.

The strong similarities between the photosynthetic apparatus and mechanism in cyanobacteria and photosynthetic eukaryotes originally led to the view that chloroplasts evolved from oxygen-producing photosynthetic prokaryotes resembling cyanobacteria. Structural and biochemical evidence, and evidence from the comparison of chromosome and gene structure, and from protein and nucleic acid sequences leave little doubt that plastids and cyanobacteria are evolutionarily related.

The cyanelles of endocyanomes have characteristics of both chloroplasts and cyanobacteria. They have been regarded variously as an endosymbiotic cyanobacterium, as a chloroplast, and as a 'bridge' (Aitken and Stanier, 1979) or a 'missing link' (Kies, 1980; Jaynes and Vernon, 1982) between the free-living cyanobacteria and the plastids from eukaryotic algae and higher plants. Jaynes and Vernon (1982) have called the cyanelle from *C. paradoxa* 'almost a cyanobacterial chloroplast'. In a strict sense the term chloroplast should be reserved for the plastids of the Spermatophyceae and the Chlorophyceae. The statement by Jaynes and Vernon implies that cyanelles can be viewed as a type of plastid which is distinguished from other types of plastids by characteristics normally associated with cyanobacteria.

In our view, the cyanelles from *Cyanophora paradoxa* are not endosymbionts, but a separate line of plastid evolution. Cyanelles function as photosynthetic organelles. Although their biochemical integration into the metabolism of the whole cell is not yet sufficiently studied, the extent of this integration appears to be as in higher plant chloroplasts. Clearly the cyanelles form an obligatory association (Trench, 1979). The overall gene content of cyanelles is as in the chloroplast. At present, the only differences in gene content between the higher plant chloroplast and the cyanelle in *C. paradoxa* are the genes for the small subunit of ribulose 1,5-bisphosphate carboxylase/oxygenase and the phycobiliprotein genes present in the cyanelle DNA. Another point of divergence is the presence of peptidoglycan in the cyanelle. It is expected, but not proven, that the genes involved in its synthesis are contained in the cyanelle DNA. A comparison of the gene organization of other plastid types and of cyanelle gene organization shows that many evolutionary lines are possible. It is interesting to note that cyanelles are not confined to a single group of eukaryotic cells, suggesting that the mechanisms by which organelles arose may have been common or that such mechanisms may be still at work.

The cyanelles in *Cyanophora paradoxa*, the plastids from *Euglena sp.* and the chloroplasts of most other eukaryotes have arrived at approximately the same genome complexity. This may argue for either a common starting point or for common constraints on the size of organelle DNAs. However, because plastids have been characterized in detail from only two evolutionary lines, the Euglenophyceae and Chlorophyceae/Spermatophyceae, it seems premature to emphasize differences in gene content as a means to establish evolutionary relationships. Already one organism, *Acetabularia sp.* (Chlorophyceae), is known to differ with respect to the complexity and possibly the organization of its plastid DNA (Tymms and Schweiger, 1985). It appears likely that within other subclasses, for example, the Cryptophyceae, Dinophyceae, Crysophyceae, Rhodophyceae, and Phaeophyceae (for a survey of the characteristics of different algal classes see Rosowski and Parker, 1982), other forms of plastid chromosome organization and differences in gene content will be found.

Acknowledgements

We wish to acknowledge support for our work by the Deutsche Forschungsgemeinschaft and the Arizona Agric. Exp. Station (HJB), by 'Fonds zur Förderung der wissenschaftlichen Forschung' (WL), and by the Department of Energy Contract No. DE-AC02-76ERO-1338 (CCW). This work was aided in part by the Arizona Agricultural Experiment Station (ARZT No. 174441). We wish to thank Ms. Helen Markes and Christine Michalowski for their help in preparing this manuscript, and Dr. Robert W. Hoshaw (University of Arizona) for his comments.

References

Aitken, A. and Stanier, R.Y. (1979) Characterization of peptidoglycan from the cyanelles of *Cyanophora paradoxa*. J. Gen. Microbiol. 212, 218–223.

Aldrich, J., Cherney, B., Merlin, E., Williams, C. and Mets, L. (1985) Recombination within the inverted repeat sequences of the *Chlamydomonas reinhardii* chloroplast genome produces two orientation isomers. Curr. Genet. 9, 233–238.

Alt, J. and Herrmann, R.G. (1984) Nucleotide sequence of the gene for pre-apocytochrome *f* in the spinach plastid chromosome. Curr. Genet. 8, 551–557.

Alt, J., Winter, P., Sebald, W., Moser, J.G., Schedel, R., Westhoff, P. and Herrmann, R.G. (1983a) Localization and nucleotide sequence of the gene for the ATP synthase proteolipid subunit on the spinach plastid chromosome. Curr. Genet. 7, 129–138.

Alt, J., Westhoff, P., Sears, B.B., Nelson, N., Hurt, E., Hauska, G. and Herrmann, R.G. (1983b) Genes and transcripts for the polypeptides of the cytochrome b_6/f complex from spinach thylakoid membranes. EMBO J. 2, 979–986.

Alt, J., Morris, J., Westhoff, P. and Herrmann, R.G. (1984) Nucleotide sequence of the clustered genes for the 44 Kd chlorophyll *a* apoprotein and the '32 Kd- like' protein of the photosystem II reaction center in the spinach plastid chromosome. Curr. Genet. 8, 597–605.

Bedbrook, T.R. and Kolodner, R. (1979) The structure of chloroplast DNA. Annu. Rev. Plant Physiol. 30, 593–620.

Boettcher, U., Brandt, P., Mueller, B. and Tischner, R. (1982) Physiological characterization of the endocyanelle *Cyanocyta korschikoffiana* Hall and Claus I. Photosynthetic and N-assimilating properties in the symbiotic association *Cyanophora-Cyanocyta*. Z. Pflanzenphysiol. 106, 167–172.

Bogorad, L. (1975) Evolution of organelles and eukaryotic genomes. Science 188, 891–898.

Bohnert, H.J. and Loeffelhardt, W. (1982) Cyanelle DNA from *Cyanophora paradoxa* exists in two forms due to intramolecular recombination. FEBS Lett. 150, 403–406.

Bohnert, H.J. and Loeffelhardt, W. (1984) Genome and gene organization of the cyanelle DNA from *Cyanophora paradoxa* in relation to the common organization in chloroplasts. In W. Wiessner, D. Robinson and R.C. Starr (eds.), Compartments in Algal Cells and Their Interaction. Springer Verlag, Berlin-Heidelberg, pp. 58–68.

Bohnert, H.J., Crouse, E.J. and Schmitt, J.M. (1982a) Organization and expression of plastid genomes. In B. Parthier and D. Boulter (eds.), Encyclopedia of Plant Physiology, New Series, Vol. 14B. Springer Verlag, Berlin-Heidelberg-New York, pp. 475–530.

Bohnert, H.J., Crouse, E.J., Pouyet, J., Mucke, H. and Loeffelhardt, W. (1982b) The subcellular localization of DNA components from *Cyanophora paradoxa*, a flagellate containing endosymbiotic cyanelles. Eur. J. Biochem. 126, 1–38.

Bohnert, H.J., Michalowski, C., Koller, B., Delius, H., Mucke, H. and Loeffelhardt, W. (1983) The cyanelle genome from *Cyanophora paradoxa*. In H.E.A. Schenk and W. Schwemmler (eds.), Endocytobiology II. de Gruyter, Berlin-New York, pp. 433–448.

Bohnert, H.J., Michalowski, C., Bevacqua, S., Mucke, H. and Loeffelhardt, W. (1985) Cyanelle DNA from *Cyanophora paradoxa*. Physical mapping and location of protein coding regions. Mol. Gen. Genet. 201, 565–575.

Bothe, H. and Floener, L. (1978) Physiological characterization of *Cyanophora paradoxa* a flagellate containing cyanelles in endosymbiosis. Z. Naturforsch. 33, 981–987.

Brandt, P. and Marten, S. (1983) Chloroplast-like thylakoid development in the endocyanelle *Cyanocyta korschikoffiana* in synchronized *Cyanophora paradoxa*. In H.E.A. Schenk and W. Schwemmler (eds.), Endocytobiology, II. de Gruyter, Berlin-New York, pp. 465–470.

Broach, J.R., Guarascio, V.R. and Jayaram, M. (1982) Recombination within the yeast plasmid 2 micron circle is site-specific. Cell 29, 227–234.

Bryant, D.A., de Lorimier, R., Lambert, D.H., Dubbs, J.M., Stirewalt, V.L., Stevens, S.E., Jr., Porter, R.D., Tam, J. and Jay, E. (1985) Molecular cloning and nucleotide sequence of the α and β subunits of allophycocyanin from the cyanelle genome of *Cyanophora paradoxa*. Proc. Natl. Acad. Sci. USA 82, 3242–3246.

Burnap, R.L. and Trench, R.K. (1984) Isolation and characterization of ribulose bisphosphate carboxylase from the cyanelle of *Cyanophora paradoxa* (Korschikoff). FEMS Microbiol. Lett. 21, 271–275.

Cavalier-Smith, T. (1975) The origin of nuclei and eukaryotic cells. Nature (London) 256, 463–468.

Cavalier-Smith, T. (1981) Eukaryote kingdoms: seven or nine? Biosystems 14, 461–481.

Chapman, D.T. (1966) Pigments of the symbiotic algae (Cyanosomes) of *Cyanophora paradoxa* and *Glaucocystis nostochinearum* and the Rhodophyceae *Porphyridium aerugineum* and *Asterocystis ranosa*. Arch. Microbiol. 55, 17–25.

Codd, G.A. and Stewart, W.D.P. (1976) Polyhedral bodies and ribulose 1,5-diphosphate carboxylase of the blue-green alga *Anabaena cylindrica*. Planta (Berlin) 130, 323–326.

Codd, G.A. and Stewart, W.D.P. (1977) Quaternary structure of the D-ribulose-1,5-diphosphate carboxylase from the cyanelles of *Cyanophora paradoxa*. FEMS Microbiol. Letts. 1, 35–38.

Crouse, E.J., Bohnert, H.J. and Schmitt, J.M. (1984) Chloroplast RNA synthesis. In R.J. Ellis (ed.),

Chloroplast Biogenesis, Seminar Series of the Society for Experimental Biology, Vol. 21. Cambridge University Press, Cambridge, pp. 83–136.

Crouse, E.J., Schmitt, J.M. and Bohnert, H.J. (1985) Chloroplast and cyanobacterial genomes, genes and RNAs: a compilation. Plant Mol. Biol. Rep. 3, 43–89.

Curtis, S.E. and Haselkorn, R. (1983) Isolation and sequences of the gene for the large subunit of ribulose-1,5-bisphosphate carboxylase from the cyanobacterium *Anabaena* 7120. Proc. Natl. Acad. Sci. USA 80, 1835–1839.

Curtis, S.E. and Haselkorn, R. (1984) Isolation, sequence and expression of two members of the 32 Kd thylakoid membrane protein gene family from cyanobacterium *Anabaena* 7120. Plant Mol. Biol. 3, 249–258.

De Lorimier, R., Bryant, D.A., Porter, R.D., Liu, W.-Y., Jay, E. and Stevens, S.E. (1984) Genes for the α and β subunits of phycocyanin. Proc. Natl. Acad. Sci. USA 81, 7946–7950.

Deno, H., Shinozaki, K. and Sugiura, M. (1983) Nucleotide sequence of tobacco chloroplast gene for the α subunit of proton-translocating ATPase. Nucl. Acids Res. 11, 2185–2191.

Drews, G. and Weckesser, J. (1982) Function, structure and composition of cell walls. In N.A. Carr and B.A. Whitton (eds.), The Biology of Cyanobacteria, University of California Press, Berkeley, pp. 333–358.

Edelman, M., Swinton, D., Schiff, J.A., Epstein, H.T. and Zelden, B. (1967) Deoxyribonucleic acid of the blue-green algae (Cyanophyta). Bacteriol. Rev. 31, 315–331.

Ellis. R.J. (1977) Protein synthesis by isolated chloroplasts. Biochim. Biophys. Acta 463, 185–215.

Ellis, R.J. (1981) Chloroplast-proteins: synthesis, transport, and assembly. Annu. Rev. Plant Physiol. 32, 111–137.

Erickson, J.M., Rahire, M. and Rochaix, J.D. (1984) *Chlamydomonas reinhardii* gene for the 32,000 mol. wt. protein of photosystem II contains four large introns and is located entirely within the chloroplast inverted repeat. EMBO J. 3, 2753–2762.

Floener, L. and Bothe, H. (1982) Metabolic activities in *Cyanophora paradoxa* and its cyanelles. II. Photosynthesis and respiration. Planta 156, 78–83.

Floener, L., Danneberg, G. and Bothe, H. (1982) Metabolic activities in *Cyanophora paradoxa* and its cyanelles. Planta 156, 70–77.

Fox, G.E., Stackebrandt, E., Hespell, R.B., Gibson, J., Mailoff, J., Dyer, T.A., Wolfe, R.S., Blach, W.E., Tanner, R.S., Magrum, L.J., Zablen, L.B., Blakemore, R., Gupta, R., Bonen, L., Lewis, B.J., Stahl, D.A., Luehrsen, K.R., Chen, K.N. and Woese, C.R. (1980) The phylogeny of prokaryotes. Science 209, 457–463.

Frederick, J.F. (ed.) (1981) Origins and evolution of eukaryotic intracellular organelles. Ann. NY Acad. Sci. 361, New York Academy of Science, New York.

Gantt, E. (1981) Phycobilisomes. Annu. Rev. Plant Physiol. 32, 327–347.

Gantt, E., Lipschultz, C.A., Grabowski, J. and Zimmerman, B.K. (1979) Phycobilisomes from blue-green and red algae. Isolation criteria and dissociation characteristics. Plant Physiol. 63, 615–620.

Geitler, L. (1959) Syncyanosen. In Ruhland's Handbuch der Pflanzenphysiologie, 11. Springer Verlag, Berlin-Goettingen-Heidelberg, pp. 530–545.

Giddings, Jr., T.H., Wasmann, C. and Staehelin, L.A. (1983) Structure of the thylakoids and envelope membranes of the cyanelles of *Cyanophora paradoxa*. Plant Physiol. 71, 409–419.

Glazer, A.N. (1984) Phycobilisome. A macromolecular complex optimized for light energy transfer. Biochim. Biophys. Acta 768, 29–51.

Gray, M.W. and Doolittle, W.F. (1982) Has the endosymbiont hypothesis been proven? Microbiol. Rev. 46, 1–42.

Hall, W.T. and Claus, G. (1963) Ultrastructural studies on the blue-green algal symbiont in *Cyanophora paradoxa* Korschikoff. J. Cell Biol. 19, 551–563.

Hall, W.T. and Claus, G. (1967) Ultrastructural studies on the cyanelles of *Glaucocystis nostochinearum* Korschikoff. J. Phycol. 3, 37–51.

Hallick, R.B. and Bottomley, W. (1983) Proposals for the naming of chloroplast genes. Plant Mol. Biol. Rep. 1, 38–43.

Hallick, R.B., Gingrich, J.C., Johaningmeier, U. and Passavant, C.W. (1985) Introns in *Euglena* and *Nicotiana* chloroplast protein genes. In L. van Vloten-Doting, G.S.P. Groot and T.C. Hall (eds.), Molecular Form and Function of the Plant Genome. Plenum Publ. Co., New York and London, pp. 211–220.

Heinemeyer, W., Alt, J. and Herrmann, R.G. (1984) Nucleotide sequence of the clustered genes for apocytochrome *b6* and subunit 4 of the cytochrome *b/f* complex in the spinach plastid chromosome. Curr. Genet. 8, 543–549.

Heinhorst, S. and Shively, J.M. (1983) Encoding of both subunits of ribulose-1,5-bisphosphate carboxylase by organelle genome of *Cyanophora paradoxa*. Nature (London) 304, 373–374.

Herdman, M. and Stanier, R.Y. (1977) The Cyanelle: Chloroplast or endosymbiotic prokaryote? FEMS Microbiol. Lett. 1, 7–12.

Herdman, M., Janvier, J., Waterbury, J.B., Rippka, R., Stanier, R.Y. and Mandel, M. (1979a) Deoxyribonucleic acid base composition of cyanobacteria. J. Gen. Microbiol. 111, 63–71.

Herdman, M., Janvier, J., Rippka, R. and Stanier, R.Y. (1979b) Genome size of cyanobacteria. J. Gen. Microbiol. 111, 73–85.

Herrmann, R.G., Alt, J., Schiller, B., Widger, W.R. and Cramer, W.A. (1984) Nucleotide sequence of the gene for apocytochrome *b*-559 on the spinach plastid chromosome: implications for the structure of the membrane protein. FEBS Lett. 176, 239–244.

Holschuh, K., Bottomley, W. and Whitfeld, P.R. (1984) Structure of the spinach chloroplast genes for the D2 and 44kd reaction-centre proteins of photosystem II and for tRNA (UGA). Nucl. Acids Res. 12, 8819–8834.

Hudspeth, M.E.S., Shumard, D.S., Bradford, D.J.R. and Grossman, L.I. (1983) Organization of *Achlya* mtDNA: a population with two orientations and a large inverted repeat containing the rRNA genes. Proc. Natl. Acad. Sci. USA 80, 142–146.

James, T.W. and Jope, C. (1978) Visualization of DNA by the fluorescent dye DAPI. J. Cell Biol. 79, 623–630.

Jaynes, J.M. and Vernon, L.P. (1982) The cyanelle of *Cyanophora paradoxa*: almost a cyanobacterial chloroplast. Trends Biochem. Sci. 7, 22–24.

Jaynes, J.M., Vernon, L.P., Klein, S.M. and Strobel, G.A. (1981) Genome size of cyanelle DNA from *Cyanophora paradoxa*. Plant Sci. Lett. 21, 345–356.

Kies, L. (1974) Elektronenmikroskopische Untersuchungen an *Paulinella chromatophora* Lauterborn, einer Thekamoebe mit blaugruenen Endosymbionten (Cyanellen). Protoplasma 80, 69–89.

Kies, L. (1976) Untersuchungen zur Feinstruktur und taxonomischen Einordnung von *Gloeochaete wittrockiana*, einer apoplastidalen capsalen Alge mit blaugruenen Endosymbionten (Cyanellen). Protoplasma 87, 419–446.

Kies, L. (1979) Zur systematischen Einordnung von *Cyanophora paradoxa*, *Gloeochaete wittrockiana* und *Glaucocystis nostochinearum*. Ber. Dtsch. Bot. Ges. 92, 445–454.

Kies, L. (1980) Morphology and systematic position of some endocyanomes. In W. Schwemmler and H.E.A. Schenk (eds.), Endocytobiology, Endosymbiosis and Cell Biology. de Gruyter, Berlin-New York, pp. 7–19.

Kies, L. (1984) Cytological aspects of blue-green algal endosymbiosis. In W. Wiessner, D. Robinson and C. Starr (eds.), Compartments in Algal Cells and Their Interaction. Springer Verlag, Berlin-Heidelberg, pp. 191–199.

Kies, L. and Kremer, B.P. (1979) Function of cyanelles in the Thecamoeba *Paulinella chromatophora*. Naturwiss. 66, 578.

Klein, S., Jaynes, J.M., Kent, S.S. and Vernon, L.P. (1981a) Properties of the photosynthetic system and DNA of *Cyanophora paradoxa* cyanelles. Plant Physiol. 68, 407–410.

Klein, S.M., Vernon, L.P. and Kent, S.S. (1981b) Photosynthetic properties and components of the endosymbiont *Cyanophora paradoxa*. In G. Akoyunoglou (ed.). Photosynthesis III. Structure and Molecular Organization of the Photosynthetic Apparatus. Elsevier, Amsterdam, pp. 175–183.

Koessel, H., Edwards, D., Fritzsche, E., Koch, W. and Schwarz, Zs. (1983) Phylogenetic significance of nucleotide sequence analysis. In U. Jensen and D.E. Fairbrothers (eds.), Proteins and Nucleic Acids in Plant Systematics. Springer Verlag, Berlin-Heidelberg, pp. 36–37.

Koller, B. and Delius, H. (1980) *Vicia faba* chloroplast DNA has only one set of ribosomal RNA genes as shown by partial denaturation mapping and R-loop analysis. Mol. Gen. Genet. 178, 261–269.

Koller, B. and Delius, H. (1982) A chloroplast DNA of *Euglena gracilis* with five complete tRNA operons and two extra 16S rRNA genes. Mol. Gen. Genet. 188, 305–308.

Korschikoff, A.A. (1924) Protistologische Beobachtungen I. *Cyanophora paradoxa*. Russ. Ark. Protistol. 3, 57–74.

Kremer, B.P. (1983) *Cyanidium caldarium*: Rhodophyte, cyanome or transitional species? In H.E.A. Schenk and W. Schwemmler (eds.), Endocytobiology, II. de Gruyter, Berlin-New York, pp. 947–962.

Kremer, B.P., Feige, G.B. and Schneider, H.A.W. (1978) A new proposal for the systematic position of *Cyanidium caldarium*. Naturwissensch. 65, 157–161.

Kremer, B.P., Kies, L. and Rostami-Rabet, A. (1979) Photosynthetic performance of cyanelles in the endocyanomes *Cyanophora*, *Glaucosphaera*, *Gloeochaete*, and *Glaucocystis*. Z. Pflanzenphysiol. 92, 303–317.

Kuntz, M., Crouse, E.J., Mubumbila, M., Bukard, G., Weil, J.-H., Bohnert, H.J., Mucke, J. and Loeffelhardt,

W. (1984) Transfer RNA gene mapping studies on cyanelle DNA from *Cyanophora paradoxa*. Mol. Gen. Genet. 194, 508–512.

Lambert, D.H., Bryant, D.A., Stirewalt, V.L., Dubbs, J.M., Stevens, S.E. and Porter, R.D. (1985) Gene map for the *Cyanophora paradoxa* cyanelle genome. J. Bacteriol. 164, 659–664.

Lemaux, P.G. and Grossman, A.R. (1984) Isolation and characterization of a gene for a major light-harvesting polypeptide from *Cyanophora paradoxa*. Proc. Natl. Acad. Sci. USA 81, 4100–4104.

Lemaux, P.G. and Grossman, A.R. (1985) Major light-harvesting polypeptides encoded in polycistronic transcripts in an eukaryotic alga. EMBO J. 4, 1911–1919.

Lemieux, C., Turmel, M., Lee, R.W. and Bellemare, G. (1985) A 21 kilobase-pair deletion/addition difference in the inverted repeat sequence of chloroplast DNA from *Chlamydomonas eugametos* and *C. moewusii*. Plant Mol. Biol. 5, 77–84.

Loeffelhardt, W., Mucke, H. and Bohnert, H.J. (1980) Cyanelle DNA from *Cyanophora paradoxa*: analogies to chloroplast DNA. In W. Schwemmler and H.E.A. Schenk (eds.), Endocytobiology. de Gruyter, Berlin-New York, pp. 532–530.

Loeffelhardt, W., Mucke, H., Crouse, E.J. and Bohnert, H.J. (1983) Comparison of the cyanelle DNA from the two different strains of *Cyanophora paradoxa*. Curr. Genet. 7, 139–144.

Loeffelhardt, W., Michalowski, C. and Bohnert, H.J. (1985) Sequence homologies between cyanobacterial and chloroplast genes. Endocyt. Cell Res. 2, 213–231.

Margulis, L. (1981) Symbiosis in Cell Evolution. W.H. Freeman and Co., San Francisco.

Marten, S., Brandt, P. and Wiessner, W. (1981) Improved culture conditions and biphasic growth of *Cyanophora paradoxa* (Korschikoff). Ber. Dtsch. Bot. Ges. 94, 727–732.

Marten, S., Brandt, P. and Robinson, D.G. (1983) Thylakoid development in the endocyanelle *Cyanocyta korschikoffiana* in synchronized *Cyanophora paradoxa*. Eur. J. Cell Biol. 30, 18–24.

Mereschkowsky, C. (1905) Ueber Natur und Ursprung der Chromatophoren im Pflanzenreiche. Biol. Centralbl. 25, 593–604.

Mignot, J.P., Joyon, L. and Pringsheim, E.G. (1969) Quelques particularités structurales de *Cyanophora paradoxa* Korsch., protozoaire flagellé, J. Protozool. 16, 138–145.

Moestrup, O. (1978) On the phylogenetic validity of the flagellar apparatus in green algae and other chlorophyll a and b containing plants. Bio Systems 10, 117–144.

Montandon, P.E. and Stutz, E. (1984) The genes for the r-proteins S12 and S7 are clustered with the gene for the EF-Tu protein on the chloroplast genome of *E. gracilis*. Nucl. Acids Res. 12, 2851–2859.

Morris, J. and Herrmann, R.G. (1984) Nucleotide sequence of the gene for the P680 chlorophyll *a* apoprotein of the photosystem II reaction center from spinach. Nucl. Acids Res. 12: 2837–2850.

Mubumbila, M., Gordon, K.H.J., Crouse, E.J., Burkard, G. and Weil, J.-H. (1983) Construction of the physical map of the chloroplast DNA of *Phaseolus vulgaris* and localization of ribosomal and transfer RNA genes. Gene 21, 257–266.

Mucke, H., Loeffelhardt, W. and Bohnert, H.J. (1980) Partial characterization of the genome of the 'endosymbiotic' cyanelles from *Cyanophora paradoxa*. FEBS Lett. 111, 347–352.

Neu-Mueller, M. and Schenk, H.E.A. (1983) On the biosynthesis of photosynthetic pigments in *Cyanophora paradoxa*. In H.E.A. Schenk and W. Schwemmler (eds.), Endocytobiology, II. de Gruyter, Berlin-New York, pp. 451–463.

Nierzwicki-Bauer, S., Curtis, S.E. and Haselkorn, R. (1984) Cotranscription of ribulose 1,5-bisphosphate carboxylase in the cyanobacterium *Anabaena* 7120. Proc. Natl. Acad. Sci. USA 81, 5961–5965.

Palmer, J.D. (1983) Chloroplast DNA exists in two orientations. Nature (London) 301, 92–93.

Palmer, J.D. (1985) Comparative organization of chloroplast genomes. Ann. Rev. Genet. 19, 325–354.

Pascher, A. (1914) Ueber Symbiosen von Spaltpilzen und Flagellaten mit Blaualgen. Ber. Dtsch. Bot. Ges. 32, 339–352.

Pascher, A. (1929) Studien ueber Symbiosen. I. Ueber einige Endosymbiosen von Blaualgen in Einzellern. Jahrb. Wiss. Bot. 71, 386–462.

Pickett-Heaps, J. (1972) Cell division in *Cyanophora paradoxa*. New Phytol. 71, 561–567.

Preston, R.D. (1971) Studies on the fine structure of *Glaucocystis nostochinearum* Itzigs. II. Membrane morphology and taxonomy. Br. Phycol. J. 6, 113–128.

Ravel-Chapuis, P., Flamant, F., Nicolas, P., Heizmann, P. and Nigon, V. (1984) Diversity of the ribosomal structures in the *Euglena gracilis* chloroplast genome: description of a mutant with two ribosomal operons and possible mechanism for its production. Nucl. Acids Res. 12, 1039–1048.

Raven, P.H. (1970) A multiple origin for plastids and mitochondria. Science 169, 641–646.

Reichelt, B.Y. and Delaney, S.F. (1983) The nucleotide sequence for the large subunit of ribulose 1,5-bisphosphate carboxylase from a unicellular cyanobacterium *Synechococcus* PCC 6301. DNA 2, 121–129.

Reisfeld, A. and Edelman, M. (1976) Ribulose diphosphate carboxylase from *Cyanophora paradoxa*. Isr. J. Bot. 25, 97–102.

Reisser, W. (1984) Endosymbiotic Cyanobacteria and Cyanellae. In H.F. Linskens and J. Heslop-Harrison (eds.), Cellular Interactions, Encyclopedia of Plant Physiology New Series, Vol. 17. Springer Verlag, Berlin-Heidelberg-New York-Tokyo, pp. 91–112.

Robinson. D.G. and Preston, R.D. (1971) Studies on the fine strucure of *Glaucocystis nostochinearum* Itzigs. I. Wall structure. J. Exp. Bot. 22, 635–643.

Rosowski, J.R. and Parker, B.C. (eds.) (1982) Selected Papers in Phycology, II. Phycol. Soc. America, Lawrence, KS.

Schenk, H.E.A. (1970) Nachweis einer lysozymempfindlichen Stuetzmembran der Endocyanellen von *Cyanophora paradoxa* (Korschikoff). Z. Naturforsch. 25B, 656.

Schenk, H.E.A. (1977) Zur osmotischen Beeinflussung der Photosynthese von *Cyanophora paradoxa* Korsch. durch Saccharose. Arch. Microbiol. 114, 261–266.

Schmidt, A. and Christen, U. (1979) A PAPS-dependent sulfotransferase in *Cyanophora paradoxa* inhibited by 5′-AMP, 5′ADP, and APS. Z. Naturforsch. 34, 222–228.

Schmidt, B., Kies, L. and Weber, A. (1979) Die Pigmente von *Cyanophora paradoxa*, *Gloeochaete wittrockiana* und *Glaucocystis nostochinearum*. Arch. Protistenk. 122, 164–170.

Schnepf, E., Koch, W. and Deichgraeber, G. (1966) Zur Cytologie und taxonomischen Einordung von *Glaucocystis*. Arch. Mikrobiol. 55, 149–174.

Schnepf, E. and Brown, R.M. (1971) On the relationships between endosymbiosis and the origin of plastids and mitochondria. In J. Reinert and H. Ursprung (eds.), The Origin and Continuity of Cell Organelles. Springer Verlag, New York-Heidelberg-Berlin, pp. 299–322.

Schwartz, R.M. and Dayhoff, M.O. (1981) Chloroplast origins: inferences from protein and nucleic acid sequences. Ann. NY Acad. Sci. 361, 260–272.

Scott, O.T., Castenholz, R.W. and Bonnett, H.T. (1984) Evidence for a peptidoglycan envelope in the cyanelles of *Glaucocystis nostochinearum* Itzigsohn. Arch. Microbiol. 139, 130–138.

Seckbach, J. (1972) On the fine structure of the acidophilic hot-spring alga *Cyanidium caldarium*. Microbios. 5, 113–142.

Seckbach, J., Fredrick, J.F. and Garbary, D.J. (1983) Auto- or exogenous origin of transitional algae: an appraisal. In H.E.A. Schenk and W. Schwemmler (eds.), Endocytobiology, II. de Gruyter, Berlin-New York, pp. 947–962.

Shinozaki, E. and Sugiura, M. (1983) The gene for the small subunit of ribulose-1,5-bisphosphate carboxylase/oxygenase is located close to the gene for the large subunit in the cyanobacterium *Anacystis nidulans* 6301. Nucl. Acids Res. 11, 6957–6964.

Shinozaki, K. and Sugiura, M. (1985) Genes for the large and small subunits of ribulose-1,5-bisphosphate carboxylase/oxygenase constitute a single operon in a cyanobacterium *Anacystis nidulans* 6301. Mol. Gen. Genet. 200, 27–32.

Siebens, H.C. and Trench, R.K. (1978) Aspects of the relation between *Cyanophora paradoxa* (Korschikoff) and its endosymbiotic cyanelles *Cyanocyta korchikoffiana* (Hall and Claus). III. Characterization of ribosomal ribonucleic acids. Proc. Roy. Soc. London B. 202, 463–472.

Sijben-Mueller, G., Hallick, R.B., Alt, J., Westhoff, P. and Herrmann, R.G. (1986) Spinach plastid genes coding for initiation factor IF-1, ribosomal protein S11 and RNA polymerase α-subunit. Nucl. Acids Res., 14, in press.

Skuja, H. (1954) Glaucophyta. In A. Melchior and E. Werdermann (eds.), Syllabus der Pflanzenfamilien, 1. Gebrüder Bornträger, Berlin, pp. 56–57.

Starnes, S.M., Lambert, D.H., Maxwell, E.S., Stevens, S.E., Porter, R.D. and Shively, J.M. (1985) Co-transcription of the large and small subunit genes of ribulose-1,5-bisphosphate carboxylase/oxygenase in *Cyanophora paradoxa*. FEMS Microbiol. Lett. 28, 165–169.

Stutz, E. (1984) The chloroplast genome of algae. Bot. Helv. 94, 145–159.

Subramanian, A.R., Steinmetz, A. and Bogorad, L. (1983) Maize chloroplast DNA encodes a protein sequence homologous to the bacterial ribosome assembly protein S4. Nucl. Acids Res. 11, 5277–5286.

Sugita, M. and Sugiura, M. (1983) A putative gene of tobacco chloroplast coding for ribosomal protein similar to *E. coli* ribosomal protein S19. Nucl. Acids Res. 11, 1913–1918.

Trench, R.K. (1979) The cell biology of plant-animal symbiosis. Annu. Rev. Plant Physiol. 485–531.

Trench, R.K. (1982a) Physiology, biochemistry, and ultrastructure of cyanellae. In F.E. Round and D.J. Chapman (eds.), Progress in Phycological Research, Vol. 1, Elsevier, Amsterdam, pp. 257–288.

Trench, R.K. (1982b) Cyanelles. In J.A. Schiff (ed.), On the Origins of Chloroplasts. Elsevier/North-Holland, New York-Amsterdam-Oxford, pp. 55–76.

Trench, R.K. and Ronzio, G.S. (1978) Aspects of the relation between *Cyanophora paradoxa* (Korschikoff) and its endosymbiotic cyanelles *Cyanocyta korschikoffiana* (Hall and Claus). II. The photosynthetic pigments. Proc. Roy. Soc. London B. 202, 445–462.

Trench, R.K. and Siebens, H.C. (1978) Aspects of the relation between *Cyanophora paradoxa* (Korschikoff) and its endosymbiotic cyanelles *Cyanocyta korschikoffiana* (Hall and Claus). IV. The effects of rifampicin, chloramphenicol and cycloheximide on the synthesis of ribosomal ribonucleic acids and chlorophyll. Proc. Roy. Soc. London B. 202, 473–482.

Trench, R.K., Pool, R.R., Logan, M. and Engelland, A. (1978) Aspects of the relation between *Cyanophora paradoxa* (Korschikoff) and its endosymbiotic cyanelles *Cyanocyta korschikoffiana* (Hall and Claus). I. Growth, ultrastructure, photosynthesis, and the obligate nature of the association. Proc. Roy. Soc. London B. 202, 423–443.

Tymms, J.J. and Schweiger, H.-G. (1985) Tandemly repeated nonribosomal DNA sequences in the chloroplast genome of an *Acetabularia mediterranea* strain. Proc. Natl. Acad. Sci. USA 82, 1706–1710.

Uzzell, T. and Spolsky, C. (1974) Mitochondria and plastids as endosymbionts: a revival of special creation? Am. Sci. 62, 334–343.

Wasmann, C.C. (1985) The Cyanelle and the Cyanelle Genome of *Cyanophora paradoxa*. Dissertation, Michigan State University.

Westhoff, P., Alt, J., Nelson, N. and Herrmann, R.G. (1985a) Genes and transcripts for the ATP synthase CF_0 subunits I and II from spinach thylakoid membranes. Mol. Gen. Genet. 199, 290–299.

Westhoff, P., Alt, J., Widger, W.R., Cramer, W.A. and Herrmann, R.G. (1985b) Localization of the gene for apocytochrome *b*-559 on the plastid chromosome of spinach. Plant Mol. Biol. 4, 103–110.

Whatley, J.M. and Whatley, F.R. (1981) Chloroplast evolution. New Phytol. 87, 233–247.

Whatley, J.M., John, P. and Whatley, F.R. (1979) From extracellular to intracellular: the establishment of mitochondria and chloroplasts. Proc. Roy. Soc. London B. 204, 165–187.

Whitfeld, P.R. and Bottomley, W. (1983) Organization and structure of chloroplast genes. Annu. Rev. Plant Physiol. 34, 279–310.

Willey, D.K., Howe, C.J., Auffret, A.D., Bowman, C.M., Dyer, T.A. and Gray, J.C. (1984) Location and nucleotide sequence of the gene for cytochrome *f* in wheat chloroplast DNA. Mol. Gen. Genet. 194, 416–422.

Woese, C.R. (1977) Endosymbionts and mitochondrial origins. J. Mol. Evol. 10, 93–96.

Wollgiehn, R. and Parthier, B. (1980) RNA and protein synthesis in plastid differentiation. In J. Reinert (ed.), Results and Problems in Cell Differentiation, Vol. 10. Springer Verlag, Berlin-Heidelberg, pp. 97–145.

Yamada, T. (1983) Characterization of inverted repeat sequences and ribosomal RNA genes of chloroplast DNA from *Chlorella ellipsoidea*. Curr. Genet. 7, 481–487.

Yamanaka, G., Glazer, A.N. and Williams, R.C. (1978) Cyanobacterial phycobilisomes. Characterization of the phycobilisomes of *Synechococcus* 6301. J. Biol. Chem. 253, 8303–8310.

Zurawski, G., Perrot, P., Bottomley, W. and Whitfeld, P.R. (1981) The structure of the gene for the large subunit of ribulose-1,5-bisphosphate carboxylase from spinach chloroplast DNA. Nucl. Acids Res. 9, 3251–3270.

Zurawski, G., Bottomley, W. and Whitfeld, P.R. (1982a) Structures of the genes for the B and E subunits of spinach chloroplast ATPase indicate a dicistronic mRNA and an overlapping translation stop-/start signal. Proc. Natl. Acad. Sci. USA 79, 6260–6264.

Zurawski, G., Bohnert, H.J., Whitfeld, P.R. and Bottomley, W. (1982b) Nucleotide sequence of the gene for the 32 kd thylakoid membrane protein from *Spinacia oleracea* and *Nicotiana debneyi* predicts a totally conserved primary translation product of M_r 38,950. Proc. Natl. Acad. Sci. USA 79, 7699–7703.

Zurawski, G., Bottomley, W. and Whitfeld, P.R. (1984) Junctions of the large single copy region and the inverted repeats in *Spinacia oleracea* and *Nicotiana debneyi* chloroplast DNA: sequence of the genes for $tRNA^{His}$ and the ribosomal proteins S19 and L2. Nucl. Acids Res. 12, 6547–6558.

P. Fay and C. Van Baalen (Eds.) The Cyanobacteria
© 1987 Elsevier Science Publishers B.V. (Biomedical Division)

12

Photomovement

DONAT-P. HÄDER

Fachbereich Biologie-Botanik, Lahnberge, D-3550 Marburg, FRG

1. Introduction

Light-dependent movement responses by microorganisms have been described as early as 1817 (Treviranus). Possibly due to their slow gliding movement, light-induced reactions in cyanobacteria have been observed only several decades later (Cohn, 1863; Engelmann, 1882). Despite of the lack of elaborate optical instruments, the basic phenomena of movement have been studied in the last century (Hansgirg, 1883; Kolkwitz, 1896). Light was found to be a major factor in orientation (Aderhold, 1888; Verworn, 1889; Phillips, 1904) in addition to chemical, thermal and mechanical stimuli (Mendelssohn, 1902; Fechner, 1915; Schmid, 1923; Crozier and Federighi, 1924; Burkholder, 1933; Haupt, 1962).

During the first half of this century both motility and light-induced responses were studied in cyanobacteria in great detail (e.g. Pringsheim, 1912; Pieper, 1915; Nienburg, 1916; Funk, 1923; Ullrich, 1929). Most of this work has been carefully reviewed and summarized (cf. Harder, 1920; Haupt, 1959, 1972; Halldal, 1962; Nultsch, 1973, 1974b, 1975, 1980; Nultsch and Häder, 1979, 1980; Häder, 1979a, 1984a), and need not to be repeated here.

All light-dependent reactions can be described by a series of events: *photoperception, sensory transduction and motor response.*

The nature of the photoreceptors utilized by cyanobacteria has been covered by Bendix (1960), Clayton (1964), Nultsch (1965), Perez-Miravete (1973), Checcucci and Weale (1973), Ferrara (1975) and Kivic and Walne (1983). The sensory transduction chain includes both a signal transmission over some distance (either intracellular or intercellular) and a signal amplification which often involves a change in the form of energy (light, chemical, electrical energy) (Jahn and Bovee, 1965; Carlile, 1980; Lenci and Colombetti, 1978; Fay, 1983). Eventually, this signal controls the activity of the motor apparatus and causes a modulation in speed, a reversal of movement or a directional change (Drews and Nultsch, 1962; Castenholz, 1973a,b; Halfen, 1979). When studied on the level of whole populations, light influences the choice of the habitat and controls the orientation in space and time (Stanier and Cohen-Bazire, 1957; Seitz, 1975; Castenholz, 1982; Rosen, 1983).

Cyanobacteria show the three major light-induced motor responses found also in many other microorganisms, collectively called photomovement. Unfortunately, the terms used to describe the light reactions have changed in the past. Instead of going into the historical use of these definitions (cf. Nultsch, 1975), the following paragraph will briefly explain the basic responses using the currently widely accepted terminology (Diehn et al., 1977; Nultsch and Häder, 1979).

1. *Phototaxis* describes an orientation of movement with respect to the direction of the incident light beam. The response can be either towards (positive phototaxis) or away from the light source (negative phototaxis) (Nultsch et al., 1983b; Schuchart and Nultsch, 1984). Thus, the vectorial component of light direction is the physical quality which defines the stimulus, but light intensity controls the sign of the response: like many other microorganisms, cyanobacteria show positive phototaxis at low fluence rates and negative at higher. (In this chapter the term 'light intensity' is used to include the term 'irradiance' measured as fluence rate of monochromatic light with the unit [Wm^{-2}] and the term 'illuminance' of white light measured in lux with the unit [lx].)

2. *Photokinesis* is defined as a steady-state dependence of the speed of movement on the incident fluence rate; it is independent of the light direction. When at a tested light intensity an organism moves with a speed higher than in the dark control, this is called a positive photokinesis. When it is slower (or stops altogether) at that intensity this is defined as negative photokinesis.

3. *Photophobic respons s* are also independent of the light direction. But unlike photokinetic reactions, they are transient phenomena and are triggered by a sudden increase or decrease in light intensity (dI/dt). In the filamentous cyanobacteria, photophobic responses can usually be observed as a sudden stop and a subsequent reversal of movement on the original track. Both temporal and spatial changes in the light intensity induce phobic responses: the organisms reverse movement when the overall light intensity is suddenly changed or when they enter a field illuminated with a different intensity (Dangeard, 1928; Nuernbergk, 1928).

Though all three photomovement responses are used by cyanobacteria to find a suitable niche in their photoenvironment, probably photophobic responses are the most effective means of orientation.

2. Mechanisms of movement: facts, models and hypotheses

To state the most important fact first: the mechanism of movement, at least for filamentous cyanobacteria, is still a mystery.

There is no microscopically visible organelle involved in gliding movement nor do the organisms show a noticeable change in shape. Therefore, it is not astonishing that a number of different models, more or less probable hypotheses and even wild speculations have been proposed. The more convincing ones are based on directed slime secretion or undulating surface structures and the speculative ones involve electrokinetic elementary processes or electrical fields (Toman, 1955; Jarosch, 1962, 1963a; Halfen, 1973). Most of the older work has been reviewed by Walsby (1968).

2.1. Slime secretion

Most cyanobacteria secrete a slime sheath which sometimes can be much thicker than the filament itself. In addition, some species produce a large amount of mucilage of a softer consistency (Wolk, 1980), though this material may simply be sheath substance in a more hydrated form. It can be visualized by suspending the organisms in a dilute indian ink suspension (Schmid, 1918; Nikitschek, 1934; Walsby, 1968). It was hypothesized that the filaments propel themselves by a directed secretion of mucilage (Prell, 1921; Pringsheim, 1968). Hosoi (1951) held *Oscillatoria* filaments in a fixed position by means of microneedles and observed the mucilage move along the surface which he interpreted as a proof that the cells propel themselves by extruding slime. The secreted mucilage swells considerably by hydration after excretion which is supposed to enhance the propelling force.

The surface structure of cyanobacteria has been studied by light and electron microscopy (Mühldorf, 1938; Johnson and Baker, 1947; Schulz, 1955). The substances forming the slime sheath may be extruded from the cells by either or both of two sets of structures revealed by electron microscopy. In several *Oscillatoria* species Schulz (1955) found large pores (30–60 nm in diameter) in the outer walls which may only represent depressions rather than open pores. The other modification of the cell wall are girdles of small pores (<20 nm) in the longitudinal walls which line the cross walls on both sides in a number of cyanobacteria (Drawert and Metzner, 1958; Frank et al., 1962; Pankratz and Bowen, 1963). Both these structures are not seen in scanning electron microscopy of *Phormidium uncinatum* after rapid freeze fixation in liquid propane (Fig. 1). Instead, the surface appears to be smooth and uniform with subtle constrictions indicating the position of the cross walls. The slime sheath is ruptured during dehydration and forms the reticulate debris covering the filaments.

The slime sheath is formed by all cells of a trichome. In *Phormidium* it is rather thin when the organism glides continuously in one direction. When it is forced to reverse repeatedly (by e.g. photophobic responses, see below), new layers are added (Fig. 2). The slime sheath collapses when the filament leaves it and it is twisted by the rotation around the longitudinal axis (Fig. 3). During this process it can be seen to be composed of about 10 or 12 individual strands when stained by

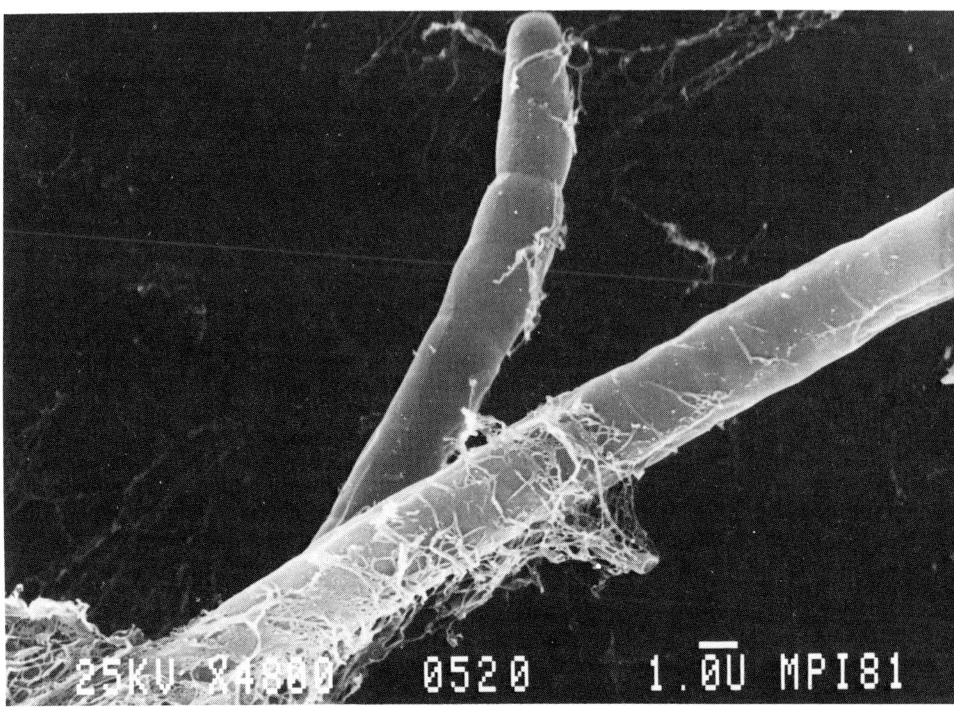

Fig. 1. Scanning electron micrograph of *Phormidium uncinatum* after fixation in liquid propanol. The bar indicates 1 μm. (Courtesy of Dr. Claviez)

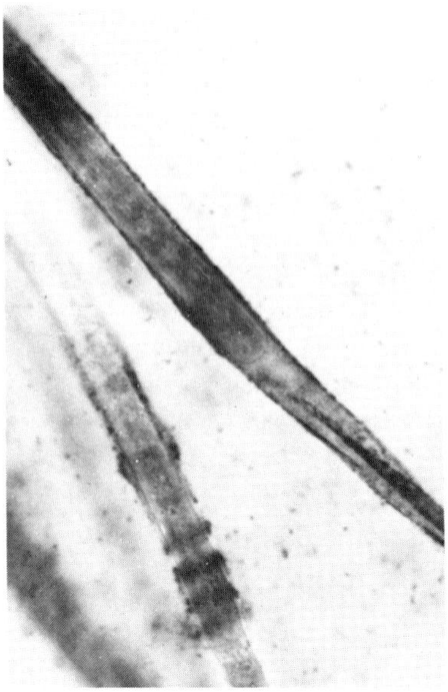

Fig. 2. (left) Trichome of *Phormidium uncinatum* in its slime sheath stained in vivo with india ink particles (0.5 %) in interference contrast. The filament is about 5 nm wide. (Häder, unpublished)

Fig. 3. (right) Slime sheath of *Phormidium uncinatum* stained in vivo with india ink particles (0.5 %) in interference contrast. (Häder, unpublished)

either india ink or polystyrene particles of selected diameter (0.125 μm). After a reversal the filament can reutilize the slime sheath or break out laterally. The naked tip of the filament immediately forms a new sheath which can be visualized four or five cells behind the tip cell and gradually increases in thickness by additional material secreted by the following cells.

Holton and Freeman (1965) have calculated that for a jet propulsion mechanism the production of mucilage necessary to power the forward locomotion would exceed the volume of the cells several times per second. Even when the slime swells considerably by uptake of water outside the cells this mechanism would require a highly uneconomical waste of available resources.

There is another conceptional difficulty with the idea that the slime secretion should directly power the forward movement: while the Nostocaceae just move along their long axis the Oscillatoriaceae additionally rotate around this axis (Burkholder, 1934). The direction of this rotation can be a left-handed or right-handed helix but it is specific for the species (Ruthsack-Künzenbach, 1967; Thomas, 1970). It is not easy to explain how an ejection of a slime stream can cause a rotational movement around the long axis even if one assumes extracellular microfibrils in the mucilage oriented at the same angle as the pitch of rotation (Lamont, 1969). It is likewise difficult to explain sudden reversals of movement on the basis of the simple slime jet expulsion.

2.2. Movement by surface undulations

The alternative explanation to a jet stream propulsion is the involvement of contractile elements which cause surface undulations. Jarosch (1963b) had postulated the existence of contracting or rotating protoplasmic fibrils. Thin sections of the outer cell wall layers show a convoluted appearance in a number of species (Echlin and Morris, 1965; Jost, 1965). Though this appearance could be an artifact resulting from shrinkage during fixation or dehydration, there is growing evidence for the presence of fibrils in the outer cell wall layers. Wildon (1965) demonstrated 15-nm fibrils in *Anabaena cylindrica* and Halfen and Castenholz (1970, 1971a,b) found similar structures in an *Oscillatoria* species. These fibrils are oriented at an angle corresponding to the pitch of the rotation observed in this organism. This observation is a strong indication that gliding motility may indeed be based on contractile elements generating surface undulations.

The propagation of waves over the surface of filaments has never been observed (with the exception of one obscure report), but it can be speculated that these undulations are either too small or too fast (or both) to be observable under the light microscope. Work is currently in progress to demonstrate the existence of such waves, using the recently developed acoustical microscope (Thaer et al., 1982). Contractile elements have also been proposed to be the basis for flexing locomotion in the gliding bacterium *Flexibacter FS-1* (Burchard, 1982). However, the filamentous structures proposed to contract seem to operate with a different mechanism than contractile elements of eukaryotic cells since gliding motility is unaffected by cytochalasin B (Spangle and Armstrong, 1973), but it is inhibited by *p*-chloromercuribenzoate (Abe, 1963).

The energy required for locomotion is thought to be derived from both photophosphorylation and oxidative phosphorylation in form of ATP (Manten, 1948; Nultsch, 1969). Recently, Glagoleva et al. (1980) have demonstrated that the proton motive force directly supports gliding in cyanobacteria. Other bacteria can even utilize the energy stored in form of Na^+ or K^+ ion gradients (Brown et al., 1983). The biophysics group of Skulachev has demonstrated that the energy requirement of even a long *Phormidium* filament can be fulfilled by a few photosynthesizing cells. The strain of *Phormidium uncinatum* isolated from Lake Baikal stops moving a few minutes after being transferred into darkness. It could be demonstrated in a film that a light spot covering no more than three or four cells induces motility in a formerly immotile filament (Chailakhyan et al., 1982). Since all cells of the filament start moving almost simultaneously there must be an efficient intercellular power transmission along the filament (Raven, 1983).

Motility in the marine cyanobacterium *Spirulina subsalsa* is dependent on Ca^{2+}. Since dicyclohexylcarbodiimide (DCCD) inhibits motility, it has been speculated that the calcium action operates via a membrane bound Ca-ATPase (Abeliovich and Gan, 1982). Acetylcholine lowers the calcium concentration necessary to induce motility.

A biochemical study by Simon (1981) indicated that all motile strains of both unicellular and filamentous cyanobacteria produce a high molecular mass glycoprotein (ca. 200 000 daltons) which non-motile strains lack. This protein was produced in very large amounts by some *Oscillatoria* and *Phormidium* species.

3. Photokinesis

Engelmann (1882) was the first to use the term 'photokinesis' to describe the dependence of speed of movement on light intensity. He observed that some purple bacteria became immotile after some time in darkness. After a short irradiation of a few seconds or minutes they started moving. He concluded that "a pool of some substance is produced by the irradiation which is necessary for movement and is consumed in the dark".

3.1. White light behavior

Photokinesis in cyanobacteria was studied extensively by Nultsch (1962a, 1968, 1975). Photokinetic effects can be analyzed either by measuring the speed of movement of individual organisms (the distance travelled in a unit time) or by determining the speed of spreading of a population from an inoculation point, e.g. on an agar surface (Nultsch, 1974a; Häder, 1979a). In a *Phormidium* strain which is motile in darkness, the speed of movement increases almost linearly with the illuminance of white light starting at a threshold of about 0.02 lx. It reaches an optimum at about 2000 lx to decline at higher light intensities. At an illuminance >30 000 lx the speed of movement is slower than in darkness which by definition is a negative photokinesis. *Anabaena variabilis*, for comparison has a much higher threshold of about 10 lx, an optimum near 1000 lx and a maximum (where the speed of movement reaches the value of the dark control) of 90 000 lx. Negative photokinesis has not been observed in *Anabaena*.

3.2. Photoreceptor pigments

The photoreceptor pigments responsible for light-induced motor responses are usually determined by the comparison of action spectra with the absorption spectra either in vivo or of extracted pigments (cf. Tevini and Häder, 1985). The photokinetic action spectrum of *Phormidium uncinatum* (Fig. 4) closely resembles the in vivo absorption spectrum only in the red and blue spectral regions.

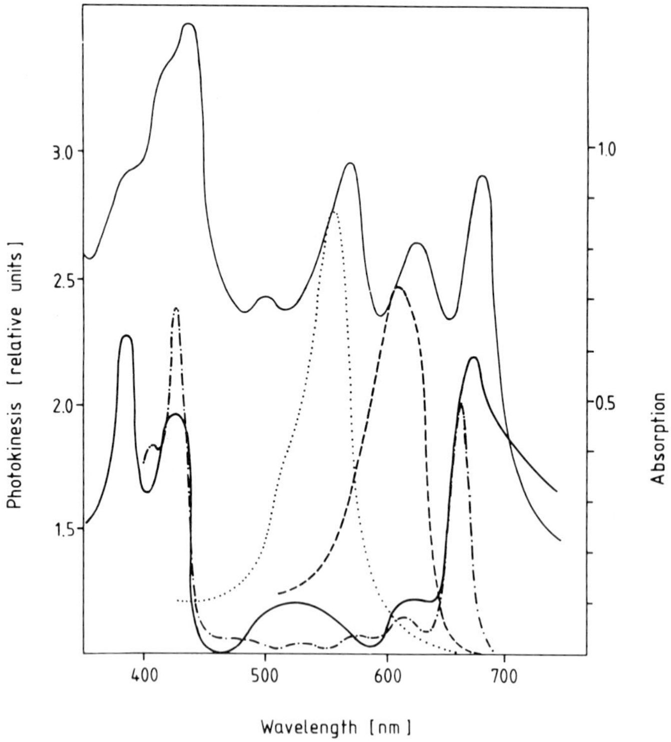

Fig. 4. Action spectrum of photokinesis (heavy solid line) and absorption spectrum (fine solid line) of *Phormidium uncinatum* in vivo (heavy lines) and of isolated pigments, chlorophyll *a* (dash-dotted line), C-phycoerythrin (dotted line), C-phycocyanin (dashed line). (After Nultsch, 1974a)

The main peak centered around 430 nm wavelength and the maximum in the red near 675 nm indicate the involvement of chlorophyll *a*. The phycobilins, C-phycoerythrin and C-phycocyanin, have maxima at about 565 nm and 625 nm, respectively. The main biliprotein of *Ph. uncinatum* is C-phycoerythrin. In contrast, *Ph. ambiguum* does not contain any detectable C-phycoerythrin, but a large amount of C-phycocyanin. (Both species contain only a small amount of allo-phycocyanin.) However, the photokinetic action spectrum of *Ph. ambiguum* closely resembles that of *Ph. uncinatum*.

The spectra suggest that only part of the photosynthetic apparatus is involved in the perception of photokinetically active radiation. Since only chlorophyll *a* contributes to photokinesis while the phycobilins do not, Nultsch (1974a) concluded that only Photosystem I is involved. Thus, only cyclic photophosphorylation drives photokinesis in these *Phormidium* species.

The photokinetic action spectrum of *Anabaena variabilis*, however, indicates a strong participation of C-phycocyanin (Nultsch and Hellman, 1972). (This organism does not produce C-phycoerythrin.) Chlorophyll *a* may as well be active in the red region, though no distinctive maximum or shoulder can be detected and the Soret band (430 nm) is totally ineffective. This result indicates that both photosystems are involved and that either the noncyclic or the pseudocyclic photophosphorylation produces the additional energy to maintain the higher speed of movement. Comparison with the action spectra of other species indicated that members of the Oscillatoriaceae (*Phormidium*, *Oscillatoria*, *Pseudanabaena* and *Spirulina*) basically utilize chlorophyll *a* (and to a certain extent also carotenoids) as photoreceptors while members of the Nostocaceae (*Anabaena* and *Nostoc*) use the accessory pigments (phycobilins) which feed their energy into photosystem II (Nultsch, 1974a).

3.3. Effects of inhibitors and uncouplers

Since the photosynthetic apparatus is involved in photokinesis, the sensory transduction chain leading to the final control of the speed of movement can be studied using inhibitors, uncouplers

TABLE I

Effects of inhibitors on photokinesis in cyanobacteria (values in per cent of the uninhibited control)

DCMU (M)	*Oscillatoria*	*Spirulina*
10^{-6}	98 %	100 %
2×10^{-6}	90 %	99 %
5×10^{-6}	80 %	100 %
10^{-5}	60 %	99 %
2×10^{-5}	35 %	85 %
5×10^{-5}	12 %	75 %
DBMIB (M)	***Anabaena***	***Phormidium***
2×10^{-7}	86 %	–
5×10^{-7}	75 %	99 %
10^{-6}	58 %	94 %
2×10^{-6}	45 %	85 %
5×10^{-6}	24 %	66 %
10^{-5}	15 %	52 %

(From Nultsch, 1974a)

and redox systems affecting the photosynthetic primary and secondary processes (Nultsch, 1965; Nultsch and Jeeji-Bai, 1966).

Photokinesis of those cyanobacteria utilizing the photosynthetic electron transport between Photosystems I and II for the photokinetic effect is extremely affected by dichlorophenyl-dimethylurea (DCMU), an inhibitor which blocks the noncyclic electron transport near the quencher of Photosystem II (Table I). Those organisms in which photokinetic effects are driven by cyclic electron transport (*Spirulina, Phormidium, Lyngbya*) show only a moderate inhibition by DCMU. Similar differences have been found for dibromomethyl-isopropyl-*p*-benzoquinone (DBMIB), which inhibits the reoxidation of plastoquinone, and other inhibitors of the photosynthetic electron transport chain (Nultsch, 1974a).

Redox substances with the appropriate redox potential are known to interfere with the photosynthetic electron transport since they can operate as artificial electron donors or acceptors. Most of 19 redox systems tested by Nultsch (1968) had an effect on photokinesis. The most effective one was methylviologen, which had the most negative redox potential ($E'_0 = -0.440$ V). Substances with less negative potentials had increasingly less effect, but those with positive potentials were again effective with a maximal inhibition by phenazinmethosulfate ($E'_0 = +0.064$ V). The interpretation of the results is that the redox systems trap electrons from the photosynthetic electron transport chain.

A second effect of artificially applied redox systems is the inhibition of movement in darkness which is probably due to the fact that they abstract electrons from the respiratory electron transport chain if their redox potential is more positive than that of NAD^+.

The results described above do not allow to distinguish between two alternative interpretations for the coupling of photokinesis to photosynthesis: either the movement response is directly coupled to the electron transport chain (as are photophobic responses, see below) or the light dependent speed of movement is linked to the photophosphorylation driven by the electron transport. Inhibitors and redox systems impair both processes. In order to distinguish between the two alternatives, uncouplers can be used which block phosphorylation without affecting the electron transport.

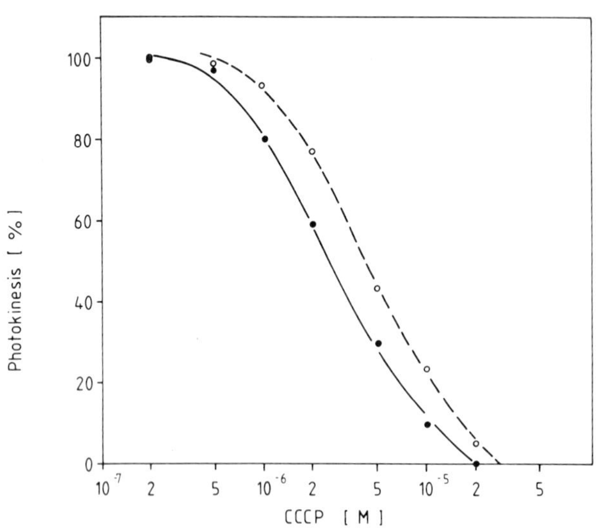

Fig. 5. Effect of CCCP on photokinesis of *Phormidium* (solid line) and *Anabaena* (broken line). (After Nultsch, 1974a)

Carbonylcyanide-*m*-chlorophenylhydrazone (CCCP) drastically impairs photokinesis both in *Phormidium* (Oscillatoriaceae) and in *Anabaena* (Nostocaceae); a concentration of 10^{-5} M almost totally abolishes any photokinetic effect (Fig. 5). Similar results have been found with other uncouplers such as dinitrophenol, sodium azide, imidazol and atebrin (Nultsch, 1974a). In addition to affecting photophosphorylation, most uncouplers also impair oxidative phosphorylation equally well which reduces the dark motility.

Summarizing the results, the photoreceptor pigments for photokinesis are subsets of the photosynthetic pigments and differ amongst families of gliding cyanobacteria. The increase in speed of movement is brought about by a higher energy supply to the motor apparatus in form of either ATP or proton motive force. This interpretation also holds for other photosynthetic prokaryotes (Clayton, 1959; Throm, 1968) and even for some eukaryotes such as red algae (Nultsch et al., 1979) and desmids (Häder, 1982).

4. Phototaxis

As mentioned above, phototaxis (formerly also called phototopotaxis) describes a directed movement with respect to the light direction. Most cyanobacteria either show positive or negative phototaxis (movement toward or away from the light source) or both (Haupt, 1959; Stanier et al., 1971; Häder, 1979a). Directed movement is not necessarily the result of an active steering.

4.1. Strategies for orientation

The more primitive mode of orientation is based on a trial-and-error mechanism and is used by members of the Oscillatoriaceae (Drews, 1957, 1959). *Phormidium* filaments reverse their direction of movement with a more or less regular rhythm (minutes to hours). When the filaments are oriented parallel to the incident light beam or deviate only by a small angle they continue their movement toward the light source for longer periods of time and reverse faster than without a stimulus when moving away from it (Fig. 6A). This behavior eventually brings the population closer to the light source (Burkart and Häder, 1980; Häder, 1981a).

The alternative mechanism is a true steering into the light direction which can be found in *Anabaena* (Fig. 6B) and other Nostocaceae (Drews, 1957; Nultsch et al., 1979; Nultsch and Wenderoth, 1983). It could be speculated that the differences in the strategy of phototactic orientation are related to the differences found in the mechanism of motility described above, i.e. the fact that the Oscillatoriaceae rotate while the Nostocaceae do not. In lateral light, *Anabaena* turns either toward or away from the light source depending on the intensity until the filament moves parallel to the incident beam (Nultsch, 1975). Often the filaments move in a U-shaped manner; in this case the leading bent zone initiates the turn. Partial irradiation of trichomes using small light fields projected onto the lateral half of a filament indicates that each cell is able to compare the light intensity in the left half versus the right half. When a low intensity light spot (1.25 Wm^{-2}) is placed on one half of a few central cells, the filament bulges out and this new front starts to glide in the direction of the light spot (Nultsch and Wenderoth, 1983). When the light spot is moved with the filament, the organism can be 'drawn' over a considerable distance. In high intensity light fields (27 Wm^{-2}), the filament moves in the opposite direction. These results indicate that *Anabaena* orients with respect to an internally established light gradient. A similar mechanism has been found in *Dictyostelium* pseudoplasmodia and amoebae (Häder and Burkart, 1983; Häder et al., 1983).

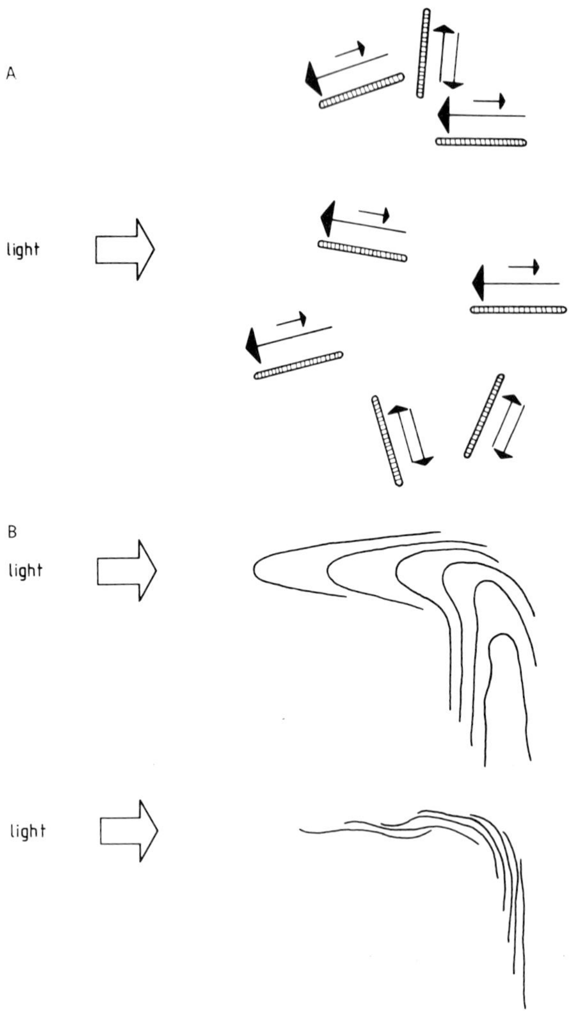

Fig. 6. Schematic representation of the mechanisms of phototactic orientation. A, trial-and-error; B, active steering.

4.2. Action spectra and transduction chain

The spectral sensitivity of both *Phormidium uncinatum* and *Ph. autumnale* extends to 640 nm (Nultsch, 1961, 1962b). The main activity peaks are at around 390 nm, 480 nm and 560 nm. While the UV and blue peaks could indicate the involvement of a flavin type photoreceptor the longer wavelengths maxima correspond with the absorption maxima of C-phycoerythrin (560 nm) and C-phycocyanin (625 nm). It is interesting to note that only the accessory photosynthetic pigments are involved while chlorophyll obviously is not active (Fig. 7). In *Oscillatoria mougeotii*, however, the action spectrum extends to 720 nm (Drews, 1959).

Tyagi (1976) measured an action spectrum of phototaxis in *Cylindrospermum alatosporum* which also extends well into the red region of the spectrum. Thus, there are not only differences in the

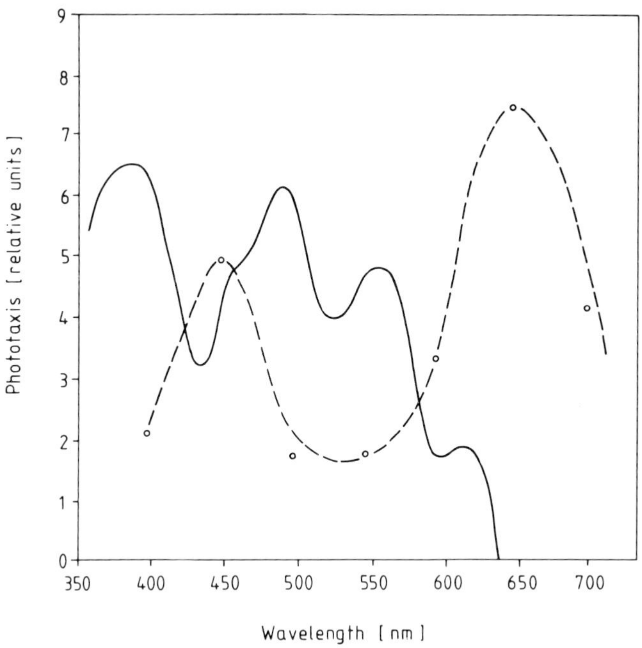

Fig. 7. Phototactic action spectrum of *Phormidium* (solid line) and *Cylindrospermum* (broken line). (From Nultsch, 1961; Tyagi, 1976)

strategy of phototactic orientation but also in the photoreceptor pigments. This was further studied by Nultsch et al. (1979, 1983a) in *Anabaena variabilis*. The action spectrum for positive phototaxis points to C-phycocyanin as the main photoreceptor pigment. But in contrast to *Phormidium* also chlorophyll *a* is represented by peaks near 440 nm and 670 nm. This is a strong indication for the photosynthetic pigments to be the phototactically active photoreceptor molecules. The same pigments are found to be responsible for negative phototaxis. However, there are two additional active bands at about 550 nm and at 730 nm. The pigments discussed to be responsible for these peaks are a *b*-type phycochrome and a phytochrome-like pigment in the far red band.

Although the photosynthetic pigments are involved in the photoperception, the inhibitors of the photosynthetic electron transport chain, DCMU and DBMIB, had no effect on photoorientation. Thus, a direct linkage between phototaxis and the photosynthetic machinery is still obscure.

4.3. Control of the reaction sense

Many microorganisms are known to show positive phototaxis in low irradiance and negative in higher. The results described above indicate that two at least partially different sets of photoreceptor pigments are responsible for the two phototactic orientations found in *Anabaena*. What is the nature of the switch which controls the sign reversal? Nultsch et al. (1983a) have shown that the one pigment system which includes the phycobiliproteins and chlorophyll *a* as photoreceptors is sensitive to sodium azide (NaN_3) while the other absorbing at 560 nm and above 700 nm is insensitive. Application of NaN_3 reverses negative responses at high fluence rates to positive ones. NaN_3 has been reported to have several different effects on plants and microorganisms. In this case it does not seem to impair the cyclic electron transport chain and photophosphorylation dramatically. Another effect of NaN_3 is that it quenches active oxygen species, especially singlet oxygen. Therefore, it was concluded that the sign reversal from positive to negative phototaxis is controlled by the level of singlet oxygen which is produced by some of the primary photosynthetic reactions. More recent experiments with carotenoids, crocetin and other $^1O_2$ quenchers support this hypothesis.

5. Photophobic responses

Photophobic responses were among the first light-induced reactions observed in cyanobacteria. The detailed and well-documented earlier observations are reviewed by Haupt (1959), Nultsch (1975), Nultsch and Häder (1979) and Häder (1979a). The more recent work has concentrated on *Phormidium uncinatum* (Oscillatoriaceae). This organism responds to either a temporal or a spatial decrease in light intensity with a stop a few seconds after the step-down and a subsequent reversal of movement, so that the filament glides back on its own trail. Repeated responses cause the accumulation of organisms in light traps (Häder and Burkart, 1978). The zero threshold was found to be at 0.03 lx (Nultsch, 1962c) and some organisms respond to a difference of as low as 4% between two light fields (Nultsch and Häder, 1970; Häder, 1973).

5.1. Linkage between phobic responses and photosynthesis

The action spectrum of photophobic responses in *Phormidium* corresponds well with the in vivo absorption spectrum and indicates the involvement of the phycobiliproteins, chlorophyll *a* and possibly carotenoids (Nultsch, 1962c; Häder, 1973, 1976). Uncoupler experiments indicate that the phobic response is not linked to photophosphorylation. Inhibitors of the photosynthetic electron transport chain, however, strongly impair phobic reactions (Häder, 1975). Both photosystems are involved in the response and the coupling site seems to be located in the linear electron transport chain (Häder, 1974). Since DCMU blocks the phobic reactions mediated by Photosystem II while DBMIB inhibits those induced by Photosystem I, plastoquinone seems to be the coupling site.

5.2. Proton gradients and electrical potentials

Plastoquinone plays a central role in the photosynthetic electron transport chain: during its redox cycle it takes up protons from the outside of the thylakoids (Mitchell, 1978) and releases them to the inside of the thylakoid vesicles (which in cyanobacteria are scattered throughout the cytoplasm and are not encapsulated in a plastid envelope; cf. Chapter 1 of this volume). This vectorial proton transport results in a pH difference and an electrical gradient. Light-dependent modulation of the gradient plays a central role in the sensory transduction of photophobic responses in *Phormidium* (Häder, 1978a,b, 1979b, 1981b; Murvanidze et al., 1982a,b).

In light, the vectorial proton transport causes a hyperpolarization of the cytoplasm since positive charges are moved into another compartment (thylakoid vesicles or extracellular space). When a filament moves into a dark area this gradient breaks down again. Electrical potential changes can be measured by means of intracellular recording using microglass electrodes with tip diameter of about 0.1–0.2 μm (Häder, 1978a,b). The action spectrum of the light-induced potential changes coincides with that of the photophobic responses and the potential changes can be suppressed by DCMU and DBMIB (Häder, 1979b). The results indicate that modulation of the proton gradient is an early step in the sensory transduction chain of photophobic responses. A further proof is that the primary photoreceptor can be bypassed using pH jump experiments. The organisms buffer their growth medium to a pH of about 7.3. When in darkness the pH is forced to drop into the range of 4.9 to 5.7, the filaments undergo a reversal of movement which closely resembles a photophobic response, though the organisms are observed in darkness (i.e. in infra red light >780 nm wavelength, with an infrared-sensitive camera) (Häder, 1981b).

Judging from the low zero and difference thresholds, the electrical potential changes are larger than those to be expected from an exclusive proton transport. These considerations suggest a mechanism for signal amplification. There are two in biological systems widespread mechanisms for this task. One uses enzyme cascading as found in vision (Stieve, 1982) and the other involves gated ionic currents (Hille, 1978). The existence of voltage-dependent, cation-specific channels in the cytoplasmic membrane could be demonstrated by the following results (Häder, 1982; Murvanidze and Glagolev, 1983):

(1) Specific calcium blockers such as lanthanum ions or ruthenium red inhibit photophobic responses (Häder, 1982).

(2) Likewise, organic calcium antagonist such as D600 (gallopamilhydrochloride) impair photophobic reactions (Häder, unpublished results).

(3) The calcium ionophore calcimycin (A23187), which causes a calcium current to bypass the natural calcium channels, decreases the photophobic reactivity (Murvanidze, 1981; Häder, 1982). Ionophores for monovalent cations such as gramicidin or valinomycin do not impair the response.

(4) Increasing the gated calcium current by the application of phosphatidic acid or caffeine alters the phobic behavior (Häder, unpublished).

(5) The Ca-ATPase inhibitor, poly-L-lysine, decreases photophobic responses while ouabain, an inhibitor of the Na/K pump, does not (Häder, 1982).

(6) During a photophobic response a massive uptake of $^{45}Ca^{2+}$ into the filaments can be detected a few seconds after a decrease in illuminance (Häder and Poff, 1982).

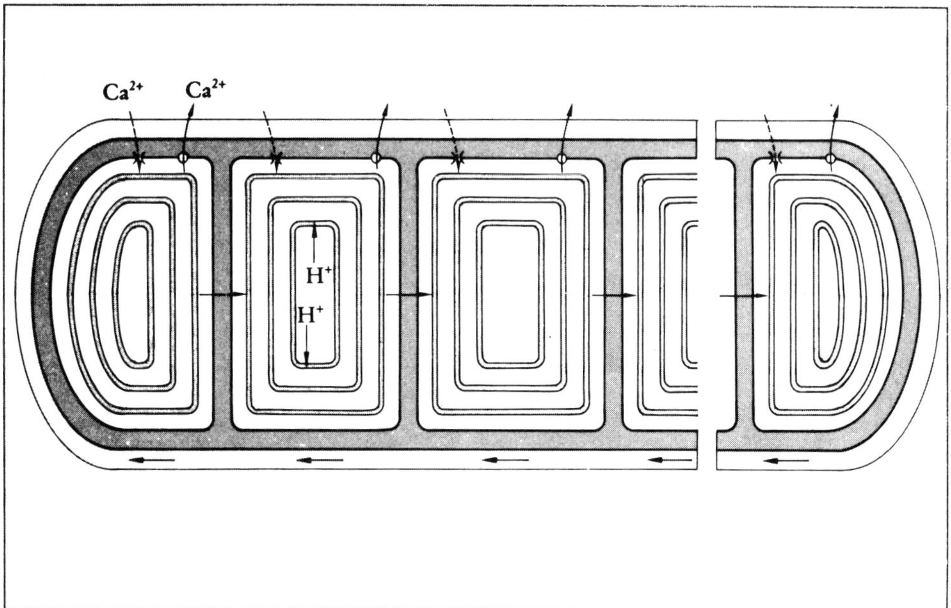

Fig. 8. Model of photophobic responses in *Phormidium*. For explanation see text. (From Häder, 1984a)

These results gave rise to a model for photophobic responses shown in Fig. 8 (Häder, 1984a). In light, protons are transported from the cytoplasmic compartment into the thylakoid vesicles which hyperpolarize the cytoplasm. When the front cells of a filament enter a dark area, the proton gradient breaks down which slightly depolarizes the cytoplasm. These small electrical potential changes trigger the voltage-dependent, calcium-specific channels in the cytoplasmic membrane to open. In constant light calcium is constantly pumped out of the cell by energy-dependent Ca-ATPases. During a photophobic response there is a massive influx of calcium into the cell along the previously established gradient. This drastically depolarizes the cell and causes a considerable signal amplification.

The next question is how the intracellular potential change causes a reversal of movement. It could be shown that the direction of movement is dictated by an electrical potential difference between the two ends of a filament (Häder, 1978b; Murvanidze and Glagolev, 1982; Murvanidze et al., 1982a,b). When the potential in the front cells breaks down upon entering a dark area, the longitudinal potential of the filament reverses its polarity which causes a reversal of movement. This process can be manipulated artificially by applying an external voltage, which inhibits photophobic responses at a light-dark boundary (Häder, 1977, 1985).

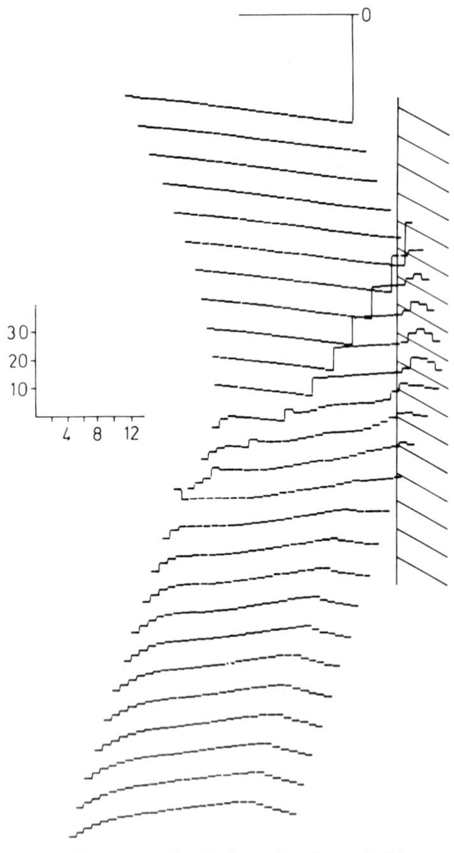

Fig. 9. Computer simulation of a photophobic reversal. The hatched area represents the dark zone. The abscissa indicates the number of cells in a filament and the ordinate the electrical potential of each cell in mV. (From Häder and Burkart, 1982a)

Photophobic reactions in *Phormidium* have been simulated by a mathematical model (Häder and Burkart, 1982a,b). The model assumes that each cell of a filament has an electrical potential, the front cells having the highest (negative) potential (Fig. 9). When the front cells enter a dark area the potential is reduced and the longitudinal gradient reverses. This causes a reversal of movement.

The next task will be to determine the molecular nature of the motor apparatus for gliding and the control of the direction of movement by electrical gradients. It could be speculated that the spatial molecular configuration of protein macromolecules is dependent on the electrical field.

6. Ecological significance of photomovement

Photosynthetic cyanobacteria need solar irradiation for their existence. This necessity may have ultimately selected for the evolution of photomovement. However, most cyanobacteria are typical shade plants and too strong an irradiation photobleaches the photosynthetic and accessory pigments. In culture, many strains are known to thrive best at illuminances of 500–1000 lx and higher illuminances are detrimental (Walsby, 1968; Häder, 1985). (See Chapter 13 for details.)

6.1. Photobleaching

Anabaena trichomes exposed to photon fluence rates exceeding 3 Wm^{-2} are bleached within 1 or 2 days. This effect has been quantified by microspectrophotometry (Nultsch et al., 1983a,b). Photosystem I is less sensitive than Photosystem II; therefore, heterocysts — which contain only a functional Photosystem I — are less affected. When kept at low fluence rates after photobleaching the cells recover and normal pigmentation reappears within a few days provided the photobleaching irradiation did not exceed 50 Wm^{-2}, which eventually kills the organisms. It is interesting to note that photobleaching can partially be prevented by addition of 1 mM NaN_3. One of the possible explanations is that singlet oxygen is produced in high fluence rates which photooxidizes the photopigments; NaN_3 quenches the $^1O_2$ production. (See also Section 4.1. of Chapter 13.)

6.2. Strategies for the choice of habitat

How do cyanobacteria optimize their need for solar irradiation versus the photodamage encountered in too high fluence rates? In *Anabaena*, the sign reversal between positive and negative phototaxis is an effective mechanism to reach this goal (cf. Section 4.1.).

Another mechanism to select a habitat of optimal light intensity is the photophobic response. *Phormidium* shows a step-down photophobic response when entering a dark zone (e.g. when gliding into the substratum or into the shade of a leaf, etc.). In contrast, the filaments show a step-up response when entering a too bright light field. This primitive, but rather effective mechanism results in an accumulation of the population in a narrow light intensity range as can be shown with a simple experiment (Fig. 10): when a photographic negative is projected into a homogeneous suspension of filaments in an agar layer, the organisms move to areas of suitable light intensities. Thus, they form a rather detailed positive image of the photographic negative. The image can be preserved when the organisms are fixed in their position by allowing the agar to dry during irradiation.

Photophobic responses have also been demonstrated to be the main mechanism for light adjustment in the natural habitat (Castenholz, 1968; Nelson and Castenholz, 1982). *Beggiatoa* shows diurnal vertical migrations into and out of sediments in warm springs. The movement out of too

Fig. 10. Accumulation of *Phormidium* trichomes in bright areas of a photographic negative producing a positive (market place in Marburg). (Häder, unpublished)

bright light intensities is caused by step-up photophobic responses. A crude action spectrum indicates the involvement of a blue light photoreceptor. Growth is inhibited by as little as 5 000 lx from fluorescence lamps (Nelson and Castenholz, 1982). *Microcoleus lyngbyaceus* also escapes higher light intensities (exceeding 2.5 klx) by step-up photophobic responses. When the organisms are buried by fine sediments they move back to the surface and form dense mats using light responses, though chemotactic responses were also observed (Pentecost, 1984).

Recently, the potential hazard of a stratospheric ozone reduction by man-made gaseous pollutants such as chlorinated fluorocarbons with a resulting increase of UV-B (280–320 nm) irradiation has been studied (Häder, 1984b). Even a moderate increase of the solar UV-B irradiation strongly inhibits photoaccumulations of *Phormidium*. This is partially due to an impaired motility, but both phototaxis and photophobic responses are equally disturbed. The organisms are not directly damaged or killed by the UV-B irradiation but die either from lack of energy when moving into dark areas (step-down phobic response and positive phototaxis impaired) or by photobleaching in too bright white light intensities (step-up phobic response inhibited). Thus, an increase of the solar UV-B level by only a few percent could have a drastic effect on the survival of terrestrial or aquatic cyanobacteria in the top layers of the water column.

Acclimation of planktonic cyanobacteria to light quantity by means of buoyancy regulation is discussed in Chapter 14 of this volume.

References

Abe, S. (1963) The effect of *p*-chloromercuribenzoate on amoeboid movement, flagellar movement and gliding movement. Biol. Bull. 124, 107–114.

Abeliovich, A. and Gan, J. (1982) Site of Ca^{2+} action in triggering motility in the cyanobacterium *Spirulina subsalsa*. Cell Motil. 2, 393–403.

Aderhold, R. (1888) Beiträge zur Kenntnis richtender Kräfte bei der Bewegung niederer Organismen. Jena. Z. Med. Naturwiss. 22, 311–342.

Bendix, S.W. (1960) Phototaxis. Bot. Rev. 26, 145–208.

Brown, I.I., Galperin, M.Y., Glagolev, A.N. and Skulachev, V.P. (1983) Utilization of energy stored in the form of Na^+ and K^+ ion gradients by bacterial cells. Eur. J. Biochem. 134, 345–349.

Burchard, R.P. (1982) Evidence for contractile flexing of the gliding bacterium *Flexibacter FS-1*. Nature 298, 663–665.

Burkart, U. and Häder, D.-P. (1980) Phototactic attraction in light trap experiments: a mathematical model. J. Math. Biol. 10, 257–269.

Burkholder, P.R. (1933) Movement in the Cyanophyceae. The effect of pH upon movement in *Oscillatoria*. J. Gen. Physiol. 16, 875–881.

Burkholder, P.R. (1934) Movement in Cyanophyceae. Quart. Rev. Biol. 9, 438–459.

Carlile, M.J. (1980) From prokaryote to eukaryote: gains and losses. In G.W. Gooday, D. Lloyd and A.P.J. Trinci (eds.), The Eukaryotic Microbial Cell. Soc. Gen. Microbiol., Symp. 30. Cambridge University Press, Cambridge, pp. 1–40.

Castenholz, R.W. (1968) The behavior of *Oscillatoria terebriformis* in hot springs. J. Phycol. 4, 132–139.

Castenholz, R.W. (1973a) Movements. In N.G. Carr and B.A. Whitton (eds.), The Biology of Blue-green Algae. Botanical Monographs, 9. Blackwell Sci. Publ., Oxford, London, Edinburgh, Melbourne, pp. 320–339.

Castenholz, R.W. (1973b) Ecology of blue-green algae in hot springs. In N.G. Carr and B.A. Whitton (eds.), The Biology of Blue-green Algae. Botanical Monographs, 9. Blackwell Sci. Publ., Oxford, London, Edinburgh, Melbourne, pp. 379–414.

Castenholz, R.W. (1982) Motility and taxes. In N.G. Carr and B.A. Whitton (eds.), The Biology of Cyanobacteria. Botanical Monographs, Vol. 19, Blackwell Sci. Publ., Oxford, London, Edinburgh, Melbourne, pp. 413–439.

Chailakhyan, L.M., Glagolev, A.N., Glagoleva, T.N., Murvanidze, G.V., Potapova, T.V. and Skulachev, V.P. (1982) Intercellular power transmission along trichomes of cyanobacteria. Biochim. Biophys. Acta 679, 60–67.

Checcucci, A. and Weale, R.A. (eds.) (1973) Primary Molecular Events in Photobiology. Elsevier Sci. Publ., Amsterdam, London, New York.

Clayton, R. (1959) Phototaxis of purple bacteria. In W. Ruhland (ed.), Encyclopedia of Plant Physiology, Vol. 17, Physiology of Movements, Part 1. Springer, Berlin, Göttingen, Heidelberg, pp. 371–387.

Clayton, R.K. (1964) Phototaxis in microorganisms. In A.C. Giese (ed.), Photophysiology, Vol. 2. Academic Press, New York, London, pp. 51–77.

Cohn, F. (1863) Verhalten der grünen mikroskopischen Pflanzen und Thiere zum Lichte. Jahresber. Schles. Gesellsch. f. Vaterl. Cultur, 41, 102–105.

Crozier, W.J. and Federighi, H. (1924) Critical thermal increment for the movement of *Oscillatoria*. J. Gen. Physiol. 7, 137–150.

Dangeard, P.A. (1928) Le déterminisme des mouvements chez les organismes inférieurs. Ann. Protistol. 1, 3–10.

Diehn, B., Feinleib, M., Haupt, W., Hildebrand, E., Lenci, F. and Nultsch, W. (1977) Terminology of behavioral responses of motile microorganisms. Photochem. Photobiol. 26, 559–560.

Drawert, H. and Metzner, I. (1958) Fluoreszenz- und elektronen-mikroskopische Untersuchungen an *Oscillatoria borneti*. Z. Bot. 46, 16–25.

Drews, G. (1957) Die phototaktischen Reaktionen einiger Cyanophyceen. Ber. Dtsch. Bot. Ges. 70, 259–262.

Drews, G. (1959) Beiträge zur Kenntnis der phototaktischen Reaktionen der Cyanophyceen. Arch. Protistenkd. 104, 389–430.

Drews, G. and Nultsch, W. (1962) Spezielle Bewegungsmechanismen von Einzellern (Bakterien, Algen). In W. Ruhland (ed.), Encyclopedia of Plant Physiology, Vol. 17, Physiology of Movements, Part 2. Springer, Berlin, Göttingen, Heidelberg, pp. 876–919.

Echlin, P. and Morris, I. (1965) The relationships between blue-green algae and bacteria. Biol. Rev. 40, 143–187.

Engelmann, T.W. (1882) Über Licht und Farbenperception niedrigster Organismen. Pflügers Arch. Ges. Physiol. 29, 387–400.

Fay, P. (1983) The Blue-greens. The Institute of Biology's Studies in Biology, 160, E. Arnold, London.

Fechner, R. (1915) Die Chemotaxis der Oscillarien und ihre Bewegungserscheinungen überhaupt. Z. Botanik 7, 289–364.

Ferrara, R. (1975) General review on phototactic action spectra. In G. Colombetti (ed.), Biophysics of Photoreceptors and Photobehaviour of Microorganisms. Lito Felici, Pisa, pp. 121–145.

Frank, H., Lefort, M. and Martin, H.H. (1962) Elektronenoptische und chemische Untersuchungen an Zellwänden der Blaualge *Phormidium uncinatum*. Z. Naturforsch. 17, 262–268.

Funk, G. (1923) Über Bewegungen der Kolonien von *Oscillatoria amphibia*. Beih. Bot. Zentralbl. 39, 257–320.

Glagoleva, T.N., Glagolev, A.N., Gusev, M.V. and Nikitina, K.A. (1980) Protonmotive force supports gliding in Cyanobacteria. FEBS Lett. 117, 49–53.

Häder, D.-P. (1973) Untersuchungen zur Photo-phobotaxis bei *Phormidium uncinatum*. Ph.D. Thesis, Marburg.

Häder, D.-P. (1974) Participation of two photosystems in the photo-phobotaxis of *Phormidium uncinatum*. Arch. Microbiol. 96, 255–266.

Häder, D.-P. (1975) The effect of inhibitors on the electron flow triggering photophobic reaction in Cyanophyceae. Arch. Microbiol. 103, 169–174.

Häder, D.-P. (1976) Phobic reactions between two adjacent monochromatic light fields. Z. Pflanzenphysiol. 78, 173–176.

Häder, D.-P. (1977) Influence of electric fields on photophobic reactions in blue-green algae. Arch. Microbiol. 114, 83–86.

Häder, D.-P. (1978a) Evidence of electric potential changes in photophobically reacting blue-green algae. Arch. Microbiol. 118, 115–119.

Häder, D.-P. (1978b) Extracellular and intracellular determination of light-induced potential changes during photophobic reactions in blue-green algae. Arch. Microbiol. 119, 75–79.

Häder, D.-P. (1979a) Photomovement. In W. Haupt and M.E. Feinleib (eds.), Encyclopedia of Plant Physiology. New Series, Vol. 7, Physiology of Movements. Springer, Berlin, Heidelberg, pp. 268–309.

Häder, D.-P. (1979b) Effect of inhibitors and uncouplers on light-induced potential changes triggering photophobic responses. Arch. Microbiol. 120, 57–60.

Häder, D.-P. (1981a) Computer-based evaluation of phototactic orientation in microorganisms. EDV Med. Biol. 12, 27–30.

Häder, D.-P. (1981b) Electrical and proton gradients in the sensory transduction of photophobic responses in the blue-green alga, *Phormidium uncinatum*. Arch. Microbiol. 130, 83–86.

Häder, D.-P. (1982) Gated ion fluxes involved in photophobic responses of the blue-green alga, *Phormidium uncinatum*. Arch. Microbiol. 131, 77–80.

Häder, D.-P. (1984a) Photomovement. In H. Senger (ed.), Blue Light Effects in Biological Systems. Springer-Verlag, Berlin, Heidelberg, pp. 435–443.

Häder, D.-P. (1984b) Effects of UV-B on motility and photoorientation in the Cyanobacterium, *Phormidium uncinatum*. Arch. Microbiol. 140, 34–39.

Häder, D.-P. (1985) Bewegungssteuerung von Blaualgen durch Licht. Biol. in unserer Zeit 15, 27–29.

Häder, D.-P. and Burkart, U. (1978) Mathematical model for photophobic accumulations of blue-green algae in light traps. J. Math. Biol. 5, 293–304.

Häder, D.-P. and Burkart U. (1982a) Mathematical simulation of photophobic responses in blue-green algae. Math. Biosci. 58, 1–17.

Häder, D.-P. and Burkart, U. (1982b) Enhanced model for photophobic responses of the blue-green alga, *Phormidium uncinatum*. Plant Cell Physiol. 23, 1391–1400.

Häder, D.-P. and Burkart, U. (1983) Optical properties of *Dictyostelium discoideum* pseudoplasmodia responsible for phototactic orientation. Exp. Mycol. 7, 1–8.

Häder, D.-P. and Poff, K.L. (1982) Dependence of the photophobic response of the blue-green alga, *Phormidium uncinatum*, on cations. Arch. Microbiol. 132, 345–348.

Häder, D.-P., Claviez, M., Merkl, R. and Gerisch, G. (1983) Responses of *Dictyostelium discoideum* amoebae to local stimulation by light. Cell Biol. Inter. Rep. 7, 611–616.

Halfen, L.N. (1973) Gliding motility of *Oscillatoria*: Ultra-structural and chemical characterization of the fibrillar layer. J. Phycol. 9, 248–253.
Halfen, L.N. (1979) Gliding movements. In W. Haupt and M.E. Feinleib (eds.), Encyclopedia of Plant Physiology, New Series, Vol. 7, Physiology of Movements. Springer, Berlin, Heidelberg, New York, pp. 250–267.
Halfen, L.N. and Castenholz, R.W. (1970) Gliding in a blue-green alga: a possible mechanism. Nature 225, 1163–1165.
Halfen, L.N. and Castenholz, R.W. (1971a) Gliding motility in the blue-green alga *Oscillatoria princeps*. J. Phycol. 7, 133–145.
Halfen, L.N. and Castenholz, R.W. (1971b) Energy expenditure for gliding motility in a blue-green alga. J. Phycol. 7, 258–260.
Halldal, P. (1962) Taxes. In R.A. Lewin (ed.), Physiology and Biochemistry of Algae. Academic Press. New York, London, pp. 583–593.
Hansgirg, A. (1883) Bemerkungen über die Bewegungen der *Oscillatorien*. Bot. Zeitung 41, 831–843.
Harder, R. (1920) Über die Reaktionen freibeweglicher Organismen auf plötzliche Änderung der Lichtintensität. Z. Bot. 12, 353–462.
Haupt, W. (1959) Die Phototaxis der Algen. In W. Ruhland (ed.), Encyclopedia of Plant Physiology, Vol. 17, Physiology of Movements, Part 1. Springer, Berlin, Göttingen, Heidelberg, pp. 318–370.
Haupt, W. (1962) Thermotaxis. In W. Ruhland (ed.), Encyclopedia of Plant Physiology, Vol. 17, Physiology of Movements, Part 2. Springer, Berlin, Göttingen, Heidelberg, pp. 29–33.
Haupt. W. (1972) Bewegungen. Fortschr. Bot. 34, 227–235.
Hille, B. (1978) Ionic channels in excitable membranes. Biophys. J. 2, 283–294.
Holton, R.W. and Freeman, A.W. (1965) Some theoretical and experimental considerations of the gliding movement of blue-green algae. Am. J. Bot. 52, 640.
Hosoi, A. (1951) Secretion of the slime substance in *Oscillatoria* in relation to its movement. Bot. Mag. Tokyo 64, 14–16.
Jahn, T.L. and Bovee, E.C. (1965) Movement and locomotion of microorganisms. Ann. Rev. Microbiol. 19, 21–58.
Jarosch, R. (1962) Gliding. In R.A. Lewin (ed.), Physiology and Biochemistry of Algae. Academic Press, New York, London, pp. 573–581.
Jarosch, R. (1963a) Grundlagen einer Schrauben-Mechanik des Protoplasmas. Protoplasma 57, 448–500.
Jarosch, R. (1963b) Gleitbewegung und Torsion von *Oscillatorien*. Öster. Bot. Z. 111, 143–148.
Johnson, F.H. and Baker, R.F. (1947) The electron and light microscopy of *Beggiatoa*. J. Cell Comp. Physiol. 30, 131–146.
Jost, M. (1965) Die Ultrastruktur von *Oscillatoria rubescens*. Arch. Mikrobiol. 50, 211–245.
Kivic, P.A. and Walne, P.L. (1983) Algal photosensory apparatus probably represent multiple parallel evolutions. Bio Systems 16, 31–38.
Kolkwitz, R. (1896) Über die Krümmungen bei den *Oscillariaceen*, Ber. Dtsch. Bot. Ges. 14, 422–431.
Lamont, H.C. (1969) Shear-oriented microfibrils in the mucilaginous investments of two motile oscillatoriacean blue-green algae. J. Bacteriol. 97, 350–361.
Lenci, F. and Colombetti, G. (1978) Photobehaviour of microorganisms: a biophysical approach. Annu. Rev. Biophys. Bioeng. 7, 341–361.
Manten, A. (1948) Phototaxis, phototropism and photosynthesis in purple bacteria and blue-green algae. Ph.D. Thesis, Utrecht.
Mendelssohn, M. (1902) Recherches sur la thermotaxie des organismes unicellulaires, J. Physiol. Pathol. Gen. 4, 393–409.
Mitchell, P. (1978) Protonmotive chemiosmotic mechanisms in oxidative and photosynthetic phosphorylation. TIBS 3, N58–N61.
Mühldorf, A. (1938) Einige Betrachtungen zur Membranmorphologie der Blaualgen. Ber. Dtsch. Bot. Ges. 56, 316–336.
Murvanidze, G.V. (1981) The role of Ca^{2+} and the membrane potential in the formation of the taxis signal in cyanobacteria. Bull. Acad. Sci. Georgian SSR 104, 173–176.
Murvanidze, G.V. and Glagolev, A.N. (1982) Electrical nature of the taxis signal in cyanobacteria. J. Bacteriol. 150, 239–244.
Murvanidze, G.V. and Glagolev, A.N. (1983) Role of Ca concentration gradient and the electrochemical potential of H^+ in the reaction of phototaxis of cyanobacteria. Biophysics 28, 887–891.

Murvanidze, G.V., Gabai, V.L. and Glagolev, A.N. (1982a) Taxis responses in *Phormidium uncinatum*. J. Gen. Microbiol. 128, 1623–1630.

Murvanidze, G.V., Gabai, V.L. and Glagolev, A.N. (1982b) The role of membrane potential in the photophobic response and certain aspects of chemotaxis in the cyanobacterium *Phormidium uncinatum*. Academica NAUK CCCP Microbiologia 51, 240–246.

Nelson, D.C. and Castenholz, R.W. (1982) Light responses of *Beggiatoa*. Arch. Microbiol. 131, 146–155.

Nienburg, W. (1916) Die Perzeption des Lichtreizes bei den Oszillarien und ihre Reaktion auf Intensitätsschwankungen. Z. Bot. 8, 161–193.

Nikitschek, A. (1934) Das Problem der Oscillatorienbewegung. Beitr. Bot. Zbl. Abt. A 52, 205–254.

Nuernbergk, E. (1928) Phototropismus und Phototaxis bei Pflanzen. Handb. Norm. Pathol. Physiol. 12, 36–59.

Nultsch, W. (1961) Der Einfluß des Lichtes auf die Bewegung der Cyanophyceen. I. Phototopotaxis von *Phormidium autumnale*. Planta 56, 632–647.

Nultsch, W. (1962a) Der Einfluß des Lichtes auf die Bewegung der Cyanophyceen. II. Photokinesis bei *Phormidium autumnale*. Planta 57, 613–623.

Nultsch, W. (1962b) Phototaktische Aktionsspektren von Cyanophyceen. Ber. Dtsch. Bot. Ges. 75, 443–453.

Nultsch, W. (1962c) Der Einfluss des Lichtes auf die Bewegung der Cyanophyceen. III. Photophobotaxis bei *Phormidium uncinatum*. Planta 58, 647–663.

Nultsch, W. (1965) Light reaction system in Cyanophyceae. Photochem. Photobiol. 4, 613–619.

Nultsch, W. (1968) Einfluss von Redox-Systemen auf die Bewegungsaktivität und das phototaktische Reaktionsverhalten von *Phormidium uncinatum*. Arch. Microbiol. 63, 295–320.

Nultsch, W. (1969) Effect of desaspidin and DCMU on photokinesis of blue-green algae. Photochem. Photobiol. 10, 119–123.

Nultsch, W. (1973) Relation between photomotion and photosynthesis. In A. Checcucci and R.A. Weale (eds.), Primary Molecular Events in Photobiology. Elsevier Science Publ., Amsterdam, London, New York, pp. 245–273.

Nultsch, W. (1974a) Der Einfluss des Lichtes auf die Bewegung phototropher Mikroorganismen. Abh. Marburger Gelehrt. Ges. 2, 143–213.

Nultsch, W. (1974b) Movement. In W.D.P. Stewart (ed.), Algal Physiology and Biochemistry. Blackwell Sci. Publ., Oxford, London, Edinburgh, Melbourne, pp. 864–893.

Nultsch, W. (1975) Phototaxis and photokinesis. In M.J. Carlile (ed.) Primitive Sensory and Communication Systems. Academic Press, New York, London, San Francisco, pp. 29–90.

Nultsch, W. (1980) Photomotile responses in gliding organisms and bacteria. In F. Lenci and G. Colombetti (eds.), Photoreception and Sensory Transduction in Aneural Organisms. Plenum Press, New York, London, pp. 69–88.

Nultsch, W. and Häder, D.-P. (1970) Bestimmung der photo-phobotaktischen Unterschiedsschwelle bei *Phormidium uncinatum*. Ber. Dtsch. Bot. Ges. 83, 185–192.

Nultsch, W. and Häder, D.-P. (1979) Photomovement of motile microorganisms. Photochem. Photobiol. 29, 423–437.

Nultsch, W. and Häder, D.-P. (1980) Light perception and sensory transduction in photosynthetic prokaryotes. Structure Bonding 41, 111–146.

Nultsch, W. and Hellmann, W. (1972) Untersuchungen zur Photokinesis von *Anabaena variabilis*. Arch. Microbiol. 82, 76–90.

Nultsch, W. and Jeeji-Bai (1966) Untersuchungen über den Einfluß von Photosynthese Hemmstoffen auf das phototaktische und photokinetische Reaktionsverhalten blau-grüner Algen. Z. Pflanzenphysiol. 54, 84–98.

Nultsch, W. and Wenderoth, K. (1983) Partial irradiation experiments with *Anabaena variabilis* (Kütz). Z. Pflanzenphysiol. 111, 1–7.

Nultsch, W., Schuchart, H. and Höhl, M. (1979) Investigation on the phototactic orientation of *Anabaena variabilis*. Arch. Microbiol. 122, 85–91.

Nultsch, W., Benedetti, P.A. and Gualtieri, P. (1983a) Microspectrophotometric investigations of photobleaching in *Anabaena variabilis* cells and heterocysts and its prevention by sodium azide. Z. Pflanzenphysiol. 111, 327–332.

Nultsch, W., Schuchart, H. and Koenig, F. (1983b) Effects of sodium azide of phototaxis on the blue-green alga *Anabaena variabilis* and consequences to the two-photoreceptor systems-hypothesis. Arch. Microbiol. 134, 33–37.

Pankratz, H.S. and Bowen, C.C. (1963) Cytology of blue-green algae. 1. The cells of *Symploca muscorum*. Am. J. Bot. 50, 387–399.

Pentecost, A. (1984) Effects of sedimentation and light intensity on mat-forming Oscillatoriaceae with particular reference to *Microcoleus lyngbyaceus* Gomont. J. Gen. Microbiol. 130, 983–990.

Perez-Miravete, A. (1973) Behaviour of Micro-organisms. Plenum Press, London, New York.

Phillips, O.P. (1904) A comparative study of the cytology and movements of the Cyanophyceae. Contribut. Bot. Lab. Univ. Penn. 2, 237–335.

Pieper, A. (1915) Die Phototaxis der Oscillarien. Ph.D. Thesis. Berlin.

Prell, H. (1921) Zur Theorie der sekretorischen Orstsbewegung. Arch. Bot. 42, 99–156.

Pringsheim, E.G. (1912) Die Reizbewegungen der Pflanzen. Springer Verlag, Berlin.

Pringsheim, E.G. (1968) Cyanophyceen-Studium. Arch. Microbiol. 63, 331–355.

Raven, J.A. (1983) Cyanobacterial motility as a test of the quantitative significance of proticity transmission along membranes. New Phytol. 94, 511–519.

Rosen, G. (1983) Space-time rescaling symmetry in the tactic migration of micrometer-size organisms. Physical Rev. A 27, 1232–1233.

Ruthsack-Künzenbach, R. (1967) Untersuchungen über die Struktur des Centroplasmas und die Drehrichtung einiger *Oscillatoria*arten. Schweiz. Z. Hydrol. 29, 185–190.

Schmid, G. (1918) Zur Kenntnis der Oscillarienbewegung. Flora 111, 327–379.

Schmid, G. (1923) Das Reizverhalten künstlicher Teilstücke, die Kontraktilität und das osmotische Verhalten der *Oscillatoria jenensis*. Jahrb. Wiss. Bot. 62, 328–419.

Schröder, B. (1902) Untersuchungen über Gallertbildungen der Algen. Verh. Nat.-Med. Ver. Heidelberg N.F. 7, 139–196.

Schuchart, H. and Nultsch, W. (1984) Possible role of singlet molecular oxygen in the control of the phototactic reaction sign of *Anabaena variabilis*. J. Photochem. 25, 317–325.

Schulz, G. (1955) Bewegungsstudien sowie elektronenmikroskopische Membranuntersuchungen an Cyanophyceen. Arch. Microbiol. 21, 335–370.

Seitz, K. (1975) Orientation in space: plants. In O. Kinne (ed.), 'Marine Ecology', II, 2. Wiley, New York, pp. 451–497.

Simon, R.D. (1981) Gliding motility in *Aphanothece halophytica*: analysis of wall proteins in mot mutants. J. Bacteriol. 148, 315–321.

Spangle, L. and Armstrong, P.B. (1973) Gliding motility of algae is unaffected by cytochalasin B. Exp. Cell Res. 80, 490–493.

Stanier, R.Y. and Cohen-Bazire, G. (1957) The role of light in the microbiol world: some facts and speculations. Microbiol. Ecol. 56–89.

Stanier, R.Y., Kunisawa, R., Mandel, M. and Cohen-Bazire, G. (1971) Purification and properties of unicellular blue-green algae (order Choococcales). Bacteriol. Rev. 35, 171–205.

Stieve, H. (1982) Biophysik der Photorezeption. In W. Hoppe, W. Lohmann, H. Markl and H. Ziegler (eds.), Biophysik. Springer, Berlin, Heidelberg, New York, pp. 734–755.

Tevini, M. and Häder, D.-P. (1985) Allgemeine Photobiologie. Georg Thieme Verlag, Stuttgart, New York.

Thaer, A., Hoppe, M. and Patzelt, W.J. (1982) Akustomikroskop Elsam. Leitz-Mitt. Wiss. u. Techn. Bd. VIII, 61–67.

Thomas, E.A. (1970) Beobachtungen über das Wandern von *Phormidium autumnale*. Schweiz. Z. Hydrol. 32, 523–531.

Throm, G. (1968) Untersuchungen zum Reaktionsmechanismus von Phobotaxis und Kinesis an *Rhodospirillum rubrum*. Arch. Protistenkd. 110, 313–371.

Toman, M. (1955) Theorie der Oscillatorienbewegung. Arb. d. II. Sektion d. Slow. Wiss. Acad. Biol. Serie 1, 5–23.

Treviranus, L.C. (1817) Die Bewegung der grünen Materie im Pflanzenreiche. Verm. Schriften Anat. Physiol. Inh. 2, 71–92.

Tyagi, V.V.S. (1976) Studies on phototopotaxis in the blue-green alga, *Cylindrospermum*. Biochem. Physiol. Pflanz. 170, 17–22.

Ullrich, H. (1929) Über die Bewegung der Beggiatoaceen und Oscillatoriaceen. Planta 9, 144–194.

Verworn, M. (1889) Psychophysiologische Protistenstudien. Verlag Gustav Fischer, Jena, pp. 25–130.

Walsby, A.E. (1968) Mucilage secretion and the movements of blue-green algae. Protoplasma 65, 223–238.

Wildon, D.C. (1965) Ultrastructure and respiratory function in blue-green algae. Ph.D. Thesis. University of Sydney, Australia.

Wolk, C.P. (1980) Cyanobacteria (blue-green algae). In Biochemistry of Plants, Vol. 1. Academic Press, New York, pp. 659–686.

P. Fay and C. Van Baalen (Eds.) The Cyanobacteria

13

Acclimation to the natural light climate

MICHAEL WYMAN and PETER FAY

Department of Biological Sciences, University of Warwick, Coventry, UK, and Department of Botany and Microbiology, University College, University of London, London, UK

1. Introduction

Like all photoautotrophs, cyanobacteria are dependent upon solar radiation as their primary source of energy in the natural environment. With the exception of those few species capable of chemoheterotrophic growth, active populations of cyanobacteria are confined to those regions of the biosphere to which sunlight penetrates for at least some proportion of the daylight hours. The natural light climate is extremely heterogeneous, however, being subject to considerable temporal and spatial variation in both quantity and quality. The ancient lineage of cyanobacteria and their continued success in contemporary environments is perhaps sufficient evidence that these microorganisms have developed mechanisms which enable them to acclimate to the changeable nature of the incident light field. Such mechanisms of acclimation have their origin in the light-mediated regulation of the expression of the cyanobacterial genome and are manifest at the structural, biochemical and physiological level. In this sense we have used the term *acclimation* in order to distinguish between short-term phenotypic responses and the longer term genotypic *adaptation* of cyanobacteria to the natural light climate. For reasons that are discussed in the following section, variation in the quantity and quality of the natural light field is at its most extreme in aquatic environments and therefore this article concentrates on the response of planktonic cyanobacteria for the most part.

2. The natural light climate

The following account draws attention to those physico-chemical and biotic factors which modify the nature of the incident solar radiation on its transmission through the atmosphere and the

Abbreviations: μ, specific growth rate; μ_{max}, maximum specific growth rate; I, irradiance; $I\mu_{max}$, irradiance at which $\mu = \mu_{max}$; α, initial slope of P vs. I curve; P_{max}, maximum rate of photosynthesis; I_k, intercept of α and P_{max}; V_{max}, maximum specific nutrient uptake rate; K^u_s, half saturation constant for nutrient uptake.

aquatic environment. A comprehensive treatment of the optical properties of the natural light climate may be found in the texts of Jerlov (1976) and Kirk (1983).

2.1. Solar radiation

Sunlight reaching the earth's atmosphere is selectively attenuated by a number of atmospheric components. Ozone present in the upper atmosphere absorbs much of the short wavelength radiation below 290 nm (Halldal, 1979) and, in addition, a fraction of the light at longer wavelengths. Both oxygen (688 and 762 nm) and water vapour (723 nm) have strong absorption bands in the red and far-red regions of the spectrum (Smith, 1982; Kirk, 1983). The direct solar beam is also subject to both Rayleigh and Mie scattering. Atmospheric molecules smaller than the wavelength of light (e.g. O_2, N_2 etc.) act as Rayleigh particles and therefore scatter the shorter wavelengths of light preferentially. This introduces a diffuse component to the incident radiation and accounts for the blue colour of skylight. Mie scattering by atmospheric particles of a diameter greater than the wavelength of light (e.g. water droplets, pollutants) although generally wavelength-independent, diminishes the intensity of terrestrial daylight (Smith and Morgan, 1981; Smith, 1982).

Three further factors which influence the nature of the light received at the earth's surface may be identified: (a) the diurnal variation in irradiance, (b) the seasonal variation in daylight, and (c) the prevailing meteorological conditions.

The irradiance received at a given geographical location is determined by the solar elevation (which varies with the seasons and throughout the course of single day) and on the solar declination (the angle through which the geographical point of measurement is tilted toward the sun) at the time of measurement (Kirk, 1983). Solar elevation and daylength are dependent upon latitude and reach their maximum values in the summer. As the solar elevation declines during the winter months and at dawn and dusk throughout the year, the contribution of the diffuse component of daylight (skylight) becomes increasingly important. At solar altitudes below 20° the direct solar beam and skylight contribute equally to the total light flux. A small amount of cloud can increase the total irradiance received at the earth's surface by 5–10% providing that the sun is not obscured. However, thick cloud cover may diminish the incident radiation by as much as 50–80% (Kirk, 1983). Hence, the spectral distribution of daylight depends on the relative contributions of direct and diffuse radiation while the quantity of light received depends on the extent of cloud cover, latitude, and the time of day.

2.2. The underwater light climate

The contribution of solar radiation to the underwater light field is determined by reflection and refraction at the water surface and by absorption and scattering of light below the surface (Smith, 1982). Although a fraction of the incident sunlight is reflected at the surface, losses to the downwelling radiation due to reflectance are probably insignificant at high solar altitudes but increase in importance at lower solar elevations. At a solar altitude of 10° as much as 35% of the incident light may be reflected; wave action, however, tends to diminish the influence of reflectance on the downwelling radiation.

The underwater light climate darkens with depth, owing to absorption and scattering of the incident radiation penetrating the water surface. Rayleigh scattering by water molecules and dissolved salts attenuates light principally in the blue region of the spectrum. Owing to the higher concentration of dissolved salts, pure seawater has a higher scattering coefficient than freshwaters and scatters light about 30 times more intensely than pure water (Kirk, 1983). Mie particles such as suspended silts, clays and phytoplankton cells may reduce the depth of the euphotic (illuminated) zone considerably, particularly in productive coastal, estuarine and inland waters.

Selective absorption radically modifies the quality of light available at depth in aquatic environments. Absorption of light by water molecules is negligible in the blue and green regions of the spectrum but increases significantly at wavelengths above 550 nm. Pure water at 1 metre depth absorbs 35 % of the radiation at 680 nm (Kirk, 1983) and an even greater proportion of the downwelling light of higher wavelengths. Dissolved and colloidal substances of organic origin (Gelbstoff) colour the water through the selective absorption of the ultraviolet and blue wavebands. This material which is derived from the decomposition of plant tissue in situ or leaches from the terrestrial environment may represent a substantial proportion of the total organic carbon present in inland waters (see Section 4 of Chapter 16). Kirk (1976) found that in a number of organically-rich freshwaters 60–80 % of the total quanta captured were absorbed by Gelbstoff rather than by water molecules, thereby extinguishing the contribution of blue light to the downwelling radiation at quite moderate depths. The contribution of Gelbstoff to light absorption in the marine environment is substantially less and diminishes in importance with distance from shore.

In productive waters the absorption of light by the photosynthetic pigments of phytoplankton cells is considerable such that these organisms may limit their own growth by 'self-shading'. The amount of light absorbed is a function of cell size, shape and pigment composition (Kirk, 1974, 1975a,b, 1976). Individual cells and filamentous forms with narrow cell diameters are much more efficient at absorbing light than large-celled or colonial phytoplankton. As the diameter of the cells/colonies increases, the specific absorption coefficient per milligram of photopigment present declines owing to the 'package effect'. Hence picoplanktonic (< 1.0 μm diameter) cyanobacteria and eukaryotic algae have significantly higher absorption coefficients than larger phytoplankton (see also Section 2.3. of Chapter 15).

2.3. Optical classification of natural waters

On the basis of their optical properties natural waters may be divided into a number of types. Talling (1971) recognised three optical classes of lakes. In the first class green light is the most penetrating component while the red waveband has the highest attenuation coefficient. The second class includes lakes in which red light penetrates further than the blue waveband while in the third class blue light also has the highest attenuation coefficient but red and green light penetrate approximately to the same extent at a somewhat lower depth. Talling (1971) noted that the lakes in his three optical classes corresponded to the general concept of oligotrophic, mesotrophic and eutrophic nutrient status, respectively.

Jerlov (1976) described three basic categories of oceanic water (I–III) and nine classes of coastal seawater (1–9). Blue light is the least attenuated component in the clearest oceanic waters (Type I) in which water molecules are responsible for the capture of the majority of quanta with depth. Coastal waters, however, have a much shallower euphotic zone owing to the higher concentrations of detritus, Gelbstoff and phytoplankton cells. Most coastal waters show a broad peak of transmittance in the green waveband (500–575 nm) but in highly turbid waters (Class 9) there is a wavelength shift in maximum transmittance to a defined peak at 574 nm.

3. The influence of light climate on growth rate

The maximum growth rate (μ_{max}) of most temperate freshwater cyanobacteria in laboratory culture is reached at a 'white' light irradiance of approximately 50–60 μE $m^{-2}{\cdot}s^{-1}$ (Van Liere et al., 1979b; Foy, 1980; Gibson and Smith, 1982; Gibson and Foy, 1983; Wyman and Fay, 1986a). While many cyanobacteria are able to maintain growth rate at irradiances considerably in excess of those required for μ_{max} (Raps et al., 1983; Wyman and Fay, 1986a), certain strains appear to be particularly light-sensitive e.g. *Microcystis aeruginosa* 7005 (Eloff, 1978), *Oscillatoria rubescens* (Meffert, 1971). Photoinhibition of growth rate has been reported in a number of other instances including *O. agardhii*, *Anabaena flos-aquae* (Zevenboom and Mur, 1980), *O. redekei* (Gibson and Foy, 1983) and *O. agardhii* var. *isothrix*, *A. solitaria* (Wyman and Fay, 1986a). Richardson et al. (1983) quote 5 μE $m^{-2}{\cdot}s^{-1}$ as the minimum irradiance supporting the growth of unicellular cyanobacteria while Gibson and Foy (1983) reported the active growth of planktonic strains at irradiances as low as 1–2 μE $m^{-2}{\cdot}s^{-1}$. The minimum irradiance supporting growth shows a significant dependence on light quality (Wyman and Fay, unpublished).

During light-limited growth, variation in the specific growth rate (μ) with light quality is largely a function of the special absorption characteristics of the cell (Wyman and Fay, 1986b). Preferential illumination of phycocyanin or phycoerythrin (when present) produces a significant enhancement of growth rate in comparison to that possible in white light of a similar quantum flux (Table I). In the chromatically-adapting cyanobacterium *Nostoc* sp. strain MAC, Kipe-Nolt et al. (1982) also showed that μ may vary with light quality (μ = 0.043 h^{-1}, red light; 0.039 h^{-1}, green light). The increase in μ noted in red light may be related to a reduction in the overall cell maintenance costs associated with the repression of phycoerythrin synthesis in this light regime. An increase in μ_{max} in red light has been observed in a chromatically-adapting isolate of *Gloeotrichia echinulata* (Wyman and Fay, 1986b).

Considerable variation in μ_{max} is observed between individual species and strains of planktonic cyanobacteria. Generally, small unicellular species and those filamentous forms with narrow cell diameters (and high surface to volume ratios) have the highest maximum growth rates. For example, Foy (1980) found that *Oscillatoria limnetica* (FBA 100) had a μ_{max} of 0.079 h^{-1} at 20°C whereas the equivalent rate for the larger celled *O. agardhii* var. *isothrix* was 0.016 h^{-1} (see also Section 2.3. of Chapter 15).

Cyanobacteria do not appear to have an obligate growth requirement for alternating light : dark cycles (see Fogg et al., 1973), although Zevenboom and Mur (1984) have shown recently that

TABLE I

Specific growth rates (divisions day^{-1} $\pm$ standard error) of freshwater planktonic cyanobacteria grown in different light qualities of a constant quantum flux (12 μE $m^{-2}{\cdot}s^{-1}$)

Species	White	Red	Green	Blue
M. aeruginosa	0.135 ± 0.016	0.319 ± 0.024	0.086 ± 0.001	0.072 ± 0.021
O. redekei	0.150 ± 0.009	0.560 ± 0.071	0.107 ± 0.012	< 0.05
A. solitaria	0.091 ± 0.007	0.263 ± 0.019	0.062 ± 0.018	< 0.05
G. echinulata	0.091 ± 0.003	0.219 ± 0.024	0.143 ± 0.015	0.092 ± 0.03

(After Wyman and Fay, 1986b)

Microcystis aeruginosa requires a dark period of about 2 h in each 24-h cycle in order to reach μ_{max}. In continuous light $\mu_{max} = 0.02\,h^{-1}$ whereas in a 22:2 h light : dark cycle $\mu_{max} = 0.03\,h^{-1}$. The minimum irradiance supporting μ_{max} in continuous light was found to be 80 $\mu E\ m^{-2}\cdot s^{-1}$ but approximately 40 $\mu E\ m^{-2}\cdot s^{-1}$ for *M. aeruginosa* grown in a 16:8 h light : dark cycle.

Temperature also has a considerable influence on the light requirements for growth and on the value of μ_{max}. Foy et al. (1976) found a linear relationship between the logarithm of μ_{max} and growth temperature in the range 10–20°C with four species of planktonic cyanobacteria. μ_{max} was linearly related to temperature (10–30°C) for the growth of *Oscillatoria agardhii* (Post et al., 1985). At 10°C $\mu_{max} = 0.01\ h^{-1}$ while at 30°C $\mu_{max} = 0.05\ h^{-1}$; the Q_{10} (the change in rate for each 10°C increase in temperature) for the growth of *O. agardhii* was found to be equivalent to approximately 0.02 h^{-1}. The minimum irradiance required to support μ_{max} in *O. agardhii* varied from 33 $\mu E\ m^{-2}\cdot s^{-1}$ at 10°C to 85 $\mu E\ m^{-2}\cdot s^{-1}$ at 20°C, although at very low light fluxes ($< 4\ \mu E\ m^{-2}\cdot s^{-1}$) μ was independent of growth temperature.

4. The influence of light climate on the synthesis of photosynthetic pigments

Variation in either the availability of light and/or its spectral quality can lead to profound changes in the organisation of the photosynthetic apparatus of cyanobacteria (see Chapters 1 and 3 of this volume). The biosynthesis of the light-harvesting pigments is subject to both quantitative and qualitative control; cyanobacteria may show marked alterations in the amounts and in the relative proportions of the photosynthetic pigments present within the cell. Individual species, however, differ in their ability to regulate the synthesis of their light-harvesting pigments and it is evident that such differences may be of considerable importance in determining the species distribution of cyanobacteria in the natural environment.

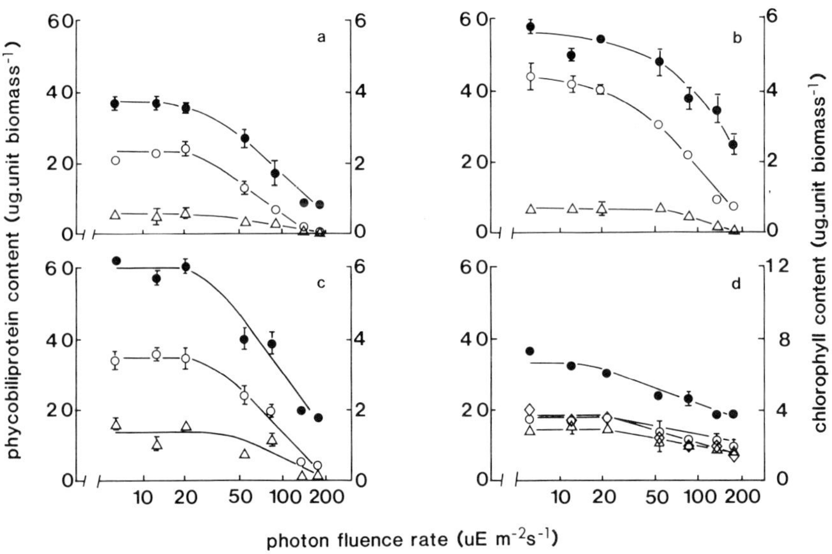

Fig. 1. The influence of photon fluence rate on the cell concentrations of chlorophyll and phycobiliproteins in freshwater planktonic cyanobacteria. (a) *Microcystis aeruginosa*; (b) *Oscillatoria redekei*; (c) *Anabaena solitaria*; (d) *Gloeotrichia echinulata*. Chlorophyll (●); phycocyanin (○); phycoerythrin (◇); allophycocyanin (△). (After Wyman and Fay, 1986a)

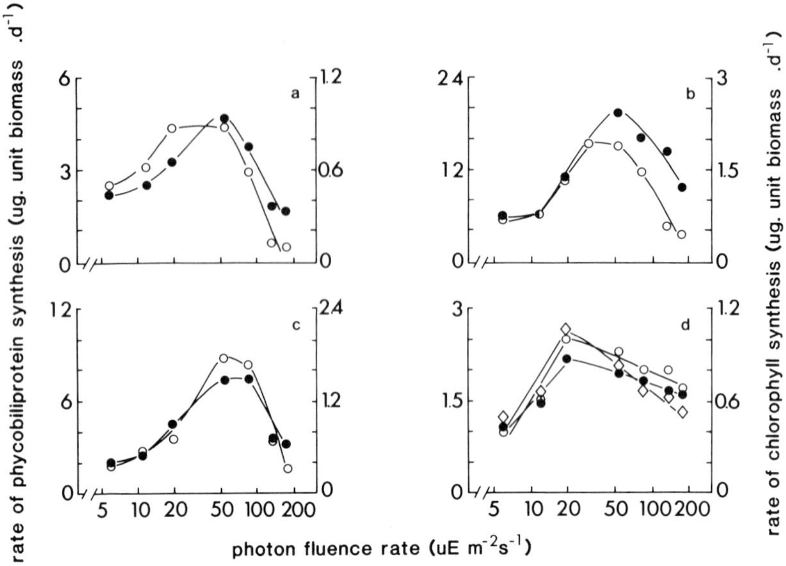

Fig. 2. The relation between the rates of chlorophyll and phycobiliprotein synthesis and photon fluence rate. Key and symbols as in Fig. 1. (After Wyman and Fay, 1986a)

4.1. Light quantity

The amounts of chlorophyll and phycobiliproteins present per cell in freshwater planktonic cyanobacteria may show a 4–5-fold variation in concentration between low and high irradiance (I) (Fig. 1). The net rates of chlorophyll and phycobiliprotein synthesis, however, show a more complex relationship with irradiance (Fig. 2). Variations in cell pigment content, therefore, are not simply the result of alterations in the rates of chlorophyll and phycobiliprotein synthesis but are the product of the net rates of synthesis and the rate of dilution through cell division (growth). In severely to moderately light-limited cells (i.e. $< 20\ \mu E\ m^{-2} \cdot s^{-1}$; $< 0.5\ \mu_{max}$), the net rates of chlorophyll and phycobiliprotein synthesis vary coordinately with μ and hence cell pigment concentrations remain relatively stable. At higher irradiance ($20\ \mu E\ m^{-2} \cdot s^{-1}$–$I\ \mu_{max}$; $> 0.5\ \mu_{max}$–$\leqslant \mu_{max}$), however, μ increases with energy input at a more rapid rate than the relative rates of chlorophyll and phycobiliprotein synthesis and results in a decline in the steady-state concentrations of these light-harvesting pigments. Beyond $I\mu_{max}$ the further decline in cell chlorophyll and phycobiliprotein content arises from the direct reduction in the net rates of synthesis where μ_{max} is maintained (for the dilution rate is constant), or remains the product of both net synthesis and μ where photoinhibition of growth occurs.

The biosynthesis and maintenance of the cyanobacterial light-harvesting apparatus represents a substantial metabolic investment; the phycobiliproteins alone may account for 50% of the total cell protein (Cohen-Bazire and Bryant, 1982). Therefore, the reduction in cellular chlorophyll and phycobiliprotein content with increases in incident irradiance, which is accompanied by a decline in the synthesis of associated thylakoid membrane components (Utkilen et al., 1983), amounts to considerable conservation of energy; energy that may be directed toward the biosynthesis of other cell constituents. Given that the rate of photon absorption determines μ, the observation that cell photosynthetic pigment content begins to decline when $I < I\ \mu_{max}$ (Fig. 1) may appear paradoxi-

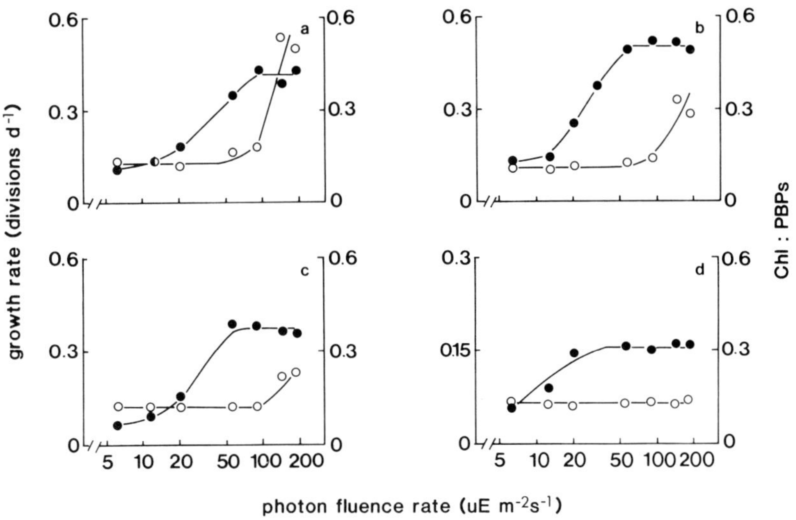

Fig. 3. The relation between growth rate (●), the quotient Chl: PBPs (○) and photon fluence rate. Key as in Fig. 1. (After Wyman and Fay, 1986a)

cal. However, the benefits of reducing the metabolic expense of chlorophyll and phycobiliprotein synthesis may exceed the energetic advantage conferred by maintaining optimal light-harvesting capacity. By definition, where $I > I\,\mu_{max}$, the positive control of chlorophyll and phycobiliprotein synthesis will not influence growth rate, except perhaps in further reducing light absorption capacity as a mechanism of avoiding potentially damaging photo-oxidative reactions. Indeed, while photoinhibition of photosynthesis has been observed frequently in cyanobacteria subjected to a sudden shift from low to high irradiance (Codd, 1981; Tytler et al., 1984; Zevenboom and Mur, 1984; Barlow and Alberte, 1985), susceptibility to photoinhibition diminishes following a period of acclimation in high light (Zevenboom and Mur, 1984). Acclimation to high irradiance is associated with a decrease in the cell chlorophyll/carotenoid ratio (Codd, 1981; Raps et al., 1983) in addition to a decline in chlorophyll and phycobiliprotein content. The photoprotective role of carotenoids has been well documented in cyanobacteria (Kellar and Paerl, 1980; Codd, 1981) and in many other photosynthetic microorganisms (Krinsky, 1978). Wyman and Fay (1986a) have proposed that qualitative as well as quantitative changes in the net rates of chlorophyll and phycobiliprotein synthesis may also be of importance during acclimation to high irradiance. During light limited growth variation in cell chlorophyll and phycobiliprotein content with μ is coordinate; i.e. the chlorophyll/phycobiliprotein ratio is constant (Fig. 3). Where $I > I\,\mu_{max}$, however, the net rate of phycobiliprotein synthesis declines more rapidly than that of chlorophyll resulting in an increase in the chlorophyll/phycobiliprotein ratio. The differential decline in phycobiliprotein content specifically affects the synthesis of that biliprotein (phycoerythrin or phycocyanin) located in the distal region of the rod substructures of the phycobilisome (see Sections 2 and 6 of Chapter 3). This leads to the production of phycobilisomes of smaller dimensions (Fig. 4), to a reduction in the cell light absorption cross-section and, in phycoerythrin-producing cyanobacteria, to a reduction in the cell light-absorption bandwidth. The phycobiliproteins are associated with Photosystem II directly and possibly exclusively (Manodori et al., 1984; see also Section 7 of Chapter 3) and hence the decline in phycobilisome size at high irradiance specifically reduces the throughput of light energy to reaction centre II chlorophyll. This form of acclimation to high light levels may be of critical

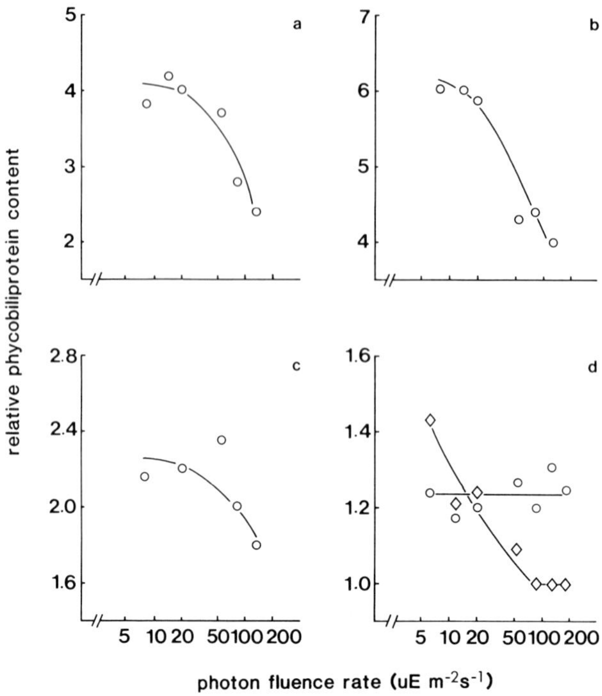

Fig. 4. The influence of photon fluence rate on mean phycobilisome size in freshwater planktonic cyanobacteria. Phycocyanin (○) and phycoerythrin (◇) content is normalised to unit allophycocyanin. Key as in Fig. 1. (Fig. 4d after Wyman and Fay, 1986a)

importance in avoiding photoinhibition and in maintaining μ_{max}, for Tytler et al. (1984) have shown in the planktonic cyanobacterium *Microcystis aeruginosa* that photoinhibition of the light reactions of photosynthesis occurs at a site close to the Photosystem II reaction centre following an abrupt shift from low to high irradiance. Under these conditions Photosystem I activity is unaffected.

The ability of cyanobacteria to acclimate to and tolerate high irradiance, however, is a complex process which involves mechanisms other than the control of light-harvesting capacity mediated through the regulation of pigment synthesis. Inherent in the foregoing discussion has been the assumption that light energy, once absorbed, has a single fate, i.e. it is utilised in photosynthesis with a constant quantum efficiency. However, depending on the availability of light and nutrients, considerable variation may occur in the efficiency with which light energy absorbed by the major accessory photosynthetic pigments (phycoerythrin, phycocyanin, allophycocyanin, light-harvesting chlorophyll) is transferred to reaction centre chlorophyll (Ghosh and Govindjee, 1966; Carthew, 1980), a phenomenon which to date appears to be at its most extreme in certain marine *Synechococcus* species (Barlow and Alberte, 1985; Wyman et al., 1985). In *Synechococcus* sp. strain WH7803 (DC2), although its biosynthesis is also light-regulated, phycoerythrin accumulates in non-light limited nitrogen-replete cells in excess of the requirement for this biliprotein as a light-harvesting pigment. Under these conditions a substantial fraction of the light energy absorbed by phycoerythrin is dissipated as autofluorescence rather than directed to reaction centre II chlorophyll (Fig. 5). The constitutive synthesis of energetically uncoupled phycoerythrin led Wyman et al. (1985) to

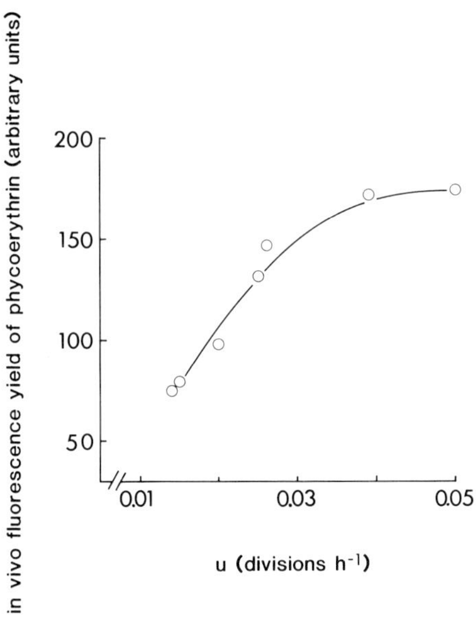

Fig. 5. The relation between the steady-state specific growth rate (μ) in light-limited continuous cultures and the relative light-harvesting efficiency of phycoerythrin in *Synechococcus* sp. strain WH7803 (DC2). (After Wyman et al., 1985)

propose that this photopigment serves two functionally distinct roles in *Synechococcus* sp. strain WH7803: as a defined reserve of cellular nitrogen as well as a collector of quanta for photosynthesis.

Having discussed the control of pigment synthesis at high irradiance it is pertinent to examine those factors which determine and possibly limit the rate of chlorophyll and phycobiliprotein production in cyanobacteria during growth under severely light-limiting conditions. While the steady-state concentrations of chlorophyll and phycobiliproteins in individual species of planktonic cyanobacteria are similar at irradiances below approximately 20 μE $m^{-2} \cdot s^{-1}$ i.e. when $\mu < 0.5\ \mu_{max}$ (Fig. 1; Wyman and Fay, 1986a), the net rates of their synthesis decline with progressive light limitation (Fig. 2). This suggests that the energy demand of pigment synthesis itself or some related process (e.g., the assimilation and incorporation of nitrogen) limits the maximum photosynthetic pigment content and thereby the maximum light absorption capacity of the cell. Indeed, many nitrogen-fixing cyanobacteria show a reduced cell complement of photosynthetic pigments (particularly phycobiliproteins), when utilizing dinitrogen rather than combined sources of this element for growth (Meeks et al., 1983), presumably because of the increased energy demand imposed by the assimilation of molecular nitrogen. In the presence of combined nitrogen, however, constraints on cell chlorophyll and phycobiliprotein content may operate at the genetic rather than at the physiological and biochemical level. In *Anacystis nidulans*, Utkilen et al. (1983) found that the specific chlorophyll content of the thylakoids was constant at irradiances below 282 μE $m^{-2} \cdot s^{-1}$ while the photosynthetic membrane area per unit cell volume was invariant below approximately 180 μE $m^{-2} \cdot s^{-1}$. Several investigations have shown that, irrespective of irradiance, a similar number of chlorophyll molecules is associated with individual photosynthetic units in *Anacystis nidulans* (Oquist, 1974; Kawamura et al., 1979; Vierling and Alberte, 1980), and hence the data of Utkilen et al. (1983) suggest that this organism, and perhaps cyanobacteria in general, has a maximum cell den-

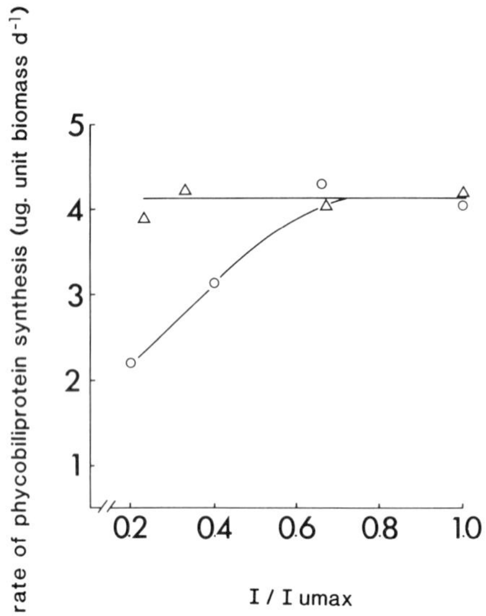

Fig. 6. The relation between growth rate and the combined rates of phycocyanin and phycoerythrin synthesis in *Gloeotrichia echinulata* during steady-state growth in nitrogen-replete cultures (○) or following an earlier period (5 days) of nitrogen starvation (△).

sity of photosynthetic units. The maximum size of individual photosynthetic units also appears to be under regulatory control. This, as Raven and Beardall (1982) have proposed, may be an evolutionary response to the loss of light energy transfer efficiency in photosynthetic units beyond an optimum size. Photosynthetic units associated with Photosystem I have been shown to be of a similar size, irrespective of irradiance (Kawamura et al., 1979; Vierling and Alberte, 1980; Raps et al., 1983), while photosynthetic units associated with Photosystem II show variation with irradiance only in the number of associated phycobiliproteins rather than in the number of light-harvesting chlorophyll molecules (Wang et al., 1977; Vierling and Alberte, 1980). Wyman and Fay (1986a) showed that the mean size of the phycobilisomes increased in eight strains of planktonic cyanobacteria exposed to progressive light limitation (Fig. 4). In the majority of the strains examined, however, neither the size nor the number of phycobilisomes present per cell showed any further significant increase below an incident irradiance of 20 μE $m^{-2} \cdot s^{-1}$ owing to a parallel decline in growth rate and the rate of phycobiliprotein synthesis. In contrast to nitrogen-replete cells, Wyman and Fay (1986b) recorded similar rates of total phycobiliprotein synthesis in light-limited cultures of *Gloeotrichia echinulata* supplied with combined nitrogen after cell phycobiliprotein content had been depleted by approximately 50% during an earlier period of nitrogen starvation (see Section 6.1. of Chapter 3). For a given growth rate the initial combined rates of phycocyanin and phycoerythrin synthesis were considerably in excess of those recorded in non-nitrogen starved cultures (Fig. 6). Therefore, the lower rates of phycobiliprotein production observed in light-limited, nitrogen-sufficient cultures appear to be the result of limitations on synthesis imposed by factors other than the availability of light and nitrogen. Given that the maximum size and density of photosynthetic units per cell appears to be subject to genotypic limitations, Wyman and Fay (1986a) put forward the hypothesis that at low irradiance the rate of chlorophyll and phycobiliprotein synthesis declines as a consequence of the limitation of cell division rather than as a result of light/nutrient limitation of pigment synthesis.

4.2. Light quality

By virtue of the diversity of their photosynthetic pigments cyanobacteria are capable of efficient light absorption over a broad spectral range. Shifts in the spectral distribution of the incident light, however, may induce remarkable alterations in the rates of pigment synthesis resulting in profound changes in light absorption capacity. The process of acclimation to variation in light quality may be analogous to or quite distinct from the responses induced by shifts in light quantity.

4.2.1. Chromatic adaptation

The role of light quality in the control of phycobiliprotein synthesis has been studied extensively in recent years in those phycoerythrin-producing cyanobacteria capable of complementary chromatic adaptation. Chromatic adaptation results in the preferential synthesis of that phycobiliprotein with an absorption spectrum complementary to the available light quality, i.e. phycocyanin in red light, phycoerythrin in green light. Tandeau de Marsac (1977) described three levels of response to chromatic light amongst phycoerythrin-producing cyanobacteria. A number of strains were found to be incapable of modulating the differential rates of phycobiliprotein synthesis and these were assigned to Group I. Those strains capable of regulating phycoerythrin synthesis only were placed in Group II, while those cyanobacteria able to modulate the synthesis of both phycoerythrin and phycocyanin were assigned to Group III. The repression of phycoerythrin synthesis in red light results in an overall reduction in phycobilisome size in Group II but not in Group III strains, since the phycoerythrin units of the phycobilisome rod substructures are replaced by stoichiometric numbers of phycocyanin hexamers (Glazer, 1982; see also Section 6.1. of Chapter 3). Investigation of the chromatic adaptation phenomenon has led to an understanding of this process at the biochemical as well as at the structural level (see Chapter 3 for details). Recent advances in molecular biology have enabled the analysis of the light regulation of phycobiliprotein synthesis at the level of transcriptional control (Conley et al., 1985). Consideration of the ecological significance of chromatic adaptation, while alluded to in much of the literature, has not been, nor was it intended to be, the central theme of such studies. Although of primary importance in determining phycobiliprotein composition in chromatically adapting cyanobacteria, illumination exclusively in the red waveband remains an exotic condition in the context of the natural environment.

Against the background of the wealth of information generated in laboratory-based studies it is surprising perhaps that reports of chromatic adaptation in field populations of cyanobacteria are comparatively rare. The development of metalimnetic populations of the red-coloured species *Oscillatoria rubescens* has been frequently regarded as a symptom of lake eutrophication (Jaag, 1972; Reynolds and Walsby, 1975; Skulberg, 1978), for the green light absorbed by phycoerythrin is the most penetrating component of the downwelling radiation under these conditions. Individual strains of *O. rubescens*, however, appear to differ in their capacity for chromatic adaptation. Konopka (1980) reported a depth-dependent increase in the phycoerythrin/phycocyanin ratio of a field population while laboratory cultures of the isolate *O. rubescens* EAWAG A19 were shown to be incapable of modulating phycobiliprotein synthesis in response to variation in the light quality provided for growth (Wyman and Fay, unpublished). Amongst other *Oscillatoriaceae* Jewson (1977) observed an increase in the cell phycoerythrin content of *O. redekei* at a time when the population was at its maximum. Kohl and Nicklisch (1981) have also reported chromatic adaptation in this species as well as in *O. agardhii*. An apparent instance of chromatic adaptation in a metalimnetic population of *Lyngbya limnetica* was reported by Reynolds et al. (1983). Chromatic adaptation, although generally of rare occurrence in unicellular cyanobacteria (Rippka et al., 1979), has been observed in a field population of *Merismopedia minima* (Dakshini and Gupta, 1981). To date none of the *Synechococcus* strains isolated from open ocean waters has been shown to be capable of chromatic adaptation (Waterbury, personal communication).

Phycoerythrin synthesis appears to be of infrequent occurrence amongst planktonic members of those genera of cyanobacteria which form heterocysts (Wyman and Fay, unpublished), although members of several heterocyst-producing genera produce phycoerythrocyanin (Bryant et al., 1976; Rippka et al., 1979), a phycobiliprotein with absorption maxima at approximately 568 and 590 nm (see Fig. 2 of Chapter 3). Füglistaller et al. (1981) reported chromatic adaptation in the thermophilic, phycoerythrocyanin-containing cyanobacterium *Mastigocladus laminosus*, although according to Bryant (unpublished, quoted by Cohen-Bazire and Bryant, 1982), induction of phycoerythrocyanin synthesis is not dependent on light quality but rather is the general result of growth at low irradiance. Chromatic adaptation in phycoerythrin-producing cyanobacteria has been shown to be largely independent of light quantity (Fujita and Hattori, 1960; Tandeau de Marsac, 1977), provided that the chromatic light regime is maintained for an extended period of time (Diakoff and Scheibe, 1973). However, Wyman and Fay (1986b) found considerable interaction between the differential rates of phycobiliprotein synthesis and the quantitative/qualitative properties of the incident light field in a chromatically adapting (belonging to Group II, sensu Tandeau de Marsac, 1977) isolate of the planktonic cyanobacterium *Gloeotrichia echinulata*. Repression of phycoerythrin synthesis by red light showed a distinct dependence on irradiance (Fig. 7) in the range 4–15 $\mu E\ m^{-2} \cdot s^{-1}$ ($I\mu_{max} = 10\ \mu E\ m^{-2} \cdot s^{-1}$ in red light at 20°C). Phycoerythrin synthesis was also repressed with increasing irradiance during growth in white light (Wyman and Fay, 1986a; Fig. 8), a response which appears to be analogous to the repression of phycoerythrin synthesis in non-chromatically adapting cyanobacteria and phycocyanin synthesis in those strains which do not produce phycoerythrin.

Although in the context of the laboratory culture exposed to red light the continued synthesis of a phycobiliprotein with absorption properties that are ill-suited to the prevailing light regime appears eccentric, the occurrence of a corresponding light regime is limited in the natural environment. Whereas green light absorbed by phycoerythrin is often the most penetrating component of the downwelling radiation in freshwater lakes, red light is most frequently encountered at shallower depths, where it forms a component of light having a broader spectral distribution (including green light) and a relatively higher quantum flux. In the natural environment, therefore, re-

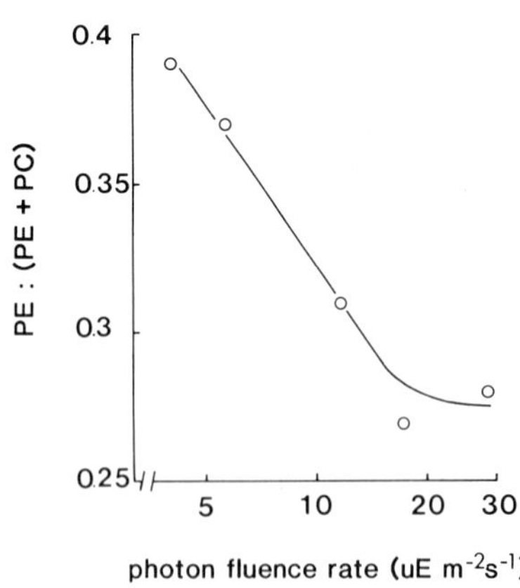

Fig. 7. The influence of photon fluence rate on the relative cell concentrations of phycocyanin (PC) and phycoerythrin (PE) in *Gloeotrichia echinulata* grown in red light. (After Wyman and Fay, 1986b)

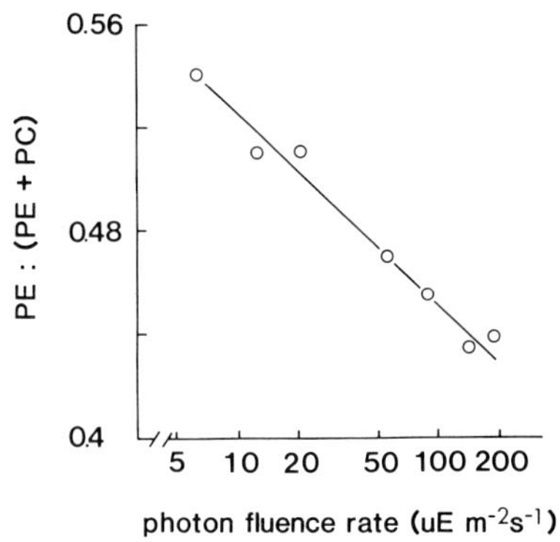

Fig. 8. The influence of photon fluence rate on the relative cell concentrations of phycocyanin (PC) and phycoerythrin (PE) in *Gloeotrichia echinulata* grown in white light. (After Wyman and Fay, 1986a)

sponses to red light will occur alongside similar changes in the relative rates of phycobiliprotein synthesis induced by increases in light quantity. In *Gloeotrichia echinulata* the primary competitive advantage gained by the facility to modulate phycobiliprotein synthesis appears to lie not in the conservation of biosynthetic expense gained by the repression of phycoerythrin synthesis but in the ability to derepress the synthesis of this photopigment; a property which under severe light limitation is largely independent of light quality (Wyman and Fay, 1986b).

4.2.2. Phycobiliprotein synthesis in non-chromatically adapting cyanobacteria

Relatively few studies have considered the influence of light quality on phycobiliprotein synthesis in those cyanobacteria, which either do not produce phycoerythrin or are incapable of chromatic adaptation. Tandeau de Marsac (1977) found similar rates of phycoerythrin, phycocyanin and allophycocyanin synthesis in twelve Group I strains of cyanobacteria grown in green or red light of a quantum flux which enabled similar growth rates to be reached in each light regime. In a survey of eight planktonic cyanobacteria grown in either white, blue, green or red light of a similar quantum flux, Wyman and Fay (1986b) attributed variation in the phycobiliprotein content of the majority of strains to differences in growth rate (a function of 'perceived' light quantity) rather than to intrinsic responses to light quality. Consistent with the relatively higher growth rates recorded in each of the eight strains grown in red light, cell phycobiliprotein concentration declined in comparison to control cultures grown in white light under otherwise similar experimental conditions. However, although relative growth rate was also enhanced in green light in two phycoerythrin-producing strains (*Oscillatoria agardhii* 7821, *Gloeotrichia echinulata*), phycobiliprotein content was comparable in all strains (including those unable to produce phycoerythrin) to that observed in white light. The pattern of response in blue light was more ambiguous although in comparison to white light, cell chlorophyll rather than phycobiliprotein content was the more variable component (see Section 4.2.3.).

Lonneborg et al. (1985) examined the kinetics of phycobiliprotein synthesis in *Anacystis nidulans* following a shift from high white light (300 μE m$^{-2}\cdot$s^{-1}) to a lower quantum flux (15 μE m$^{-2}\cdot$s^{-1}) of red light. During acclimation to red light chlorophyll and phycocyanin content per cell increased, while the steady-state growth rate declined from a mean generation time of 4 h in white light to 33 h in red light. Changes in chlorophyll content were complete within 20 h. The increase in phycocyanin content was biphasic in nature: a rapid (10–15 h) phase of synthesis followed by a slower second phase which reached completion at 80 h. On the basis of parallel increases in phycocyanin and in the relative amounts of two phycobilisome linker polypeptides (see Section 6 of Chapter 3), Lonneborg et al. (1985) suggested that the first phase of acclimation was due to an increase in the size of existing phycobilisomes while the second slower phase was the result of an increase in the numbers of phycobilisomes present per cell.

4.2.3. Chlorophyll synthesis

As is the case with phycobiliprotein content in non-chromatically adapting cyanobacteria, cell chlorophyll content also appears to show a dependence on the perceived light quantity (and, hence, the rate of cell division) rather than on the light quality experienced during growth (Tandeau de Marsac, 1977; Wyman and Fay, 1986b). An exception, however, has been noted in a number of cyanobacteria grown in either blue or far-red light. Jones and Myers (1965) found that while phycobiliprotein content was largely unaffected, the cell chlorophyll content of *Anacystis nidulans* declined to as little as 25 % of that found in white light during growth in these chromatic light regimes. It was proposed that the response to blue and far-red light (principally absorbed by chlorophyll *a*) was due to an imbalance in the photon distribution to the two photosystems. Preferential illumination of chlorophyll results in Photosystem I receiving energy in excess to that available to Photosystem II (for which the phycobiliproteins are the principal light-harvesting photopigments). The decline in chlorophyll content was interpreted as a mechanism for reducing Photosystem I activity to a rate compatible with the rate of Photosystem II activity under these light conditions. In a later investigation Myers et al. (1978) showed that while the maximum rate of photosynthesis (P_{max}) declined, *A. nidulans* was found to have a higher photosynthetic quantum efficiency following acclimation to far-red light. More recently Myers et al. (1982) demonstrated the accumulation of protochlorophyll(ide) in cells of *A. nidulans* and *Agmenellum quadruplicatum* grown in far-red light, and they suggested that an anomaly in chlorophyll synthesis forms part of the general low chlorophyll syndrome in far-red light. Although growth rate was severely limited in blue light, Pulich and Van Baalen (1974) showed little variation in either chlorophyll or phycobiliprotein content in four cyanobacteria including the marine forms *Agmenellum quadruplicatum* and *Coccochloris elabens*. However, Wyman and Fay (1986b) noted a significant decline in chlorophyll concentration in a number of freshwater planktonic cyanobacteria grown in blue light, although this decline was not as extreme as in the case of *A. nidulans*.

5. The influence of light climate on metabolism

5.1. Photosynthesis

Vertical profiles of photosynthetic activity in freshwater and marine habitats usually show depth-dependent variations in the rate of carbon fixation (Talling, 1971; Harris, 1978; see also Section 2.3. of Chapter 15 and Section 1.1. of Chapter 14). Maximum photosynthetic activity is generally recorded at some depth below the water surface, there being an exponential decline in activity with depth below the photosynthetic maximum and often a reduction in carbon fixation in surface waters due to photoinhibition. The reproducibility of photosynthesis-depth profiles from a diverse range of water bodies and phytoplankton assemblages has stimulated the development of a number of theoretical analyses which primarily seek to describe phytoplankton productivity in terms of temporal changes in the underwater light climate (Talling, 1971; Bannister, 1974). The basis for such productivity models is generally the relationship between photosynthesis and incident irradiance (PvI) and the factors which influence this essentially curvilinear function. While it has been the practice in the field to consider light as a single quantitative variable, the possible influence of depth-dependent changes in light quality on the shape of PvI curves has received warranted attention recently (Lewis et al., 1985; Wood, 1985; Glover et al., 1985).

5.1.1. Light quantity

Vierling and Alberte (1980) examined the influence of irradiance (10 or 100 μE $m^{-2} \cdot s^{-1}$) applied during growth on the photosynthetic characteristics of *Anacystis nidulans*. Little variation was found in the initial slope (α) of oxygen evolution v. irradiance in cells taken from either light regime. However, high light-grown cells showed a more than twofold higher rate of maximum photosynthetic activity (P_{max}) per unit chlorophyll although similar saturated rates of O_2 evolution per cell were recorded for low and high light-grown cultures. The irradiance at which light saturation of photosynthesis was observed decreased in low light-grown cells. This decrease was correlated with an increase in photosynthetic unit size at low irradiance as a result of the enhanced production of phycocyanin. Similar changes in photosynthetic characteristics were noted by Foy and Gibson (1982a) in 20 freshwater strains of planktonic cyanobacteria grown in culture at either 30 or 150 μE $m^{-2} \cdot s^{-1}$. In comparison to the *Oscillatoria* strains examined (which showed a mean reduction in P_{max} of 42 % at low irradiance) the depression of P_{max} was less pronounced (13 %) in *Anabaena* and *Aphanizomenon* strains. The reduction in I_k (intercept of α and P_{max}) observed at low irradiance was primarily the result of a decline in P_{max} although in a number of instances (particularly amongst *Anabaena* and *Aphanizomenon* strains) α decreased in high light grown cells. In *Oscillatoria redekei* the increase in P_{max} (per unit chlorophyll) with culture irradiance was more pronounced in cultures maintained in continuous light than under a 16 : 8 h light : dark cycle (Foy and Gibson, 1982b). Similarly, α declined more rapidly with growth irradiance in cultures grown under continuous light. Foy and Gibson (1982b) found that variation in α with either culture irradiance or photoperiod length was closely correlated with changes in the cell phycocyanin/chlorophyll ratio: α increased when the relative cell phycocyanin content was high (i.e. under low irradiance or short light periods). Despite similar light-dependent changes in the phycocyanin/chlorophyll ratio, however, α (per unit chlorophyll) did not show any clear relationship with culture irradiance in phosphate-limited chemostat cultures of *Oscillatoria agardhii* (Riegman et al., 1985). Photosynthetic affinity also appeared to be independent of the cell phycocyanin/chlorophyll ratio in two strains of *Microcystis aeruginosa* exposed to a range of illumination conditions (Raps et al., 1983; Zevenboom and Mur, 1984). This suggests that variation in the affinity for light is dependent on not only the relative proportions of the photosynthetic pigments within the cell but also on the degree of energy coupling between the light-harvesting antennae. Variation in the efficiency of light energy transfer from phycobiliproteins to reaction centre chlorophyll has been described in *Anacystis nidulans* (Ghosh and Govindjee, 1966), *Oscillatoria limnetica* (Carthew, 1980) and most recently in the marine *Synechococcus* sp. strain WH7803 (Barlow and Alberte, 1985; Wyman et al., 1985). In the latter example increases in both α and P_{max} in low light grown cells were correlated with an increase in the number of photosynthetic units present per cell and with a reduction in the loss of absorbed light energy as phycoerythrin autofluorescence (Barlow and Alberte, 1985).

The dependence of photosynthetic parameters on previous light history has also been described in field populations of planktonic cyanobacteria. In a mixed assemblage of *Oscillatoria agardhii, O. redekei* and *Aphanizomenon flos-aquae*, α and I_k were correlated with the mean daily irradiance experienced during the 5 days prior to sampling, rather than with the incident irradiance recorded at the time of measurement (Jones, 1978). At low irradiance α increased and I_k declined while during periods of sustained fine weather α decreased and I_k increased. Jones (1978) suggested that variation in the populations photosynthetic characteristics was the result of changes in the relative cell phycobiliprotein content. Increases in the photosynthetic efficiency of marine *Synechococcus* present in deep water in comparison to the population present in surface waters have also been related to an increase in cell phycoerythrin concentrations (Glover et al., 1985). Increased photosynthetic efficiency in bloom-forming freshwater cyanobacteria, however, has been related to the enhanced production of carotenoid pigments rather than to increases in phycobiliprotein synthesis (Paerl et al., 1983; Paerl, 1984; Paerl et al., 1985), and it was proposed that in addition to their photoprotective role photosynthetically-active carotenoids may make a considerable contribution to light-harvesting in surface populations.

Light quantity has been reported to influence the primary route of CO_2 fixation in a number of planktonic cyanobacteria. At low light levels aspartate, glutamate and citrate were detected as early photosynthetic products in cultures of *Microcystis aeruginosa* which led Krüger and Jüttner (1980) to suggest that under these circumstances CO_2 fixation was mediated principally by phosphoenolpyruvate (PEP) carboxylase. At irradiances above the compensation point for photosynthesis, however, the principal route of carbon incorporation was via the reductive pentose phosphate pathway (see Section 2 of Chapter 4). Similar conclusions were drawn from experiments with cultures of *Merismopedia tennuissima* and *Oscillatoria rubescens*: low light favoured protein synthesis while incorporation of carbon into polysaccharide was more pronounced at higher irradiances (Konopka and Schnur, 1980). A consistent effect of previous light history on protein synthesis was not observed in field populations of *O. rubescens*, however, which Konopka (1982) suggested may be the result of parallel responses to transitions in nutrient availability. No effect of light quantity on the pattern of carbon incorporation was observed in a recent study of a field population of marine *Synechococcus* (Glover et al., 1985).

5.1.2. Light quality

Photosynthetic action spectra of cyanobacteria closely resemble their spectral absorption properties, particularly in the region of phycobiliprotein absorption. Hence quantitative/qualitative variation in pigmentation has a profound influence on the wavelength dependent efficiency of carbon fixation. In nutrient-sufficient cultures of *Aphanocapsa* 6701 peaks in the photosynthetic action spectrum occurred at 570, 625 and 652 nm which corresponded to the absorption maxima of phycoerythrin, phycocyanin and allophycocyanin, respectively (Lemasson et al., 1973). While chlorophyll *a* accounted for a substantial fraction of light absorption, oxygen evolution in the corresponding spectral region made a relatively minor contribution to the overall rate of photosynthetic activity. Profound changes were observed in the photosynthetic action spectrum of nitrogen-starved cells depleted in phycobiliproteins, however, in which chlorophyll *a* was found to be the principal light-harvesting pigment. The impact of depth-dependent changes in light quality on the photosynthetic efficiency of natural populations of marine cyanobacteria and other phytoplankton has been described recently (Lewis et al., 1985; Glover et al., 1985; Wood, 1985; Glover et al., 1986). Light in the blue and green regions of the visible spectrum was utilised for photosynthesis by phycoerythrin-rich *Synechococcus* spp. with higher efficiency than red light of a similar quantum flux (Glover et al., 1985). Wood (1985) demonstrated that whereas in artificial white light the photosynthetic response to variation in light quantity was similar for both phycoerythrin-rich and phycocyanin-rich clones of *Synechococcus*, phycoerythrin-rich strains utilised the natural underwater light climate of the open oceans with higher efficiency over a greater depth integral. In addition to a reduction in α, cells of the phycocyanin-rich *Microcystis aeruginosa* incubated in blue light showed an apparent reduction in P_{max} (per cell and per unit chlorophyll) when compared to the value obtained for the same cells incubated in either green or red light (Table II; Wyman and Fay, unpublished). In contrast, the value of P_{max} for the phycoerythrin-rich *Oscillatoria rubescens* was independent of the light

TABLE II

The influence of the wavelength of measurement on the photosynthetic characteristics of *Microcystis aeruginosa* (phycocyanin-rich) and *Oscillatoria rubescens* (phycoerythrin-rich)

Species	Wavelength (nm)			
	494	575	625	675
M. aeruginosa				
P_{max}[a]	80 ± 15	183 ± 16	200 ± 33	198 ± 24
α[b]	0.52 ± 0.04	1.12 ± 0.04	1.97 ± 0.09	1.49 ± 0.01
O. rubescens				
P_{max}[a]	300 ± 36	355 ± 45	326 ± 30	318 ± 53
α[b]	2.4 ± 0.1	2.7 ± 0.04	3.1 ± 0.17	2.9 ± 0.3

[a]μmol O_2 mg chlorophyll$^{-1}\cdot$h^{-1}.
[b]μmol O_2 mg chlorophyll$^{-1}\cdot$h^{-1} μE m$^{-2}\cdot$s^{-1}.

quality experienced during the determination of photosynthetic activity. Measurements of the partial reactions of photosynthesis showed no evidence of an impairment in either Photosystem I or Photosystem II activity in blue light which suggested that the observed decline in photosynthetic potential may be the result of changes in the rate of electron transport between the two photosystems.

Exposure to blue light has also been shown to influence the activity of the primary carboxylating enzyme ribulose 1,5-bisphosphate (RuBP) carboxylase in *Microcystis aeruginosa* (Codd, 1981). Blue light photoinactivation of RuBP carboxylase was stimulated in vitro by the presence of oxygen and the addition of flavin mononucleotide (FMN). No significant loss in RuBP carboxylase activity was detected in cells illuminated with an equal flux of red light or in cells that were incubated in darkness.

In investigations of the effect of light quality on the relative rates of synthesis of protein and carbohydrate in *M. aeruginosa* and *Oscillatoria rubescens* a significant reduction in the amount of label incorporated into the polysaccharide fraction was observed in blue light (Wyman and Fay, unpublished). However, as the overall rate of carbon incorporation in blue light was reduced substantially in comparison to cultures incubated in red light, specific responses to light quality were difficult to separate from the known effect of variation in light quantity on the pattern of carbon assimilation in cyanobacteria (Krüger and Jüttner, 1980; Konopka and Schnur, 1980). Glover et al. (1985) found little evidence for an effect of light quality on the pattern of carbon fixation in a natural population of *Synechococcus* present in the Gulf of Maine.

5.2. Respiration

In addition to the regular nocturnal dark periods, natural populations of planktonic cyanobacteria may also spend a considerable fraction of the daylight hours in dark or in dimly illuminated water depending on the depth of circulation. Basal metabolism is supported in the dark by energy derived from the mobilisation of reserves of carbohydrate (glycogen) accumulated during periods of active photosynthesis (see Chapter 5).

Although influenced by other environmental factors such as temperature, pH and nutrient concentration, previous light history has a profound effect on the rate of dark respiration. Kratz and Myers (1955) found that in comparison to cultures maintained in prolonged darkness, a short pre-incubation in the light led to a five-fold stimulation of respiratory oxygen uptake in a number of cyanobacteria. The stimulatory effect of light on the subsequent rate of dark respiration has also been observed in field populations of *Oscillatoria redekei* (Gibson, 1975) and *Anabaena flos-aquae* (Yallop, 1982). Gibson (1975) found that the gradual decline in the rate of respiration observed in *O. redekei* maintained in prolonged darkness was associated with a parallel decline in cellular carbohydrate content. In cultures entrained to regular light : dark cycles, mobilisation of glycogen provided sufficient energy to support the continued synthesis of proteins, DNA and other macromolecules during the dark period in *Oscillatoria agardhii* (Van Liere et al., 1979b) and *O. agardhii* and *O. redekei* (Foy and Smith, 1980).

The rate of glycogen accumulation in the light and its subsequent mobilisation during the hours of darkness was shown to depend on the irradiance experienced during growth (Stone and Ganf, 1981) and on the duration of the light period (Foy and Smith, 1980; Loogman et al., 1982). At an irradiance sufficient to support μ_{max} in *O. agardhii* the rate of glycogen synthesis declined as the length of the light period in a 24-h light : dark cycle was increased (Loogman et al., 1982). Maximal cell glycogen content was achieved in light periods of 8 h or more while the dark rate of glycogen mobilization increased in proportion to the decline in the number of hours of darkness experienced during the 24-h cycle.

Little is known of the influence of the spectral distribution of light supplied for autotrophic growth on the dark respiratory activity of cyanobacteria. However, Ohki et al. (1980) reported that short periods of red light administered to heterotrophic cultures of *Fremyella diplosiphon* stimulated dark respiration whereas green light reversed this effect. Although blue light is known to stimulate respiratory activity in green algae (reviewed by Kowallik, 1982), Pulich and Van Baalen (1974) found a 50% inhibition of respiration in cultures of *Agmenellum quadruplicatum* and *Coccochloris elabens* grown in blue light in comparison to material incubated in white light.

The magnitude of respiration in the light (particularly at low light levels and temperatures) is of considerable importance in determining the productivity of natural populations of cyanobacteria. Generally, respiration is inhibited to some extent at low irradiance (Imafuku and Katoh, 1976; Scherer and Böger, 1982), although at higher light levels the rate of oxygen consumption may exceed the rate of dark respiration considerably. The frequently observed inhibition of respiratory activity at low light levels has been interpreted as a consequence of the prokaryotic cell organisation of cyanobacteria in which the respiratory and photosynthetic electron transport systems share common intermediates (Jones and Myers, 1963; Lockau, 1981). Scherer et al. (1981), however, proposed that photosynthetic and respiratory electron transport components were located in different membrane regions in *Anabaena variabilis*. At low light levels, CO_2 evolution in the light was less than 10% of the equivalent rate measured in darkness in each of nine strains of cyanobacteria (Scherer and Böger, 1982). The light inhibition of respiratory activity was found to be only 50–85% of the dark rate, however, when allowance was made for the immediate re-fixation of evolved CO_2 (see also Sections 6 and 7 of Chapter 5).

5.3. Nitrogen assimilation

Under photoautotrophic conditions light is required for the assimilation and utilisation of inorganic (Losada and Guerrero, 1979), organic (Lee-Kaden and Simonis, 1982) and atmospheric nitrogen (Fay, 1970), although a certain amount of nitrogen assimilation may be possible in the dark at the expense of photochemically generated products. Nitrogen uptake shows a distinct dependence on irradiance, although the spectral requirements for the assimilation of different nitrogenous compounds may differ considerably.

5.3.1. Inorganic and organic nitrogen

In nitrogen-limited batch or chemostat cultures of *Oscillatoria agardhii*, Van Liere et al. (1979a) found that the half-saturation constant (K^u_s) for nitrate assimilation was independent of the incident irradiance in the range 0.1–5.5 Wm^{-2}. During light-limited growth K^u_s was found to be similarly independent of μ (and hence irradiance) although V_{max} (maximum specific uptake rate) for NO^-_3 assimilation increased with increases in μ (Zevenboom and Mur, 1980). The rate of NO^-_3 assimilation increased six-fold from its low level in the dark during the first 2 h of the light period in cultures of *Synechococcus leopoliensis* (*Anacystis nidulans*) grown in a 12 : 12 h light : dark cycle (Tischner and Schmidt, 1984). Subsequently, the rate of assimilation declined and reached approximately 50% of its maximum rate at 5 h into the light period. Changes in the rate of NO^-_3 uptake were attributed to the light activation of the assimilatory enzymes nitrite reductase, glutamine synthetase and isocitrate dehydrogenase. The similarity in vivo to the activation kinetics of these enzymes in vitro in the presence of thioredoxin and dithioerythreitol was strongly suggestive that nitrate assimilation was regulated by a ferredoxin/thioredoxin system in intact cells (see Sections 2.2. and 2.5. of Chapter 6).

Although Tischner and Schmidt (1984) found no evidence for light regulation of nitrate reductase activity, Stevens and Van Baalen (1973) had earlier reported that the rate of nitrate reduction showed a similar response to irradiance as μ in a nitrite reductase deficient mutant of *Agmenellum quadruplicatum*. Although nitrate reduction was light-stimulated, a low but constant rate of NO^-_3 assimilation occurred in the dark. Foy and Smith (1980) emphasised the potential ecological importance of dark nitrate assimilation in natural populations of planktonic cyanobacteria. In *Oscillatoria redekei* grown in a 6 : 18 h light : dark cycle 39% of the NO^-_3 assimilated occurred during the dark period. As the light-dependent assimilation of NO^-_3 is directly dependent upon photosynthesis and is sensitive to uncouplers of photosynthetic electron transport (Ortega et al., 1976; Losada and Guerrero, 1979), it is likely that the two processes share similar spectral requirements. However, Stevens and Van Baalen (1973, 1974) found that the action spectrum for the induction of nitrate reductase in *Agmenellum quadruplicatum* resembled the absorption spectrum of chlorophyll *a* indicative of a dependence of Photosystem I activity alone. Blue light was found to have a slight inhibitory effect on nitrate reduction, however, during the first 24 h of incubation in the presence of nitrate.

Many cyanobacteria can photoassimilate amino acids and other organic nitrogen sources, although relatively few systematic investigations of their utility as sole nitrogen sources have been undertaken. Assimilation of amino acids was dependent on photosynthesis and stimulated by increases in irradiance in cultures of the photoautotroph *Anacystis nidulans* (Lee-Kaden and Simonis, 1982). In monochromatic light, uptake was sensitive to dichlorophenyl-dimethylurea (DCMU) at 630 nm and dibromothymoquinone (DBMIB) at 717 nm which suggested that both partial reactions of photosynthesis were involved in assimilation. The energy generated by Photosystem I alone, however, was found to be sufficient to support the assimilation of amino acids in the absence of Photosystem II. In *Oscillatoria rubescens*, Wyman and Fay (unpublished) found that the action spectrum for arginine incorporation was similar to the absorption spectrum of cell suspensions but differed from the action spectrum for photosynthetic oxygen evolution in showing a distinct stimulation in uptake in the regions of chlorophyll absorption.

5.3.2. Nitrogen fixation

The close relationship between photosynthesis and nitrogen-fixation in heterocyst-forming cyanobacteria was established in a number of early studies (Fogg and Than-Tun, 1958, 1960; Fay and Fogg, 1962). Cox (1966) and Cox and Fay (1969) found that nitrogen fixation in *Anabaena cylindrica* was dependent on light quantity and it is now established that nitrogenase activity is light-stimulated in all N_2-fixing cyanobacteria (Bothe, 1982). N_2 fixation was found to be far less sensitive than CO_2 fixation to chlorophenyl dimethylurea (CMU) in *A. cylindrica* and led Cox and Fay (1969) to suggest that cyclic photophosphorylation was sufficient to support the ATP requirement of nitrogenase. This suggestion was supported by Fay (1970) who found that the action spectrum for nitrogenase activity was similar to that of Photosystem I, an observation confirmed by the finding that N_2 fixation was sensitive to DBMIB (Bothe and Loos, 1972). Consistent with the absence of an oxygen-evolving Photosystem II, phycobiliproteins were shown to be absent or only present in minor amounts in isolated heterocysts (Fay et al., 1968; Fay, 1969b; Alberte et al., 1980). More recently, several investigations have demonstrated that phycobiliproteins may be retained in mature, functional heterocysts of some species (Peterson et al., 1981; Yamanaka and Glazer, 1983; Kumar et al., 1983), and that light absorbed by these pigments is transferred to Photosystem I exclusively (see Sections 7 and 8 of Chapter 7). In a group III chromatically adapting isolate of *Nostoc*, phycoerythrin-rich cultures showed an enhanced capacity to utilise green light for nitrogen fixation in comparison to cultures grown in red light in which the synthesis of phycoerythrin was repressed (Wyman and Fay, unpublished). The presence of phycoerythrin in heterocysts was demonstrated by fluorescence microscopy, thereby providing a possible explanation for the observed stimulation of nitrogenase activity in green light.

In natural populations of planktonic cyanobacteria, N_2 fixation rates may show considerable variation with season, throughout the course of a single day and with depth (Horne and Goldman, 1972; Horne, 1975, 1979; Vanderhoef et al., 1975; see Section 4 of Chapter 21). Such variation is undoubtedly due in part to changes in water chemistry, temperature and pH but the dependence of nitrogenase on light as a source of energy and reductant (Stewart, 1980) accounts for a proportion of the observed variability. In active populations of *Anabaena* spp. a light-mediated recovery in N_2 fixation rates was observed after noon when CO_2 fixation rates were depressed owing to the accumulation of photosynthetically-generated oxygen (Paerl, 1978; Kellar and Paerl, 1980). Maximum CO_2 fixation rates preceded maximum N_2 fixation rates; an observation which led to the suggestion that temporal separation of these two processes may reduce the competition between them for light-generated reductant. Levine and Lewis (1984) found that the rate of N_2 fixation declined dramatically during the night hours ($< 6\%$ of the maximum daytime rate), although appreciably higher levels of nocturnal N_2 fixation have been recorded in other field populations (Horne and Fogg, 1970; Horne, 1975). Nitrogenase activity declined to negligible rates 3 h after transfer to darkness of photoautotrophically grown cultures of *Anabaena* ATCC 33047 (= *Anabaena* CA) but recovered rapidly (within 60 min) following re-illumination (Ramos et al., 1985). The redevelopment of N_2 fixing capacity was found to be dependent upon the de novo synthesis of both messenger RNA and protein.

Nitrogen fixation in the non-heterocystous cyanobacteria *Gloeothece* sp. and *Microcoleus chthonoplastes* grown in light : dark cycles occurs predominantly during the dark period (Mullineaux et al., 1981; Pearson et al., 1981). A similar temporal separation of carbon and nitrogen fixation was reported recently in a marine *Oscillatoria* sp. (Stal and Krumbein, 1985). In cultures incubated in continous light, nitrogenase activity showed diurnal variation while in 24-h light : dark cycles N_2 fixation was confined to the dark period exclusively when the length of the light period was shorter than 16 h. Dark nitrogenase activity was not observed, however, when chloramphenicol was added to cultures 3 h before the start of the dark period; an observation which supported the conclusion that de novo nitrogenase synthesis was required during each diel cycle.

6. The influence of light on structure and differentiation

Cyanobacteria may exhibit considerable morphological plasticity in response to changes in the physico-chemical nature of their environment. Variations in the light climate may result in marked changes in cell size and shape and in trichome morphology. Whereas such morphological responses may arise from differences in growth rate due to changes in the availability of light energy, light appears to be an essential trigger in the control of heterocyst, akinete and hormogonium differentiation.

6.1. Morphology

Mann and Carr (1974) found a logarithmic relationship between the cell volume of *Anacystis nidulans* and the organism's division rate during the exponential phase of growth in light-limited batch cultures. Cell volume increased from approximately 1.2 μm^3 at a growth rate of 0.004 h^{-1} to 2.3 μm^3 when $\mu = 0.3\,h^{-1}$ and was accompanied by increases in cell DNA and RNA content. In contrast, mean cell volume was found to increase at low irradiance in cultures of *Microcystis aeruginosa* (Kruger et al., 1981). Bennett and Bogorad (1973) reported striking variations in cell size and shape in *Fremyella diplosiphon*, induced by changes in light quality. In green light, cells were relatively large, cylindrical in shape and had less pronounced cross wall constrictions, while in red light cells became spherical and formed bead-like filaments. It is not clear, however, to what extent differences in growth rate between each light regime had influenced cell morphology. Transfer of *F. diplosiphon* to red light also resulted in a rapid decline in mean trichome length, owing to filament fragmentation. A similar effect of red light on filament length has been noted in *Tolypothrix tenuis* (Haury and Bogorad, 1977). In nitrogen-fixing cultures of *Gloeotrichia echinulata* grown under severe light limitation, Wyman and Fay (unpublished) noted a considerable increase in mean trichome length. The length of the terminal hairs (see Section 2.5 of Chapter 21) increased several-fold, which, in view of the known effect of nitrogen limitation on hair formation in other members of the *Rivulariaceae* (Sinclair and Whitton, 1977), suggested that energy limitation of nitrogenase activity may have been the underlying cause.

Light also appears to influence the coiling of trichomes in certain cyanobacteria. Bai and Seshadri (1980) found three distinct morphological variants in cultures of *Spirulina fusiformis*. High light flux and low nutrient availability favoured the growth of a highly coiled trichome variant whereas at low light a more loosely coiled form predominated. Nutrient enrichment of high light grown cultures favoured the growth of a variant of intermediate helix frequency. The growth of a highly coiled trichome variant of *Arthrospira* 8005 was found to be promoted by either red light or white light of a higher quantum flux (Wyman and Fay, unpublished). Helical trichomes are frequently encountered in natural populations of certain planktonic *Anabaena* spp., although this character is often lost in culture (Gorham et al., 1964; Booker and Walsby, 1979).

6.2. Differentiation

6.2.1. Heterocysts

Heterocyst differentiation invariably follows a stepdown in the availability of combined nitrogen.

Although heterocysts may be produced in the dark in cyanobacteria capable of heterotrophic growth, under photoautotrophic conditions light is required to initiate the differentiation process. Following the removal of inorganic nitrogen from the growth medium, a minimum light period of 3 h was required before a small proportion of the vegetative cells was capable of completing subsequent differentiation into heterocysts in cultures of *Anabaena cylindrica* (Bradley and Carr, 1977). Further increases in the light period produced a higher frequency of heterocysts such that the full complement was formed after illumination for a period of 9 h or more. In addition to photoperiod, light quantity may also influence the course of heterocyst differentiation. Fogg (1949) found a quantum requirement corresponding to about 3–4 μE $m^{-2} \cdot s^{-1}$ (240 lx) for heterocyst development in *A. cylindrica* whereas Kale (1972) reported that maximum heterocyst frequency in a tropical strain of *Anabaena ambigua* occurred at an irradiance of 1300 lx (approximately 20 μE $m^{-2} \cdot s^{-1}$) and that below this light flux the number of heterocysts produced was directly proportional to the incident irradiance. Short periods of high irradiance (approximately ten-fold that required to support μ_{max} initiated the formation of double heterocysts in *A. cylindrica* at a frequency which was proportional to the length of the period of irradiation (Adams and Carr, 1981).

Kale and Talpasayi (1969) and Kale (1972) reported that whereas blue or green light were ineffective, white or red light induced heterocyst differentiation in *A. ambigua*. Parallel measurements of filament length made during the experimental period showed that growth was arrested in blue and green light, and, therefore, the reported effect of light quality on heterocyst differentiation appears to be difficult to separate from its effect on cell division in this instance. However, at somewhat higher light fluxes Tyagi and Ahluwalia (1978) found that the action spectrum for heterocyst formation corresponded to the absorption spectrum of the major light-harvesting pigments (chlorophyll and phycocyanin) of *Anabaena doliolum*. Heterocyst differentiation was inhibited in the absence of CO_2 which indicated a requirement for photosynthetic products. The need for CO_2 was circumvented, however, by the addition of glucose or sucrose, indicating that reduced carbon rather than photosynthesis itself was essential for differentiation. Incubation of the cyanobiont *Anabaena azollae* in green light led to a dramatic enhancement of heterocyst differentiation (heterocyst frequency 40 %); a response which was interpreted as an adaptation to this organism's natural microenvironment in which the chlorophyll present in the overlying fern cells presumably depletes light in the red and blue wavebands (Wu-Guo-Liang et al., 1982).

6.2.2. Akinetes

6.2.2.1. Akinete differentiation. Many heterocystous cyanobacteria produce akinetes under environmental conditions which are unfavourable for the maintenance of cell division and growth (see Chapter 9 for details). Although a wide range of physico-chemica factors have been shown to promote akinete production, energy limitation appears to be a central factor in the initiation of the differentiation process. Fay (1969a) reported that akinete differentiation was associated with the termination of the logarithmic growth phase in light-limited cultures of *Anabaena cylindrica*. The association of akinete differentiation with the termination of exponential growth in batch culture was confirmed by Simon (1977) and Nichols et al. (1980), and has been observed also in *Aphanizomenon flos-aquae* (Rother and Fay, 1979), *Nostoc* 7524 (Sutherland et al., 1979), *Anabaena circinalis* (Fay et al., 1984) and *Gloeotrichia echinulata* (Wyman and Fay, 1986c). Akinete differentiation during logarithmic growth has been reported, however, in cultures of *Cylindrospermum licheniforme* (Hirosawa and Wolk, 1979).

The influence of light quantity on akinete differentiation was investigated in *Nostoc* 7524 (Sutherland et al., 1979) and *Anabaena cylindrica* (Nichols et al., 1980). Both organisms showed a similar

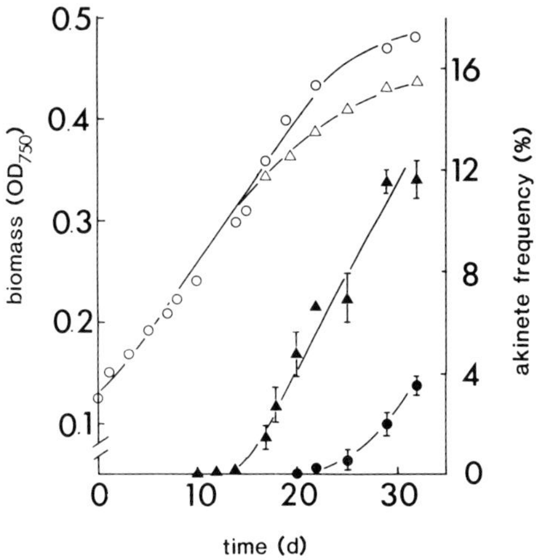

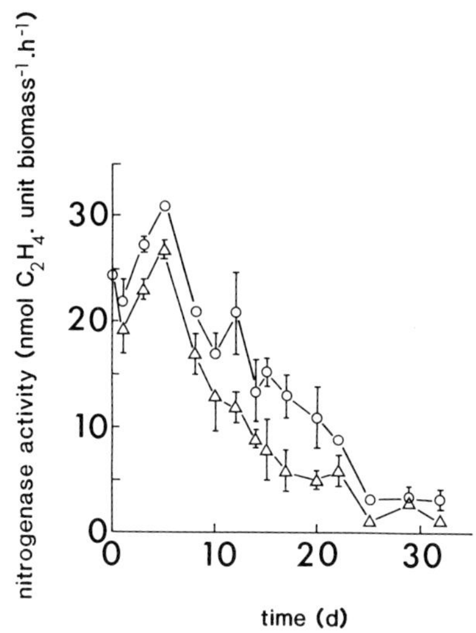

Fig. 9. The relation between culture age, biomass (open symbols) and akinete frequency (closed symbols) in *Gloeotrichia echinulata* grown in white (○, ●) or green (△, ▲) light. (After Wyman and Fay, 1986c)

Fig. 10. The relation between culture age and the rate of acetylene reduction (nitrogenase activity) in *Gloeotrichia echinulata* grown in white (○) or green (△) light. (After Wyman and Fay, 1986c)

response; at low irradiance the initiation of akinete differentiation occurred at a lower culture density than that reached by cultures exposed to higher light fluxes (see Section 2.6.2. of Chapter 9).

Less attention has been directed to the effect of variation in light quality on the course of akinete differentiation. However, Wyman and Fay (1986c) found that green light stimulated akinete production in a phycoerythrin-producing strain of *Gloeotrichia echinulata*. The effect of light quality was particularly evident in cultures grown in the absence of combined nitrogen. Under nitrogen-fixing conditions, akinete differentiation was initiated in green light at an earlier time and at a lower culture density than in comparable cultures grown in white light at a similar initial growth rate (Fig. 9). In both light regimes nitrogenase activity was greatest in young cultures but as culture density increased, activity subsequently declined more rapidly in green light (Fig. 10). It was concluded that akinete differentiation was the consequence of light energy limitation of nitrogen fixation. Cultures grown in green light became nitrogen-limited at a lower culture density, owing to the higher quantum requirement to satisfy the energy demand of nitrogenase in this light regime (Fay, 1970). That cultures incubated in green light were nitrogen-limited was supported by the observations that (a) cell phycobiliprotein content declined more rapidly than in white light; and (b) the addition of $NaNO_3$ to post-exponential phase cultures resulted in a resumption of cell division and growth, the resynthesis of phycobiliproteins and a cessation of akinete differentiation.

6.2.2.2. Akinete germination. A number of studies have shown a specific light requirement for the germination of akinetes. However, Kaushik and Kumar (1970) reported that akinete germination was independent of photosynthesis in *Anabaena doliolum* and *Fischerella mucicola* while Cmiech et al. (1984) observed germination of *Anabaena solitaria* akinetes in the absence of light. Yamamoto (1976) found that DCMU blocked akinete germination, although, in comparison to photosynthetic electron transport, akinete germination in *A. variabilis* was found to be one order of magnitude less sensitive to DCMU (Braune, 1979). Germination was dependent upon the incident irradiance and showed an action spectrum similar to the absorption spectrum of phycocyanin. A similar wavelength dependence was noted in germinating akinetes of *Anabaena vaginicola* (Rai and Pandey, 1981). A detailed account of the structural and biochemical aspects of akinete germination in *Nostoc* 7524 has been published recently by Sutherland et al. (1985a,b). Earlier, Chauvat et al. (1982) had shown that cyclic photophosphorylation alone was insufficient to support akinete germination in this organism and required the participation of both Photosystems I and II to achieve 100% germination (see Section 2.5.1. of Chapter 9 for details).

In natural populations of *Anabaena circinalis* Reynolds (1972) suggested that the seasonal increase in irradiance and daylength stimulated akinete germination during the early summer months. Similar observations have been made in natural populations of *Gloeotrichia echinulata* (Roelofs and Ogelsby, 1970). However, Rother and Fay (1977) found that akinetes germinated soon after maturation, following the collapse of mixed blooms of planktonic cyanobacteria, a phenomenon which was also noted by Cmiech et al. (1984). Akinete germination following a decrease in culture turbidity, owing to the collapse of the aged vegetative cell population, has also been observed in *Anabaena cylindrica* (Fay, 1969a).

6.3. Hormogonia

Hormogonia, which consist of short chains of undifferentiated cells, are produced by many filamentous cyanobacteria and in the natural environment evidently aid in population dispersal. Hormogonia may be capable of gliding motility and may be positively buoyant, owing to the presence of gas vacuoles. Although produced by many members of the *Oscillatoriales* and *Nostocales*, the influence of light on hormogonium differentiation has been studied in most detail in species of the genus *Nostoc*.

The development of hormogonia was found to require a short period of illumination in the facultative heterotroph *Nostoc muscorum* (Lazaroff and Vishniac, 1961). Lazaroff (1966) found that the light requirement was not absolute, for hormogonia also developed in cultures maintained for prolonged periods (2 months) in darkness. However, light (particularly red light) greatly accelerated the differentiation process. Robinson and Miller (1970) found a similar stimulation of hormogonium formation by red light in *Nostoc commune* but noted that differentiation was blocked by exposure to green light. In photoautotrophic cultures of *Nostoc muscorum*, hormogonium formation was stimulated by exposure to high irradiance and accompanied by the induction of gas vacuole formation (Waaland et al., 1971; Armstrong et al., 1983).

References

Adams, D.G. and Carr, N.G. (1981) Heterocyst differentiation and cell division in the cyanobacterium *Anabaena cylindrica:* effect of high light intensity. J. Cell Sci. 49, 341–352.

Alberte, R.S., Tel-Or, E., Packer, L. and Thornber, J.P. (1980) Functional organization of the photosynthetic apparatus in heterocysts of the nitrogen-fixing cyanobacteria. Nature (London) 284, 481–483.

Armstrong, R.E., Hayes, P.K. and Walsby, A.E. (1983) Gas vacuole formation in hormogonia of *Nostoc muscorum*. J. Gen. Microbiol. 128, 263–270.

Bai, N.J. and Seshadri, C.V. (1980) On coiling and uncoiling of trichomes in the genus *Spirulina*. Arch. Hydrobiol., Suppl. 60, 32–47.

Bannister, T.T. (1974) Production equations in terms of chlorophyll, quantum yield, and upper limit of production. Limnol. Oceanogr. 19, 1–12.

Barlow, R.G. and Alberte, R.S. (1985) Photosynthetic characteristics of phycoerythrin-containing marine *Synechococcus* spp. Mar. Biol. 86, 63–74.

Bennett, A. and Bogorad, L. (1973) Complementary chromatic adaptation in a filamentous blue-green alga. J. Cell Biol. 58, 419–435.

Booker, M.J. and Walsby, A.E. (1979) The relative form resistance of straight and helical blue-green algal filaments. Br. Phycol. J. 14, 141–150.

Bothe, H. (1982) Nitrogen fixation. In N.G. Carr and B.A. Whitton (eds.), The Biology of Cyanobacteria. Blackwell, Oxford, pp. 87–104.

Bothe, H. and Loos, E. (1972) Effect of far-red light and inhibitors on nitrogen fixation and photosynthesis in the blue-green alga *Anabaena cylindrica*. Arch. Mikrobiol. 86, 241–254.

Bradley, S. and Carr, N.G. (1977) Heterocyst development in *Anabaena cylindrica*: the necessity for light as an initial trigger and sequential stages of commitment. J. Gen. Microbiol. 101, 291–297.

Braune, W. (1979) C-phycocyanin – the main photoreceptor in the light dependent germination process of *Anabaena* akinetes. Arch. Microbiol. 122, 289–295.

Bryant, D.A., Glazer, A.N. and Eiserling, F.A. (1976) Characterization and structural properties of the major biliproteins of *Anabaena* sp. Arch. Microbiol. 110, 61–75.

Carthew, R.W. (1980) The thermodynamics of photosynthetic adaptation to photon fluence rate in the blue-green alga *Oscillatoria limnetica*. Photosynthetica 14, 204–212.

Chauvat, F., Corre, B., Herdman, M. and Joset-Espardellier, F. (1982) Energetic and metabolic requirements for the germination of akinetes of cyanobacterium *Nostoc* PCC 7524. Arch. Microbiol. 133, 44–49.

Cmiech, H.A., Reynolds, C.S. and Leedale, G.F. (1984) Seasonal periodicity, heterocyst differentiation and sporulation of planktonic Cyanophyceae in a shallow lake, with special reference to *Anabaena solitaria*. Br. Phycol. J. 19, 245–257.

Codd, G.A. (1981) Photoinhibition of photosynthesis and photoinactivation of ribulose bisphosphate carboxylase in algae and cyanobacteria. In H. Smith (ed.), Plants and the Daylight Spectrum. Academic Press, London, pp. 315–337.

Cohen-Bazire, G. and Bryant, D.A. (1982) Phycobilisomes. Composition and structure. In N.G. Carr and B.A. Whitton (eds.), The Biology of Cyanobacteria. Blackwell, Oxford, pp. 143–190.

Conley, P.B., Lemaux, P.G. and Grossman, A.R. (1985) Cyanobacterial light harvesting complex subunits encoded in two red light-induced transcripts. Science (Wash.) 230, 550–553.

Cox, R.M. (1966) Physiological studies on nitrogen fixation in the blue-green alga *Anabaena cylindrica*. Arch. Mikrobiol. 53, 263–276.

Cox, R.M. and Fay, P. (1969) Special aspects of nitrogen fixation by blue-green algae. Proc. Roy. Soc. Lond. B 172, 357–366.

Dakshini, K.M.M. and Gupta, S.K. (1981) Chromatic adaptation in *Merismopedia minima* at two depths in Damdama Lake. Curr. Sci. 50, 36–38.

Diakoff, S. and Scheibe, J. (1973) Action spectra for chromatic adaptation in *Tolypothrix tenuis*. Plant Physiol. 51, 382–385.

Eloff, J.N. (1978) The photooxidation of laboratory cultures of *Microcystis* under low light intensities. Environ. Biogeochem. Geomicrobiol. 1, 95–100.

Fay, P. (1969a) Metabolic activities of isolated spores of *Anabaena cylindrica*. J. Exp. Bot. 20, 100–109.

Fay, P. (1969b) Cell differentiation and pigment composition in *Anabaena cylindrica*. Arch. Mikrobiol. 67, 62–70.

Fay, P. (1970) Photostimulation of nitrogen fixation in *Anabaena cylindrica*. Biochim. Biophys. Acta 216, 353–356.

Fay, P. and Fogg, G.E. (1962) Studies on nitrogen fixation by blue-green algae. III. Growth and nitrogen fixation in *Chlorogloea fritschii*. Arch. Mikrobiol. 42, 310–321.

Fay, P., Lynn, J.A. and Majer, S.C. (1984) Akinete development in the planktonic blue-green alga *Anabaena circinalis*. Br. Phycol. J. 19, 163–173.

Fay, P., Stewart, W.D.P., Walsby, A.E. and Fogg, G.E. (1968) Is the heterocyst the site of nitrogen fixation in blue-green algae. Nature (London) 220, 810–812.

Fogg, G.E. (1949) Growth and heterocyst production in *Anabaena cylindrica*. II. In relation to carbon and nitrogen metabolism. Ann. Bot. 13, 241–259.

Fogg, G.E., Stewart, W.D.P., Fay, P. and Walsby, A.E. (1973) The Blue-Green Algae. Academic Press, London.

Fogg, G.E. and Than-Tun (1958) Photochemical reduction of elementary nitrogen in the blue-green alga *Anabaena cylindrica*. Biochim. Biophys. Acta 30, 209.

Fogg, G.E. and Than-Tun (1960) Interrelations of photosynthesis and assimilation of elementary nitrogen in a blue-green alga. Proc. Roy. Soc. Lond. B 153, 111–127.

Foy, R.H. (1980) The influence of surface to volume ratio on the growth rates of planktonic blue-green algae. Br. Phycol. J. 15, 279–289.

Foy, R.H. and Gibson, C.E. (1982a) Photosynthetic characteristics of planktonic blue-green algae: the response of twenty strains grown under high and low light. Br. Phycol. J. 17, 169–182.

Foy, R.H. and Gibson, C.E. (1982b) Photosynthetic characteristics of planktonic blue-green algae: changes in photosynthetic capacity and pigmentation of *Oscillatoria redekei* van Goor under high and low light. Br. Phycol. J. 17, 183–193.

Foy, R.H. and Smith, R.V. (1980) The role of carbohydrate accumulation in the growth of planktonic *Oscillatoria* species. Br. Phycol. J. 15, 139–150.

Foy, R.H., Gibson, C.E. and Smith, R.V. (1976) The influence of daylength, light intensity and temperature on the growth rates of planktonic blue-green algae. Br. Phycol. J. 11, 151–163.

Füglistaller, P., Widmer, H., Sidler, W., Frank, G. and Zuber, H. (1981) Isolation and characterisation of phycoerythrocyanin and chromatic adaptation in the thermophilic cyanobacterium *Mastigocladus laminosus*. Arch. Microbiol. 129, 268–274.

Fujita, Y. and Hattori, A. (1960) Effect of chromatic lights on phycobilin formation in a blue-green alga *Tolypothrix tenuis*. Plant Cell Physiol. 1, 293–303.

Ghosh, A.K. and Govindjee (1966) Transfer of excitation energy in *Anacystis nidulans* grown to obtain different pigment ratios. Biophys. J. 6, 611–617.

Gibson, C.E. (1975) A field and laboratory study of oxygen uptake by planktonic blue-green algae. J. Ecol. 63, 867–880.

Gibson, C.E. and Foy, R.H. (1983) The photosynthesis and growth efficiency of a planktonic blue-green alga, *Oscillatoria redekei*. Br. Phycol. J. 18, 39–45.

Gibson, C.E. and Smith, R.V. (1982) Freshwater plankton. In N.G. Carr and B.A. Whitton (eds.), The Biology of Cyanobacteria. Blackwell, Oxford, pp. 463–489.

Glazer, A.N. (1982) Phycobilisomes: structure and dynamics. Annu. Rev. Microbiol. 36, 173–198.

Glover, H.E., Phinney, D.A. and Yentsch, C.S. (1985) Photosynthetic characteristics of picoplankton compared with those of larger phytoplankton populations, in various water masses in the Gulf of Maine. Biol. Oceanogr. 3, 223–248.

Glover, H.E., Keller, M.D. and Guillard, R.R.L. (1986) Light quality and oceanic ultraphytoplankters. Nature (London) 319, 142–143.

Gorham, P.R. (1964) Isolation and culture of toxic strains of *Anabaena flos-aquae*. Verh. Internat. Verein. Limnol. 15, 796–804.

Halldal, P. (1979) Biological energy capture: Special features of marine plants. In D.S. Simonett (ed.), Marine Sciences and Ocean Policy Symposium. University of California Press, Santa Barbara.

Harris, G.P. (1978) Photosynthesis, productivity, and growth: the physiological ecology of phytoplankton. Erg. Limnol. H10, 1–171.

Haury, J.F. and Bogorad, L. (1977) Action spectra for phycobiliprotein synthesis in a chromatically adapting cyanophyte, *Fremyella diplosiphon*. Plant Physiol. 60, 835–839.

Hirosawa, T. and Wolk, C.P. (1979) Factors controlling the formation of akinetes adjacent to heterocysts in the cyanobacterium *Cylindrospermum licheniforme* Kutz. J. Gen. Microbiol. 114, 423–432.

Horne, A.J. (1975) Algal nitrogen fixation in Californian streams: Diel cycles and nocturnal fixation. Freshwat. Biol. 5, 471–477.

Horne, A.J. (1979) Nitrogen fixation in Clear Lake, California. 4. Diel studies on *Aphanizomenon* and *Anabaena* blooms. Limnol. Oceanogr. 24, 329–341.

Horne, A.J. and Fogg, G.E. (1970) Nitrogen fixation in some English lakes. Proc. Roy. Soc. Lond. B 175, 351–366.

Horne, A.J. and Goldman, C.R. (1972) Nitrogen fixation in Clear Lake, California. I. Seasonal variation and the role of heterocysts. Limnol. Oceanogr. 17, 678–692.

Imafuku, H. and Katoh, T. (1976) Intracellular ATP level and light-induced inhibition of respiration in a blue-green alga, *Anabaena variabilis*. Plant Cell Physiol. 17, 515–524.

Jaag, O. (1972) *Oscillatoria rubescens* D.C. In T.V. Desikachary (ed.), Taxonomy and Biology of Blue-Green Algae. University of Madras, Madras, pp. 296–299.

Jerlov, N.G. (1976) Marine Optics. Elsevier, Amsterdam.

Jewson, D.H. (1977) Light penetration in relation to phytoplankton content of the euphotic zone of Lough Neagh, N. Ireland. Oikos 28, 74–83.

Jones, L.W. and Myers, J. (1963) A common link between photosynthesis and respiration in a blue-green alga. Nature (London) 199, 670–672.

Jones, L.W. and Myers, J. (1965) Pigment variations in *Anacystis nidulans* induced by light of selected wavelengths. J. Phycol. 1, 7–14.
Jones, R.I. (1978) Adaptations to fluctuating irradiance by natural phytoplankton communities. Limnol. Oceanogr. 23, 920–926.
Kale, S.R. (1972) In T.V. Desikachary (ed.), Taxonomy and Biology of Blue-Green Algae. University of Madras Press, Madras, pp. 203–208.
Kale, S.R. and Talpasayi, E.R.S. (1969) Heterocysts: a review. Indian Biologist 1, 19–29.
Kaushik, M. and Kumar, H.D. (1970) The effect of light on growth and development of two nitrogen-fixing blue-green algae. Arch. Mikrobiol. 74, 52–57.
Kawamura, M., Mimuro, M. and Fujita, Y. (1979) Quantitative relationship between two reaction centers in the photosynthetic system of blue-green algae. Plant Cell Physiol. 20, 697–705.
Kellar, P.E. and Paerl, H.W. (1980) Physiological adaptations in response to environmental stress during an N_2-fixing *Anabaena* bloom. Appl. Environ. Microbiol. 40, 587–595.
Kipe-Nolt, J.A., Stevens, S.E. and Bryant, D.A. (1982) Growth and chromatic adaptation of *Nostoc* sp. strain MAC and the pigment mutant R-MAC. Plant Physiol. 70, 1549–1553.
Kirk, J.T.O. (1974) The contribution of phytoplankton to the attenuation of light within natural waters: a theoretical analysis. In M. Avron (ed.), Proceedings of the Third International Congress on Photosynthesis. Elsevier, Amsterdam, pp. 245–253.
Kirk, J.T.O. (1975a) A theoretical analysis of the contribution of algal cells to the attenuation of light within natural waters. I. General treatment of suspensions of pigmented cells. New Phytol. 75, 11–20.
Kirk, J.T.O. (1975b) A theoretical analysis of the contribution of algal cells to the attenuation of light within natural waters. II. Spherical cells. New Phytol. 75, 21–36.
Kirk, J.T.O. (1976) Yellow substance (Gelbstoff) and its contribution to the attenuation of photosynthetically active radiation in some inland and coastal South-Eastern Australian waters. Aust. J. Mar. Freshwat. Res. 27, 61–71.
Kirk, J.T.O. (1983) Light and Photosynthesis in Aquatic Ecosystems. Cambridge University Press, Cambridge.
Kohl, J.G. and Nicklisch, A. (1981) Chromatic adaptation of the planktonic blue-green alga *Oscillatoria redekei* and its ecological significance. Int. Rev. Ges. Hydrobiol. 66, 83–94.
Konopka, A. (1980) Physiological changes within a metalimnetic layer of *Oscillatoria rubescens*. Appl. Environ. Microbiol. 40, 681–684.
Konopka, A. (1982) Physiological ecology of a metalimnetic *Oscillatoria rubescens* population. Limnol. Oceanogr. 27, 1154–1161.
Konopka, A. and Schnur, M. (1980) Effect of light intensity on macromolecular synthesis in cyanobacteria. Microbiol. Ecol. 6, 291–301.
Kowallik, W. (1982) Blue light effects on respiration. Annu. Rev. Plant Physiol. 33, 51–72.
Kratz, W.A. and Myers, J. (1955) Photosynthesis and respiration of three blue-green algae. Plant Physiol. 30, 275–280.
Krinsky, N.I. (1978) Non-photosynthetic functions of carotenoids. Phil. Trans. Roy. Soc. Lond. B 284, 581–590.
Krüger, G.H.J. and Jüttner, F. (1980) Carbon metabolism in *Microcystis* under low light intensities. J. Gen. Microbiol. 119, 27–30.
Krüger, G.H.J., Eloff, J.N. and Pretorius, J.A. (1981) Morphological changes in toxic and non-toxic *Microcystis* isolates at different irradiance levels. J. Phycol. 17, 52–56.
Kumar, A., Tabita, F.R. and Van Baalen, C. (1983) High endogenous nitrogenase activity in isolated heterocysts of *Anabaena* sp. strain CA after nitrogen starvation. J. Bact. 155, 493–497.
Lazaroff, N. (1966) Photoinduction and photoreversal of the Nostocacean developmental cycle. J. Phycol. 2, 7–17.
Lazaroff, N. and Vishniac, W. (1961) The effect of light on the developmental cycle of *Nostoc muscorum*, a filamentous blue-green alga. J. Gen. Microbiol. 25, 365–374.
Lee-Kaden, J. and Simonis, W. (1982) Amino acid uptake and energy coupling dependent on photosynthesis in *Anacystis nidulans*. J. Bacteriol. 151, 229–236.
Lemasson, C., Tandeau de Marsac, N. and Cohen-Bazire, G. (1973) Role of allophycocyanin as a light harvesting pigment in cyanobacteria. Proc. Natl. Acad. Sci. USA 70, 3130–3133.
Levine, S.N. and Lewis, W.M. (1985) Diel variation of nitrogen fixation in Lake Valencia, Venezuela. Limnol. Oceanogr. 29, 887–893.
Lewis, M.R., Warnock, R.E. and Platt, T. (1985) Absorption and photosynthetic action spectra for natu-

ral phytoplankton populations: Implications for production in the open ocean. Limnol. Oceanogr. 30, 794–806.

Lockau, W. (1981) Evidence for a dual role of cytochrome *c*-553 and plastocyanin in photosynthesis and respiration of the cyanobacterium *Anabaena variabilis*. Arch. Microbiol. 128, 336–340.

Lonneborg, A., Lind, L.K., Kalla, S.R., Gustafsson, P. and Oquist, G. (1985) Acclimation processes in the light-harvesting system of the cyanobacterium *Anacystis nidulans* following a light shift from white to red light. Plant Physiol. 78, 110–114.

Loogman, J.G., Van Liere, L. and Mur, L.R. (1982) The hoaxing of the *Oscillatoria*. In Abstracts, IVth International Symposium on Photosynthetic Prokaryotes, Bombannes, France, pp. A39.

Losada, M. and Guerrero, M.G. (1979) The photosynthetic reduction of nitrate and its regulation. In J. Barber (ed.), Photosynthesis in Relation to Model Systems. Elsevier, Amsterdam.

Mann, N. and Carr, N.G. (1974) Control of macromolecular composition and cell division in the blue-green alga *Anacystis nidulans*. J. Gen. Microbiol. 83, 399–405.

Manodori, A., Alhadeff, M., Glazer, A.N. and Melis, A. (1984) Photochemical apparatus organization in *Synechococcus* 6301 (*Anacystis nidulans*). Effect of phycobilisome mutation. Arch. Microbiol. 139, 117–123.

Meeks, J.C., Wycoff, K.L., Chapman, J.S. and Enderlin, C.S. (1983) Regulation of expression of nitrate and dinitrogen assimilation by *Anabaena* species. Appl. Environ. Microbiol. 45, 1351–1359.

Meffert, M.E. (1971) Cultivation and growth of two planktonic *Oscillatoria* species. Mitt. Int. Ver. Theor. Angew. Limnol. 19, 189–205.

Mullineaux, P.M., Gallon, J.R. and Chaplin, A.E. (1981) Acetylene reduction (nitrogen fixation) by cyanobacteria grown under alternating light-dark cycles. FEMS Microbiol. Lett. 10, 245–247.

Myers, J., Graham, J.R. and Wang, R.T. (1978) On spectral control of pigmentation in *Anacystis nidulans* (Cyanophyceae). J. Phycol. 14, 513–518.

Myers, J., Graham, J.R. and Wang, R.T. (1982) Protochlorophyll(ide) in a blue-green alga. Plant Physiol. 69, 549–550.

Nichols, J.M., Adams, D.G. and Carr, N.G. (1980) Effect of canavanine and other amino acid analogues on akinete formation in the cyanobacterium *Anabaena cylindrica*. Arch. Microbiol. 127, 67–75.

Ohki, K., Isono, T. and Fujita, Y. (1980) Role of the light of shorter wavelength in photobiological phenomena in blue-green algae. In H. Senger (ed.), The Blue Light Syndrome. Springer-Verlag, Berlin.

Oquist, G. (1974) Distribution of chlorophyll between the two photosystems of the blue-green alga *Anacystis nidulans* grown at two different light intensities. Physiol. Plant. 30, 38–44.

Ortega, T., Castillo, F. and Cardenas, J. (1976) Photolysis of water coupled to nitrate reduction by *Nostoc muscorum* subcellular particles. Biochem. Biophys. Res. Commun. 71, 885–891.

Paerl, H.W. (1978) Light-mediated recovery of N_2-fixation in the blue-green algae *Anabaena* spp. in O_2 supersaturated waters. Oecologia 32, 135–139.

Paerl, H.W. (1984) Cyanobacterial carotenoids: their roles in maintaining optimal photosynthetic production among aquatic bloom forming genera. Oecologia 61, 143–149.

Paerl, H.W. and Kellar, P.E. (1979) Nitrogen-fixing *Anabaena:* physiological adaptations instrumental in maintaining surface blooms. Science (Wash.) 204, 620–622.

Paerl, H.W., Tucker, J. and Bland, P.T. (1983) Carotenoid enhancement and its role in maintaining blue-green algal (*Microcystis aeruginosa*) surface blooms. Limnol. Oceanogr. 28, 847–857.

Paerl, H.W., Bland, P.T., Bowles, N.D., and Haibach, M.E. (1985) Adaptation to high intensity, low wavelength light among surface blooms of the cyanobacterium *Microcystis aeruginosa*. Appl. Environ. Microbiol. 49, 1046–1052.

Pearson, H.W., Malin, G. and Howsley, R. (1981) Physiological studies on in vitro nitrogenase activity by axenic cultures of the blue-green alga *Microcoleus chthonoplastes*. Br. Phycol. J. 16, 139.

Peterson, R.B., Dolan, E., Calvert, H.E. and Ke, B. (1981) Energy transfer from phycobiliproteins to photosystem I in vegetative cells and heterocysts of *Anabaena variabilis*. Biochim. Biophys. Acta 634, 237–248.

Post, A.F., De Wit, R. and Mur, L.R. (1985a) Interactions between temperature and light intensity on growth and photosynthesis of the cyanobacterium *Oscillatoria agardhii*. J. Plankton Res. 7, 487–495.

Post, A.F., Loogman, J.G. and Mur, L.R. (1985b) Regulation of growth and photosynthesis by *Oscillatoria agardhii* grown with a light/dark cycle. FEMS Microbiol. Lett. 31, 97–102.

Pulich, W.M. and Van Baalen, C. (1974) Growth requirements of blue-green algae under blue light conditions. Arch. Microbiol. 97, 303–312.

Rai, A.K. and Pandey, G.P. (1981) Influence of environmental stress on the germination of *Anabaena* akinetes. Ann. Bot. 48, 361–370.

Ramos, J.L., Madueno, F. and Guerrero, M.G. (1985) Regulation of nitrogenase levels in *Anabaena* sp.

ATCC 33047 and other filamentous cyanobacteria. Arch. Microbiol. 141, 105–111.

Raps, S., Wyman, K., Siegelmal, H.W. and Falkowski, P. (1983) Adaptation of the cyanobacterium *Microcystis aeruginosa* to light intensity. Plant Physiol. 72, 829–832.

Raven, J.A. and Beardall, J. (1982) The lower limit of photon fluence rate for phototrophic growth: The significance of 'slippage' reactions. Plant Cell Environ. 5, 117–124.

Reynolds, C.S. (1972) Growth, gas vacuolation and buoyancy in a natural population of a planktonic blue-green alga. Freshwat. Biol. 87–106.

Reynolds, C.S. and Walsby, A.E. (1975) Water blooms. Biol. Rev. 50, 437–481.

Reynolds, C.S., Tundisi, J.G. and Hino, K. (1983) Observations on a metalimnetic *Lyngbya* population in a stably stratified tropical lake (Lagoa Carioca, Eastern Brasil). Arch. Hydrobiol. 97, 7–17.

Richardson, K., Beardall, J. and Raven, J.A. (1983) Adaptation of unicellular algae to irradiance: an analysis of strategies. New Phytol. 93, 157–191.

Riegman, R., Rutgers, M. and Mur, L.R. (1985) Effects of photoperiodicity and light irradiance on phosphate-limited *Oscillatoria agardhii* in chemostat cultures. I. Photosynthesis and carbohydrate storage. Arch. Microbiol. 142, 66–71.

Rippka, R., Deruelles, J., Waterbury, J.B., Herdman, M. and Stanier, R.Y. (1979) Generic assignments, strain histories and properties of pure cultures of cyanobacteria. J. Gen. Microbiol. 111, 1–61.

Robinson, B.L. and Miller, J.H. (1970) Photomorphogenesis in the blue-green alga *Nostoc commune* 584. Physiol. Plant 23, 461–472.

Roelofs, T.D. and Oglesby, R.T. (1970) Ecological observations on the planktonic cyanophyte *Gloeotrichia echinulata*. Limnol. Oceanogr. 15, 224–229.

Rother, J.A. and Fay, P. (1977) Sporulation and the development of blue-green algae in two Salopian meres. Proc. Roy. Soc. Lond. B 196, 317–332.

Rother, J.A. and Fay, P. (1979) Some physiological-biochemical characteristics of planktonic blue-green algae during bloom formation in three Salopian meres. Freshwat. Biol. 9, 369–379.

Scherer, S. and Böger, P. (1982) Respiration of blue-green algae in the light. Arch. Microbiol. 132, 329–332.

Scherer, S., Sturzl, E. and Boger, P. (1981) Arrhenius plots indicate localization of photosynthetic and respiratory electron transport in different membrane regions of *Anabaena*. Z. Naturforsch. 36, 1036–1040.

Simon, R.D. (1977) Sporulation in the filamentous cyanobacterium *Anabaena cylindrica*. The course of spore formation. Arch. Microbiol. 111, 283–288.

Sinclair, C. and Whitton, B.A. (1977) Influence of nutrient deficiency on hair formation in Rivulariaceae. Br. Phycol. J. 12, 297–313.

Skulberg, O.M. (1978) Some observations on red-coloured species of *Oscillatoria* (Cyanophyceae) in nutrient-enriched lakes of southern Norway. Verh. Int. Ver. Limnol. 20, 776–787.

Smith, H. (1982) Light quality, photoperception and plant strategy. Annu. Rev. Plant Physiol. 33, 481–518.

Smith, H. and Morgan, D.C. (1981) The spectral characteristics of the visible radiation incident upon the surface of the Earth. In H. Smith (ed.), Plants and the Daylight Spectrum. Academic Press, London, pp. 3–20.

Stal, L.J. and Krumbein, W.E. (1985) Nitrogenase activity in the non-heterocystous cyanobacterium *Oscillatoria* sp. grown under alternating light-dark cycles. Arch. Microbiol. 143, 67–71.

Stevens, S.E. and Van Baalen, C. (1973) Characteristics of nitrate reduction in a mutant of the blue-green alga *Agmenellum quadruplicatum*. Plant Physiol. 51, 350–356.

Stevens, S.E. and Van Baalen, C. (1974) Control of nitrate reductase in a blue-green alga. The effect of inhibitors, blue light and ammonia. Arch. Biochem. Biophys. 161, 146–152.

Stewart, W.D.P. (1980) Some aspects of structure and function in N_2-fixing cyanobacteria. Annu. Rev. Microbiol. 34, 497–536.

Stone, S. and Ganf, G. (1981) The influence of previous light history on the respiration of four species of freshwater phytoplankton. Arch. Hydrobiol. 91, 435–462.

Sutherland, J.M., Herdman, M. and Stewart, W.D.P. (1979) Akinetes of the cyanobacterium *Nostoc* PCC 7524: macromolecular composition, structure and control of differentiation. J. Gen. Microbiol. 115, 273–287.

Sutherland, J.M., Stewart, W.D.P. and Herdman, M. (1985a) Akinetes of the cyanobacterium *Nostoc* PCC 7524: morphological changes during synchronous germination. Arch. Microbiol. 142, 269–274.

Sutherland, J.M., Reaston, J., Stewart, W.D.P. and Herdman, M. (1985b) Akinetes of the cyanobacterium *Nostoc* PCC 7524: macromolecular and biochemical changes during synchronous germination. J. Gen. Microbiol. 131, 2855–2863.

Tandeau de Marsac, N. (1977) Occurrence and nature of chromatic adaptation in cyanobacteria. J. Bacteriol. 130, 82–91.
Talling, J.F. (1971) The underwater light climate as a controlling factor in the production ecology of freshwater phytoplankton. Mitt. Int. Ver. Limnol. 19, 214–243.
Tischner, R. and Schmidt, A. (1984) Light mediated regulation of nitrate assimilation in *Synechococcus leopoliensis*. Arch. Microbiol. 137, 151–154.
Tyagi, V.V.S. and Ahluwalia, A.S. (1978) Heterocyst formation in the blue-green alga *Anabaena doliolum* – a study of some aspects of photoregulation. Ann. Bot. 42, 1333–1341.
Tytler, E.M., Whitelam, G.C., Hipkins, M.F. and Codd, G.A. (1984) Photoinactivation of photosystem II during photoinhibition in the cyanobacterium *Microcystis aeruginosa*. Planta 160, 229–234.
Utkilen, H.C., Briseid, T. and Eriksson, B. (1983) Variation in photosynthetic membrane and pigment content with light intensity for *Anacystis nidulans* grown in continuous culture. J. Gen. Microbiol. 129, 1717–1720.
Vanderhoef, L.N., Huang, C.Y., Musil, R. and Williams, J. (1975) Diurnal variation in algal acetylene reduction (nitrogen fixation) in situ. Plant Physiol. 55, 273–276.
Van Liere, L., De Groot, G.J. and Mur, L.R. (1979a) Pigment variation with irradiance in *Oscillatoria agardhii* Gomont. in nitrogen (nitrate)-limited chemostat cultures. FEMS Microbiol. Lett. 6, 337–340.
Van Liere, L., Mur, L.R., Gibson, C.E. and Herdman, M. (1979b) Growth and physiology of *Oscillatoria agardhii* Gomont. cultivated in continuous culture with a light-dark cycle. Arch. Microbiol. 123, 315–318.
Vieling, E. and Alberte, R.S. (1980) Functional organization and plasticity of the photosynthetic unit of the cyanobacterium *Anacystis nidulans*. Physiol. Plant. 50, 93–98.
Waaland, J.R., Waaland, S.D. and Branton, D. (1970) Gas vacuoles: light shielding in blue-green algae. J. Cell Biol. 48, 212–215.
Wang, R.T., Stevens, S.E. and Myers, J. (1977) Action spectra for photoreactions I and II of photosynthesis in the blue-green alga *Anacystis nidulans*. Photochem. Photobiol. 25, 103–108.
Wood, A.M. (1985) Adaptation of photosynthetic apparatus of marine ultraphytoplankton to natural light fields. Nature (London) 316, 253–255.
Wu Guo-liang, Zhong Ze-pu, Bai Ke-zhi, Wang Fa-zhu and Cui Cheng (1982) The effects of light quality on the growth and development of *Anabaena azollae*. Acta Bot. Sin. 24, 46–53.
Wyman, M. and Fay, P. (1986a) Underwater light climate and the growth and pigmentation of planktonic blue-green algae (Cyanobacteria). I. The influence of light quantity. Proc. Roy. Soc. Lond. B, 227, 367–380.
Wyman, M. and Fay, P. (1986b) Underwater light climate and the growth and pigmentation of planktonic blue-green algae (Cyanobacteria). II. The influence of light quality. Proc. Roy. Soc. Lond. B, 227, 381–393.
Wyman, M. and Fay, P. (1986c) Interaction between light quality and nitrogen availability in the differentiation of akinetes of the blue-green alga *Gloeotrichia echinulata*. Br. Phycol. J., 21, 147–153.
Wyman, M., Gregory, R.P.F. and Carr, N.G. (1985) A novel role for phycoerythrin in the marine cyanobacterium *Synechococcus* sp. DC2. Science (Wash.) 230, 818–820.
Yallop, M.L. (1982) Some effects of light on algal respiration and the validity of the light and dark bottle technique for measuring primary productivity. Freshwat. Biol. 12, 427–433.
Yamamoto, Y. (1976) Effect of some physical and chemical factors on the germination of akinetes of *Anabaena cylindrica*. J. Gen. Appl. Microbiol. 22, 311–323.
Yamanaka, G. and Glazer, A.N. (1983) Phycobiliproteins in *Anabaena 7119* heterocysts. In G.C. Papageorgiou and L. Packer (eds.), Photosynthetic Prokaryotes: Cell Differentiation and Function. Elsevier, Amsterdam, pp. 69–90.
Zevenboom, W. and Mur, L.R. (1980) N_2-fixing cyanobacteria: why they do not become dominant in Dutch hypertrophic lakes. In J. Barica and L.R. Mur (eds.), Hypertrophic Ecosystems. Junk, Den Haag, The Netherlands, pp. 123–130.
Zevenboom, W. and Mur, L.R. (1984) Growth and photosynthetic response of the cyanobacterium *Microcystis aeruginosa* in relation to photoperiodicity and irradiance. Arch. Microbiol. 139, 232–239.

P. Fay and C. Van Baalen (Eds.) The Cyanobacteria

14

Mechanisms of buoyancy regulation by planktonic cyanobacteria with gas vesicles

A.E. WALSBY

Department of Botany, University of Bristol, Bristol BS8 1UG, UK

1. The need for buoyancy

Planktonic cyanobacteria, like other phytoplankton, are obligate photoautotrophes that need a certain minimum average light intensity for growth. The light intensity in a water body diminishes with depth owing to attenuation by the water itself and matter suspended in it, and below a certain depth, the bottom of the euphotic zone, there is insufficient light to support net photosynthesis and growth (Yentsch, 1980; see Section 2 of Chapter 13). The cell constituents of cyanobacteria – principally protein, carbohydrates, glycolipids and minerals – are all denser than water and without special buoyancy devices these organisms will settle out of suspension (Hutchinson, 1967; Walsby and Reynolds, 1980). The speed of sinking, which varies as the square of diameter of a particle, may be very slow in the case of very tiny cells, like the picoplanktonic cyanobacteria (Waterbury et al., 1979; see Sections 1 and 2.3. of Chapter 15), and in turbulent water of the open ocean the losses of these cells by settling out of suspension may be much less than the gains through growth. Nevertheless, losses by sinking do persist even when the water column is thoroughly mixed (see Reynolds, 1984, p. 48) and an organism that is positively buoyant will more effectively remain in the euphotic zone.

Gas vacuoles, which are present in many planktonic cyanobacteria (Walsby, 1981a) and also in large numbers of planktonic bacteria (Walsby, 1981b), are the most effective buoyancy aids. They are made up of gas vesicles which have a density of about a tenth that of water (see below); the only structures that would provide more buoyancy are air bubbles, but for reasons of surface tension these structures are unstable in very small cells (Walsby, 1972a) and are thus ruled out.

While the suspension of non-buoyant cells is favoured under turbulent conditions, the flotation of buoyant ones is favoured in conditions of calm, when populations of buoyant cyanobacteria may accumulate in waterblooms at the water surface (Reynolds, 1971). Varying conditions of turbulence may benefit gas-vacuolate cyanobacteria in different ways. In lakes that suffer alternating periods of turbulence and calm, buoyant forms will be distributed throughout the mixed depth in rough weather but will float upwards to the surface when mixing stops. Near the surface not only will they benefit from the higher light intensity for photosynthesis but they will shade out potential competitors below them. Note that such alternating stable and unstable conditions will particularly benefit the larger colonial cyanobacteria which will float more quickly to the surface (Reynolds and Walsby, 1975). Even during periods of mixing buoyant cells that are carried down to the bottom of the mixed layer or the lake bottom will remain in suspension, whereas non-buoyant ones may be permanently lost to the hypolimnion or bottom sediment.

1.1. The need for buoyancy regulation

The light intensity at the water surface may be so high as to cause inhibition of photosynthesis, damage to cell structure and, ultimately, cell death. These inhibiting and damaging effects are usually time-dependent and accrue with exposure durations of several hours. On a timescale of days, dense populations of cyanobacteria crowded into the top millimeter or so of the water column may locally deplete dissolved nutrients so that growth is prevented, and they may even consume carbon dioxide so voraciously that the rate of supply by diffusion from the surrounding water and overlying atmosphere becomes rate limiting to photosynthesis (see Booker and Walsby, 1981). The inhibiting effects of high light intensity are inextricably linked with the limitations of nutrient and inorganic carbon supply.

There is no doubt that different species of planktonic cyanobacteria respond to high light intensities at the surface in different ways: for example, water blooms of *Aphanizomenon* may collapse within a few hours through cell lysis, releasing a milky-blue mixture of gas vesicles and phycocyanin, while surface blooms of *Microcystis* may persist for many weeks without noticeable degradation of the cells (observations made by A.J. Horne and the author on Clear Lake in California). Eloff et al. (1976) showed that *Microcystis* taken from natural populations survived exposure to conditions of high light intensity that caused photooxidative death of several non-adapted strains of cyanobacteria; survival was correlated with the ability of strains to maintain high levels of the enzyme superoxide dismutase (see Van Liere and Walsby, 1982).

If photoinhibition sets in at or near the water surface there are clear advantages in moving down the water column to depths where the light intensity is lower; a cyanobacterium which has floated up in the first place must lose its buoyancy in order to sink down, and ideally, once it has reached an optimal intensity it should stop sinking. This is the essence of buoyancy regulation according to the light intensity. It is not possible for an organism to regulate its buoyancy rapidly enough to stay precisely at the optimal depth, but depending on a number of hydrodynamic factors, it may oscillate around its preferred depth. The rest of this article analyses the different problems attending buoyancy regulation in different water bodies and different cyanobacteria, and the various types of gas vesicles and behaviour that have evolved to provide the means of buoyancy regulation.

2. The gas vesicle

The gas vesicle in cyanobacteria is a hollow cylindrical structure, usually less than 0.1 μm wide, up to 1 μm (exceptionally 2 μm) long and closed by conical ends (Bowen and Jensen, 1965). The wall is made of protein (Walsby and Buckland, 1969; Jones and Jost, 1970) and is very thin, about 2 nm wide (Jost and Jones, 1970; Walsby and Armstrong, 1979). The protein is arranged in 4.5 nm wide ribs running at right angles to the cylinder axis. It is not definitely known whether the ribs form a shallow continuous spiral or a stack of hoops.

Only one type of protein has been found in the gas vesicle (Jones and Jost, 1971; Falkenberg

```
                                            10                            20                            30
Anabaena flos-aquae          A V E K T N S S S S L A E V I D R I L D K G I V I D A W V R V S L V G
Aphanizomenon flos-aquae     A V E K T N S S S S L A E V I D R I L D K G I V I D A W V R V S L V G
Calothrix sp.                A V E K T N S S S S L A E V I D R I L D K G I V V D/N A W V R V S L V G
Oscillatoria agardhii        A V E K V N S S S S L A E V I D R I L D K G I V I D A W V R V S L V G
Microcystis sp.              A V E K T N S S S S L A E V I D R I L D K G I V I D A W A R V S L V G
Dactylococcopsis salina      A V E K T N S S S S L G E V V D R I L D K G V V V D L W V R V S L V G
Halobacterium halobium       A - - Q P D - S S G L A E V L D R V L D K G V V V D V W A R V S L V G
Halobacterium salinarium     A - - Q P D - S S G V A E M L D R V L D K G V V V D V W A

                   40                            50                            60                            70
I Q L L A I E A R I V I A S V E T Y L K Y A E A V G L T Q S A V P A A
I P/E L L A I P A R I V I A S V P T...
I E L L A I E A R I V I A X V E T Y L K Y A...
I P L L S I P A R I V...
I E L L A I E A R V V I A S V E T Y L K Y A E A V G L T Q X A X X A..
I E L L A L E A R V V I...
I P/E I L T V E A R V V...
```

Fig. 1. The complete amino acid sequence of *Anabaena flos-aquae* GVP, and partial amino acid sequences of GVPs from five other cyanobacteria and two halobacteria. (1) Boxed areas show where the sequence differs from *Anabaena* GVP; (2) – denotes a deletion (no residue present); X denotes a residue present but not identified; (3) single letter amino acid code (Biochem. J. (1969) 113, 1–4); (4) data for GVPs of *Anabaena*, *Calothrix* and *Microcystis*, from Hayes et al. (1986); *Aphanizomenon*, *Oscillatoria*, *Dactylococcopsis* and both *Halobacterium* species, from Walker et al. (1984).

et al., 1972). It is referred to as 'gas vesicle protein' or GVP (Weathers et al., 1977). The complete amino acid sequence has been determined for GVP from *Anabaena flos-aquae* (Walker et al., 1984; Hayes et al., 1986); it is a protein of 70 amino acid residues and molecular mass of 7 388 daltons. The amino acid sequence is shown in Fig. 1 together with partial sequences for GVPs from five other cyanobacteria and two halobacteria determined by Walker et al. (1984) and Hayes et al. (1986). Comparison of these sequences establishes that while the cyanobacterial GVPs and halobacterial GVPs form distinct clusters the two are highly homologous. The GVP sequences of the three heterocystous cyanobacteria *Anabaena*, *Aphanizomenon* and *Calothrix* differ from one another in only two of the first 45 residues while those of *Oscillatoria*, a non-heterocystous filamentous form, and of *Microcystis*, a unicellular form, differ in only three residues from that of *Anabaena* GVP. The sequence of *Dactylococcopsis* GVP shows eight differences over the same region; it is interesting that the GVP sequence of *Dactylococcopsis*, a halophilic cyanobacterium isolated from Solar Lake, a brine pool in the Sinai Peninsula (Walsby et al., 1983a), is intermediate between those of the other cyanobacteria investigated, which are all freshwater forms, and those of halobacteria.

Most of the differences in the amino acid sequences shown in Fig. 1 represent conservative changes, that is, the replacement of one amino acid by another of similar properties. There are several instances, for example, where one residue with an aliphatic side chain is replaced by another that differs by a single -CH_2- group in the side chain (Walker et al., 1984), viz. G for A (residue 12), A for V (residues 27, 29) V for L or I (residues 11, 15, 18, 25, 27, 41, 45) and also L for I (residues 15, 38, 41). Such replacements will not bring about any essential changes in the gross properties of the protein but they may bring about small changes in volume or shape of the GVP molecule. The gas vesicles of the different cyanobacteria vary in size (see below, Table I). In each case the diameter of the circular rib, which stacks to form the cylindrical portion of the vesicle, must be determined by the size and shape of the individual GVP molecules, and, of course, by the number of molecules in the rib. In this way the amino acid sequence of each GVP may indirectly specify the size of the gas vesicle it forms. It is shown below that there must have been natural selection for gas vesicles of different size in cyanobacteria of different habitats. Evolution may thus have worked on a single gene specifying the GVP, to determine all the properties of the gas vesicle. Probably the story is more complex than this; it is known, for example, that in *Nostoc*

TABLE I

Widths and critical collapse pressures of gas vesicles in different cyanobacteria

Cyanobacterium	Cylinder width (nm)	Critical pressure (bar)	Reference to critical pressure
Dactylococcopsis	107	5.0	Walsby et al., 1983
Anabaena	85	6.0	Walsby, 1980
Aphanizomenon	79	6.2	Hayes and Walsby, 1986
Microcystis	67	7.6	Thomas and Walsby, 1985
Oscillatoria	64	9.0	Utkilen et al., 1985a

Reproduced from Hayes and Walsby, 1986

muscorum the formation of gas vesicles is affected by light intensity and nutrient conditions (Waaland and Branton, 1969; Waaland et al., 1971; Armstrong et al., 1983). Here the synthesis must be regulated by other proteins, and it is not ruled out that still others may be required in the assembly of a gas vesicle and perhaps also in shaping the structure.

2.1. Gas vesicle genes

The determination of the amino acid sequence of GVP has provided a means of isolating the structural gene for this protein. Tandeau de Marsac et al. (1985) (and see Section 5.3. of Chapter 10 of this book) have made a synthetic 29-base oligonucleotide with a sequence that coded for amino acid residues 19–27 of the *Anabaena* GVP. They used this oligonucleotide to probe a Southern blot of the total genomic DNA from a species of *Calothrix*. A fragment of this DNA, formed by digestion with the restriction enzyme Hind III, was found to hybridise strongly with the probe. Sequencing of the DNA fragment and an adjacent fragment demonstrated an open reading frame that coded for a 7 375 dalton (70 residue) protein nearly identical with *Anabaena* GVP. A second, very similar gene, coding for an identical amino acid sequence, has now been located about a hundred base pairs downstream from the first. A second cistron was predicted from mutant studies by Walsby (1977).

The isolation of *gvp* genes should open up new areas of gas vesicle research. One of the first problems to be investigated will be the high level of expression of the gene. In many of the planktonic cyanobacteria where gas vesicles are produced constitutively, there may be a highly active promoter to the gene and its mRNA transcription product may be highly stable. In the red-coloured *Oscillatoria agardhii* that ceases gas vesicle production in high-light intensity (see Section 5.2. below), it may be expected that the transcription of the *gvp* genes is regulated directly or otherwise, by light. In those cyanobacteria that produce gas-vacuolate hormogonia but non-vacuolated vegetative filaments, it may turn out that the *gvp* genes and other genes whose products are involved in the differentiation of hormogonia are under the control of a common factor. Perhaps the most important question to answer is whether there are proteins other than GVP involved in gas vesicle formation. Investigations with gas vesicle defective mutants (e.g. Walsby, 1977) may resolve these problems.

2.2. Molecular structure of the gas vesicle

The X-ray crystallographic studies of Blaurock and Walsby (1976) showed that there is a 1.15 nm repeating subunit along the rib of the gas vesicle, the ribs are 4.57 nm wide and the walls stack with a periodicity which indicates that their overall thickness is 1.95 nm. The product of these three values ($1.15 \times 4.57 \times 1.95$ nm^3) defines the volume of the repeating unit cell, 10.25 nm^3. Multiplying by the density of the protein, 1.32 g·cm^{-3}, and the Avogadro number gives the molecular weight of the subunit, 8 150. Using the average thickness of the gas vesicle wall, 1.8 nm, instead of the stacking periodicity, which includes some unfilled space (Hayes et al., 1986), gives a slightly smaller MW estimate, 7 520. This value agrees closely with the molecular weight of GVP (7 388), which must, therefore, correspond with the repeating unit cell.

The crystallographic studies also indicate that the protein structure is at least partly made up of β-sheet within the thickness of the wall and in each layer the β-chains run at an angle of 55° to the long axis of the rib (i.e. 35° to the long axis of the cylindrical gas vesicle). This angle results in the amino acid residues of adjacent chains being in horizontal register (Walsby, 1978a). The precise angle generated is arctan $2\,a/2\,h = 36.16°$ (Atkins and Walsby, in preparation) where a is the perpendicular distance between adjacent β-chains (0.472 nm) and $2\,h$ is the dipeptide repeat along the chain (0.69 nm). This angle is, therefore, generated by the particular way the polypeptide β-chain is folded on itself. Hydrogen bonds between residues in adjacent chains would lie at right angles to the chains themselves, i.e. at 35° to the axis of the rib or 55° to the axis of the gas vesicle cylinder. As is discussed below, this is of importance in the mechanical strength of the structure.

Exactly how the GVP polypeptide chain is folded up to generate this structure has yet to be resolved. One possibility is of runs of β-chains folded back to form antiparallel β-sheet; alternating runs of 8 and 7 dipeptide residues would cross the rib at the 55° angle (Atkins and Walsby, in preparation). It is not known if the β-sheet in the two layers of the wall run in the same direction or opposite directions, but the latter arrangement would give an overlapping structure that might explain why the gas vesicle wall is so difficult to break down into monomers: sodium dodecylsulphate (SDS) polyacrylamide gel electrophoresis shows that the GVP tends to remain aggregated as dimers, trimers or even pentamers in detergent suspensions of gas vesicles from different species (Hayes et al., 1986).

Eventually it is hoped to reconcile the amino acid sequence data with the crystallographic structure of the gas vesicle protein. As Hayes et al. (1986) have pointed out there are features of the sequence that could generate a polypeptide chain folded in the manner proposed.

3. The natural selection of gas vesicles

Two factors, economy and strength, are known to conflict in the natural selection of gas vesicle size (Walsby, 1972b).

3.1. Economy

In order to float, a cell must accumulate sufficient gas vesicle gas space to negate the ballast provided by protein, carbohydrate and other components. The wider is the gas vesicle the more gas space is provided per unit of GVP in the enclosing wall. The cost of providing the space is, therefore, equivalent to the cost of synthesizing the required amount of GVP; as indicated by the following example, buoyancy is not a very cheap commodity.

Oliver and Walsby (1984, Table I) found that 1 μl of buoyant *Anabaena flos-aquae* cells cultured at low light intensity contained 168.4 μg of protein and 57.1 μg of carbohydrate, which together accounted for 80 % of the dry mass (283 μg·μl^{-1} of cell volume); from the density of these components it was calculated that their ballast (mass minus the mass of water each displaced) was 39.1 μg for protein and 21.5 μg for the denser carbohydrate, in total 60.6 μg. (Much of the remaining 20% of the dry matter was glycolipid which, with a density (1.05 g cm^{-3}) close to that of water, would have contributed little more to the ballast). To counteract this ballast requires the displacement of 60.6 μg of water occupying 60.7 nl of space by gas. Walsby and Armstrong (1979) showed by direct measurements on pure gas vesicles from *Anabaena flos-aquae* that 1 μg of GVP encloses 7.67 nl of gas space and hence 60.7 nl of gas requires 7.9 μg of GVP. This, then, is the amount of GVP that must be synthesised to provide the cell with neutral buoyancy: it represents 4.7% of the cell protein and 2.8% of the total dry mass of the cell. The actual amount of gas vesicle space present in the positively buoyant cells grown at low light intensity was 99.6 nl, requiring 12.95 μg of GVP to enclose it or 7.7% of the cell protein, a substantial investment to make the cells float.

Measurements by Hayes and Walsby (1986) have shown that the gas vesicles of other cyanobacteria are of different widths. It is probable that in each case the wall thickness is similar since they are all made of homologous proteins and X-ray diffraction studies indicate an identical stacking periodicity of collapsed walls from gas vesicles of *Anabaena* and *Halobacterium* (cf. Blaurock and Walsby, 1976; Blaurock and Wober, 1976), which in other respects differ markedly. The probable ratio of gas volume to wall volume in the cylindrical part of the gas vesicle therefore varies with the radius (see Table I). The overall ratio, which can be calculated from the shape equations of Walsby and Armstrong (1979), requires knowledge of the length distribution of the gas vesicles and the height (or end angle) of the conical end caps.

It is clear that the gas vesicles of the halophilic *Dactylococcopsis* are much more efficient in providing buoyancy than those of *Anabaena*, while those of *Trichodesmium* and *Oscillatoria* are much less so. If natural selection works to produce the optimum why do we find such a profligate use of GVP in these last organisms? The answer lies in the pressures the gas vesicles must withstand.

3.2. Strength

In order to maintain the gas-filled space that actually provides buoyancy, a gas vesicle must prevent the space from filling with water and from disappearing by collapse under pressure. Walsby (1971) has calculated that pressures of up to 40 bar would be required to overcome surface tension and force water inside a gas vesicle if the inner gas-facing surface of the wall were hydrophobic. The location of amino acid residues with hydrophobic side chains at the inner surface would provide the required waterproofing. Indirect evidence for such hydrophobicity from neutron diffraction studies (Worcester, 1975) and electron diffraction studies (Blaurock and Wober, 1976) is reviewed by Walsby (1978a).

The collapse of the gas vesicle by pressure is prevented by the rigidity of the protein wall itself, rather than the pressure of gas inside it. For example, if the gas vesicles of *Anabaena flos-aquae* are suspended in water equilibrated with air, the average hydrostatic pressure required to collapse them is 6 bars above atmospheric pressure (Walsby, 1980a). The gas vesicles are highly permeable to gases, which diffuse in from the surrounding water (Walsby, 1969, 1984) and they generate an internal pressure equal only to the atmospheric pressure (Walsby, 1971): the pressure difference that can be borne by the gas vesicle wall itself before it collapses therefore averages 6 bar in *Anabaena*. At subcritical pressures, the gas vesicle maintains its form indefinitely and with only a slight, elastic (reversible) decrease in volume – it shrinks in volume by only 1 part in 650 per bar. This was demonstrated by placing a concentrated suspension of gas vesicles in a glass tube connected to a fine capillary and measuring the contraction on the suspension in the capillary when the suspension was exposed to pressure (Walsby, 1982). The measurements are so delicate that due allowance has to be made for the elastic compressibility of the water itself (1 part in 22000 bar^{-1}) and the walls of the glass vessel.

With macroscopic structures of homogeneous construction the compressibility of a hollow cylinder with closed ends is related to the elastic (or Young's) modulus (Y) of the material from which it is constructed by the equation

$$dV/V = Pr\,(4\sigma - 5)/2\,Yt$$

(see Newman and Searle, 1957). For the gas vesicle, where a pressure P of 1 bar produces a relative volume change dV/V of 1/650, the cylinder radius r is 84 nm, its wall thickness t is 1.8 nm, and the Poisson ratio, σ, (the ratio of lateral to longitudinal strain when the material is stretched) is taken to be 0.33, then Y is calculated to be 28000 bar.

If we were to make a smooth-walled macroscopic cylinder of the same shape and relative dimensions as the *Anabaena* gas vesicle, from homogeneous material with the same Young's modulus, we should expect that, if it collapsed by instability failure in the way that a gas vesicle does, its critical collapse pressure would be

$$p_c = \frac{Y}{4(1-\sigma^2)}\frac{t^3}{r^3}$$

Substituting the values given above, this equation yields a p_c of 0.6 bar. This is 10 times less than the actual average critical pressure. The reason for the structure being so much stronger must lie in the fact that the gas vesicle wall material is not homogeneous, with bonds of equal strength in all directions, but is laid down in ribs oriented in such a way that they would postpone buckling of the cylinder.

Another feature that may be of importance in determining the strength of the structure is the orientation at 55° to the cylinder axis of hydrogen bonds that tie together the adjacent β-chains of the GVP polypeptide. R.G.C. Arridge has pointed out to me that this is very close to the angle used by engineers when reinforcing pressure bottles or hose pipes with fibres. For geometrical reasons the ratio of wall area to total cross-sectional area of a cylinder when cut at right angles to its long axis ($2\,\pi rt/\pi r^2 = 2\,t/r$) is twice that when cut parallel to its long axis ($2\,th/2\,rh = t/r$). It, therefore, follows that when the cylinder is subject to pressure the longitudinal stress in the wall ($pr/2\,t$) is half the transverse stress (pr/t). The stresses from these two directions are equalised in fibres oriented at arctan $\sqrt{2} = 54.7°$. The similarity between this angle and that of the hydrogen bonds can hardly be fortuitous. In engineering structures a second set of fibres is wound in the opposite sense (i.e., $-54.7°$) to prevent it from untwisting. It is not possible to determine from the X-ray crystallographic data from intact gas vesicles whether the two layers of β-sheet within the thickness of the wall lie in the same or the opposite sense (see Blaurock and Walsby, 1976) but these mechanical considerations would support the second possibility.

Thus, in two ways – by the oblique orientation of the polypeptide chains in the GVP molecules, and by the transverse orientation of ribs in the cylindrical structure – a stiffened structure results which is able to postpone instability failure and resist collapse. Nevertheless, the critical pressure is ultimately determined by the stress in the wall in relation to the yield strength of the wall material. As we have seen the stress in the wall is dependent on the ratio of r/t and will therefore be greater in a wide cylinder than in a narrow one.

The mean critical pressures of gas vesicles of different cyanobacteria are also given in Table I. It demonstrates the expected inverse correlation between width and strength. Now we have to ask why gas vesicles of different strength should be needed in different cyanobacteria.

4. Critical pressure requirements of gas vesicles in different microorganisms

The pressure that a gas vesicle must withstand is effectively the sum of the hydrostatic pressure from the overlying water column and the turgor pressure of the cell. The hydrostatic pressure (Pa) is generated by the product of depth (m), water density ($kg{\cdot}m^{-3}$) and the gravitational acceleration ($9.8\ m{\cdot}s^{-2}$): it increases by about 10^5 Pa (1 bar) per 10 m depth. The turgor pressure of a cell is generated by the difference in osmotic pressure of the fluids inside it and the medium outside it. Turgor pressure can be measured as the difference in hydrostatic pressure required to collapse gas vesicles in cells suspended in hypertonic sucrose (which removes turgor) and in the growth medium (see Walsby, 1971, 1980); in cyanobacteria of freshwater turgor pressure is usually within the range of 2 to 5 bar.

In broad terms gas-vacuolate microorganisms have gas vesicles that are just strong enough to withstand the sum of the hydrostatic and turgor pressures they are likely to experience. This idea can be illustrated by reference to the following examples.

4.1. Halobacteria

The extreme halophiles live in sabkras and brine pools with salinities from 3 M to 6 M NaCl (Friedman and Krumbein, 1985). The pools are typically one metre or less in depth so that the water column will generate a hydrostatic pressure usually not exceeding 0.1 bar. The cells of these organisms accumulate high concentrations of K^+ and Cl^- internally to balance the high external concentrations of NaCl (see Larsen, 1967). They do not have measurable cell turgor pressure; the gas vesicles, which collapse at pressure of only 0.9 bar, the weakest in any microbial group, are therefore quite adequate for these organisms. The square halophilic bacteria isolated from a sabkra in southern Sinai have gas vesicles which are similarly very weak (Walsby, 1980b). The gas vesicles of the extreme halophiles are by far the widest known and are thus the most efficient in providing buoyancy.

4.2. Halophilic cyanobacteria

The presence of gas vesicles in this group was first demonstrated in Solar Lake, Sinai (Walsby et al., 1983a). The pool is up to 5 m deep during the winter; a 1-m layer of 8 % salinity brine floats over a denser, 15 to 19 % salinity hyperlimnion that becomes heated to 60°C or more by solar insolation. *Dactylococcopsis salina* was the principal gas-vacuolate cyanobacterium present in 1979–1981. It regulated its buoyancy according to light intensity, maintaining station at a depth of about 3.5 m during midwinter (but was eliminated from these depths in spring when the water there reached lethal temperatures). The critical pressures of the gas vesicles varied from about 2 to 7 bar with a median value at about 5 bar. This was sufficient to withstand the sum of hydrostatic pressure, 0.6 bar at the lake bottom, turgor pressure, about 0.5 bar, and a possible difference between gas pressure in the overlying atmosphere (1 bar) and that in the gas vesicles (0.2 bar if nitrogen gas is depleted by nitrogen fixing activity in the lake hypolimnion). The gas vesicles of this organism are the weakest encountered in cyanobacteria (see Table I) but are also the widest and hence most efficient of those found in this group. *Aphanothece halophytica* isolated from the same lake has gas vesicles of similar strength but their dimensions have not yet been investigated.

4.3. Cyanobacteria of freshwater lakes

The commonly encountered gas-vacuolate cyanobacteria of temperate freshwater lakes include the ubiquitous species of *Anabaena*, *Aphanizomenon*, *Microcystis* and *Oscillatoria*, as well as *Nodularia*, *Coelosphaerium*, and *Gloeotrichia*. Tropical lakes contain these and other genera such as *Spirulina*, which is also found in soda lakes of high pH (see Walsby, 1981a, for a more complete list).

The gas vesicles of *Anabaena flos-aquae* have a critical pressure range from about 4.5 to 7.5 with a median at 6.0 bars (Walsby, 1971, 1980a). This organism is frequently found in the epilimnion of thermally stratified lakes, where it will experience a hydrostatic pressure of 0.5 to 1 bar in depths of 5 to 10 m. Added to this are cell turgor pressures which may vary from 2.5 to 5 bar (Walsby, 1971; Dinsdale and Walsby, 1972; Grant and Walsby, 1977). This implies that the gas vesicles of this organism are not strong enough to withstand the pressure it is likely to encounter. This apparent defect is understood when it is seen that this organism is actually able to regulate its buoyancy by modulating its turgor pressure to control its gas vesicle content (see below). The same mechanism is found in species of *Gloeotrichia* (Allison, unpublished), *Aphanizomenon* (Konopka et al., 1978) and *Nostoc* (Armstrong et al., 1983).

The gas vesicles of *Microcystis* have a critical pressure distribution that is about 1.5 bar higher than those of *Anabaena*, and this seems to give the required protection from collapse. The critical pressure distribution is narrower; although 50% of the gas vesicles collapse at 7.5 bar, and the strongest survive over 10 bars, hardly any collapse below 7.0 bars. Oliver et al. (1985) calculated that no significant loss of gas vesicles (or buoyancy) would be caused by the combination of the maximum hydrostatic pressure (1.2 bar) and turgor pressure (nearly 5 bar) encountered by *Microcystis* cells in Blelham Tarn (English Lake District). The gas vesicles here, being that much stronger than those of the previously considered groups of cyanobacteria, appear to be endowed with just sufficient strength to survive. This is, again, bought at the expense of efficiency in buoyancy, for the gas vesicles of *Microcystis* are 15% narrower than those of *Anabaena* (see Table I).

4.4. Cyanobacteria of deeper lakes

The gas vesicles of a red-pigmented strain of *Oscillatoria agardhii* in Lake Gjersjøen, Southern Norway, have a critical pressure distribution of between 8 and 12 bar with a median at 9 bar. The cells have a turgor pressure of up to 5 bar (Walsby et al., 1983b). This organism, therefore, has the capacity to go to a depth of 30 m (where the hydrostatic pressure is about 3 bar) without losing any of its gas vesicles or the buoyancy they provide. In the summer this *Oscillatoria* stratifies at a depth of 6 m, though the population extends down to 10 m. At first sight this would suggest that the gas vesicles are stronger than they need to be. However, in the autumn the lake overturns and much of the water is circulated to a depth of 30 m or more. The *Oscillatoria* filaments, with their strong gas vesicles would be able to float back towards the lake surface when an ice cover forms for the winter (Walsby et al., 1983b). Local aggregations do form under the ice where they can be seen as red discolorations (T. Källqvist, personal communication). The buoyancy of the organisms would again survive circulation to depth during spring holomixis, so that the filaments would be able to float up to the epilimnion during the following summer stratification. Utkilen et al. (1985b) have made a small survey of *Oscillatoria* strains isolated from various Norwegian lakes. It indicates that strong gas vesicles, with median critical pressures up to 10.5 bar, are associated with red-coloured strains. Green-coloured strains tend to have weaker gas vesicles and it is suggested that they may inhabit shallower lakes. Examples of these green-coloured strains include that investigated by Walsby and Klemer (1974) in Lake Deming, Minnesota, of mean depth about 14 m, and that investigated by Konopka (1982) in Crooked Lake, Indiana, in which the median critical pressures are 7.5 bar.

4.5. Cyanobacteria of the tropical ocean

At least three distinct species of *Trichodesmium* (*Oscillatoria*) are found in the tropical ocean, where under calm conditions they may form extensive surface blooms (see Fogg, 1982 and Section 3 of Chapter 15). Carpenter and Price (1976, 1977) showed that the populations may form substantial subsurface maxima at depths of 15 or 25 m and that some *Trichodesmium* colonies could be recovered from depths exceeding 175 m. The critical pressure distribution of gas vesicles in *T. erythraeum* was found to range between 7 and 20 bars with a median at 12 bar; in *T. contortum* between 20 and 45 bars with a median at 34 bar, and in *T. thiebautii* over a similar range with a median at 37 bar (Walsby, 1978b). These very high critical pressures must enable the colonies to remain gas-vacuolate and buoyant after being mixed down to great depths by storms, and then to float back to the water surface. These organisms formed large colonies which, when highly gas-vacuolate, were found to float up with a velocity of between 0.5 and 3 $mm \cdot s^{-1}$ (40 to 250 $m \cdot day^{-1}$). Gantt et al. (1984) found that the vesicles of *T. thiebautii* are only 40 nm wide.

In summary, it is seen that in each group of cyanobacteria gas vesicles are produced that are just strong enough to withstand the pressures they are likely to encounter in their natural environment. The requirement for extra strength carries with it the penalty of decreased width, and less buoyancy provided by a given amount of GVP. It is, therefore, proposed that in each of these organisms buoyancy is provided with the greatest economy allowed by the constraint of pressure.

5. Mechanisms of buoyancy regulation

A cell may alter its buoyancy by changing the respective rates at which intact gas vesicles and other cell materials are produced or destroyed. If these changes are correlated with physicochemical conditions that form vertical gradients in natural waters they may provide cells with the means of regulating their depth in the water column (Walsby, 1972b).

Experiments under controlled growth conditions have shown that various gas-vacuolate cyanobacteria float when incubated at low light intensity and sink when shifted to high intensity. This response has been demonstrated in *Anabaena flos-aquae* (Walsby, 1969, 1971; Dinsdale and Walsby, 1972), *Microcystis aeruginosa* (Dinsdale and Walsby, 1972; Kromkamp and Mur, 1984; Thomas and Walsby, 1985), *Aphanizomenon flos-aquae* (Konopka et al., 1978), *Dactylococcopsis salina* (Walsby et al., 1983a), and *Oscillatoria agardhii* of both green (Walsby and Klemer, 1974; Konopka, 1982; Klemer, 1976; Van Rijn and Shilo, 1983) and red coloured (Walsby et al., 1983b) varieties. In some instances the reverse has been demonstrated, regain of buoyancy when the cyanobacteria are shifted back to low light intensity (Dinsdale and Walsby, 1972) or darkness (Kromkamp and Mur, 1984). Such responses to light intensity would enable these organisms either to stratify on a vertically decreasing light intensity gradient or to migrate up and down during night and day (see Reynolds and Walsby, 1975).

Other factors have been shown to affect buoyancy, including availability of inorganic carbon (Booker and Walsby, 1981), nitrogen (Klemer, 1978; Klemer et al., 1982; Spencer and King, 1985), phosphate (Booker and Walsby, 1981) and temperature (Thomas and Walsby, 1986). In each case these factors appear to act by affecting the buoyancy responses to light intensity (Konopka, 1984).

It has now been demonstrated that there are at least three different mechanisms that can bring about a buoyancy decrease at high light intensity. These can be summarised as (1) destruction of gas vesicles, (2) regulation of gas vesicle formation, and (3) increase of cell ballast. A given organism may use more than one of these mechanisms at once. In order to determine the relative importance of each mechanism it is necessary to measure the amount of each of the major components that determine a cell's buoyant density (Oliver and Walsby, 1984). As outlined above (Section 3.1.) from measurements of mass and density, the amount of ballast weight that each component contributes can be calculated. The sum of these ballast contributions (expressed as μg per μl of cell volume) should be equal to the overall excess buoyant density of the cell (cell density less the density of the suspending water) which can be determined independently by centrifuging the cells into an isopycnic layer on a Percoll density gradient (Oliver et al., 1981). The negative ballast or buoyant lift provided by gas vesicles can be determined from direct measurements of gas volume in the capillary compression tube (Walsby, 1982); if this is set against the positive ballast from the other cell components, it will determine whether the cell sinks or floats. By this quantitative approach the mechanisms of buoyancy regulation have been elucidated in the following cyanobacteria.

5.1. Destruction of gas vesicles by turgor pressure in *Anabaena*

It was first demonstrated in *Anabaena flos-aquae* that the cell turgor pressure increased when the light intensity was raised and that the increase might cause collapse of the weaker gas vesicles originally present (Walsby, 1971). Dinsdale and Walsby (1972) found that turgor pressure could increase by nearly 1 bar within an hour of the light intensity rise and they showed that a loss of buoyancy accompanied the resulting loss of gas vesicles. Both the turgor rise and buoyancy loss were dependent on photosynthesis under the increased illumination, since depriving cultures of CO_2 or adding the photosynthetic inhibitor DCMU eliminated the response. Grant and Walsby (1977) showed that the osmotic pressure generated by low molecular weight products of photosynthesis dissolved in the cytoplasmic fluids accounted for up to half of the turgor rise, and Allison and Walsby (1981) found that the other half of the rise could be generated by light-dependent uptake of K^+ ions. Oliver and Walsby (1984) measured the changes in carbohydrate, protein and gas vesicle space that accompanied light-stimulated loss of buoyancy, and found that the decrease in gas vesicle space from 99 to 33 nl per μl of cell volume could account completely for the buoyancy change. There was an increase in ballast provided by carbohydrate (from 22 to 35 μg per μl of cell volume) but it was not by itself sufficient to cause buoyancy loss, and was anyway partly offset by a decrease in ballast due to protein.

Hayes and Walsby (1984) have investigated the fate of gas vesicles collapsed by pressure. The results of a ^{14}C-labelling experiment suggested that 60 % of the protein in the collapsed membranes found its way into the new gas vesicles that were subsequently assembled. The rate of formation of new gas vesicles increased by nearly 40 % after collapsing pre-existing ones.

Dinsdale and Walsby (1972) pointed out that the turgor pressure rise that collapses gas vesicles could result from the amount of carbon fixed in photosynthesis exceeding that which can be assimilated by cells whose net growth rate is limited by other factors, such as temperature or the availability of particular nutrients. With this response the cyanobacterium might sink to a depth where light intensity would support just sufficient photosynthesis to sustain maximal growth.

As commented above (p. 385) the ability of *Anabaena* to regulate its buoyancy by gas vesicle destruction depends on its producing gas vesicles whose range of critical pressure overlaps the range of turgor pressure that the cells can generate. The same mechanism of regulation has been demonstrated in natural populations of *Aphanizomenon* (Konopka et al., 1978), and cultures of *Gloeotrichia* (Allison, unpublished) and *Nostoc muscorum* (Armstrong et al., 1983).

5.2. Regulation of gas vesicle formation in *Oscillatoria*

The red-coloured *Oscillatoria agardhii*, which stratified in Lake Gjersjøen, Norway, was found to become buoyant when cultured at a photon flux density of less than 5 μmol $m^{-2} \cdot s^{-1}$, and to lose buoyancy when transferred to more intense illumination. It was clear, however, that buoyancy loss could not be brought about by turgor collapse of gas vesicles as the gas vesicles were far too strong (Walsby et al., 1983b). Quantitative analysis of the major cell constituents showed that while those providing positive ballast increased after the increase in light intensity, gas vesicle formation ceased. Thus, while there was no loss of gas vesicles from the cells, the proportion of cell volume occupied by gas vesicle space decreased (from 106 to 83 nl per μl of cell volume) owing to dilution out by growth. It was calculated that this dilution effect was sufficient to account for buoyancy loss though there was an additional contribution to cell density from accumulation of carbohydrate (see Section 5.3. below). These results suggest that gas vesicle formation in *Oscillatoria* can be modulated in response to light intensity, perhaps by control of expression of the GVP gene.

5.3. Increase of cell ballast in *Microcystis*

The gas vesicles of certain *Microcystis* strains that have been investigated, while not as strong as those of the red *Oscillatoria*, are still too strong to be regulated by turgor pressure (Oliver et al., 1985; Thomas and Walsby, 1985). With one strain, Kromkamp and Mur (1984) showed that the buoyant density of cells increased in parallel with the gain in cell carbohydrate content in the light, and decreased in parallel with the loss of carbohydrate in the dark; they suggested that such changes might be implicated in buoyancy regulation. Quantitative analyses on another strain of *Microcystis* by Thomas and Walsby (1985) demonstrated that such changes in carbohydrate caused the cells to become negatively or positively buoyant. In this strain there was no evidence that the organism modulated its gas vesicle formation (see Section 5.2.); it appeared that gas vesicles provided background buoyancy against which changes in ballast, provided mainly by carbohydrate and protein, would determine whether the cell floated or sank. Carbohydrate has a proportionately greater effect because its buoyant density ($\sim$ 1.6 $g \cdot cm^{-3}$) is greater than that of protein ($\sim$ 1.3 $g \cdot cm^{-3}$). Carbohydrate accumulates in the cell when the amount of carbon fixed by photosynthesis exceeds the demands of growth. In low light intensity or darkness the carbohydrate is depleted by respiratory activity or by conversion into protein (Konopka and Schnur, 1980; Foy, 1983; Post et al., 1985). Such changes in the cell's store of carbohydrate probably occur in all types of cyanobacteria (and in other photosynthetic organisms), but they have been put to work in the regulation of buoyancy by certain of the planktonic strains in such a way that they will sink down when they have had too much light, and will float up when they have had too little of it.

6. Conclusions

It is fascinating that three mechanisms should have evolved to regulate the buoyancy of planktonic cyanobacteria and produce broadly the same result.

The mechanism of gas vesicle collapse by turgor pressure provides the most rapid response. Substantial turgor rise may occur within a few minutes, causing loss of gas vesicles and buoyancy (Grant and Walsby, 1977). Recovery of buoyancy takes longer as it requires formation of new gas vesicles when the turgor has fallen in low light intensity. The fact that collapse of gas vesicles stimulates the formation of new ones (Hayes and Walsby, 1984) suggests homeostasis in this mechanism of regulation. Such a method of regulation appears to make use of a rise in turgor pressure that may be a general consequence of increased light intensity in photosynthetic tissue (Utkilen et al., 1985a), but the method can only operate where the range of cell turgor pressure overlaps the range of gas vesicle critical pressure. It is, therefore, not possible in cyanobacteria of deep lakes or oceans where strong gas vesicles are required to recover cells from deep water after mixing, and it is not possible in halophilic cyanobacteria where only low turgor pressures are generated.

The mechanism of regulation by carbohydrate accumulation likewise makes use of a universal response of photosynthetic tissue in high light intensity. This mechanism should be universally applicable in all planktonic cyanobacteria and all lake types. Carbohydrate accumulation was found in each of the cyanobacteria investigated, though its quantitative importance varied.

What both these mechanisms provide is the automatic and simple solution to an inherently complex problem, the positioning of an organism at the appropriate light intensity on a vertically decreasing light gradient. The 'appropriate' intensity does not have a fixed absolute value; it depends on the amount of light needed to support photoautotrophic growth. For a cell well supplied with all the necessary mineral nutrients the appropriate light intensity will be higher than for a cell starved of one or more nutrients. In the case of the mechanism described in Section 5.1., it is proposed that the rise in turgor, which collapses gas vesicles and destroys buoyancy, results from excess photosynthetic activity. In the case of the mechanism described in Section 5.3., the accumulation of carbohydrate, which raises cell density and destroys buoyancy, similarly results from photosynthesis providing fixed carbon in excess of the demands for growth.

The third mechanism, regulation of gas vesicle formation, probably by the regulation of GVP synthesis, might appear to be less cumbersome and more sophisticated than the other two (and it is more appealing to molecular biologists). Such a mechanism requires, however, a complex control system responding not just to light intensity but also to other factors that determine growth.

6.1. Buoyancy regulation in natural habitats

It is not my intention here to discuss in detail the ecological consequences of buoyancy regulation but, since buoyancy has relevance only in the natural habitats of planktonic cyanobacteria, its regulation should be put into context.

The prediction that cyanobacteria depleted of nutrients should lose their buoyancy at a lower light intensity than those replete with nutrients (Dinsdale and Walsby, 1972) has been supported by several observations. For example, Klemer (1976, 1978) found that enrichment of natural populations of *Oscillatoria* with combined nitrogen caused them to float up in Deming Lake, Minnesota. Booker and Walsby (1981) found that phosphate concentration similarly influenced the buoyancy of *Anabaena* in cultures, and the depth at which the same organism stratified in a 2-m deep experimental water column. Spencer and King (1985) have recently shown that enrichment of *Anabaena* cultures with combined nitrogen also increases buoyancy.

The depletion of inorganic carbon has the opposite effect on buoyancy, of course, because without a ready supply of it neither carbohydrate accumulation nor the turgor rise can take place (Dinsdale and Walsby, 1972). Under nutrient-rich conditions in the experimental water column Booker and Walsby (1981) demonstrated that the *Anabaena* population quickly exhausted dissolved inorganic carbon and then floated to the water surface forming a 'waterbloom'. Diffusion of CO_2 from the overlying atmosphere failed to keep up with the photosynthetic demands of the cyanobacterium and it continued to float at the water surface, even though the light intensity was 10-fold higher than that needed to cause buoyancy loss when carbon was readily available (Walsby and Booker, 1980). Depletion of inorganic carbon is probably one of the major contributory factors in the formation of summer waterblooms in natural waters (see Reynolds and Walsby (1975) for a more detailed discussion).

Cyanobacterial waterblooms may persist through until late autumn but then sediment out to the lake bottom where they may overwinter (Preston et al., 1980; Reynolds et al., 1981). We have recently found that a population of *Microcystis*, in Abbots Pool near Bristol, regulates its buoyancy in the usual way during early autumn, sinking from the surface by day and regaining buoyancy to float back at night. As the water temperature drops, however, it takes progressively longer for colonies to dispose of their excess carbohydrate and the population gradually becomes less buoyant (Thomas and Walsby, 1986). Below 11°C buoyancy regain effectively ceases and the colonies are lost to the bottom mud. Oliver et al. (1985) found another and unexpected cause of *Microcystis* sedimentation in Blelham Tarn, English Lake District. The still-buoyant colonies became entrapped in a dense iron-containing colloid that precipitated in the lake water when ferrous salts in the anaerobic hypolimnion became oxidised on mixing with aerobic water of the epilimnion at the autumnal overturn.

References

Allison, E.M. and Walsby, A.E. (1981) The role of potassium in the control of turgor pressure in a gas-vacuolate blue-green alga. J. Exp. Bot. 32, 241–249.

Armstrong, R.E., Hayes, P.K. and Walsby, A.E. (1983) Gas vacuole formation in hormogonia of *Nostoc muscorum*. J. Gen. Microbiol. 128, 263–270.

Blaurock, A.E. and Walsby, A.E. (1976) Crystalline structure of the gas vesicle wall from *Anabaena flos-aquae*. J. Mol. Biol. 105, 183–199.

Blaurock, A.E. and Wober, W. (1976) Structure of the wall of *Halobacterium halobium* gas vesicles. J. Mol. Biol. 106, 871–888.

Booker, M.J. and Walsby, A.E. (1981) Bloom formation and stratification by a planktonic blue-green alga in an experimental water column. Br. Phycol. J. 14, 141–150.

Bowen, C.C. and Jensen, T.E. (1965) Blue-green alga: fine structure of the gas vacuoles. Science NY 147, 1460–1462.

Carpenter, E.J. and Price, C.C. (1976) Marine *Oscillatoria* (*Trichodesmium*): explanation for aerobic nitrogen fixation without heterocysts. Science (NY) 191, 1278–1280.

Carpenter, E.J. and Price, C.C. (1977) Nitrogen fixation, distribution, and production of *Oscillatoria* (*Trichodesmium*) spp. in the western Sargasso and Caribbean Seas. Limnol. Oceanogr. 22, 60–72.

Dinsdale, M.T. and Walsby, A.E. (1972) The interrelations of cell turgor pressure, gas-vacuolation, and buoyancy in a blue-green alga. J. Exp. Bot. 23, 561–570.

Eloff, J.N., Steinitz, Y. and Shilo, M. (1976) Photooxidation of cyanobacteria in natural conditions. Appl. Environ. Microbiol. 31, 119–126.

Falkenberg, P., Buckland, B.A. and Walsby, A.E. (1972) Chemical composition of gas vesicles isolated from *Anabaena flos-aquae*. Arch. Mikrobiol. 85, 304–309.

Fogg, G.E. (1982) Marine plankton. In N.G. Carr and B.A. Whitton (eds.), The Biology of Cyanobacteria. Blackwell, Oxford, pp. 491–519.

Foy, R.H. (1983) Interaction of temperature and light on the growth rates of two planktonic *Oscillatoria* species under a short photoperiod regime. Br. Phycol. J. 18, 267–273.

Friedman, G.M. and Krumbein, W.E. (1985) Hypersaline Ecosystems. Ecological studies, Vol. 53. Springer Verlag, Berlin.

Gantt, E., Ohki, K. and Fujita, Y. (1984) *Trichodesmium thiebautii*; structure of a nitrogen-fixing blue-gree alga (*Cyanophyta*). Protoplasma 119, 188–196.

Grant, N.G. and Walsby, A.E. (1977) The contribution of photosynthate to turgor pressure rise in the planktonic blue-green alga *Anabaena flos-aquae*. J. Exp. Bot. 28, 409–415.

Hayes, P.K. and Walsby, A.E. (1984) An investigation into the recycling of gas vesicle protein derived from collapsed gas vesicles. J. Gen. Microbiol. 130, 1591–1596.

Hayes, P.K. and Walsby, A.E. (1986) The inverse correlation between width and strength of gas vesicles in cyanobacteria. Br. Phycol. J. 21, 191–197.

Hayes, P.K., Walsby, A.E. and Walker, J.E. (1986) Complete amino acid sequence of cyanobacterial gas vesicle protein indicates a 70 residue molecule that corresponds in size to the crystallographic unit cell. Biochem. J., 236, 31–36.

Hutchinson, G.E. (1967) A Treatise on Limnology, Vol. II, Introduction to Lake Biology and the Limnoplankton. Academic Press, London, pp. 1115.

Jones, D.D. and Jost, M. (1970) Isolation and chemical characterization of gas vacuole membranes from *Microcystis aeruginosa* Kuetz. emend. Elenkin. Arch. Mikrobiol. 70, 43–64.

Jones, D.D. and Jost, M. (1971) Characterization of the protein from gas vacuole membranes of the blue-green alga *Microcystis aeruginosa*. Planta 100, 277–287.

Jost, M. and Jones, D.D. (1970) Morphological parameters and macromolecular organisation of gas vacuole membranes of *Microcystis aeruginosa* Kuetz. emend. Elekin. Can. J. Microbiol. 16, 159–164.

Klemer, A.R. (1976) The vertical distribution of *Oscillatoria agardhii* var. *isothrix*. Arch. Hydrobiol. 78, 343–362.

Klemer, A.R. (1978) Nitrogen limitations of growth and gas vacuolation in *Oscillatoria rubescens*. Verh. Int. Verein. Limnol. 20, 2293–2297.

Klemer, A.R., Feuillade, J. and Feuillade, M. (1982) Cyanobacterial blooms: carbon and nitrogen limitations have opposite effects on the buoyancy of *Oscillatoria*. Science (NY) 215, 1629–1631.

Konopka, A.E. (1982) Buoyancy regulation and vertical migration by *Oscillatoria rubescens* in Crooked Lake, Indiana. Br. Phycol. J. 17, 427–442.

Konopka, A.E. (1984) Effect of light-nutrient interactions on buoyancy regulation by planktonic cyanobacteria. In M.J. Klug and C.A. Reddy (eds.), Current Perspectives in Microbial Ecology. American Society for Microbiology, pp. 41–48.

Konopka, A. and Schnur, M. (1980) Effect of light intensity on macromolecular synthesis in cyanobacteria. Microb. Ecol. 6, 291–301.

Konopka, A., Brock, T.D. and Walsby, A.E. (1978) Buoyancy regulation by planktonic blue-green algae in Lake Mendota, Wisconsin. Arch. Hydrobiol. 83, 524–537.

Kromkamp, J. and Mur, L. (1984) Buoyant density changes in the cyanobacterium *Microcystis aeruginosa* due to changes in the cellular carbohydrate content. FEMS Microbiol. Lett. 25, 105–109.

Larsen, H. (1967) Biological aspects of extreme halophilism. Adv. Microbial Physiol. 1, 97–132.

Newman, F.H. and Searle, V.H.L. (1957) The General Properties of Matter. Edward Arnold, London.

Oliver, R.L. and Walsby, A.E. (1984) Direct evidence for the role of light-mediated gas vesicle collapse in the buoyancy regulation of *Anabaena flos-aquae* (Cyanobacteria). Limnol. Oceanogr. 29, 879–886.

Oliver, R.L., Kinnear, A.J. and Ganf, G.G. (1981) Measurements of cell density of three freshwater phytoplankters by density gradient centrifugation. Limnol. Oceanogr. 26, 285–294.

Oliver, R.L., Thomas, R.H., Reynolds, C.S. and Walsby, A.E. (1985) The sedimentation of buoyant *Microcystis* colonies caused by precipitation with an iron-containing colloid. Proc. Roy. Soc. Lond. B. 223, 511–528.

Post, A.F., Loogman, J.G. and Mur, L.R. (1985) Regulation of growth and photosynthesis by *Oscillatoria agardhii* grown with a light/dark cycle. FEMS Microbiol. Ecol. 31, 97–102.

Preston, T., Stewart, W.D.P. and Reynolds, C.S. (1980) Bloom-forming cyanobacterium *Microcystis aeruginosa* overwinters on sediment surface. Nature (Lond.) 288, 365–367.

Reynolds, C.S. (1971) The ecology of the planktonic blue-green algae in the North Shropshire meres. Field Stud. 3, 409–432.

Reynolds, C.S. (1984) The Ecology of Freshwater Phytoplankton. Cambridge University Press, pp. 384.

Reynolds, C.S. and Walsby, A.E. (1975) Water-blooms. Biol. Rev. 50, 437–481.

Reynolds, C.S., Jaworski, G.H.M., Cmiech, H.A. and Leedale, G.F. (1981) On the annual cycle of the blue-green alga *Microcystis aeruginosa* Kütz. emend. Elenkin. Phil. Trans. Roy. Soc. B 293, 419–477.

Spencer, C.N. and King, D.L. (1985) Interactions between light, NH_4^+, and CO_2 in buoyancy regulation of *Anabaena flos-aquae* (Cyanophyceae). J. Phycol. 21, 194–199.

Tandeau de Marsac, N., Mazel, D., Bryant, D.A. and Houmard, J. (1985) Molecular cloning and nucleotide sequence of a developmentally regulated gene from the cyanobacterium *Calothrix* PCC 7601: a gas vesicle protein gene. Nucl. Acids Res. 13, 7223–7236.

Thomas, R.H. and Walsby, A.E. (1985) Buoyancy regulation in a strain of *Microcystis*. J. Gen. Microbiol. 131, 799–809.

Thomas, R.H. and Walsby, A.E. (1986) The effect of temperature on recovery of buoyancy by *Microcystis*. J. Gen. Microbiol, 132, 1665–1672.

Utkilen, H.C., Oliver, R.L. and Walsby, A.E. (1985a) Buoyancy regulation in a red *Oscillatoria* unable to collapse gas vacuoles by turgor pressure. Arch. Hydrobiol. 102, 319–329.

Utkilen, H.C., Skulberg, O.M. and Walsby, A.E. (1985b) Buoyancy regulation and chromatic adaptation in planktonic *Oscillatoria* species: alternative strategies for optimising light absorption in stratified lakes. Arch. Hydrobiol. 104, 407–417.

Van Liere, E. and Walsby, A.E. (1982) Interactions of cyanobacteria with light. In N.G. Carr and B.A. Whitton (eds.), The Biology of the Cyanobacteria. Oxford, Blackwell Sci. Publ., pp. 237–262.

Van Rijn, J. and Shilo, M. (1983) Buoyancy regulation in a natural population of *Oscillatoria* spp. in fishponds. Limnol. Oceanogr. 28, 1034–1037.

Waaland, R. and Branton, D. (1969) Gas vacuole development in a blue-green alga. Science 163, 1339–1341.

Waaland, J.R., Waaland, S.D. and Branton, D. (1971) Gas vacuoles. Light shielding in blue-green algae. J. Cell Biol. 48, 212–215.

Walker, J.E., Hayes, P.K. and Walsby, A.E. (1984) Homology of gas vesicle proteins in cyanobacteria and halobacteria. J. Gen. Microbiol. 130, 2709–2715.

Walsby, A.E. (1969) The permeability of blue-green algal gas-vacuole membranes to gas. Proc. Roy. Soc. Lond. B 173, 233–255.

Walsby, A.E. (1971) The pressure relationships of gas vacuoles. Proc. Roy. Soc. Lond. B 178, 301–326.

Walsby, A.E. (1972a) Gas-filled structures providing buoyancy in photosynthetic organisms. In M.A. Sleigh and A.G. MacDonald (eds.), The Effect of Pressure on Microorganisms. Symp. Soc. Exp. Biol. 26. University Press, Cambridge, pp. 233–250.

Walsby, A.E. (1972b) Structure and function of gas vacuoles. Bacteriol. Rev. 36, 1–32.

Walsby, A.E. (1977) Absence of gas vesicle protein in a mutant of *Anabaena flos-aquae*. Arch. Microbiol. 114, 167–170.

Walsby, A.E. (1978a) The gas vesicles of aquatic prokaryotes. In R.Y. Stanier, H.J. Rogers and J.B. Ward (eds.), Relations between Structure and Function in the Prokaryotic Cell, Symp. Soc. Gen. Microbiol. 28. University Press, Cambridge, pp. 327–358.

Walsby, A.E. (1978b) The properties and buoyancy-providing role of gas vacuoles in *Trichodesmium* Ehrenberg. Br. Phycol. J. 13, 103–116.

Walsby, A.E. (1980a) The water relations of gas-vacuolate prokaryotes. Proc. Roy. Soc. Lond. B 208, 73–102.

Walsby, A.E. (1980b) A square bacterium. Nature (Lond.) 283, 69–71.

Walsby, A.E. (1981a) Cyanobacteria: planktonic gas-vacuolate forms. In M. Starr, H. Stolp, H. Truper, A. Balows and H.G. Schlegel (eds.), The Prokaryotes. Springer Verlag, New York, pp. 224–235.

Walsby, A.E. (1981b) Gas-vacuolate bacteria (apart form cyanobacteria). In M. Starr, H. Stolp, H. Truper, A. Balows and H.G. Schlegel (eds.), The Prokaryotes. Springer Verlag, New York, pp. 441–447.

Walsby, A.E. (1982) Cell-water and cell-solute relations. In N.G. Carr and B.A. Whitton (eds.), The Biology of the Cyanobacteria. Blackwell Sci. Publ., Oxford, pp. 237–262.

Walsby, A.E. (1984) Lower limit of the permeability coefficient of gas vesicles. Proc. Roy. Soc. Lond. B 223, 117–196.

Walsby, A.E. and Armstrong, R.E. (1979) Average thickness of the gas vesicle wall. J. Mol. Biol. 129, 279–285.

Walsby, A.E. and Booker, M.J. (1980) Changes in buoyancy of a planktonic blue-green alga in response to light intensity. Br. Phycol. J. 15, 311–319.

Walsby, A.E. and Buckland, B.A. (1969) Isolation and purification of intact gas vesicles from a blue-green alga. Nature (Lond.) 224, 716–717.

Walsby, A.E. and Klemer, A.R. (1974) The role of gas vacuoles in the microstratification of a population of *Oscillatoria agardhii* var. *isothrix* in Deming Lake, Minnesota. Arch. Hydrobiol. 74, 375–392.

Walsby, A.E. and Reynolds, C.S. (1981) Sinking and floating. In I. Morris (ed.), The Physiological Ecology of Phytoplankton. Blackwell Sci. Publ., Oxford, pp. 371–412.

Walsby, A.E., Van Rijn, J. and Cohen, Y. (1983a) The biology of a new gas-vacuolate cyanobacterium, *Dactylococcopsis salina* sp. nov., in Solar Lake. Proc. Roy. Soc. Lond. B 217, 417–447.

Walsby, A.E., Utkilen, H.C. and Johnsen, I.J. (1983b) Buoyancy changes of a red coloured *Oscillatoria agardhii* in Lake Gjersjøen, Norway. Arch. Hydrobiol. 97, 18–38.

Waterbury, J.B., Watson, S.W., Guillard, R.R.L. and Brand, L.E. (1979) Widespread occurrence of a unicellular, marine, planktonic, cyanobacterium. Nature (Lond.) 277, 293–294.

Weathers, P.H., Jost, M. and Lamport, D.T.A. (1977) The gas vacuole membrane of *Microcystis aeruginosa*. A partial amino acid sequence. Arch. Biochem. Biophys. 178, 226–244.

Worcester, D.L. (1975) Neutron diffraction studies of biological membranes and membrane components. In Neutron Scattering for the Analysis of Biological Structures. Brookhaven Symp. Biol. 27. Brookhaven National Library, Upton, NY, pp. 37–57.

Yentsch, C.S. (1980) Light attenuation and phytoplankton photosynthesis. In I. Morris (ed.), The Physiological Ecology of Phytoplankton. Blackwell, Oxford, pp. 95–127.

P. Fay and C. Van Baalen (Eds.) The Cyanobacteria

15

Marine planktonic cyanobacteria

G.E. FOGG

School of Ocean Sciences, Marine Science Laboratories, Menai Bridge, Gwynedd LL59 5EH, UK

1. Introduction

The two greatest constraints on the growth of phytoplankton are that sufficient light for photosynthesis is available only near the water surface and that some essential nutrients may be present only in low concentrations. The more familiar kinds of phytoplankton – those retained by the traditional fine silk net – rely on motility, positive buoyancy or exploitation of turbulence in the water to remain in the photic zone and on relative motion between cell and water to accelerate uptake or to reach regions where nutrients are present in higher concentrations. However, the same ends can be achieved simply by being very small, in which case rates of sinking become negligible and rates of diffusion greatly accelerated. The familiar type – having linear dimensions between 20 and 200 μm and now usually referred to as microplankton – is represented by only a few cyanobacterial genera in the sea, and these are virtually absent from most temperate and polar waters. The other includes nanoplankton (2–20 μm) and picoplankton (0.2–2 μm), this latter class including minute coccoid cyanobacteria which appear to be ubiquitous and abundant in most seas. It is the recent discovery that picoplanktonic cyanobacteria probably make a major contribution towards oceanic primary productivity that justifies the writing of this chapter so soon after the publication of another review of the same topic (Fogg, 1982a). An important discussion on picoplankton (Platt, in preparation) occurred too recently for anything more than cursory mention here of the material presented.

2. Picoplanktonic cyanobacteria

2.1. Occurrence and distribution

It was not realised that a substantial proportion of the numerous minute cells usually found in seawater were cyanobacteria until a technique for distinguishing them by means of their characteristic fluorescence was employed by Waterbury et al. (1979) and by Johnson and Sieburth (1979).

Illuminated through an excitation filter passing wavelengths shorter than 500 nm and observed through a barrier filter transmitting wavelengths longer than 510 nm, phycoerythrin-containing cyanobacteria can be readily differentiated under the microscope from bacteria sensu stricto, which do not fluoresce, and eukaryotic algae which fluoresce red, by their yellow to orange fluorescence (c. 575 nm). Electron microscopy shows that such cells are, in fact, cyanobacteria and not cryptophyceans or unicellular rhodophyceans, which also contain phycoerythrin. Not all unicellular cyanobacteria contain phycoerythrin (Chapman, 1973) and since the other biliprotein pigment, phycocyanin, fluoresces at a wavelength (647–660 nm) near to that at which chlorophyll *a* fluoresces (670–685 nm), it is not easy to distinguish these from eukaryotic algae which are also reported in the picoplankton fraction (Johnson and Sieburth, 1982; Joint and Pipe, 1984). El Hag (1986), for example, noted that a marine coccoid cyanobacterium isolated by him in axenic culture contained a preponderance of phycocyanin and fluoresced a deeper orange than other strains. It is possible, therefore, that counts made by the epifluorescence technique may sometimes underestimate numbers of cyanobacteria. Nevertheless, Johnson and Sieburth (1979) reported reasonable correspondence between counts made by this auto-fluorescence method and those made by transmission electron microscopy, and strains with phycocyanin predominating seem to be only a minor component in both coastal (Campbell et al., 1983) and oceanic waters (Murphy and Haugen, 1985). Haas (1982) has described an improved technique using proflavine hemi-sulphate as a fluorochrome which enables phototrophic and non-phototrophic pico- and nano-plankton to be enumerated in the same sample.

Using the autofluorescence technique, counts have been made in many different sea areas. The immediately striking feature (Table I) is that, allowing for seasonal periodicity and the effects of nutrient enrichment to be discussed below, the observations of different workers show great consistency. Coccoid cyanobacteria appear to be ubiquitous in temperate and tropical seas in concentrations in surface waters within an order of magnitude on either side of 10^4 cells $\cdot$ ml^{-1}. They appear to be less numerous, or possibly absent, in polar waters. Murphy and Haugen (1985) found that going north in the Atlantic from 36°51′N to 61°35′N, cyanobacterial numbers decreased by a factor of 10 to 100, whilst eukaryotic ultraplankton (5–10 μm) increased 5-fold. Smith et al. (1985) found an indication of the presence of cyanobacteria in the waters of the northern Foxe Basin (68°N 80°W), but the abundance of these organisms in such waters needs to be further investigated.

The vertical distribution of cyanobacteria has been examined in several studies (Waterbury et al., 1979; Krempin and Sullivan, 1981; Davis et al., 1985; Murphy and Haugen, 1985; El Hag and Fogg, 1986). The findings agree with a generalisation made by Murphy and Haugen (1985) for oceanic waters that in well-mixed situations numbers are relatively constant with depth within the euphotic zone, whereas in stratified waters they reach a maximum on the thermocline. This maximum corresponded in the North Atlantic with an irradiance level of 1% of that incident at the surface, whereas eukaryotic ultraplankton showed a maximum below this at an irradiance level of 0.5% (Murphy and Haugen, 1985). At depth, 400 m or more, cyanobacterial cells are sparse (Waterbury et al., 1979; Davis et al., 1985). More of the effect of hydrographical conditions is shown in the observations of El Hag and Fogg (1986), who made counts of cyanobacterial numbers in the vicinity of a shallow-sea tidal mixing front in the western Irish Sea. The distribution at the surface between the main areas of stratified and mixed waters on either side was not perceptibly different, but vertical sections approximately normal to the front showed that when it was well established and the cyanobacteria near their summer maximum, there was a distinct relation between distribution and hydrographic structure (Fig. 1). Apart from a concentration in the region of the pycnocline, such as already noted, there was a concentration at the surface adjacent to the

TABLE I

Some representative estimates of coccoid cyanobacterial numbers reported from various sea areas

Station	Date	Water type	Depth (m)	Cell numbers (10^4 cells · ml^{-1})	Reference
Peru					Waterbury et al., 1979
13°52′S 76°34′W	March 78	Upwelling	10	2.6	
13°57′S 77°10′W	March 78		40	0.024	
13°57′S 77°10′W	March 78	Upwelling	0	8.8	
			40	0.076	
Arabian Sea					,,
21°16′N 68°05′E	Jan 78	Upwelling	10	12.5	
			50	6.7	
			400	0.01	
24°12′N 64°44′E	Jan 78	Upwelling	0	5.3	,,
			50	3.95	
			400	0.001	
Gulf Stream		Slope	0	0.13	,,
Woods Hole Harbor	May 78	Coastal	–	0.2	,,
	June 78	Coastal	–	36.0	
Sargasso Sea					Johnson and Sieburth, 1979
29°00′N 65°36′W	April 78	Oceanic	50	0.13	
			100	0.19	
22°00′N 64°07′W	April 78	Oceanic	50	0.19	
			100	0.22	
Barbados Deep					,,
12°26′N 60°26′W	April 78	Oceanic	50	0.66	
			100	0.06	
Grenada Deep					,,
12°28′N 61°58′W	April 78	Oceanic	50	1.35	
			100	1.29	
York River					Exton et al., 1983
Virginia		Estuarine	0	1.7	
Chesapeake Bay		Coastal	0	13.0	,,
Sargasso Sea					Murphy and Haugen, 1985
34°57′N 67°07′W	June/July 81	Oceanic	75	1.5	
38°18′N 69°38′W	,,	Slope	50	4.3	

TABLE I *continued)*

Station	Date	Water type	Depth (m)	Cell numbers (10^4 cells·ml^{-1})	Reference
Gulf of Maine					Murphy and Haugen, 1985
40°41′N 65°59′W	Oct 81	Slope	1–25	2.5	
43°33′N 69°29′W	,,	Coastal	25	3.5	
42°30′N 67°59′W	,,	,,	10	8.7	
Puerto Rico					,,
18°30′N 66°33′W	Feb 82	Coastal	35	2.4	
19°30′N 66°30′W	,,	Oceanic	75	0.5	
North Atlantic					,,
36°53′N 60°31′W	April 82	Oceanic	1	1.9	
38°00′N 48°35′W	,,	,,	5	18.0	
58°32′N 33°41′W	May 82	,,	5	0.2	
63°00′N 06°57′W	,,	,,	25	0.87	
43°30′N 21°02′W	,,	,,	2.25	1.1	
Gulf of Maine					,,
39°59′N 66°00′W	Aug 82	Slope	30	2.7	
43°48′N 68°00′W	,,	Coastal	1	15.0	
42°40′N 70°00′W	,,	,,	20	7.7	
Sargasso Sea					,,
34°29′N 67°30′W	Sept 82	Oceanic	2.40	0.5	
39°55′N 70°00′W	,,	Slope	50	2.1	
Costa Rica Dome					Li et al., 1983
9°25′N 89°30′W	March 81	Upwelling	0.50	150	
'Biostat'					,,
9°45′N 93°45′W	March 81	Oceanic	0.75	1	
Southern California	Nov 78	Coastal	0	4.3	Krempin and Sullivan, 1981
Isthmus Reef off Santa Catalina Island			30	2	,,
Menai Straits	Aug 82	Coastal	0	15	El Hag and Fogg, 1986
Irish Sea	July 81	Shelf	0	8.1	,,

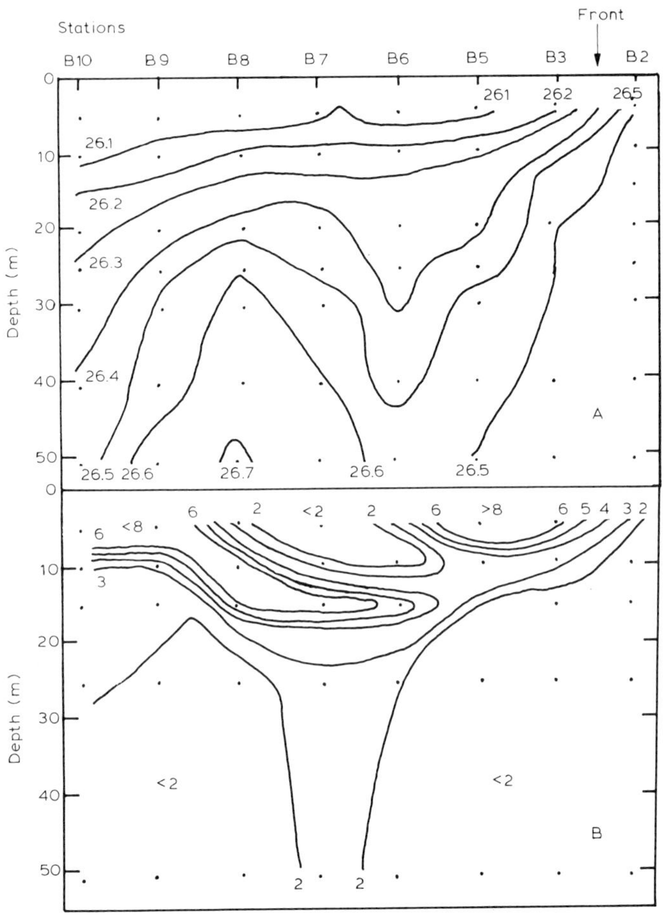

Fig. 1. Distribution of coccoid cyanobacteria in relation to hydrographic structure in the vicinity of a shallow-sea tidal mixing front in the western Irish Sea. The transect was about 45 km long, approximately normal to the front and was made on 16 April, 1982. (A) density (σ_t), (B) cyanobacterial numbers (10^3 cells·ml^{-1}). (Data of El Hag and Fogg, 1986)

front on the stratified side. A chlorophyll maximum in this position has been reported frequently and various explanations for it have been put forward. A likely possibility is that eddies on the front pick up the phytoplankton populations from the thermocline and redistribute it through the water column by vertical circulation within the eddy (Fogg et al., 1985).

Two trends, in relation to nutrient availability and temperature, respectively, seem discernible in geographical distribution. In Table I coastal, estuarine and upwelling areas have, on the whole, higher cyanobacterial numbers than oceanic areas. Temporal and spatial effects may be confused here, but El Hag and Fogg (1986), who followed changes over 2 years in inshore and offshore waters, found that during the summer period of increase numbers were consistently higher in the more nutrient-rich inshore waters of the Menai Straits than in the open western Irish Sea. The decrease in numbers with increasing latitude in the North Atlantic Ocean (Murphy and Haugen, 1985), already noted, seems most likely to reflect a positive correlation of cyanobacterial numbers with temperature, and we shall see that there is a similar correlation in time. No marked effects

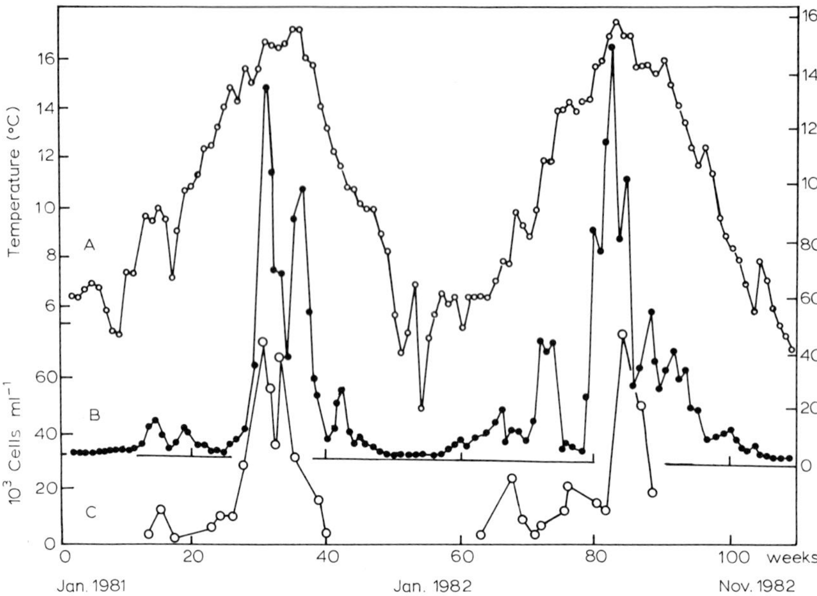

Fig. 2. Seasonal variations of surface water temperatures (°C) in the Menai Straits (A), and numbers of coccoid cyanobacteria (10^3 cells·ml^{-1}) in the surface waters of (B) the Menai Straits and (C) the western Irish Sea. Correlation coefficients: between A and B, $r = 0.77$ ($n = 50$, $p<0.001$); between B and C, $r = 0.77$ ($n = 60$, $p<0.001$). (Data of El Hag and Fogg, 1986)

of salinity were observed by El Hag and Fogg (1986), and high cyanobacterial numbers have been reported in estuarine waters (Exton et al., 1983).

In temperate, fully marine, regions the seasonal cycle is perhaps a simple one with a maximum just after mid-summer and an amplitude of about a 100-fold (Krempin and Sullivan, 1981; El Hag and Fogg, 1986; Fig. 2). It is striking that similar cycles were observed in the shallow vigorously mixed waters of the Menai Straits and in an area of the western Irish Sea subject to summer stratification, there being in both cases a strong correlation with water temperature (El Hag and Fogg, 1986). It cannot be said, however, that this correlation has a direct causal basis and, since the populations appeared not to be nutrient-limited, it is more likely that it is differential effects on growth and predation causing standing stock to rise with temperature. In an estuarine situation off Long Island, New York, Campbell et al. (1983) found a spring maximum as well as a late summer maximum in numbers of *Synechocystis* but attributed the trough between them to wash-out following high precipitation. In the same study a phycoerythrin-containing *Synechococcus* was found to persist in high numbers into the winter. Much the same pattern was found by Davis et al. (1985) in the shallow waters of Narragansett Bay, Rhode Island, in which an irregular plateau in total picoplankton numbers persisted from May to the end of November.

2.2. Taxonomy

On the basis of light microscope studies representatives of 13 genera of chroococcalean cyanobacteria have been reported in marine plankton (Sournia, 1970). However, the most abundant forms are too small to be differentiated from each other in this way, and information about their taxonomy is correspondingly sparse. In isolating clones for detailed study one runs the risk of selecting strains amenable to culture but not really representative of natural populations. Transmission electron microscopy of sections of pelleted natural populations indicates the relative abundance of structural types but gives no physiological information.

Most of the strains isolated (Waterbury et al., 1979; Perkins et al., 1981; Campbell et al., 1983; El Hag, 1986) have the typical cell morphology and ultrastructure of *Synechococcus*, with peripheral thylakoids, a size around 1×2 μm, and no sheath, as described by Rippka et al. (1979). Johnson and Sieburth (1979) found two smaller *Synechococcus*-like forms 0.7×0.9 μm and 0.4×0.5 μm in dimensions, the former with peripheral thylakoids and the latter with intracytoplasmic membranes present as a band of three or four in the centre or to one side of the cell, but they did not isolate them in culture. Two *Synechococcus* isolates studied by Perkins et al. (1981) possessed spinae, 0.044 to 0.065 μm in diameter and up to 2.7 μm in length, the function of which is not known, but this was a variable feature. Spinae have also been described for another *Synechococcus*-like form by Sarokin and Carpenter (1981). A most remarkable finding is that some oceanic *Synechococcus* strains have a capacity for swimming motility (Waterbury et al., 1985). These strains do not have flagella and the mechanism by which they swim, at rates up to 25 $\mu m \cdot s^{-1}$, is unknown.

Campbell et al. (1983) studied a strain designated as *Synechocystis* sp. because it divides in three planes but recognised that assignment to this genus may be premature. Johnson and Sieburth (1979) obtained in culture a sheathed *Gloeocapsa*-like form, which they described as ubiquitous. Its cells were 0.8×1.0 μm with thylakoids arranged peripherally.

The *Synechococcus* isolated by Waterbury et al. (1979) had a predominance of phycoerythrin and appeared red in culture, whereas that of El Hag (1986) was a bright blue-green with phycocyanin dominating. Although the two strains studied by Perkins et al. (1981) were morphologically identical, one was red and the other green. Campbell et al. (1983) showed, using an indirect immunofluorescence technique, that phycoerythrin and phycocyanin-containing strains were serologically different. Cross reactions occurred within each of these groups between strains isolated from similar oceanic environments. Seasonal succession of strains seemed to occur in Great South Bay, New York, in which *Synechocystis* populations were present only during spring and summer, phycocyanin-containing *Synechococcus* strains were only a minor component in the spring and summer, and phycoerythrin-containing *Synechococcus* populations became significant in summer, remaining so until late autumn or winter.

El Hag and Fogg (1986) classified coccoid cyanobacteria into three size groups and found that the largest of these (more than 1×1.4 μm in size) comprised up to half of the total population at the inshore site in the Menai Straits, whereas in the western Irish Sea it only made up 20 to 25% of the total. It is not known, however, whether these size groups have any taxonomic significance and it may be that larger cells simply reflect greater availability of nutrients in inshore waters (see Floodgate et al., 1981).

2.3. Physiology and biochemistry

The physical consequences of small size may be briefly stated. Applying Stokes's equation and assuming the density difference between cyanobacterial cells and sea water to be 74 $kg \cdot m^{-3}$ and the coefficient of viscosity of seawater (10°C, 35‰ salinity) to be 1.39×10^{-3} $kg \cdot m^{-1} \cdot s^{-1}$, then a spherical cell 1 μm in diameter sinks at a rate of 2.5 $mm \cdot day^{-1}$ as compared with a rate of 25 $cm \cdot day^{-1}$ for one of 10 μm diameter. As its linear dimensions decrease, so the diffusion gradients around a cell steepen and this can be an important means of reducing diffusion transport limitation of nutrient uptake (Gavis, 1976). There is also an increase in efficiency of light absorption with decreasing size, this effect being most marked at strongly absorbed as compared with weakly absorbed wavelengths. The specific absorption coefficient at the chlorophyll absorption maximum in the red increases some six-fold as the diameter of the cells into which a given amount of chlorophyll is packaged decreases from 100 μm to 1 μm (Kirk, 1983). Thus, small size gives picoplankton considerable competitive advantages against which the disadvantages appear only to be a possible greater vulnerability to predators and an inability to migrate to regions of more favourable light intensity or nutrient concentration.

Although the small size of picoplankton reduces limitation by molecular diffusion, this is not entirely eliminated at low nutrient concentrations. For the freshwater *Synechococcus leopoliensis* which is about the same size as the forms we are considering, Mierle (1985) showed that there is a deviation at low concentrations of phosphate from the Michaelis-Menten relationship describing uptake which can be attributed to limitation by diffusion of phosphate ion through water. This becomes appreciable at concentrations less than 0.1 g $P \cdot l^{-1}$ such as are often encountered in temperate seas in the summer and in oligotrophic tropical sea waters. At higher concentrations, uptake will depend on the characteristics of the carrier in the cell membrane, i.e. on V_{max}, the maximum rate of uptake and K, the half saturation constant. These do not appear to have been determined for marine coccoid cyanobacteria in respect of any nutrient. Wymer and Thake (1980) have demonstrated that with *Selenastrum gracile* and *Anabaena variabilis*, both microplanktonic freshwater algae, it is the latter, the cyanobacterium, which has the higher V_{max} for phosphate, and for this reason it is successful in competition with the former in phosphorus-limited chemostat cultures. Other workers (Lang and Brown, 1981) have obtained similar results. A freshwater form, *Synechococcus* sp. has a lower K and higher V_{max} for phosphate than larger cyanobacteria (Healey, 1982). However, it remains to be seen whether it is justifiable to infer from these findings that picoplanktonic cyanobacteria are more effective in nutrient uptake than their eukaryotic counterparts and microphytoplankton generally. An indication that this may be so is provided by Fisher's (1985) conclusion that *Synechococcus* is a major route for movement of particle-reactive metals in the sea.

Studies on the utilisation of different nitrogen sources by marine coccoid cyanobacteria are few. *Agmenellum quadruplicatum*, reclassified by Rippka et al. (1979) as a *Synechococcus*, and which is sometimes found in coastal plankton (Voltolina, 1983), has been shown to utilise glycine in phototrophic growth (Bonin et al., 1982). An interesting feature was that it seemed capable of using this substrate only within a narrower temperature range, 29°C to 40°C, than that within which growth on nitrate was possible, i.e. 17°C to 40°C. Luxury consumption of nitrogen may occur: Barlow and Alberte (1985) considered that a large part of the phycoerythrin in *Synechococcus* cells is not contained in phycobilisomes and is uncoupled from photosynthesis. Such phycoerythrin may serve as a reservoir of nitrogen and in agreement with this they found that it tends to accumulate at growth-inhibiting light intensities. Correspondingly Wyman (in Platt, in preparation) has reported that in a phycoerythrin-containing *Synechococcus* strain the fluorescence yield was greater, indicating more of the pigment uncoupled from photosynthesis, in nitrogen-replete cells as compared with nitrogen-starved cells (see Section 4.1. of Chapter 13).

An ability to fix molecular nitrogen has been reported for an axenic *Synechococcus* strain isolated from coastal water (Duerr and Mitsui, 1980). However, this preliminary report has not yet been confirmed and El Hag (1986) failed to find any evidence, using the acetylene reduction assay, for nitrogenase activity in his strain of *Synechococcus* under either aerobic or anaerobic conditions. Waterbury (in Platt, in preparation) has recently reported a strain of *Synechocystis* from tropical waters as nitrogen-fixing. It is not unlikely that marine chroococcoid cyanobacteria should fix nitrogen since it has been definitely established that a *Gloeothece* of freshwater origin can do so, although it is not yet clear how in this photosynthetic unicell the inactivation of nitrogenase by oxygen is avoided (Kallas et al., 1983). At high light intensities *Gloeothece* grows poorly on molecular nitrogen, evidently because the capacity of an oxygen-scavenging system is overloaded at high rates of photosynthesis. The preference of the marine picoplanktonic cyanobacteria for low light intensities might thus be conducive to nitrogen fixation, and, in view of the possible ecological importance of nitrogen fixation by these forms, it is desirable that more strains should be tested for nitrogenase activity. In light utilisation the physical advantages of small size are reinforced in the picoplanktonic cyanobacteria by biological features. It may be noted first of all that Waterbury et al. (1979) found that the *Synechococcus* which they isolated did not adapt chromatically (see Section 4.2. of Chapter 13) and that this seems to be true of all strains examined. From a survey of data available for phytoplankton in general, Richardson et al. (1983) concluded that cyanobacteria generally photosynthesise and grow at low photon flux densities better than all other algal classes except the dinoflagellates, and studies with picoplanktonic forms bear this out (see Section 3 of Chapter 13). Brand and Guillard (1981) compared *Synechococcus bacillaris* with 21 other species of marine phytoplankton. They found that in continuous light it saturated at 0.023 langley·min^{-1} (= 80 $\mu E \cdot m^{-2} \cdot s^{-1}$) achieving almost two divisions per day at 24°C, and was inhibited above 0.1 langley·min^{-1}. In a 14:10 light/dark regime it saturated at 0.1 langley·min^{-1} and showed no inhibition up to 0.23 langley·min^{-1}. Morris and Glover (1981) investigated two *Synechococcus* strains, one with phycoerythrin predominating, the other with phycocyanin predominating. There was little difference in their photosynthetic characteristics, the former saturating at 130, the latter at 110 $\mu E \cdot m^{-2} \cdot s^{-1}$. Maximum growth rates were achieved at lower irradiances than these, division times being 13 h at 45 $\mu E \cdot m^{-2} \cdot s^{-1}$ and 11 h at 55 $\mu E \cdot m^{-2} \cdot s^{-1}$, respectively. Photosynthesis was oxygen-sensitive, but this effect was diminished with increasing bicarbonate concentration in the medium. *Agmenellum quadruplicatum* (*Synechococcus* sp.?) required a minimum irradiance of 7.6 $\mu E \cdot m^{-2} \cdot s^{-1}$ for growth and at a temperature of around 25°C achieved maximum growth rate at 25 $\mu E \cdot m^{-2} \cdot s^{-1}$ (Bonin et al., 1982). However, at 40°C this thermophilic strain required at least 547 $\mu E \cdot m^{-2} \cdot s^{-1}$ for maximum growth (8 divisions per day). Barlow and Alberte (1985) grew two clones of *Synechococcus*, both with phycoerythrin predominating, in a range of photon flux densities. Both achieved maximum growth, 0.59 divisions per day at 18°C, at 25 $\mu E \cdot m^{-2} \cdot s^{-1}$ and photosynthesis saturated at between 66 and 112 $\mu E \cdot m^{-2} \cdot s^{-1}$ in cells grown at low irradiance. Determination of the pigment contents of cells grown at different irradiances showed that growth at low irradiances was optimised by increasing the number and decreasing the size of the photosynthetic units (see Section 4.1. of Chapter 13). This response is exactly the opposite to that shown by diatoms and dinoflagellates (Richardson et al., 1983). One strain showed marked photoinhibition at high irradiances, e.g. P_{max} of cells grown at 10 to 50 $\mu E \cdot m^{-2} \cdot s^{-1}$ was reduced by 20 to 30% by exposure to more than 150 $\mu E \cdot m^{-2} \cdot s^{-1}$. However, for brief periods photosynthesis by such cells could be sustained at higher levels than could be achieved by cells grown at these relatively high irradiances. No photoinhibition occurred in the other strain under comparable conditions. El Hag (1986) found that a phycocyanin-rich *Synechococcus* attained its maximum division rate at 20° of one division in 18 h at 30 $\mu E \cdot m^{-2} \cdot s^{-1}$ and required a minimum of 2.5 to 3 $\mu E \cdot m^{-2} \cdot s^{-1}$ to sustain growth. All these results point to these cyanobacteria being adapted to grow at exceptionally low irradiances – one may compare the values just given with the average required by diatoms for maximum growth, 84 $\mu E \cdot m^{-2} \cdot s^{-1}$ (Richardson et al., 1983) and with full sunlight, about 2000 $\mu E \cdot m^{-2} \cdot s^{-1}$. (For more details on acclimation of cyanobacteria to the natural light climate see Chapter 13 of this volume.)

A preliminary study of the biochemical composition of a *Synechococcus* strain and short-term patterns of incorporation of ^{14}C supplied to it as bicarbonate was made by Cuhel and Waterbury (1984). They found the elemental ratio, C:N:P:S, in exponentially growing cells, to be 95:16:3:1. Compared with the Redfield ratio (C:N:P, 106:16:1) this suggests a rather higher nitrogen content and a much higher phosphorus content than those for eukaryotic phytoplankton. In tracer experiments it was found that acetate and adenine were assimilated, but without contributing significantly to cell carbon, whereas glucose and thymidine were not. These findings suggest, but do not, of course, prove, that this cyanobacterium is not capable of heterotrophic growth.

In agreement with the tendency found in the other marine cyanobacteria the *Synechococcus* isolated by El Hag (1986) has been found to accumulate glycosylglycerol as an organic osmoticum (Reed et al., 1984).

Unusual phycoerythrins, of a type found in Rhodophyceae but not in freshwater or soil cyanobacteria, have been reported from marine *Synechococcus* spp. Phycoerythrin purified from a *Synechococcus* by Ong et al. (1984) had molecular properties typical of rhodophycean phycoerythrins and a high phycourobilin content. It was suggested that this phycoerythrin could be of value as a reagent in fluorescence-activated cell sorting and analysis and fluorescence immunoassay because its conjugates have higher fluorescence intensity than those of fluorescein. Alberte et al. (1984) found two spectral types of phycoerythrin, distinct from other known phycoerythrins, in marine *Synechococcus* strains, one, like that described by Ong et al. (1984), containing phycourobilin (see Section 4 of Chapter 3). The biochemical features of these phycoerythrins suggested to them a possible evolutionary link between the marine *Synechococcus* spp. containing them and the red algal chloroplast.

2.4. The role of picoplankton cyanobacteria in the pelagic ecosystem

Apart from the direct counts obtained with the epifluorescence microscope there is much evidence to show that picoplankton cyanobacteria are an important component of the marine plankton community. Yentsch (1983) found that the spectral characteristics of fluorescence from material in sea water samples passing through a glass fibre filter of 1 to 2 μm pore size suggested the presence of cyanobacteria and other algae, which were not, however, diatoms or dinoflagellates. Laser-stimulated fluorescence of bulk samples of water from Chesapeake Bay corresponded to phycoerythrin in cyanobacteria and cryptophytes (Exton et al., 1983). Fluorescence associated with the latter decreased to zero after estuarine samples had passed through a pump, evidently as a result of cell breakage and cessation of energy transfer, but coastal water with a preponderance of cyanobacteria showed no perceptible change on pumping. Such observations may provide a basis for remote sensing of picoplankton. Barlow and Alberte (1985) went further in suggesting that an inverse correlation between phycoerythrin fluorescence and maximum specific photosynthetic fluorescence and maximum specific photosynthetic and relative growth rates of cyanobacterial clones might be used to obtain from remote sensing data an indication of the in situ physiological state of the picoplankton and the hydrographical history of the water column. Guillard et al. (1985) found evidence that particles of less than 1 μm diameter associated with zeaxanthin in water samples from the North Atlantic were cyanobacterial. A latitudinal gradient in zeaxanthin concentrations correlated with one in *Synechococcus* numbers but with the occurrence of no other phytoplankton types. Such evidence points to picoplanktonic cyanobacteria usually being a major component of the total phytoplanktonic biomass. Correspondingly it seems that they contribute a significant proportion of the total planktonic photosynthesis.

Platt et al. (1983) measured photosynthesis in samples from the oligotrophic tropical North Atlantic after passage through a 1 μm filter and concluded that this fraction, which contained around 10^4 cyanobacterial cells$\cdot$ml^{-1}, carried out 60 % of the total primary productivity. In the eastern tropical Pacific Ocean from 25 to 90 % of the phytoplankton biomass, measured in terms of chlorophyll *a*, passed a filter with pores of 1 μm diameter and this contributed 20 to 80 % of the photosynthesis (Li et al., 1983). Again, cyanobacteria formed a substantial component of the picoplankton. In subtropical Hawaiian waters, cells less than 3 μm accounted for some 80% of the photosynthesis (Takahashi and Bienfang, 1983). In the North Pacific and South China Sea 70% of the subsurface chlorophyll maximum was contributed by picoplankton in which was identified by transmission electron microscopy a *Chlorella*-like eukaryote (1.2–1.5 μm) and a cyanobacterium (0.5–2 μm) (Takahashi and Hori, 1984). In the temperate waters of the Celtic sea 20 to 30% of primary production in the summer was attributable to cells below 1 μm in diameter (Joint and Pomroy, 1983). Off Nova Scotia, Douglas (1984) found photosynthesis by this fraction to amount to 12.7 to 65% of total particulate production. At higher latitudes 10 to 25% of photosynthesis was carried out by fraction less than 3 μm in size, which included yellow fluorescing particles, presumably cyanobacteria (Smith et al., 1985). Herbland et al. (1985) carried out transects covering the entire equatorial Atlantic Ocean which showed that everywhere the bulk of the chlorophyll *a* was in organisms of less than 3 μm. Cells of less than 1 μm contained 71% of the total chlorophyll on average in nitrate-depleted waters, falling to 25% when nitrate was in ample supply at the surface.

Li et al. (1983) obtained evidence that photosynthesis in waters put through filters with 1 μm pores is carried out by intact cells and not cell fragments, and this was substantiated by Douglas (1984) who in observations by microautoradiography found ^{14}C-labelling in cells showing a consistent morphology. No apportionment of photosynthesis between cyanobacteria and other autotrophic picoplankton seems to have been reported as yet. Johnson and Sieburth (1982) found, by transmission electron microscopy of concentrated samples of estuarine and coastal waters, that eukaryotic phototrophs are abundant in the picoplankton. 'Chorella-like green algae', with cells in the size range 0.85–2 μm, were reported by Joint and Pipe (1984) as abundant in the Irish Sea. Nevertheless, from numbers of cyanobacteria and the chlorophyll content per cell it appears that they constitute a substantial proportion of the total phototrophic biomass in the plankton and contribute accordingly to primary productivity (El Hag and Fogg, 1986).

Both their photosynthetic characteristics and their distribution in the water column suggest these cyanobacteria as members of the planktonic 'shade flora'. This flora comprises a great variety of forms (Sournia, 1982). Furuya and Marumo (1983), who studied the taxonomic composition of the flora of the sub-surface in the western North Pacific Ocean, found that high species diversity characterised the flora. *Micromonas* and *Ochromonas* were dominant in the sub-surface maximum and cyanobacteria, coccolithophorids, silicoflagellates and diatoms were present in a markedly different assemblage from that in the surface layer. The physiological properties of picoplanktonic cyanobacteria appear to be such as to enable them to compete successfully in this community. The light intensities which saturate their photosynthesis represent about 2 to 7% of that at the surface on a sunny day. At the depths at which such intensities are attained the predominating penetrating wavelengths are in the green (see Section 2.3. of Chapter 13), and it is in the green that their characteristic pigment, phycoerythrin, shows a maximum absorption. Glover (1982–1983) has shown that the ratio of photosynthetic rates in green and white light was up to twice as much in the plankton fraction passing a 3-μm filter as in that retained by it. This effect presumably includes a size factor (see Section 2.3. of this chapter) as well as any due to differences in absorption maxima. However, as we have seen, there is doubt about the extent to which light absorption by phycobilins is effective in the photosynthesis of cyanobacterial picoplankton. This doubt is reinforced by the finding of Lewis et al. (1985) that action spectra of natural phytoplankton assemblages show no peaks corresponding to phycobilin absorption. It would nevertheless be expected that picoplanktonic cyanobacteria would contribute substantially to primary productivity in the lower part of the euphotic zone. In the upper part their greater propensity to photorespiration and photoinhibition than other kinds of phytoplankton would render their contribution much less (see Section 5.1.2. of Chapter 13).

Since some picoplanktonic cyanobacteria are able to fix nitrogen, their contribution to the combined nitrogen income of the oceans may be considerable. However, assays on seawater by both the acetylene reduction and ^{15}N techniques have so far given negative results unless microplanktonic nitrogen fixers such as *Trichodesmium* or *Richelia* were present (e.g. Mague et al., 1974; Maruyama, 1975). Bryceson and Fay (1981) found in coastal waters off Tanzania that nitrogen fixation assayed by ^{15}N incorporation was associated with *Trichodesmium* rather than the 0.2 to 2 μm fraction of the plankton. It might be argued that their ability to grow at low light intensities enables picoplanktonic cyanobacteria to occupy a position in the water column where supply of combined nitrogen is more favourable than elsewhere so that they have no need to fix nitrogen. Uptake of combined nitrogen by picoplankton was found to be 27 % of the total in oceanic waters as compared with circa 10 % in coastal waters in the southern Benguela upwelling system (Probyn, 1985). Turnover rates of particulate nitrogen in the 1 μm fraction were faster in oceanic than in shelf or inshore waters. Picoplankton, it was concluded, rely mostly on ammonium and urea as nitrogen sources and in oceanic waters are supported mainly by regenerated nitrogen. From their study of Great South Bay, New York, which has a high nutrient loading, Kaufman et al. (1983) suggested that small cyanobacteria may have a preference for urea as against ammonium nitrogen.

Forms such as *Synechococcus* are an important component of what Sieburth and Davis (1982) call the 'microlitersphere' – an assemblage of phototrophic, chemotrophic, heterotrophic and phagotrophic micro-organisms that is responsible for most of the recycling activity in the pelagic ecosystem. Turn-over times for nitrogen by this fraction of the plankton are a few days and it is estimated that 90 % of primary productivity in the oceans is dependent on the regenerated nitrogen it produces (for references, see Fogg, 1982b). The picoplanktonic cyanobacteria appear to be grazed mainly by nanoplanktonic protozoa. The helioflagellate *Actinomonas* and the ciliate *Uronema* have been maintained on them in the laboratory for over a year. *Uronema* feeds preferentially on chroococcoid cyanobacteria rather than heterotrophic bacteria. This protozoan can achieve 4.6 doublings a day, but its numbers in the sea seem distinctly less than that which could be supported by the observed cyanobacterial populations. This suggests that there must be other major predators. Chroococcoid cyanobacteria have been found in the gut contents and faeces of calanoid copepods but persist intact and Johnson et al. (1982) suggest that they may provide an appreciable food source for abyssal and benthic protozoa when they are transported in faecal pellets to deep water. An interesting suggestion arising from the work of Turley (1985) is that urea, which occurs in concentrations of up to 5 $\mu m \cdot l^{-1}$ and shows rapid turn-over in shelf sea waters, may be derived from the cyanobacterial reserve product, cyanophycin (see Section 2 of Chapter 8), which is particularly rich in arginine, as a result of grazing activity.

3. Free-living microplanktonic cyanobacteria

In this group the main forms are *Trichodesmium* (*Oscillatoria*) spp. largely confined to tropical and subtropical seas, and *Nodularia*, *Aphanizomenon* and *Anabaena* spp. in some brackish waters. Since these have been the subject of a recent review (Fogg, 1982a) and *Trichodesmium* has received special attention in articles by Carpenter (1983a,b), it is only necessary here to discuss the more recent publications relating to them.

3.1. Occurrence and distribution

The distribution of *Trichodesmium* (the old name, *Trichodesmium*, rather than *Oscillatoria*, saves confusion and it does not seem worthwhile for present purposes to distinguish between species) has been considered thoroughly by Carpenter (1983b) in his assessment of its contribution to the combined nitrogen income of the oceans. From this it appears to be generally present, subject to temperature and hydrographic conditions being suitable, between 50°N and 35°S, in concentrations between 10^2 and 10^7 trichomes·m^{-3} (a trichome was assumed to contain 100 cells and a colony to have 200 trichomes). Although it has been recorded as far north as 64°N near Iceland, Carpenter concluded that 20°C is the minimal temperature for physiologically active *Trichodesmium*. The hydrographic conditions conducive to its growth are still not clearly defined. Blooms normally only attract attention when the population accumulates at the water surface and hence are associated with calm conditions and a stable water column but it seems clear that upwelling areas do not favour growth of *Trichodesmium*. Steven and Glombitza (1972), who observed a more or less regular periodicity in the occurrence of *Trichodesmium* off Barbados, proposed that there was a 'seed population' near the bottom of the photic zone where nutrients were more available. This idea was elaborated by Fogg (1982a) who pointed out that as a single filament grows to form a colony, the rate of its rise to the surface – and consequently its rates of photosynthesis and nutrient uptake – is accelerated so that surface blooms form suddenly. Once formed, blooms shade out the deep water population of single filaments from which they arose and only after their dissipation or decay can the cycle restart. Borstad (1982) has drawn attention to a horizontal dimension to the problem, adducing changes in salinity and plankton composition in phase with the roughly 3-monthly occurrence of *Trichodesmium* blooms off Barbados as evidence for the advection of *Trichodesmium*-rich water from the productive regions off the Guiana Shelf. Here there would be a sufficient nutrient supply and it seems significant that off Barbados *Trichodesmium* is associated with diatoms rather than the dinoflagellates and coccolithophorids characteristic of oligotrophic waters. It is postulated that the timing of the blooms off Barbados is controlled by the complex movements of the Guiana Current. The idea that *Trichodesmium* favours eutrophic rather than oligotrophic conditions is borne out by the observation of Revelante and Gilmartin (1982) that standing stocks of phytoplankton, of which *Trichodesmium* accounts for a large fraction, in the Great Barrier Reef lagoon, are strongly influenced by nutrient input from land drainage. Dense blooms were, however, more closely correlated with calm conditions than with nutrient concentrations. Marshall (1985), too, found that highest concentrations of *Trichodesmium* in the north-eastern coastal waters of the United States were associated with the flow patterns of the major estuarine systems along the coast.

In a temperate sea area, Liverpool Bay, affected by river inflow, Voltolina (1982) recorded eight species of cyanobacteria among which *Lyngbya limnetica*, usually described from fresh or brackish waters, dominated the phytoplankton in fully marine stations in June. *Isocystis infusionum*, a taxonomically dubious form, perhaps to be placed in the *Nostocaceae*, was, however, the most widespread of the cyanobacteria encountered. In more brackish waters it is often species of *Nodularia*, *Aphanizomenon* and *Anabaena* which are prominent. In recent years, probably as a result of increasing eutrophication, they have become particularly abundant in the Baltic, satellite images (Horstmann et al., 1978) showing blooms to be widespread. This was attributed by Niemi (1979) to the low N:P ratio, 3 to 5:1 in Baltic sea water as compared with the usual value of 16:1, which favours these nitrogen-fixing species. Close to centres of pollution the non-nitrogen-fixing species *Oscillatoria agardhii* contributes most of the phytoplankton biomass. In a study of phytoplankton succession in the Baltic, Elder (1979) found *Aphanizomenon* to be present all the year round, but more abundant in summer, whereas *Nodularia*, although present sporadically at other times, was judged to be the only true summer species. Both, in their season, were dominant components of the phytoplankton, and blooms seemed to be associated with a water temperature of more than 17°C. Another situation in which *Nodularia* blooms have been studied is the Peel-Harvey Estuary, Western Australia, where, again, they are associated with high phosphate inputs (Huber, 1984a). The cyanobacterium occurs in small numbers in the water column throughout the year, but tends to produce massive blooms in the austral spring. The sudden occurrence of these blooms is attributed to the mass germination of propagules, the majority of which seem to be akinetes, which over a 4-year period have been found to increase 100-fold in concentration in the sediments of this shallow estuary (Huber, 1984b). Akinetes are also produced by *Aphanizomenon* and *Anabaena* and, as has been found for their freshwater counterparts by Rother and Fay (1977), these no doubt play a role in enabling the species to survive unfavourable conditions at the water surface and to overwinter in bottom deposits (see Section 2.6.2. of Chapter 9). Apart from ample supply of phosphate it is likely that hydrographic conditions play an important part in determining the occurrence of these blooms. In the Baltic the stable water column enables the biological advantage of buoyancy control with gas-vesicles to be exploited (Fogg, 1982a), whereas in the very shallow waters of the Peel-Harvey Estuary (mean depth 1 m) mixing does not exceed the critical depth even for a dense population.

3.2. Physiology

All the microplanktonic cyanobacteria just mentioned appear to be gas-vacuolate (Walsby, 1981; see Chapter 14 of this volume). The gas vesicles of *Trichodesmium* spp. are more robust than those of freshwater species, in keeping with their exposure to a wide range of hydrostatic pressures in the turbulence of the open sea (Walsby, 1978; see also Section 4.5. of Chapter 14). With such strong vesicles, buoyancy regulations cannot be achieved by vesicle collapse through increase in turgor pressure, but may occur by preferential accumulation of carbohydrate, which, being denser than other cell components such as protein, accelerates buoyancy loss (Thomas and Walsby, 1985). Freshwater strains of *Anabaena* and *Aphanizomenon* have the turgor pressure mechanism (Thomas and Walsby, 1985), but it is not known whether the brackish water forms possess it. In the brackish water cyanobacterial osmoregulation must be particularly important and one species, *Nodularia harveyana*, has been examined in detail from this point of view by Warr et al. (1984). The major low molecular weight carbohydrate which accumulated in response to increased salinity was sucrose. Intracellular concentrations of this substance, which was evidently produced from glycogen, varied rapidly in response to salinity changes. In short-term experiments photosynthesis remained at a high level at salinities up to 52.5‰ and nitrogenase activity remained high at salinities up to 35‰. Growth, as measured by final yields of chlorophyll *a*, was highest within the range 0 to 35‰ salinity. This species at least seems well adapted to conditions in the Baltic.

Investigation of the physiology of *Trichodesmium* is still hampered by difficulty in growing it in laboratory culture (Walsby, 1981; Fogg, 1982a; Borstad and Borstad, 1984). However, Mingren et al. (1984) have recently reported the successful culture in f/2 medium of Guillard and Ryther (1962) of a new species, *T. maccii*, from the Qingdao coast of China. Possibly success would be achieved with other strains if care was taken to use actively growing single filaments from the bottom of the photic zone, rather than photo-inhibited colonies, as starting material. Experiments with natural populations of *Trichodesmium* have perhaps often been carried out with photo-inhibited unhealthy material. Observations of this kind have indicated a high half-saturation constant for orthophosphate uptake but a high alkaline phosphatase activity, so that there should be a high potential for utilising phosphomonoesters (McCarthy and Carpenter, 1979). Uptake rates of ammonium- and nitrate-nitrogen are low (see Carpenter, 1983a). Whether estimated in terms of cell nitrogen or cell carbon in situ, doubling times appear lengthy, estimates lying in the range 18 to 180 days (see Carpenter, 1983a). Although, presumably, there are no particular difficulties in obtaining the brackish water forms of *Anabaena*, *Aphanizomenon* and *Nodularia* in culture, their nutrition and growth physiology do not appear to have been studied.

The physiology of nitrogen fixation by marine planktonic cyanobacteria has been reviewed by Fogg (1982a) and Carpenter (1983a). There is still discussion of how a photosynthetic form, such as *Trichodesmium*, in which nitrogen fixation is not partitioned off in heterocysts, avoids inactivation of nitrogenase by oxygen when the medium is usually saturated with this gas (see also Section 6 of Chapter 7). Bryceson and Fay (1981) support the theory of Carpenter and Price (1976) that a central region in the trichome bundle contains cells incapable of photosynthesis, in which sufficiently reducing conditions for nitrogen fixation can be maintained. Like Carpenter and Price, they found that nitrogenase activity is dependent on the trichome bundles being intact, and they went on to show that the paler, non-granulated, central regions of bundles lack carboxysomes, show greater reducing activity towards tetrazolium than other regions, and contain more bacteria, the respiration of which would presumably lower the local oxygen concentration. However, Saino and Hattori (1982), who confirmed that nitrogenase activity is *Trichodesmium* is sensitive to oxygen and diminishes when colonies are disintegrated, nevertheless found that separate trichomes still showed some light-dependent acetylene reduction. Since they were able to show that colonies evolved hydrogen when placed under argon in the light, they suggested that the uptake hydrogenase of the cyanobacterium, by recombining hydrogen and oxygen, protects the nitrogenase from damage by oxygen. Scranton (1983, 1984), while tentatively attributing the supersaturation with hydrogen which is commonly observed in surface oceanic waters, to *Trichodesmium*, considered that Saino and Hattori (1982) over-estimated its hydrogen production capacity. Her results indicated rates of hydrogen production (0.002 to 0.05 nmol$\cdot$colony$^{-1}\cdot$h^{-1}), which were low compared with those of nitrogen fixation (0.3 to 1.5 nmol$\cdot$colony$^{-1}\cdot$h^{-1}). It begins to seem as if *Trichodesmium* achieves the reducing environment needed for the high rates of nitrogen fixation of which it is capable by combination of a number of different mechanisms, none of which is effective by itself. There is no such problem with the brackish water forms *Anabaena*, *Aphanizomenon* and *Nodularia*, all of which possess heterocysts (although, interestingly, *Aphanizomenon*, which forms bundles of trichomes like those of *Trichodesmium*, sometimes produces heterocysts only sparsely), and there seems to be no suggestion as yet that the non-heterocystous littoral species of *Lyngbya* and *Oscillatoria* are capable of fixing nitrogen.

Photosynthesis by *Trichodesmium* has been considered in the reviews of Fogg (1982a) and by Carpenter (1983a). Li et al. (1980) in an investigation of carbon assimilation by *T. thiebautii* in the Caribbean Sea found an I_k value of 358 $\pm$ 166 $\mu E\cdot m^{-2}\cdot s^{-1}$ which is of the same order as that found by McCarthy and Carpenter (1979) and in a much higher range than the values for *Synechococcus* strains appear to be. Thus *Trichodesmium* seems not to be a shade form; however, considerable self-shading must occur within its large colonies and the situation is likely to be different for solitary trichomes. It was inhibited by irradiances greater than 600 $\mu E\cdot m^{-2}\cdot s^{-1}$, an inhibition which increased with oxygen concentration and amounted to 100% in full sunlight. The phycoerythrin which is its predominant phycobiliprotein pigment has unusual features, resembling the phycoerythrin with a predominance of phycourobilin chromophores which has been found in a *Synechococcus* strain (Alberte et al., 1984) mentioned earlier. Possession of such a major accessory pigment gives the impression that *Trichodesmium* is basically adapted to maximise photosynthesis at low light intensities. The brackish water cyanobacteria are blue-green in colour, having phycocyanin as their predominant accessory pigment, but no studies appear to have been carried out on their photosynthetic behaviour.

Carpenter (1983a) summarised information on the chemical composition of *Trichodesmium* and there seems to be nothing to add to his account. No specific information on composition appears to be available for the brackish water cyanobacterial microplankton.

3.3. Role in the oceanic ecosystem

Here, interest centres on nitrogen fixation, a topic which has been discussed by Fogg (1982a,b), Capone and Carpenter (1982) and Carpenter (1983b). There seems little doubt that *Trichodesmium* is the principal biological agency fixing nitrogen in the marine pelagic ecosystem, contributing an estimated 4.8 megatonnes (Mt) of fixed nitrogen per year as against 15.4 Mt fixed in oceanic benthic environments (Capone and Carpenter, 1982). Two-thirds of the pelagic fixation is attributed to the Indian Ocean, and to the Pacific only 0.34 Mt but for this area there are little data on which to base estimates. The highest rates on a global basis were found in the northern winter and spring, principally because of the favourable conditions provided for fixation in the Indian Ocean by the Northern Monsoon (Carpenter, 1983b). It is generally concluded (see Fogg, 1982a) that fixation by *Trichodesmium* supplies only a small fraction of the combined nitrogen required by the phytoplankton as a whole in most situations.

The contribution of nitrogen fixation by the brackish water cyanobacteria is negligible on the global scale (Fogg, 1982a; Carpenter, 1983b) but in restricted areas it is important. Some estimates put the contribution of fixed nitrogen by these forms in the Baltic as of about the same order as the annual discharge of combined nitrogen by rivers into the sea (see Fogg, 1982b). Hübel and Hübel (1980), however, emphasise that there is great local variation; in a backwater, the Kleiner Jasmunder Bodden, a rate of 374 g $N_2 \cdot l^{-1} \cdot h^{-1}$ was observed – some 10^3 times higher than elsewhere in the Baltic. Since it is generally supposed that nitrogen is the principal limiting nutrient in the sea, the question arises as to why nitrogen-fixing cyanobacteria are not more prominent in the marine phytoplankton than they are. Fogg (1982a) suggested that one of the cyanobacterium's great biological advantages – buoyancy control by means of gas vesicles – is less effective in the turbulent conditions of the open sea than it is in sheltered and more stable waters. This would be a handicap to the heterocystous nitrogen fixers, the multicellular nature of which necessitates a size such that buoyancy control is an advantage. The idea of Stewart (in press) that the osmoregulatory mechanisms of cyanobacteria are unable to operate in heterocysts in hypertonic medium, seems a more convincing explanation for these forms being absent from the open sea. It does not, of course, explain why non-heterocystous nitrogen-fixers such as *Trichodesmium* are not more abundant than they are. However, it has been suggested by Doremus (1982) that it is the high phosphate requirement of cyanobacterial nitrogen-fixers, coupled to the low phosphate levels in the oceans, which limits their activity. Mee et al. (1984) recorded active nitrogen fixation by *Trichodesmium* in Mazatlan Bay, Pacific coast of Mexico, in which supplies of both combined nitrogen and phosphate were ample. It was concluded that nitrogen fixation by the cyanobacterium did not contribute appreciably to eutrophication of the Bay. Since available combined nitrogen represses nitrogenase synthesis, it may seem surprising that *Trichodesmium* fixed nitrogen under these conditions. However, the much greater concentration of N_2 in seawater, 414 $\mu mol \cdot l^{-1}$ at 35 % salinity and 20°C as compared with a usual concentration of nitrate of 5 $\mu mol \cdot l^{-1}$, would mean that in the interior of the *Trichodesmium* bundle, where the nitrogenase is supposed to be situated, N_2 would be the only available form. Another suggestion as to why nitrogen-fixing organisms are not more abundant in the sea has been put forward by Howarth and Cole (1985), who found that uptake of molybdenum, an essential element in nitrogenase, is inhibited by sulphate, which is, of course, present in relatively high concentrations in sea water.

Microplanktonic cyanobacteria are locally the main primary producers and *Trichodesmium* supports a characteristic community which has been described by Fogg (1982a) and Carpenter (1983a). Knowledge of this community has been added to by Nair et al. (1980) who, in investigating blooms in the nearshore waters off Goa, India, found a correlation between total zooplankton biomass and *Trichodesmium* biomass and a periodicity in the succession of groups of zooplankton which was regularly associated with the *Trichodesmium* bloom. Nair et al. (1980) say that there is no direct grazing by zooplankton on *Trichodesmium*, but they record substantial populations of the copepod *Macrosetella* sp., one species of which genus, *M. gracilis*, is well known from the Caribbean area as a consumer of *Trichodesmium* (see Fogg, 1982a; and Carpenter, 1983a). There are also reports of extensive grazing of *Trichodesmium* by fish (see Carpenter, 1983a). Study of the natural abundance of ^{15}N in the plankton of the North Pacific Ocean indicates a close trophic connexion between *Trichodesmium* and its associated zooplankton (Wada and Hattori, 1976). Nair et al. (1980) regard the zooplankton population as subsisting mainly on diatoms and other phytoplankton, growth of which is promoted by the decay of *Trichodesmium* blooms. As already noted, Borstad (1982) also found an association between *Trichodesmium* and diatoms, but Revelante and Gilmartin (1982) recorded an inverse relationship between the two in the Great Barrier Reef lagoon. Here much would probably depend on the stability of the water column, and it would be expected that calm conditions would result in a separation between positively buoyant *Trichodesmium* and negatively buoyant diatoms. Clearly, much remains to be learnt about the ecology of the *Trichodesmium* community, and little appears to be known, except by analogy with their freshwater counterparts, about that of the brackish water cyanobacteria. Horstmann et al. (1978) summarised observations on the ecology of blooms in the Baltic Sea. A quantitative method for the study of sparse cyanobacterial populations in estuarine waters suggested by Huber (1984a) may help to extend our knowledge of these forms further.

4. Symbiotic cyanobacteria in the plankton

There seems to have been little recent progress in the study of these at either the experimental or oceanographic levels, but the comprehensive review by Taylor (1982) should be noted. Endosymbiotic *Richelia intracellularis* (see Fogg, 1982a) infects up to 98% of populations of the diatom *Rhizosolenia* spp. in tropical and subtropical waters, although its host's ranges may extend beyond this. *Richelia* is a nitrogen fixer but its contribution to the ocean nitrogen budget is uncertain (Carpenter, 1983b). Martinez et al. (1983) found nitrogen fixation in floating mats of two species of *Rhizosolenia* occurring in abundance in various oligotrophic oceanic areas, but *Richelia* was absent from their cells and fixation appeared to be due to associated bacteria. *Synechococcus* and *Synechocystis* spp., of larger size than the picoplankton forms, have been observed as epibionts on non-photosynthetic dinophysoids (Taylor, 1982). There is presumably some nutritional benefit to the host, possibly by phagotrophy, but the dependence is not absolute since the dinophysoids may sometimes be without them. The chloroplast-like bodies of *Amphisolenia globifera*, *Dictyocha speculum* and *Streptotheca thamensis* have been described as cyanelles (see Chapter 11), but this has not been substantiated (Taylor, 1982).

5. Conclusions

Whereas at one time (Fogg, 1973) it was difficult to understand why the cyanobacteria, with their considerable tolerance of saline conditions and frequent dominance in pelagic freshwater systems, are not more conspicuous in the marine plankton, it is now seen that they have in fact developed extremely successfully along two different lines in this ecosystem. The discovery of the abundance of coccoid cyanobacteria and their importance as primary producers in the sea, coming together with a realization of the intense activity in recycling shown by plankton in the smallest size range, has opened up a new vista in marine ecology. The prokaryotic type of cell structure appears to allow the essential vital functions to be packaged into smaller units than does the eukaryotic type. Thus, the picoplanktonic cyanobacteria have been able fully to exploit the advantages conferred by small size in remaining in suspension, in rapid uptake of nutrients and in efficient light absorption. Carried passively by water movements, they multiply rapidly wherever conditions are favourable, and being adapted to utilise low irradiance levels efficiently in photosynthesis, they develop densest populations at the bottom of the euphotic zone where nutrients diffuse up from below. They thus predominate in oligotrophic waters and are exceeded by microphytoplankton only when nutrients are in more ample supply. These larger forms compensate for the slower uptake of nutrients attendant on their small surface volume ratios and the less steep diffusion gradients surrounding them, by motion relative to the water giving forced convection. The cyanobacteria achieve this by sinking and floating. The colonies produced by *Trichodesmium*, *Nodularia*, *Aphanizomenon* and *Anabaena* are among the largest found in the marine phytoplankton. Cyanobacteria typically have high nitrogen contents on a dry weight basis, both their accessory pigments and gas-vesicle membrane being proteinaceous, calling for nitrogen for synthesis above that required in eukaryotic cells. Nitrogen limitation can be overcome by fixation of molecular nitrogen, high concentrations of which are dissolved in sea water. At present it appears that the picoplankton forms in general do not fix nitrogen and, considering the rapid regeneration of combined nitrogen sources which takes place at the picoplanktonic level and the high rates of uptake of which they are capable, it seems it is not necessary for them to do so. In contrast, nearly all the microplanktonic cyanobacteria fix nitrogen. This seems to leave phosphate as the major limiting nutrient for them and it begins to appear that their distribution and abundance is, in fact, controlled by availability of phosphate.

References

Alberte, R.S., Wood, A.M., Kursar, T.A. and Guillard, R.R.L. (1984) Novel phycoerythrins in marine *Synechococcus* spp. Plant Physiol. 75, 732–739.

Barlow, R.G. and Alberte, R.S. (1985) Photosynthetic characteristics of phycoerythrin-containing marine *Synechococcus* spp. I. Responses to growth photon flux density. Mar. Biol. 86, 63–74.

Bonin, D.J., Antia, N.J. and Pelaez-Hudlet, J. (1982) Influence of temperature and light intensity on the utilization of glycine as nitrogen source for phototrophic growth of a marine unicellular cyanophyte (cyanobacterium). Bot. Mar. 25, 493–499.

Borstad, G.A. (1982) The influence of the meandering Guiana Current on surface conditions near Barbados: temporal variations of *Trichodesmium* (*Cyanophyta*) and other plankton. J. Mar. Res. 40, 435–452.

Borstad, G.A. and Borstad, L.E. (1984) Needlelike projectiles observed issuing from lysing *Trichodesmium* (*Cyanophyta*) cells. Limnol. Oceanogr. 29, 1097–1099.

Brand, L.E. and Guillard, R.R.L. (1981) The effects of continuous light and light intensity on the production rates of twenty-two species of marine phytoplankton. J. Exp. Mar. Biol. Ecol. 50, 119–132.

Bryceson, I. and Fay, P. (1981) Nitrogen fixation in *Oscillatoria* (*Trichodesmium*) *erythraea* in relation to bundle formation and trichome differentiation. Mar. Biol. 61, 159–166.

Campbell, L., Carpenter, E.J. and Iacono, V.J. (1983) Identification and enumeration of marine chroococcoid cyanobacteria by immunofluorescence. Appl. Environ. Microbiol. 46, 553–559.

Capone, D.G. and Carpenter, E.J. (1982) Nitrogen fixation in the marine environment. Science 217, 1140–1142.

Carpenter, E.J. (1983a) Physiology and ecology of marine planktonic *Oscillatoria* (*Trichodesmium*). Mar. Biol. Lett. 4, 69–85.

Carpenter, E.J. (1983b) Nitrogen fixation in *Oscillatoria* (*Trichodesmium*) in the world's oceans. In E.J. Carpenter and D.J. Capone (eds.), Nitrogen in the Marine Enviroment. Academic Press, New York, pp. 65–103.

Carpenter, E.J. and Price, C.C. (1976) Marine *Oscillatoria* (*Trichodesmium*): explanation for aerobic nitrogen fixation without heterocysts. Science 191, 1278–1280.

Chapman, D.J. (1973) Biliproteins and bile pigments. In N.G. Carr and B.A. Whitton (eds.), The Biology of Blue-Green Algae. Blackwell Sci. Publ., Oxford, pp. 162–185.

Cuhel, R.L. and Waterbury, J.B. (1984) Biochemical composition and short term nutrient incorporation patterns in a unicellular marine cyanobacterium, *Synechococcus* (WH7803). Limnol. Oceanogr. 29, 370–374.

Davis, P.G., Caron, D.A., Johnson, P.W. and Sieburth, J. McN. (1985) Phototrophic and apochlorotic components of picoplankton and nanoplankton in the North Atlantic: geographic, vertical, seasonal and diel distributions. Mar. Ecol. Prog. Ser. 21, 15–26.

Doremus, C. (1982) Note. Geochemical control of dinitrogen fixation in the open ocean. Biol. Oceanogr. 1, 429–436.

Douglas, D.J. (1984) Microautoradiography-based enumeration of photosynthetic picoplankton with estimates of carbon-specific growth rates. Mar. Ecol. Prog. Ser. 14, 223–228.

Duerr, E.O. and Mitsui, A. (1980) Aerobic growth and nitrogenase activity of a marine unicellular blue-green alga, *Synechococcus* sp. Plant. Physiol. 65, 160.

Elder, L. (1979) Phytoplankton succession in the Baltic Sea. Acta Bot. Fenn. 110, 75–78.

El Hag, A.G.D. (1986) Physiological studies on a coccoid marine blue-green alga (*Cyanobacterium*). Br. Phycol. J., 21, 315–319.

El Hag, A.G.D. and Fogg, G.E. (1986) The distribution of coccoid blue-green algae (*Cyanobacteria*) in the Menai Straits and the Irish Sea. Br. Phycol. J., 21, 45–54.

Exton, R.J., Houghton, W.M., Esaias, W., Haas, L.W. and Hayward, D. (1983) Spectral differences and temporal stability of phycoerythrin fluorescence in estuarine and coastal waters due to the domination of labile cryptophytes and stabile cyanobacteria. Limnol. Oceanogr. 28, 1225–1231.

Fisher, N.S. (1985) Accumulation of metals by marine picoplankton. Mar. Biol. 87, 137–142.

Floodgate, G.D., Fogg, G.E., Jones, D.A., Lochte, K. and Turley, C.M. (1981) Microbiological and zooplankton activity at a front in Liverpool Bay. Nature (Lond.) 290, 133–136.

Fogg, G.E. (1973) Physiology and ecology of marine blue-green algae. In N.G. Carr and B.A. Whitton (eds.), The Biology of the Blue-green Algae. Blackwell Sci. Publ., Oxford, pp. 368–378.

Fogg, G.E. (1982a) Marine plankton. In N.G. Carr and B.A. Whitton (eds.), The Biology of Cyanobacteria. Blackwell, Oxford, pp. 491–513.

Fogg, G.E. (1982b) Nitrogen cycling in sea waters. Phil. Trans. Roy. Soc. London. B 296, 511–520.

Fogg, G.E., Egan, B., Floodgate, G.D., Jones, D.A., Kassab, J.Y., Lochte, K., Rees, E.I.S., Scrope-Howe, S. and Turley, C.M. (1985) Biological studies in the vicinity of a shallow-sea tidal mixing front. VII. The frontal ecosystems. Phil. Trans. Roy. Soc. London. B, 310, 555–571.

Furuya, K. and Marumo, R. (1983) The structure of the phytoplankton community in the subsurface chlorophyll maxima in the western North Pacific Ocean. J. Plankton Res. 5, 393–406.

Gavis, J. (1976) Munk and Riley revisited: nutrient diffusion transport and rates of phytoplankton growth. J. Mar. Res. 34, 161–179.

Glover, H.E. (1982–83) Photosynthetic characteristics of small, coccoid, marine cyanobacteria. Bigelow Laboratory for Ocean Sciences, Annual Report 1982–83, 22–23.

Guillard, R.R.L. and Ryther, J.H. (1962) Studies of marine planktonic diatoms I. *Cyclotella nana Hustedt* and *Detonula confervacea* (Cleve) Gran. Can. J. Microbiol. 8, 229–239.

Guillard, R.R.L., Murphy, L.S., Foss, P. and Liaaen-Jensen, S. (1985) *Synechococcus* spp. as likely zeaxanthin-dominant ultraphytoplankton in the North Atlantic. Limnol. Oceanogr. 30, 412–414.

Haas, L.W. (1982) Improved epifluorescence microscopy for observing planktonic micro-organisms. Ann. Inst. Oceanogr. (Paris) (58) (S), 261–266.

Healey, F.P. (1982) Phosphate. In N.G. Carr and B.A. Whitton (eds.), The Biology of Cyanobacteria. Blackwell Sci. Publ., Oxford, pp. 105–124.

Herbland, A., Le Bouteiller, A. and Raimbault, P. (1985) Size structure of phytoplankton biomass in the equatorial Atlantic Ocean. Deep-Sea Res. 32, 819–836.

Horstmann, U., Ulbricht, K.A. and Schmidt, D. (1978) Detection of eutrophication processes from air and space. Proceedings of 12th International ERIM Symposium on Remote Sensing of Environment, Manila, Philippines, April 20–26, 1978, pp. 1379–1389.

Howarth, R.W. and Cole, J.J. (1985) Molybdenum availability, nitrogen limitation, and phytoplankton growth in natural waters. Science 229, 653–655.

Hübel, H. and Hübel, M. (1980) Nitrogen fixation during blooms of *Nodularia* in coastal waters and backwaters of the Arkona Sea (Baltic Sea) in 1974. Int. Rev. Ges. Hydrobiol. 65, 793–808.

Huber, A.L. (1984a) A simple method for the isolation and enumeration of sparse populations of cyanobacteria in estuarine waters. J. Phycol. 20, 619–621.

Huber, A.L. (1984b) *Nodularia* (*Cyanobacteriaceae*) akinetes in the sediments of the Peel-Harvey Estuary, Western Australia: potential inoculum source for *Nodularia* blooms. Appl. Environ. Microbiol. 47, 234–238.

Johnson, P.W. and Sieburth, J.McN. (1979) Chroococcoid cyanobacteria in the sea: a ubiquitous and diverse phototrophic biomass. Limnol. Oceanogr. 24, 928–935.

Johnson, P.W. and Sieburth, J.McN. (1982) In situ morphology and occurrence of eukaryotic photoautotrophs of bacterial size in the picoplankton of estuarine and oceanic waters. J. Phycol. 18, 318–327.

Johnson, P.W., Xu, H. and Sieburth, J.McN. (1982) The utilization of chroococcoid cyanobacteria by marine protozooplankters but not by calanoid copepods. Ann. Inst. Oceanogr. (Paris) 58 (S) 297–308.

Joint, I.R. and Pipe, R.K. (1984) An electron microscope study of a natural population of picoplankton from the Celtic Sea. Mar. Ecol. Prog. Ser. 20, 113–118.

Joint, I.R. and Pomroy, A.J. (1983) Production of picoplankton and small nanoplankton in the Celtic Sea. Mar. Biol. 77, 19–27.

Kallas, T., Rippka, R., Coursin, T., Rebiere, M.-C., Tandeau de Marsac, N. and Cohen-Bazire, G. (1983) Aerobic nitrogen fixation by non-heterocystous cyanobacteria. In G.B. Papageorgiou and L. Packer (eds.), Photosynthetic Prokaryotes: Cell Differentiation and Function. Elsevier Sci. Publ. pp. 281–302.

Kaufman, Z.G., Lively, J.S. and Carpenter, E.J. (1983) Uptake of nitrogenous nutrients by phytoplankton in a barrier island estuary; Great South Bay, New York. Estuar. Coast Shelf Sci. 17, 483–493.

Kirk, J.T.O. (1983) Light and photosynthesis in aquatic ecosystems. Cambridge University Press, Cambridge.

Krempin, D.W. and Sullivan, C.W. (1981) The seasonal abundance, vertical distribution and relative microbial biomass of chroococcoid cyanobacteria at a station in southern California coastal waters. Can. J. Microbiol. 27, 1341–1344.

Lang, P.S. and Brown, E.J. (1981) Phosphorus-limited growth of a green and a blue-green alga. Appl. Environ. Microbiol. 42, 1002–1009.

Lewis, M.R., Warnock, R.E. and Platt, T. (1985) Absorption and photosynthetic action spectra for natural phytoplankton populations: implications for production in the open ocean. Limnol. Oceanogr. 30, 794–806.

Li, W.K.W., Glover, H.E. and Morris, I. (1980) Physiology of carbon photoassimilation by *Oscillatoria thiebautii* in the Caribbean Sea. Limnol. Oceanogr. 25, 447–456.

Li, W.K.W., Subba Rao, D.V., Harrison, W.G., Smith, J.C., Cullen, J.J., Irwin, B. and Platt, T. (1983) Autotrophic picoplankton in the tropical ocean. Science 219, 292–295.

Mague, T.H., Weare, N.M. and Holm-Hansen, O. (1974) Nitrogen fixation in the North Pacific Ocean. Mar. Biol. 24, 109–119.

Marshall, H.G. (1985) The seasonal composition and distribution patterns of *Cyanophyceae* in the north eastern coastal waters of the United States. Arch. Hydrobiol. Suppl. 71 (Algological Studies 38/39), 305–306.

Martinez, L.A., Silver, M.W., King, J.M. and Alldredge, A.L. (1983) Nitrogen fixation by floating diatom mats: a new source of nitrogen to oligotrophic ocean waters. Science 221, 152–154.

Maruyama, Y. (1975) Nitrogen fixation in the open sea. In H. Takahashi (ed.), JIBP Synthesis: Nitrogen Fixation and Nitrogen Cycle. University of Tokyo Press, Tokyo, pp. 79–84.

McCarthy, J.J. and Carpenter, E.J. (1979) *Oscillatoria* (*Trichodesmium thiebautii*) (*Cyanophyta*) in the central North Atlantic Ocean. J. Phycol. 15, 75–82.

Mee, L.D., Cortes-Altaminano, R. and Garcia-de-la-Parra, L.-M. (1984) Di-nitrogen fixation in a eutrophic tropical bay. Estuarine, Coastal and Shelf Science 19, 477–483.

Mierle, G. (1985) Kinetics of phosphate transport by *Synechococcus leopoliensis* (*Cyanophyta*): evidence for diffusion limitation of phosphate uptake. J. Phycol. 21, 177–181.

Mingren, L., Bozhong, B., Lewin, R.A., Cheng, L. and Shenoiu, P. (1984) Laboratory culture of a new species of planktonic blue-green alga, *Trichodesmium maccii*, from the Qingdao Coast, China. J. Shandong College Oceanol. 14, 93–101.

Morris, I. and Glover, H. (1981) Physiology of photosynthesis by marine coccoid cyanobacteria – some ecological implications. Limnol. Oceanogr. 26, 957–961.

Murphy, L.S. and Haugen, E.M. (1985) The distribution and abundance of phototrophic ultraplankton in the North Atlantic. Limnol. Oceanogr. 30, 47–58.

Nair, V.R., Devassy, V.P. and Qasim, S.Z. (1980) Zooplankton and *Trichodesmium* phenomenon. Indian J. Mar. Sci. 9, 1–6.

Niemi, A. (1979) Blue-green algal bloom and N:P ratio in the Baltic Sea. Acta Bot. Fenn. 110, 57–61.

Ong, L.J., Glazer, A.N. and Waterbury, J.B. (1984) An unusual phycoerythrin from a marine cyanobacterium. Science 224, 80–83.

Perkins, F.O., Haas, L.W., Phillips, D.E. and Webb, K.L. (1981) Ultrastructure of marine *Synechococcus* possessing spinae. Can. J. Microbiol. 27, 318–329.

Platt, T. (ed.) NATO Advanced Study Institute: Physiological Ecology of Photosynthetic Picoplankton. Can Bull. Fish Aquat. Sci., in preparation.

Platt, T., Subba Rao, D.B. and Irwin, B. (1983) Photosynthesis of picoplankton in the oligotrophic ocean. Nature 301, 702–704.

Probyn, T.A. (1985) Nitrogen uptake by size-fractionated phytoplankton populations in the southern Benguela upwelling system. Mar. Ecol. Prog. Ser. 22, 249–258.

Reed, R.H., Richardson, D.L., Warr, S.R.C. and Stewart, W.D.P. (1984) Carbohydrate accumulation and osmotic stress in cyanobacteria. J. Gen. Microbiol. 130, 1–4.

Revelante, N. and Gilmartin, M. (1982) Dynamics of phytoplankton in the Great Barrier Reef lagoon. J. Plankton Res. 4, 47–76.

Richardson, K., Beardall, J. and Raven, J.A. (1983) Adaptation of unicellular algae to irradiance: an analysis of strategies. New Phytol. 93, 157–191.

Rippka, R., Deruelles, J., Waterbury, J.B., Herdman, M. and Stanier, R.Y. (1979) Generic assignments, strain histories and properties of pure cultures of cyanobacteria. J. Gen. Microbiol. 111, 1–61.

Rother, J.A. and Fay, P. (1977) Sporulation and the development of planktonic blue-green algae in two Salopian meres. Proc. Roy. Soc. Lond. B 196, 317–332.

Saino, T. and Hattori, A. (1982) Aerobic nitrogen fixation by the marine non-heterocystous cyanobacterium *Trichodesmium* (*Oscillatoria*) spp: its protective mechanism against oxygen. Mar. Biol. 70, 251–254.

Sarokin, D.J. and Carpenter, E.J. (1981) Cyanobacterial spinae. Bot. Mar. 24, 389–392.

Scranton, M.I. (1983) The role of the cyanobacterium *Oscillatoria* (*Trichodesmium*) *thiebautii* in the marine hydrogen cycle. Mar. Ecol. Prog. Ser. 11, 79–87.

Scranton, M.I. (1984) Hydrogen cycling in the waters near Bermuda: role of the nitrogen fixer, *Oscillatoria thiebautii*. Deep Sea Res. 31, 133–143.

Sieburth, J.McN. and Davis, P.G. (1982) The role of heterotrophic nanoplankton in the grazing and nurturing of planktonic bacteria in the Sargasso and Caribbean Seas. Ann. Inst. Oceanogr. (Paris) 58 (S), 285–296.

Smith, J.C., Platt, T., Li, W.K.W., Horne, E.P.W., Harrison, W.G., Subba Rao, D.V. and Irwin, B.D. (1985) Arctic marine photoautotrophic picoplankton. Mar. Ecol. Prog. Ser. 20, 207–220.

Sournia, A. (1970) Les cyanophycées dans le plancton marin. Ann. Biol. 9, 63–76.

Sournia, A. (1982) Is there a shade flora in the marine plankton? J. Plankton Res. 4, 391–399.

Steven, D.M. and Glombitza, R. (1972) Oscillatory variation of a phytoplankton in a tropical ocean. Nature (Lond.) 237, 105–107.

Stewart, D.E. and Farmer, F.H. (1984) Extraction, identification and quantitation of phycobiliprotein pigments from phototrophic plankton. Limnol. Oceanogr. 29, 392–397.

Stewart, W.D.P. (1986) Leeuwenhoek Lecture, 1984. Proc. Roy. Soc. Lond. B, in press.

Takahashi, M. and Bienfang, P.K. (1983) Size structure of phytoplankton photosynthesis in subtropical Hawaiian waters. Mar. Biol. 76, 203–211.

Takahashi, M. and Hori, T. (1984) Abundance of picophytoplankton in the sub-surface chlorophyll maximum (SCM) layer in subtropical and tropical waters. Mar. Biol. 79, 177–186.

Taylor, F.J.R. (1982) Symbioses in marine microplankton. Ann. Inst. Oceanogr. (Paris) 58 S, 61–90.

Thomas, R.H. and Walsby, A.E. (1985) Buoyancy regulation in a strain of *Microcystis*. J. Gen. Microbiol. 131, 799–809.

Turley, C.M. (1985) Biological studies in the vicinity of a shallow-sea tidal mixing front. IV. Seasonal and spatial distribution of urea and its uptake by phytoplankton. Phil. Trans. Roy. Soc. Lond. B., 310, 471–500.

Voltolina, D. (1983) The phytoplankton of Liverpool Bay (1977–1978): an annotated checklist. Nova Hedwigia 37, 403–434.

Wada, E. and Hattori, A. (1976) Natural abundance of ^{15}N in particulate organic matter in the North Pacific Ocean. Geochem. Cosmochem. Acta 40, 249–251.

Walsby, A.E. (1978) The properties and buoyancy-providing role of gas vacuoles in *Trichodesmium* Ehrenberg. Br. Phycol. J. 13, 103–116.

Walsby, A.E. (1981) Cyanobacteria: planktonic gas-vacuolate forms. In M.P. Starr, N. Stolp, H.G. Truper, A. Balows and H.G. Schlegel (eds.), The Prokaryotes, I. Springer-Verlag, Berlin, pp. 224–235.

Warr, S.R.C., Reed, R.H. and Stewart, W.D.P. (1984) Physiological responses of *Nodularia harveyana* to osmotic stress. Mar. Biol. 79, 21–26.

Waterbury, J.B., Watson, S.W., Guillard, R.R.L. and Brand, L.E.E. (1979) Widespread occurrence of a unicellular, marine, planktonic cyanobacterium. Nature 277, 293–294.

Waterbury, J.B., Willey, J.M., Franks, D.G., Valois, F.W. and Watson, S.W. (1985) A cyanobacterium capable of swimming motility. Science NY 230, 74–76.

Wymer, P.E.O. and Thake, B. (1980) The importance of phosphorus in microalgal growth and species composition in mixed populations: experiments and simulations. Proc. Roy. Soc. Lond. B 209, 333–353.

Yentsch, C.S. (1983) A note on the fluorescence characteristics of particles that pass through glass-fiber filters. Limnol. Oceanogr. 28, 597–599.

P. Fay and C. Van Baalen (Eds.) The Cyanobacteria

16

Iron chelation and uptake

GREGORY L. BOYER[a], ANDREW H. GILLAM[b,]* and CHARLES TRICK[c]

[a]*Department of Chemistry, SUNY-College of Environmental Science and Forestry, Syracuse, NY 13210, USA,* [b]*Institute of Offshore Engineering, Heriot-Watt University, Edinburgh EH14 4AS, UK and* [c]*Institute of Environmental Studies, University of Toronto, Toronto, Ontario M5S 1A4, Canada*

1. Introduction

Iron, behind oxygen, silicon and aluminum, is the fourth most abundant element by weight in the earth's crust. However, due to its poor solubility in oxic aqueous solution, it is an element with a low biological availability. Iron is normally found in one of two ionic forms; ferrous (Fe^{2+}) and ferric (Fe^{3+}). In the presence of oxygen, aqueous Fe^{2+} is rapidly oxidized to Fe^{3+}, and precipitates at physiological pH as its insoluble ferric hydroxide ($K_s = 10^{-39}$). While recent reports (Anderson and Morel, 1980; Finden et al., 1984) have suggested that the iron requirement of some algae can be met by Fe^{2+} species, ferric iron (Fe^{3+}) is the principal form of 'biologically available iron' utilized by microorganisms. Living organisms are constantly faced with the problem of how to overcome the low solubility of the ferric ion to meet their cellular needs of iron.

Iron is involved in a wide variety of biochemical processes, both as a cofactor and catalyst. These include photosynthesis, chlorophyll production, nitrate reduction, dinitrogen fixation, and other biological oxidations. A complete list of iron-containing proteins common to many microbial species can be found in Neilands (1974). While these enzymes will also apply to cyanobacteria, a critical analysis of the metal composition and the influence of iron availability on the activity and production of the cyanobacterial enzymes is lacking in most instances. Enzymes known to be under the influence of iron metabolism in cyanobacteria are summarized in Section 3 of this chapter.

In nature, cyanobacteria (and phytoplankton in general) are often limited by the supply of available nutrients. Iron has been shown to limit primary production in both marine (Menzel and Ryther, 1961; Barber and Ryther, 1969; Glover, 1978; Entsch et al., 1983; Yamochi, 1983), and freshwater (Schelske, 1962; Wetzel, 1966; Murphy et al., 1983; Goldman, 1964; Sakamoto, 1971) systems. This role of iron as a critical determinant of primary production is widely recognized but little is known about the mechanisms by which algae acquire their required iron. Studies have

*Present address: Dodrocky Seatech, 9865 W. Saanich Road, P.O. Box 6500, Sidney, British Columbia V8T 2E5, Canada.

shown that a large percentage of the iron present in the natural environment is not biologically available to support phytoplankton growth. This fraction can be increased by the addition of chelators. Wetzel (1972), Sakamoto (1971), and Barber and Ryther (1969) all report a stimulation of $^{14}C\text{-}HCO_3^-$ uptake by the addition of EDTA. The mechanism of this stimulation is not clear. The addition of EDTA is capable of solubilizing iron in freshly prepared culture media or seawater, but is not very effective in promoting growth after the culture has been allowed to age (Lewin and Chen, 1971; Wells et al., 1983). EDTA may also depress inhibitory Cu^{2+} activity (Jackson and Morgan, 1978). However for the marine diatom, *Ditylum brightwellii*, the observed increase in growth was not due to a decreased trace metal toxicity but rather to an increase in soluble iron (Canterford, 1979).

In addition to artificial chelators such as EDTA, phytoplankton are also known to respond to a wide range of natural chelators. Humic acids, amorphous materials which arise from bacterial action and the degradation of plant and animal materials, are important contributors to the dissolved organic carbon (DOC) present in natural waters. Generally acidic in nature, these substances chelate a wide variety of trace metals including iron (see Section 4.4.). Reports have correlated the occurrence of phytoplankton blooms with the level of humic materials in the water column (Ingle and Martin, 1971), though it has not been established if this is due to an increased availability of essential trace elements or an increased complexation (hence making them biologically unavailable) of toxic heavy metals.

Besides responding to general chelators, blue-green algae, certain bacteria, and fungi are known to produce specific high affinity iron chelators known as siderophores. These low molecular weight compounds are highly specific in their ability to chelate iron with affinity constants (K_f) often greater than 10^{-30}. Their production and uptake mechanism in cyanobacteria have been the subject of detailed investigation and are further described in Section 5.

2. Iron requirements in cyanobacteria

2.1. General considerations

There is little argument that iron is an essential trace metal for growth in cyanobacteria. However, the level of iron required is a subject of much debate. Experimentally, this problem has been approached from two directions. One is to determine the amount of iron necessary to support maximal cell growth rates. This approach is also dependent on the levels of other nutrients present in the system. The second approach is to determine the iron quota for the cell, e.g. the minimal amount of iron needed to support cell growth. While little information is available specifically on cyanobacteria, several attempts have been made to estimate the cellular iron quota for certain eukaryotic algae.

Both approaches are affected by the previous iron history of the cell. Phytoplankton are known to show luxury consumption of nutrients, and mechanisms exist in a large number of organisms to store iron in times of iron excess (see Section 2.4. below). Iron history also has a strong influence on cell size and morphology in cyanobacteria. McLachlan et al. (1963) found that the accumulation of iron and other trace metals stimulated the formation of aggregates of *Aphanizomenon flos-aquae*. O'Flaherty and Phinney (1970) noted that flake size of *Aphanizomenon* was related to the available iron concentration, with the largest flakes formed at 0.18 $mg \cdot l^{-1}$ Fe^{3+}. *Anabaena flos-aquae* showed pronounced filament coiling and cell size variation in iron-deficient cultures (Gorham et al., 1964). Similarly, individual cells of iron-deficient *Anacystis nidulans* R2 cultures were 1/2 to 1/3 shorter than their iron-sufficient controls (Sherman and Sherman, 1983).

2.2. Iron requirements for maximal growth rates

The effect of limiting iron on growth rate has been studied in several species. *Agmenellum quadruplicatum* PR-6 showed a slight reduction in growth rate (80% of maximum rate) when cultured under iron-deficient conditions sufficient to induce the production of iron-chelating organics (Armstrong and Van Baalen, 1979). Hutber et al. (1977) noted only a small reduction in the growth rate of *Nostoc*, though production of flavodoxin was indicative of an iron-deficient cell (see Section 3.3. below and Section 2.1. of Chapter 2).

In contrast, Sherman and Sherman (1983) noted no difference in the growth rate of *Anacystis nidulans* R2 cells in iron-sufficient and iron-deficient media. Examination of the physical composition of the cells verified that the cells were truly iron-stressed.

The iron requirement for maximal growth rates in culture has been estimated for a few algae (Table I). Iron requirements for cyanobacteria are similar to those observed for other algal species. Yet, in spite of this similarity, it is suggested that cyanobacteria have a higher cellular iron requirement than eukaryotic algae. Increased iron concentrations are reported to selectively stimulate cyanobacterial growth over competing chlorophytes (Morton and Lee, 1974; Parr and Smith, 1976).

This perceived higher demand for iron by cyanobacteria may be due to the small number of isolates that have been tested under carefully controlled metal concentrations. Alternatively, high

TABLE I

Iron requirement for maximal growth rate in culture

Organism	Iron concentration (μM)	Reference[a]
Bacillariophyta		
Thalassiosira pseudonana	0.002	1
Chlorophyta		
Ankistrodesmus braunii	11.0	2
Chlorella vulgaris	1.2	3
C. pyrenoidosa (heterotrophic)	0.001	2
C. pyrenoidosa (autotrophic)	18.0	2
Cyanophyta		
Agmenellum quadruplicatum	1.75	4
Aphanocapsa 6714	0.3–3.0	5
Nostoc muscorum	0.2–4.0	6
Rhodophyta		
Asparagopsis armata	0.5–1.0	7
Phaeophyta		
Macrocystis pyrifera	0.34	8

[a](1) Anderson and Morel, 1982; (2) O'Kelly, 1974; (3) Pirt and Walach, 1978; (4) Armstrong and Van Baalen, 1979; (5) Sandmann and Malkin, 1983; (6) Carnahan and Castle, 1958; (7) Segot and Codomier, 1981; (8) Kuwabara, 1981.

cyanobacterial iron requirements may reflect a difference in physiology. Iron requirements may increase when the cells are fixing nitrogen. Carnahan and Castle (1958) have shown that *Nostoc muscorum* required from 1–20 mg $Fe \cdot l^{-1}$ to achieve maximum growth rates (depending on the nitrogen source) but 10 mg $Fe \cdot l^{-1}$ for maximal rate of nitrogen fixation. At low (0.06 $mg \cdot l^{-1}$) iron concentrations, *Nostoc* exhibited normal growth on nitrate or urea but little growth on molecular nitrogen. At optimal iron concentrations (10 $mg \cdot l^{-1}$), cultures achieved equal densities regardless of the nitrogen source.

2.3. Cellular iron quotas

Determination of cellular iron quotas for microorganisms is hampered by contamination from trace levels of iron present in the salts used to make up the media. This may supply sufficient iron to lead to large errors in determination of the iron quota (Q_{Fe}) (for discussion see Lankford, 1973; Morel et al., 1979). Attempts to estimate cellular iron quotas for select eukaryotic algae give values ranging from 0.04–0.23 pg Fe per cell (Droop, 1973; Sakshaug and Holm-Hansen, 1977; Pirt and Walach, 1978). *Aerobacter indologenes* is reported to contain 0.3% iron (wt/dry wt) in iron-sufficient cells and 0.003% in iron-stressed cells. The equivalent experiments with cyanobacteria have not been conducted but similar values may be expected for cells grown with nitrate or ammonium. It is unknown if increased cellular demand for iron during nitrogen fixation will translate into a significant increase in cell quota. Iron-containing enzymes comprise only a small (10 to 18%) fraction of the total cell iron quota (Lankford, 1973; Hewitt, 1983) and a large portion of the iron in bacteria and higher plants is unaccountable.

2.4. Iron storage and iron excess

In nature cells will be exposed to both periods of excess iron as well as iron-limiting conditions. There is little information on iron storage in cyanobacteria. Algae (Pueschel and Cole, 1980), bacteria (Yariv et al., 1981), higher plants (Van der Mark et al., 1982), fungi (David and Easterbrook, 1971) and higher animals (Bienfait and Van der Mark, 1983) are able to store iron in a large molecular weight protein-enveloped form called ferritin. While ferritin compounds have not been demonstrated for cyanobacteria, Evans et al. (1979) reported on the presence of potential iron storage proteins in the membranes of *Chlorogloea fritschii*, and Sandmann and Malkin (1983) on the presence of large amounts of membrane-bound nonheme iron in iron sufficient *Aphanocapsa*. Such compounds may serve a dual role, in the storage of iron and as a protective sink to buffer the cell from excess free iron, as has been proposed for higher plants (Van der Mark et al., 1982).

3. The function of iron in cellular metabolism

In addition to its overall effect on cell growth, there is increasing information about the effects of iron on cell metabolism. Iron is a necessary cofactor for many enzymes and often crucial for catalytic activity. Metabolic effects of iron stress often reflect the loss of biochemical pathways, rather than the decrease in substrate observed under carbon, nitrogen and phosphorus limitation. Pathways involved in nitrogen metabolism, photosynthesis, and electron transport are extremely sensitive to iron limitation. These pathways, and the specific enzymes known to be influenced by iron, are discussed in the following sections. In addition, several other enzymes may be under iron control by inference with other organisms. Catalase activity is reported to be influenced by the amount of available iron in *Arthrobacter* JG-9 (Burnham and Neilands, 1961) and a similar situation is expected in cyanobacteria. A list of iron-dependent enzymes in different organisms can be found in the comprehensive treatise edited by Neilands (1974).

3.1. The effect on the photosynthetic apparatus

The chlorotic effect of iron deficiency in cyanobacteria is well known. Initial work by Öquist (1971, 1974a,b) described the changes in chlorophyll content of iron-deficient *Anacystis nidulans*. This was manifested as a shift in the 679-nm absorption peak of chlorophyll *a* and a change in the photosynthetic action spectrum. This was confirmed by others for *Anacystis nidulans* (Peschek, 1979; Sherman and Sherman, 1983; Guikema and Sherman, 1984) and for *Agmenellum quadruplicatum* (Hardie et al., 1983a) (see also Section 3.3. of Chapter 1). Under iron stress alteration of the photosynthetic apparatus resulted in a decreased phycocyanin and chlorophyll content per cell (Hardie et al., 1983a; Peschek, 1979) and an increased phycocyanin/chlorophyll ratio (Guikema and Sherman, 1984). The degradation of chlorophyll and phycocyanin was also reflected in reduced photosynthetic activity. Field analysis of iron-deficient natural cyanobacterial populations also showed a rapid drop in chlorophyll (to 1/3 of the non-iron-stressed level) upon becoming iron limited (Murphy and Lean, 1975).

Changes in the photosynthetic unit size have been measured by Hardie et al. (1983a) for *A. quadruplicatum*. The quotient chlorophyll/P700 reaction centers increases as cells become iron stressed. Since the chlorophyll content per cell decreased, the P700 content must decrease at an even faster rate. These results are similar to those obtained with higher plants (Spiller and Terry, 1980).

Sandmann and Malkin (1983) examined the influence of iron stress on the components of the photosynthetic electron transport chain of *Aphanocapsa* 6714, and have assumed that instead of the number of photosynthetic units decreasing, all components of the photosynthetic electron transport chain have decreased, including those components which did not contain iron. Under moderate iron stress (1/10 normal iron), *Aphanocapsa* cells exchanged their iron-containing ferredoxin with a non-iron-containing flavodoxin (see Section 3.3. below and Section 2.1. of Chapter 2). Other photosynthetic redox proteins were not affected and this adaptation had little effect on the rate of photosynthetic electron transport or on growth. Further iron stress (1/100 normal iron) resulted in a decrease of all iron-containing photosynthetic redox proteins, as well as in the levels of other non-iron-containing redox proteins, causing a corresponding reduction in electron transport and growth.

Ultrastructural changes induced by iron deficiency in *A. quadruplicatum* and *A. nidulans* have also been reported (Hardie et al., 1983b; Sherman and Sherman, 1983). The initial stages of iron deficiency have been correlated with a rearrangement of the thylakoids and an accumulation of glycogen storage granules. With prolonged exposure to low iron conditions, a gradual deterioration of the cytoplasmic membrane and the thylakoids was observed. Cytoplasmic changes included the depletion of ribosomes and of peripheral lipid bodies, and also the loss of carboxysomes in *A. nidulans* but not in *A. quadruplicatum*. Guikema and Sherman (1984) noted the formation of two unique membrane proteins in *A. nidulans* when grown under iron stress. One, a 34-kD membrane protein, was similar in size to the inducible siderophore transporter/receptor system characterized in *E. coli* (Neilands, 1982) though its role in *A. nidulans* is not known.

3.2. The effect on nitrogen metabolism

Many of the major enzymes of nitrogen metabolism are iron-containing proteins (nitrate reductase, nitrite reductase, glutamine synthetase, and nitrogenase) and thus nitrogen metabolism is extremely sensitive to iron stress. The work of Hardie et al. (1983b) and of Sherman and Sherman (1983) provides experimental evidence in support of the observations by Glover (1977) that iron deficiency in eukaryotic algae particularly affects nitrogen metabolism and manifests itself in symptoms characteristic of nitrogen-deficient cells. The degradation of phycocyanin in cyanobacteria under iron stress (Boussiba and Richmond, 1980), the depletion of ribosomes, and the accumulation of glycogen are all possible symptoms of nitrogen stress induced during low iron conditions. Glycogen may function as a storage mechanism for carbon normally incorporated into nitrogen-containing proteins (Allen and Smith, 1969; Lehmann and Wober, 1976).

The specific effects of iron limitation on key enzymes in nitrogen metabolism have also been investigated. Peschek (1979) examined the kinetics of nitrate reductase and nitrite reductase in iron-deficient cultures of *A. nidulans*. These cultures grew at a much reduced rate when compared to the iron-sufficient controls and the activities of the enzymes were also reduced. Hardie et al. (1983a) reported a gradual increase in nitrate and nitrite reductase activities in *A. quadruplicatum* followed by a similar gradual decrease in enzyme activities 40 h after transfer to low iron media. Enzyme activities in the iron-sufficient controls remained constant through the experiment. Verstreate et al. (1980) found a significant decrease in the activities of nitrate reductase, nitrite reductase, and glutamine synthetase in iron-stressed *Anabaena flos-aquae*. High levels of iron (ca. 40 μM) stimulated the activities of all three enzymes in *A. flos-aquae*, but not in the green alga *Scenedesmus bijugatus*.

3.3. The effect on ferredoxin levels

Iron is a primary component in the electron transport system, both as an integral part of heme in cytochromes, and in the non-heme iron-sulfur proteins. Ferredoxins (see Chapter 2 of this volume) transfer electrons by allowing the iron molecule to oscillate between the oxidized (Fe^{3+}) and reduced (Fe^{2+}) state. They can serve as electron acceptors in photosynthesis, mitochondrial respiration, nitrate reduction, nitrite reduction, hydrogenase reactions and dinitrogen fixation.

Under iron-limiting conditions, cells adjust their ferredoxin levels. In an early study, Smillie (1965) noted changes in the ferredoxin levels in iron-stressed *A. nidulans*. Subsequently, two proteins which would transfer electrons to NADP:ferredoxin oxidoreductase were isolated. One was the iron-containing ferredoxin, the second a flavoprotein, termed phytoflavin or flavodoxin, that contained no iron group (see Section 2.1. of Chapter 2).

Several groups examined the relative proportion of ferredoxins to flavodoxins under iron limitation. Hutber et al. (1977) reported the influence of iron deficiency on the production of two ferredoxins and a phytoflavin. At higher iron levels, the two ferredoxins were present at their highest level. Phytoflavin was not present. This arrangement remained constant until the iron content of the medium was reduced to 2 μM. Below this level there was a rapid loss of the two ferredoxins and a similarly sharp increase in the amount of flavodoxin present.

The replacement of the ferredoxins with flavodoxin is significant in that it allows physiological processes, such as dinitrogen fixation, to proceed even under periods of iron stress. The production of an iron-free phytoflavin in response to low iron stress has been used as a biochemical marker of cells under iron stress (Hutber et al., 1977; Entsch et al., 1983).

4. The role of humics in iron chelation

4.1. General considerations

The study of naturally occurring organic materials in terrestrial soils and aquatic sediments has long been an area of interest (Kononova, 1966; Schnitzer and Khan, 1972, 1978). In contrast, knowledge of the organic materials present in aqueous environments is much less detailed (for a recent review, see Boggs et al., 1985). Several terms have been suggested for the humic fraction of aquatic matter, e.g. water humus, Gelbstoff, and gilvin. More important than a label for these compounds is an operational definition of aquatic humic substances analogous to the simple definitions adopted for soil humic substances.

Humic substances are operationally divided into three major groups (Kononova, 1966): (1) humic acids, the alkali (0.5 N NaOH) extractable material from soil which precipitates when the solution is acidified to pH 2; (2) fulvic acids, the alkali-extractable material which remains soluble after the solution is acidified; and (3) humin, the organic material in soil which is not alkali extractable. These definitions have been used for studies on humic substances obtained from both soils and aquatic sediments (Schnitzer and Khan, 1972; Kononova, 1966; Boggs et al., 1985). Humic substances are high molecular weight, polycondensed organic materials with important ecological properties. Of interest to this review is their ability to complex certain trace metals (e.g. Fe) in the aquatic environment.

4.2. Techniques for the characterization of humic substances

Due to the low concentrations of humic matter in natural waters (1–20 $mg \cdot l^{-1}$), humic samples must be enriched prior to study. Column enrichment techniques, using ion exchange resins or hydrophobic adsorbants, provide a convenient and inexpensive technique for the concentration of humic substances. The fraction of dissolved organic matter adsorbed using a specified column (generally XAD-8) when an acidified (pH 2) and filtered sample of water is passed through the resin is used as an operational definition of humic substances.

A variety of approaches are utilized to investigate the structure of humic substances. Physical techniques, such as ultraviolet-visible, infra-red, and Fourier transform nuclear magnetic resonance spectroscopy have been coupled with more classical oxidative and reductive degradations. Molecular weight determinations are made by means of gel filtration and combined with direct measurements by electron microscopy. Additional techniques include pyrolysis-gas chromatography-mass spectrometry, X-ray analysis, and potentiometric titrations. Although these techniques provide an increasing amount of information on the structural organization of humic substances, the complexity of the molecular framework of these materials has hindered attempts to define these compounds in great detail.

4.3. Humic substances in aquatic systems

Recent studies suggest that 10% of the dissolved organic carbon (DOC) in natural waters consists of identifiable compounds, i.e., amino acids, carbohydrates and volatile fatty acids (Christman and Gjessing, 1983). The balance consists of polymeric, macromolecular carbon forms. About 50% of the total DOC is defined operationally (by XAD-8 retention) as humic substances. While aquatic humic acids are thought to originate in part from soil humic substances, Nissenbaum and Kaplan (1972), using isotopic labeling, found that a significant fraction of marine humic substances were formed in situ from the breakdown and lysis of plankton. In addition, Craigie and McLachlan (1964) reported that several species of brown algae released a yellow, ultraviolet-absorbing substance into the surrounding seawater. They postulated that these substances (believed to be polyphenolic in nature) were an important contributor to the humic substances present in coastal seawater.

4.4. Role of humic acids in metal chelation

The role of humic and fulvic acids in the chelation of metal ions and the techniques utilized for studying metal-humic interactions have been extensively reviewed (Schnitzer and Khan, 1978; Boggs et al., 1985). Kerndorff and Schnitzer (1980) examined the binding of humic acids to 11 different metals at varying pH. On the basis of their stability constants, they divided the metals into the following groups: strong binders: Fe^{3+}, Hg^{2+}; intermediate binders: Cu^{2+}, Pb^{2+}, Al^{3+}; and weak binders: Ni^{2+}, Cr^{3+}, Zn^{2+}, Mn^{2+}, Co^{2+}, Cd^{2+}.

Since humic and fulvic acids do not represent defined compounds, classical stability constants for metal-humic interactions as determined by Scatchard analysis may not be applicable. Constants, determined under laboratory conditions, may not correlate well with those in the natural environment. Factors, such as pH, ionic strength, the presence of competing ions, and the type of organic material present are known to influence humic binding to metals (Schnitzer and Hansen, 1970; Boggs et al., 1985).

Humics seem to have a variable effect on the availability of iron. Prakash et al. (1973) and Lange (1970) have reported that high levels of humic acids stimulate $^{14}C\text{-}HCO_3^-$ uptake. However Sakamoto (1971) and Wetzel (1972) noted that humic substances appeared to restrict photosynthesis by making iron unavailable. As mentioned earlier, several reports have correlated the occurrence of phytoplankton blooms with the levels of humic acids in the water (Slobodkin, 1953; Ingle and Martin, 1971; Prakash, 1971), though it is not clear whether this was due to increased availability of trace metals or to increased complexation of toxic elements.

5. Role of siderophores in iron uptake

Siderophores are defined as low molecular mass (500–1000 daltons) microbial iron transport compounds (Greek: sider = iron, phore = bearer). They are produced by a wide range of prokaryotic and eukaryotic microorganisms under iron-limiting conditions and serve as virtually iron-specific extracellular ligands to aid in the solubilization and assimilation of Fe^{3+}. Their chemistry and physiology has been the subject of a number of recent reviews (Neilands, 1974, 1981, 1982, 1984; Hider, 1984; Sugiura and Nomoto, 1984).

5.1. Classification of siderophores

Microbial siderophores can be broadly divided into three major categories: catechols, hydroxamates, and phytosiderophores. There is also a small group of compounds with novel structural features which do not fit readily into these major categories. The structural formulas of representative siderophores are shown in Fig. 1.

5.1.1. Catechol-type siderophores

The phenolate-catechol siderophores are exemplified by the structure of enterobactin (also named

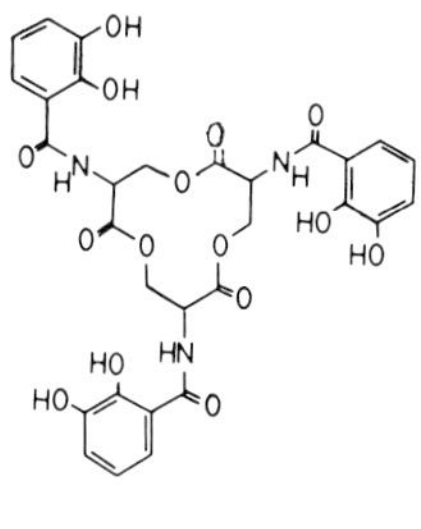

1

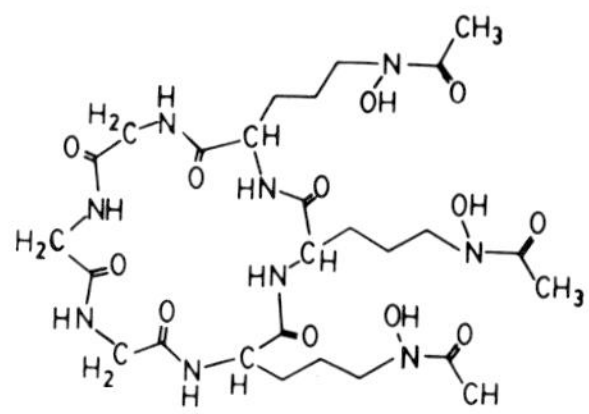

2

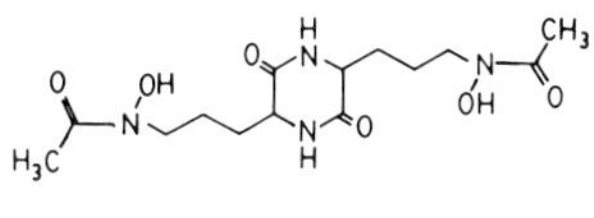

3

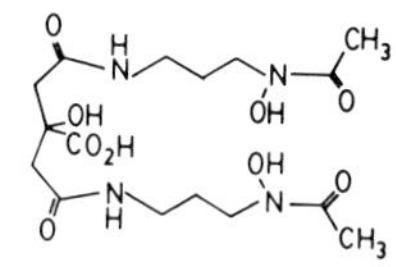

4

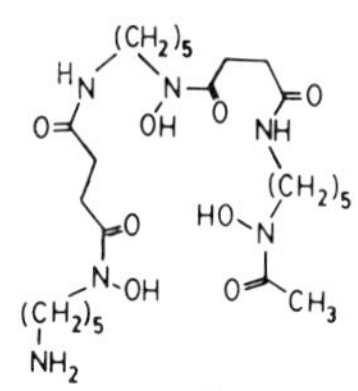

5

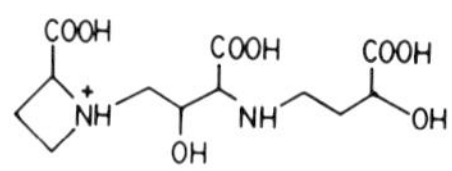

6

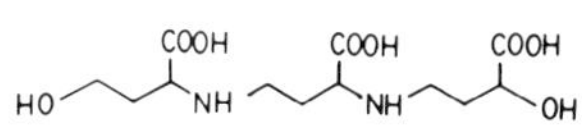

7

Fig. 1. The structure of several representative siderophores: [*1*] enterobactin, [*2*] deferriferrichrome, [*3*] rhodotorulic acid, [*4*] schizokinen, [*5*] desferrioxamine B, [*6*] mugeneic acid, [*7*] avenic acid.

enterochelin, see structure *1*). It can be isolated from low iron cultures of *Salmonella typhimurium* and *Escherichia coli* and is the common siderophore of the enteric bacteria (Pollack and Neilands, 1970; O'Brien and Gibson, 1970; Neilands, 1981). The monomeric unit of enterobactin, 2,3-dihydroxy-*N*-benzylserine, was isolated from iron-deficient cultures of *E. coli* K_{12} (Brot et al., 1966). Other catechol siderophores, 2,3-dihydroxy-*N*-benzylglycine from *B. subtilis* (Ito and Neilands, 1958), the threonine derivative from *Klebsiella oxytoca*, and a di-lysine derivative from *Azotobacter vinelandii* (Corbin and Bulen, 1969) have also been described. The identification of 2,3-dihydroxybenzoic acid and other phenolic acids in a wide range of bacterial filtrates suggests that amino acid-catechol type siderophores may have a broad distribution (Lankford, 1973). In addition to the amino acid linked catechols, two spermidine-linked catechol siderophores, agrobactin and parabactin, have been isolated from low iron cultures of *Agrobacterium tumefaciens* and *Paracoccus* (*Micrococcus*) *denitrificans* (Ong et al., 1979; Tait, 1974).

5.1.2. Hydroxamate-type siderophores

Ferrichrome (structure *2*) was the first of a series of naturally occurring trihydroxamate siderophores to be isolated (Neilands, 1952) and has since been isolated from a number of fungi (Neilands, 1984) and bacteria (Peterson et al., 1980). Ferrichromes are cyclic hexapeptides, containing a tripeptide sequence of *N*-acyl-*N*-hydroxyornithine and a tripeptide composed of small, neutral amino acids such as glycine, serine, or alanine. A detailed description of the ferrichrome series of siderophores is given in Hider (1984).

Rhodotorulic acid (structure *3*), the parent compound in a second class of hydroxamate siderophores, is a cyclic dipeptide of *N*-acetyl-*N*-hydroxyornithine and produced in gram quantities by low iron cultures of *Rhodotorula pilimanae* and related yeasts (Atkin and Neilands, 1968; Atkin et al., 1970). Additional members of this class include dimerumic acid (isolated from *Fusarium dimerum*; Widmer and Keller-Schierlein, 1974), coprogen (a growth factor isolated from *Neurospora* and other fungal species; Keller-Schierlein and Diekmann, 1970), and triornicin and isotriornicin (isolated from the soil fungus *Epicoccum purpurascens;* Frederick et al., 1981, 1982).

Three members of the citrate family of hydroxamate siderophores have been characterized; schizokinen (structure *4*), arthrobactin, and aerobactin, isolated from *Bacillus megaterium*, *Arthrobacter* sp., and *Aerobacter aerogenes* 62-1 (*Klebsiella pneumoniae*), respectively (Mullis et al., 1971; Linke et al., 1972; Gibson and Magrath, 1969). Schizokinen is also the only siderophore that has been isolated from a cyanophyte (*Anabaena* sp.) that has been structurally identified (Simpson and Neilands, 1976).

Other hydroxamate siderophores include the fusarinines (isolated from species of *Fusarium* and *Penicillium*) and ferrioxamines (isolated from *Streptomyces* and *Actinomyces*). The mesylate salt of the linear desferrioxamine B (structure *5*) is marketed under the commercial trade name Desferal for use in deferration therapy.

The mycobactin siderophores produced by mycobacteria contain both the hydroxamate and phenolate-catecholate functionalities and may be viewed as a hybrid of both types (Snow, 1970). The mycobactins are unusual in that both they and their iron complexes are highly lipid soluble. They are thought to function with a second (structurally unknown) extracellular chelating agent termed exochelins (Stephenson and Ratledge, 1980).

5.1.3. Diaminobutyrate-type of siderophores (phytosiderophores)

In higher plants a novel class of iron-binding compounds are secreted from the roots which seem to aid in iron uptake (Mino et al., 1981; Takagi et al., 1984; Sugiura and Nomoto, 1984). These are based on 2,4-diaminobutyrate and the amino acid, azetidine. They constitute the third and newest class of siderophores. Members of the group include mugeneic acid (structure *6*) and distichonic acid A from barley (Takemoto et al., 1978; Sugiura and Nomoto, 1984), avenic acid (structure *7*) from oats (Fushiya et al., 1980), 2′-deoxymugeneic acid from wheat (Nomoto et al., 1981), 3-hydroxymugeneic acid from rye (Sugiura and Nomoto, 1984), and nicotianamine from a number of different species (Noma and Noguchi, 1976; Kristensen and Larsen, 1974; Budesinsky et al., 1980). Azetidine-2-carboxylic acid, a possible precursor to both mugeneic acid and nicotianamine, has been found in a wide variety of plants (Kristensen et al., 1974; Sugiura and Nomoto, 1984).

In addition to the three major types of siderophores, there exist several compounds that do not readily fit into the above classes. Rhizobactin, recently isolated from *Rhizobium meliloti*, resembles the phytosiderophores in that it lacks hydroxamate and catechol ligands for its iron binding (Smith et al., 1985). Low molecular weight iron (III) binding compounds have been isolated from *Corynebacterium diphtheriae* and *Aeromonas salmonicida*, which also apparently lack these ligands, though their detailed structure is unknown (Chart and Trust, 1983; Russell and Holmes, 1983). Other iron-binding compounds isolated from microbial sources include aspergillic acid, mycelianamide, and pulcherriminic acid, but it is unknown if these compounds act as true siderophores.

5.2. Isolation and characterization of siderophores

In some cases, e.g. ferrichrome, it is advantageous to isolate the siderophore as its ferric complex. For others, e.g. rhodotorulic acid, the free ligand is more easily extracted and purified. Solvent extractions with benzyl alcohol or a 1:1 (w/w) mixture of phenol and chloroform have been widely used by Neilands and co-workers. Enterobactin, a catechol-type siderophore, was isolated from a concentrated acidified cell-free supernatant by ether extraction, crystallization from hot benzene-dioxane, and recrystallization from ethanol-water (Pollack and Neilands, 1970). Crude rhodotorulic acid was precipitated from a chilled supernatant of *Rhodotorula rubra* and further purified using phenol/chloroform extraction (Atkin and Neilands, 1968). Procedures such as column chromatography may be necessary. Schizokinen was isolated from cultures of *Bacillus megaterium* using ion exchange chromatography followed by gel permeation chromatography to concentrate and purify the material (Mullis et al., 1971).

Modern techniques of instrumental analysis have been utilized for the structural analysis of siderophores. Both ^{1}H- and ^{13}C-NMR (nuclear magnetic resonance) have been applied with success to the iron-free siderophores, though the paramagnetic line broadening due to Fe^{3+} prevents their use with the ferric compounds. In the past, mass spectrometry of siderophores has not proved of great utility due to the low volatility of these compounds. However, recent applications using soft ionization techniques such as field desorption and fast atom bombardment have been successfully used on a wide range of siderophores to determine their molecular ions (Dell et al., 1982). Such techniques show considerable promise for the characterization of novel compounds.

The stability constants of a wide range of siderophores have been measured using standard methods of inorganic and coordination chemistry. The catechol siderophore enterobactin was shown to form the most stable Fe^{3+} complex with a formation constant of 10^{52}. The majority of bacterial siderophores have formation constants in the range of 10^{30} (Neilands, 1984) while the phytosiderophores generally range around 10^{18} (Sugiura and Nomoto, 1984).

5.3. The occurrence of siderophores in cyanobacteria

Several different techniques have been utilized to investigate the occurrence of siderophores in cyanobacteria (Table II). While no single method is a universal indicator for the presence or absence of these compounds, their occurrence can be investigated using a combination of approaches. Some species (e.g. *Agmenellum quadruplicatum* PR-6) produce siderophores detectable using a variety of analytical methods, while others (e.g. *Gomphosphaeria aponina*) do not produce compounds detectable by any technique. They may still produce siderophores, but current techniques, both biological and chemical, are inadequate to detect them.

TABLE II

Siderophore activity in cyanobacteria

Species (isolate)	Growth[a]	JG-9[b]	Cu[c]	Csaky[d]	Ref.[e]
Agmenellum quadruplicatum (PR6)	+ (−Fe−C)	+		+	1
A. quadruplicatum (BG1)	+ (−Fe−C)	−		−	1
Anabaena circinalis (W1038)	− (−C)				2
Anabaena cylindrica (UCB 629)	+ (−Fe)			+	3
A. cylindrica (IU 629)	+ (−C)				2
A. cylindrica (UTEX 1611)	+ (−C)			+	4, 5

TABLE II (*continued*)

Species (isolate)	Growth[a]	JG-9[b]	Cu[c]	Csaky[d]	Ref.[e]
A. cylindrica (UTEX 144)	+ (–C)		+		6
Anabaena flos-aquae	– (?)	+		+	7
A. flos-aquae	+ (–Fe–C)	+		+	8
A. flos-aquae (UTEX)	+				9
A. flos-aquae (UTEX 1444)	+ (–C)		+	+	4, 5
Anabaena sp. (No. 6411)	+ (–Fe)			+[f]	3
Anacystis marina (WH-20)		–		–	1
Anacystis nidulans (No. 6301)	+ (–C)		+	+	4
Anacystis sp. (UCB No. 69)	+ (–Fe)			–	3
Aphanizomenon flos-aquae (W1062)	–				2
Chroococcus sp. (No. 6501)	+ (–Fe)			–	3
Coccochloris elabens (17A)	+ (–Fe–C)	+		+	1
C. elabens (Di)	+ (–Fe–C)	–		–	1
Coccochloris peniocystis (No. 6307)	+ (–Fe)			–	3
Coccomyxa chodatii (UTEX 279)	(+) (–C)			+	4
Gloeocapsa alpicola (No. 6308)	+ (–Fe)			–	3
G. alpicola (No. 6308)	+ (–C)		+	+	4
Gloeotrichia echinulata (W1052)	–				2
Gomphosphaeria aponina	+ (–Fe)	–		–	10
Lyngbya lagerheimii (Mont)		–		–	1
Lyngbya sp. (IU 487)	+ (–C)				2
Microcoleus chthonoplastes (Ba-1)		–		–	1
Microcystis aeruginosa (W1036)	+ (–C)				2
M. aeruginosa	– (–Fe)	–		+	7, 8
M. aeruginosa (UTEX 2061)	+ (–C)		+	+	4
Nostoc muscorum (W1013)	+ (–C)				2
N. muscorum (UTEX 1832)	+ (–C)			+	4
Oscillatoria rubescens (clone 51)	+ (–C)				2
Oscillatoria williamsii (Mev)		–		–	1
Oscillatoria sp. (JCM)		–		–	1
Phormidium autumnale	– (–Fe)	–		+	7
Phormidium faveolarum (IU427)	+ (–C)				2
Plectonema terebrans (Cal sp-31)		–		–	1
Schizothrix calcicola (Man)		–		–	1
Synechococcus leopoliensis (UTEX 625)	+ (–C)		+	+	5
Synechococcus sp. (WHOI DC-2)				+	11
unidentified filamentous (A-2)		–		–	1

[a]Growth on media deficient in iron (–Fe) or EDTA (–C).
[b]*Arthrobacter* JG-9 bioassay.
[c]Chelators also exhibited strong copper binding activity.
[d]Csaky test for bound hydroxylamines [Csaky, 1948].
[e]References: (1) Armstrong and Van Baalen, 1979; (2) Lange, 1974; (3) Simpson and Neilands, 1976; (4) McKnight and Morel, 1979; (5) McKnight and Morel, 1980; (6) Van den Berg et al., 1979; (7) Murphy et al., 1976; (8) Murphy, 1976; (9) Murphy et al., 1983; (10) Eng-Wilmont and Martin, 1979; (11) Gillam, unpublished.
[f]Structure was determined to be schizokinen [structural formula *4*].

5.3.1. Biological tests

The first test for siderophore production is the ability to grow on iron- and/or chelator-deficient media. The assumption is that siderophores increase the bioavailability of residual iron in the media. Lange (1974) examined 10 species of cyanobacteria for the ability to grow in the absence of external chelator (EDTA) and found only six species which where able to condition the media to support their own growth. The preconditioned medium would also support the growth of the four other 'non-chelator' producing cyanobacterial species, thus supporting the hypothesis that an extracellular metabolite was responsible for the observed growth increase. The organic chelators produced by these six species of cyanobacteria in response to iron or chelator depletion need not necessarily be high affinity siderophores. Several species of algae, including at least three species of cyanobacteria produce only low affinity chelators in response to growth on chelator-deficient media (McKnight and Morel, 1979).

A second test for siderophore activity is the ability to promote growth in siderophore auxotrophs. The organism commonly employed in this bioassay is the soil bacterium *Arthrobacter flavescens* JG-9 (ATCC 29091), an obligate auxotroph for hydroxamate siderophores (Neilands, 1984). Though it does not respond to the catechol siderophores, JG-9 is quite non-specific in its response to a wide range of hydroxamate siderophores (including ferrichrome, ferrichrome A, rhodotorulic acid, and schizokinen) and some non-siderophore compounds (haemin, aspergillic acid, kojic acid, and 2,3-dihydroxybenzoyl serine) and synthetic chelators (Burnham and Neilands, 1961; Morrison et al., 1965; Emery and Emery, 1973). Estep et al. (1975) utilized this system to show blue-green algal mats from various near-shore marine ecosystems contained JG-9 growth promoting activity. Other mutant test organisms (e.g. *Salmonella typhimurium* LT-2) exist that respond to a wide range of siderophores (including catechols, Luckey et al., 1972), however they have yet to be applied to work with cyanobacteria.

A third biological approach is the ability to ameliorate copper toxicity. The free cupric ion is toxic to phytoplankton at very low metal concentrations (Sunda and Guillard, 1976; Sunda and Lewis, 1978; Anderson and Morel, 1978; Morel et al., 1978). Phytoplankton, including several species of blue-green algae, produce strong copper-complexing agents that reduce this free metal ion concentration (Van den Berg et al., 1979; McKnight and Morel, 1979). The production of strong copper-complexing compounds in four species of cyanobacteria was greatly stimulated by iron limitation but was not affected by increased cupric ions. The iron-chelator complex is significantly more stable than the copper-chelator complex, hence it was hypothesized that these strong copper-complexing compounds were hydroxamate siderophores (McKnight and Morel, 1980).

5.3.2. Chemical tests

Two major chemical tests have been utilized in search of siderophore production in cyanobacteria; the Arnow test for catechols (Arnow, 1937) and the Csáky method for bound hydroxylamines (Csáky, 1948; Gillam et al., 1981). While the Arnow test was utilized to screen numerous species of blue-green algae (*Anabaena* sp., *A. cylindrica*, *A. spiroides*, *Chroococcus* sp., *Coccochloris peniocystis*, *Gloeocapsa alpicola*, *Anacystis* sp.), the results have been consistently negative, providing no evidence in support of the presence in cyanobacteria of catechol-type siderophores common in enteric bacteria (Simpson and Neilands, 1976; Murphy, 1976).

Several investigators have used the Csáky test to screen both cell extracts and media filtrates for the presence of hydroxamate-type siderophores. The results of these tests are summarized in Table II. Murphy et al. (1976) reported that iron-stressed *Anabaena flos-aquae* produced a bound hydroxamate whose peak of natural abundance corresponded with the peak period of iron uptake. Isolation of the active chelate by column chromatography showed that bound iron, hydroxamate content as determined by the Csáky test, and JG-9 bioassay activity all co-occurred in the same fraction. Gillam (unpublished) has shown that filtrates from the marine blue-green algae *Synechococcus* sp. grown in Fe-free medium contained 3.6 $\mu g \cdot l^{-1}$ Desferal equivalents of hydroxamate.

Armstrong and Van Baalen (1979) found a high level of Csáky-positive material (1 mg Desferal equiv.$\cdot l^{-1}$) in filtrates of *Agmenellum quadruplicatum* PR-6 when grown on low iron media (0.1 $\mu g \cdot ml^{-1}$) lacking EDTA. The production of Csáky-positive material was strain specific in that a related strain, *A. quadruplicatum* (BG-1) did not produce Csáky-positive material when grown under similar conditions. Hydroxamate production was sensitive to both N and Fe levels in the medium; it increased about 100-fold when 2 mM NH_4Cl was added to the medium and was totally repressed by the addition of 50 μM iron. Siderophore production in *Agmenellum* (as measured by the JG-9 bioassay) occurred in a pulse, with maximal activity 48 h after the start of iron depletion. Extracellular siderophore was removed from the medium as the algae showed increased growth, an observation qualitatively similar to that observed by Trick et al. (1983) in marine phytoplankton. The active material from *Agmenellum* was purified by chloroform-phenol extraction and LH-20 column chromatography to yield 2.5 $mg \cdot l^{-1}$ of a compound whose ultraviolet absorption spectra showed a broad peak at 430 nm, typical of ferric hydroxamates. Only a small absorbance change was observed with changing pH (pH 2.0–9.5), suggesting the compound was a trihydroxamate (Neilands, 1967). This UV behavior, coupled with the observations that the biosynthesis of the compound was strongly de-repressed by the depletion of iron, and the positive bioassay results using both iron-depleted *Arthrobacter* JG-9 and *Agmenellum quadruplicatum* (PR-6), suggested that the compound acts and functions as a siderophore.

Similarly, Simpson and Neilands (1976) found Csáky-positive material in two isolates of *Anabaena* grown in iron and EDTA-deficient media. In *A. cylindrica*, the Csáky reaction was strongly dependent on the level of combined nitrogen in the medium and was not repressed by the addition of 10 μM ferrous sulfate. It is unlikely that this compound was a high affinity siderophore. In contrast, the production of Csáky-positive material in cultures of *Anabaena* sp. was completely repressed by the addition of 0.5 μM Fe. After extraction into chloroform-phenol, ion exchange chromatography followed by gel chromatography yielded approximately 60 mg purified siderophore from a 20-l culture. The compound was identified as schizokinen on the basis of the following evidence: (1) the iron-free compound exhibited the same relative mobility on paper electrophoresis as authentic schizokinen, (2) hydrolysis of both schizokinen and the *Anabaena* siderophore released 1-amino-3-*N*-hydroxyaminopropane, and (3) the ^{1}H-NMR spectrum of the purified material was essentially identical with that reported by Mullis et al. (1971) for schizokinen isolated from *Bacillus megaterium*. In addition, the siderophore from *Anabaena* was accompanied by variable amounts of a neutral substance. Though this material was not characterized further, it is similar to what is observed in *B. megaterium* and may correspond to schizokinen A.

It is also possible that cyanobacterial siderophores contain neither hydroxamate nor phenolate ligands. The phytosiderophores (e.g. mugeneic acid), rhizobactin, and the iron-binding compounds isolated from the bacteria *Corynebacterium diphtheriae* and *Aeromonas salmonicida* all give negative Arnow and Csáky tests (Chart and Trust, 1983; Russell and Holmes, 1983; Sugiura and Nomoto, 1984; Smith et al., 1985). *Anabaena cylindrica* and the green alga *Scenedesmus basiliensis* were both reported to excrete small molecular weight peptides that bind iron. These peptides required acid hydrolysis prior to reaction with ninhydrin, suggesting that they might be cyclic in structure (Fogg and Westlake, 1955; Murphy, 1976). Thus, the possibility exists that undetected siderophores are present in cyanobacteria. Work is needed to develop new chemical and biological assays that detect a wider range of iron-binding compounds.

5.4. Siderophore-mediated iron uptake

5.4.1. *General considerations*

In view of the large number of different siderophores now recognized, more than one uptake system for siderophore-mediated iron transport may exist. In *Escherichia coli*, four different iron uptake systems have been characterized, each involved in the transport of the iron complex into the cell. These include: (1) the iron (III)-enterobactin receptor, (2) the ferrichrome receptor, (3) the iron (III)-citrate receptor, and (4) the iron (III)-aerobactin receptor (Hider, 1984). All four receptors have separate outer membrane receptor proteins which can be distinguished on the basis of their binding of cell toxins and antibiotics. In the case of the iron (III)-aerobactin uptake system, the outer membrane receptor is plasmid encoded and has been cloned (Bindereif and Neilands, 1983). Similar multiple siderophore transport systems have been observed for the related species *Salmonella typhimurium* (Hider, 1984).

There also exist a number of mechanisms to increase the availability of iron that are not siderophore dependent. Higher plants are reported to utilize a number of approaches to increase iron absorption including the release of hydrogen ions, reducing compounds, and/or organic acids (particularly citrate) from the root. Reduction of Fe^{3+} to form the more soluble Fe^{2+} at the root membrane has also been reported for higher plants and fungi (Brown, 1978; Ecker et al., 1982). Some algae are able to absorb free iron directly. Anderson and Morel (1982) have examined the uptake rates of ^{59}Fe in the diatom *Thalassiosira weissflogii* in the presence of different chelators, and concluded that the uptake was dependent on the free Fe^{3+} concentration. They have postulated that iron uptake in *Thalassiosira* is an energy-dependent process requiring the presence of a phytotransferrin molecule located on the membrane surface (Anderson and Morel, 1980, 1982).

Eng-Wilmont and Martin (1979) studied iron uptake kinetics in continuous cultures of the marine blue-green *Gomphosphaeria aponina*, and found that iron (as complexed with EDTA) was growth limiting for *G. aponina*. Uptake of iron exhibited biphasic kinetics. The initial rate was fast, light independent, and similar in iron-deficient and iron-sufficient cells. This was interpreted as an initial adsorption of iron on to the cell surface. It was followed by a slower uptake rate, presumably dependent on iron utilization. An identical pattern of iron uptake has been reported for *Anabaena flos-aquae*.

Both *A. flos-aquae* and *G. aponina* have been examined for the presence of siderophores as a component of their iron uptake systems. *A. flos-aquae* grown under iron limitation produced Csáky-positive bound hydroxamate and gave positive results with the *Arthrobacter* JG-9 bioassay (Murphy, 1976). *G. aponina* did not appear to produce detectable amounts of a hydroxamate siderophore when assayed by either the Csáky test or the *Arthrobacter* JG-9 bioassay. Care must be therefore taken to avoid the generalization that, because a cyanobacterium contains or can utilize an available siderophore, it also provides a physiologically important route of iron uptake in the organism. In all but a few cases, our knowledge of siderophores and cyanobacterial ecology is insufficient to make such claims.

5.4.2. *Siderophore-mediated transport in cyanobacteria*

While mechanisms of siderophore uptake in cyanobacteria are much less characterized than those in the enteric bacteria, siderophore uptake in *Anabaena* species has been examined in a series of papers by Sanders-Loehr and coworkers (Lammers and Sanders-Loehr, 1982a, 1982b; Goldman et al., 1983; Plowman et al., 1984). *Anabaena* sp., grown in a medium containing less than 1 μM $FeCl_3$, showed a marked increase in the production of schizokinen. There was also an increase in an iron transport system that allowed for increased uptake of iron (III)-schizokinen. Transport activity showed a high specificity for iron (III)-schizokinen with an apparent K_m of 0.04 μM, similar to the K_m's reported for the high affinity uptake systems of ferrienterobactin (0.10–0.36 μM) and ferrichrome (0.15–0.25 μM) (Negrin and Neilands, 1978; Lammers and Sanders-Loehr, 1982a) in *E. coli*. Iron (III)-aerobactin and iron (III)-acetylschizokinen were transported by the same uptake system but with a lower affinity than observed for iron (III)-schizokinen. Desferal (desferrioxamine B) showed little, if any, affinity for the transport receptor. The uptake of siderophore displayed typical carrier-mediated kinetics and was saturated at concentrations of Fe-schizokinen greater than 0.1 μM.

Studies with metabolic inhibitors have shown that ferrischizokinen uptake is an energy-dependent process (Table III). Two possible sources of energy, the direct involvement of ATP or the use of the transmembrane electrical potential, are available for nutrient uptake. Work with *E. coli* and *Neurospora* has implicated the membrane potential as the driving force for siderophore uptake (Negrin and Neilands, 1978; Winkelmann and Huschka, 1984). In contrast, siderophore uptake in *Anabaena* is believed to involve ATP as the primary energy source (Lammers and Sanders-Loehr, 1982a). Transport of ferrischizokinen was blocked in both light- and dark-grown cells by disruption of the electrochemical proton gradient with the uncouplers carbonylcyanide-*m*-chlorophenylhydrazone (CCCP) or 2,4-dinitrophenol (DNP). Addition of *N,N'*-dicyclohexylcarbodiimide (DCCD), which blocks the proton translocating ATPase in bacteria, chloroplasts and mitochondria (and hence the formation of ATP), also obviates the uptake of ^{55}Fe-labeled siderophore. That ATP is required as the energy source in *Anabaena* is supported by the finding that siderophore uptake is inhibited by sodium arsenate, an ATP analog. Inhibition of iron uptake by cyanide in the dark indicates that the ATP requirement can be met by means of oxidative phosphorylation, while lack of inhibition of iron uptake by cyanide in the light strongly suggests that the required ATP is generated by photophosphorylation.

TABLE III

Effect of inhibitors on ferrischizokinen uptake in *Anabaena* sp.

Inhibitor	Light uptake[a]	Dark uptake[a]
Uncouplers:		
CCCP, 8 μM	10%	0%
DNP, 100 μM[b]	20%	–
ATPase inhibitors:		
DCCD, 40 μM	5%	5%
arsenate, 20 mM	50%	15%
Respiration inhibitors:		
KCN, 100 μM	90%	20%

[a]Uptake activity relative to light or dark control in the absence of an inhibitor.
[b]From Lammers and Sanders-Loehr, 1982b.
Abbreviations: CCCP, carbonylcyanide-*m*-chlorophenylhydrazone; DNP, 2,4-dinitrophenol; DCCD, *N,N'*-dicyclohexylcarbodiimide.
(from Lammers and Sanders-Loehr, 1982a)

6. Summary; iron and cyanobacterial ecology

It has been increasingly recognized that iron, which is an important component of a great number of enzymes, is a major determinant of primary production in a wide variety of ecosystems. Because of the low solubility of ferric iron, organisms do not respond to the total iron concentration but rather to the 'biologically available iron', the ratio of which is greatly affected by the level of chelators present. Blue-green algae in nature can make use of a wide variety of chelators to solubilize their needed iron. These range from the relatively non-specific (but widespread) humic substances to the production of highly specific siderophores that can be actively transported into the cell.

The addition of humic substances has been shown to increase growth of a wide variety of phytoplankton. Several efforts have been made to predict blooms of algae on the basis of an iron index related to the level of humic acids in the river run-off. However, these compounds are relatively non-specific chelators, hence it is difficult to determine if the increased growth is due to increased solubility of the essential trace metals such as iron, or to the decreased levels of inhibitory elements such as copper.

In contrast, the primary function of siderophores is to increase the solubility and uptake of iron. Cyanobacteria have been shown to produce siderophore-like compounds though only one, schizokinen from *Anabaena* sp., has been rigorously characterized. Production of the siderophore is induced along with a specific membrane uptake system under conditions of low iron availability. This ability to tie up iron in a form that is available only to the cyanobacterium may play an important part in the establishment and domination of cyanobacterial populations.

References

Allen, M.M. and Smith, A.J. (1969) Nitrogen chlorosis in blue-green algae. Arch. Mikrobiol. 69, 114–120.

Anderson, D.M. and Morel, F.M.M. (1978) Copper sensitivity of *Gonyaulax tamarensis*. Limnol. Oceanogr. 23, 283–295.

Anderson, M.A. and Morel, F.M. (1980) Uptake of Fe(III) by a diatom in oxic culture medium. Mar. Biol. Lett. 1, 263–268.

Anderson, M.A. and Morel, F.M.M. (1982) The influence of aqueous iron chemistry on the uptake of iron by the coastal diatom *Thalassiosira weissflogii*. Limnol. Oceanogr. 27, 789–813.

Armstrong, J.E. and Van Baalen, C. (1979) Iron transport in microalgae: the isolation and biological activity of a hydroxamate siderophore from the blue-green algae *Agmenellum quadruplicatum*. J. Gen. Microbiol. 111, 253–262.

Arnow, L.E. (1937) Colorimetric determination of the components of 3,4-dihydroxyphenylalanine-tyrosine mixtures. J. Biol. Chem. 118, 531–537.

Atkin, C.L. and Neilands, J.B. (1968) Rhodotorulic acid, a diketopiperazine dihydroxamic acid with growth-factor activity. I. Isolation and characterization. Biochemistry 7, 3734–3739.

Atkin, C.L., Neilands, J.B. and Phaff, H.J. (1970) Rhodotorulic acid from species of *Leucosporidium*, *Rhodosporidium*, *Rhodotorula*, *Sporidiobolus*, and *Sporobolomyces*, and a new alanine-containing ferrichrome from *Cryptococcus melibiosum*. J. Bacteriol. 103, 722–733.

Barber, R.T. and Ryther, J.H. (1969) Organic chelators: factors affecting primary production in the Chromwell current upwelling. J. Exp. Mar. Biol. Ecol. 3, 191–199.

Bienfait, H.F. and Van der Mark, F. (1983) Phytoferritin and its role in iron metabolism. Ann. Proc. Phytochem. Soc. 21, 111–123.

Bindereif, A. and Neilands, J.B. (1983) Cloning of the aerobactin-mediated assimilation system of plasmid ColV. J. Bacteriol. 153, 1111–1113.

Boggs, S.Jr., Livermore, D.G. and Seitz, M.G. (1985) Humic macromolecules in natural waters. J. Macromol. Sci.–Rev. Macromol. Chem. Phys. C25, 599–657.

Boussiba, S. and Richmond, A.E. (1980) C-Phycocyanin as a storage protein in the blue-green alga *Spirulina platensis*. Arch. Microbiol. 125, 143–147.

Brot, N., Goodwin, J. and Fales, H. (1966) In vitro and in vivo formation of 2,3-dihydroxybenzoylserine by *Escherichia coli* K12. Biochem. Biophys. Res. Comm. 25, 454–461.

Brown, J.C. (1978) Mechanism of iron uptake by plants. Plant Cell Environ. 1, 249–257.

Budešínsky, M., Budzikiewicz, H., Procházka, Z., Ripperger, H., Römer, A., Scholz, G. and Schreiber, K. (1980) Nicotianamine, a possible phytosiderophore of general occurrence. Phytochemistry 19, 2295–2297.

Burnham, B.F. and Neilands, J.B. (1961) Studies on the metabolic function of the ferrichrome compounds. J. Biol. Chem. 236, 554–559.

Canterford, G.S. (1979) Effect of EDTA on growth of the marine diatom *Ditylum brightwellii* (West) Grunow. Aust. J. Mar. Freshwater Res. 30, 765–772.

Carnahan, J.E. and Castle, J.E. (1958) Some requirements of biological nitrogen fixation. J. Bacteriol. 75, 121–124.

Chart, H. and Trust, T.J. (1983) Acquisition of iron by *Aeromonas salmonicida*. J. Bacteriol. 156, 758–764.

Christman, R.F. and Gjessing, E.T. (eds.) (1983) Aquatic and Terrestrial Humic Materials. Ann Arbor Sci. Publ., Ann Arbor, Michigan, 538p.

Corbin, J.L. and Bulen, W.A. (1969) The isolation and identification of 2,3-dihydroxybenzoic acid and 2-*N*,6-*N*-di-(2,3-dihydroxybenzoyl)-L-lysine formed by iron deficient *Azotobacter vinelandii*. Biochemistry 8, 757–762.

Craigie, J.S. and McLachlan, J. (1964) Excretion of colored ultraviolet absorbing substances by marine algae. Can. J. Bot. 42, 23–33.

Csáky, T.Z. (1948) On the estimation of bound hydroxylamine in biological materials. Acta Chem. Scand. 2, 450–454.

David, C.N. and Easterbrook, K. (1971) Ferritin in the fungus *Phycomyces*. J. Cell Biol. 48, 15–28.

Dell, A., Hider, R.C., Barber, M., Bordoli, R.S., Sedgwick, R.D., Tyler, A.N., and Neilands, J.B. (1982) Field desorption and fast atom bombardment mass spectrometry of hydroxamate containing siderophores. Biomed. Mass Spect. 9, 158–161.

Droop, M.R. (1973) Some thoughts on nutrient limitation in algae. J. Phycol. 9, 264–272.

Ecker, D.J., Lancaster, J.R., Jr. and Emery, T. (1982) Siderophore iron transport followed by electron paramagnetic resonance spectroscopy. J. Biol. Chem. 257, 8623–8626.

Emery, T. and Emery, L. (1973) The biological activity of some siderophore derivatives. Biochem. Biophys. Res. Comm. 50, 670–675.

Eng-Wilmont, D.L. and Martin, D.F. (1979) Uptake of iron by *Gomphosphaeria aponina*, a possible control agent for the Florida red tide *Pytochodiscus brevis*. Microbios 26, 103–113.

Entsch, B., Sim, R.G. and Hatcher, B.G. (1983) Indications from photosynthetic components that iron is a limiting nutrient in primary producers on coral reefs. Mar. Biol. 73, 17–30.

Estep, M., Armstrong, J.E. and Van Baalen, C. (1975) Evidence for the occurrence of specific iron (III)-binding compounds in near-shore marine ecosystems. Appl. Microbiol. 30, 186–188.

Evans, E.H., Rush, J.D., Johnson, C.E. and Evans, M.C.W. (1979) Mossbauer spectra of photosystem-I reaction centres from the blue-green alga, *Chlorogloea fritschii*. Biochem. J. 182, 861–865.

Finden, D.A.S., Tippings, E., Jaworski, G.H.M. and Reynolds, C.S. (1984) Light-induced reduction of natural iron (III) oxide and its relevance to phytoplankton. Nature 309, 783–784.

Fogg, G.E. and Westlake, D.F. (1955) The importance of extracellular products of algae in freshwater. Verh. Int. Verein. Limnol. 12, 219–232.

Frederick, C.B., Bentley, M.D. and Shive, W. (1981) Structure of triornicin, a new siderophore. Biochemistry 20, 2436–2438.

Frederick, C.B., Bentley, M.D. and Shive, W. (1982) The structure of the fungal siderophore, isotriornicin. Biochem. Biophys. Res. Comm. 105, 133–138.

Fushiya, S., Sato, Y., Nozoe, S., Komoto, K., Takemoto, T. and Takagi, S. (1980) Avenic acid A, a new amino acid possessing an iron chelating activity. Tetrahedron Lett. 21, 3071–3072.

Gibson, F. and Magrath, D.I. (1969) The isolation and characterization of a hydroxamic acid (aerobactin) formed by *Aerobacter aerogenes* 62-I. Biochim. Biophys. Acta 192, 175–184.

Gillam, A.H., Lewis, A.G. and Andersen, R.J. (1981) Quantitative determination of hydroxamic acids. Anal. Chem. 53, 841–844.

Glover, H.E. (1977) Effects of iron deficiency on *Isochrysis galbana* (Chrysophyceae) and *Phaeodactylum tricornutum* (Bacillariophyceae). J. Phycol. 13, 208–212.

Glover, H.E. (1978) Iron in Maine coastal waters; seasonal variation and its apparent correlation with a dinoflagellate bloom. Limnol. Oceanogr. 23, 534–537.

Goldman, C.R. (1964) Primary productivity and micronutrient limiting factors in some North American and New Zealand lakes. Verh. Int. Verein. Limnol. 15, 365–374.

Goldman, S.J., Lammers, P.J., Berman, M.S. and Sanders-Loehr, J. (1983) Siderophore-mediated iron uptake in different strains of *Anabaena* sp. J. Bacteriol. 156, 1144–1150.

Gorham, P.R., McLachlan, J.L., Hammer, U.T. and Kim, U.K. (1964) Isolation and culture of toxic strains of *Anabaena flos-aquae* (Lyngb.). Verh. Int. Verein. Theor. Angew. Limnol. 15, 796–804.

Guikema, J.A. and Sherman, L.A. (1984) Influence of iron deprivation on the membrane composition of *Anacystis nidulans*. Plant Physiol. 74, 90–95.

Hardie, L.P., Balkwill, D.L. and Stevens, S.E. (1983a) Effects of iron starvation on the physiology of the cyanobacterium *Agmenellum quadruplicatum*. Appl. Environ. Microbiol. 45, 999–1006.

Hardie, L.P., Balkwill, D.L. and Stevens, S.E. (1983b) Effects of iron starvation on the ultrastructure of the cyanobacterium *Agmenellum quadruplicatum*. Appl. Environ. Microbiol. 45, 1007–1017.

Hewitt, E.J. (1983) A perspective on mineral nutrition: essential and functional metals in plants. Ann. Proc. Phytochem. Soc. 21, 277–323.

Hider, R.C. (1984) Siderophore mediated absorption of iron. Struct. Bonding 58, 25–88.

Hutber, G.N., Hutson, K.G. and Rogers, L.J. (1977) Effect of iron deficiency on levels of two ferredoxins and flavodoxin in a cyanobacterium. FEMS Microbiol. Lett. 1, 193–196.

Ingle, R.M. and Martin, D.F. (1971) Prediction of the Florida red tide by means of the iron index. Environ. Lett. 1, 69–74.

Ito, T. and Neilands, J.B. (1958) Products of 'low-iron fermentation' with *Bacillus subtilis*: isolation, characterization, and synthesis of 2,3-dihydroxybenzoylglycine. J. Am. Chem. Soc. 80, 4645–4647.

Jackson, G.A. and Morgan, J.J. (1978) Trace metal-chelator interactions and phytoplankton growth in seawater media: Theoretical analysis and comparison with reported observations. Limnol. Oceanogr. 23, 268–282.

Keller-Schierlein, W. and Diekmann, H. (1970) Stoffwechselprodukte von Mikroorganismen: zur Konstitution des Coprogens. Helv. Chim. Acta 53, 2035–2044.

Kerndorff, H. and Schnitzer, M. (1980) Sorption of metals on humic acid. Geochim. Cosmochim. Acta 44, 1701–1708.

Kononova, M.M. (1966) Soil Organic Matter. Pergamon Press, Oxford, 544p.

Kristensen, I. and Larsen, P.O. (1974) Azetidine-2-carboxylic acid derivatives from seeds of *Fagus silvatica* L. and a revised structure for nicotianamine. Phytochemistry 13, 2791–2798.

Kristensen, I., Larsen, P.O. and Sorensen, H. (1974) Free amino acids and γ-glutamyl peptides in seeds of *Fagus silvatica*. Phytochemistry 13, 2803–2811.

Kuwabara, J.S. (1981) Gametophytic growth by *Microcystis pyrifera* (Phaeophyta) in response to various iron and zinc concentrations. J. Phycol. 17, 417–419.

Lammers, P.J. and Sanders-Loehr, J. (1982a) Active transport of ferric schizokinen in *Anabaena* sp. J. Bacteriol. 151, 288–294.

Lammers, P.J. and Sanders-Loehr, J. (1982b) Active transport of ferrischizokinen in cyanobacteria. In P. Saltman and Hegenauer, J. (eds.), The Biochemistry and Physiology of Iron. Elsevier/North-Holland, pp. 720–721.

Lange, W. (1970) Cyanophyta-bacteria systems: effects of added carbon compounds or phosphate on algal growth at low nutrient concentrations. J. Phycol. 6, 230–234.

Lange, W. (1974) Chelating agents and blue-green algae. Can. J. Microbiol. 20, 1311–1321.

Lankford, C.E. (1973) Bacterial assimilation of iron. CRC Crit. Rev. Microbiol. 2, 273–331.

Lehmann, M. and Wober, G. (1976) Accumulation, mobilization and turn over of glycogen in the blue-green bacterium *Anacystis nidulans*. Arch. Microbiol. 111, 93–97.

Lewin, J. and Chen, C.-H. (1971) Available iron: a limiting factor for marine phytoplankton. Limnol. Oceanogr. 16, 670–675.

Linke, W.-D., Crueger, A. and Diekmann, H. (1972) Stoffwechselprodukte von Mikroorganismen: zur Konstitutions des Terregen-Faktors. Arch. Mikrobiol. 85, 44–50.

Luckey, M., Pollack, J.R., Wayne, R., Ames, B.N. and Neilands, J.B. (1972) Iron uptake in *Salmonella typhimurium*: utilization of exogenous siderochromes as iron carriers. J. Bacteriol. 111, 731–738.

McKnight, D.M. and Morel, F.M.M. (1979) Release of weak and strong copper-complexing agents by algae. Limnol. Oceanogr. 24, 823–837.

McKnight, D.M. and Morel, F.M.M. (1980) Copper complexation by siderophores from filamentous blue-green algae. Limnol. Oceanogr. 25, 62–71.

McLachlan, J.L., Hammer, U.T. and Gorham, P.R. (1963) Observations on the growth and colony habits of ten strains of *Aphanizomenon flos-aquae*. Phycologia 2, 157–168.

Menzel, D.W. and Ryther, J.H. (1961) Nutrients limiting the production of phytoplankton in the Sargasso Sea, with special reference to iron. Deep Sea Res. 7, 276–281.

Mino, Y., Ishida, T., Ota, N., Inoue, M., Nomoto, K., Yoshioka, H., Takemoto, T., Sugiura Y. and Tanaka, H.
(1981) Molecular structure and spectroscopic properties of a copper (II) complex with mugineic acid, a novel amino acid from graminaceous plants. Inorg. Chem. 20, 3440–3444.

Morel, N.M.L., Rueter, J.G. and Morel, F.M.M. (1978) Copper toxicity to *Skeletonema costatum* (Bacillariophyceae). J. Phycol. 14, 43–48.

Morel, F.M.M., Rueter, J.G., Anderson, D.M. and Guillard, R.R.L. (1979) Aquil: a chemically defined phytoplankton culture medium for trace metal studies. J. Phycol. 15, 135–141.

Morrison, N.E., Antoine, A.D. and Dewbrey, E.E. (1965) Synthetic metal chelators which replace the natural growth of *Arthrobacter terregens*. J. Bacteriol. 89, 1630.

Morton, S.D. and Lee, T.H. (1974) Algal blooms – possible effects of iron. Environ. Sci. Tech. 8, 673–674.

Mullis, K.B., Pollack, J.R. and Neilands, J.B. (1971) Structure of schizokinen, an iron-transport compound from *Bacillus megaterium*. Biochemistry 10, 4894–4898.

Murphy, T.P. (1976) Some Aspects of Iron Metabolism of Algae. Ms. Thesis, University of Toronto, 141p.

Murphy, T.P. and Lean, D.R.S. (1975) The distribution of iron in a closed ecosystem. Verh. Int. Verein. Limnol. 9, 258–266.

Murphy, T.P., Lean, D.R.S. and Nalewajko, C. (1976) Blue-green algae: their excretion of iron-selective chelators enables them to dominate other algae. Science 192, 900–902.

Murphy, T.P., Hall, K.J. and Yesaki, I. (1983) Biogenic regulation of iron availability in a eutrophic hard-water lake. Science of the Total Environment 28, 37–50.

Negrin, R.S. and Neilands, J.B. (1978) Ferrichrome transport in inner membrane vesicles of *Escherichia coli* K12. J. Biol. Chem. 253, 2339–2342.

Neilands, J.B. (1952) A crystalline organo-iron pigment from a rust fungus (*Ustilago sphaerogena*). J. Am. Chem. Soc. 74, 4846–4847.

Neilands, J.B. (1967) Hydroxamic acids in nature. Science 156, 1443–1447.

Neilands, J.B. (ed.) (1974) Microbial Iron Metabolism, a Comprehensive Treatise. Academic Press, NY, 597p.

Neilands, J.B. (1981) Microbial iron compounds. Ann. Rev. Biochem. 50, 715–731.

Neilands, J.B. (1982) Microbial envelope proteins related to iron. Ann. Rev. Microbiol. 36, 285–309.

Neilands, J.B. (1984) Methodology of siderophores. Struct. Bonding 58, 1–24.

Nissenbaum, A. and Kaplan, I.R. (1972) Chemical and isotopic evidence for the in situ origin of marine humic substances. Limnol. Oceanogr. 17, 570–582.

Noma, M. and Noguchi, M. (1976) Occurrence of nicotianamine in higher plants. Phytochemistry 15, 1701–1702.

Nomoto, K., Yoshioka, H., Arima, M., Fushiya, S., Takagi, S. and Takemoto, J. (1981) Structure of 2′-deoxymugineic acid, a novel amino acid possessing an iron-chelating activity. Chimia 35, 249–250.

O'Brien, I.G. and Gibson, F. (1970) The structure of enterochelin and related 2,3-dihydroxy-*N*-benzoylserine conjugates in *Escherichia coli*. Biochim. Biophys. Acta 215, 393–402.

O'Flaherty, L.M. and Phinney, H.K. (1970) Requirements for the maintenance and growth of *Aphanizomenon flos-aquae* in culture. J. Phycol. 6, 95–97.

O'Kelly, J.C. (1974) Inorganic nutrients. In W.D.P. Stewart (ed.), Algal Physiology and Biochemistry, University of California Press, Berkeley, pp. 610–635.

Ong, S.A., Peterson, T. and Neilands, J.B. (1979) Agrobactin, a siderophore from *Agrobacterium tumefaciens*. J. Biol. Chem. 254, 1860–1865.

Öquist, G. (1971) Changes in plant composition and photosynthesis induced by iron-deficiency in the blue-green alga *Anacystis nidulans*. Physiol. Plant. 25, 188–191.

Öquist, G. (1974a) Iron deficiency in the blue-green alga *Anacystis nidulans*: changes in pigmentation and photosynthesis. Physiol. Plant. 30, 30–37.

Öquist, G. (1974b) Iron deficiency in the blue-green alga *Anacystis nidulans*: fluorescence and absorbance spectra recorded at 77°K. Physiol. Plant. 31, 55–58.

Parr, M.P. and Smith, R.V. (1976) The identification of phosphorus as a growth-limiting nutrient in Lough Neagh using bioassays. Water Res. 10, 1151–1154.

Peschek, G.A. (1979) Nitrate and nitrite reductase and hydrogenase in *Anacystis nidulans* grown in Fe- and Mo-deficient media. FEMS Microbiol. Lett. 6, 371–374.

Peterson, T., Falk, K.E., Leong, S.A., Klein, M.P. and Neilands, J.B. (1980) Structure and behavior of spermidine siderophores. J. Am. Chem. Soc. 102, 7715–7718.

Pirt, S.J. and Walach, M. (1978) Biomass yields of *Chlorella* from iron (Y x/Fe) in iron-limited batch cultures. Arch. Microbiol. 116, 293–296.

Plowman, J.E., Loehr, T.M., Goldman, S.J. and Sanders-Loehr, J. (1984) Structure and siderophore activity of ferric schizokinen. J. Inorg. Biochem. 20, 183–197.

Pollack, J.R. and Neilands, J.B. (1970) Enterobactin, an iron transport compound from *Salmonella typhimurium*. Biochem. Biophys. Res. Comm. 38, 989–991.

Prakash, A. (1971) Terrigenous organic matter and coastal phytoplankton fertility. In J.D. Costlow (ed.), Fertility of the Sea, Vol. 2. Gordon and Breach, London, pp. 351–368.

Prakash, A., Rashid, M.A., Jensen, A. and Subba Rao, D.V. (1973) Influence of humic substances on the growth of marine phytoplankton: diatoms. Limnol. Oceanogr. 18, 516–524.

Pueschel, C.M. and Cole, K.M. (1980) Phytoferritin in the red alga *Constantinea* (Cryptonemiales). J. Ultrastruct. Res. 73, 282–287.

Russell, L.M. and Holmes, R.K. (1983) Initial characterization of the ferric iron transport system of *Corynebacterium diphtheriae*. J. Bacteriol. 155, 1439–1442.

Sakamoto, M. (1971) Chemical factors involved in the control of phytoplankton production in the experimental lakes area, Northwestern Ontario. J. Fish. Res. Bd. Canada 28, 203–213.

Sakshaug, E. and Holm-Hansen, O. (1977) Chemical composition of *Skeletonema costatum* (Grev.) Cleve and *Pavlova (Monochrysis) lutheri* (Droop) Green as a function of nitrate-, phosphate-, and iron-limited growth. J. Exp. Mar. Biol. Ecol. 29, 1–34.

Sandmann, G. and Malkin, R. (1983) Iron-sulfur centers and activities of the photosynthetic electron transport chain in iron-deficient cultures of the blue-green alga *Aphanocapsa*. Plant Physiol. 73, 724–728.

Schelske, C.L. (1962) Iron, organic matter, and other factors limiting primary productivity in a marl lake. Science 136, 45–46.

Schnitzer, M. and Hansen, E.H. (1970) Organo-metallic interactions in soils. 8. An evaluation of methods for the determination of stability constants of metal-fulvic acid complexes. Soil Sci. 109, 333–340.

Schnitzer, M. and Khan, S.U. (1972) Humic Substances in the Environment. Marcel Dekker, Inc., New York, 327pp.

Schnitzer, M. and Khan, S.U. (1978) Soil Organic Matter. Elsevier, New York, 319 pp.

Segot, M. and L. Codomier (1981) Culture media with enriched seawater: optimal and inhibitory concentrations of some chemical compounds added to seawater for the culture of *Asparagopsis armata* (Rhodophyceae, Bonnemaisoniales) Bot. Mar. 24, 63–67.

Sherman, D.M. and Sherman, L.A. (1983) Effect of iron deficiency and iron restoration on ultrastructure of *Anacystis nidulans*. J. Bacteriol. 156, 393–401.

Simpson, F.B. and Neilands, J.B. (1976) Siderochromes in cyanophyceae: isolation and characterization of schizokinen from *Anabaena* sp. J. Phycol. 12, 44–48.

Slobodkin, L.B. (1953) A possible initial condition for red tides on the coast of Florida. J. Mar. Res. 12, 148–155.

Smillie, R.M. (1965) Isolation of phytoflavin, a flavoprotein with chloroplast ferredoxin activity. Plant Physiol. 40, 1124–1128.

Smith, M.J., Shoolery, J.N., Schwyn, B., Holden, I. and Neilands, J.B. (1985) Rhizobactin, a structurally novel siderophore from *Rhizobium meliloti*, J. Am. Chem. Soc. 107, 1739–1743.

Snow, G.A. (1970) Mycobactins: iron-chelating growth factors from mycobacteria. Bacteriol. Rev. 34, 99–125.

Spiller, S. and Terry, N. (1980) Limiting factors in photosynthesis. II. Iron stress diminishes photochemical capacity by reducing the number of photosynthetic units. Plant Physiol. 65, 121–125.

Stephenson, M.C. and Ratledge, C. (1980) Specificity of exochelins for iron transport in three species of Mycobacteria. J. Gen. Microbiol. 116, 521–523.

Sugiura, Y. and Nomoto, K. (1984) Phytosiderophores – structures and properties of mugineic acids and their metal complexes. Struct. Bonding 58, 107–135.

Sunda, W. and Guillard, R.R.L. (1976) The relationship between cupric ion activity and the toxicity of copper to phytoplankton. J. Mar. Res. 34, 511–529.

Sunda, W.G. and Lewis, J.A.M. (1978) Effect of complexation by natural organic ligands on the toxicity of copper to a unicellular alga, *Monochrysis lutheri*. Limnol. Oceanogr. 23, 870–876.

Tait, G.H. (1974) Iron-binding compounds formed by *Micrococcus denitrificans*. Biochem. Soc. Trans. 2, 657–658.

Takagi, S., Nomoto, K. and Takemoto, T. (1984) Physiological aspects of mugineic acid, a possible phytosiderophore of graminaceous plants. J. Plant Nutr. 7, 469–477.

Takemoto, T., Nomoto, K., Fushiya, S., Ouchi, R., Kusano, G., Hikino, H., Takagi, S., Matsuura, Y. and Kakudo, M. (1978) Structure of mugineic acid, a new amino acid possessing an iron-chelating activity from root washing of water-cultured *Hordeum vulgare* L. Proc. Japan Acad. 54B, 469–73.

Trick, C.G., Andersen, R.J., Price, N.M., Gillam, A. and Harrison, P.J. (1983) Examination of hydroxamate-siderophore production by neritic eukaryotic marine phytoplankton. Mar. Biol. 75, 9–17.

Van den Berg, C.M.G., Wong, P.T.S. and Chau, Y.K. (1979) Measurement of complexing materials excreted from algae and their ability to ameliorate copper toxicity. J. Fish Res. Bd. Can. 36, 901–905.

Van der Mark, F., Van den Briel, M.L., Van Oers, J.W.A.M. and Bienfait, H.F. (1982) Ferritin in bean leaves with constant and changing iron stress. Planta 156, 341–344.

Verstreate, D.R., Storch, T.A. and Dunham, V.L. (1980) A comparison of the influence of iron on the growth and nitrate metabolism of *Anabaena* and *Scenedesmus*. Physiol. Plant. 50, 47–51.

Wells, M.L., Zorkin, N.G. and Lewis, A.G. (1983) The role of colloid chemistry in providing a source of iron to phytoplankton. J. Mar. Res. 41, 731–746.

Wetzel R.G. (1966) Productivity and nutrient relationships in marl lakes of northern Indiana. Verh. Int. Verein. Limnol. 16, 331–332.

Wetzel, R.G. (1972) The role of carbon in hardwater marl lakes. In G.E. Likens (ed.) Nutrients and Eutrophication. Am. Soc. Limnol. Oceanogr. Spec. Symp., pp. 84–91.

Widmer, J. and Keller-Schierlein, W. (1974) Stoffwechselprodukte von Mikroorganismen; Synthesen in der Sideramin-Reiche: Rhodotorulasäure und Dimerumsäure. Helv. Chim. Acta 57, 1904–1912.

Winkelmann, G. and Huschka, H. (1984) A study on the mechanism of siderophore transport – a proton symport. J. Plant Nutr. 7, 479–487.

Yamochi, S. (1983) Mechanisms for an outbreak of *Heterosigma-akashiwo* red tide in Osaka Bay, Japan 1. Nutrient factors involved in controlling the growth of *Heterosigma-akashiwo*. J. Oceanogr. Soc. Jpn. 39, 310–316.

Yariv, J., Kalb, A.J., Sperling, R., Bauminger, E.R., Cohen, S.G. and Ofer, S. (1981) The composition and structure of bacterioferritin of *Escherichia coli*. Biochem. J. 197, 171–175.

P. Fay and C. Van Baalen (Eds.) The Cyanobacteria

17

Secondary metabolism in the cyanobacteria

FLORENCE K. GLEASON and JOHN M. WOOD

Gray Freshwater Biological Institute, University of Minnesota, P.O. Box 100, Navarre, MN 55392, USA

1. Introduction

Most plants and microorganisms, from the simplest unicellular algae to complex angiosperms, produce a variety of unusual compounds collectively referred to as secondary metabolites. Thousands of these compounds have been chemically characterized and classified on the basis of their functional groups and/or chemical properties (Swain, 1977). Some of these unusual metabolites have useful pharmacological properties, for example, the antibiotics, penicillin and erythromycin and drugs such as morphine and vinblastin. We have only begun to tap the scientific and economic potential of secondary metabolites.

One question which is often ignored by researchers in this field is the possible role of secondary metabolites in the producing organism. Since they serve no basic metabolic function and vary in concentration and structure, even in closely related species, they were once considered metabolic waste products (Muller, 1969). However, most of these compounds are structurally very complex and, therefore, metabolically expensive. In addition, they have a rapid turnover rate in the cell (Seigler and Price, 1976). Such properties are hardly those of waste products.

Advances in plant physiology and biochemistry have shown that many secondary metabolites facilitate interactions between plants and their environment. Examples of these are the plant hormones, auxins and gibberellins, which govern plant growth and development in response to environmental parameters such as light and temperature. The synthesis of some secondary metabolites, the phytoalexins, is induced by parasitic invasion and may enable plants to fight off infections (Deverall, 1977). It has also been suggested that many secondary metabolites are allelopathic substances; i.e., compounds produced by one organism which inhibit the growth and development of competing species. Although difficult to prove in an ecological setting, this is apparently the rationale behind bacterial and fungal production of antibiotics. It has also been shown that similar substances produced by higher plants can be used as 'natural herbicides' (Putnam, 1983).

Abbreviations: Adda, 3-amino-9-methoxy-2,6,8-trimethyl-10-phenyldeca-4,6-dienoic acid; DCMU, 3-(3,4-dichlorophenyl)-1,1-dimethyl urea; NMR, Nuclear Magnetic Resonance; NOE, Nuclear Overhauser Enhancement; sp. act., specific activity in mCi/mmol; UTEX, University of Texas at Austin – Culture Collection of Algae; LD_{50}, LD_{100}, dosage which was lethal to 50% and 100%, respectively, of the experimental animals.

While the secondary metabolism of terrestrial species has received the greatest attention, aquatic organisms, particularly those in the marine environment, are also known to produce a complex array of unusual metabolites (see Barbier, 1981, for a general review). Although the function of most of these compounds is unknown, some have pharmacological uses such as the anthelmintic agent, kainic acid from the red alga, *Digenea simplex*, or possible antitumor activity as demonstrated by a number of halogenated metabolites from marine algae (Fenical, 1975). Aquatic algae and bacteria are also known to synthesize compounds with functions similar to their terrestrial counterparts, such as hormones and antibiotics. Among these, the secondary metabolism of the cyanobacteria (blue-green algae) has received relatively little attention. However, like other marine algae, these prokaryotic algae produce a variety of secondary metabolites, many of them halogenated (see Moore, 1981, for a review).

2. Natural products from marine cyanobacteria and algae

The biosynthesis of halogenated secondary metabolites in the marine environment is much more widespread than in both terrestrial and freshwater ecosystems. Biological halogenation requires the involvement of specific peroxidases. Peroxidases utilize hydrogen peroxide in a variety of reactions, but perhaps the most unusual one is that of biological halogenation by the enzymes chloroperoxidase and bromoperoxidase. These enzymes are responsible for the synthesis of a large number of halogenated secondary metabolites produced exclusively by marine organisms (Fenical, 1982; Suida et al., 1975; Fowden, 1968; Hager et al., 1975).

In this reaction pathway it is believed that halide ion reacts with highly reactive 'Compound I' to give coordinated hypochlorite (OCl^-), thus creating an electrophilic reagent which reacts rapidly with a wide variety of organic substrates.

As early as 1940, Clutterbuck et al. (1940) showed that certain fungi were capable of using halide ions in the biosynthesis of haloorganic compounds as secondary metabolites. Later Shaw and Hager (1960) isolated an enzyme from the fungus *Caldariomyces fumago* which catalyzed the oxida-

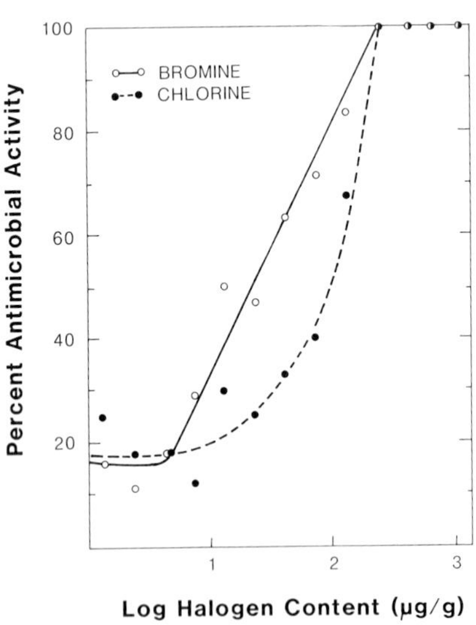

Fig. 1. The correlation between organo-halogen content of marine organisms and antimicrobial activity.

tion of Cl^-, Br^- and I^- to give an enzyme-bound electrophilic halogenating agent. This halogenating reagent was shown to react with a variety of nucleophiles to give halogenated reaction products. Since the enzyme has a heme prosthetic group, and since it catalyzes a peroxidase reaction, it was given the name chloroperoxidase.

In the marine environment significant evolutionary pressures have led to the utilization of halide ions. Hager et al. (1975) have conducted a survey of over 900 marine animal and plant species for the presence of halogenated organic compounds. Approximately 25% of species tested had lipid extracts which contained greater than 10 μg of organic halogen per gram wet weight of tissue. The Rhodophyta (red algae) were found to be particularly rich in halogenated organic compounds. Also most of these halocarbons were shown to be cytotoxic to microorganisms. In all cases antimicrobial activity was found to be a function of the halogen content of these lipid extracts (Fig. 1). Therefore it is likely that organisms in the marine environment have evolved mechanisms to detoxify these halogenated organic compounds through special metabolic pathways. Studies by Suida et al. (1975) with extracts of the red alga *Bonnemaisonia hamifera* have demonstrated the presence of a number of halogenated heptanones which arise by direct halogenation with a bromoperoxidase. By analogy with chloroperoxidase the following reaction sequence in Scheme I explains the biosynthesis of brominated heptanones from 3-oxo-octanoic acid.

I SYNTHESIS OF BROMINATED HEPTANONES

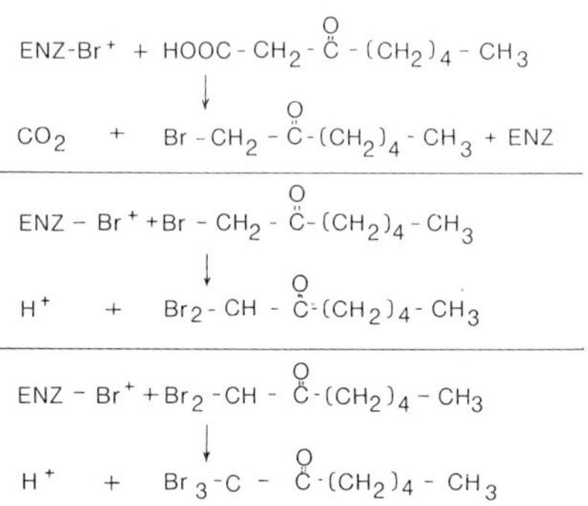

Hewson and Hager (1980) have surveyed 72 different species of marine algae for the presence of bromoperoxidase, and 55 species were found to have high levels of this enzyme. The Rhodophyta were found to be better brominators, having the highest levels of bromoperoxidase and the greatest lipid halogen content. The Phaeophyta were the poorest halogenators. Besides containing a great variety of halogenated aliphatic compounds, marine organisms synthesize a multitude of halogenated aromatic compounds. For example, red algae of the genus *Laurencia* synthesize a great variety of brominated and iodinated aromatic compounds.

Several of these halogenated natural products are cytotoxic, and some of them resemble synthetic medicaments, insecticides, or pesticides. Brominated pyrrole derivatives as well as unique chlorinated amides (malyngamide) have been isolated from marine cyanobacteria (Fig. 2) (Faulkner, 1978; Cardellina et al., 1979).

Since these halogenated natural products are synthesized by marine organisms, it follows that such organisms must have dehalogenation mechanisms to degrade them. Most halogenated synthetic compounds of industrial origin are regarded as 'persistent' or even 'recalcitrant' in the terres-

A

B

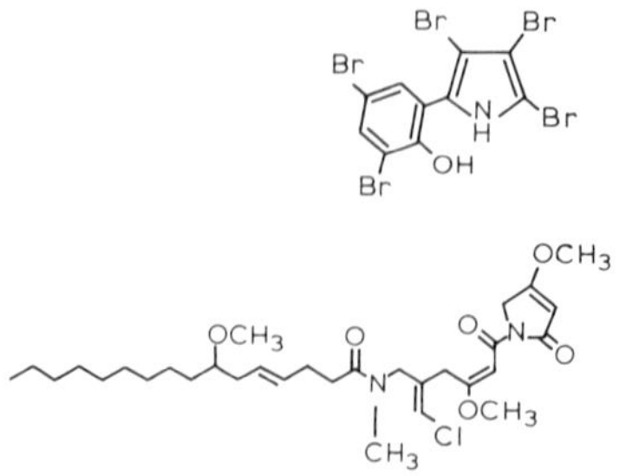

Fig. 2. The structures of a brominated pyrrole (A) and malyngamide (B). Both halogenated secondary metabolites were isolated from marine cyanobacteria.

trial environment. Can the same be assumed when compounds of industrial origin contaminate the marine environment?

Dehalogenation reactions can occur either as a result of volatilization of low molecular weight halocarbons or by biological elimination of halide ions from organohalogen metabolic intermediates. Probably the most surprising discovery is the volatilization of dibromomethane, bromoform, and pentylbromide by the red alga *Bonnemaisonia hamifera*. Tribromo-heptanone is a likely substrate for the biosynthesis of bromoform (Scheme II.1.). Pentylbromide probably arises by the bromination of oxo-octanoic and by bromoperoxidase followed by hydrolysis of the brominated intermediate (Scheme II.2.).

II.1 **BIOSYNTHESIS OF BROMOFORM**

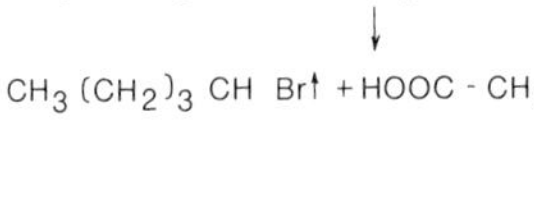

II.2 **BIOSYNTHESIS OF PENTYL-BROMIDE**

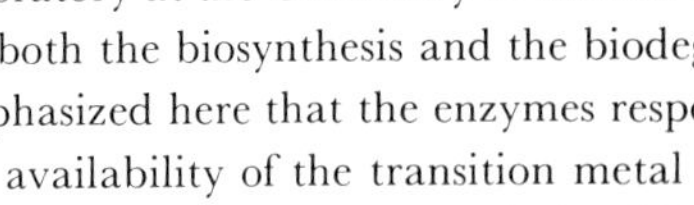

The research described above is largely the product of almost 20 years of work in Lowell Hager's laboratory at the University of Illinois. Peroxidative halogenation provides us with one mechanism for both the biosynthesis and the biodegradation of halogenated organic compounds. It should be emphasized here that the enzymes responsible contain heme, are classical peroxidases, and rely on the availability of the transition metal iron for their synthesis. The availability of, and resistance to, metal ion toxicity will be dealt with later in this chapter.

3. Natural products from freshwater cyanobacteria

In contrast to the marine algae, the secondary metabolism of freshwater cyanobacteria has been relatively ignored. Although these organisms might be expected to be as metabolically versatile as their marine counterparts, research on freshwater cyanobacteria has been directed to those species which are known to synthesize toxins. Since many cyanobacteria give rise to blooms in drinking water supplies, problems related to their growth and metabolism have become a public health concern. Although relatively few cases of human poisoning can be directly attributed to the freshwater cyanobacterial toxins, the potential problem has become more acute with the continuing eutrophication of many urban water supplies. In addition, the loss in fish, wildlife and domestic animals to cyanobacterial toxicosis is more than a minor nuisance. This field has been reviewed most recently in a monograph edited by Carmichael (1982).

3.1. Anatoxin

One of the best characterized cyanobacterial toxins is anatoxin-*a* produced by the heterocystous cyanobacterium, *Anabaena flos-aquae*. The earliest reports of blooms which were associated with animal poisoning are from the American Midwest and Canada (see Gorham and Carmichael, 1979, for a review). The Canadian workers isolated several strains of *A. flos-aquae* from a toxic bloom on Lake Burton, Saskatchewan (Gorham et al., 1964). Some of these are currently available in axenic culture. The toxic metabolite was purified from a unialgal clone designated NRC-44h. The active component was assayed by intraperitoneal injection into mice. The initial observations suggested that anatoxin-*a* was a small, relatively polar molecule. On final purification, the compound was found to have a molecular weight of 165 and a molecular formula of $C_{10}H_{15}NO$. The acetyl derivative of anatoxin-*a* was subsequently shown to be a bicyclic amine by X-ray crystallographic analysis (Huber, 1972). The structure of anatoxin-*a* is illustrated in Fig. 3.

Anatoxin-*a* can be prepared from mass cultures of *Anabaena flos-aquae* NRC-44h by extraction of freeze-dried cells with acidic methanol. Milligram quantities are obtained by subsequent preparative thin-layer chromatography on silica gel. A final toxicity of 0.25 mg/kg (intraperitoneal injection into mice) has been reported (Devlin et al., 1977).

Partial chemical synthesis of anatoxin-*a* by ring expansion of the 8-azabicyclo octane system of cocaine has been reported (Campbell et al., 1977). A more general synthetic method, utilizing an intramolecular cyclization between a keto iminium salt and a nucleophilic carbon atom, also yielded synthetic anatoxin with an LD_{50} of 2.5 mg/kg in mice (Bates and Rapoport, 1979).

Intraperitoneal injections of anatoxin-*a* are toxic to mammals, birds and fish. Injection of the minimum lethal dose results in death within a few minutes from respiratory failure. Since anatoxin-*a* is a small, water-soluble molecule it is also readily absorbed by ingestion which accounts for the many cases of animal toxicity from blooms of *Anabaena flos-aquae*. In contrast, the toxin is apparently not transported across the gill membranes in fish (Carmichael et al., 1975). Symptoms observed after ingestion indicated that anatoxin-*a* acts as a neuromuscular inhibitor. Work done on isolated nerve-muscle preparations subsequently demonstrated that anatoxin-*a* acts as a depolarizing agent, blocking nerve transmission at the neuromuscular junction (Carmichael et al., 1979). Its activity somewhat resembles that of the known depolarizing agent, decamethonium bromide, although there is no structural resemblance between the two molecules to substantiate a similar

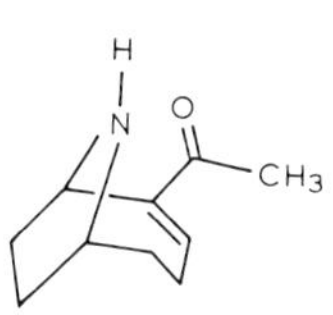

Fig. 3. Structure of anatoxin-*a*, isolated from a toxic clone of *Anabaena flos-aquae* (Huber, 1972).

mode of action. Anatoxin-*a* does bear some structural resemblance to cocaine which is an 8-membered bicyclic ring. However, unlike cocaine, anatoxin-*a* does not affect norepinephrine reabsorption.

Nothing is known about the biosynthesis of anatoxin nor the parameters which control its synthesis. Fortunately most blooms of *Anabaena flos-aquae*, which are quite common in mildly eutrophic freshwater ecosystems, are non-toxic. Among the known toxic strains, anatoxin concentration can vary widely depending on environmental conditions (Gorham, 1964). In laboratory grown cultures, toxin content is dependent on parameters such as light intensity, temperature, bacterial contamination and also on the stage of growth. Anatoxin-*a* production is maximal in mature cultures and can only be found in the medium in late exponential to stationary phase (Devlin et al., 1977). This finding raises the usual question as to its function in *Anabaena flos-aquae*. Defense against predators seems a bit contrived, especially since the toxin is released only when the cyanobacterial bloom is well established and possibly degenerating. In addition, there is no evidence of an effect of anatoxin-*a* on protozoan or cladoceran grazers, nor on the most immediate cyanobacterial competitors, other species of blue-green algae and bacteria.

Further work on isolated strains of *Anabaena flos-aquae* has shown that these organisms are capable of producing a variety of anatoxins. While all are lethal, the symptoms induced by these other compounds vary slightly (Carmichael and Gorham, 1978). No structural characterization of these toxins has been reported as yet and their resemblance, if any, to anatoxin-*a* is unknown.

3.2. Microcystis toxin

Probably the most common bloom-forming cyanobacterium in fresh-water ecosystems is the unicellular organism, *Microcystis aeruginosa*. Animal poisonings due to toxins produced by this organism have been reported world-wide (Gorham and Carmichael, 1979). Canadian workers were again the first to isolate a toxic strain of *M. aeruginosa* (designated NRC-1) which grew well under laboratory conditions (Hughes et al., 1958). The toxic substance was isolated from NRC-1 and shown to be a cyclic polypeptide consisting of seven amino acids. The LD_{100} on intraperitoneal injection into mice was 0.47 mg/kg. Death was apparently due to extensive liver damage and internal hemorrhage (Bishop et al., 1959). There are obvious experimental difficulties in isolating and characterizing small peptides and, in addition, *Microcystis aeruginosa* strains produce a variety of peptide toxins. Therefore, several other peptides with varying composition were subsequently isolated and partially characterized (Gorham and Carmichael, 1979). A strain of *M. aeruginosa*, isolated from a toxic bloom which occurred in Witbank Dam in the Transvaal, was found to produce four different peptide toxins when grown under laboratory conditions. The toxins all appear to be somewhat water soluble and were separated by DEAE column chromatography in aqueous solvents. Final toxin purification was by electrophoresis. All four purified toxins were cyclic heptapeptides containing β-methyl aspartic acid, glutamic acid and alanine. The other residues in each fraction were variable (Botes et al., 1982).

One of these toxic variants was characterized further and designated cyanoginosin-LA, LA indicating that it contained the variable amino acids, leucine and alanine. Cyanoginosin-LA has an amino acid composition of erythro-β-D-aspartic acid, D-alanine, L-alanine, D-glutamic and L-leucine. Fast-atom bombardment mass spectra indicated a mass of 909. A partial structure containing the above residues was deduced from 500 MHz ^{1}H NMR spectra of cyanoginosin. ^{13}C NMR data combined with electron impact mass spectra of permethylated and perdeuterio-permethylated cyanoginosin demonstrated the presence of an unusual 3-amino-9-methoxy-2,6,8-trimethyl-10-phenyldeca-4,6-dienoic acid residue (Adda). Partial degradation of the peptide and NMR analysis

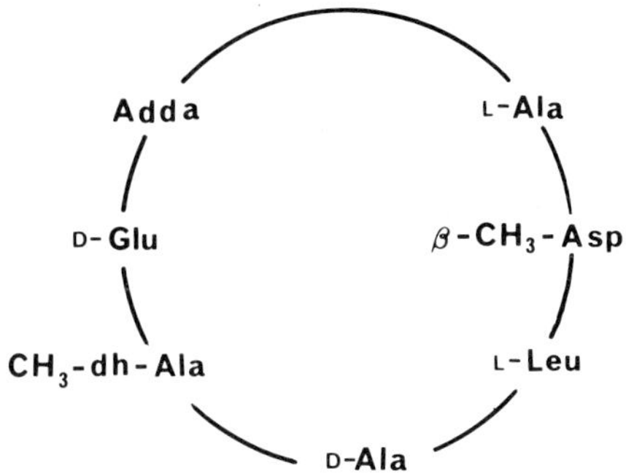

Fig. 4. Partial structure of cyanoginosin-LA, a cyclic heptapeptide from *Microcystis aeruginosa* (Botes et al., 1984).

of the fragments generated a sequence of the toxic peptide as shown in Fig. 4 (Botes et al., 1984). Structural data generated by these workers (Botes et al., 1982) and an Australian group (Elleman et al., 1978) indicate that the peptide toxins from various *M. aeruginosa* strains are similar i.e., all are cyclic peptides with a D-Ala, D-Glu and β-methyl-Asp residues and the remainder, variable amino acids and/or blocking groups.

An LD_{100} as low as 0.070 mg/kg (intraperitoneal injection in mice) has been reported for *M. aeruginosa* toxin purified from a natural bloom (Elleman et al., 1978). This is considerably more toxic than laboratory grown strains and indicated that not only strain difference but growth conditions also influence toxicity. A systematic study of *M. aeruginosa* strains in axenic culture showed that toxin production was maximal in exponentially growing cells and slowly declined in stationary cultures. Alkaline pH (above 8.5) also promoted toxin production. The general conclusion was that slowly growing cells produced more toxin than cultures in which growth was optimized (Van der Westhuizen and Eloff, 1983).

The toxin from *M. aeruginosa* is lethal to a number of mammalian species and birds, both after intraperitoneal injection or ingestion. Death occurs in 1–3 h and is apparently due to internal bleeding. Pathological examinations show extensive liver damage (Hughes et al., 1958). However, microcystis toxins have relatively hydrophobic properties, due to blocking groups and the presence of the unusual aromatic residue, Adda. It has been proposed that these compounds may be general membrane disruptors and that liver damage in higher organisms is a consequence of the general detoxification function of this organ. Runnegar and coworkers showed that low levels of microcystis toxin caused a relatively rapid deformation of the cell membrane in isolated hepatocytes. It was suggested that microcystis toxin may act in a manner analogous to the fungal toxin, phalloidin, which is also a cyclic peptide. However, these workers found no evidence that microcystis toxins bind to actin which is the major mechanism of action of the fungal toxin (Runnegar et al., 1981). Slatkin and others reported the occurrence of pulmonary thrombi after lethal injections of microcystis toxin. These clots precede and may lead to the subsequent liver damage noted by others. Microcystis toxin is a potent thrombogenic agent which is not counteracted by the usual anticoagulants (Slatkin et al., 1983). This same group recently reported that microcystis toxin may have to be activated by the liver before it becomes lethal. They confirmed an earlier observation that the toxin has no effect on young mice with incomplete liver function. In addition, mice with livers previously damaged by carbon tetrachloride, were not immediately affected by microcystis toxin. The lethal effects of the toxin could be counteracted by hydrocortisone which may be due to the effect of the drug on platelet aggregation. They found no direct effect of microcystis toxin on platelets which also suggests that an activated form is functioning in vivo (Adams et al., 1985).

Microcystis toxin is often referred to as an 'endotoxin', i.e. it is not found in the water column until the bloom begins to deteriorate. Thus, its toxic activity is apparently unrelated to any physiological role it may have in the cyanobacterium. Given the ubiquitous nature of *Microcystis aeruginosa*, it is fortunate that the majority of strains do not produce toxins, at least under most environmental conditions. No clearly documented cases of human poisoning have been attributed to this organism. However, the possibility of diarrhea due to growth of these algae in drinking water supplies, particularly in third world countries where water purification systems are often inadequate, is considered a possible human health problem (Aziz, 1974).

3.3. Uncharacterized toxins

Although numerous other cases of toxicity have been attributed to cyanobacteria, few of the offending organisms have been cultured in the laboratory, nor have the toxic metabolites been characterized. Blooms of the heterocystous, *Aphanizomenon flos-aquae* in Kezar Lake, New Hampshire, were toxic to fish and other aquatic animals (Sawyer et al., 1968). A toxic strain was cultured under laboratory conditions and an extract was partially purified by column chromatography. This fraction contained several toxic compounds, two of which were chromatographically identical to the *Gonyaulax* metabolites, saxitoxin and neosaxitoxin which were extracted from shellfish by a similar procedure (Ikawa et al., 1982). This identity of secondary metabolites between organisms as taxonomically remote as cyanobacteria and dinoflagellates, if confirmed, would be most unusual. Like the other toxic cyanobacteria, most strains of the widespread *Aphanizomenon flos-aquae* are apparently not toxic or do not produce toxin under prevailing environmental conditions (Gentile and Maloney, 1969).

Deleterious effects have also been ascribed to the contamination of water supplies with cyanobacterial endotoxins, compounds similar to the cell wall fractions from other gram-negative bacteria (see, e.g., Keleti and Sykora, 1982). The cyanobacteria have a cell wall which structurally and chemically resembles that of typical gram-negative bacteria. The structural component, designated lipid A, is comparable in some cyanobacterial species to that of non-photosynthetic bacteria (Weckesser et al., 1979). The lipid A is a complex molecule consisting of variable fatty acids, phosphate and attached sugars. In species such as *Salmonella*, a variety of toxic reactions are attributed to ingestion or even inhalation of this cell wall fraction (Galanos et al., 1977). However, lipopolysaccharide from the unicellular cyanobacterium, *Synechococcus* 73109 was shown to be significantly less toxic than *Salmonella* endotoxin (Buttke and Ingram, 1975). A corresponding fraction from several of the filamentous *Phormidium* species had no typical endotoxin activity (Mikheyskaya et al., 1977). Although only a small number of cyanobacterial lipopolysaccharides have been characterized, it seems that most do not have significant endotoxin activity and thus pose a minor human health hazard. This is apparently true even of lipid A fractions from toxic *Microcystis* species (Raziuddin et al., 1983). Further work in this area is warranted to generate additional information on cyanobacterial endotoxins and their activity.

The biological function of the cyanobacterial toxins, in general, remains obscure. It seems unlikely that they play any role in growth and development of the organisms or perform an antibiotic or allelopathic function which would account for the ecological dominance of a particular strain. However, it is likely that blue-green algae produce compounds with this latter property. Many of the marine algal metabolites, particularly halogenated compounds, are believed to discourage grazers or inhibit the growth of competing species (Hellebust, 1974). The best documented example of algal allelopathy in freshwater ecosystems was described by Keating in a study of species succession in a eutrophic lake, Linsley Pond in Connecticut. Lake water collected during a cyanobacterial bloom and filter sterilized, contained substances which were specifically toxic to other algae isolated from the same lake (Keating, 1977). Metabolites excreted by blue-green algae which bloom in late Fall also proved inhibitory to diatom species which normally dominate the pond in Spring (Keating, 1978). Laboratory analysis showed that the antibiotic substances were small, heat-labile molecules which were apparently excreted into the water column by the dominant cyanobacteria. The problems involved in isolating a metabolite from a dilute and chemically complex natural system precluded further characterization of these algal antibiotics. (See also Section 9 of Chapter 18.)

3.4. Cyanobacterin

Research in the area of algal antibiotics using laboratory cultures of cyanobacteria has been disappointing. Potential antibiotic properties have been generally attributed to competition for limiting nutrients or to release of excess primary metabolites such as fatty acids. One exception is the filamentous cyanobacterium, *Scytonema hofmanni*. This slow-growing benthic organism easily maintains itself in two-species culture with even the most prolific planktonic cyanobacteria. Filaments of *S. hofmanni* were shown to excrete a small, hydrophobic metabolite which was toxic to most cyanobacteria and also to some green algae (Mason et al., 1982). This active metabolite, named cyanobacterin, was isolated from *S. hofmanni* (UTEX B1581, now available in axenic culture as UTEX 2349) by extraction with ethyl ether or other non-polar organic solvents. The compound was purified by preparative thin-layer chromatography and high-performance liquid chromatography on silica gel. High resolution mass spectra showed a molecular weight of 430 and a molecular formula of $C_{23}H_{23}ClO_6$. The combined data from mass, infrared and ^1H and ^{13}C nuclear magnetic resonance spectra indicated that cyanobacterin is a diaryl substituted γ-ylidene, γ-butyrolactone (Pignatello et al., 1983). The structure is illustrated in Fig. 5. The γ-ylidene, γ-butyrolactone structure is unusual in that cyanobacterin does not contain any additional α,β-unsaturation as is generally found in other natural products. In addition, it is the only halogenated metabolite known to be produced by a freshwater alga. Cyanobacterin contains two asymmetric centers at C-2 and C-3. Nuclear magnetic resonance (NMR) experiments, using lanthanide shift reagents and nuclear overhauser enhancement (NOE) effect, indicated a *cis* configuration of the H and OH substituents. However, the absolute configuration of the natural product could not be assigned. Additional proof of the structure of cyanobacterin was provided by the synthetic work of Jong and coworkers (1984). Essentially, the two aromatic halves of the molecule were synthesized and subsequently coupled to yield an acetylenic carboxylic acid. Cyclization of this intermediate with silver nitrate in methanol yielded the γ-butyrolactone with the proper *cis* configuration at C-2 and C-3. However, the material synthesized by this procedure is a mixture of optical isomers and X-ray diffraction analysis did not allow assignment of the absolute configuration. In addition, biological analysis in-

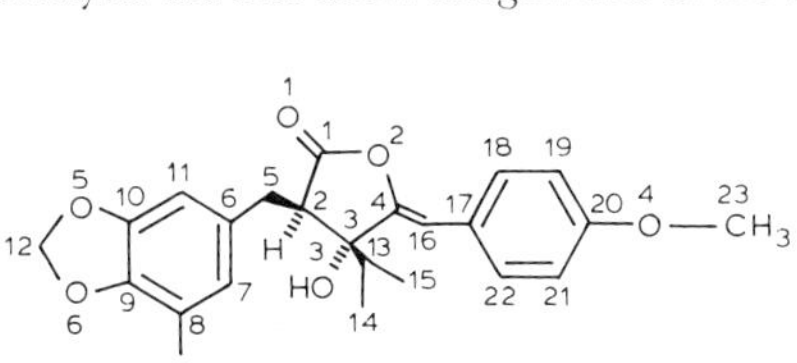

Fig. 5. Structure of the naturally occurring isomer of cyanobacterin, isolated from *Scytonema hofmanni* UTEX 2349 (Gleason et al., 1986).

dicated that the synthetic material had approximately half the inhibitory activity of the natural product, suggesting that the unnatural isomer is not active (Gleason and Paulson, 1984). Later crystallization and X-ray analysis of the natural cyanobacterin demonstrated that the substituents have the 'R' configuration about both asymmetric carbons as indicated in Fig. 5. (Gleason et al., 1986).

Cyanobacterin is toxic to most cyanobacteria at a concentration of approximately 5 μM. The compound has no effect on eubacteria, the photosynthetic bacterium *Rhodospirillum rubrum*, nor on *Scytonema hofmanni* itself (Gleason and Paulson, 1984). The compound is also toxic at the same level to eukaryotic algae but has no effect on protozoans or fungi (Gleason and Baxa, 1986). Cyanobacterin will inhibit the growth of the aquatic angiosperm, *Lemna gibba* (duckweed) when present in as little as 1 μM in the medium. Cyanobacterin is toxic to higher plants when sprayed on the leaves but has no effect when applied to the soil (Gleason and Case, 1986). The compound has been shown to inhibit photosynthetic electron transport in spheroplasts of *Synechococcus*, at a concentration as low as 23 nM (Gleason and Paulson, 1984). Cyanobacterin also inhibits the Hill reaction in isolated pea chloroplasts in the presence of the artificial electron acceptors, ferricyanide, dichlorophenolindophenol, *p*-benzoquinone and silicomolybdate (Gleason and Case, 1986). These data indicate that the site of action is in Photosystem II. Thus cyanobacterin is the natural counterpart of the well-known Photosystem II herbicides, 3-(3,4-dichlorophenyl)-1,1-dimethyl urea (DCMU) and atrazine. These herbicides act by binding to a protein associated with the thylakoid membranes in cyanobacteria and chloroplasts. This protein, the quinone-binding or 32 kD protein, mediates electron transport from a primary quinone acceptor to the quinone pool. Plants which are resistant to these herbicides often have a mutation in the quinone-binding protein (Hirschberg and McIntosh, 1983).

Cyanobacterin bears no obvious structural resemblance to these known Photosystem II inhibitors, but like them, is a hydrophobic, halogenated molecule with an affinity for a site on the thylakoid membranes (Baxa and Gleason, 1985). A DCMU-resistant mutant of *Anacystis nidulans* was reported to have an altered quinone-binding protein (Golden and Sherman, 1984). Thylakoids prepared from this organism were not able to carry out photosynthetic electron transport in the presence of cyanobacterin, although they were unaffected by DCMU. These data indicate that cyanobacterin does not bind to the same protein site as DCMU and atrazine (Gleason et al., 1986).

Relatively little is known about the biosynthesis of cyanobacterin or any secondary metabolite in the blue-green algae. Cyanobacterin is found in the culture medium of *Scytonema hofmanni* at all stages of growth although the concentration of the compound increases dramatically in old, densely populated cultures. Several attempts were made to incorporate radioactive precursors into cyanobacterin by adding labeled compounds to *S. hofmanni* medium. Methionine, labeled in the methyl group, was readily taken up by the cells. Incorporation into purified cyanobacterin was 0.002% (sp. act. purified cyanobacterin/sp. act. methionine). Given the small amount of cyanobacterin produced per cell mass, the data suggest a direct incorporation into the methoxy (C-23) and methylene dioxy (C-12) groups by way of *S*-adenosyl methionine. Uniformly labeled phenylalanine and tyrosine were also taken up by *S. hofmanni* with over 90% of the label associated with the cell mass after harvest. However, after purification, no radioactivity was detected in cyanobacterin. It can be concluded that the aromatic rings in cyanobacterin do not arise from these amino acids via the phenylalanine- or tyrosine-ammonia lyase reaction. Phenylalanine-ammonia lyase activity has been reported to occur in only one alga, the Rhodophyte, *Porphyridium* (Camm and Towers, 1973). Although this enzyme (and associated tyrosine-ammonia lyase) is a key enzyme in the production of secondary metabolites in higher plants, it appears to be rather rare in microorganisms. It has been proposed that in the course of evolution, this enzyme did not appear before the development of an oxygenic atmosphere (Chapman and Ragan, 1980). If true, the cyanobacteria may also lack such activity since they are an ancient group of organisms which presumably made the original and major contribution to oxygen in the atmosphere (Schopf and Walter, 1983). These incorporation experiments do not exclude the possibility of cyanobacterin synthesis from intermediates of the shikimic acid pathway.

An alternative pathway for the synthesis of aromatic secondary metabolites is from acetate through intermediates called polyketides. This pathway is found in higher plants and is especially common in fungi. Radioactive acetate is incorporated into cyanobacterin at the level of 0.001%. Unfortunately, acetate is a precursor of many metabolites and its incorporation into cyanobacterin was not high enough to establish any pathway. Labeled ferulic acid, a key intermediate in polyketide synthesis, was incorporated at the 0.01% level into cyanobacterin, which suggests possible synthesis of the chlorine-containing ring of cyanobacterin from this intermediate (Gleason, unpublished). Our preliminary data indicate that elucidation of pathways of secondary metabolism in cyanobacteria by means of classical 'feeding' procedures is feasible.

Labeling experiments with ^{36}Cl demonstrated that *S. hofmanni* extracts contain several halogenated compounds. Some of these may be biosynthetic precursors or degradation products of cyanobacterin (Gleason et al., 1983). One of these is anhydrocyanobacterin which results from loss of water in the lactone ring. Anhydrocyanobacterin has no biological activity (Gleason and Paulson, 1984). Other compounds show slight algicidal activity but have not been characterized. Although cyanobacterin is (thus far) a unique molecule, it is quite probable that other cyanobacteria produce antibiotic compounds with algicidal or bactericidal activity.

3.5. Chelating agents

It has been proposed that cyanobacteria in nature can often out-compete other algae and even bacteria by excretion of iron-chelating compounds (Murphy et al., 1976). Iron-chelating metabolites have been isolated from culture medium of the filamentous cyanobacterium, *Anabaena* sp. 6411. These compounds resemble the hydroxamate siderophores described in other eubacterial species. Some of these compounds are highly species-specific while others can be utilized by bacteria other than the producing strain (Goldman et al., 1983). (See Chapter 16 of this volume for details.)

4. Resistance mechanisms in the cyanobacteria

The biosynthesis of secondary metabolites with antibiotic properties is useful to slow growing cyanobacteria as they compete for both bulk nutrients and essential trace elements. However, in addition to investing energy in active transport processes for essential nutrients, energy must also be expended to keep unwanted, non-essential and toxic elements outside the cell. The rates at which elements are cycled in the aqueous environment depends on the availability of the bulk nutritional elements C, N, O, P and S (Solomans and Baccini, 1986).

The cycling of trace elements follows the cycling of the bulk elements, and the uptake of trace elements requires the input of considerable energy by living cells and, therefore, the biochemical processes for transport are carefully regulated at the gene level (Chakrabarty, 1985). Consequently, living cells are not at equilibrium with their external environment (Wood and Wang, 1983; Wood, 1984).

For those elements which are non-essential to the life of cyanobacteria a variety of resistance mechanisms have evolved to prevent toxicity. These resistance mechanisms require energy also, and they represent non-equilibrium processes (Wood and Wang, 1983; Wood, 1984; Chakrabarty, 1985). Resistance mechanisms have evolved for the removal of toxic elements by the processes of precipitation, complexation, volatilization, alkylation, hydrolysis, oxidation, reduction and by specific efflux mechanisms (Brinckman and Bernhard, 1985). These strategies for microbial resistance to toxic elements differ greatly between microbial cells (prokaryotes) and more advanced forms of life such as green algae (eukaryotes).

This point can be illustrated by a comparative study of the transport and toxicity of the transition metal nickel. Nickel becomes toxic at relatively low concentrations and therefore bacteria and algae have evolved energy-dependent mechanisms to regulate the uptake of nickel. Nickel is also of interest in three other respects: (1) it is soluble as Ni^{2+} at ambient pH and temperature (e.g., at pH 8.0 nickel in aqueous solution largely exists as Ni^{2+} with $NiOH^{+}$ and insoluble $Ni(OH)_2$ only becoming significant at pH 9.0 and above); (2) insoluble nickel complexes such as nickel subsulfide are known to be potent carcinogens in laboratory animals and in humans (Sunderman, 1981); and (3) $^{63}Ni^{2+}$ is a stable isotope (β-emitter) and it provides for extremely precise measurements for nickel uptake by microbiological microtechniques (Wang and Wood, 1984).

Table I presents a study of the uptake of nickel by two eukaryotes (*Scenedesmus* and *Chlamydomonas*) and two prokaryotes (*Oscillatoria* and *Synechococcus*). The green algae are much more tolerant to nickel than the cyanobacteria. *Chlamydomonas* will tolerate up to 150 μg/ml of nickel, *Scenedesmus* 5 μg/ml, and *Oscillatoria* 1 μg/ml with *Synechococcus* showing extreme sensitivity at 0.02 μg/ml. In

TABLE I

Concentration factor[a] for $^{63}Ni^{II}$ by different strains of cyanobacteria and green algae at different pH

Strain designation	pH				
	4	5	6	7	8
Scenedesmus ATCC 11460	NS	NS	7.9×10^2	1.8×10^3	2.2×10^3
Synechococcus ATCC 17146	NS	NS	4.2×10^2	3.3×10^3	3.3×10^3
Oscillatoria UTEX 1270	NS	2.9×10^1	9.5×10^2	1.1×10^3	1.1×10^1
Chlamydomonas UTEX 89	2.7×10^1	3.8×10^1	5.4×10^2	NS	NS

[a]Concentration factor (CF) = $\dfrac{\text{AC}}{\mu\text{g of } ^{63}\text{Ni/ml of culture medium}}$, where accumulation coefficient (AC) = $\dfrac{\mu\text{g of Ni removed}}{\text{g of alga (dry weight)}}$.

$[^{63}Ni^{II}] = 0.02$ μg/ml (0.34 μM), incubated for 6 h at 20°C in the light.
NS, no significant uptake at the 95% confidence level.

all cultures Ni^{2+} was found to be primarily complexed at the cell surfaces which accommodated a 3×10^3 fold increase in concentration over the external environment. Nickel transport in *Chlamydomonas* was found to be dependent on the concentration of Mg^{2+} in the culture medium. Competition experiments between Mg^{2+} and Ni^{2+} showed that Mg^{2+} prevents Ni^{2+} toxicity. Similar competition between Ca^{2+} and Ni^{2+} could not be shown, indicating that Ni^{2+} is probably transported in *Chlamydomonas* by way of the ATP-dependent Mg^{2+} channel. In the absence of Mg^{2+}, but at equivalent concentration of Ni^{2+}, growth inhibition occurs for a period of time until cells develop an efflux mechanism which pumps out the intracellular nickel. Similar efflux mechanisms have been discovered in prokaryotes for the specific removal of Cd^{2+} (*Staphylococcus aureus*) and for the removal of arsenicals and antimonials (*Escherichia coli*).

Synechococcus was found to be the most sensitive organism to Ni^{2+} in this series of experiments. Therefore, attempts were made to select for more tolerant mutants. This was accomplished by selecting surviving colonies from solid culture media containing 10^{-5} M nickel sulfate. After approximately 2 weeks colonies appeared which resisted nickel poisoning by the selection of two different mechanisms. One class of mutants was found to synthesize large quantities of an intracellular polymer which complexed Ni^{2+} approximately 300-fold over the internal cytoplasm concentration. The second class of mutants did not accumulate nickel as in the case with the parent culture *Synechococcus* ATCC 27146. This ability to resist both nickel complexation at the cell surface, and nickel uptake, was brought about by the synthesis and excretion of a low molecular weight polar ligand which complexes nickel outside the cell and prevents its uptake.

The structure of this ligand has not yet been determined but its presence indicates that cyanobacteria have the capability to excrete small metabolites to prevent the uptake of toxic metal ions. The biosynthesis of chelating agents such as the siderophores for the complexation and uptake of essential elements such as iron, which is not too available under aerobic conditions, is well documented in the cyanobacteria (see Chapter 16 of this volume). Therefore it is understandable that cellular energy may be invested also for the biosynthesis of secondary metabolites which have the function of keeping unwanted toxic elements from being transported into the cytoplasm.

Acknowledgements

F.K. Gleason would like to acknowledge research support from NOAA, through the Sea Grant College Program, Grant No. DOC/NA83AA-D-00056 R/NP-1 and also thanks her colleagues for critical reading of the manuscript.

J.M. Wood acknowledges research support from the National Institutes of Health Grant AM18101 and from Atlantic Richfield Company.

References

Adams, W.H., Stoner, R.D., Adams, D.G., Slatkin, D.N. and Siegelman, H.W. (1985) Pathophysiologic effects of a toxic peptide from *Microcystis aeruginosa*. Toxicon. 23, 441–447.

Aziz, K.M.S. (1974) Diarrhea toxin obtained from a water bloom-producing species, *Microcystis aeruginosa*. Kützing, Science 183, 1206–1207.

Barbier, M. (1981) Marine chemical ecology: the roles of chemical communication and chemical pollution. In P. Scheurer (ed.), Marine Natural Products, Vol. IV. Academic Press, New York, pp. 148–186.

Bates, H.A. and Rapoport, H. (1979) Synthesis of anatoxin-*a* via intra-molecular cyclization of iminium salts. J. Am. Chem. Soc. 101, 1259–1265.

Baxa, C.B. and Gleason, F.K. (1985) Algicidal properties of cyanobacterin in the eukaryotic alga *Euglena gracilis* Z. Abst. Am. Soc. Microbiol., p. 191.

Bishop, C.T., Anet, E.F.L.J. and Gorham, P.R. (1959) Isolation and identification of the fast-death factor in *Microcystis aeruginosa* NRC-1. Can. J. Biochem. Physiol. 37, 453–471.

Botes, D.P., Kruger, H. and Viljoen, C. (1982) Isolation and characterization of four toxins from the blue-green alga, *Microcystis aeruginosa*. Toxicon. 20, 945–954.

Botes, D.P., Tuinman, A.A., Wessels, P.L., Viljoen, C.C., Kruger, H., Williams, D.H., Santikarn, S., Smith, R.J. and Hammond, S.J. (1984) The structure of cyanoginosin-LA, a cyclic heptapeptide toxin from the cyanobacterium *Microcystis aeruginosa*. J. Chem. Soc. Perkin Trans. 1, 2311–2318.

Brinckman, F. and Bernhard, M. (eds.) (1987) Dahlem Konferenzen on Chemical Speciation. Springer-Verlag, Berlin, in press.

Buttke, T.M. and Ingram, L.O. (1975) Comparison of lipopolysaccharides from *Agmenellum quadruplicatum* to *Escherichia coli* and *Salmonella typhimurium* by using thin-layer chromatography. J. Bacteriol. 124, 1566–1573.

Camm, E.L. and Towers, G.H.N. (1973) Phenylalanine ammonia lyase. Phytochemistry 12, 961–973.

Campbell, H.F., Edwards, O.E. and Kolt, R. (1977) Synthesis of noranatoxin-*a* and anatoxin-*a*. Can. J. Chem. 55, 1372–1379.

Cardellina, J.H. II, Mamer, F.J. and Moore, R.E. (1979) The structure of malyngamide. J. Am. Chem. Soc. 101, 240–242.

Carmichael, W.W. (ed.) (1982) The Water Environment: Algal Toxins and Health. Plenum Press, New York.

Carmichael, W.W. and Gorham, P.R. (1978) Anatoxins from clones of *Anabaena flos-aquae* isolated from lakes of western Canada. Mitt. Int. Ver. Limnol. 21, 285–295.

Carmichael, W.W., Biggs, D.F. and Gorham, P.R. (1975) Toxicology and pharmacological action of *Anabaena flos-aquae* toxin. Science 187, 542–544.

Carmichael, W.W., Biggs, D.F. and Peterson, M.A. (1979) Pharmacology of anatoxin-*a*, produced by the freshwater cyanophyte *Anabaena flos-aquae* NRC-44-1. Toxicon 17, 229–236.

Chakrabarty, A.M. (1987) In F. Brinkman and M. Bernhard (eds.), Proceedings of the Dahlem Konferenzen on Chemical Speciation. Springer-Verlag, Berlin, in press.

Chapman, D.J. and Ragan, M.A. (1980) Evolution of biochemical pathways: evidence from comparative biochemistry. Ann. Rev. Plant Physiol. 31, 639–678.

Clutterbuck, P.W., Mukhopahyay, S.C., Oxford, A.E. and Raistrick, H. (1940) Microbial halogenation. Biochem. J. 34, 664.

Deverall, B.J. (1977) Defense Mechanisms of Plants. Cambridge University Press, Cambridge.

Devlin, J.B., Edwards, O.E., Gorham, P.R., Hunter, N.R., Pike, R.K. and Stavric, B. (1977) Anatoxin-a, a toxic alkaloid from *Anabaena flos-aquae* NRC-44h. Can. J. Chem. 55, 1367–1371.

Elleman, T.C., Falconer, I.R., Jackson, A.R.B. and Runnegar, M.T. (1978) Isolation, characterization and pathology of the toxin from a *Microcystis aeruginosa* bloom. Aust. J. Biol. Sci. 31, 209–218.

Faulkner, D.J. (1978) In P.S. Sammes (ed.), Topics in Antibiotic Chemistry, Vol. 2. Ellis and Huranne, Chichester, England, pp. 9–58.

Fenical, W. (1975) Halogenation in the Rhodophyta – a review. J. Phycol. 11, 245–259.

Fenical, W. (1982) Natural products chemistry in the marine environment. Science 215, 923–928.

Fowden, L. (1968) Halogenated organic compounds from the marine environment. Proc. Roy. Soc. Lond. Series B 171, 5–18.

Galanos, G., Lüderitz, O., Rietschel, E.T. and Westphal, O. (1977) Newer aspects of the chemistry and biology of bacterial lipopolysaccharides, with special reference to their lipid a component. In J.W. Goodwin (ed.), Int. Rev. Biochem., 14, 239–335.

Gentile, J. and Maloney, T. (1969) Toxicity and environmental requirements of a strain of *Aphanizomenon flos-aquae* (L.) Ralfs. Can. J. Microbiol. 15, 165–173.

Gleason, F.K. and Baxa, C.A. (1986) Activity of the natural algicide, cyanobacterin, on eukaryotic microorganisms. FEMS Microbiol. Lett. 33, 85–88.

Gleason, F.K. and Case, D.E. (1986) Activity of the natural algicide, cyanobacterin, on angiosperms. Plant Physiol. 80, 834–838.

Gleason, F.K. and Paulson, J.P. (1984) Site of action of the natural algicide, cyanobacterin in the blue-green alga, *Synechococcus* sp. Arch. Microbiol. 138, 273–277.

Gleason, F.K., Michurski, S. and Wood, J.M. (1983) Biosynthesis of the antibiotic, cyanobacterin. Fed. Proc. 42, 2640.

Gleason, F.K., Case, D.E., Sipprell, K.D. and Magnuson, T.S. (1986) Effect of the natural algicide, cyanobacterin, on a herbicide-resistant mutant of *Anacystis nidulans* R2. Plant Sci., 46, 5–10.

Gleason, F.K., Porwoll, J., Flippen-Anderson, J.L. and George, C. (1986) X-ray structure determination of the naturally occurring isomer of cyanobacterin. J. Org. Chem., 51, 1615–1616.

Golden, S.S. and Sherman, L.A. (1984) Biochemical and biophysical characterization of herbicide-resistant mutants of the unicellular cyanobacterium, *Anacystis nidulans* R2. Biochim. Biophys. Acta 764, 239–246.

Goldman, S.J., Lammers, P.J., Berman, M.S. and Sanders-Loehr, J. (1983) Siderophore-mediated iron uptake in different strains of *Anabaena* sp. J. Bacteriol. 156, 1144–1150.

Gorham, P.R. (1964) Toxic algae. In D. Jackson (ed.), Algae and Man. Plenum Publ. Corp., New York, pp. 307–336.

Gorham, P.R. and Carmichael, W.W. (1979) Phycotoxins from blue-green algae. Pure Appl. Chem. 52, 165–174.

Gorham, P.R., McLachlan, J., Hammer, U.T. and Kim, W.K. (1964) Isolation and culture of toxic strains of *Anabaena flos-aquae* (Lyngb.) de Breb. Verh. Int. Ver. Limnol. 15, 796–804.

Hager, L.P., Hollenberg, P.F., Rand-Meir, T., Chaing, R. and Doubek, D. (1975) Mechanism of action of chloroperoxidase. Ann. NY Acad. Sci. 244, 80–93.

Hellebust, J.A. (1974) Extracellular products. In W.D.P. Stewart (ed.) Algal Physiology and Biochemistry. University of California Press, Berkeley CA, pp. 838–863.

Hewson, W.D. and Hager, L.P. (1980) Bromoperoxidase and halogenated lipids in marine algae. J. Phycol. 16, 340–345.

Hirschberg, J. and McIntosh, L. (1983) Molecular basis of herbicide resistance in *Amaranthus hybridus*. Science 222, 1346–1349.

Huber, C.S. (1972) The crystal structure and absolute configuration of 2,9-diacetyl-9-azabicyclo (4,2,1) non-2,3-ene. Acta Crystallogr. B28, 2577–2582.

Hughes, E.O., Gorham, P.R. and Zehnder, A. (1958) Toxicity of a unialgal culture of *Microcystis aeruginosa*. Can. J. Microbiol. 4, 225–236.

Ikawa, M., Wegener, K., Foxall, T.L. and Sasner, J.J. (1982) Comparison of the toxins of the blue-green alga *Aphanizomenon flos-aquae* with the *Gonyaulax* toxins. Toxicon 20, 747–752.

Jong, T-T., Williard, P.G. and Porwoll, J.P. (1984) Total synthesis and x-ray structure determination of cyanobacterin. J. Org. Chem. 49, 735–736.

Keating, K.I. (1977) Allelopathic influence on blue-green bloom sequence in a eutrophic lake. Science 196, 885–887.

Keating, K.I. (1978) Blue-green algal inhibition of diatom growth: transition from mesotrophic to eutrophic community structure. Science 199, 971–973.

Keleti, G. and Sykora, J.L. (1982) Production and properties of cyanobacterial endotoxins. Appl. Environ. Microbiol. 43, 104–109.

Mason, C.P., Edwards, K.R., Carlson, R.E., Pignatello, J., Gleason, F.K. and Wood, J.M. (1982) Isolation of chlorine-containing antibiotic from the freshwater cyanobacterium, *Scytonema hofmanni*. Science 215, 400–402.

Mikheyskaya, L.V., Ovodova, R.G. and Ovodov, Yu.S. (1977) Isolation and characterization of lipopolysaccharides from cell walls of blue-green algae of the genus *Phormidium*. J. Bacteriol. 130, 1–3.

Moore, R.E. (1981) Constituents of blue-green algae. In P. Scheuer (ed.), Marine Natural Products, Vol. IV. Academic Press, New York, pp. 1–52.

Muller, C.H. (1969) Coevolution. Science 165, 415–416.

Murphy, T.P., Lean, D.R.S. and Nalewajko, C. (1976) Blue-green algae: their excretion of iron selective chelators enables them to dominate other algae. Science 192, 900–902.

Pignatello, J.J., Porwoll, J., Carlson, R.E. Xavier, A., Gleason, F.K. and Wood, J.M. (1983) Structure of the antibiotic cyanobacterin, a chlorine-containing γ-lactone from the freshwater cyanobacterium *Scytonema hofmanni*. J. Org. Chem. 48, 4035–4038.

Putnam, A.R. (1983) Allelopathic chemicals. Chem. Eng. News 61, 34–45.

Raziuddin, S., Siegelman, H.W. and Tornabene, T.G. (1983) Lipopolysaccharides of the cyanobacterium *Microcystis aeruginosa*. Eur. J. Biochem. 137, 333–336.

Runnegar, M.T., Falconer, I.R. and Silver, J. (1981) Deformation of isolated rat hepatocytes by a peptide hepatotoxin from the blue-green alga *Microcystis aeruginosa*. Naunyn-Schmiedeberg's Arch. Pharmakol. 317, 268–272.

Sawyer, P.J., Gentile, J.H. and Sasner, J.J. (1968) Demonstration of a toxin from *Aphanizomenon flos-aquae*

(L.) Ralfs. Can. J. Microbiol. 14, 1199–1204.
Schopf, J.W. and Walter, M.R. (1983) Archean microfossils: new evidence of ancient microbes. In J.W. Schopf (ed.), Earth's Earliest Biosphere. Princeton University Press, New Jersey, pp. 214–239.
Seigler, D. and Price, P.W. (1976) Secondary compounds in plants: primary functions. Am. Nat. 110, 101–105.
Shaw, P.D. and Hager, L.P. (1960) The peroxidation of halogens. J. Biol. Chem. 236, 1626–1630.
Slatkin, D.N., Stoner, R.D., Adams, W.H., Kycia, J.H. and Siegelman, H.W. (1983) Atypical pulmonary thrombosis caused by a toxic cyanobacterial peptide. Science 220, 1383–1385.
Solomans, W. and Baccini, P. (1987) Chemical speciation in the environment. In F. Brinckman and M. Bernhard (eds.), Dahlem Konferenzen on Chemical Speciation. Springer-Verlag, Berlin, in press.
Suida, J.F., Van Blaricum, G.R., Shaw, P.D., Johnson, R.D., White, R.H., Hager, L.P. and Rinehart, K.L. Jr. (1975) The isolation of brominated natural products from marine organisms. J. Am. Chem. Soc. 97, 937.
Sunderman, F.W. (1981) The carcinogenicity of nickel subsulfide. Environ. Health Persp. 40, 131–141.
Swain, T. (1977) Secondary compounds as protective agents. Ann. Rev. Plant Physiol. 28, 479–501.
Van der Westhuizen, A.J. and Eloff, J.N. (1983) Effect of culture age and pH of culture medium on the growth and toxicity of the blue-green alga *Microcystis aeruginosa*. Z. Pflanzenphysiol. 110, 157–163.
Wang, Hong Kang and Wood, J.M. (1984) Bioaccumulation of nickel by algae. Environ. Science Technol. 18, 106, 109.
Weckesser, J., Drews, G. and Mayer, H. (1979) Lipopolysaccharides of photosynthetic prokaryotes. Ann. Rev. Microbiol. 33, 215–239.
Wood, J.M. (1984) Evolutionary aspects of metal ion transport through cell membranes. Chapter 7, pp. 223–237, and Microbial strategies in resistance to metal ion toxicity, Chapter 10, pp. 333–349. In H. Sigel (ed.), Metal Ions in Biological Systems, Vol. 18, Circulation of Metals in the Environment. Marcel Dekker, New York, Basel.
Wood, J.M. and Wang, Hong Kang (1983) Microbial resistance to heavy metals. Environ. Science Technol. 17, 12, 582A–590A.

P. Fay and C. Van Baalen (Eds.) The Cyanobacteria

18

Volatile organic substances

FRIEDRICH JÜTTNER

Institut für Chemische Pflanzenphysiologie der Universität Tübingen, Corrensstrasse 41, D-74 Tübingen, FRG

1. Introduction

Volatile organic biogenic substances (VOBS) are liberated by a large number of microorganisms and many are well known as typical products of fermentation. The VOBS of the cyanobacteria differ from the latter by being synthesized in the presence of oxygen. They represent a group of substances characterized by the physical feature of volatility regardless of the class of organic compound to which its members belong. The most important chemical classes in which VOBS are found are those of the saturated and unsaturated hydrocarbons, alcohols, aldehydes, ketones, carboxylic acids, amines, aromatic and sulfur compounds. Both the molecular weight and the functional groups of a compound contribute to its volatility. In homologous series of compounds it is primarily the number and type of functional groups that determine the upper molecular weight limit of volatility. For instance in the unsubstituted alkane and alkene series the upper limit for volatility is around 18 C-atoms (octadecane). Introduction of a single functional group (e.g. hydroxy-, aldehyde-, carboxy-) lowers the upper limit of the molecular weight of each homologous series. Di- and multi-substituted compounds (succinic acid, glycerol, glycolaldehyde, pyruvic acid) do no longer belong to the VOBS.

While alkanes and alkenes of low molecular weight which occur as gases are in general treated separately, the upper molecular weight limit for volatility is to a certain degree determined by the limits set by current detection methodology. As methods are improved, a steady increase in the molecular weight of volatile compounds which may be detected can be observed. The method of choice for studying the VOBS is gas chromatography, usually combined with mass spectrometry. A prerequisite for the application of this technique is a potent concentration method that enables sufficient amounts of high purity VOBS to be separated from the site of their occurrence (e.g. growth media, biomass, mud, natural waters) for gas chromatography and mass spectrometry. A method of concentration often used until recently was the static head space technique (Hachenberg and Schmidt, 1977; Drozd and Novák, 1981). Though good yields of highly volatile substances were obtained by this method, it failed to adequately extract even those members of homologous series that are intermediate in volatility. The introduction of the stripping technique in a closed loop system by Grob (1973), and application of thermal desorption techniques with Tenax (Bellar and Lichtenberg, 1974; Jüttner and Wurster, 1979), Chromosorb 105 (Murray, 1977) and other sorbants was essential for extending the range of compounds which could be analyzed by the gas chromatography/mass spectrometry (GC/MS) technique.

The separation of all VOBS can be achieved without derivation on capillary columns by gas chromatography. Since individual compounds often occur in low amounts and are contaminated with large amounts of ubiquitous pollution products, high performance chromatographic separation is a basic requirement for their analysis. Since other separation techniques of the same resolution as capillary gas chromatography can only be achieved at much greater expense, combined GC/MS is the standard method for the elucidation of the structure of individual VOBS. An essential prerequisite for unambiguous identification of a particular structure is the availability of an authentic reference sample. The lack of pure and sufficient material usually prevents the application of more powerful analytical techniques such as nuclear magnetic resonance spectrometry. Some strains, however, liberate considerable quantities of a particular product to enable the isolation of sufficient material for a more detailed analysis.

The distinction between VOBS and contaminants is often not easy, especially in those cases where a particular compound may have a biogenic origin as well as being derived from other sources. Though laboratory air is suspected to be a major source of contaminants, distilled water, which is often considerably enriched with pollution products (with seasonal variation when its precursor was surface water), is much more critical. Other important sources of contamination are autoclaving, flaming of glassware and aeration gases. Means to reduce the introduction of non-biogenic volatile organic compounds and ways to distinguish between VOBS and contaminants are discussed by Jüttner (1983).

VOBS observed in cyanobacteria have also been reported under the term excretion products. A distinction between 'true' excretion and other mechanisms involved in the liberation of cell constituents is in many cases questionable. The major part of certain substances synthesized by cyanobacteria under photoautotrophic growth conditions has been found to be released immediately into the medium. This is true for mesityloxide in *Anabaena cylindrica* and isopropyl thiol in *Microcystis* PCC 7806 and may be equally valid for acetone, butanone and dimethylallylic alcohol, though no experimental evidence has been put forward for the latter compounds.

The use of the term excretion product is more difficult to justify when oxygenases, which are most effectively activated under conditions of stress, are involved in the formation of compounds like β-cyclocitral and crocetindial in *Microcystis* (Jüttner and Höflacher, 1985). Low amounts of these substances produced by oxygenase activity are found under optimum growth conditions. It is not clear whether oxygenase cleavage products are liberated at a low rate from undisturbed cells, as a result of transcellular trichome breakage, which is a wide-spread feature of filamentous cyanobacteria (Rippka et al., 1981), or by stress conditions which may occur at the meniscus between liquid medium and gas phase as the result of cell deposition on the glass walls of the culture vessel. The lytic processes which are involved in the normal life cycle of many cyanobacteria (e.g. hormogonium formation, Geitler, 1960) may also contribute to the above effect. Under natural conditions, various chemical and biological factors could cause the disintegration of cyanobacterial cells. Cyanophages, lytic bacteria, fungi and feeding by protozoa and other zooplankton organisms are well-documented agents (Fogg et al., 1973). In addition, photooxidation (Eloff et al., 1976) and drying of cell matter during scum deposition may contribute to cell lysis and the subsequent release of oxygenase products.

The known VOBS of cyanobacteria can be classified in (1) nor-carotenoids, which represent degradation products of carotenoids, (2) lipoxygenase products and their derivatives which are regarded as degradation products of fatty acids, (3) metabolites which are released into the medium during the growth phase, (4) sulfur compounds, (5) terpenoids, (6) aromatic compounds and (7) amines.

2. Nor-carotenoids

The designation nor-carotenoid is used for compounds which have an unequivocal terpenoid structure but which, in most cases, do not fit the rule of Ruzicka in being built-up of isoprene units. Instead, their C-skeleton can readily be derived from carotenoids by assuming cleavage reactions at double bonds. This feature applies to those nor-carotenoids which have hitherto been detected in cyanobacteria. They are listed in Table I. Their structures indicate that they may be derived equally from cyclic and acyclic carotenes. However, in growing cultures of cyanobacteria the latter

TABLE I

Nor-carotenoids isolated from different cyanobacterial strains

Nor-carotenoid	Organism	Reference
β-Cyclocitral (1)	*Microcystis aeruginosa*	Jüttner, 1976
	M. wesenbergii	
	Anabaena cylindrica	Jüttner et al., 1983
β-Ionone (2)	*Anabaena cylindrica*	Jüttner, 1981
	Synechococcus PCC 6911	Jüttner, 1981
Hydroxy-β-cyclocitral (3)	*Microcystis* PCC 7806	Jüttner and Höflacher, 1985
β-Ionone-5,6-epoxide (4)	*Synechococcus* PCC 6301	Henatsch and Jüttner, 1983
Dihydroactinidiolide (5)	*Synechococcus* PCC 6301	Henatsch and Jüttner, 1983
	Synechococcus PCC 6911	Henatsch and Jüttner, 1983
2,6,6-Trimethylcyclohex-2-en-1,4-dione (6)	*Synechococcus* PCC 6301	Henatsch and Jüttner, 1983
6-Methylhept-5-en-2-one (7)	*Anabaena cylindrica*	Jüttner, 1981
	Synechococcus PCC 6301	Henatsch and Jüttner, 1983
	Synechococcus PCC 6911	Henatsch and Jüttner, 1983
6-Methylhept-5-en-2-ol (8)	*Synechococcus* PCC 6911	Henatsch and Jüttner, 1983
	Anabaena cylindrica	Jüttner et al., 1983
Tetrahydrogeranyl acetone (9)	*Synechococcus* PCC 6911	Henatsch and Jüttner, 1983
Hexahydrofarnesyl acetone (10)	*Oscillatoria splendida*	Tsuchiya et al., 1981

The numbers refer to the structural formulas below:

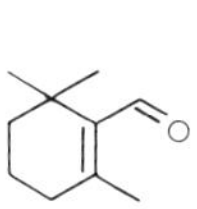
(1)

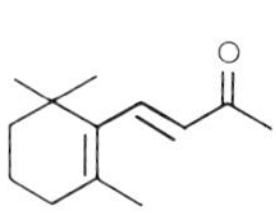
(2)

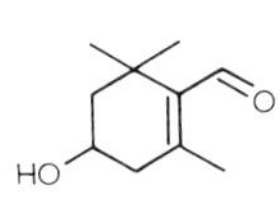
(3)

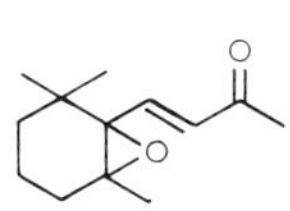
(4)

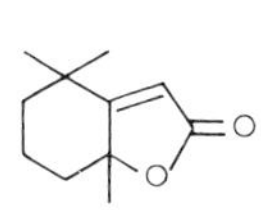
(5)

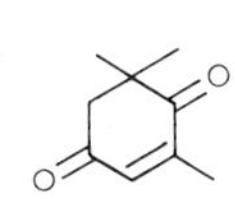
(6)

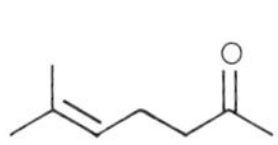
(7)

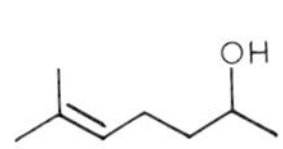
(8)

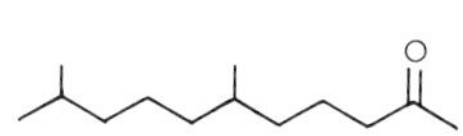
(9)

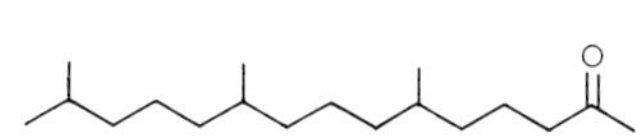
(10)

which are intermediates in carotene synthesis, are generally below the detection limit. Direct proof of the origin of nor-carotenoids is difficult to obtain since they are in most cases liberated in trace amounts and usually occur in complex mixtures. This holds for both higher plants (Enzell et al., 1977) and cyanobacteria. The first exception to this rule are members of the genus *Microcystis* which produce large amounts of the nor-carotenoid β-cyclocitral (Jüttner, 1976). *Microcystis* PCC 7806 liberates as much as 0.6 μg/mg dry weight of this compound and turned out to be a suitable organism for physiological and biosynthetic studies on nor-carotenoids.

Recent experiments with *Microcystis* (Jüttner and Höflacher, 1985) have demonstrated that the enzyme system responsible for the oxidative cleavage of carotenoids in *Microcystis* can be activated in any way which leads to the disintegration of the cells. French press treatment, sonification, incubation with 20% ethanol, 20% NaCl or freezing result equally well in the activation of this enzymatic reaction, though freezing has been found to be the most convenient and effective method.

More detailed studies on the reaction mechanism revealed that β-carotene and zeaxanthin are oxidatively cleaved at the double bonds 7,8 and 7′,8′ (Fig. 1). The cleavage of β-carotene leads to the formation of β-cyclocitral (Formula 1) and crocetindial (Formula 11). A stoichiometry of about 2 mol β-cyclocitral formed per 1 mol β-carotene degraded was measured. β-Apo-8′-carotenal which has lost only one β-ionone ring and which was detected in trace amounts may be an intermediate in the reaction which leads ultimately to the formation of crocetindial. In a similar reaction hydroxy-β-cyclocitral (Formula 3) and crocetindial are formed from zeaxanthin. Echinenone and myxoxanthophyll which are also major carotenoids in *Microcystis* are not degraded in detectable amounts. An explanation may be that the specificity of the enzyme does not allow their cleavage or that a spatial separation of substrate and enzyme in the cell does not enable them to come into contact with each other.

The cleavage enzyme has the feature of a β-carotene 7,8 (7′,8′) oxygenase. The oxygen introduced into the aldehyde group of β-cyclocitral and hydroxy-β-cyclocitral is derived from dioxygen. No formation of β-cyclocitral can be observed under strictly oxygen-free conditions. This is consistent with the labelling, up to 86% of the aldehyde group in β-cyclocitral with 18-oxygen after application of $^{18}O_2$. The difference between this figure and the 98.7% purity of the $^{18}O_2$ applied can be explained by a rapid exchange of the labelled oxygen in the aldehyde group with that of unlabelled water during the experimental procedure.

The β-carotene oxygenase is membrane bound and has not yet been solubilized. Essential cofactors have not been found: a loss of activity was not observed when the particulate fraction was washed, nor was an activation measured when the unreacted soluble fraction was added. A co-

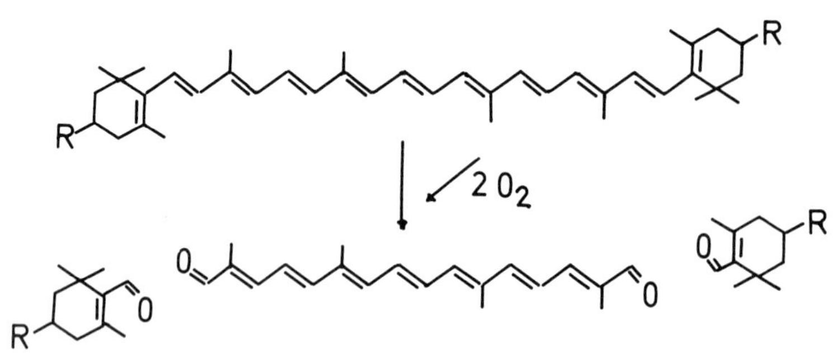

Fig. 1. Oxidative cleavage of β-carotene (R = H) and zeaxanthin (R = OH) in *Microcystis* PCC 7806. The products obtained are crocetindial (11), β-cyclocitral (R = H) and hydroxy-β-cyclocitral (R = OH).

oxidation reaction with unsaturated fatty acids such as has been shown for higher plants (Grosch and Laskawy, 1979) can be ruled out because the addition of various unsaturated fatty acids was found to be inhibitory in all cases. A similarity to known lipoxygenases is indicated by the strong inhibitory effects of *o*-phenanthroline and by the stimulatory effects of iron ions which suggest that iron may be an essential component of the enzyme system. Antioxidants are effective inhibitors of the reaction. Complete inhibition was observed with 0.5 mM methyl gallate, nordihydroguaiaretic acid and *tert*-butylhydroquinone, slightly less with 2-aminoethylisothiouronium bromide hydrobromide. Though the occurrence of radicals during inhibitor studies points to likely intermediates of the oxygenation reaction, it is still only possible to speculate on the reaction mechanism.

Studies of the structure of other nor-carotenoids isolated from cyanobacteria suggest that further, as yet uncharacterized, carotene oxygenases may be involved in their formation and that further modifications of the structures entail the reduction of C-C and C-O double bonds. Feeding *Synechococcus* PCC 6911 with 6-methylhept-5-en-2-one results in ready formation of the corresponding alcohol. Reduction of double bonds in nor-carotenoids derived from acyclic carotenes would lead to such molecules as tetrahydrogeranyl acetone and hexahydrofarnesyl acetone reported as excretion products of *Synechococcus* PCC 6911 (Henatsch and Jüttner, 1983) and *Oscillatoria splendida* (Tsuchiya et al., 1981), respectively.

3. Lipoxygenase products

The term lipoxygenase products defines substances that are produced during the cleavage of fatty acids. Compounds from cyanobacteria for which such an origin is postulated are listed in Table II. It has been shown that thylakoid-bound lipoxygenases are responsible for the formation in a light-mediated reaction of pentane and a pentene isomer from linoleic and γ-linolenic acid (Sandmann and Böger, 1982), though there is no direct proof that related reactions do take place in other cyanobacteria. However, comparative studies on fungi and higher plants suggest by analogy that similar reactions are very likely to occur also in cyanobacteria. Preferential formation of C_5-, C_8- and C_9-cleavage products is observed in cyanobacteria while the C_6-compounds, which are typically found in higher plants, are rather an exception. No strain has hitherto been detected that liberates C_6-compounds as major products. In most cases a palette of differently substituted compounds with an oxygen function in the 1 or 3 position is observed. The vinylketones or α-unsaturated aldehydes that are typically formed are reduced to the corresponding alcohols in various ratios depending on the strain. In a group of *Anabaena* strains, represented by *A. randhawae* UTEX 1823, the major product is oct-1-en-3-ol, in another, represented by *A. oscillarioides* CCAP 1403-9, octa-1,5-dien-3-ol and oct-1-en-3-ol are produced in nearly equal amounts.

Reductions can also affect C-C double bonds resulting in saturated alcohols and ketones though these reactions seem to be of minor importance in the strains so far studied. However, the reactions do offer an explanation for the origin of small amounts of pentanol-1, octanol-1 and octanone-3 which can often be detected in culture media from various cyanobacteria. The biotransformation of octanone-3 into octanol-3, of octanal into octanol-1 and *trans*-2-octenal into *trans*-2-octen-1-ol has been successfully demonstrated in shaken cultures of *A. oscillarioides*. While introduction of an oxygen function into the cleavage products is the major reaction route, olefines may also be formed to a certain degree. Formation of 1,3-octadiene, the stereochemistry of which has not yet been established, and *trans,cis* 1,3,5-octadiene are examples of olefinic substances which are formed during the lipoxygenase reaction in *A. oscillarioides* (Möhren and Jüttner, 1983).

TABLE II

Lipoxygenase products and their occurrence in different strains of cyanobacteria

Compound	Organism	Reference
Pentane	*Anabaena variabilis*	Sandmann and Böger, 1982
Pentanol-1	*Anabaena cylindrica*	Jüttner et al., 1983
Pent-1-en-3-ol		
Penten-3-one		
trans 2-Pentenal		
cis 2-Pentenal		
Hexanal	*Anabaena cylindrica*	Jüttner et al., 1983
	Anabaena oscillarioides	Möhren and Jüttner, 1983
Hexanol-1	*Anabaena cylindrica*	Jüttner et al., 1983
	Anabaena oscillarioides	Möhren and Jüttner, 1983
1,3-Octadiene	*Anabaena oscillarioides*	Möhren and Jüttner, 1983
trans,cis 1,3,5-Octatriene		
Octanol-1		
Octanone-3		
Octanol-3		
Oct-1-en-3-ol		
Octa-1,5-dien-3-ol		
Oct-2-en-1-ol		
Octa-2,5-dien-1-ol		
trans,cis-2,6-Nonadienal	*Anabaena utermöhli*	Von Tasch, 1985

4. Metabolites of cyanobacteria

A limited number of metabolites was found to be excreted into the medium by cyanobacteria. They resemble the fermentation products of heterotrophic bacteria, but are synthesized under aerobic conditions. Substances belonging to this group are acetone and butanone both of which have been isolated from cultures of *Nostoc* PCC 6314 and 6310 (Möhren and Jüttner, 1983) and mesityloxide (4-methylpent-3-en-2-one) which is produced in large quantities by *Anabaena cylindrica* (Jüttner, 1981; Jüttner et al., 1983).

Some preliminary investigations have been made on the biochemistry of the latter compound, which is unique for this cyanobacterial strain. Though the production of mesityloxide in *A. cylindrica* only occurs in the light, a direct connection with photosynthesis is not likely since dichlorophenyl dimethylurea (DCMU) applied at concentrations which completely and immediately inhibited the photosynthetic oxygen evolution had no immediate effect on its formation. Only after another hour of incubation with DCMU has the rate of mesityloxide synthesis decreased (Jüttner et al., 1983). In a light-dark cycle (Fig. 2) mesityloxide formation was observed only during the light phase. Cultures grown under continuous light conditions for 4 days displayed mesityloxide concentrations twice as high (90 μg/l) as those held in a 12:12 h light-dark cycle (Von Tasch, 1985).

Though the biosynthetic origin is not known, a connection with steps of the fatty acid synthesis can be assumed. Formally, this molecule can easily be constructed by condensation reactions from acetate moieties (Formula 12).

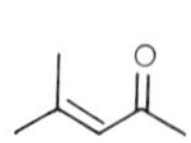

(12)

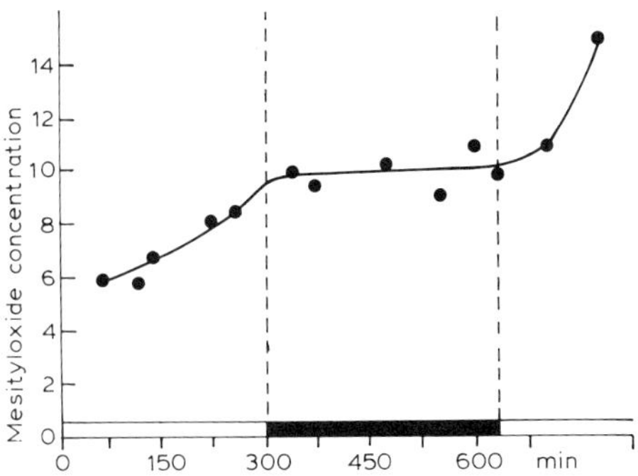

Fig. 2. Formation of mesityloxide in a 2 l-fermenter culture of *Anabaena cylindrica* during a 5:5 h light-dark cycle. The aeration gas of the culture supplemented with 20 mM $NaHCO_3$ was stopped at the beginning of the experiment. (From Jüttner et al., 1983)

Studies with inhibitors (Von Tasch, 1985) indicate that formation of mesityloxide is most likely accomplished in a branch reaction during early steps in the pathway of fatty acid synthesis. Iodoacetamide in the range of 30–500 μM partially inhibits (60%) the formation of mesityloxide. Under the same conditions a partial inhibition of fatty acid synthesis and a much stronger inhibition of carotenoid synthesis has also been recorded. The sensitivity of isopentenyl pyrophosphate isomerase to iodoacetamide, reported by Agranoff et al. (1960), may be the reason for the inhibition of carotenoid synthesis. A competitive reaction for a common precursor could explain the weaker inhibition of mesityloxide production observed at very high concentrations (0.5 mM) of iodoacetamide at which the synthesis of carotenoids is strongly inhibited.

A competitive reaction involving a precursor, which gives rise to amino acids of the glutamate family, may also exist. Monofluoroacetate (1 mM), which inhibits the flow of carbon into the amino acids of the glutamate family, is the only effector that has been found to increase the formation of mesityloxide (30%). The increase is accompanied by a complete inhibition of chlorophyll *a* synthesis and a strong inhibition of protein synthesis, presumably caused by the lack of glutamate family amino acids.

The most effective inhibitor of mesityloxide formation is cerulenin. Low concentrations (0.3 μg/ml) are sufficient to entirely inhibit mesityloxide synthesis in *A. cylindrica*. A 50% inhibition was found at 0.1 μg/ml, but growth (as determined by chlorophyll *a* and carotenoid concentrations) was not affected even in the presence of much higher concentrations (0.6 μg cerulenin/ml). Neither was the synthesis of fatty acids severely impaired. A slight decrease of linoleic and linolenic acid was accompanied by a 5-fold increase of palmitoleic acid. Though effects of cerulenin on growth (Herczeg et al., 1979), and on fatty acid metabolism (Ōmura, 1976; D'Agnolo et al., 1973; Oesterhelt et al., 1977) have been reported with respect to specific enzymes, the studies of Funabashi et al. (1983) demonstrate that, in principle, all enzymes with thiol groups can be inhibited by the irreversible binding of cerulenin to cysteine. The experimental data can be interpreted as showing that the enzymes catalyzing the formation of mesityloxide contain sensitive thiol groups which are more accessible to binding than those in the enzymes of the fatty acid and isoprenoid pathway.

Mesityloxide is an unusual metabolite which has, up to now, only been found in *A. cylindrica*. Several experiments have been conducted in which no induction or repression of mesityloxide synthesis could be observed. In addition, no proof has yet been obtained that this product can be remetabolized by the organism. Although the growth rate of cyanobacterial cells (*Synechococcus*) is not affected by concentrations up to 100 μg/l mesityloxide (Jüttner and Bogenschütz, 1983), such treated cells showed significantly higher protein values than the controls.

5. Sulfur and nitrogenous compounds

Though sulfur compounds of freshwater cyanobacteria were reported as early as 1967 (Jenkins et al., 1967), a clear distinction between bacterial and cyanobacterial products could not be achieved before axenic cultures of cyanobacteria were available. Two different groups of alkyl sulfides have been observed up to now, a series of methyl derivatives (dimethyl sulfide, dimethyl disulfide, dimethyl trisulfide, dimethyl tetrasulfide) which seem to be more widespread in and typical for cyanobacteria and isopropyl sulfides which have been detected in *Microcystis* PCC 7806. The major compound in the latter strain is diisopropyl disulfide accompanied by methylisopropyl disulfide and diisopropyl trisulfide (Jüttner, 1984a).

Under anaerobic conditions the major sulfur excretion product for the *Microcystis* strain PCC 7806 is isopropyl thiol (Hofbauer and Jüttner, unpublished). This finding, obtained with axenic fermenter cultures, supports the early observation of Jenkins et al. (1967) on isopropyl thiol obtained with bacterized cultures of *M. flos-aquae* and provides definite proof that isopropyl thiol is of cyanobacterial origin. Alkyl sulfides so far isolated from axenic cultures of cyanobacteria, are listed in Table III. No studies have been reported on the biosynthesis of these compounds in cyanobacteria.

A much more complex but not yet fully formulated structure has been put forward for an akinete-stimulating substance isolated by Hirosawa and Wolk (1979a, 1979b) from phosphate-deficient cultures of *Cylindrospermum licheniforme* ATCC 29412. It consists of a γ-lactone fused by a five-membered ring that contains the very rarely found thioketone group (see Section 2.6.1. of Chapter 9).

It has not yet been shown with certainty whether any volatile nitrogenous compounds are synthesized by cyanobacteria. The occurrence in culture media of *Microcystis aeruginosa* of propyl amine has been reported by Herrmann and Jüttner (1977), and indole and 3-methylindole have been isolated from a strain of *Oscillatoria* (Tsuchiya et al., 1979). However, since both cyanobacterial strains used in these studies were bacterized, definite proof that cyanobacteria are able to liberate amines has yet to be obtained.

TABLE III

Volatile alkyl sulfides isolated from axenic cultures of cyanobacteria

	MSH	MSM	MS_2M	MS_3M	MS_4M	ISH	IS_2I	IS_3I	IS_2M
Synechococcus cedrorum (1)	+	+	+						
Anacystis nidulans (1)		+	+						
Anabaena cylindrica (2)			+						
Microcystis PCC 7005 (3)			+	+	+				
Microcystis PCC 7806 (3,4)						+	+	+	+

M, methyl; I, isopropyl.

The numbers in parentheses indicate the references: (1) Bechard and Rayburn, 1973; (2) Jüttner et al., 1983; (3) Jüttner, 1984a (4) Hofbauer and Jüttner, unpublished.

6. Aromatic compounds

To date, very few aromatic compounds have been reported to be released from cyanobacteria. Pedersén and DaSilva (1973) isolated 2,3,5-tribromo-4-hydroxybenzaldehyde (Formula 13), 2,3,5-tribromo-4-hydroxybenzyl alcohol (Formula 14) and traces of lanosol (2,3-dibromo-4,5-dihydroxybenzyl alcohol) (Formula 15) from the medium of axenically cultured *Calothrix brevissima*. These compounds, however, which were produced upon addition of NaBr (0.075%) to the growth medium cannot be regarded in a strict sense as volatile. A phenol isolated from *Oscillatoria* has been identified by Tsuchiya et al. (1979) to be *p*-cresol (Formula 16). However, as the culture was not axenic, the cyanobacterial origin of the product requires further confirmation.

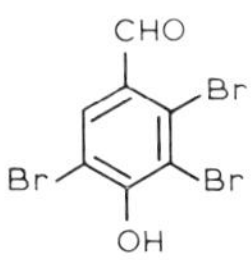

(13)

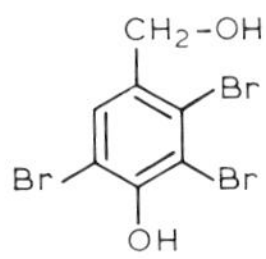

(14)

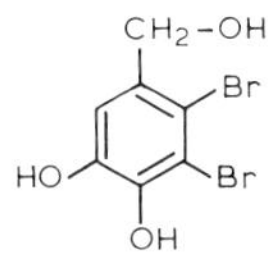

(15)

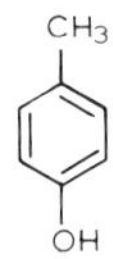

(16)

7. Terpenoid compounds

Terpenoid compounds do not have the same importance as products of secondary metabolism in cyanobacteria as they do in higher plants. Their existence was first observed by Sirenko and Sakevich (1967) by thin-layer chromatography but a structural determination was not achieved. Cyclic monoterpenes have not yet been reported, though this may be due to the limited number of strains which have been investigated thoroughly. Acyclic monoterpenes have repeatedly been observed. From crude cyanobacterial material (*Microcystis aeruginosa*, *Aphanizomenon*) collected in the Dnepr river, Mel'nichenko et al. (1974) isolated geraniol (Formula 17), nerol (Formula 18), geranyl acetate, citronellol (Formula 19), citronellal (Formula 20) and citronellyl formate. Henatsch and Jüttner (1983) isolated nerol and geraniol from the growth medium of *Synechococcus* PCC 6911. One assumption is that they are derived from classical terpene metabolism.

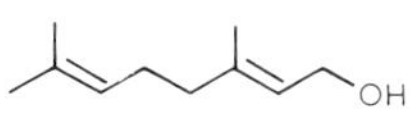

(17)

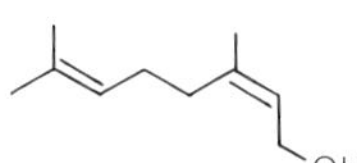

(18)

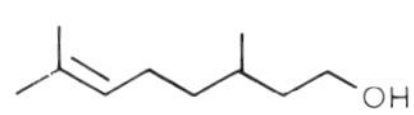

(19)

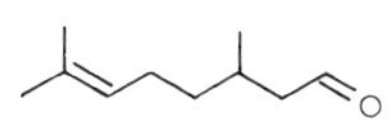

(20)

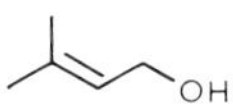

(21)

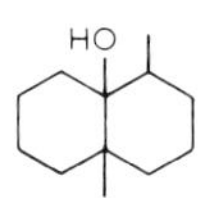

(22)

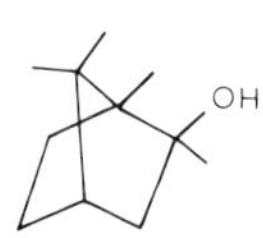

(23)

The only hemiterpene so far observed is dimethylallylic alcohol (Formula 21) which is excreted in appreciable amounts into the medium by axenically grown *Anabaena cylindrica* (Jüttner et al., 1983). Its genesis is similar to that of geraniol and nerol, both of which are probably liberated by the action of phosphatases upon phosphorylated intermediates of cell terpene metabolism.

No doubt the most important terpenoid substances isolated from cyanobacteria are geosmin (Formula 22) and 2-methylisoborneol (Formula 23). An intense search for the producers of these compounds has been undertaken worldwide as humans have extremely low odor thresholds for these compounds (Persson, 1980), with the consequence that these odorous compounds are often responsible for complaints on malodor and taste of drinking water. Both compounds were earlier isolated from actinomycetes (Gerber, 1968; Medsker et al., 1969), and they were only later de-

TABLE IV

Occurrence of geosmin and 2-methylisoborneol (MIB) in different axenic and non-axenic strains of cyanobacteria.

Organism	Geosmin	MIB	Axenic	Non-axenic	Reference
Anabaena macrospora	+			+	Yagi et al., 1983
A. scheremetievi	+			+	Izaguirre et al., 1982
A. spiroides	+			+	Yagi, 1983
Aphanizomenon flos-aquae	+			+	Yagi, 1983
A. sp.	+			+	Tsuchiya et al., 1981
A. gracile	+			+	Jüttner et al., 1986
Lyngbya aestuarii	+		+		Tabachek and Yurkowski, 1976
L. cryptovaginata		+		+	Tabachek and Yurkowski, 1976
Oscillatoria agardhii	+		+		Tabachek and Yurkowski, 1976
O. amoena	+			+	Tsuchiya et al., 1981
O. animalis	+			+	Yagi, 1983
O. bornetii	+			+	Skulberg and Skulberg, 1985
O. brevis	+	+	+		Berglind et al., 1983
O. chlorina	+			+	Yagi, 1983
O. cortiana	+		+		Tabachek and Yurkowski, 1976
O. curviceps		+		+	Izaguirre et al., 1982
O. geminata		+		+	Tsuchiya et al., 1981
O. limnetica		+		+	Tsuchiya et al., 1981
O. prolifica	+		+		Tabachek and Yurkowski, 1976
O. simplicissima	+			+	Izaguirre et al., 1982
O. splendida	+			+	Tsuchiya et al., 1981
O. splendida	+			+	Tabachek and Yurkowski, 1976
O. tenuis	+		+		Tabachek and Yurkowski, 1976
O. tenuis		+		+	Izaguirre et al., 1982
O. variabilis	+		+		Tabachek and Yurkowski, 1976
Phormidium autumnale	+		+		Tabachek and Yurkowski, 1976
P. tenue		+		+	Kikuchi et al., 1983
Schizothrix muelleri	+			+	Kikuchi et al., 1973
Symploca muscorum	+			+	Safferman et al., 1967
Symploca muscorum	+			+	Medsker et al., 1968
Symploca muscorum	+			+	Tabachek and Yurkowski, 1976

tected in cyanobacteria. Geosmin was first reported in cyanobacteria (*Symploca muscorum*) by Safferman (1967) and 2-methylisoborneol by Tabachek and Yurkowski (1976) in *Lyngbya cryptovaginata*.

Typical for these two compounds is the exclusive occurrence of either one or the other in any particular strain of cyanobacteria. An exception was reported by Berglind et al. (1983), who isolated geosmin and 2-methylisoborneol from the same strain of *Oscillatoria brevis*. The occurrence of geosmin is widespread. In addition to cyanobacteria, streptomycetes, myxobacterales and fungi (Wood et al., 1983) are potent producers of geosmin. Within the cyanobacteria it is mainly members of the genera *Oscillatoria, Lyngbya, Phormidium* that synthesize geosmin, but it has also been reported (Tsuchiya et al., 1981; Izaguirre et al., 1982; Jüttner et al., 1986) that heterocyst-forming cyanobacteria of the genera *Anabaena* and *Aphanizomenon* are potent producers of geosmin. Among the unicellular Chroococcales no representative has so far been found that is capable of producing either geosmin or 2-methylisoborneol. Cyanobacteria that have been reported to produce either geosmin or 2-methylisoborneol are listed in Table IV. Though in most cases axenic cultures were not available, care was taken that actinomycetes were absent.

Both geosmin and 2-methylisoborneol are irregular terpenes. Labelling experiments with [*methyl*-^{14}C]methionine in streptomycetes indicated that 2-methylisoborneol can be regarded as a methylated monoterpene. Since no labelling was observed in geosmin under these conditions, a pathway via a sesquiterpene and subsequent loss of an isopropyl group seems to be most likely (Bentley and Meganathan, 1981). It has been demonstrated that label from acetate is incorporated in geosmin, but the arrangement of the acetate moieties has not yet been determined. ^{13}C NMR spectrometry, which has become possible after determination of the chemical shift values of geosmin (Gerber and Denney, 1977), could not yet been used because of the difficulty to obtain sufficient labelling in the molecule. There is an urgent need for physiological and biochemical studies on the formation of geosmin and 2-methylisoborneol.

Several regular sesquiterpenes have been identified in cultures of a non-axenic strain of *Oscillatoria*. Tsuchiya et al. (1981) determined two structures as γ-cadinene (Formula 24) and germacrene D (Formula 25). Data of additional members of this chemical group are expected in due course.

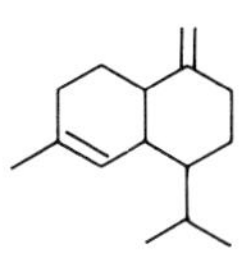

(24)

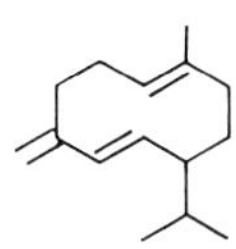

(25)

8. Occurrence and dynamics of VOBS in freshwater ecosystems

Though numerous VOBS have been detected which are liberated in cyanobacterial cultures, evidence of their occurrence in natural aquatic ecosystems is limited to only a few compounds. Difficulties arise when the specificity of a compound for a particular species or group of organisms is not sufficient to rule out the possibility that other sources could be responsible for the synthesis of the compound in question (Slater and Blok, 1983). In addition, the difficulties of collecting and determining trace amounts of a single component in complex mixtures restrict our knowledge of their occurrence and distribution in natural waters.

The most frequently studied VOBS are geosmin and 2-methylisoborneol. They have been reported for many freshwater and brackish water ecosystems throughout the world, but appear to be absent in marine ecosystems. Because they were first detected in actinomycetes their occurrence was attributed initially to the presence of these microorganisms (see Tabachek and Yurkowski, 1976; Wood et al., 1983). Most studies of surface waters have, however, indicated that the number of actinomycetes in the water body is usually very low and they are regarded to be of allochthonous origin introduced from the surrounding land by surface drainage. Highly significant correlations between cell numbers and threshold odor numbers of water and fish (Persson, 1982) which were used as a measure of geosmin, were found for cyanobacteria rather than for actinomycetes and were reported for the reservoir De Grote Rug in Holland (Cees et al., 1974) and the Tuusulanjärvi lake in Finland (Vakkuri, 1980). Seasonal variations of the musty odor of geosmin reached their maximum values in Japanese lakes when cyanobacterial populations were at their peak development (Hachimoto et al., 1978).

Few quantitative chemical analyses have been performed in lakes. Izaguirre et al. (1982) studied the distribution of 2-methylisoborneol in Lake Matheus (California). A depth profile indicated that 2-methylisoborneol was concentrated in the epilimnion and negligible in the hypolimnion. Similar results were obtained for geosmin in the Silverwood lake (McGuire et al., 1981), where only very low concentrations were measured in the hypolimnion.

Besides variations in spatial distribution, pronounced seasonal dynamics have been found for 2-methylisoborneol in Lake Matheus (Izaguirre et al., 1982). In late summer the levels increased to about 25 ng/l and they dropped in late November to 3 ng/l. The sources of 2-methylisoborneol in this lake, however, were not planktonic but rather benthic forms of cyanobacteria. Mats of *Oscillatoria curviceps*, which grew in significant amounts in the relatively shallow water around the shoreline, were regarded the major producers. In the immediate vicinity of these algal mats concentrations of 2-methylisoborneol surpassing 1 μg/l were measured (McGuire et al., 1981).

Different behavior has been reported for geosmin in the Federsee lake, a shallow eutrophic lake in S.W.-Germany (Jüttner et al., 1984; Jüttner et al., 1986). In this lake (maximum depth 2.4 m) a seasonal succession of algal and cyanobacterial blooms occurs every year. Geosmin, which was

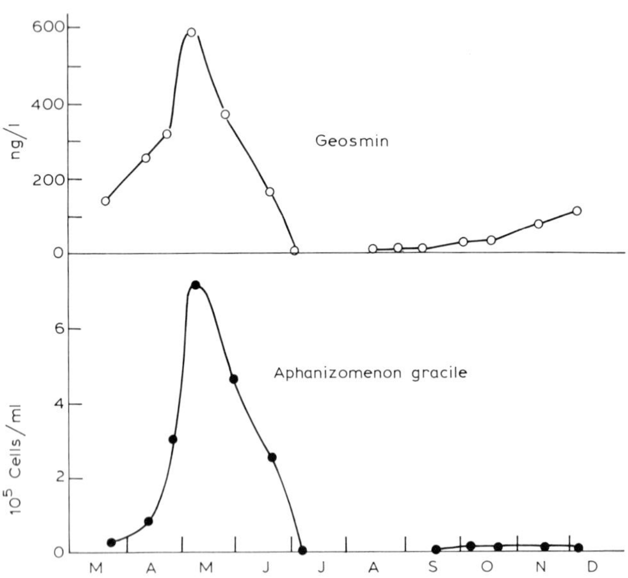

Fig. 3. Seasonal dynamics of geosmin and *Aphanizomenon gracile* in a shallow eutrophic lake (Federsee) in S.W.-Germany.

measured quantitatively with other VOBS over the growth season, in 1983 reached its maximum in mid-April as did the cell numbers of *Aphanizomenon gracile*, both exhibiting rather similar dynamics (Fig. 3). The much more moderate autumnal increase of geosmin concentrations could, however, not be attributed to the occurrence of the small population of *Aphanizomenon gracile*. (A possible source for this autumnal increase of geosmin may have been another cyanobacterium, *Lyngbya limnetica*, though definite proof for this could not be obtained.) In the previous year, concentrations of geosmin as high as 1.3 μg/l were estimated at the end of May (Jüttner, 1984b), yet geosmin concentrations have dropped to the level of the detection limit only 2 weeks after. Since the residence time of water in this lake is 0.2 a, effective mechanisms for the elimination of geosmin must be present in the lake.

Another compound which can clearly be recognized as a cyanobacterial product by its unique chemical structure is β-cyclocitral. So far, only members of the genus *Microcystis* have been shown to be able to synthesize this compound in appreciable amounts (Jüttner, 1984a). Good correlations between the cell numbers of *Microcystis* and concentrations of β-cyclocitral have been obtained (Fig. 4) in the Federsee lake (Jüttner et al., 1986). In August a bloom of *M. aeruginosa* superseded that of *M. wesenbergii*. This second maximum is indicated by a shoulder in the seasonal β-cyclo-

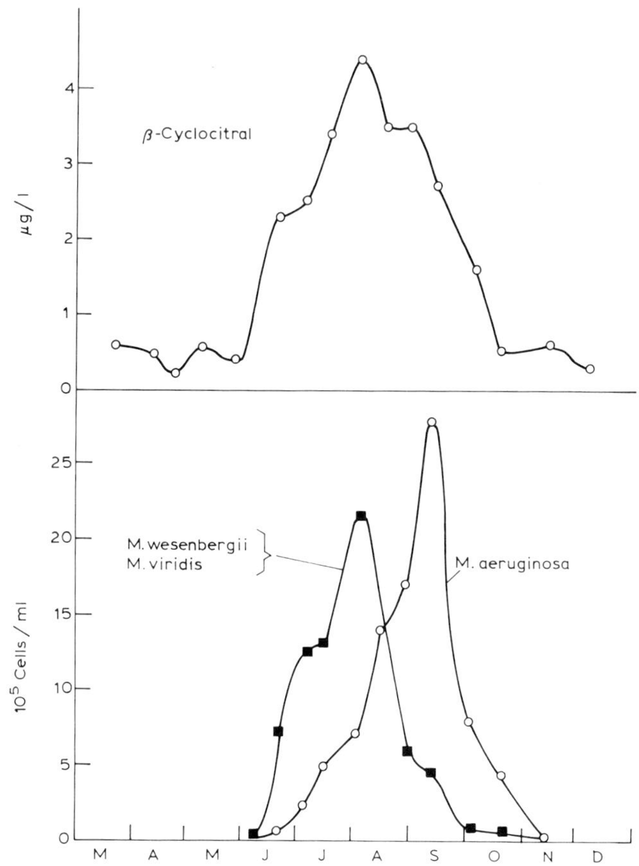

Fig. 4. Seasonal dynamics of β-cyclocitral and different species of *Microcystis* in Federsee lake, S.W.-Germany.

citral concentration curve. However, it should be noted that the concentrations of β-cyclocitral represent the total amount of the compound dissolved in the water and the particulate matter and are formed by the cells under the stress conditions applied.

9. Allelopathic and intraspecies regulation

The classical locality for allelopathic effects is the microenvironment in soil and mud (Brock, 1966). It is well known that most antibiotic substances have been isolated from soil microorganisms. Similar compounds may act in the aquatic microenvironments by the concentration of cells within a benthic population, inside cyanobacterial colonies or within a dense bloom of planktonic cyanobacteria (see also Chapter 17). Although interactions between cyanobacteria themselves and between cyanobacteria and eukaryotic algae have been frequently reported (Keating, 1977; Maestrini and Bonin, 1981; Jüttner and Faul, 1984), no definitive proof has been put forward for the influence of any particular compound under natural conditions. The effects of some VOBS have been tested on aquatic bacteria (Reichardt, 1981) and cyanobacteria (Jüttner and Bogenschütz, 1983). The concentrations necessary to obtain measurable growth inhibition, change in the pigment composition or a decline of glucose uptake and respiration, grossly exceeded those which have so far been measured in the natural environment. However, it has been shown that extremely high concentrations of certain VOBS may be present during short periods in the macroenvironment, as has been demonstrated for certain VOBS in a freshwater lake (Jüttner and Wurster, 1984). The analytical difficulties of recording very short periods of production in the natural environment render the systematic study of such events extremely difficult.

References

Agranoff, B.W., Eggerer, H., Henning, U. and Lynen, F. (1960) Biosynthesis of terpenes. VII. Isopentenyl pyrophosphate isomerase. J. Biol. Chem. 235, 326–332.

Bechard, M.J. and Rayburn, W.R. (1979) Volatile organic sulfides from freshwater algae. J. Phycol. 15, 379–383.

Bellar, T.A. and Lichtenberg, J.J. (1974) Determining volatile organics at microgram-per-litre levels by gas chromatography. J. Am. Wat. Works Ass. 66, 739–744.

Bentley, R. and Meganathan, R. (1981) Geosmin and methylisoborneol biosynthesis in streptomycetes. Evidence for an isoprenoid pathway and its absence in non-differentiating isolates. FEBS Lett. 125, 220–222.

Berglind, L., Johnsen, I.J., Ormerod, K. and Skulberg, O.M. (1983) *Oscillatoria brevis* (Kütz.) Gom. and some other especially odouriferous benthic cyanophytes in Norwegian inland waters. Wat. Sci. Technol. 15, 241–246.

Brock, T.D. (1966) Principles of Microbial Ecology. Prentice Hall, Inc., Englewood Cliffs, New Jersey.

Cees, B., Zoeteman, J. and Piet, G.J. (1974) Cause and identification of taste and odour compounds in water. Sci. Total Environ. 3, 103–115.

D'Agnolo, G., Rosenfeld, I.S., Awaya, J., Omura, S. and Vagelos, P.R. (1973) Inhibition of fatty acid synthesis by the antibiotic cerulenin. Specific inactivation of β-ketoacyl-acyl carrier protein synthetase. Biochim. Biophys. Acta 326, 155–166.

Drozd, J. and Novák, J. (1981) Headspace gas analysis by gas chromatography. J. Chromatogr. Libr. 19, 141–165.

Eloff, J.N., Steinitz, Y. and Shilo, M. (1976) Photooxidation of cyanobacteria in natural conditions. Appl. Environ. Microbiol. 31, 119–126.

Enzell, C.R., Wahlberg, I. and Aasen, A.J. (1977) Isoprenoids and alkaloids of tobacco. Fortschr. Chem. Org. Naturst. 34, 1–79.

Fogg, G.E., Stewart, W.D.P., Fay, P. and Walsby, A.E. (1973) The Blue-Green Algae. Academic Press, London-New York, pp. 281–297.

Funabashi, H., Iwasaki, S., Okuda, S. and Omura, S. (1983) A model study on the mechanism of fatty acid synthetase inhibition by antibiotic cerulenin. Tetrahedron Lett. 24, 2673–2676.

Geitler, L. (1960) Schizophyzeen. Gebr. Borntraeger, Berlin-Nikolassee.

Gerber, N.N. (1968) Geosmin, from microorganisms, is *trans*-1,10-dimethyl-*trans*-9-decalol. Tetrahedron Lett. 2971–2974.

Gerber, N.N. and Denney, D.Z. (1977) The carbon-13 nuclear magnetic resonance spectra of four eudesmane sesquiterpenols. Phytochemistry 16, 2025–2027.

Grob, K. (1973) Organic substances in potable water and in its precursor. Part I. Methods for their determination by gas-liquid chromatography. J. Chromatogr. 84, 255–273.

Grosch, W. and Laskawy, G. (1979) Co-oxidation of carotenes requires one soybean lipoxygenase isoenzyme. Biochim. Biophys. Acta 575, 439–445.

Hachenberg, H. and Schmidt, A.P. (1977) Gaschromatographic Headspace Analysis. Heyden and Son Ltd., London.

Hachimoto, T., Ichinoe, M., Tsuneizumi, Y., Yagi, M., Sayado, Y., Hori, H. and Onozawa, T. (1978) Musty odor problems in Japanese lakes and reservoirs as sources of water supply. J. Jap. Wat. Works Ass. 531, 30–47.

Henatsch, J.J. and Jüttner, F. (1983) Volatile odorous excretion products of different strains of *Synechococcus* (Cyanobacteria). Wat. Sci. Techn. 15, 259–266.

Herczeg, T., Lehoczki, E., Farkas, T., Rojik, I. and Szalay, L. (1979) Cerulenin-induced modification of structural and photosynthetic characteristics in *Chlorella*. Z. Pflanzenphysiol. 94, 55–64.

Herrmann, V. and Jüttner, F. (1977) Excretion products of algae. Identification of biogenic amines by gas-liquid chromatography and mass spectrometry of their trifluoroacetamides. Analyt. Biochem. 78, 365–373.

Hirosawa, T. and Wolk, C.P. (1979a) Factors controlling the formation of akinetes adjacent to heterocysts in the cyanobacterium *Cylindrospermum licheniforme* Kütz. J. Gen. Microbiol. 114, 423–432.

Hirosawa, T. and Wolk, C.P. (1979b) Isolation and characterization of a substance which stimulates the formation of akinetes in the cyanobacterium *Cylindrospermum licheniforme* Kütz. J. Gen. Microbiol. 114, 433–441.

Izaguirre, G., Hwang, C.J., Krasner, S.W. and McGuire, M.J. (1982) Geosmin and 2-methylisoborneol from cyanobacteria in three water supply systems. Appl. Environ. Microbiol. 43, 708–714.

Jenkins, D., Medsker, L.L. and Thomas, J.F. (1967) Odorous compounds in natural waters. Some sulfur compounds associated with blue-green algae. Environ. Sci. Technol. 1, 731–735.

Jüttner, F. (1976) β-Cyclocitral and alkanes in *Microcystis* (Cyanophyceae). Z. Naturforsch. 31c, 491–495.

Jüttner, F. (1981) Biologically active compounds released during algal blooms. Verh. Int. Ver. Limnol. 21, 227–230.

Jüttner, F. (1983) Volatile odorous excretion products of algae and their occurrence in the natural aquatic environment. Wat. Sci. Technol. 15, 247–257.

Jüttner, F. (1984a) Characterization of *Microcystis* strains by alkyl sulfides and β-cyclocitral. Z. Naturforsch. 39c, 867–871.

Jüttner, F. (1984b) Dynamics of the volatile organic substances associated with cyanobacteria and algae in a eutrophic shallow lake. Appl. Environ. Microbiol. 47, 814–820.

Jüttner, F. and Bogenschütz, O. (1983) Geranyl derivatives as inhibitors of the carotenogenesis in *Synechococcus* PCC 6911 (Cyanobacteria). Z. Naturforsch. 38c, 387–392.

Jüttner, F. and Faul, H. (1984) Organic activators and inhibitors of algal growth in water of a eutrophic shallow lake. Arch. Hydrobiol. 102, 21–30.

Jüttner, F. and Höflacher, B. (1985) Evidence of β-carotene 7,8 (7′,8′) oxygenase (β-cyclocitral, crocetindial generating) in *Microcystis*. Arch. Microbiol. 141, 337–343.

Jüttner, F. and Wurster, K. (1979) Einfache Anordnung zur Adsorption von Geruchsstoffen aus Algen an Tenax GC und deren Überführung in Gaschromatographie-Systeme. J. Chromatogr. 175, 178–182.

Jüttner, F. and Wurster, K. (1984) Evidence of ectocarpene and dictyopterenes A and C′ in the water of a freshwater lake. Limnol. Oceanogr. 29, 1322–1324.

Jüttner, F., Leonhardt, J. and Möhren, S. (1983) Environmental factors affecting the formation of mesityloxide, dimethylallylic alcohol and other volatile compounds excreted by *Anabaena cylindrica*. J. Gen. Microbiol. 129, 407–412.

Jüttner, F., Höflacher, B. and Wurster, K. (1986) Seasonal analysis of volatile organic biogenic substances (VOBS) in freshwater phytoplankton populations dominated by *Dinobryon*, *Microcystis* and *Aphanizomenon*. J. Phycol., 22, 169–175.

Keating, K.I. (1977) Allelopathic influence on blue-green bloom sequence in a eutrophic lake. Science 196, 885–887.

Kikuchi, T., Mimura, T., Harimaya, K., Yano, H., Arimoto, T., Masada, Y. and Inoue, T. (1973) Odorous metabolite of blue-green alga: *Schizothrix muelleri* Nägeli collected in the southern basin of lake Biwa. Identification of geosmin. Chem. Pharm. Bull. 21, 2342–2343.

Kikuchi, T., Kadota, S., Suehara, H., Nishi, A., Tsubaki, K., Yano, H. and Harimaya, K. (1983) Odorous metabolites of fungi. *Chaetomium globosum* Kinze ex Fr. and *Botrytis cinerea* Pers. ex Fr., and a blue-green alga, *Phormidium tenue* (Meneghini) Gomont. Chem. Pharm. Bull. 31, 659–663.

Maestrini, S.Y. and Bonin, D.J. (1981) Allelopathic relationships between phytoplankton species. Can. Bull. Fish. Aquat. Sci. 210, 323–338.

McGuire, M.J., Krasner, S.W., Hwang, C.J. and Izaguirre, G. (1981) Closed-loop stripping analysis as a tool for solving taste and odor problems. J. Am. Wat. Works Ass. 73, 530–537.

Medsker, L.L., Jenkins, D. and Thomas, J.F. (1968) Odorous compounds in natural waters. An earthy-smelling compound associated with blue-green algae and actinomycetes. Environ. Sci. Technol. 2, 461–464.

Medsker, L.L., Jenkins, D., Thomas, J.F. and Koch, C. (1969) Odorous compounds in natural waters. 2-Exo-hydroxy-2-methylbornane, the major odorous compound produced by several actinomycetes. Environ. Sci. Technol. 3, 476–477.

Mel'nichenko, L.A., Goronovski, I.T. and Sakevich, A.I. (1974) Development of methods for the analysis and identification of organoleptic substances in a technological process of water preparation. Khim. Tekhnol. (Kiev) 4, 61–63, in Russian.

Möhren, S. and Jüttner, F. (1983) Odorous compounds of different strains of *Anabaena* and *Nostoc* (Cyanobacteria). Wat. Sci. Tech. 15, 221–228.

Murray, K.E. (1977) Concentration of headspace, airborne and aqueous volatiles on Chromosorb 105 for examination by gas chromatography and gas chromatography-mass spectrometry. J. Chromatogr. 135, 49–60.

Oesterhelt, D., Bauer, H., Kresze, G.B., Steber, L. and Lynen, F. (1977) Reaction of yeast fatty acid synthetase with iodoacetamide. 1. Kinetics of inactivation and extent of carboxamidomethylation. Eur. J. Biochem. 79, 173–180.

Ōmura, S. (1976) The antibiotic cerulenin, a novel tool for biochemistry as an inhibitor of fatty acid synthesis. Bacteriol. Rev. 40, 681–697.

Pedersén, M. and DaSilva, E.J. (1973) Simple brominated phenols in the bluegreen alga *Calothrix brevissima* West. Planta 115, 83–86.

Persson, P.E. (1980) Sensory properties and analysis of two muddy odour compounds, geosmin and 2-methylisoborneol, in water and fish. Water Res. 14, 1113–1118.

Persson, P.E. (1982) Muddy odour: a problem associated with extreme eutrophication. Hydrobiologia 86, 161–164.

Reichardt, W. (1981) Influence of methylheptenone and related phytoplankton norcarotenoids on heterotrophic aquatic bacteria. Can. J. Microbiol. 27, 144–147.

Rippka, R., Waterbury, J.B. and Stanier, R.Y. (1981) Provisional generic assignments for cyanobacteria in pure culture. In M.P. Starr, H. Stolp, H.G. Trüper, A. Balows and H.G. Schlegel (eds.), The Prokaryotes. Springer-Verlag, Berlin-Heidelberg-New York, pp. 247–256.

Safferman, R.S., Rosen, A.A., Mashni, C.I. and Morris, M.E. (1967) Earthy-smelling substance from a blue-green alga. Environ. Sci. Technol. 1, 429–430.

Sandmann, G. and Böger, P. (1982) Volatile hydrocarbons from photosynthetic membranes containing different fatty acids. Lipids 17, 35–41.

Sirenko, L.A. and Sakevich, A.I. (1967) Formation of terpenes in blue-green algae. Dokl. Akad. Nauk SSSR 177, 959–960, in Russian.

Skulberg, O.M. and Skulberg, R. (1985) Planktic species of *Oscillatoria* (Cyanophyceae) from Norwey. Arch. Hydrobiol. Suppl. 71, 157–174.

Slater, G. and Blok, V.C. (1983) Isolation and identification of odorous compounds from a lake subject to cyanobacterial blooms. Water Sci. Technol. 15, 229–240.

Tabachek, J.L. and Yurkowski, M. (1976) Isolation and identification of blue-green algae producing muddy odor metabolites, geosmin, and 2-methylisoborneol, in saline lakes in Manitoba. J. Fish. Res. Board Can. 33, 25–35.

Tsuchiya, Y., Shudo, K. and Okamoto, T. (1979) The odorous compounds in the blue-green algae, *Oscillatoria* sp. and the river water. Identification of 2-methylisoborneol, geosmin, *p*-cresol, indole and 3-methylindole. Eisei Kagaku 25, 216–220.

Tsuchiya, Y., Matsumoto, A. and Okamoto, T. (1981) Identification of volatile metabolites produced by blue-green algae, *Oscillatoria splendida, O. amoena, O. geminata* and *Aphanizomenon* sp. Yakugaku Zasshi 101, 852–856.

Vakkuri, T. (1980) Observations on the connection between blue-green algae and the odour of water leaving Lake Tuusulanjärvi. Vesitalous 21, 30–34.

Von Tasch, S. (1985) Flüchtige organische Verbindungen der Gattung *Anabaena* (Cyanobacteria). Biochemische und chemotaxonomische Untersuchungen. Thesis, Universität Tübingen.

Wood, S., Williams, S.T. and White, W.R. (1983) Microbes as a source of earthy flavours in potable water- A review. Int. Biodeterior. Bull. 19, 83–97.

Yagi, M. (1983) Odor produced by blue green algae. Eisei Kagaku 29, 16–22.

Yagi, M., Kajino, M., Matsuo, U., Ashitani, K., Kita, T. and Nakamura, T. (1983) Odour problems in Lake Biwa. Wat. Sci. Technol. 15, 311–321.

P. Fay and C. Van Baalen (Eds.) The Cyanobacteria

19

Organic geochemical aspects of cyanobacterial mats

JAAP J. BOON and JAN W. DE LEEUW

FOM-Institute for Atomic and Molecular Physics, Kruislaan 407, 1098 SJ Amsterdam, The Netherlands and Delft University of Technology, Organic Geochemistry Unit, De Vries van Heystplantsoen 2, 2628 RZ Delft, The Netherlands

1. Introduction: aims of organic geochemistry and the geochemical fate of cyanobacteria

Organic compounds present in recent and ancient sediments of all sorts reflect the biochemical composition of the overlying water column and the in situ benthic activity at the time of deposition. Changes in the compositions of sedimentary organic compounds also reflect diagenetic pathways and inform us about the geological fate of biosynthesized compounds. This geochemical fate is highly dependent on the nature of the organic compounds and on the environmental conditions prevailing at the early stages of sedimentation.

Three major groups of organic compounds can be discriminated in this respect: compounds that are completely mineralized and thus disappear from the record, compounds that survive microbiological action and are preserved, and compounds that are microbially or chemically transformed to other compounds.

By studying well-defined sedimentary environments and organisms, knowledge has been accumulating regarding the survival potential of numerous organic compounds and about their species specificity (Ourisson et al., 1979; Mackenzie et al., 1982; Schenck and De Leeuw, 1982). There is still very little understanding, however, about the dynamics of these processes.

Environments dominated by prokaryotic organisms are attractive for investigation because of their relative simplicity. In the case of many cyanobacterial mat systems a superposition occurs of photosynthetically active layers on top of older inactive layers that are liable to processes of decomposition and preservation of organic matter. Dramatic changes occur in the composition of organic matter of an overgrown photosynthetic community as the newly developing decomposer communities transform deposited organic matter, and their own specific compounds, and induce the formation of mineral substances.

We will focus here on aspects of the geochemical fate of cyanobacterial organic matter. The discovery of remains of cyanobacteria in Precambrian stromatolitic rocks has stimulated organic geochemical studies and promoted the interest in the chemistry of recent cyanobacterial ecosystems on the tacit assumption that the chemistry of recent and Precambrian cyanobacteria is similar. Interdisciplinary studies have recently been published on fossil cyanobacterial systems (Schopf, 1983) and recent microbial mats that are viewed as potential stromatolites (Cohen et al., 1984).

Extrapolation from the present to the past has its limitations. Studies of present-day cyanobacterial mats indicate that, strictly speaking, we can no more consider them as model systems for the study of 'the geochemical fate of cyanobacteria', because significant contributions of organic compounds from other prokaryotes and even eukaryotic microorganisms are invariably present in these natural systems.

The study of recent cyanobacterial mats has become a goal in itself, since early stage diagenetic pathways can be investigated and characteristic compounds can be related to specific organisms or groups of organisms.

It is not our intention to present a complete review on the organic geochemistry of recent and ancient cyanobacterial mat systems. We selected a number of studies which were critically evaluated and we also present some new data. In this way an attempt is made to describe some interesting aspects of the organic geochemical approach and to give an impression of the scope and limitations. It was decided to concentrate on the Solar Lake cyanobacterial mats since these have been analyzed in great detail by a range of analytical techniques and by different research groups. Attention will be given to compounds with hydrocarbon skeletons and to macromolecules. Other sites will be shortly discussed and the results will be compared with the Solar Lake data.

2. Molecular signature of functionalized hydrocarbon skeletons

Historically, the organic geochemistry of functionalized hydrocarbons such as hydrocarbons, fatty acids, alcohols, steroids, hopanoids and carotenoids goes hand in hand with the development of analytical mass spectrometry and high resolution gas chromatography. Thus a gradual change can be observed from first discoveries of certain compounds 20 years ago to the qualitative and quantitative analysis of complete extract mixtures nowadays. These developments have their impact on the quality of the data and on the possibilities to compare the molecular signatures of the various microbial mat systems studied during the last two decades. Nevertheless, a comparison will be made between organic geochemistry of microbial mats from Solar Lake (Sinai, Egypt), shallow coastal areas from Abu Dhabi, shallow lagoons in Texas (USA) and sabkha-like lagoons in Baja California (Mexico).

The joint effort in the period of 1980–1983, known under the acronym ORGAST, brought together organic geochemists from the universities of Jerusalem (Israel), Delft (NL), Bristol (UK) and Berkeley (USA), and microbiologists from laboratories in Eilat (Israel) and Oldenburg (FRG). As a result of their cooperation, detailed knowledge is now available on organic geochemistry of the Solar Lake (Cohen et al., 1984) and the Gavish Sabkha (Friedman and Krumbein, 1985), both microbial mat systems on the eastern border of Sinai (Egypt). The efforts of the research groups in Delft and Bristol were directed to the analysis of extractable compounds from the mats. This has led to a unique experiment in organic geochemistry in which both groups analysed similar samples from the same relative depth in different cores for similar substances. Results from these two groups, indicated as Delft and as Bristol group, will be reviewed in the following chapters.

2.1. Solar Lake

The Solar Lake is a small heliothermal hypersaline lake bordering the Red Sea. Limnology and microbiology of the lake and its sediments have been extensively studied (Cohen et al., 1977a,b,c; Jørgenson and Cohen, 1977; Krumbein et al., 1977; Cohen, 1984). The shallow flats at a water depth of approximately 70 cm harbor extensive *Microcoleus chtonoplastes* mats in certain periods of the year, alternated by *Aphanothece* and other coccoid cyanobacteria and patchwise benthic diatoms. The organic matter from these cyanobacterial mats has accumulated to a continuous sequence of about 70 cm thickness. Practically all this organic matter is of prokaryotic origin and is biosynthesized in situ. This unique sequence of microbial mat material allows the comparison of chemical data from recent layers with those of older fossil layers synthesized under similar conditions in the past. The rationale of the importance of geochemical studies on Solar Lake cyanobacterial mats has been presented by Boon (1984). Organic geochemical data have been published by Aizenshtat et al. (1984), Boon et al. (1983), Boon (1984), Edmunds and Eglinton (1984) and Klok et al. (1984). As yet unpublished data of the Delft group dealing with investigations of the 0–0.5 mm and 0.5–3 mm layers are incorporated in this review.

2.1.1. Straight-chain and monomethyl branched compounds

2.1.1.1. Hydrocarbons. The results obtained by both the Bristol and Delft groups for this compound class were very similar. Comparison of the hydrocarbons present in the 0–0.5 mm mat layer and the 0.5–3.0 mm layer did not reveal considerable differences. We conclude therefore, that the hydrocarbon patterns published for the 0–3 mm layer typically reflects the standing crop and perhaps the first stages of decay (Boon et al., 1983; Edmunds and Eglinton, 1984). Up to at least 10 mm depth the major hydrocarbon encountered is *n*-heptadecene. *n*-Octadecene as well as *n*-eicosene and *n*-heneicosadiene are also present in the top layers. The saturated hydrocarbons range from C_{14} to C_{20} with *n*-heptadecane as the major component in the 0–3 and 3–10 mm layers. 8-Methylhexadecane (Formula I) is the only monomethyl branched hydrocarbon observed. The re-

I 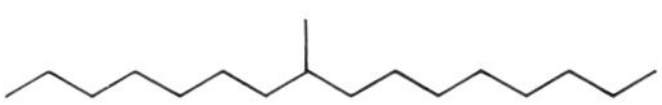

latively high abundance of these types of hydrocarbons in the top layer drops very sharply with depth. In the lower core samples (10–20 mm, 30 cm, 60 cm) these hydrocarbons are minor components representing less than a few ppm of the sediment dry weight. No unsaturated or monomethyl branched hydrocarbons occur in these lower layers. Only *n*-alkanes ranging from *n*-C_{14} to *n*-C_{28} are observed with heptadecane as a major component. The Bristol group observed also relative high abundances of the C_{20} and C_{22} components at the 25–30 cm and 65–70 cm level.

Although it is sometimes believed that hydrocarbons are relatively stable sedimentary compounds, we must conclude from the very drastic decrease (600 ppm → 2 ppm) observed in the upper 20 mm of the sediment that apparently a high degree of catabolism occurs in these upper layers. This is not only true for the major component, *n*-heptadecene, but also for the *n*-alkanes.

The much higher temperatures in the Solar Lake environment (a heliothermal lake), compared to other sedimentary environments, may be a key factor in hydrocarbon metabolism. We assume that, at least in the top layers of Solar Lake sediment, the hydrocarbons described mainly reflect the presence of living organisms. Indeed, heptadecene and *n*-alkanes ranging from C_{14} to C_{20} have been reported to occur in cyanobacteria and algae (Paoletti et al., 1976). The 8-methylheptadecane has only been reported to occur in several cyanobacteria (Gelpi et al., 1970). The *n*-alkane distribution patterns in the lower sections might – on the contrary – be indicative of diagenetic processes by which *n*-alkanes formed from functionalized precursors accumulate. This view is supported by the absence of long-chain alkanes in the upper layers and the absence of an odd over even predominance – indicative for biogenetic hydrocarbons of terrestrial plants (Kolattukudy et al., 1976).

2.1.1.2. Fatty acids. The straight-chain and monomethyl branched fatty acid data as reported by both the Delft and Bristol group for comparable core samples are basically similar, although some differences were observed. A comparison of the fatty acid patterns in the 0.0–0.5 mm mat layer and 0.5–3.0 mm layers showed, surprisingly, considerable differences (see Fig. 1). The i- and ai-C_{15} and -C_{17} fatty acids were minor if not trace components in the 0.0–0.5 mm layer, whereas the cyclopropyl C_{19}-fatty acid was completely absent in this sample. In the 0.5–3.0 mm layer, however, especially the i- and ai-$C_{15:0}$ fatty acids showed a major increase and the cyclopropyl C_{19} fatty acid was also clearly present.

The other major fatty acids (n-$C_{16:1}$, n-$C_{16:0}$, n-$C_{18:2}$, n-$C_{18:1}$ and n-$C_{18:0}$) revealed similar distribution patterns. The differences observed on the millimeter level in the top layers of the sediment as well as the sharp quantitative decrease of fatty acids with depth (300 ppm → 3 ppm) again very clearly support the idea that especially in the top mat layers the fatty acids represent the living biota. The low quantities of i- and ai- fatty acids in the living cyanobacterial mat support literature data stating that these types of fatty acids do not occur or are virtually absent in cyanobacteria and in the photosynthetic sulfur (green and purple) bacteria (Kenyon et al., 1972). The relative abundance of the cyclopropyl C_{19} fatty acid in the 0.5–3.0 mm, 3–10 mm and 10–20 mm layer samples is an indication of the presence of living heterotrophic anaerobic bacteria (Perry et al., 1979). The fatty acid data reported for the integral 0–3.0 mm layer are very similar to the 0.5–3.0 mm layer data (Boon et al., 1983).

Down the core, the fatty acid pattern becomes very simple: mainly saturated straight-chain fatty acids are present, with the C_{16} and C_{18} components dominating. A relative increase of longer chain fatty acids ranging from C_{20} to C_{30}, with a dominant C_{24} component, is observed with depth. This phenomenon has been reported for other sedimentary environments as well (see hereafter). Their origin, however, remains obscure. The n-$C_{24:0}$ fatty acid in sediments may be derived from microorganisms such as microeukaryotes (Nickels et al., 1981) and yeasts (Boon et al., 1978). However, a diagenetic pathway by which these longer chain fatty acids are released from other precursor molecules could be an alternative explanation. Their relative abundance in the deeper core samples supports such a diagenetic origin. It should be noted that the lipid data from both the Delft and the Bristol groups are limited as far as their mode of occurrence is concerned: only extractable lipids have been analyzed. Analysis of residually bound lipids might give clues to possible origin of this group of fatty acids.

2.1.1.3. Alcohols. Data for the straight-chain and monomethyl branched alcohols as reported by the Bristol group do differ greatly from the corresponding data of the Delft group; virtually no mono- or diunsaturated alcohols were reported by Bristol whereas the same components were found to be relatively abundant in the upper mat layers investigated by the Delft group (Fig. 2). In the 0–3 mm layer the diunsaturated C_{16} and C_{18} straight-chain components are the most abundant ones. This discrepancy between the Delft and Bristol data is most probably due to somewhat different TLC-separation procedures. The major alcohols observed throughout the core are the straight-chain C_{14}, C_{16} and C_{18} components. The unsaturated ones are dominant in the upper layers and are still present in the deepest sample. Longer chain saturated alcohols ranging from C_{20} to C_{28}, with C_{24} and C_{26} as the most abundant components, are mainly observed in the deeper samples. Contrary to the fatty acids and hydrocarbons, no sharp decrease is observed in the alcohol

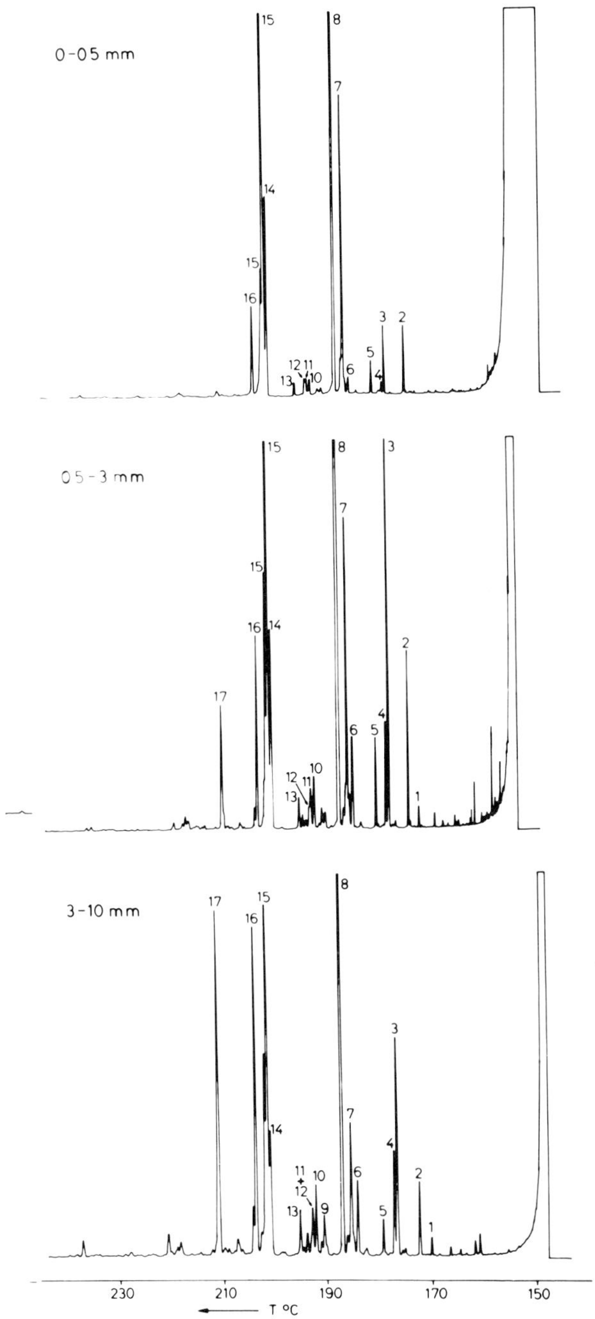

Fig. 1. Distributions of fatty acids at several depths in Solar Lake. Identifications: 1. i:$C_{14:0}$; 2. n-$C_{14:0}$; 3. i-$C_{15:0}$; 4. ai-$C_{15:0}$; 5. *n*-$C_{15:0}$; 6. i-$C_{16:0}$; 7. $C_{16:1}$; 8. *n*-$C_{16:0}$; 9. 10-Me-$C_{17:0}$; 10. i-$C_{17:0}$; 11. ai-$C_{17:0}$; 12. $C_{17:1}$; 13. *n*-$C_{17:0}$; 14. $C_{18:2}$; 15. $C_{18:1}$; 16. *n*-$C_{18:0}$; 17. cycloprop. C_{19}.

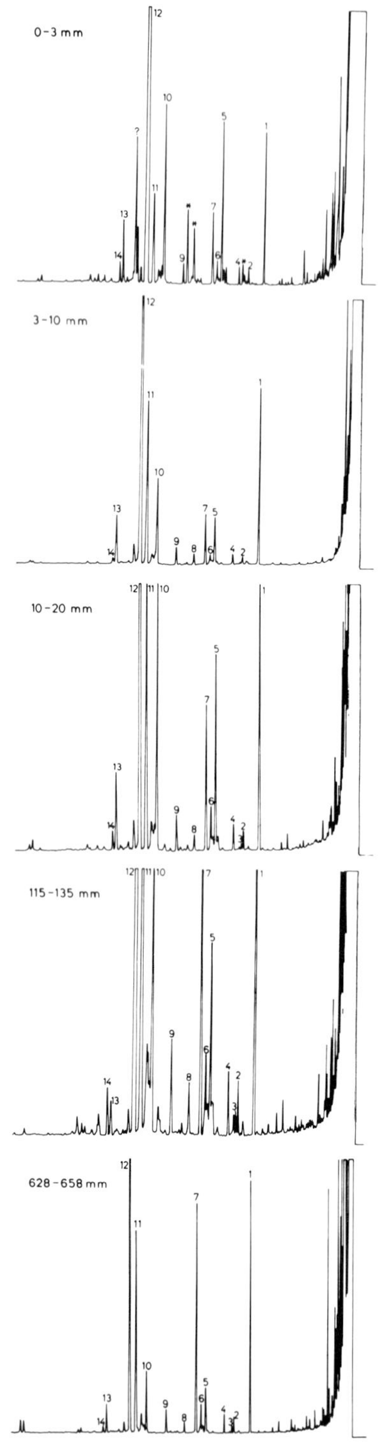

Fig. 2. Distribution of alcohols at several depths in Solar Lake. Identifications: 1. *n*-$C_{14:0}$; 2. i-$C_{15:0}$; 3. ai-$C_{15:0}$; 4. *n*-$C_{15:0}$; 5. $C_{16:2}$; 6. $C_{16:1}$; 7. *n*-$C_{16:0}$; 8. 10-Me-$C_{17:0}$; 9. *n*-$C_{17:0}$; 10. $C_{18:2}$; 11. *n*-$C_{18:0}$; 12. phytol; 13. geranylgeraniol; 14. *n*-$C_{19:0}$.

quantities. This indicates that the alcohols are much less affected during early diagenesis and are mainly associated with organisms living in the top layers. Not very much is known about the occurrence of saturated and diunsaturated alcohols in cyanobacteria or other bacteria and algae. Therefore, a more detailed interpretation of the alcohol data has to await further research.

2.1.2. Acyclic isoprenoids

The number of acyclic isoprenoids is rather limited. Phytol is very abundantly present in the top mat layers (up to 576 ppm). In the deeper samples phytol is still relatively dominant (54 ppm). Undoubtedly, the larger part of phytol is derived from chlorophyll *a*. Pristane and phytane are – with one exception – not encountered. Phyt-1-ene is abundantly present in the 0–3 mm layer. A 100-fold decrease, however, is already observed in the 3–10 mm layer. Although it was proposed that phyt-1-ene originates from phytol (De Leeuw et al., 1977), such a diagenetic relationship is not supported by the Solar Lake data since the relative quantities of phytol and phyt-1-ene greatly vary with depth.

An archaebacterial origin of phyt-1-ene seems unlikely, since no 'typical' C_{25} or C_{40} acyclic isoprenoid compounds have been encountered in Solar Lake (De Rosa et al., 1983), but the Solar Lake data suggest that phyt-1-ene does occur in certain eubacteria. Squalene is another acyclic isoprenoid encountered that reflects a microbial input. The 6,10,14-trimethylpentadecan-2-one is present in all samples, and probably originates from the diagenesis of phytol. The absence of other acyclic isoprenoids, such as farnesol and $\Delta^{2,10}$-phytanol, indicates that photosynthetic sulfur bacteria are virtually absent in Solar Lake.

2.1.3. Steroids

The 4-desmethyl sterols encountered in the core sample by the Bristol and theDelft groups are quite similar. A discrepancy in the relative intensities of C_{27}, C_{28} and C_{29} components was observed for the 0–3 mm samples. Such a difference can be rationalized by assuming geographical dissimilarities as observed in the surface mats. The $\Delta^{5,22}$-, Δ^5- and 5αH,Δ^{22}- and 5α-H-C_{27}, C_{28} and C_{29} sterols were present throughout the core. Although the quantities of these sterols did not vary very much with depth, the 5αH/Δ^5 ratios change dramatically within the first 2 cm (see Table I). It is assumed that the 4-desmethyl Δ^5-sterols are mainly biosynthesized by the diatoms (Ehrlich, 1978), and to some extent by cyanobacteria and other aerobic eubacteria. The 5αH-sterols are thought to be derived from the corresponding Δ^5-sterols by microbiological hydrogenation. Large changes in the 5αH/Δ^5 ratios are observed even within the first millimeter.

TABLE I

Ratios of 5αH/Δ^5 sterols in Solar Lake

Depth (mm)	C_{27}	Δ^{22}-C_{28}	C_{28}	C_{29}
0 –0.5	0.5	0.5	0.9	0.6
0.5–3	0.8	0.8	1.2	0.7
3 –10	4.6	5.1	3.8	1.2
10 –20	0.6	13.2	11.4	4.3
115 –135	4.3	12.0	2.3	0.5
415 –445	1.4	18.0	2.7	1.1
628 –658	2.8	6.0	1.9	0.6

Since the absolute quantities of 5αH sterols do not vary very much, it is tempting to conclude that a very effective hydrogenation is taking place by the action of bacteria living in the uppermost part of the sediments. Cyanobacteria might be responsible for this hydrogenation since the high ratios observed (up to 10–20) are not reported for other recent sediments. However, no literature data exist about the potential capability of cyanobacteria to saturate Δ^5-sterols. It cannot be excluded that some specific heterotrophic bacteria are capable of such a hydrogenation (Fauve and Kergomard, 1981). It should be noted that in the case of Solar Lake sediments no 5β(H)-stanols which result from Δ^5-sterol hydrogenation by strictly anaerobic bacteria, have been detected (Eyssen et al., 1973).

It can be seen from the Solar Lake data that somewhat different ratios are obtained for the C_{27}, C_{28} and C_{29} components. We have to keep in mind, however, that microbial hydrogenation of Δ^5-sterols probably only occurs with the free sterol as a substrate (Parmentier and Eyssen, 1974). Since we have not discriminated between free and esterified Δ^5-steroids, the different ratios per C-number can be due to differences between esterified and free sterols per C-number. In this respect the relative increase of the Δ^5-sterols in the deeper samples might be interpreted as a release, possibly by hydrolysis, of free sterols during diagenesis from more polar sterol derivatives. Since at these depths bacterial activities are strongly reduced, no rapid conversion of Δ^5-sterols to 5α(H)-stanols can occur (Jørgensen and Cohen, 1977).

The observation that 5α(H)-stanols can become very dominant within a few centimeters of sediment depth may have far reaching consequences. Numerous organic geochemical investigations of recent sediments reveal the relatively high abundance of 5α(H)-stanols, and many suggestions have been put forward to explain this phenomenon (De Leeuw et al., 1983; Robinson et al., 1984). Selected enrichment of 5α(H)-stanols and specific organisms biosynthesizing 5α(H)-sterols are two main mechanisms mentioned. When, however, certain photosynthetic (cyano)bacteria are capable of Δ^5-hydrogenation the origin of 5α(H)-stanols is to be found in the photic zones of oceans and lakes.

A number of other – less common – sterols and steroid ketones were observed by the Delft and the Bristol groups. Cycloartenol (Formula II), cycloartanol (Formula III) and 9,19-cyclo-4,14-di-

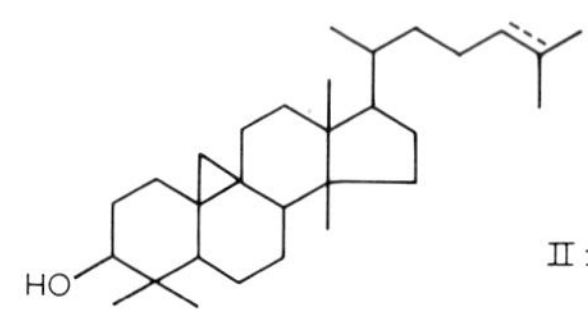

II: Δ^{24} III: saturated

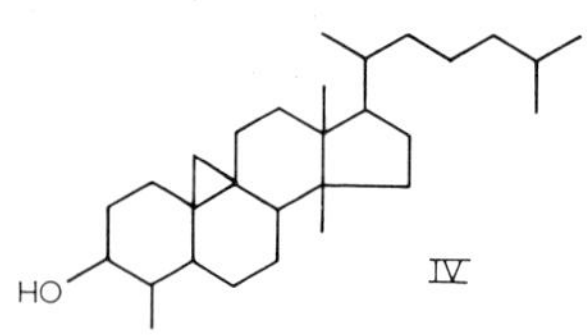

IV

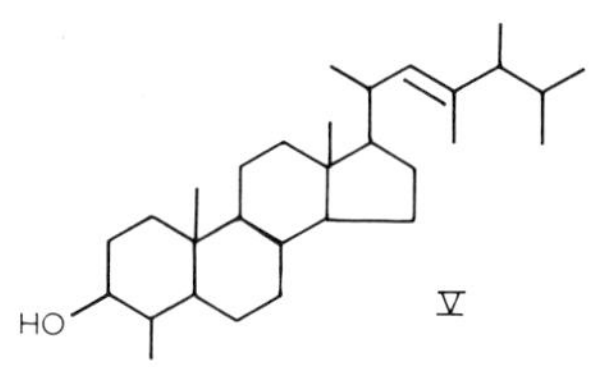

V

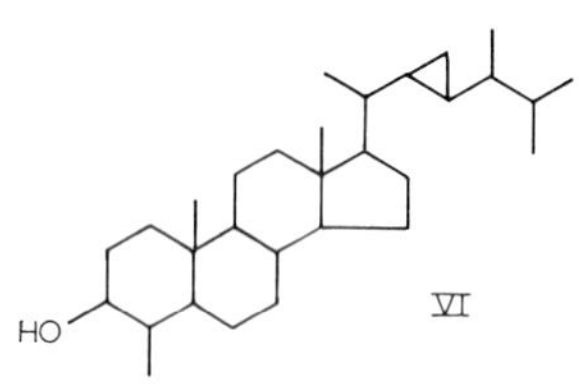

VI

methyl-cholestanol (Formula IV) as well as their corresponding ketones have been encountered in several core samples and may derive from various photosynthetic bacteria. Dinosterol (Formula V) and its Δ^5-analog as well as 4-Me-gorgosterol and -gorgostanol (Formula VI) were present occasionally and are thought to orginate from free-living or symbiotic dinoflagellates (Boon et al., 1979; De Leeuw et al., 1983). A few Δ^7-sterols were also present as well as a compound tentatively identified as 14-methyl-cholestanol.

Besides the very rapid $\Delta^5 \rightarrow 5\alpha$H conversion in the top layers another diagenetic pathway is clearly observed from the steroid data. Δ^2-Sterenes and to some extent $\Delta^{3,5}$-steradienes, thought to result from dehydration of 5α(H)- and Δ^5-sterols, respectively, were encountered throughout the core with exception of the 0–3 mm layer. This diagenetical dehydration of sterols and especially 5αH stanols is very well known to occur in recent sediments (Mackenzie et al., 1982). Basically, as expected, the sterene patterns correlate well with the sterol and stanol patterns.

2.1.4. Hopanoids

Extended hopanoids are biosynthesized by a number of aerobic bacteria (Rohmer et al., 1984; Taylor, 1984). It is, therefore, not surprising that several hopanetetrols were encountered in the top mats. Their presence in the bottom mats, however, was somewhat unexpected and indicates that these molecules can survive early diagenetic processes to some extent. The other hopanoid compounds, 22-hydroxyhomohopanol (Formula VII), homohopan-31-ol (Formula VIII) and

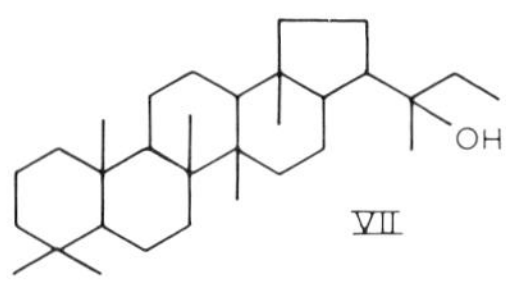

VII

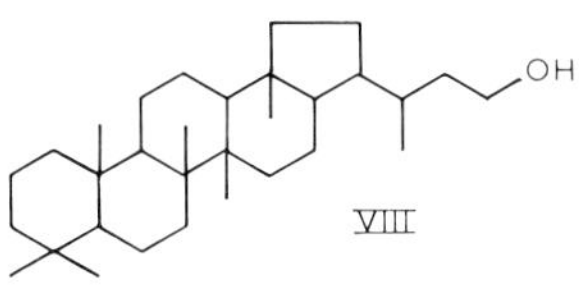

VIII

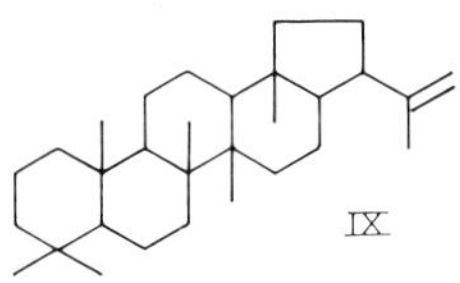

IX

hop-22(29)ene (Formula IX), were also encountered in the 0–3 mm samples, indicating that these compounds might occur as such in living organisms. The tentatively identified 22-hydroxyhomohopanol is relatively abundant in all samples and probably reflects a specific source organism. Hop-22(29)ene is widespread among bacteria (Taylor, 1984). The presence of hop-17(21)-ene and hop-21-ene in the deeper samples probably indicates a diagenetically induced isomerization of the 22(29) double bond (Brassell et al., 1980). Although not present in the 0–3 mm layer, $\beta\beta$-bishomohopanoic acid (Formula X) is the most abundant hopanoid acid in the other samples. Much

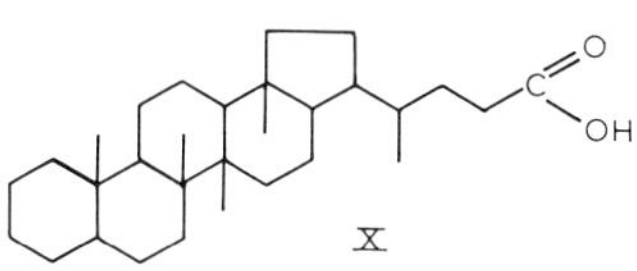

X

smaller amounts of C_{31} and C_{33} hopanoic fatty acids are also present and they too reflect an early diagenetic pathway from the polyhydroxytetrols. 3-Methylbishomohopanoic acids were detected in some of the deeper samples. An origin from specific bacteria is assumed (Rohmer et al., 1984). A diagenetically produced suite of hopanoid ketones was encountered in the deeper samples.

2.1.5. Carotenoids

A number of carotenoids were analysed in the 0–3 mm sample and in the deepest sample by the Bristol group. β-Carotene, echinone, rhodopin, dehydrorhodopin, canthaxanthin, hydroxyechinone, zeaxantin and myxoxanthophyll were identified. Rhodopin and dehydrorhodopin are believed to be indicative of purple photosynthetic bacteria and myxoxanthophyll is indicative of cyanobacteria. Other carotenoids specific for diatoms like fucoxanthin were not encountered, although diatoms make a contribution to the top mat layers (Ehrlich, 1978). It is likely that diatom carotenoids are much less resistant and therefore disappear very rapidly (Repeta and Gagosian, 1983). It is noteworthy that most of the carotenoids present in the top layer are also encountered in the bottom mats, suggesting the ability to survive to some extent a burial of 2 500 yrs.

2.2. Abu Dhabi

In Abu Dhabi intertidal mats of the sabkha type have developed in hypersaline pools (Cardoso et al., 1978; Watts et al., 1977). Unfortunately, the 20 cm core of very nicely laminated cyanobacterial mats from Abu Dhabi was subjected to lipid and carotenoid analysis without previous sectioning. Thus, no advantage was taken from the microbial stratigraphy present in the core and the results can only be considered in a general context.

The extractable lipids were analysed after saponification and subsequent fractionation.

2.2.1. Straight-chain and monomethyl branched compounds

2.2.1.1. Hydrocarbons. The *n*-alkanes ranged from C_{17} to C_{31}. *n*-Heptadecane was the most abundant component. The *n*-alkanes > C_{20} were of relatively minor importance. They exhibit odd-even predominance, probably indicating a minor contribution from terrigenous material. Small amounts of 7- and 8-methylhexadecanes were reported and the presence of branched C_{17}, C_{20} and C_{23} alkenes (monocyclic?) was noted.

2.2.1.2. Fatty acids. *n*-Alkanoic and *n*-alkenoic acids maximized at C_{16} and C_{18}. No components > C_{20} were present. Two relatively abundant cyclopropyl C_{19} fatty acids were also identified. Relatively high amounts of i- and ai-C_{15} and C_{17} fatty acids and of i-C_{16} fatty acid were present.

2.2.1.3. Alcohols. A series of *n*-alkanols was observed with only even number carbon chains in the range between C_{18}–C_{26}. The C_{22} component showed the highest intensity. Because of the small abundance of *n*-alkanes > C_{20} and the absence of fatty acids > C_{20} an origin from higher plants is not likely.

2.2.2. Acyclic isoprenoids

Phytol, as in Solar Lake, was the most abundant component identified. Pristane and phytane, were also present. These isoprenoid hydrocarbons were not ascribed to petroleum contamination. Phytanic acid was tentatively identified.

2.2.3. Steroids

The common 4-desmethylsterols were detected together with their 5α(H) and 5β(H)-analogs. The 5α(H)-stanols are much more abundant than the Δ^5-sterols. A series of Δ^2-sterenes was also encountered indicating the diagenetical conversion of 5α(H)-stenols. The specific sterols cycloartanol and cycloartenol were also reported.

2.2.4. Hopanoids

17β-Bishomohopanoic acid was definitely present as well as 17β-bishomohopanol. Hop-17(21)ene and hop-22(29)ene were encountered in the hydrocarbon fraction.

2.2.5. Carotenoids

Myxoxanthophyll, specific to cyanobacteria, was the major characteristic carotenoid. Other carotenoids present were β-carotene, torulene, echinenone, spirilloxanthin, rhodopin, 3,4-dehydrorhodopin and zeaxanthin. Echinenone is also thought to be of cyanobacterial origin, whereas spirilloxanthin, rhodopin and 3,4-dehydrorhodopin are thought to originate from purple bacteria. Torulene may indicate the presence of fungi.

2.2.6. Comparison with Solar Lake

A variety of characteristic lipids probably related to cyanobacteria (e.g. 7- and 8-Me-hexadecane, cycloartanol, myxoxanthophyll and hopenes), to purple bacteria (e.g. rhodopin) and to heterotropic bacteria (i-, ai- and cyclopropyl fatty acids) are present in both the Solar Lake and the Abu Dhabi ecosystems.

Several other series of compounds (bishomohopanoic acid and -alcohol, Δ^2-sterenes, 5α(H)-sterols) reflect early stage diagenetic pathways. The occurrence of 5β(H)-sterols in Abu Dhabi suggests the presence of strictly anaerobic bacteria (Parmentier and Eyssen, 1974); 5β(H)-sterols were not present in Solar Lake. The relative abundance of the n-$C_{17:1}$-alkene and a C_{31}-homohopanol in Solar Lake and their absence in Abu Dhabi could be an indication of different photosynthetic bacterial communities. It is not known whether the common 4-desmethylsterols present in Abu Dhabi can be ascribed to diatoms.

2.3. Texas, USA

In Texas, USA, extensive cyanobacterial mats have developed in shallow lagoons and mud flats. The best studied site in this area is the Kleberg lagoon (Baffin Bay) in which the sediments consist of cyanobacterial mats, muds and marls (Dalrimple, 1957).

2.3.1. Fatty acids

The major fatty acids ranging from C_{12} to C_{20} present in the extract after saponification have been investigated (Parker and Leo, 1965). They consisted mainly of saturated, mono-unsaturated and di-unsaturated C_{16} and C_{18} fatty acids. Samples taken at different depths have shown selective enrichment in saturated fatty acids. This phenomenon was also observed in Solar Lake. From a historical point of view it is interesting to note that the first discovery of i- and ai-fatty acids in sediments was made by Parker and Leo (1965) in the cyanobacterial mats of Baffin Bay. Some of the Baffin Bay samples were saponified after extraction (Brown et al., 1980). The residual bound lipids thus released consisted mainly of fatty acids. I- and ai-fatty acids were present throughout the core, unsaturated fatty acids decreased strongly and the longer chain fatty acids maximizing at C_{24} increased considerably with depth. The changes with depth reflect in situ processes such as microbial decomposition, selective preservation and changes in input of organic matter by winds and floods due to hurricanes.

2.3.2. Sterols

Four subfractions (0–6, 6–13, 13–18 and 18–22.5 cm depth) taken from a core of laminated cyanobacterial mats at Baffin Bay were investigated for extractable sterols (Huang and Meinschein,

1978). The common Δ^5- and $\Delta^{5,22}$-sterols (C_{27}, C_{28}, C_{29}) were encountered as well as their 5α(H)-counterparts. A number of other sterols present were not further identified. The $5\alpha/\Delta^5$ ratios had values greater than 1 in the 0–6 cm core section. In the other sections the $5\alpha H/\Delta^5$ ratios were similar and much larger than 1 (2–10). These data again demonstrate that the reduction of Δ^5-sterols occurs at very shallow depths.

A comparison of the Baffin Bay data with those of Solar Lake is hardly possible, due to the very limited data on specific compounds and to the heterogeneity of the Baffin Bay sediments.

2.4. Baja California

In Baja California (Mexico) two 'cyanobacterial sites' have been investigated: Laguna Mormona (Cardoso et al., 1976; Tibbetts et al., 1978; Horodyski et al., 1977; Margulis et al., 1980) and Laguna Guerrero (Philp et al., 1978a).

2.4.1. Laguna Mormona

This sediment represents cyanobacterial mats present in a hypersaline coastal pond. Two sections of a core were investigated for lipids. The surface (± 0–2 cm depth) consisted of filamentous cyanobacteria, diatoms and photosynthetic bacteria, the lower sample, referred to as ooze (± 10 cm depth), was partly layered and consisted of cyanobacterial sheaths, mucilages, bacterial debris and plant debris.

2.4.1.1. Hydrocarbons. The surface sample was dominated by *n*-heptadecane. Smaller amounts of *n*-alkane homologs maximizing at *n*- C_{27} with a high CPI-value were also present. In the ooze sample the most abundant *n*-alkanes range from C_{23}–C_{29} (CPI 4.5). *n*-Heptadecane is present in relatively low amounts. The 7- and 8-methylheptadecanes as well as a monocyclic $C_{20}H_{40}$-component were shown to be present in both the surface and the ooze sample. Their concentrations in the ooze sample, however, were very low. Mono- and di-unsaturated hydrocarbons ranging from C_{17}–C_{20} were present in both samples but were not further identified.

2.4.1.2. Fatty acids. In the mat sample the *n*-C_{16} and *n*-C_{18} saturated and mono-unsaturated fatty acids are dominantly present. Higher homologs of *n*-alkanoic acids ranging between C_{20}–C_{32} were present in relatively small amounts in the surface sample. These higher homologs maximizing at C_{24} were, however, major components in the ooze sample. The unsaturated C_{16} and C_{18} fatty acids were relatively less abundant in the ooze sample. Iso- and anteiso fatty acids (C_{14} and C_{15}) and cyclopropyl fatty acids (C_{17} and C_{19}) were shown to be present in the mat sample.

2.4.1.3. Alcohols. Only straight chain alcohols ranging from C_{18} to C_{26} were encountered in both the surface and the ooze sample.

2.4.1.4. Acyclic isoprenoids. Phytol was by far the most abundant component in both the surface and the ooze sample. Pristane and phytane as well as 6,10,14-trimethylpentadecan-2-one and phytanic acid were present in relatively low amounts in both the surface and the ooze sample. The authors (Cardoso et al., 1976) assume that phytane and pristane are indigenous and are not due to oil pollution.

2.4.1.5. Steroids. Mixtures of the common Δ^5-sterols and their 5α(H)-counterparts were present in both the surface and the ooze sample. 5β(H)-Sterols were only observed in the surface sample. The

$5\alpha/\Delta^5$ ratio was determined for the mat sample and was 1 (C_{29}) or greater than 1 (C_{27}, C_{28}). The ratio of $5\alpha/5\beta$-sterols was approximately 4. Cycloartenol and -artanol were also observed in the surface sample. Cholest-2-ene and other mono- and diunsaturated sterenes ranging from C_{27}–C_{29} were present only in the ooze sample and point to a diagenetic dehydration process of the original sterols.

2.4.1.6. Hopanoids. Hop-22(29)-ene, 17β(H)-bishomohopanoic acid, 17β-bishomohopanol were present in both the surface and the ooze sample.

2.4.1.7. Carotenoids. The same suites of characteristic carotenoids were observed – though exhibiting different distribution patterns – in both the surface and the ooze sample: torulene (characteristic of fungi), myxoxanthophyll, canthaxanthin, hydroxyechinenone, echinenone and zeaxanthin (probably all of cyanobacterial origin); tetrahydrospirilloxanthin, okenone, dehydrorhodopin, rhodopin and spirilloxanthin (from purple photosynthetic bacteria); and spheroidenone (from non photosynthetic bacteria).

2.4.2. Laguna Guerrero

In many respects the cyanobacterial mats in Laguna Guerrero, which is situated 300 km south of Laguna Mormona, resemble those of Laguna Mormona. The extractable lipids of three subsamples taken from the upper layer (± 1.3 cm thick; composed almost solely of *Microcoleus chthonoplastes*), the middle layer (± 0.7 cm thick contains purple photosynthetic and sulfate-reducing bacteria), and the lower layer (± 7.5 cm thick, ooze with anaerobic bacteria) were investigated.

The results are comparable with those from the Laguna Mormona sample (see also Philp, 1980). Some additional compounds were found. *n*-Heptadecene was a dominant alkene in Laguna Guerrero as were several phytene isomers. Four not earlier reported hydrocarbons with molecular formulas $C_{20}H_{32}$ and $C_{20}H_{38}$ were encountered in the lower layer.

In general, the lipid and carotenoid data of Laguna Mormona, Laguna Guerrero and Solar Lake show great similarity. Virtually the same characteristic components are encountered, indicating similar early diagenetical pathways in these cyanobacterial ecosystems. The long chain alkanes, alcohols and fatty acids, which are more conspicuous in Laguna Mormona and Laguna Guerrero than in Solar Lake, indicate an additional input of higher plants which surround the ponds in Baja California. The relative abundance of 5β-stanols in Laguna Mormona probably indicates a more anaerobic environment. The occurrence of tetrahydrospirilloxanthin, okenone and spirilloxanthin in the Baja California mats may reflect a more dominant occurrence in these mats of purple photosynthetic and non-photosynthetic bacteria than in Solar Lake sediments.

2.5. Gavish Sabkha (Southern Sinai)

In Sabkha environments, cyanobacterial communities make important contributions to the sedimentary ecosystem. These cyanobacteria generally occur in association with many other photosynthetic and heterotrophic bacteria and halobacteria, and the lipid composition of the mats reflects these various groups of organisms in the few samples analyzed. Composition of the lipids is therefore quite different when compared with Solar Lake (De Leeuw et al., 1985).

The organic geochemical investigations were part of an excellent interdisciplinary study of the Gavish Sabkha which has been published recently (Friedman and Krumbein, 1985).

3. Molecular signatures in macromolecular organic matter

The polymeric framework of cyanobacteria consists of intracellular material with mainly metabolic functions and extracellular material, which functions as structural matter and acts as barrier against the environment outside the organism. The intracellular materials play a dynamic role in the organism and their high turnover requires rapid synthesis and easy degradation. Hence the preservation potential of these substances is likely to be limited. The extracellular materials have a protective and supportive function and play a role in adhesion to mineral matter. They are less easily biodegraded and are likely to have a higher preservation potential. Little is known yet about the in situ biochemical composition of mat building cyanobacteria. According to Golubic (personal communications), the external layers of the cyanobacteria from natural samples are very well developed and thick, whereas the morphology of the organism under natural, usually nutrient poor conditions, differs from that in laboratory cultures. It is therefore by no means certain that findings from culture experiments, which have a tendency to be rather selective, are representative for what is found in the natural environment. Chemical analyses of in vitro cultures should, therefore, be used with caution.

Analytical methods en vogue for the study of mixtures of macromolecules are either non-destructive spectroscopic techniques, such as nuclear magnetic resonance spectroscopy (NMR), Fourier transform infrared spectroscopy (FTIR) and X-ray diffraction spectrometry, or destructive methods, in which the polymeric material is depolymerized via hydrolysis or thermolysis prior to the analysis using chromatography and mass spectrometry. The use of the microscope is instrumental in the selection of samples and has been used advantageously in analytical strategies (Boon et al., 1983). Small scale sampling and even the use of micromanipulators should be encouraged because often geochemists tend to overlook the existence of microenvironments and apply relatively coarse sampling schemes. Boon (1984) has pioneered analytical pyrolysis methods and mass spectrometry for characterization of microsamples from the cyanobacterial mat environment. These methods provide information on a wide range of biochemicals and require only microgram quantities of sample. Aside from these relatively novel approaches, the classical methods of biochemistry using hydrolysis and chromatography remain of great value if scaled down sufficiently to very small sample size. Very little has been done up till now with non-destructive methods such as solid state NMR or FTIR. Apart from these approaches, immunological techniques for recognition of specific structural units in recent and fossil cyanobacterial mats are now being tested and have great potential (Muyzer et al., 1985). Especially the use of monoclonal antibodies in connection with microscopic detection of antigens should be of great value in the study of cyanobacterial mats.

3.1. Hydrolysis studies

Carbohydrates are important constituents in cyanobacterial mats and may constitute up to 30% of the carbon (Boon et al., 1983). Monosaccharide composition in hydrolysates of cyanobacterial mats taken from Solar Lake have been studied in detail by Klok et al. (1984). The major sugars found were rhamnose, fucose, arabinose, ribose, xylose, mannose, galactose and glucose. Glucose is predominant in the living cyanobacterial mat which points to reserve starch, that is readily metabolized. The concentration of glucose drops to low values below a depth of 10 mm. Xylose, which becomes relatively important in lower sediment layers has been postulated as an essential constituent of the preserved sheath material. The relative abundance of ribose in the 3 to 10 mm layer of the Solar Lake *Microcoleus* mats is thought to be indicative of nucleic acids and nucleotides. The sugars in the lower sediment layers probably originate from the omnipresent sheath material as well as from bacterial decomposers. The large number of minor pentose and hexose components often with one or two methoxy groups are thought to be released from bacterial cell walls. The 4-*O*-Me-Xyl, 2-*O*-Me-Man, 6-*O*-Me-Man, 2-*O*-Me-Glu and 6-*O*-Me-Glu are major methylated sugars in the living *Microcoleus* mat (first 0.5 mm of the sediment). Only the 6-*O*-Me-glucose drops to low concentrations in the 0.5–3 mm layer and in lower sediment layers. Klok et al. (1984) suggest a metabolic role for the 6-*O*-Me-Glu in living cyanobacteria.

Aizenshtat et al. (1984) have estimated that about 30% of the organic matter in the top centimeters of the Solar Lake mats consists of hydrolysable proteins. The amino acid distribution profile from the photic zone to the sulfate reduction zones does not show drastic changes. The percentage nitrogen in the core varies from 6 to 8% with no apparent trend. Galactoseamine peaks at 2 to 3 cm. The γ-butyric acid becomes especially abundant below about 2 cm.

Polyhydroxyalkanoates, characteristic carbon storage products of *Bacilli* and several other eubacteria (Gottschalk, 1979), have been shown in *Aphanothece* sp. from Lake Joondalup in Australia and in halophilic cyanobacteria from mats in Shark Bay, Australia (Capon et al., 1983). Indicators for the presence in Solar Lake and Baffin Bay (Texas) cyanobacterial mats have been found by pyrolysis gas chromatography/mass spectrometry (PyGCMS). The preservation potential of these polymers was found to be very low and comparable to the fate of the starches in the photosynthetic zone.

3.2. Pyrolysis studies

Analytical pyrolysis has been used for studies of cyanobacterial mats from Baffin Bay (Texas) (Boon, unpublished results), Shark Bay (Boon and De Leeuw, 1980), Solar Lake (Boon et al., 1983; Boon, 1984), Gavish Sabkha (Southern Sinai) (Boon et al., 1985) and several Pacific atols (Boon and Trichet, unpublished results). Pyrolysis mass spectrometry (PyMS) combined with factor-discriminant analysis is used for comparative studies and core profiles, whereas PyGCMS is applied for identification of the pyrolysis products. Contrary to hydrolysis studies of mats, which often have a target analysis character, the thermal dissociation (pyrolysis) studies yield a broad array of structural units from the various biopolymers and modified fossilized polymer systems in cyanobacterial sediments. These methods allow the chemical characterization of mat material on a microscale.

Characteristic markers in microbial mats for proteins, peptidoglycan, polysaccharides, polyhydroxyalkanoates, lipids, chlorophyll-phytolesters and compounds, which probably act as compatible solutes, have been shown by PyGCMS (Boon, 1984). These markers are expressed as molecular ion and fragment ion distributions in low voltage pyrolysis mass spectra. A great deal of similarity exists between these spectra from several environments. Fig. 3 shows pyrolysis mass spectra from mats of Baffin Bay (Texas), which are fairly similar to the top mats and the fossil mats of Solar Lake (compare Boon et al., 1983). Comparison of the spectra of the top mats with those of the preserved mats points to a number of changes which relate to the rapid metabolism of storage polysaccharides (marker peak m/z 126) and polyalkanoates (marker peaks m/z 86 and 100). These changes in composition agree with the data from hydrolysis studies and are to be expected in the upper cm of a microbial mat system. Below a few centimeters, the bulk organic matter composition changes much less and appears under the microscope to consist of mucilages and sheath (Cohen, 1984). An unpublished PyMS survey by Boon, on a series of Solar Lake mat layers which were carefully dissected under a binocular microscope, reveals some of the chemical changes which take place. The discriminant function map shown in Fig. 4, which is based on PyMS data of 12 layers, collected (as part of the cooperative ORGAST project) from the top 21 mm of the Solar Lake

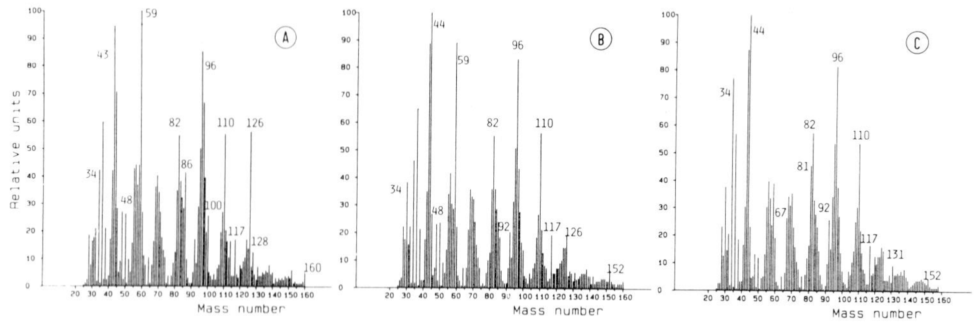

Fig. 3. Curie point pyrolysis low voltage mass spectra of Baffin Bay cyanobacterial mats taken at a depth of 0–3 mm (A), 3–7 mm (B) and 625 mm (C) in the Kleberg lagoon (Texas, USA). Algal mat material was kept frozen until analysis, homogenised and suspended in 0.7 M sodium phosphate buffer and analysed immediately thereafter with ferromagnetic wires of 510°C.

sediment, illustrates the relative differences in composition between the samples. The 0–2 mm top mat is different from all other samples and discriminated in the first discriminant space (D_1) by PyMS markers for polyhydroxyalkanoates, polysaccharides, proteins and several volatiles. An elongated cluster of samples coded 2 to 8 (2 to 14.5 mm depth) is separated by the second discriminant function (D_2) from a cluster with samples 9 to 12 (14.5 to 21 mm depth). Samples 9 to 12 all show relatively intense reduced organic sulfur compounds in their PyMS and mark the beginning of the permanent sulfuretum. Chemical differences interpreted as differences in sugar composition are reflected in the third discriminant function (D_3) and were found to be most prominent in samples 6, 7 and 11 (indicated by black dots in Fig. 4). Samples 6 and 7 taken in a relatively

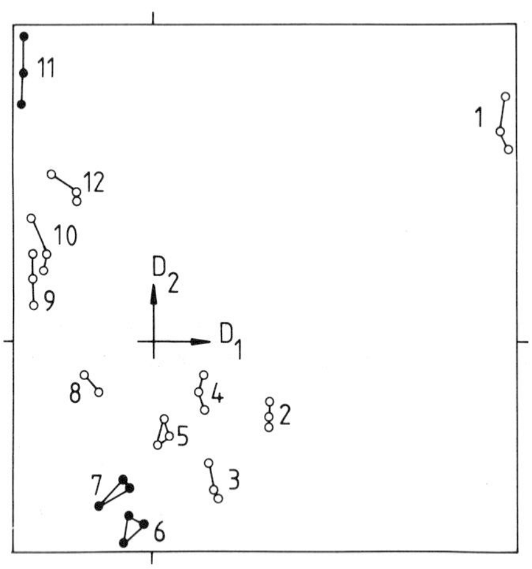

Fig. 4. Discriminant function score map calculated for pyrolysis mass spectra from a Solar Lake *Microcoleus chthonoplastes* top mat sampled in April 1981. The relative distances in this map indicate relative differences in chemical composition. The black dots indicate samples which are significantly discriminated by the third discriminant function and which lie in fact above the D_1D_2 plane. Cyanobacterial mat material was analysed at a Curie point temperature of 510°C. The sample codes in this map relate to the following depths in the top mat: 1, 0–2.1 mm; 2, 2.1–4 mm; 3, 4–6.3 mm; 4, 6.3–8 mm; 5, 8–9.4 mm; 6, 9.4–10.3 mm; 7, 10.3–11.8 mm; 8, 11.8–14.5 mm; 9, 14.5–16.5 mm; 10, 16.5–18 mm; 11, 18–19.5 mm and 12, 19.5–21 mm.

pure fossil *Microcoleus* mat system, are discriminated in the D_3 space by *m/z* 114 and 128, which were confirmed as the markers for pentoses and deoxyhexoses by PyGCMS. This correlates with a relatively abundant xylose just below the top mat (Klok et al., 1984). No drastic changes were apparent in the protein pyrolysis products, which is confirmed by the amino acid data of Lipiner (unpublished master thesis, Hebrew University, Jerusalem) on similar samples. A study of the diurnal cycle of organic matter in the top mats (upper 2 mm) of Solar Lake by pyrolysis mass spectrometry (Boon, 1984) showed a clear correlation between light intensity, carbohydrate fraction and dimethylsulfide, postulated as marker for dimethylpropriothetin, whereas markers for other biopolymers remained relatively constant. It is possible, however, that the thickness of the mat system which was sampled in the 24-h period was too great to allow the observation of significant changes in macromolecular composition.

In exceptional cases, like the one in the Gavish Sabkha (Krumbein et al., 1979), a living top mat may extend over several centimeters. Careful dissection leads to samples in which certain microorganisms predominate. PyMS and multivariant analysis were able to show a high correlation between classes of microorganisms in the zonations and the PyMS characteristics (Boon et al., 1985). Layers with cyanobacteria were clearly different in composition from those with purple bacteria. Clustering was found for a number of layers with coccoid cyanobacteria and for layers with filamentous cyanobacteria, however coccoid and filamentous bacteria have clearly different compositions in this microbial mat system. The PyMS of the coccoid cyanobacteria showed very intense peaks at *m/z* 96 and 110, indicative of their glycocalyx. Polyhydroxyalkanoates (*m/z* 86 and 100) were detected in the layers with purple bacteria. The spectrum of this top mat when covered by a mud flow inducing decomposition in the period of about 1 year, showed great resemblance to the PyMS of other fossil mat systems elsewhere in the world.

Pyrolysis mass spectra and GCMS data of the fossil mats from many different environments show markers for a carbohydrate fraction and for proteins. PyGCMS not only demonstrates general carbohydrate characteristics such as furan derivates but also 3-hydroxy-2-penteno-lactone and a methylated isomer, which points to xylanes and deoxyhexoses. PyGCMS data also suggest a relatively high abundance of aliphatic amino acids in proteins present in the sheath material (Boon, 1984). Muyzer et al. (1985) have recently been able to extract several proteins from sheath, which were found to be immunologically active. Thermogravimetric data (Ioselis et al., 1985) also point to a relative abundance of protein in fossil mat systems.

3.3. Degradation methods

Relative harsh degradation methods such as chromic acid oxidation in 3 M sulfuric acids, alkaline potassium permanganate oxidation and hydrogen bromide treatment have been performed on residues of cyanobacterial mats from the Laguna Mormona (Baja California, Mexico) and Baffin Bay (Texas) obtained after prolonged reflux in 6 M HCl (Philp et al., 1978b; Philp and Calvin, 1977). The rationale for this approach was the idea that kerogen, insoluble organic matter in rock, might have its precursors called 'protokerogen' in recent sediments. Indeed, insoluble residues were obtained from microbial mats after 6 M HCl reflux for 6 h and solvent extraction of the residues. Lipid extracts after chromic acid oxidation of these residues yielded normal fatty acids ranging from C_{12} to C_{27}, diacids ranging from C_8 to C_{12} and isoprenoid acids ranging from C_{14} to C_{20}. The authors suggest that this points to a system of cross-linked polymethylene chains in these 'protokerogen' fractions. Pyrolysis studies of similar residues from Baffin Bay mats (Philp et al., 1978b; Philp, 1980) revealed a regular pattern of alkanes and alkenes, which are now considered as markers for a resistant biopolymer in plant cuticles (Nip et al., 1985). In contrast, these patterns have not been found in pyrolysates of the Solar Lake mats, concordant with the known absence of higher plant material in these sediments. It is likely that the residues obtained from the mat systems of Laguna Mormona and Baffin Bay after 6 M HCl are resistant fractions from higher plants.

3.4. Non-destructive analyses

Antisera have been directed against top mats and lower cyanobacterial mat sections from the Solar Lake *Microcoleus* mat sediments (Muyzer et al., 1985). All antisera were found to give positive reactions in a dot-blot immunobinding assay. Antisera were further applied against recent and subfossil sheaths from *Microcoleus* sp. using immunohistochemical techniques under the microscope. The tests have shown that antigenic determinants on the sheaths are preserved over more than 2 000 years. Two proteins with an estimated size of 25 000 daltons were demonstrated after extraction with strong buffers, electrophoresis and immunoblotting. No evidence was found for new macromolecules with antigenic properties, only a decrease in antigenic activity was seen from top to bottom.

4. Molecular aspects of Precambrian stromatolites

Precambrium rocks yield rather small quantities of organic carbon. The organic matter is classified broadly into extractable molecules and insoluble macromolecular material (kerogen). Kerogen is the organic residue after dissolution of the rock matrix with strong acids (HF/HCl) and extraction with organic solvents. Kerogens from rocks of Archean age are heavily dehydrogenated and thus structurally more altered than the younger Proterozoic rocks, but even those are dehydrogenated to such an extent that very little preservation of biosynthetic carbon skeletons can be expected (Hayes et al., 1981). Only kerogens can be considered as autochthonous original organic matter, because mobile phases can percolate even through the densest silicate rocks and leave materials of much younger age (Oehler, 1977; Nagy, 1976). Pyrolysis methods have been used for thermal dissociation of Precambrian stromatolitic rock kerogen, but only low yields of pyrolysis products are obtained (McKirdy et al., 1980). The aliphatic component of these kerogen pyrolysates is marked by odd/even predominance of C_{10} + *n*-alkanes, which are speculated by the authors to be of bacterial cell wall origin. A number of these Precambrian kerogens also yield aromatic compounds. Furfuraldehyde and benzonitrile were discovered in the pyrolysate of kerogen from Bulawayan stromatolite (2.8 G yr old) (Sklarew and Nagy, 1975). Alkylated furans, aliphatic nitriles and pyrrole were released by pyrolysis of kerogen from Transvaal stromatolite (2.3 G yr old) (Nagy, 1976). These furans and the nitrogen compounds suggest the preservation of dehydrated carbohydrate polymers, possibly in the form of melanoidins, which have a furan polymer backbone (Rubinsztain et al., 1984). Furans and also tetrahydrofuran derivatives were also released by ozonolysis, which confirms their presence as such in a polymeric structure (Zumberge and Nagy, 1975). The implication of cyanobacteria in the deposition of these polymers is strongly indicated by the morphology of the microfossils (Nagy, 1974).

Despite these interesting data, we agree with Hayes et al. (1981) that "it is difficult to conclude that much has been learned thus far from these degradative investigations, in spite of their laudable goal". It remains to be seen whether much more could be learned considering the enormous time that has elapsed since the formation of these stromatolites. Unfortunately there is a paucity of analytical alternatives, although carbon isotopic measurement of the pyrolysis products relative to those of the kerogens has been suggested as a helpful further step in this difficult field. Also an improved version of solid state ^{13}C NMR may help to discover further clues. The absence of organic geochemical studies of stromatolites from the Phanerozoic is a further handicap to span the gap between the Precambrian and the present time. A systematic study of stromatolite rocks ranging in age from the late Precambrian to the Holocene should be encouraged. Such studies should concentrate on the fate of sheath material which appears to be one of the most abundant and resistant remains of cyanobacteria in subrecent deposits.

5. Summary and conclusions

– (1) Lipid marker molecules in cyanobacterial mat systems provide information on cyanobacteria, other photosynthetic and heterotrophic bacteria and on eukaryotic microorganisms such as diatoms and dinoflagellates. Most of the lipid markers of cyanobacteria are relatively easily metabolized. Free fatty acids and steroids appear to be more easily transformed and metabolized than the equivalent bound moieties. Alkanes have a low preservation potential in the Solar Lake mats, which may be related to the relatively high temperatures in the sediments.

– (2) Carbohydrates and proteins form the major part of the organic matter in the cyanobacterial mats of Solar Lake. Although storage polysaccharides are rapidly lost in the top mats, a significant fraction remains in lower layers. Proteinaceous material is preserved as well in the fossil mats. Both appear to be associated with sheath material from filamentous cyanobacteria, which is highly resistant to degradation in the sediment. Immunological methods, applied to samples from a mat profile in the Solar Lake, have shown the preservation of antigenic properties in sheath material sampled from different depths. A detailed study of the molecular architecture of this material is essential.

– (3) A great number of methylated sugars points to the presence and perhaps the preservation of bacterial cell wall material. Although these sugars are highly specific even after hydrolysis, it would be worthwhile to investigate their polysaccharide precursors, using immunological techniques.

– (4) Detailed sampling in the centimeter and millimeter range of well-defined microbial mat systems has been of great importance in detecting the changes in the distribution of many organic compounds. Correlating these data with microscopy and microbiological assays marks the beginning of a molecular stratigraphy of sediments.

– (5) A detailed organic geochemical survey of fossilized microbial mat systems in post-Precambrian rocks could provide a sound basis for the extrapolation of observations on recent mat systems to stromatolites. There is a serious lack of knowledge of the organic geochemistry of Phanerozoic stromatolites.

– (6) Microbial mat systems are suitable for studies of the fate of organic matter from photosynthetic communities, because organic synthesis and decomposition occurs, both spatially and temporarily, in separate ecological compartments. Their smaller scale allows dynamic studies with labelled compounds, essential for the detailed understanding of cycles and turnover rates of the sedimentary organic compounds.

References

Aizenshtat, Z., Lipiner, G. and Cohen, Y. (1984) Biogeochemistry of carbon and sulfur cycle in the microbial mats of the Solar Lake (Sinai). In Y. Cohen, R.W. Castenholz and H.O. Halverson (eds.), Microbial Mats: Stromatolites. Alan Liss. Inc., New York, pp. 281–313.

Boon, J.J. (1984) Tracing the origin of chemical fossils in microbial mats: biogeochemical investigations of Solar Lake cyanobacterial mats using analytical pyrolysis methods. In Y. Cohen, R.W. Castenholz and H.O. Halverson (eds.), Microbial Mats: Stromatolites. Alan Liss Inc., New York, pp. 313–343.

Boon, J.J. and De Leeuw, J.W. (1982) Early stromatolite lithification – organic geochemical aspects. In P. Westbroek and E.W. De Jong (eds.), Biomineralisation and Biological Metal Accumulation. Reidel Publ. Comp., Dordrecht, pp. 327–334.

Boon, J.J., De Leeuw, J.W. and Burlingame, A.L. (1978) Organic geochemistry of Walvis Bay diatomaceous ooze. III. Structural analysis of the monoenoic and polycyclic fatty acids. Geochem. Cosmochim. Acta 42, 631–644.

Boon, J.J., Rijpstra, W.I.C., De Lange, F., De Leeuw, J.W., Yoshioko, M. and Shimizu, Y. (1979) Black Sea sterol – a molecular fossil for dinoflagellate blooms. Nature 277, 125–127.

Boon, J.J., Hines, H., Burlingame, A.L., Klok, J., Rijpstra, W.I.C., De Leeuw, J.W., Edmunds, K.E. and Eglinton, G. (1983) Organic geochemical studies of Solar Lake cyanobacterial mats. In M. Bjorøy et al. (eds.), Advances in Organic Geochemistry – 1981. John Wiley and Sons, New York, pp. 207–227.

Boon, J.J., De Leeuw, J.W. and Krumbein, W.E. (1985) Biogeochemistry of Gavish Sabkha sediment. II. Pyrolysis mass spectrometry of the laminated microbial mats in the permanently water-covered zone before and after the desert sheet flood of 1979. In G.M. Friedman and W.E. Krumbein (eds.), Hypersaline Ecosystems: the Gavish Sabkha. Springer-Verlag, Berlin-Heidelberg-New York, pp. 368–381.

Brassell, S.C., Comet, P.A., Eglinton, G., Isaacson, P.J., McEvoy, J., Maxwell, J.R., Thomson, I.D., Tibbetts, P.J.C and Volkman, J.K. (1980) The origin and fate of lipids in the Japan Trench. In A.G. Douglas and J.R. Maxwell (eds.), Advances in Organic Geochemistry – 1979. Pergamon Press, Oxford, pp. 375–392.

Brown, S., Baillie, T.A. and Burlingame, A.L. (1980) Analytical approaches to the investigation of kerogen structure by mass spectrometric techniques. In A.G. Douglas and J.R. Maxwell (eds.), Advances in Organic Geochemistry – 1979. Pergamon Press, Oxford, pp. 475–484.

Capon, R.J., Dunlop, R.W., Ghisalberti, E.L. and Jefferies, P.R. (1983) Poly-3-hydroxyalkanoates from marine and freshwater cyanobacteria. Phytochemistry 22, 1181–1184.

Cardoso, J., Brooks, P.W., Eglinton, G., Goodfellow, R., Maxwell, J.R. and Philp, R.P. (1976) In J.O. Nriagu (ed.), Environmental Biogeochemistry, Vol. 1. Ann Arbor Science, Ann Arbor, pp. 149–173.

Cardoso, J., Watts, C.D., Maxwell, J.R., Goodfellow, R., Eglinton, G. and Golubic, S. (1978) A biogeochemical study of the Abu Dhabi algal mats: a simplified ecosystem. Chem. Geol. 23, 273–291.

Cohen, Y. (1984) The Solar Lake cyanobacterial mats: strategies of photosynthetic life under sulphide. In Y. Cohen, R.W. Castenholz and H.O. Halverson (eds.), Microbial Mats: Stromatolites. Alan Liss Inc., New York, pp. 133–149.

Cohen, Y., Krumbein, W.E., Goldberg, M. and Shilo, M. (1977a) Solar Lake (Sinai). 1. Physical and chemical limnology. Limnol. Oceanogr. 22, 597–608.

Cohen, Y., Krumbein, W.E., Goldberg, M. and Shilo, M. (1977b) Solar Lake (Sinai). 2. Distribution of photosynthetic microorganisms and primary production. Limnol. Oceanogr. 22, 609–620.

Cohen, Y., Krumbein, W.E., Goldberg, M. and Shilo, M. (1977c) Solar Lake (Sinai). 3. Bacterial distribution and production. Limnol. Oceanogr. 22, 621–635.

Cohen, Y., Castenholz, R.W. and Halverson, H.O. (1984) The interdisciplinary approach to the study of microbial mats: perspectives for future research – discussion. In Y. Cohen, R.W. Castenholz and H.O. Halverson (eds.), Microbial Mats: Stromatolites. Alan R. Liss Inc., New York, pp. 471–477.

Dalrimple, D.W. (1957) Calcium carbonate deposition associated with blue-green algal mats, Baffin Bay, Texas. Contrib. Mar. Sc. 4, 187–200.

De Leeuw, J.W., Simoneit, B.R.T., Boon, J.J., Rijpstra, W.I.C., De Lange, F., Van der Leeden, J.C.W., Correia, V.A., Burlingame, A.L. and Schenck, P.A. (1977) Phytol-derived compounds in the geosphere. In R. Campos and J. Goni (eds.), Advances in Organic Geochemistry – 1975. Enadimsa, Madrid, pp. 61–79.

De Leeuw, J.W., Rijpstra, W.I.C., Schenck, P.A. and Volkman, J.K. (1983) Free, esterified and residual bound sterols in Black Sea Unit I sediment. Geochim. Cosmochim. Acta 47, 455–465.

De Leeuw, J.W., Sinninghe Damsté, J.S., Klok, J., Schenck, P.A. and Boon, J.J. (1985) Biogeochemistry of Gavish Sabkha sediment. I. Studies on neutral reducing sugars and lipid moieties by gas chromatography-mass spectrometry. In G.M. Friedman and W.E. Krumbein (eds.), Hypersaline Ecosystems. Springer-Verlag, Berlin-Heidelberg-New York, pp. 350–367.

De Rosa, M., Gambacorta, A., Nicolaus, B., Chappe, B. and Albrecht, P. (1983) Isoprenoid ethers; backbone of complex lipids of the archaebacterium *Sulfolobus solfataricus*. Biochim. Biophys. Acta 753, 249–256.

Edmunds, K.L.H. and Eglinton, G. (1984) Microbial lipids and carotenoids and their early diagenesis in the Solar Lake laminated microbial mat sequence. In Y. Cohen, R.W. Castenholz and H.O. Halverson (eds.), Microbial Mats: Stromatolites. Alan Liss Inc., New York, pp. 343–389.

Ehrlich, A. (1978) The diatoms of the hypersaline Solar Lake (Sinai). Isr. J. Bot. 27, 1–13.

Eyssen, H.J., Parmentier, G.G., Compernolle, F.C., De Pauw, G. and Piessens-Denef, M. (1973) Biohydrogenation of sterols by *Eubacterium* ATCC 21, 408-Nova species. Eur. J. Biochem. 36, 411–421.

Fauve, A. and Kergomard, A. (1981) Réductions microbiologiques stéréoselectives des Δ^4-4 céto-3-stéroides. Tetrahedron 37, 899–902.

Friedman, G.M. and Krumbein, W.E. (1985) Hypersaline Ecosystems. The Gavish Sabkha. Springer-Verlag, Berlin-Heidelberg-New York, pp. 484.

Gelpi, E., Schneider, H., Mann, J. and Oro, J. (1970) Hydrocarbons of geochemical significance in microscopic algae. Phytochemistry 9, 603–612.

Gottschalk, G. (1979) Bacterial Metabolism. Springer-Verlag, Berlin-Heidelberg-New York, pp. 281.

Hayes, J.M., Kaplan, I.R. and Wedeking, K.W. (1983) Precambrian organic geochemistry preservation of the record. In J.W. Schopf (ed.), The Earth's Earliest Biosphere, its Origin and Evolution. Princeton University Press, Princeton, pp. 93–135.

Horodyski, R.J., Bloeser, B. and Van der Haar, S. (1977) Laminated algal mats from a coastal lagoon, Laguna Mormona, Baja California, Mexico. J. Sed. Petrol. 47, 680–696.

Huang, W.Y. and Meinschein, W.G. (1978) Sterols in sediments from Baffin Bay, Texas. Geochim. Cosmochim. Acta 42, 1391–1396.

Ioselis, P., Rubinsztain, Y., Ikan, R., Aizenshtat, Z. and Frankel, M. (1985) Thermal characterization of natural and synthetic humic acids. Org. Geochem. 8, 95–101.

Jørgensen, B.B. and Cohen, Y. (1977) Solar Lake (Sinai). 5. The sulfur cycle of the benthic cyanobacterial mats. Limnol. Oceanogr. 22, 657–666.

Kenyon, C.N., Rippka, R. and Stanier, R.Y. (1972) Fatty acid composition and physiological properties of some filamentous blue-green algae. Arch. Mikrobiol. 83, 216–236.

Klok, J., Cox, H.C., Baas, M., De Leeuw, J.W. and Schenck, P.A. (1984) Carbohydrates in recent marine sediments. II. Occurrence and fate of carbohydrates in a recent stromatolitic deposit: Solar Lake, Sinai. Org. Geochem. 7, 101–109.

Kolattukudy, P.E., Croteau, R. and Buckner, J.S. (1976) In P.E. Kolattukudy (ed.), Chemistry and Biochemistry of Natural Waxes. Elsevier, Amsterdam-Oxford-New York, pp. 289–312.

Krumbrein, W.E., Cohen, Y. and Shilo, M. (1977) Solar Lake (Sinai). 4. Stromatolitic cyanobacterial mats. Limnol. Oceanogr. 22, 635–656.

Krumbein, W.E., Buchkolz, H., Franke, P., Giani, D., Giele, C. and Wonneberger, K. (1979) O_2 and H_2S coexistence in stromatolites. A model for the origin of mineralogical lamination in stromatolites and banded iron formations. Naturwissenschaften 66, 381–389.

Mackenzie, A.S., Brassell, S.C., Eglinton, G. and Maxwell, J.R. (1982) Chemical fossils: the geological fate of steroids. Science 217, 491–504.

Margulis, L., Barghoorn, E.S., Ashendorf, D., Banerjee, S., Chase, D., Francis, S., Giavannoni, S. and Stolz, J. (1980) The microbial community in the layered sediment at Laguna Figueroa, Baja California, Mexico: does it have Precambrian analogues? Precambrian Res. 11, 93–123.

McKirdy, D.M., McHugh, D.J. and Tardif, J.W. (1980) Comparative analysis of stromatolitic and other microbial kerogens by pyrolysis hydrogenation-gas chromatography (PH-GC). In P.A. Trudinger, M.R. Walter and B.J. Ralph (eds.), Biogeochemistry of Ancient and Modern Environments. Australian Academy of Science, Canberra, pp. 187–200.

Muyzer, G., De Koster, S., Van Zijl, Y., Boon, J.J. and Westbroek, P. (1986) Immunological studies on microbial mats from Solar Lake (Sinai), a contribution to the organic geochemistry of sediments. Org. Geochem., 10, 697–704.

Nagy, B. (1976) Organic chemistry of the young earth. Naturwissenschaften 63, 499–505.

Nagy, L.A. (1974) Transvaal stromatolite first evidence for the diversification of cells about 2.2 billion years ago. Science 183, 514–516.

Nickels, J.S., Robbie, R.J., Lott, D.F., Martz, R.F., Benson, P.H. and White, D.C. (1981) Effect of manual brush cleaning on the biomass and community structure of the microfouling film formed on aluminum and titanium surfaces exposed to rapidly flowing seawater. Appl. Environ. Microbiol. 41, 1442–1553.

Nip, M., Tegelaar, E.W., De Leeuw, J.W., Schenck, P.A. and Holloway, P.J. (1986) A new, highly aliphatic and resistant biopolymer in plant cuticles; evidence from pyrolysis and ^{13}C-NMR analysis of modern and fossil plants. Naturwissenschaften, 73, 579–585.

Oehler, J.H. (1977) Irreversible contamination of Precambrian kerogen by ^{14}C-labelled compounds. Precambrian Res. 4, 221–227.

Ourisson, G., Albrecht, P. and Rohmer, M. (1979) The hopanoids. Paleochemistry and biochemistry of groups of natural products. Pure Appl. Chem. 51, 709–729.

Paoletti, C., Pushparaj, B., Florenzano, G., Capella, P. and Lercker, G. (1976) Unsaponifiable matter of green and blue-green algal lipids as a factor of biochemical differentiation of their biomass. I. Total unsaponifiable and hydrocarbon fraction. Lipids 11, 258–265.

Parker, P.L. and Leo, R.F. (1965) Fatty acids in blue-green algal mat communities. Science 148, 873.

Parmentier, G.G. and Eyssen, H.J. (1974) Mechanism of biohydrogenation of cholesterol to coprostanol by *Eubacterium* ATCC 21, 408. Biochim. Biophys. Acta 348, 279–284.

Perry, G.J., Volkman, J.K. and Johns, R.B. (1979) Fatty acids of bacterial origin in contemporary marine sediments. Geochim. Cosmochim. Acta 43, 1715–1725.

Philp, R.P. (1980) Comparative organic geochemical studies of recent algal mats and sediments of algal origin. In P.A. Trudinger, M.R. Walter and B.J. Ralph (eds.), Biochemistry of Ancient and Modern Environments. Austr. Acad. Science, Canberra, pp. 173–185.

Philp, R.P. and Calvin, M. (1977) Kerogenous material in recent algal mats in Laguna Mormona, Baja California. In R. Campos and J. Goni (eds.), Advances in Organic Geochemistry – 1975. Enadimsa, Madrid, pp. 735–752.

Philp, R.P., Brown, S., Calvin, M., Brassell, S.C. and Eglinton, G. (1978a) Hydrocarbon and fatty acid distributions in recent deposited algal mats at Laguna Guerrero, Baja California. In W.E. Krumbein (ed.), Environmental Biogeochemistry and Geomicrobiology, Vol. 1. Ann Arbor Science, Ann Arbor, pp. 255–271.

Philp, R.P., Calvin, M., Brown, S. and Yang, E. (1978b) Organic geochemical studies on kerogen precursors in recently deposited algal mats and oozes. Chem. Geol. 22, 207–231.

Repeta, D.J. and Gagosian, R.B. (1983) Carotenoid transformation products in the upwelled water off the Peruvian coast: suspended particulate matter, sediment trap material, and zooplankton fecal pellet analyses. In M. Bjorøy et al. (eds.), Advances in Organic Geochemistry – 1981. John Wiley and Sons, Chichester, pp. 380–388.

Robinson, N., Cranwell, P.A., Finlay, B.J. and Eglinton, G. (1984) Lipids of aquatic organisms as potential contributors to lacustrine sediments. Org. Geochem. 6, 143–152.

Rohmer, M., Bouvier-Nave, P. and Ourisson, G. (1984) Distribution of hopanoid triterpanes in prokaryotes. J. Gen. Microbiol. 130, 1137–1150.

Rubinsztain, Y., Ioselis, P., Ikan, R. and Aizenshtat, Z. (1984) Investigations on the structural units of melanoidins. Org. Geochem. 6, 791–805.

Schenck, P.A. and De Leeuw, J.W. (1982) Molecular organic geochemistry. In O. Hutzinger (ed.), Handbook of Environmental Chemistry, Vol. 1B. Springer-Verlag, Berlin-Heidelberg-New York, pp. 111–129.

Schopf, J.W. (1983) Earth's earliest biosphere – its origin and evolution. Princeton University Press, Princeton.

Sklarew, D.S. and Nagy, B. (1975) 2,5-Dimethylfuran from the 2.7×10^9-year-old Rupemba-Belingwe stromatolite, Rhodesia: potential evidence for remnants of carbohydrates. Proc. Natl. Acad. Sci. USA 76, 10–14.

Taylor, R.F. (1984) Bacterial triterpenoids. Microbiol. Rev. 48, 181–198.

Tibbetts, P.J.C., Maxwell, J.R. and Golubic, S. (1978) Carotenoids in an algal mat and ooze from Laguna Mormona, Baja California. In W.E. Krumbein (ed.), Environmental Biogeochemistry and Geomicrobiology, Vol. 1. Ann Arbor Science, Ann Arbor, pp. 271–283.

Watts, C.D., Maxwell, J.R. and Kjøsen, H. (1977) The potential of carotenoids as environmental indicators. In R. Campos and J. Goni (eds.), Advances in Organic Geochemistry – 1975. Enadimsa, Madrid, pp. 391–413.

Zumberge, J.E. and Nagy, B. (1975) Alkyl substituted cyclic ethers in 2 300 Myr old Transvaal algal stromatolite. Nature 255, 695–696.

P. Fay and C. Van Baalen (Eds.) The Cyanobacteria

20

Petroleum toxicity and the oxidation of aromatic hydrocarbons

MARTHA L. NARRO

Department of Genetics, Stanford University School of Medicine, Stanford, CA 94305, USA

1. Introduction

In the past 15 years a large number of observations from both field and laboratory studies have accumulated which clearly demonstrate the toxicity of petroleum to natural assemblages of microalgae and to pure cultures. The toxicity of crude oils, fuel oils and the water-soluble fractions of these oils, together with a variety of pure compounds known to occur in petroleum, have been examined (for a recent review see Oil in the Sea, 1985). It is now clear that petroleum contains many types of compounds that are remarkably algalcidal, as low as the ppb range for some compounds, and an even greater number that exhibit algalstatic properties. What is largely missing from these studies on the toxic effects of petroleum on algae are mechanistic studies aimed at elucidating the biochemical basis for the toxicity of any given compound that is a component of petroleum. This has not proven an easy area to approach; however, it has now been established that the cyanobacteria can oxidize aromatic compounds. This finding may provide partial insight into the question of the biochemical basis for petroleum toxicity in cyanobacteria. In addition these observations suggest that the cyanobacteria possess a heretofore unrecognized metabolic capability. A brief discussion of the toxicity of petroleum, petroleum fractions and pure compounds found in petroleum is presented at the beginning of this chapter. The major part of the chapter is a summary of studies of the metabolites formed by cyanobacteria from aromatic compounds found in petroleum. The postulated pathways and mechanisms of formation of these metabolites are discussed.

2. Toxicity

The deleterious effects that petroleum hydrocarbons can have on phytoplankton in general, or selectively upon a specific class of phytoplankton, or even upon a single species, are becoming increasingly apparent as studies accumulate in the literature. Such studies are divided into laboratory work with pure or unialgal cultures and work with freshly collected, mixed populations tested in situ or in some type of incubator. Both of these approaches are valid and each can contribute to our understanding of the phytoplankton-petroleum conundrum. Studies on the effects of hydrocarbons on phytoplankton and cyanobacteria have mainly dealt with photosynthesis and growth of the organisms, and, in a few instances, with nitrogen fixation (Gordon and Prouse, 1973; Jordan et al., 1978; O'Brien and Dixon, 1976; Soto et al., 1975; Vandermeulen and Ahern, 1976). A short-time reversible inhibition of photosynthesis has been reported (Vandermeulen and Ahern, 1976; Winters et al., 1977) as well as a presumably irreversible inhibition which is necessarily lethal (Batterton et al., 1978a). The reasons for the inhibition of photosynthesis are obscure. A word of caution must be added; there has been little uniformity in materials and methods employed in these types of studies and the interested reader is well advised to examine the original paper in some detail.

Photosynthesis is generally thought of in terms of the reduction of carbon. However, during growth appreciable electron flow, amounting to approximately 20% of that of CO_2, must go to nitrogen assimilation primarily for nitrate reduction or, in some cyanobacteria, for the more specialized process of nitrogen fixation. There are no studies of petroleum hydrocarbon effects on nitrogen assimilation via nitrate reduction, and only limited comments on nitrogen fixation. In an Arctic mesotrophic meadow, nitrogen fixation by epiphitic cyanobacteria (*Nostoc* sp.) failed to recover in a year after a single dose of ASA 90 crankcase oil (Jordan et al., 1978). If nitrogen fixation proves particularly sensitive to petroleum hydrocarbons then there could be important consequences for shallow marine environments where nitrogen fixation by cyanobacteria is thought to contribute significantly to maintaining the combined nitrogen level.

Batterton et al. (1978b) tested four crude oils and five fuel oils for toxic effects with three microalgae including the cyanobacterium, *Agmenellum quadruplicatum* PR-6. The oils were absorbed on filter paper pads and the pads were submerged in the growth medium. The effects of the oils on the growth rate of *A. quadruplicatum* PR-6 are shown in Table I. The four crude oils were not particularly lethal. In contrast, the fuel oils, with the exception of the Montana fuel oil, were quite toxic. Although one expects to find similar classes of compounds in both crude and fuel oils, the enrichment of fuel oils with certain classes of compounds may provide a partial explanation for the differential toxicities. While the toxicity of the fuel oils could not be attributed to a specific compound or compounds, the toxic effects of the fuel oils were found to be in the higher boiling aromatic fraction and may, therefore, persist in the environment.

The water-soluble fractions from four fuel oils showed considerably different inhibitory effects to the growth of six microalgae including two cyanobacteria, *A. quadruplicatum* PR-6 and *Coccochloris elabens* 17a (Winters et al., 1976). The water solubles of two of the fuel oils, Baytown and Montana, were lethal to the cyanobacteria tested (Table II). This toxicity was traceable, in part, to the *p*-toluidine content of these oils (see below).

Studies on the toxicity of crude and fuel oils to algae led to an examination of the toxicity of aromatic compounds that were known to be components of oils. The algal lawn technique, wherein a test material is presented to algal cells held in agarized medium in petri dishes, has provided a rapid and covenient initial screening method. Numerous pure compounds that are found in petroleum have been tested for toxicity using the algal lawn technique with *A. quadruplicatum* as the cyanobacterial test organism (Table III). When tested against cyanobacteria, the following compounds showed very strong inhibition of growth and they merit detailed investigation as possible contributors to the petroleum toxicity problem: dimethylnaphthalenes (mixture of isomers), aniline (mono-, di-, trimethyl and ethyl derivatives), di- and trimethylphenols (various isomers), 2,5-dimethylpyrrole, indole (also mono- and dimethyl derivatives), quinolines (mono- and dimethyl

TABLE I

Growth rates (generations per day) of *Agmenellum quadruplicatum* PR-6 at 30°C in the presence of filter paper pads containing crude oils or fuel oils (doubling times calculated from turbidimetric measurements of cell number during exponential growth phase; the inoculum size was approximately 10^5 cells $\cdot ml^{-1}$)

Oil	μl oil/20 ml medium			
	1 μl[b]	5 μl	10 μl	30 μl
Control[a] 5.8 ± 0.5	–	–	–	–
Alaska crude	6.1	5.7	5.7	5.7
Kuwait crude	6.1	5.9	5.3	4.9(9)[c]
South Louisiana crude	5.6	5.1	6.2	5.2(5)
Venezuela crude	5.5	5.7	5.7	5.4(9)
API fuel oil	6.1(6)	NG-5[d]		
Baton Rouge fuel oil	6.1	4.7(6)	NG-5	
Baytown fuel oil	6.0	4.2(19)	NG-5	
Montana fuel oil	6.1	6.1	5.1	5.1
New Jersey fuel oil	5.9	3.3(12)	NG-5	

[a]Control cultures contained three oil-free pads.
[b]Oil was absorbed in paper pads up to 10 μl oil/pad. For example, for 30 μl/20 ml medium, three pads, each with 10 μl oil, were added to the culture tube.
[c]Numbers in parentheses indicate lag time in hours in initiation of growth as compared with the zero or short lag times seen in control growth curves.
[d]NG-5, no growth in 5 days.
(Adapted from Batterton et al., 1978b)

derivatives), toluidine (dimethyl and ethyl derivatives) and *p*-anisidine. *p*-Toluidine was found to be toxic to *A. quadruplicatum* PR-6 at the levels of 1 μg in the algal lawn assay (Table III) and 100 μg/l in liquid culture. Further studies by Batterton et al. (1978a) indicated that marine cyanobacteria were extremely sensitive to aniline and *p*-toluidine in an algal lawn assay. The growth of eight cyanobacteria was inhibited by 1 to 100 μg of *p*-toluidine per disk. *A. quadruplicatum* strain PR-6 was shown to be very sensitive to aniline and *p*-toluidine as evidenced by growth inhibition by as little as 1 μg of *p*-toluidine per disk (equivalent to 50 $\mu g \cdot l^{-1}$ in liquid culture). Under comparable conditions, 0.5 mg or more of *p*-toluidine was required to inhibit the growth of a green alga, a diatom and two species of bacteria. In these studies anilines appeared to be selectively toxic to cyanobacteria.

3. Metabolism

It is generally assumed that degradation of aromatic compounds in terrestrial and aquatic ecosystems is brought about by heterotrophic bacteria and fungi with animals or higher plants playing a secondary role. Little attention has been given to the microalgae in the degradation of aromatic compounds. Since many polycyclic aromatic hydrocarbons or their metabolites exhibit toxic, mutagenic or carcinogenic properties (McCann et al., 1975; Miller and Miller, 1974), there has been much interest in the metabolism of this group of compounds. Bacteria initiate attack on the aro-

TABLE II

Growth rates (generations/day) of cyanobacteria grown at 30°C in presence of 25 or 50% water solubles from fuel oils (doubling times calculated from turbidimetric measurements of cell number during exponential phase of growth)

Fuel oils	*Agmenellum quadruplicatum*	*Coccochloris elabens*
Controls	4.6 ± 0.2	3.6 ± 0.2
Baytown		
25[a]	NG-7[b]	NG-9
50	NG-7	NG-9
Baton Rouge		
25	3.4(14)[c]	3.8
50	3.9(31)	3.5
Montana		
25	NG-7	NG-9
50	NG-7	NG-9
New Jersey		
25	2.3(90)	2.5(43)
50	1.9(150)	2.3(65)

[a]Number indicates percent of ASP-2 medium equilibrated against a given fuel oil (8 parts medium, 1 part oil) added to plain medium ASP-2; e.g. 25% means 5 ml of ASP-2 containing water solubles from a fuel oil plus 15 ml of ASP-2 medium.

[b]NG: No growth, followed by number of days before termination of experiment; e.g. NG-7 means no growth in 7 days.

[c]Number in parentheses indicates lag time in hours in initiation of growth as compared to the zero or short lag times seen in control growth curves.

(Adapted from Winters et al., 1976)

matic ring via dioxygenase-catalyzed reactions which incorporate both atoms of oxygen from molecular oxygen into the ring resulting in the formation of dihydrodiols with a *cis* relative stereochemistry (Gibson, 1982). In contrast mammals and fungi possess monooxygenases which incorporate one atom of oxygen from molecular oxygen into the aromatic nucleus to form arene oxides. These oxides may then undergo isomerization to yield hydroxylated products, addition of water by epoxide hydrolase to form *trans* dihydrodiols or enzymatic or non-enzymatic addition of glutathione to form conjugated derivatives (Jakoby and Keen, 1977; Lu and Miwa, 1980; Sato and Omura, 1978).

While bacteria have a widespread capability to metabolize petroleum compounds (Gibson, 1984), there was little evidence to indicate that cyanobacteria possess such biochemical pathways. Ellis (1977) showed that six species of fresh-water algae, including two cyanobacteria, degraded uniformly ^{14}C-labeled phenol and catechol. While the recovery of $^{14}CO_2$ from ^{14}C-phenol was only 0.1% or less from *Anabaena cylindrica* and *Phormidium foveolarum*, the recovery from ^{14}C-catechol was 1.4 and 2.0%, respectively.

TABLE III

Effect of pure compounds on growth of *A. quadruplicatum* PR-6 using algal lawn technique

Compound	Amount on pad (mg)				Ref.
	10	2	1	0.5	
1,2,4-Trimethylbenzene	22	16	10	NT	h
1,3,5-Trimethylbenzene	0(15)[b]	0	0	0	h
Isopropylbenzene	NT[c]	0	0	NT	h
1,2,3,5-Tetramethylbenzene	1	0	0	0	h
Triethylbenzene[a]	3	2	0	NT	h
Diisopropylbenzene[a]	NT	12	6	NT	h
Diethylbenzene	17	6	4	2	i
Benzaldehyde	36	0	0	0	i
2,4-Dimethylbenzaldehyde	36	3	1	0	i
2,5-Dimethylbenzaldehyde	36	2	1	0	i
Indan	NT	1	0	NT	h
Naphthalene	36	2	0	0	h
1-Methylnaphthalene	36	10	4	NT	h
2-Methylnaphthalene	NT	5	2	0	e
Dimethylnaphthalene[a]	36	23	7	NT	h
1,4-Dimethylnaphthalene	36	36	0(36)	0(36)	h
1,5-Dimethylnaphthalene	3	1	1	0.5	h
2,3-Dimethylnaphthalene	0	0	0	0	h
2,6-Dimethylnaphthalene	0	0	0	0	h
1,3-Dimethylnaphthalene	36	4	3	2	i
2,3,6-Trimethylnaphthalene	NT	0	0	0	i
2-Ethylnaphthalene	14	3	2	1	h
2,6-Di-t-butylnaphthalene	0	0	0	0	i
1-Naphthol	NT	36	34	26	e
2-Naphthol	NT	35	33	28	e
trans-1,2-Dihydroxy-1,2-dihydronaphthalene	0	0	0	0	e
cis-1,2-Dihydroxy-1,2-dihydronaphthalene	0	0	0	0	e
Naphthoquinone	36	36	36	30	i
1-Hydroxymethylnaphthalene	NT	36	33(36)	17	e
2-Hydroxymethylnaphthalene	NT	28	19	13	e
1-Naphthoic acid	0	0	0	0	e
2-Naphthoic acid	0	0	0	0	e
1-Hydroxy-2-naphthoic acid	0	0	0	0	e
3-Hydroxy-2-naphthoic acid	0	0	0	0	e
Acenaphthylene	36	36	10(15)	2	i
Biphenyl	36	21	4	0	h
3,3-Dimethylbiphenyl	36	15(36)	5(8)	1	i
Fluorene	0	0	0	0	h
Phenanthrene	5(15)	4(9)	4	1	h
m-Cresol	36	6	0	0	h
o-Cresol	36	14	1	0	h
p-Cresol	36	15	8	0	h
p-Thiocresol	36	36	36	22	i
m-Thiocresol	36	36	28	15(20)	i

TABLE III *(continued)*

Compound	Amount on pad (mg)				Ref.
	10	2	1	0.5	
p-Amino-*m*-cresol	36	33	25	23	i
4-Ethylphenol	36	36	15	4	i
2,3-Dimethylphenol	36	17	1	0	h
2,4-Dimethylphenol	36	36	0(10)	0	h
2,5-Dimethylphenol	36	32	11(36)	0	h
2,6-Dimethylphenol	36	0(36)	0	0	h
3,4-Dimethylphenol	36	20(36)	8	0(5)	h
3,5-Dimethylphenol	36	20(36)	0(8)	0	h
2,3,5-Trimethylphenol	36	36	20(36)	2	h
2,3,6-Trimethylphenol	36	36	25(36)	3(36)	h
2,4,6-Trimethylphenol	36	36	36	1(36)	h
p-*n*-Propylphenol	36	30(36)	20(28)	9	h
N-Methylaniline	36	36	36	15	h
N,*N*-Dimethylaniline	36	0	0	0	h
2,3-Dimethylaniline	36	36	36	36	h
2,5-Dimethylaniline	36	36	36	36	h
2,6-Dimethylaniline	36	36	5(36)	1	h
3,4-Dimethylaniline	36	36	36	36	h
3,5-Dimethylaniline	36	36	36	36	h
2,4,6-Trimethylaniline	36	36	36	36	h
N-Ethylaniline	36	36	10	2	h
o- Ethylaniline	36	36	36	36	h
p-Ethylaniline	36	36	36	36	i
p-Isopropylaniline	36	36	36	36	i
N,*N*-Dimethyl-*p*-toluidine	36	36	36	0(1)	i
6-Ethyl-*o*-toluidine	36	36	20(36)	11	h
N-Methylpyrrole	1	0	0	0	h
2,5-Dimethylpyrrole	36	36	36	36	h
2,6-Dimethylpyridine	36	0	0	0	h
2,4,6-Trimethylpyridine	36	0(36)	0	0	h
Indole	36	25(36)	14	2(6)	h
2-Methylindole	36	25	11	6	i
3-Methylindole	36	28(36)	20(36)	4	i
7-Methylindole	36	20(36)	10(36)	3	i
5-Methylindole	36	36	36	36	i
1-Methylindole	9	9	5	4	i
2,5-Dimethylindole	12	2	1	0	h
1,2-Dimethylindole	36	36	8	1	i
2,3-Dimethylindole	18	17	9	7	i
2,3,5-Trimethylindole	6	6	6	4	i
5-Hydroxyindole	36	30(36)	20(36)	20(25)	i
Quinoline	36	36	13	0(5)	h
2-Methylquinoline	36	36	36	36	h
(Lepidine) 4-Methylquinoline	36	36	36	23	i
6-Methylquinoline	36	20(36)	2	2	i

TABLE III *(continued)*

Compound	Amount on pad (mg)				Ref.
	10	2	1	0.5	
7-Methylquinoline	36	36	36	15(20)	i
8-Methylquinoline	36	36	10	2	i
2,4-Dimethylquinoline	36	36	20(30)	3	i
2,6-Dimethylquinoline	36	23(36)	12(15)	1(5)	h
2,7-Dimethylquinoline	36	34(36)	25(36)	9(15)	h
Carbazole	0	0	0	0	h
Cyclohexylmercaptan	2	1	0	0	h
Thiophene	0	0	0	0	h
2-Methylthiophene	1	0	0	0	h
Thionaphthene	30	5	1	0	h
9-Fluorenone	11(20)	16(20)	17(36)	15(36)	h
Phthalide	10	0	0	0	i
Acenaphthene	15	3	3	1	i
Dibenzofuran	3	4	2	2	i
n-Decane	1	1	1	1	i
Undecane	2	3	3	3	i
p-Phenylenediamine	36	36	29	25	i
m-Phenylenediamine	36	36	36	36	i
o-Phenylenediamine	36	32	26	19	i
p-Anisidine	36	36	36	36	i
	Amount on pad (μg)				
	500	100	10	1	
m-Toluidine	36	36	0	0	g
o-Toluidine	36	36	12(17)	0	g
p-Toluidine	36	36	36	6(10)	g
Aniline	NT	36	36	0(8)	d
2,4-Dimethylaniline	36	36	10(15)	0	g
Phenylhydroxylamine	NT	36	36	0	d
Azobenzene	NT	1	0	0	d
p-Aminophenol	NT	14	8	1	d
m-Aminophenol	NT	36	12	0	d
o-Aminophenol	NT	36	14	0(6)	d
Formanilide	NT	8	0	0	d
Acetanilide	NT	0	0	0	d
4-Hydroxyanilide	NT	0	0	0	d
1,2-Naphthoquinone	NT	5	0	0	e
1,4-Naphthoquinone	NT	26	8	0	e
2-Methyl-1-naphthol	NT	35	3	0	e
9-Phenanthrol	33	30	12	NT	f
9,10-Phenanthrenequinone	30	24	12	NT	f
9-Methoxyphenanthrene	1	0	0	NT	f
trans-9,10-Dihydroxy-9,10- dihydrophenanthrene	0	0	0	NT	f

TABLE III *(continued)*

Note: Numbers represent zone of inhibition (mm) out from edge of filter pad.
Complete killing results in a zone of inhibition of 36 mm on the plate.
[a]Mixture of isomers.
[b]Colonies much reduced in size this distance from pad.
[c]Not tested.
[d]Cerniglia et al., 1981.
[e]Cerniglia et al., 1983b.
[f]Narro, 1985.
[g]Winters et al., 1976.
[h]Winters et al., 1977.
[i]This paper.

Collaborative efforts between the laboratories of Van Baalen and Gibson led to an extensive series of studies with marine cyanobacteria, aimed at determining whether this group of organisms had the ability to metabolize aromatic hydrocarbons that are found in petroleum and, if so, by what mechanism. These studies, which are described in the remainder of this chapter, indicate that cyanobacteria can indeed metabolize aromatic hydrocarbons and have given some insight into the potential that cyanobacteria have for detoxifying and for activating xenobiotic compounds in marine environments.

In the studies of aromatic hydrocarbon metabolism, cultures were grown photoautotrophically at 30°C in cotton-plugged Pyrex test tubes containing 20 ml of modified ASP-2 medium containing any necessary vitamins (Provasoli et al., 1957; Van Baalen, 1962). Air containing 1% CO_2 was bubbled continuously through the cultures, and illumination was provided by two fluorescent lamps (F2OT12-D) placed 8.5 cm from the tubes. When cultures reached a density of 0.3 to 0.8 $mg \cdot ml^{-1}$ dry weight (depending on the organism), they were transferred to a closed growth system consisting of a 500-ml Nephalo flask as described by Cerniglia et al. (1979, 1980c). The flask was clamped to an oscillating bar and incubated in a 30°C water bath. Cultures were illuminated by two fluorescent lamps (Westinghouse F48T12/CWX/HO) placed 29 cm below the flasks. A yellow Plexiglas filter (Rohm and Haas No. 2422, 3.18 mm thick), having no transmission below 530 nm, was placed between the cultures and the light source to prevent any direct photochemical effects on the substrate or metabolites. Organisms were grown for 24 h in the presence of aromatic substrate. CO_2 was introduced into the flask via a hole in the side-arm cap and the CO_2 concentration was maintained between 0.5 to 1.0% throughout the experiment. Control experiments with radiolabeled substrates were conducted in the absence of organisms and with boiled cultures. Metabolites were extracted from the culture medium with ethyl acetate, purified by high-pressure liquid chromatography and identified on the basis of their ultraviolet, infrared, mass and proton magnetic resonance spectral properties.

3.1. Naphthalene

Initially naphthalene was chosen as a substrate since it was a major component in the water-soluble fraction of crude and certain fuel oils (Winters et al., 1976) and since its metabolism by mammals, fungi and bacteria had been examined in detail (Cerniglia and Gibson, 1977; Davies and

Evans, 1964; Jeffrey et al., 1975; Jerina et al., 1970a). In the first studies on the metabolism of naphthalene by marine cyanobacteria, *Oscillatoria* sp., strain JCM, and *Agmenellum quadruplicatum*, strain PR-6, were shown to oxidize naphthalene under photoautotrophic growth conditions to form 1-naphthol as the major ethyl acetate soluble product (Cerniglia et al., 1979, 1980c). Preliminary evidence was also obtained for the formation of *cis*-1,2-dihydroxy-1,2-dihydronaphthalene (*cis*-1,2-naphthalene dihydrodiol). In addition, *Oscillatoria* sp. JCM formed 4-hydroxy-1-tetralone. The formation of both 1-naphthol, characteristic of the monooxygenase-catalyzed oxidation of naphthalene, and *cis*-1,2-naphthalene dihydrodiol, characteristic of dioxygenase-catalyzed oxidations, suggested that these cyanobacteria possessed more than one pathway for naphthalene oxidation. *Oscillatoria* sp. JCM oxidized 4.8% of the added ^{14}C-naphthalene in 24 h under photoautotrophic growth conditions (Cerniglia et al., 1980c). The ratio of organic-soluble to water-soluble metabolites was 41:59. The identity of these water-soluble metabolites is not known.

The major metabolite, 1-naphthol, could be formed by two possible mechanisms: (a) the formation of an arene oxide intermediate that isomerizes with a concomitant NIH shift or, (b) direct hydroxylation of the aromatic ring at the 1-position. Naphthalene-1,2-oxide is unstable and spontaneously isomerizes via a mechanism called the NIH shift (Boyd et al., 1972; Daly et al., 1972). As illustrated in Fig. 1, an NIH shift is an *ortho* migration of an aryl substituent during hydroxylation of the aromatic ring. By using deuterated naphthalenes as substrates and analyzing the metabolically formed 1-naphthol by mass spectrometry or proton magnetic resonance spectroscopy, it is possible to determine whether the 1-naphthol is being formed by an NIH shift mechanism or by direct hydroxylation.

Oscillatoria sp. JCM oxidized [1-^{2}H]- and [2-^{2}H]naphthalene to form 1-naphthol with 68 and

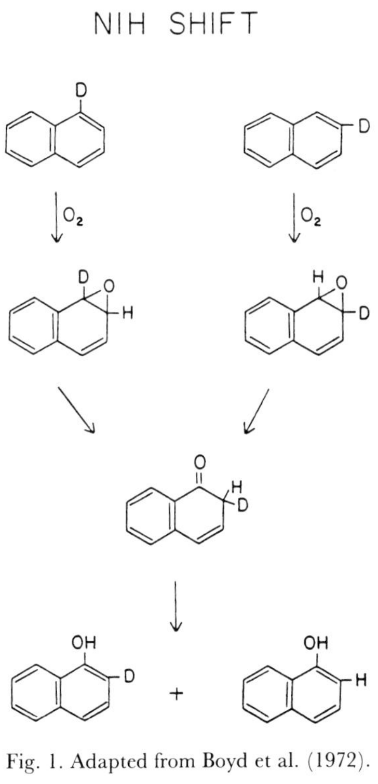

Fig. 1. Adapted from Boyd et al. (1972).

74 % retention of deuterium, respectively (Narro, 1985). These values are comparable to those reported for mammalian, fungal and bacterial systems (Boyd et al., 1972; Cerniglia et al., 1983a, 1984a). Experiments using a 1:1 mixture of unlabeled and uniformly labeled $^2H_8$-naphthalene as substrate showed that k_H/k_D was 1.05 indicating that there was no significant isotope effect during 1-naphthol formation from naphthalene. These results indicate that 1-naphthol is formed via an arene oxide intermediate with a concomitant NIH shift.

When *Oscillatoria* sp. JCM was incubated with naphthalene in the presence of $^{18}O_2$, 57 % incorporation of $^{18}O_2$ into 1-naphthol was observed (Cerniglia et al., 1980c). $^{18}O_2$ from the gas phase was readily incorporated into 1-naphthol. However, as a result of probable dilution of $^{18}O_2$ inside the cell by photosynthetically produced $^{16}O_2$, the data cannot be used quantitatively to support a specific pathway. The amount of dihydrodiol formed was insufficient for $^{18}O_2$ incorporation analysis.

Figure 2 shows the mechanisms that have been postulated for the formation of 1-naphthol from naphthalene by *Oscillatoria* sp. JCM (Narro, 1985). The lack of a primary kinetic isotope effect during the oxygenation of naphthalene suggests that *Oscillatoria* sp. JCM oxidizes naphthalene primarily via naphthalene-1,2-oxide rather than by direct hydroxylation. The retention of 68 % of the deuterium in the 1-naphthol formed from 1-^{2}H-naphthalene indicates that naphthalene-1,2-oxide isomerizes primarily with a concomitant NIH shift to a keto intermediate. Enolization of the keto intermediate occurs with a reported isotope effect (k_H/k_D) of 4, forming 1-naphthol in which the deuterium is retained (Boyd et al., 1972). Theoretically, the maximum deuterium retention would be 80 %. Since the 1-naphthol formed by *Oscillatoria* sp. JCM retained only 68 % of the deuterium, some direct loss of deuterium from the 1-position may be occurring. In addition, the possibility that a small amount of direct hydroxylation occurs cannot completely be ruled out.

Cerniglia et al. (1980a) showed that the ability to oxidize naphthalene is widely distributed throughout the algae. Eighteen algal cultures, including nine cyanobacteria, were screened for the ability to oxidize naphthalene under photoautotrophic growth conditions. The extent of naphthalene metabolism to organic-soluble products ranged from 0.1 to 2.4 %.

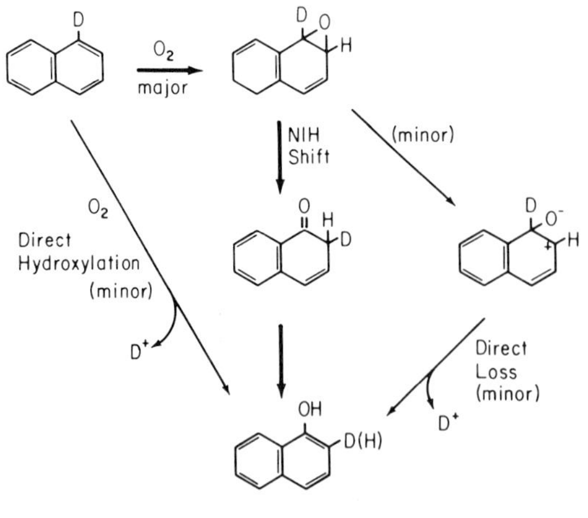

Fig. 2. Postulated mechanisms for the formation of 1-naphthol from naphthalene by the marine cyanobacterium *Oscillatoria* sp. JCM.

3.2. Phenanthrene

Considerable information about the initial mechanism of oxidative attack on the aromatic ring can be gained by examining the relative stereochemistry of the hydroxyl groups in the metabolically formed dihydrodiols and the origin of their oxygen atoms (see Section 3). The amount of dihydrodiol formed from naphthalene by *Agmenellum quadruplicatum* PR-6 and *Oscillatoria* sp. JCM was not sufficient for such studies. Other aromatic hydrocarbons were screened to determine whether these cyanobacteria oxidized them to form greater amounts of dihydrodiol. Phenanthrene, which is widely distributed in the environment as a result of pyrolytic processes (Blumer, 1976), was oxidized to dihydrodiol in sufficient amounts for rigorous chemical characterization.

Under photoautotrophic growth conditions, the marine cyanobacterium *A. quadruplicatum* PR-6 metabolized phenanthrene to form *trans*-9,10-dihydroxy-9,10-dihydrophenanthrene (*trans*-9,10-phenanthrene dihydrodiol) and 1-methoxyphenanthrene as the major ethyl acetate-extractable metabolites (Narro, 1985). Small amounts of phenanthrols were also formed. These products represented 2.2% of the added phenanthrene.

The origin of the oxygen in the metabolically formed dihydrodiol was examined by growing *A. quadruplicatum* PR-6 in the presence of phenanthrene under an atmosphere of $^{18}O_2$. The isotopic abundance of the dihydrodiol was calculated from the relative intensities of the molecular ions having mass to charge ratios of 212 (two ^{16}O's), 214 (one ^{16}O and one ^{18}O), and 216 (two ^{18}O's). The values were corrected for the average $^{18}O_2$ content of the gas phase in the flask. The mass spectral data from the $^{18}O_2$ incorporation experiment (Fig. 3) indicated that one atom of ^{18}O was incorporated into the *trans*-9,10-phenanthrene dihydrodiol formed by *A. quadruplicatum* PR-6. This result

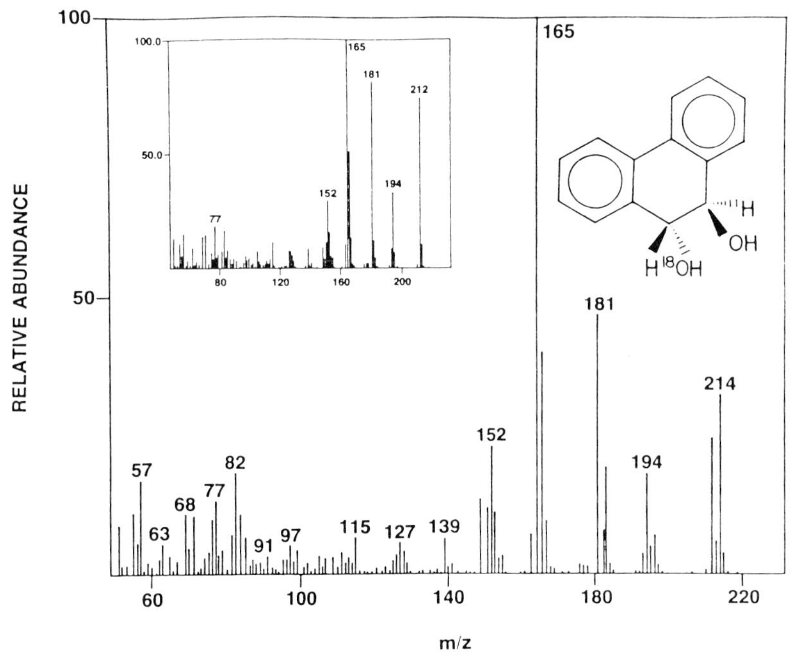

Fig. 3. Mass spectrum of *trans*-9,10-dihydroxy-9,10-dihydrophenanthrene formed by *Agmenellum quadruplicatum* PR-6 incubated in the presence of $^{18}O_2$. The inset shows the spectrum of the dihydrodiol formed in the presence of air.

indicates that the oxygenation of the aromatic ring is catalyzed by a monooxygenase rather than by a dioxygenase.

Additional evidence for the formation of the *trans* stereoisomer of 9,10-phenanthrene dihydrodiol was obtained from the ^{1}H nuclear magnetic resonance (NMR) spectrum of the metabolite. The vicinal coupling constant for the carbinol hydrogens ($J_{9,10}$) was determined to be 10 Hz from the ^{13}C-satellites of the 360 MHz ^{1}H NMR spectrum. This value is consistent with the value reported for *trans*-9,10-phenanthrene dihydrodiol by Cobb et al. (1983). The vicinal coupling constant for the *cis* stereoisomer would be 4 Hz or less.

The enantiomeric purity of the metabolically formed dihydrodiol was determined by derivatizing ^{14}C-labeled (±)-*trans*-9,10-phenanthrene dihydrodiol formed by *A. quadruplicatum* PR-6 with (−)-α-methoxy-α-trifluoromethylphenyl acetic acid [(−)-MTPA]. The resulting diastereomeric bis MTPA diesters of the dihydrodiol were separated by high pressure liquid chromatography (HPLC) and the enantiomeric purity of the metabolically formed *trans*-9,10-phenanthrene dihydrodiol was determined from the radioactivity in each diastereomer. The dihydrodiol was formed with a 22% enantiomeric excess of the (−)-9S,10S-enantiomer. Although the optical purity of the dihydrodiol formed by *A. quadruplicatum* PR-6 was relatively low, it is comparable to optical purity

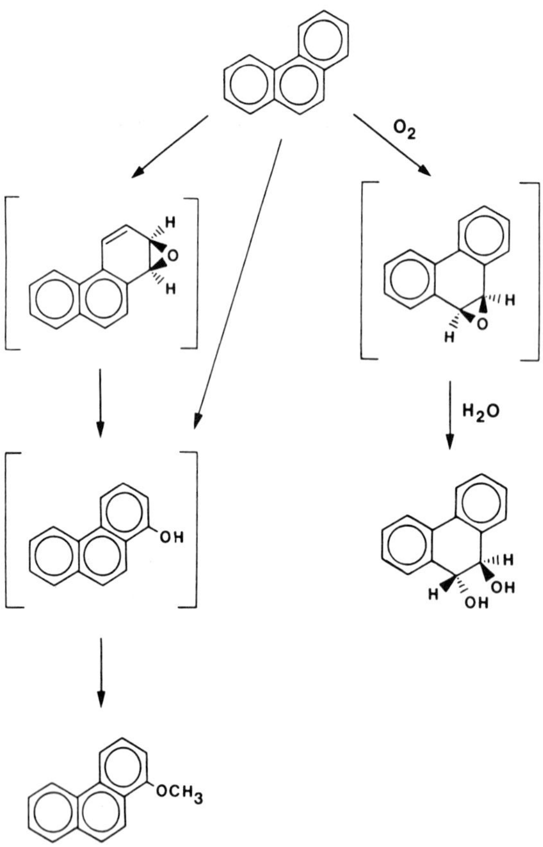

Fig. 4. Postulated pathways for phenanthrene metabolism by *Agmenellum quadruplicatum* PR-6. Brackets indicate hypothetical intermediates that have not been isolated.

values reported from mammalian systems (Jerina et al., 1970b; Nordqvist et al., 1981; Thakker et al., 1977).

The pathways postulated for the oxidation of phenanthrene by *A. quadruplicatum* PR-6 are shown in Fig. 4. A monooxygenase-catalyzed reaction could form 9,10- or 1,2-phenanthrene oxide. The 9,10-oxide is relatively stable and could undergo enzymatic hydration to yield optically active *trans*-9,10-phenanthrene dihydrodiol. In contrast the 1,2-oxide is unstable and would undergo rapid isomerization to form mainly 1-phenanthrol (Bruice et al., 1976). The latter compound could then serve as the substrate for a methyl transferase to yield 1-methoxyphenanthrene. Alternatively, 1-phenanthrol could be formed by a direct hydroxylation of the aromatic ring.

This was the first report of a prokaryotic organism metabolizing an aromatic hydrocarbon to a dihydrodiol with a *trans* relative stereochemistry. *A. quadruplicatum* PR-6 must possess enzymes that catalyze the enzymatic hydration of phenanthrene-9,10-oxide since it forms optically active *trans*-9,10-phenanthrene dihydrodiol as a metabolite of phenanthrene. However, the possibility that some non-enzymatic hydration of metabolically formed phenanthrene-9,10-oxide may occur cannot be completely ruled out (Bruice et al., 1976). The formation of an optically active *trans* dihydrodiol in which only one atom of oxygen originates from O_2 strongly indicates that *A. quadruplicatum* PR-6 oxidizes phenanthrene via a monooxygenase-catalyzed reaction.

The metabolic formation of 1-methoxyphenanthrene from phenanthrene had not been reported previously. Daly et al. (1972) reported that catechols may be methylated by catechol-*O*-methyltransferase in mammalian systems. The formation of 1-methoxyphenanthrene by *A. quadruplicatum* may serve to detoxify any 1-phenanthrol that is formed. 1-Phenanthrol is probably toxic since several hydroxylated aromatic compounds (9-phenanthrol, 1-naphthol, 2-naphthol) are known to be toxic to this organism (Cerniglia et al., 1983b). The algal lawn bioassay indicated that the phenolic and quinone derivatives of phenanthrene were considerably more toxic to *A. quadruplicatum* PR-6 than phenanthrene.

3.3. Biphenyl

Experiments with ^{14}C-biphenyl showed that over a 24-h period *Oscillatoria* sp. JCM oxidized 2.9% of the added biphenyl to ethyl acetate soluble products. 4-Hydroxybiphenyl was the major metabolite with small amounts of 4,4′-dihydroxybiphenyl also being formed (Cerniglia et al., 1980b). 4-Hydroxybiphenyl could be formed by an arene oxide mechanism. For example, a monooxygenase-catalyzed reaction could form biphenyl 3,4-epoxide which then would isomerize to 4-hydroxybiphenyl. The formation of 4-hydroxybiphenyl could also be explained by a light-dependent direct oxygen insertion mechanism. The metabolism of biphenyl by *Oscillatoria* sp. JCM was analogous to biphenyl metabolism in fungi and mammals which also form 4-hydroxybiphenyl and 4,4′-dihydroxybiphenyl (Dodge et al., 1979; Meyer and Scheline, 1976; Smith and Rosazza, 1974; Wiebkin et al., 1976). In contrast, bacteria oxidize biphenyl at the 2,3-positions to form 2,3-dihydro-2,3-dihydroxybiphenyl (Catelani et al., 1971; Gibson et al., 1973).

3.4. Methylnaphthalene

Naphthalene and methylnaphthalenes are the most toxic components in the water-soluble fraction of crude oils (Lee et al., 1974) and they have been identified as major ingredients in petroleum oil fractions of mosquito repellants (Wirtz et al., 1981) and in the aromatic hydrocarbon solvent, Aerotex 3470 (Sparling et al., 1978). In addition, these compounds have been shown to be concentrated from the water by vertebrate and invertebrate marine organisms (Anderson et al., 1974;

Melancon and Lech, 1978). Thus, Cerniglia et al. (1983b) conducted experiments to determine whether cyanobacteria could metabolize methylnaphthalenes and to test the toxicity of methylnaphthalenes and their derivatives.

Agmenellum quadruplicatum PR-6, *Oscillatoria* sp. JCM and *Anabaena* sp. CA grown photoautotrophically in the presence of either 1- or 2-methylnaphthalene oxidized both compounds at the methyl group to form 1-hydroxymethylnaphthalene and 2-hydroxymethylnaphthalene, respectively (Cerniglia et al., 1983b). The mass spectra of these metabolites are shown in Fig. 5. Ring hydroxylated metabolites were not detected. After 24 h growth in the presence of either 1- or 2-meth-

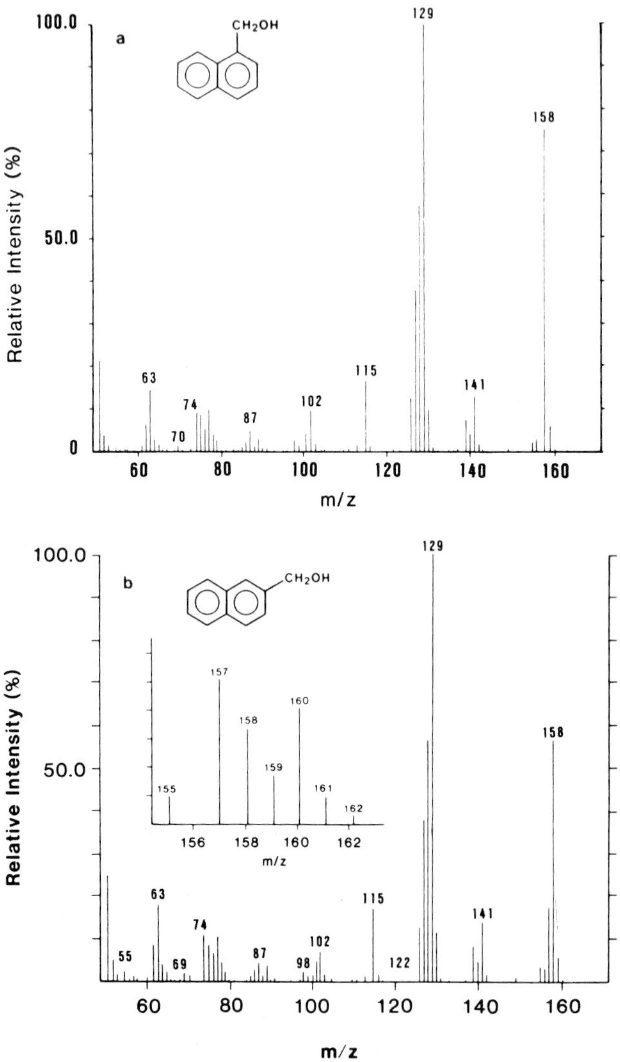

Fig. 5. Mass spectrum of 1-hydroxymethylnaphthalene (a) and 2-hydroxymethylnaphthalene (b) formed from 1-methylnaphthalene and 2-methylnaphthalene, respectively, by *A. quadruplicatum* PR-6. Inset shows the mass spectrum of 2-hydroxymethylnaphthalene formed from 2-methylnaphthalene by *A. quadruplicatum* PR-6 grown in the presence of a mixture of 20 % $^{16}O_2$ and 80 % $^{18}O_2$.

ylnaphthalene, the extent of metabolism to ethyl acetate soluble metabolites ranged from 0.7 to 3.2%.

When *A. quadruplicatum* PR-6 was grown in the presence of 2-methylnaphthalene under an atmosphere containing $^{18}O_2$, 75% of the metabolically formed 2-hydroxymethylnaphthalene contained ^{18}O. One would expect 100% incorporation of ^{18}O if the reaction was catalyzed by a monooxygenase (Cerniglia et al., 1983b). The 25% incorporation of ^{16}O may result from alternate enzymatic pathways or from incorporation of $^{16}O_2$ formed as a result of oxygenic photosynthesis.

The toxicity of naphthalene, 1- and 2-methylnaphthalene and their derivatives on strain PR-6 was investigated using the algal lawn technique. Alkyl side chain hydroxylated products, such as the hydroxymethyl derivatives of 1- and 2-methylnaphthalene were more toxic than the parent methylnaphthalenes.

The cyanobacterial metabolism of 1- and 2-methylnaphthalene to hydroxymethylnaphthalene derivatives is similar to results obtained from mammalian and fungal (Cerniglia et al., 1984b) enzyme systems. Mammals oxidize 1- and 2-methylnaphthalene primarily at the methyl group to form 1- and 2-naphthoic acids, respectively (Kaubisch et al., 1972). However, a more recent report by Breger et al. (1981) indicated that rat liver microsomes and rainbow trout oxidized 2-methylnaphthalene to three isomeric dihydrodiols. In contrast, pseudomonads oxidize 1- and 2-methylnaphthalene to 3-methylcatechol and 4-methylcatechol, respectively, which serve as substrates for ring fission (Dean-Raymond and Bartha, 1975; Williams et al., 1975).

Although 1-naphthol was the major metabolite formed from naphthalene by cyanobacteria (Cerniglia et al., 1980a,c), there was no detectable aromatic ring hydroxylation of the methylnaphthalenes. This preference for hydroxylation of the alkyl side chain rather than the aromatic ring may be a result of the ease with which a benzylic carbon, due to the electron withdrawing effects of the aromatic ring system can be oxidized (Lee, 1982). The results may also be of toxicological significance since 1- and 2-hydroxymethylnaphthalene are less toxic to *A. quadruplicatum* PR-6 than their ring hydroxylated derivatives (Table III). However, the hydroxymethylnaphthalenes are more toxic than the parent methylnaphthalenes and thus we are hesitant to ascribe the cyanobacterial oxidation of 1- and 2-methylnaphthalene solely to a detoxification pathway. The oxidation of aromatic compounds by these cyanobacteria may prove to be an unavoidable consequence of some unknown constitutive pathway rather than a specialized detoxification pathway.

3.5. Aniline

Aniline and derivatives of this aromatic amine are used in the manufacture of dyes, herbicides, drugs and other organic chemicals. The presence of these compounds in the effluent from dye-manufacturing plants (Games and Hites, 1977) prompted a study of the ability of cyanobacteria to metabolize such compounds (Cerniglia et al., 1981).

Agmenellum quadruplicatum PR-6 and *Oscillatoria* sp. JCM grown photoautotrophically in the presence of aniline metabolized the aromatic amine to formanilide, acetanilide and *p*-aminophenol. The *N*-formyl and *N*-acetyl derivatives of aniline were the predominant metabolites. The formation of formanilide as a major metabolite suggests that cyanobacteria have the ability to *N*-formylate aromatic amines. While this metabolic pathway had previously been reported for the metabolism of aromatic amines by bacteria (Alexander, 1981) and mammals (Gothoskar et al., 1979), this study represented the first example of aromatic amine metabolism by cyanobacteria.

N-Acetylation of aromatic amines by mammals is generally regarded as a detoxification reaction. In this respect, it is interesting to note that cyanobacteria produced a significant amount of acetanilide. Since cyanobacteria metabolized aniline primarily to formanilide and acetanilide, both of which are less toxic than the parent compound (Table III), it seems that the enzymatic capacity to detoxify aromatic amines may have evolved in prokaryotes.

Both cyanobacteria formed small amounts of *p*-aminophenol. *A. quadruplicatum* PR-6 was extremely sensitive to *o*-, *m*- and *p*-aminophenols (Table III). *p*-Aminophenol could be formed via initial epoxidation followed by isomerization to the phenol. These reactions have been described for the fungal (Ferris et al., 1973, Smith and Rosazza, 1974) and mammalian (Daly et al., 1972) metabolism of aniline. Alternatively, it is possible that *p*-aminophenol could be formed via direct hydroxylation of aniline or by rearrangement of phenylhydroxylamine (Booth and Boyland, 1964). *A. quadruplicatum* was also very sensitive to phenylhydroxylamine.

4. Summary

Recent work with cyanobacteria has concentrated on the isolation and identification of biotransformation products formed when various organisms were incubated with aromatic substrates that are components of petroleum. The accumulated data strongly indicate that microalgae can metabolize aromatic compounds. The list of substrates metabolized and the products identified show that the cyanobacteria studied utilize many types of reactions to metabolize aromatic compounds: ring hydroxylation, O-methylation, side-chain hydroxylation, acetylation and formylation.

Use of the algal lawn assay in conjunction with the metabolism experiments has begun to give some insight into the biochemistry of the toxicity of petroleum and aromatic hydrocarbons to cyanobacteria. It is apparent that specific compounds in crude or fuel oils, such as *p*-toluidine, dimethylanilines, 2,5-dimethylpyrrole and others can be extremely toxic to certain cyanobacteria while other compounds such as 2,3-dimethylnaphthalene, carbazole or fluorene are not. It is also quite clear that metabolism plays a crucial role in the toxicity of some compounds. For example, 1-naphthol, the major metabolite formed from naphthalene by the cyanobacteria that have been studied, is considerably more toxic than naphthalene (Table III). Only small amounts on the non-toxic naphthalene dihydrodiols were formed. The major metabolites formed from phenanthrene were *trans*-9,10-phenanthrene dihydrodiol and 1-methoxyphenanthrene. The dihydrodiol and methoxyl derivatives of phenanthrene were not toxic or were only slightly toxic under the test conditions. Only small amounts of the toxic phenanthrols were formed. The formation of more of the toxic phenolic metabolites from naphthalene than from phenanthrene may provide a partial explanation for the observation that naphthalene was more toxic than phenanthrene in the algal lawn assay. The effect of differences in water solubility and partitioning into cellular lipids are not known and may also contribute to differential toxicities of these compounds. The major metabolites formed from 1- and 2-methylnaphthalene, 1- and 2- hydroxymethylnaphthalene, were also more toxic than their respective parent compounds. From these studies it is clear that cyanobacteria have the potential to both detoxify and to activate xenobiotic compounds in the marine environment. There is essentially no understanding at the biochemical level of the mode of action for any petroleum compounds demonstrated to be toxic to a microalga.

In an environmental context it now seems that the phytoplankton should be reckoned with in any balance sheet of the degradation of aromatic compounds in the sea. To help answer this question we have begun short-time light and dark bottle experiments with mixed phytoplankton hauls incubated in situ with ^{14}C-napthalene, in order to determine if there is an appreciable light-driven oxidation of naphthalene. These primary producers may prove to be a major sink for aromatic compounds in the marine environment and their aromatic oxidation products may have important consequences for other organisms.

It is important to keep in mind that there is no such thing as a model petroleum or petroleum product nor is there a model test organism or ecosystem. The toxicity or lack of it varies depending on the petroleum, aromatic compound, organism and test conditions. It is all too easy to come up with the answer that neither photosynthesis nor growth of a particular cyanobacterium or mixed phytoplankton sample was inhibited by a certain type of petroleum presented in a certain way. By the same token, it is disturbingly easy to support the alternate notion that petroleum contains individual compounds that are remarkably toxic, in the ppb range, to individual phytoplankters. The argument that concentrations of petroleum hydrocarbons in nature will never reach the levels clearly needed to see effects on the microalgae in enclosed ecosystems or with pure cultures, assumes that the inhibition or lethality will simply be a linear function of concentration. There may be no safe level of man-induced petroleum hydrocarbon pollution of the sea. It seems best therefore, to strive to prevent a significant increase in petroleum hydrocarbon introduction to the marine environment and in particular to the photic zone.

As regards to the evolution of metabolic pathways there is an interesting speculation which follows from the finding that cyanobacteria can oxidize aromatic compounds. The cyanobacteria are generally viewed as the first group of organisms to develop the capacity for an O_2-evolving photosynthesis (Schopf, 1975). This suggests that the first oxygen-requiring enzymes capable of the hydroxylation of an aromatic ring may have evolved in an ancestral cyanobacterium.

Acknowledgements

This research was supported in part by Grant CA19078 from the National Cancer Institute to Dr. David T. Gibson.

References

Alexander, M. (1981) Biodegradation of chemicals of environmental concern. Science 211, 132–138.

Anderson, J.W., Neff, J.M., Cox, B.A., Tatem, H.E. and Hightower, G.M. (1974) The effects of oil on estuarine animals; toxicity, uptake and depuration, respiration. In F.J. Vernberg and W. Vernberg (eds.), Pollution and Physiology of Marine Organisms. Academic Press, New York, pp. 285–310.

Batterton, J., Winters, K. and Van Baalen, C. (1978a) Anilines: selective toxicity to blue-green algae. Science 199, 1068–1070.

Batterton, J., Winters, K. and Van Baalen, C. (1978b) Sensitivity of three microalgae to crude oils and fuel oils. Marine Environ. Res. 1, 31–41.

Blumer, B. (1976) Polycyclic aromatic hydrocarbons in nature. Sci. Am. 234, 34–44.

Booth, J. and Boyland, E. (1964) The biochemistry of aromatic amines. Enzymatic N-hydroxylation of arylamines and conversion of arylhydroxylamines into *o*-aminophenols. Biochem. J. 91, 362–369.

Boyd, D.R., Daly, J.W. and Jerina, D.M. (1972) Rearrangement of [1-^{2}H]- and [2-^{2}H]naphthalene 1,2-oxides to 1-naphthol. Mechanism of the NIH shift. Biochemistry 11, 1961–1966.

Breger, R.K., Franklin, R.B. and Lech, J.J. (1981) Metabolism of 2-methylnaphthalene to isomeric dihydrodiols by hepatic microsomes of rat and rainbow trout. Drug Metab. Dispos. 9, 88–93.

Bruice, P.Y., Bruice, T.C., Dansette, P.M., Selander, H.G., Yagi, H. and Jerina, D.M. (1976) Comparison of the mechanisms of solvolysis and rearrangement of K-region vs. non-K-region arene oxides. J. Am. Chem. Soc. 98, 2965–2973.

Catelani, D., Sorlini, C. and Treccani, V. (1971) The metabolism of biphenyl by *Pseudomonas putida*. Experientia 27, 1173–1174.

Cerniglia, C.E. and Gibson, D.T. (1977) Metabolism of naphthalene by *Cunninghamella elegans*. Appl. Environ. Microbiol. 34, 363–370.

Cerniglia, C.E., Gibson, D.T. and Van Baalen, C. (1979) Algal oxidation of aromatic hydrocarbons: for-

mation of 1-naphthol from naphthalene by *Agmenellum quadruplicatum*, strain PR-6. Biochem. Biophys. Res. Comm. 88, 50–58.
Cerniglia, C.E., Gibson, D.T. and Van Baalen, C. (1980a) Oxidation of naphthalene by cyanobacteria and microalgae. J. Gen. Microbiol. 116, 495–500.
Cerniglia, C.E., Van Baalen, C. and Gibson, D.T. (1980b) Oxidation of biphenyl by the cyanobacterium, *Oscillatoria* sp., strain JCM. Arch. Microbiol. 125, 203–207.
Cerniglia, C.E., Van Baalen, C. and Gibson, D.T. (1980c) Metabolism of naphthalene by the cyanobacterium *Oscillatoria* sp., strain JCM. J. Gen. Microbiol. 116, 485–494.
Cerniglia, C.E., Freeman, J.P. and Van Baalen, C. (1981) Biotransformation and toxicity of aniline and aniline derivatives in cyanobacteria. Arch. Microbiol. 130, 272–275.
Cerniglia, C.E., Althaus, J.R., Evans, F.E., Freeman, J.P., Mitchum, R.K. and Yang, S.K. (1983a) Stereochemistry and evidence for an arene oxide-NIH shift pathway in the fungal metabolism of naphthalene. Chem.-Biol. Interactions 44, 119–132.
Cerniglia, C.E., Freeman, J.P., Althaus, J.R. and Van Baalen, C. (1983b) Metabolism and toxicity of 1- and 2-methylnaphthalene and their derivatives in cyanobacteria. Arch. Microbiol. 136, 177–183.
Cerniglia, C.E., Freeman, J.P. and Evans, F.E. (1984a) Evidence for an arene oxide-NIH shift pathway in the transformation of naphthalene to 1-naphthol by *Bacillus cereus*. Arch. Microbiol. 138, 283–286.
Cerniglia, C.E., Lambert, K.J., Miller, D.W. and Freeman, J.P. (1984b) Transformation of 1- and 2-methylnaphthalene by *Cunninghamella elegans*. Appl. Environ. Microbiol. 47, 111–118.
Cobb, D.I., Lewis, D.A. and Armstrong, R.N. (1983) Solvent dependence of the conformation and chiroptical properties of *trans*-9,10-dihydroxy-9,10-dihydrophenanthrene and its monoglucuronides. J. Org. Chem. 48, 4139–4141.
Daly, J.W., Jerina, D.M. and Witkop, B. (1972) Arene oxides and the NIH shift: The metabolism, toxicity and carcinogenicity of aromatic compounds. Experientia 28, 1129–1149.
Davies, J.I. and Evans, W.C. (1964) Oxidative metabolism of naphthalene by soil pseudomonads. The ring-fission mechanism. Biochem. J. 91, 251–261.
Dean-Raymond, D. and Bartha, R. (1975) Biodegradation of some polynuclear aromatic petroleum components by marine bacteria. Dev. Ind. Microbiol. 16, 97–110.
Dodge, R.H., Cerniglia, C.E. and Gibson, D.T. (1979) Fungal metabolism of biphenyl. Biochem. J. 178, 223–230.
Ellis, B.E. (1977) Degradation of phenolic compounds by fresh-water algae. Plant Sci. Lett. 8, 213–216.
Ferris, J.P., Fasco, M.J., Stylianopoulou, F.L., Jerina, D.M., Daly, J.W. and Jeffrey, A.M. (1973) Monooxygenase activity in *Cunninghamella bainieri*. Evidence for a fungal system similar to liver microsomes. Arch. Biochem. Biophys. 156, 97–103.
Games, L.M. and Hites, R.A. (1977) Composition, treatment efficiency and environmental significance of dye manufacturing plant effluents. Ann. Chem. 49, 1433–1440.
Gibson, D.T. (1982) Microbial degradation of hydrocarbons. Toxicol. Environ. Chem. 5, 237–250.
Gibson, D.T. (1984) Microbial degradation of aromatic hydrocarbons. In D.T. Gibson (ed.), Microbial Degradation of Organic Compounds. Marcel Dekker, New York, pp. 181–251.
Gibson, D.T., Roberts, R.L., Wells, M.C. and Kobal, V.M. (1973) Oxidation of biphenyl by a *Beijerinckia* species. Biochem. Biophys. Res. Commun. 50, 211–219.
Gordon, D.C. and Prouse, N.J. (1973) The effects of three oils on marine phytoplankton photosynthesis. Mar. Biol. 22, 329–333.
Gothoskar, S.V., Benjamin, T., Roller, P.P. and Weisburger, E.K. (1979) N-Formylation of an aromatic amine as a metabolic pathway. Xenobiotica 9, 533–537.
Jacoby, W.B. and Keen, J.H. (1977) A triple threat in detoxification: the glutathione *S*-transferases. Trends Biochem. Sci. 2, 229–231.
Jeffrey, A.M., Yeh, H.J.C., Jerina, D.M., Patel, T.R., Davey, J.F. and Gibson, D.T. (1975) Initial reactions in the oxidation of naphthalene by *Pseudomonas putida*. Biochemistry 14, 575–584.
Jerina, D.M., Daly, J.W., Witkop, B., Zaltzman-Nirenberg, P. and Udenfriend, S. (1970a) 1,2-Naphthalene oxide as an intermediate in the microsomal hydroxylation of naphthalene. Biochemistry 9, 147–156.
Jerina, D.M., Ziffer, H. and Daly, J.W. (1970b) The role of arene oxide-oxepin system in the metabolism of aromatic substrates. IV. Stereochemical considerations of dihydrodiol formation and dehydrogenation. J. Am. Chem. Soc. 92, 1056–1061.
Jordan, D.C., McNicol, P.J. and Marshall, M.R. (1978) Biological nitrogen fixation in the terrestrial environment of a high Arctic ecosystem. Can. J. Microbiol. 24, 643–649.
Kaubisch, N., Daly, J.W. and Jerina, D.M. (1972) Arene oxides as intermediates in the oxidative metabo-

lism of aromatic compounds. Isomerization of methyl-substituted arene oxides. Biochemistry 11, 3080–3088.
Lee, C.C., Craig, W.K. and Smith, P.J. (1974) Water-soluble hydrocarbons from crude oil. Bull. Environ. Contam. Toxicol. 12, 212–216.
Lee, D.G. (1982) Phase transfer assisted permanganate oxidation. In W. Trahanovsky (ed.), Oxidations in Organic Chemistry, Chapter 2. Academic Press, New York, pp. 186–190.
Lu, A.Y.H. and Miwa, G.T. (1980) Molecular properties and biological functions of microsomal epoxide hydrase. Ann. Rev. Pharmacol. Toxicol. 20, 513–531.
Lunt, D. and Evans, W.C. (1970) The microbial metabolism of biphenyl. Biochem. J. 118, 54P.
McCann, J., Choi, E., Yamasaki, E. and Ames, B.N. (1975) Detection of carcinogens as mutagens in the *Salmonella* microsome test: assay of 300 chemicals. Proc. Natl. Acad. Sci. USA 72, 5135–5139.
Melancon, M.J. Jr. and Lech, J.T. (1978) Distribution and elimination of naphthalene and 2-methylnaphthalene in rainbow trout during short- and long-term exposures. Arch. Environ. Contam. Toxicol. 7, 207–220.
Meyer, T. and Scheline, R.R. (1976) The metabolism of biphenyl. II. Phenolic metabolites in the rat. Acta Pharmacol. Toxicol. 39, 419–441.
Miller, E.C. and Miller, J.A. (1974) Biochemical mechanisms of chemical carcinogenesis. In H. Bush (ed.), Molecular Biology of Cancer. Academic Press, Inc., New York, pp. 377–402.
Narro, M.L. (1985) Oxidation of aromatic hydrocarbons by marine cyanobacteria. Thesis, The University of Texas at Austin, Austin, TX 78712.
Nordqvist, M., Thakker, D.R., Vyas, K.P., Yagi, H., Levin, W., Ryan, D.E., Thomas, P.E., Conney, A.H. and Jerina, D.M. (1981) Metabolism of chrysene and phenanthrene to bay-region diol epoxides by rat liver enzymes. Mol. Pharmacol. 19, 168–178.
O'Brien, P.Y. and Dixon, P.S. (1976) The effects of oil and oil components on algae: a review. Br. Phycol. J. 11, 115–142.
Oil in the Sea. Inputs, Fates, and Effects. (1985) National Research Council, National Academy Press, Washington, D.C.
Provasoli, L., McLaughlin, J.J. and Droop, M.R. (1957) The development of artificial media for marine algae. Arch. Microbiol. 25, 392–428.
Sato, R. and Omura, T. (1978) Introduction: short history of cytochrome *P*-450. In R. Sato and T. Omura (eds.), Cytochrome *P*-450. Academic Press, New York, pp. 16–18.
Schopf, J.W. (1975) The age of microscopic life. Endeavour 122, 51–58.
Smith, R. and Rosazza, J. (1974) Microbial models of mammalian metabolism. Aromatic hydroxylation. Arch. Biochem. Biophys. 161, 551–558.
Soto, C., Hellebust, J.A., Hutchinson, T.C. and Sawa, T. (1975) Effect of naphthalene and aqueous crude oil extracts on the green flagellate *Chlamydomonas angulosa*. I. Growth. Can. J. Bot. 53, 109–117.
Sparling, J., Chittim, B., Clegg, B.S., Safe, S. and Crocker, J.F.S. (1978) The tissue distribution and clearance of Aerotex 3470, an aromatic hydrocarbon solvent. Chemosphere 7, 607–614.
Thakker, D.R., Yagi, H., Levin, W., Lu, A.Y.H., Conney, A.H. and Jerina, D.M. (1977) Stereospecificity of microsomal and purified epoxide hydrase from rat liver. J. Biol. Chem. 252, 6328–6334.
Van Baalen, C. (1962) Studies on marine blue-green algae. Bot. Mar. 4, 129–139.
Vandermeulen, J.H. and Ahern, T.P. (1976) Effect of petroleum hydrocarbons on algal physiology: review and progress report. In A.P.M. Lockwood (ed.), Effects of Pollutants on Aquatic Organisms. Cambridge University Press, London, pp. 107–125.
Wiebkin, P., Fry, J.R., Jones, C.A., Lowing, R. and Bridges, J.W. (1976) The metabolism of biphenyl by isolated viable rat hepatocytes. Xenobiotica 6, 725–743.
Williams, P.A., Catterall, F.A. and Murray, K. (1975) Metabolism of naphthalene, 2-methylnaphthalene, salicylate and benzoate by *Pseudomonas* PG: regulation of tangential pathways. J. Bacteriol. 124, 679–685.
Winters, K., O'Donnell, R., Batterton, J.C. and Van Baalen, C. (1976) Water-soluble components of four fuel oils: chemical characterization and effects on growth of microalgae. Mar. Biol. 36, 269–276.
Winters, K., Batterton, J.C., O'Donnell, R. and Van Baalen, C. (1977) Fuel oils: Chemical characterization and toxicity to microalgae. In C.S. Giam (ed.), Pollutant Effects on Marine Organisms. Lexington Books, D.C. Heath, Lexington, MA, pp. 167–189.
Wirtz, R.A., Turrentine, J.D. Jr. and Fox, R.C. (1981) Area repellents for mosquitoes (Diptera: Culicidae): identification of the active ingredients in a petroleum oil fraction. J. Med. Entomol. 18, 126–128.

The cyanobacteria–a comprehensive review P. Fay and C. an Baalen, editors

21

The biology of Rivulariaceae

B.A. WHITTON

Department of Botany, University of Durham, Durham DH1 3LE, UK

1. Introduction

The Rivulariaceae are among the morphologically more complex cyanobacteria. Desikachary (1959) gives a succinct definition, which includes most of the features listed in other taxonomic accounts: "Trichomes with a single row of cells, apices generally attenuated or tapering in a hair, unbranched or false-branched, sometimes with a distinct intercalary meristematic zone and trichothallic growth; hair with elongated more or less vacuolated cells; heterocysts present or absent, when present basal, intercalary heterocysts also present in some; hormogonia present; akinetes present or absent, when present single or in series".

Geitler (1932), who used essentially the same definition, included 12 genera, some of which are widespread and others which have rarely been recorded (Table I). No one feature is common to all, nor is there any obvious feature restricted to the family. The Rivulariaceae of classical taxonomists presumably do not represent a single genetic grouping, although certain features are common within the family. The present review focuses largely on *Calothrix*, *Rivularia* and *Gloeotrichia*, which have received the most study.

2. Morphology and ultrastructure

2.1. Range of form

It may be helpful to start with an overview of the types of filament and the changes they undergo, though generalizations should be treated with caution until studies have been made on a much wider range of forms. All species can presumably form hormogonia, though these are not always illustrated in the original accounts. Such hormogonia may be much narrower and with a different cell shape. The hormogonia develop typically into tapered trichomes, which are usually aggregated in some way and in *Rivularia* and *Gloeotrichia* form well-ordered spherical or hemispherical colonies. A hair may eventually develop at the end of a tapered trichome (Fig. 1) and in *Rivularia* and *Gloeotrichia* in nature is usually present at the end of the great majority of trichomes within a col-

TABLE I

Genera of Rivulariaceae and their principal features; status of genera in brackets doubtful

	Heterocyst	Mature trichome tapered at both ends	Cellular hair	Akinete	Filaments in a common gelatinous mucilage	Hemispherical or spherical colony	Notes
Amphithrix	–	–	? +	–	–	–	'Upright' trichomes originate from basal disc of cells
Calothrix	+ + +	–	+ +	+	+	+	One of the largest genera of cyanobacteria with great range of forms
Dichothrix	+ + +	–	+ + +	–	–	+	Trichomes false branched, with many trichomes in common sheath at base of branch
Gardnerula[a]	+ + +	–	+ + +	–	–	–	Colony up to 6 mm high, with dichotomizing thallus, each part of which consists of numerous ± parallel trichomes, themselves originating from sub-dichotomous branching

Gloeotrichia	+ + +	–	+ + +	+ + +	+ + +	+ + +	
Hammatoidea	–	+ + +	? –	–	–	–	
Homoeothrix	–	–	+ +	–	–	+ +	
Isactis	+ + +	–	+ + +	–	+ + +		Rather like a crustose expanse of *Rivularia*
(*Leptochaete*)	–	–	+	–	–	–	At base of trichome, separate cells which fall apart
Rivularia	+ + +	–	+ + +		+ + +	+ +	Species (especially in freshwaters) ill-distinguished
Sacconema	+ + +	–	+ + +	+	+ + +	–	Soft colony
(*Tapinothrix*)	–	–	–	–	–	–	Trichome ends in a mucilaginous non-cellular 'hair'

[a]*Gardnerula* is a synonym of *Polythrix* (see Friedmann, 1956).
+ + +, feature of genus: requires observation for identification.
+ +, widespread feature in genus, but sometimes absent; absence in at least some cases probably genetic.
+, occasional feature in genus.
–, feature absent in genus.

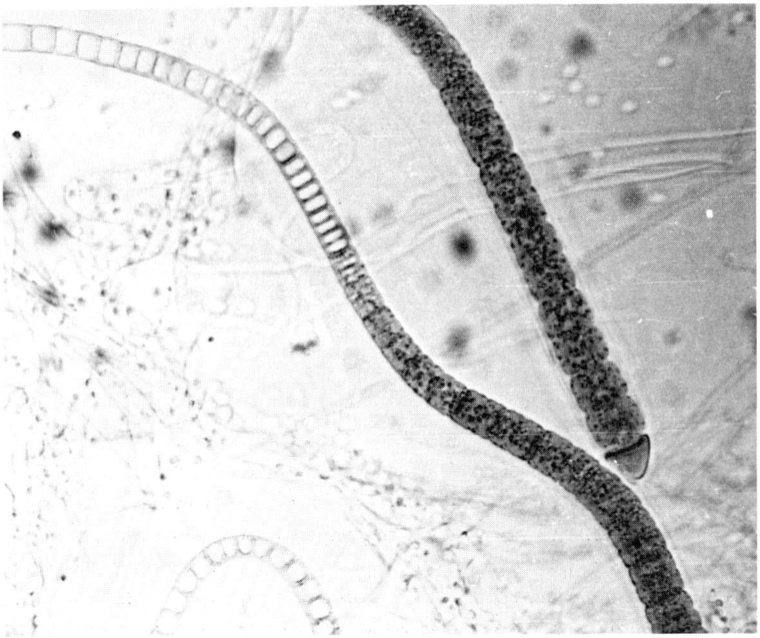

Fig. 1. *Calothrix parietina* D184, showing basal heterocyst, numerous cyanophycin granules and part of hair. (Widest part of trichome is 7 μm.)

ony. Population growth involving the formation of new trichomes appears in some species to take place through the release of hormogonia from the apical part of a mature trichome, but in the forms with spherical or hemispherical colonies, some form of false branching leads to the production of a number of generations of tapered trichomes before hormogonia are again released.

The presence of a hair at the end of a trichome is the nearest to a distinctive feature within the family. However, hairs are also present in some or all species of *Mastigocoleus*, *Nostochopsis*, *Brachytrichia* and *Kyrtuthrix* and, at least in *Brachytrichia*, appear similar to well-developed hairs of *Calothrix* and other Rivulariaceae. In contrast, the genus *Microchaete* (syn. *Fremyella*) is often considered together with *Calothrix* and *Gloeotrichia*, since there is a basal heterocyst and apical attenuation of the trichome, although no hair.

2.2. Hormogonia

Hormogonia are in some cases similar to short lengths from the narrower part of a normal trichome, but in others (e.g. many *Calothrix*) they are quite different and, in field populations, unlikely to be recognised as belonging to the same species without critical study. Hormogonia are usually relatively transient structures. After their release from the parent trichome, few, if any, cell divisions occur before further differentiation of the trichome. However, Maxwell (1974) noted elongation of hormogonia and their subsequent fragmentation by separation discs in a natural population of *Gloeotrichia echinulata*. In *Calothrix* and *Gloeotrichia* gas-vacuoles are often present in young hormogonia, although absent in other developmental stages of the same species.

Hormogonia of a population of *Calothrix parietina* D550 isolated from a stream (Livingstone and Whitton, 1983) are highly gas-vacuolate (see Chapter 14) and possess polyphosphate granules, but not cyanophycin granules (see Chapter 8). Gas-vacuoles are obvious some time before a hormogonium is released from the apex of the parent trichome, but their loss again is usually the first visually obvious change in the free hormogonium. Hormogonia of this strain can glide (as presumably all hormogonia can), but the rates are relatively slow. The maximum net rate of movement on agar was 0.03 $\mu m \cdot s^{-1}$. Hormogonia in cultures of this strain aggregate to form long ropes, with the youngest hormogonia towards the ends of the ropes (author's data), much as do those of *Gloeotrichia* (see Section 2.8 of this chapter), but unlike *Gloeotrichia* the hormogonia in these ropes do not reorganise themselves subsequently to produce a subspherical arrangement.

2.3. Trichome growth

Growth is typically meristematic, with cell division restricted to a particular part of the trichome. In forms with heterocysts and hairs division occurs somewhere between these two structures, often just below the hair (trichothallic growth – Fritsch, 1945; Bornet and Thuret, 1880; Schwenender, 1894; Weber, 1933; Darley, 1968); it is in a comparable position in species of *Homoeothrix* with hairs (Komarek and Kann, 1973). The cells of the meristematic region in some forms are markedly wider than those at the base of the trichome, giving the appearance of a spindle-shaped swelling below the hair (Schwenender, 1894; Polyanskii, 1930; Friedmann, 1956). Even in the absence of hairs the cells of trichomes of heterocystous Rivulariaceae may be shorter than those below, and thus presumably meristematic, e.g. in *Calothrix prolifera* and *C. pulvinata* (Darley, 1968).

Cell division may also be localised much nearer the base of the trichome (*Rivularia angulosa* – De Bary, 1863; *Gloeotrichia pisum* – Palla, 1893; *Calothrix*, *Gloeotrichia* – Fay et al., 1968), although there are usually a few non-dividing cells adjacent to the heterocyst except during akinete formation. Growth of *G. natans* is trichothallic in trichomes without akinetes, but, as the akinete starts to develop, the meristem shifts to the basal region (Polyanskiĭ, 1930; Serbanescu, 1966). Several examples have been reported of a secondary development of meristematic activity in an unusual part of the trichome; in *G. raciborskii* this apparently occurred as a response to attack by a parasite (Desikachary, 1945), and illustrations by Serbanescu (1966) suggest that a similar phenomenon can occur in *G. natans*, although she interpreted the results differently.

Morphological polarity largely or entirely disappears when heterocystous Rivulariaceae are transferred to a medium with combined nitrogen. Thirty three out of thirty four strains studied by Sinclair and Whitton (1977b) no longer formed heterocysts under such conditions. The trichomes of 19 of these almost all lost their taper; in the remaining non-heterocystous forms there were both untapered trichomes and tapered ones resembling *Homoeothrix* or *Hammatoidea*. Hair frequency and length were both reduced considerably in three strains which formed many long hairs in medium lacking combined nitrogen. Mutants of *Gloeotrichia ghosei* defective in nitrogen-fixing ability showed three morphological phenotypes when grown with combined nitrogen (Tiwari et al., 1979): short trichomes tapering at both ends, unusally long trichomes tapering at one end, and trichomes with complete loss of polarity.

2.4. Vegetative cell

The 16 strains (10 *Calothrix*, 5 *Gloeotrichia*, 1 *Microchaete*) studied by Fliesser and Jensen (1982) showed the typical features of a cyanobacterial cell. In addition, a number of structures were found which have been described for other cyanobacteria only occasionally, if at all. Five strains had membrane-bound crystalline inclusions and in two of them these were seen in every section. Plasma-membrane-associated microtubules, 10–14 nm in diameter, were seen in eight isolates of *Calothrix* and in one of *Microchaete*; they occur only in actively growing cultures. The tubules all appear to be perpendicular to the plasma membrane, projecting towards the centre of the cell. Another microtubule-like structure 18–22 nm in diameter was observed in representatives of all three genera; this is quite similar to structures previously reported for several species of *Anabaena* (Jensen and Ayala, 1976). Jensen (1984) gives a list of all inclusions found irregularly during previous studies of Rivulariaceae by his research group.

2.5. Hair

According to Bornet and Flahault (1886a,b), the hair is a series of narrow, elongated cells containing very little protoplasm and incapable of further growth. As this definition mixes features easy to recognise with simple light microscopy and ones which are not, Sinclair and Whitton (1977a) redefined the hair as a region of the trichome where the cells are narrow, elongated, highly vacuolated and usually apparently colourless. The hair cells of Rivulariaceae are distinct from the vacuolated cells occurring in other forms as a pathological symptom (Geitler, 1932).

Cytological and ultrastructural studies of the changes taking place along the trichome from typical vegetative to typical hair cell indicate a progressive loss of many constituents, although the sequence of changes may not always be quite the same. A study (Ueda, 1971) along trichomes of *Calothrix braunii* using fluorescence microscopy showed that 'transition' cells immediately below the hair and the hair cells themselves had from two to five times less DNA than the normal vegetative cell. An ultrastructural study of a laboratory culture of *Gloeotrichia* sp. (Miller and Lang, 1971) showed that cells near the tips of attenuated trichomes had only a very small proportion of the original thylakoid system with the two membrane components appressed. Much of the cell was intrathylakoidal space, with the remaining part of the thylakoid system existing as single membranes bounding the remnants of the interthylakoidal cytoplasm. With increasing age of vegetative cell, polyglucoside granules decreased, but at least over some distance cyanophycin granules increased. However, the only granular inclusions left in the oldest cells were polyhedral bodies or carboxysomes and lipid bodies; it is unclear whether these were fully formed hair cells. In *Rivularia mesenterica* (Feldmann and Guglielmi, 1973), cyanophycin granules were lost before polyglucoside granules and lipid bodies. Hair cells in other marine Rivulariaceae have been illustrated by Guglielmi (1975): they all show the development of intrathylakoidal spaces and progressive loss of granular inclusions.

All cytoplasmic inclusions other than membranes were absent in hair cells of a range of *Calothrix* and *Rivularia* populations described by Douglas (1979) and Wood (1984). However, gas-vesicles tend to persist in hairs of *Gloeotrichia echinulata* (Smith and Peat, 1967). Cmiech et al. (1984) reported that in a plankton population of *G. echinulata* the plasmalemma of the hair cells eventually disrupted. This does not appear to be the case in other Rivulariaceae reported; perhaps this phenomenon is restricted to trichomes with developing akinetes.

Of 36 strains of Rivulariaceae in culture (Sinclair and Whitton, 1977a), only 13 were found to produce hairs. In all cases abundant hair increase occurred under conditions of phosphate deficiency. Fe deficiency led to hair formation in eight strains and Mg deficiency in one strain, but other deficiencies had no such effect. Many of the strains lacking the ability to form hairs were described as *Calothrix* in the collections from which they originated. The question arises as to what extent the lack of hairs in *Calothrix* populations in nature is merely an environmental effect. Of 78 species with quite detailed descriptions in the taxonomic literature, hairs are described as present in 52, absent in 16 and data are inadequate for comment in 10 (Kirkby and Whitton, 1976). A population of *Calothrix* in a small freshwater pool on Aldabra Atoll has been studied in some detail (Donaldson and Whitton, 1977, unpublished data). Samples taken at different seasons and in different years have never shown the presence of a hair, neither was there any evidence for hair formation during studies of a laboratory isolate (strain D202).

2.6. Heterocyst

The presence of a heterocyst is a diagnostic feature at the generic level (Table I), though a very high concentration of combined nitrogen in the medium almost always suppresses its formation (Sinclair and Whitton, 1977b; Pattnaik and Singh, 1978). In contrast, the omission of either combined or gaseous nitrogen led to a very high (20 %) heterocyst frequency in *Calothrix parietina* D184. The fact that in some *Calothrix* strains the form developing under combined nitrogen looks much like *Hammatoidea* (see 2.3.), raises the possibility that records for the latter are merely stages of *Calothrix* lacking a heterocyst. However, at least one site is known (in the Swiss Alps) where repeated sampling has shown the presence of characteristic *Hammatoidea* (A. Zehnder, personal communication).

Where heterocysts are present, they are mostly 'terminal' i.e. with a polar nodule on only one side. The terminal heterocyst developing at the end of a hormogonium in several strains of *Calothrix parietina* differentiates from the cell which was nearest the mother trichome before the hormogonium was released (author's data). The polarity of the trichome is thus established at a stage before it is evident with the light microscope. In hormogonia of *C. parietina* D184 (Livingstone and Whitton, 1983) the characteristic shape of the heterocyst develops at the same time as other heterocyst features. The suggestion (Rippka and Herdman, 1985) that only the broad basal cell of a *Calothrix* forms a heterocyst gives a false impression. Any change in cell shape takes place during heterocyst differentiation.

Intercalary heterocysts occur occasionally in some *Calothrix* species (Geitler, 1932; Jeeji-Bai, 1977). In *C. brevissima* (Rai et al., 1978) bipolar heterocysts developed in trichomes which had been grown in the presence of ammonia and then deprived of it. Subsequently one of the polar nodules flattened and the adjoining cell died, forming a separation disc, and hence causing fragmentation of the filament. It is not at present clear that this is a general phenomenon in *Calothrix*, as suggested by Fay (1983). The terminal heterocyst in many species occurs only at one end of a filament, but in others it also occurs elsewhere, usually as a result of the development of a false branch. In comparison with natural populations of *Anabaena* and some other heterocytous genera, heterocysts tend to form a smaller proportion of the total cells in Rivulariaceae. Floristic accounts (e.g. Bornet and Flahault, 1886a,b) also suggest that heterocysts tend to be relatively small in relation to the size of vegetative cell, although there are many exceptions.

A feature which is perhaps quite widespread among Rivulariaceae is the presence of some heterocysts which are green or even blue-green (cf. Section 7 of Chapter 7). This is common among *Calothrix* and *Gloeotrichia* from rice-fields, both in nature and in the laboratory (author's data). Heterocyst ultrastructure has been studied in most detail in a lake population of *G. echinulata* (Cmiech et al., 1984). The subpolar region of the cell showed great diversity, ranging from unusual membrane configurations to no elaboration at all. The complex honeycomb lattices, which have been recorded in various *Anabaena* strains (e.g. Kulasooriya et al., 1972) and which have been attributed to increased reducing activity (Stewart et al., 1969), were not found. Cmiech et al. (1984) suggested that the massive sheath in which the heterocysts of their *Gloeotrichia* were embedded might give some protection against oxygen inactivation of nitrogenase (cf. Section 6 of Chapter 7). Later in development, closure of the intrathylakoidal lumen caused the protoplast to appear relatively homogeneous, a feature which also occurs in many other *Gloeotrichia* populations (author's data). During the early stages of akinete differentiation the polar connection became severed by the thickened sheath. Gas vesicles are retained throughout the life of the heterocyst in *G. echinulata* (Forest and Khan, 1972; Cmiech et al., 1984).

Other indications of differences occur among heterocysts of the Rivulariaceae. In a Bangladesh *Calothrix* isolate (D603) carboxysomes often occur in apparently mature heterocysts (S. L. J. Grainger, N. Harris and B. A. Whitton, unpublished data; cf. Section 4.1. of Chapter 8). There are a number of records of heterocyst germination, though this has also been recorded for *Anabaena*, *Brachytrichia*, *Nostoc* and *Tolypothrix* (see Desikachary, 1946). In populations of *Calothrix weberi* (Steinecke, 1931), *Gloeotrichia raciborskii* and *Rivularia mangini* (Desikachary, 1946), the heterocyst divided to produce a two-celled germling, but *Calothrix weberi* differed from the others in that the lower cell degenerated. Germination has been described in most detail in a fascinating study of *Gloeotrichia ghosei* (Singh and Tiwari, 1970), though one which should be repeated because the isolates were bacterized. This organism still forms heterocysts when grown in the presence of combined nitrogen. The heterocysts appear faintly coloured (using light microscopy) and homogeneous. A non-sporulating (presumed mutant) clone showed germination of up to 83% heterocysts in the presence of ammonia, but much less with nitrate and hardly any in medium free of combined nitrogen. Most of the heterocysts germinated while still attached to the healthy growing filament. The first sign of germination was an increase in size and appearance of a central vacuole. Subsequently a division wall formed; the lower of the two cells then developed into a new heterocyst, while the upper one gave rise to the rest of the trichome; the vacuole had disappeared by this stage. The whole process could be repeated a number of times.

When considered together, these studies indicate that the heterocysts of some Rivulariaceae differ considerably from the heterocyst model developed from laboratory studies of a relatively small number of *Anabaena* strains. In at least some cases differentiation of the heterocyst of Rivulariaceae appears to involve fewer changes.

2.7. Akinete

The ability to form akinetes (see Chapter 9) is a feature of *Gloeotrichia*, several species of *Calothrix* (Geitler, 1932) and *Dichothrix gelatinosa* (Böcher, 1946). The akinete develops next to the basal heterocyst in most cases, but intercalary ones have been reported for *D. gelatinosa*, *Calothrix weberi* (Steinecke, 1931) and *Gloeotrichia andreanskyana* (Claus, 1957). The akinete may develop from a single cell (Desikachary, 1959) or a number of cells with breakdown of cross-walls (Geitler, 1925; Claassen, 1973). The factors leading to akinete formation have received little study among the Rivulariaceae (Whitton, in press). Roelofs and Oglesby (1970) reported that in *G. echinulata* akinetes seemed to be produced when the colony reached a particular size, even when conditions in nature permitted logarithmic population growth (but see Section 6.2.2. of Chapter 13). In *Calothrix stagnalis* (Gomont, 1895) akinete formation was restricted to trichomes with hairs, as opposed to trichomes lacking hairs, but forming hormogonia at their apex. The germination of akinetes in *Gloeotrichia* has been described by De Bary (1863), Bornet and Thuret (1880), Palla (1893) and Polyanskiĭ (1930), all of whom noted the release of a long gas-vacuolate hormogonium from the akinete case (see also Section 2.5. of Chapter 9).

2.8. Colony

Rivularia, *Isactis*, *Gloeotrichia* and some species of *Homoeothrix* and *Dichothrix* have a colonial form. Many species of *Calothrix* also form a macroscopic thallus with a characteristic appearance e.g. *C. pulvinata*. The hormogonia of *C. pulvinata* tend to attach themselves to parent filaments, giving a tufted, bushy appearance. De Bary's (1863) account for *Rivularia* is still the most detailed available. Here the first sign of colony formation was the aggregation of motile hormogonia into small non-motile groups. The initial stages of colony formation have also been shown to involve the aggregation of motile hormogonia in populations of *Gloeotrichia echinulata* (Maxwell, 1974) and *G. pisum* D613 (A. Aziz and B.A. Whitton, unpublished data). Darley (1968), in a study of the behaviour of marine Rivulariaceae on transfer to the laboratory, found that hormogonia would develop only if a large quantity was inoculated simultaneously. However, clonal cultures of colonial Rivulariaceae are held in culture collections and these have presumably originated from single hormogonia. Although not all can still form colonies (e.g. Chang, 1983), there are examples where they do. It is unclear whether growth of the hormogonia used in establishing cultures has led to the production of further hormogonia straight away (see Section 2.2.), or whether additional hormogonia have formed only as the result of the usual sequence of morphological changes involving differentiation into a tapered trichome and eventual production of new hormogonia at the apical end of the trichome. Careful field studies are needed to establish whether the aggregation of many hormogonia is a normal requirement for colony formation. Three different morphological types of Rivulariaceae have been isolated from a single colony of a pond *Rivularia* (Sinclair and Whitton, 1977a), showing that even within one colony there may be genetic diversity. The extent to which hormogonia from different populations can aggregate together also needs investigation.

In heterocystous forms the basal heterocyst is directed towards the centre of the colony and the tapered part towards the outside. Where hairs are present they may (e.g. *Gloeotrichia echinulata*: Cmiech et al., 1984) or may not (e.g. calcified freshwater *Rivularia mesenterica*: Frémy, 1933) protrude from the mucilaginous surface of the colony. The detailed structure of the colony depends on the events associated with false branching.

3. Biochemistry

There may be chemical features which distinguish at least some Rivulariaceae from other cyanobacteria. The four strains of cyanobacteria referred to the '*Calothrix*-type' by Kenyon et al. (1972) all possess a tetra-unsaturated fatty acid (octadecatetraenoic acid), a feature found in only one strain belonging to other groups. Brominated phenolic substances have been reported from two different Rivulariaceae, an axenic culture of a freshwater *Calothrix* and a field population of a marine *Rivularia*. *Calothrix brevissima* was found to produce two simple brominated phenols, 2,3-dibromo-4,5-dihydroxybenzyl alcohol (lanosol) and 3,5-dibromo-*p*-hydroxybenzyl alcohol (Pedersen and DaSilva, 1973); the authors discussed the possible biological effects of these substances on other organisms. Six novel polybrominated biindole derivatives reported from *Rivularia firma* by Norton and Wells (1982) were termed rivularins by Maehr and Smallheer (1984), who made preliminary studies on their synthesis.

4. Physiology

While no detailed comparisons have been made on growth rates of Rivulariaceae in the laboratory or the field versus those of other cyanobacteria, the studies which have been reported suggest that most Rivulariaceae do not grow as fast as the faster strains of non-tapered filamentous cyanobacteria like *Oscillatoria* and *Anabaena*. (Compare, for instance, the data given by Sinclair and Whitton, 1977a, with those in Table 17.1 of Whitton, 1973.) However, relatively fast growth can occur, for *Gloeotrichia pisum* D613 shows a specific growth rate (k') of 0.573 (doubling time of 12.6 h) at 200 μmol photon $m^{-2} \cdot s^{-1}$ photosynthetically active radiation (PAR) and 32° (Aziz, 1985). A comparison (Carpenter and Cox, 1974) of carbon turnover rate of *Dichothrix* epiphytic on *Sargassum* in the Sargasso Sea showed that the average for the former was four times that for the latter.

The majority of laboratory physiological studies have been made on *Calothrix* and *Gloeotrichia*. All of those reported can apparently show continued growth with dinitrogen as the only nitrogen source, at least under optimum conditions, and grow in the absence of vitamins. The stimulatory effect of fulvic acid on growth of *G. echinulata* (and other cyanobacteria) was ascribed (Lange, 1970a,b) to its chelating activity. Zehnder (1963) reported that the optimum pH for (another) *G. echinulata* strain was from 8 to 9. Examples have been reported for *Calothrix* of chromatic adaptation (Ahluwalia et al., 1980) and of the ability to grow heterotrophically in the dark (Khoja and Whitton, 1971), though a strain of *Gloeotrichia echinulata* studied by Chang (1979b) could not grow heterotrophically. Tiwari et al. (1985) described studies on the action of methionine sulphoximine, an inhibitor of glutamine synthesis, on *G. ghosei*. Its inhibitory activity on uptake of nitrate or ammonium is reversible, possibly because a mechanism is developed by which the inhibitor is transported out of the cells.

Fogg (1969) reported that even (quite) low concentrations of inorganic phosphate in artificial media appear to be inhibitory for some genera, including *Gloeotrichia*. A number of subsequent studies have indicated that the phosphorus status of the environment may be a particularly important factor in the growth of Rivulariaceae. The sequence of changes taking place in the morphology of *C. parietina* D550 during batch culture in medium lacking combined nitrogen, and under conditions leading eventually to phosphorus deficiency, has been followed by Livingstone and Whitton (1983). Four different stages were recognised for descriptive purposes. Stage I commences with the release of a hormogonium and ends with the development of a heterocyst. Stage II commences with formation of a mature heterocyst; this is followed rapidly by tapering of the trichome; under phosphorus-rich conditions the apical part of this tapered trichome can continue to give rise to a hormogonium, but with increasing phosphorus deficiency, the terminal cells start to differentiate into a hair (Stage III). Hairs are formed when the average value for cellular P falls to about 1 % dry weight, a concentration found in filaments which can still be regarded as relatively healthy (Livingstone et al., 1983). When further phosphate is added to a population with hairs, a characteristic series of events takes place (Stage IV): several hormogonia start to differentiate at the apical end of the chlorophyll-containing part of the trichome; the hair falls off; the hormogonia are released by gliding from the sheaths.

Light is not essential for the formation of polyphosphate granules when phosphate is added to *C. parietina* D184 with hairs, even when the alga has been kept in the dark for 4 days, though the rate of formation is slightly slower than for material maintained in the light the whole time (Sinclair, 1977; see also Section 3.2.1. of Chapter 8). Hormogonia do not form in material kept in the dark subsequent to phosphate addition, even when the organism has previously been in the light.

Inducible phosphatase activity commences at about the same time as hair formation when *C. parietina* D550 is grown in batch culture under conditions with decreasing phosphate (Livingstone et al., 1983). Phosphomonoesterase activity occurs on walls of the hair cells, the chlorophyll-containing cells adjacent to the hair cells, and the part of the sheath furthest from the heterocyst. Phosphodiesterase and phytase activity also occur. This ability may be particularly important for this alga in nature, since it was isolated from a stream where most of the soluble phosphate is organic (Livingstone and Whitton, 1984). *Rivularia* colonies taken directly from this and other similar streams showed marked phosphatase activity; in one 24-h study, phosphate uptake by colonies from β-glycerophosphate was 86% of that from inorganic phosphate under otherwise similar conditions. A marine *Rivularia* (*atra*) population was found (Khoja et al., 1984) to be associated with a bay where dissolved phosphate was relatively high (typically 50–150 $\mu g \cdot l^{-1}$ P), but about half of it organic.

Assays for nitrogen fixation (usually measured as nitrogenase activity) have shown positive results in almost all heterocystous Rivulariaceae tested. Laboratory studies (Stewart, 1967; Chang and Blauw, 1980; Al-Mousawi, 1984; Aziz, 1985) have usually shown much higher rates in the light than the dark. Chang and Blauw (1980) found for *Gloeotrichia echinulata* that rates are sometimes higher under anaerobic as opposed to aerobic conditions. Rates change markedly during growth in batch culture, typically reaching their maximum several days after subculturing to fresh medium. In *G. echinulata* (Chang and Blauw, 1980) membrane-like colonies containing mostly short trichomes showed much higher rates than cluster-like colonies with long trichomes, when incubated at low, but not high, values of light flux.

The majority of studies have been carried out in the field with materials in vessels incubated at the same site. Diel studies of *Calothrix scopulorum* and *Rivularia* sp. showed a marked dependence of nitrogen fixation on light (Hubel and Hubel, 1974; see also Section 5.3.2. of Chapter 13). Rates were usually six to ten times higher at midday than at night; the difference was, however, less marked in July and August, presumably due to a greater daytime accumulation of carbohydrate during these months. The ratio between midday and midnight nitrogen fixation by a stream *Rivularia* (Livingstone et al., 1984) was similar, with 8–16% fixation occurring at night. Although the rates of nitrogen fixation were much higher by day, there was a relatively low correlation between fixation and light flux when individual 1-h periods were considered, in contrast to carbon dioxide fixation and light flux, where there was a significant correlation. Overall, about 400 mol of CO_2 to 1 mol of N_2 was fixed during daylight, so nitrogen fixation may supply only a small percentage of the nitrogen requirements of this cyanobacterium. *Calothrix scopulorum* (Jones and Stewart, 1969a) and *Rivularia atra* (Reed and Stewart, 1983) from the upper part of the intertidal zone both show greater inhibition of nitrogenase activity at high salinity than of photosynthesis; this may reflect the channelling of fixed carbon into an osmoticum.

Fixation appears to take place in field populations over the whole temperature range at which *Calothrix* grows. Obvious activity has been recorded in terrestrial Antarctic (Davey, 1983) and supralittoral marine (Wärmling, 1973) populations, where ground and air temperatures, respectively, were below 0°C. Tropical studies have been reported for coral reefs (Mague and Holm-Hansen, 1975; Webb et al., 1975; Wiebe et al., 1975) and the intertidal region (Potts and Whitton, 1977). There have been a number of studies of seasonal changes in nitrogen fixation by littoral *Calothrix* of temperate regions (Stewart, 1962, 1963; Lannergren, 1980), but so far fixation has not been measured simultaneously per unit area and per unit biomass. Stewart (1965) showed for a Scottish population that fixation was rapid in spring and autumn, but slower in summer, due to desiccation, and in winter, due to low temperatures. In areas of uniform cover, nitrogen fixation could be as high as 2.5 $g \cdot m^{-2} \cdot yr^{-1}$ N. A population of *C. scopulorum* on the Baltic coast showed the highest fixation rates in summer (Savela, 1983). Fixation rates per unit area in the littoral zone near Bergen were highest in July–August, with the highest activities in decaying matter at protected sites in association with *C. confervicola* and *Anabaena torulosa*. Other genera studied from various habitats include *Dichothrix* (Carpenter, 1972), *Rivularia* (Pearson and Taylor, 1978) and *Gloeotrichia* (Kostyaev, 1984).

5. Genetics

A study (Herdman et al., 1979a) of the DNA base composition of a wide range of cyanobacteria showed that all 11 *Calothrix* strains fell within a narrow span (39.8 to 44.4 mol% GC). The base span for Nostocaceae was quite similar, in spite of the considerable morphological differences. The genome sizes of the ten of these strains studied by Herdman et al. (1979b) fell into two groups: seven in the range 5.07×10^9 to 5.46×10^9 daltons and three in the range 7.75×10^9 to 8.58×10^9 daltons. The subdivision of the genus in this way is not consistent with any other known phenotypic properties (Rippka et al., 1979). Hybridisation of three different *nif* genes from *Anabaena* sp. to DNA of *Gloeothece* sp. and *Calothrix* sp. PCC 7601 showed that the *Calothrix* was more like the *Anabaena* than was the *Gloeothece* (Kallas et al., 1983). Although mutants have been used in various studies of Rivulariaceae (e.g. Tiwari et al., 1979), possible gene exchange has apparently not been investigated. Lange (1975) speculated as to whether the double helix formed by pairs of trichomes in cultures of *Gloeotrichia echinulata* U. Wisc. 1052 might be a means of aiding genetic exchange. The aggregation of trichomes during the early stages of colony formation (see Section 2.8.) also provides a situation where trichomes are in close contact, and one where the trichomes are probably often derived from different parental stocks.

6. Ecology

6.1. Occurrence

Rivulariaceae grow in such a wide range of environments that it is at first difficult to see any pattern in their occurrence. Heterocystous forms occur, for instance, in lake plankton, ponds, fast-flowing streams, hot springs, (Mann and Schlichting, 1967), seepages, soils, salt marshes, the supralittoral and intertidal regions of rocky shores (Womersley, 1946; Umezaki, 1961) and, especially in the tropics, in the shallower parts of the sublittoral (Tsuda and Kami, 1973). In moist calcareous freshwater environments there is a complex of communities in which heterocystous Rivulariaceae are particularly well-represented, termed the *Hymenostylium curvirostre-Rivularia haematites-Scytonema myochrous* microassociation in the phytosociological literature (Braun, 1968); there are many records for such communities from Europe and the USA and scattered ones from elsewhere (e.g. Argentina – Prosperi, 1983; Saudi Arabia – Whitton et al., 1986). Among reports for the marine sublittoral, examples of *Calothrix* growing on the leaves of marine angiosperms are frequent (e.g. Goering and Parker, 1972; Capone et al., 1979; Goldner, 1980).

Apart from *Gloeotrichia echinulata* in lake plankton, floating colonies of *G. natans* (Fig. 2) and the transient gas-vacuolate hormogonia of many forms, Rivulariaceae are almost always associated with surfaces, where their sheaths presumably play an important role in attachment. Another feature which appears to be widespread is their association with environments which are subject to marked changes, such as rock surfaces at pond edges or the marine intertidal. Although physical changes are usually most obvious (Fig. 3), a shifting chemical environment may perhaps be even more important. In an area of upland England, *Rivularia* is restricted to parts of streams influenced at various times by differing proportions of peaty and calcareous waters (Livingstone and Whitton, 1984).

Calothrix and *Dichothrix* occur in both very soft (Heuff and Horkan, 1983) and highly calcareous (Geitler, 1932) waters. The majority of records for freshwater *Rivularia* are from highly calcareous waters, but in Jämtland, Sweden, and probably elsewhere in Scandinavia (Johansson, 1979; C. Johansson, personal communication), this genus is also widespread in non-calcareous waters. There

Fig. 2. Abundant growth of *Gloeotrichia natans* developing on surface of drainage ditch at International Rice Research Institute, Philippines, 2 weeks after most of submerged macrophytes had been killed with a herbicide.

Fig. 3. *Rivularia* colonies in calcareous stream, W. Yorkshire, England. Due to the wide variations in flow some colonies may remain out of water for a month or more. Survival under such conditions at this particular site is apparently due to the retention of moisture in the inner part of the highly calcified colony.

are a number of records of sites where floristic comparisons or inspection of calcareous deposits show that Rivulariaceae were once much more abundant than at present. Examples include the loss of *Homoeothrix janthina* in a USA stream (Golubic, 1973), *Rivularia* and *Calothrix* at the edges of two Austrian lakes (Kann, 1982), *Rivularia* in calcareous streams in southern France (Casanova and Lafont, 1985) and *Gloeotrichia echinulata* in an English lake (Griffiths, 1925). It seems probable that these changes are associated with some aspect of eutrophication, perhaps increased levels of inorganic phosphate. There are so many records of heterocystous Rivulariaceae in freshwater and slightly brackish environments occurring together with charophytes and/or some hair-forming eukaryotes such as *Chaetophora* and *Pleurocladia*, that it seems likely that they have quite similar chemical requirements for competitive success (e.g. Claassen, 1973; Hudson et al., 1971; Kirkby et al., 1972; Mahajan and Patel, 1984; Pankow et al., 1971).

Surveys of the algal vegetation of streams in relation to water chemistry have shown (Shehata and Whitton, 1981; Whitton et al., 1981) several records of *Calothrix* at zinc concentrations just below 10 $mg \cdot l^{-1}$ Zn, a concentration well above that for other heterocystous genera. This may be another example of an environment where an ability to use organic phosphate effectively is important, since inorganic phosphate is relatively insoluble under such conditions, where the concentrations of not only zinc, but also lead, are high.

6.2. Influence of animals

Calothrix was one of three ricefield cyanobacterial genera grazed by the ostracod *Cypris* sp. in flooded soil (Osa-Afiana and Alexander, 1981), whereas *Anabaena* was little grazed. The developmental stage of the cyanobacterium may, however, have an influence on grazing ability, since there was much less effect when established populations of *Calothrix* (or *Tolypothrix tenuis*) were tested. Both *Calothrix* and *Anabaena* were grazed by mosquito larvae during a study carried out in Sri Lanka (Theivendirarajah and Jeyaseelan, 1977). In contrast to these observations, nodules of mixed *Calothrix* and *Pleurocapsa* survive in part of an Oregon hot spring with a dense population of *Potamocypris* sp., which exerts a particularly heavy grazing pressure on *Oscillatoria terebriformis*, the species which dominates in the part of the stream immediately upstream of the thermal range of the ostracod (Castenholz, 1973). The hairs of the *Calothrix*, which protrude from the nodules, are however grazed by the ostracod (R.W. Castenholz, personal communication), as is the whole organism when supplied on agar discs without the associated *Pleurocapsa* (Wickstrom and Castenholz, 1975). The grazing pressure probably accounted not only for coexistence of the nodule species, but also for their spatial relationship in the nodules, with *Calothrix* predominating towards the outside (Wickstrom and Castenholz, 1978; Wickstrom, 1980).

Cattaneo (1983) suggested that differences between *Gloeotrichia pisum* populations of a Canadian lake in exclosures and in the control environment might be due to grazing by snails, which were abundant in the former. Alkaline phosphatase activity increased markedly in the controls and decreased to almost as great an extent in the exclosures, so another explanation for the differing success of *Gloeotrichia* in the two environments might be a differences in the chemical environment. Both *G. pisum* and *G. natans* in deepwater rice-fields of Bangladesh are subject to obvious grazing by snails (author's data). The rotifer *Lindia euchromatica* appeared to have a particularly close dependence on *Gloeotrichia* in the littoral zone of Linsley Pond, Connecticut (Edmondson, 1938). It lives among the mucilage of the colonies, feeding on the filaments and depositing its eggs there. Clearly there is a great deal to be learned about possible selective effects of grazing on cyanobacteria.

6.3. Symbiotic associations

Among lichens, the phototroph of *Lichina*, *Calotrichopsis* and *Porocyphus* is *Calothrix* and of *Placynthium* is *Dichothrix* (Ahmadjian, 1967), though its characteristic form can often be seen only after isolation and growth in culture. Both Geitler (1934) and De Puymaly (1949) described the resemblance of the organism in the lichen to a species of *Nostoc*: heterocysts were present, but there was no basal-apical differentiation of the trichomes and growth was not localised in a meristem. When, however, cultured in a free-living state, growth was normal, with basal heterocyst, tapered trichome, meristem and hair. The phototroph of *Lichina rosulans* (Henssen, 1969) consisted of short, contorted chains and basal heterocysts were rare. Basal heterocysts were found (Henssen, 1973) in *L. polycarpa*, but not tapering; a normal tapered morphology was, however, retained in *L. tasmanica* and *L. willeyi*. Welsh (1965) observed *Calothrix* trichomes, often with a hair, in an unidentified lichen, but also many otherwise similar trichomes lacking heterocysts. Fungal haustoria in *Lichina pygmaea* penetrate the *Calothrix* sheath, but not the cell wall (Boissière, 1977).

The cyanobacterium *Richelia*, which occurs intracellularly in several marine centric diatoms as well as sometimes free-living, has a basal heterocyst, but lacks tapering (see Section 4 of Chapter 15). It has apparently not been obtained in axenic or even persistent impure culture (Fogg, 1982), so it is uncertain whether the presumed endosymbiont has any close resemblance to *Calothrix*. *C. chapmanii* does, however, live inside the alga *Enteromorpha*, though extracellularly (Lami and Meslin, 1959). The nitrogenase activity of another marine alga, *Codium* (from New Zealand), was due to *Calothrix* epiphytic or embedded in the thallus (Dromgoole et al., 1978). *Calothrix* was one of three cyanobacteria responsible for nitrogenase activity in a reduced microzone within the thallus of a USA species, *Codium decorticatum* (Rosenberg and Paerl, 1981).

6.4. Parasites

There are apparently no reports of phage attacking Rivulariaceae. An account of attack by what is probably a chytrid has been mentioned in Section 2.4. Chang (1983) has reported an example of the cells of *Gloeotrichia echinulata* being invaded and destroyed by another cyanobacterium, *Pseudanabaena catenata*.

7. Geological record

There is a number of older reports of supposed fossil Rivulariaceae (mostly referred to *Palaeorivularia*) from the Precambrian and Cambrian eras (e.g. Macgregor et al., 1974), but Golubic and Campbell (1981) doubt whether any are Rivulariaceae. However, Zhao-liang (1984a,b) has described from the late Precambrian a new genus, *Paleocalothrix*, with basal heterocyst, large akinete, trichome tapered towards the apex and a sheath; these seem to fulfil all the requirements of a member of the Rivulariaceae. Further, Batten and Van Geel (1985) are almost certain that *Celyphus rallus*, a form described in 1973, is an Early Cretaceous representative of the Rivulariaceae.

There are many records of modern freshwater calcareous deposits where Rivulariaceae play an important role. One of the types of calcareous nodule (oncolite) on the floor of the Untersee of Lake Constance is formed by *Phormidium* and *Calothrix* and/or *Dichothrix* (Schäfer and Stapf, 1978). These spongy nodules occur in loose carbonate sands influenced by the Rhine current. Casanova (1982, 1984) has described how various Pliocene and Pleistocene deposits appear to have been formed by a similar flora. A 20-m core of Peyre travertine (Middle Pleistocene) showed incrustations of *Phormidium*, *Dichothrix* and *Rivularia* (Casanova, 1984). Several authors have suggested explanations for the banded arrangement of carbonate deposited in colonies of modern *Rivularia haematites* (see Caudwell, 1983). What are clearly the remains of *Gloeotrichia echinulata* have been found in postglacial lake sediments. A study (Van Geel et al., 1984) of a late glacial deposit at Usselo, The Netherlands, showed that the sequence started with a (presumed) oligotrophic pool, with *Gloeotrichia* and Characeae playing an important role on the initially sandy substrate. *G. echinulata* forms a thick deposit in 14 000 B.P. lake sediments at Cedar Mountain Bog, Washington, USA (M. Tsukada and J. Turner, personal communication). Not only do the sheaths look much like those in a live colony, but the heterocyst walls also persist.

In the marine environment modern carbonate-trapping stromatolites can be formed by *Dichothrix bornetiana* (Monty, 1967) and *Gardnerula corymbosa* (Golubic and Awramik, 1974). Carbonate deposition appears to be rare among marine Rivulariaceae, but Golubic and Campbell (1981) describe aragonite concretions formed by *Rivularia mesenterica* and *R. polyotis*, although the examples were all for thalli already growing on carbonate rocks.

8. Concluding comments

Although this review covers many topics, it is far from comprehensive, especially on morphological and ecological aspects. Further physiological and molecular research may eventually make it easier to review the early literature, by providing more understanding of the causal and functional significance of morphological features long recognised by taxonomists. It will be apparent from Section 4 that I believe that this applies to the hair, a feature present in the majority of Rivulariaceae. The following hypotheses are offered because a theoretical framework is needed. The reader may of course remain sceptical: Cmiech et al. (1984), for instance, suggest that the appearance of the hair is more indicative of degeneration than of differentiation.

It is suggested that the hairs of many species, though almost certainly not all, may have an important role in maximising phosphate uptake because of their ability to hydrolyse organic phosphates. Other possibilities that should be considered are that hairs may be important in making use of phosphate presented in pulses or at very low concentrations. However, no direct evidence has been produced so far to show that hairs are a site of phosphate uptake. Within the hair-forming Rivulariaceae there appears to be a considerable variety of strategies with respect to use of nitrogen source. Some species, such as those *Homoeothrix* which do have well-developed hairs, probably depend entirely on combined nitrogen. Some heterocystous forms, while able to fix (di)nitrogen, probably still depend largely on combined nitrogen. It might be that nitrogen fixation is of critical importance in such organisms only at a particular growth stage or environmental condition. Other heterocystous forms fix nitrogen at a sufficiently high rate that it seems likely that this is their main nitrogen source. The behaviour of heterocysts in some *Gloeotrichia* strains suggests that they might be particularly well suited to environments where the relative importance of the two main nitrogen sources shifts frequently.

Acknowledgements

I am especially grateful to Dr. C. Sinclair for her thorough search through the literature in diverse languages and for permission to make use of the review in her thesis, but many other people have provided helpful conversation discussion, including Prof. A. Aziz, Prof. R.W. Castenholz, Dr. A. Donaldson, Dr. D. Douglas, Prof. G.E. Fogg, Dr. M.T. Gibson, S.L.J. Grainger, Dr. S.M.

Kirkby, Prof. T.M. Khoja, Dr. D. Livingstone, A. Peat, Dr. A. Pentecost, Prof. M. Potts, Prof. Z.N. Tahmida(-Khan), Prof. A.E. Walsby and Dr. P. Wood.

References

Ahluwalia, A.S., Rai, R.K. and Kumar, H.D. (1980) Chromatic adaptation and photoreversal in blue-green alga *Calothrix clavata* West. J. Biosci. 2, 63–68.

Ahmadjian, V. (1967) A guide to the algae occurring as lichen symbionts: isolation, culture, cultural physiology, and identification. Phycologia 6, 127–160.

Al-Mousawi, A.H.A. (1984) Biological studies on algae in ricefield soil from the Iraqi marshes. Ph.D. Thesis, University of Durham, England.

Aziz, A. (1985) Blue-green algal nitrogen fixation associated with deepwater rice in Bangladesh. Ph.D. Thesis, University of Durham, England.

Batten, D.J. and Van Geel, B. (1985) *Celyphus rallus*, probably early Cretaceous Rivulariacean blue-green alga. Rev. Paleobot. Palynol. 44, 233–241.

Böcher, T.W. (1946) *Dichothrix gelatinosa* sp. n., its structure and resting organs. K. Danske Vidensk. Selsk. Skr. IV, 4, 1–15.

Boissière, M.C. (1977) Un méchanisme possible d'absorption des glucides d'origine cyanophytique par les hyphes de quelques lichens. Rev. Bryol. Lichenol. 42, 617–635.

Bornet, E. and Flahault, C. (1886a,b) Revision des Nostocacées heterocystées contenues dans les principaux herbiers de France. Ann. Sci. Nat. Bot. VII, 3, 323–381; 4, 343–373.

Bornet, E. and Thuret, G. (1880) Notes Algologiques. G. Masson, Paris.

Braun, W. (1968) Die Kalkflachmoore und ihre wichtigsten Kontaktgesellschaften im bayerischen Alpenvorland. (Dissertationes Botanicae 1), J. Cramer, Germany, 134 pp.

Capone, D.G., Penhale, P.A., Oremland, R.S. and Taylor, B.F. (1979) Relationship between productivity and N_2 (C_2H_2) fixation in a *Thalassia testudinum* community. Limnol. Oceanogr. 24, 117–125.

Carpenter, E.J. (1972) Nitrogen fixation by a blue-green epiphyte on pelagic *Sargassum*. Science 178, 1207–1209.

Carpenter, E.J. and Cox, J.L. (1974) Production of pelagic *Sargassum* and a blue-green epiphyte in the western Sargasso Sea. Limnol. Oceanogr. 19, 429–436.

Casanova, J. (1982) Morphologie et biolithogenèse des barrages de travertins. In J. Nicod (ed.), Formations Carbonatées Externes, Tufs et Travertins. Aix-en-Provence, pp. 45–54.

Casanova, J. (1984) Genèse des carbonates d'un travertin pléistocène: interpretation paléoécologique du sondage Peyre I (Comprégnac, Aveyron). Geobios 8, 219–225.

Casanova, J. and Lafont, R. (1985) Les cyanophycées encroutantes des eaux courantes du Var (France). Verh. Int. Ver. Theor. Angew. Limnol. 22, 2805–2810.

Castenholz, R.W. (1973) Ecology of blue-green algae in hot springs. In N.G. Carr and B.A. Whitton (eds.), The Biology of Blue-green Algae. Blackwell, Oxford, pp. 379–414.

Cattaneo, A. (1983) Grazing on epiphytes. Limnol. Oceanogr. 28, 124–132.

Caudwell, C. (1983) Les Rivulariacées actuelles: interprétation possible de la structure zonée des concrétions stromatolitiques a *Rivularia haematites*. Geobios 16, 169–177.

Chang, T.-P. (1979a) Growth and acetylene reduction by *Gloeotrichia echinulata* (Smith) Richter in axenic culture. Br. Phycol. J. 14, 207–210.

Chang, T.-P. (1979b) Zur Morphologie von *Gloeotrichia echinulata* (Smith) Richter in axenischer Kultur. Nova Hedwigia Kryptogamenk. 21, 265–283.

Chang, T.-P. (1983) Interaction of water-blooming cyanophyte *Gloeotrichia echinulata* and its endophytic blue-green alga *Pseudanabaena catenata*. Arch. Hydrobiol. 97, 320–328.

Chang, T.-P. and Blauw, T.S. (1980) Nitrogen fixing capacity of two colonial types of *Gloeotrichia echinulata* (Smith) Richter. Arch. Hydrobiol. 89, 382–386.

Claassen, M.I. (1973) Freshwater algae of Southern Africa. I. Notes on *Gloeotrichia ghosei*. R.N. Singh. Br. Phycol. J. 8, 325–331.

Claus, G. (1957) *Gloeotrichia andreanszkyana*, eine neue, interessante Blaualge. Verh. Zool.-Bot. Ges. Wien 97, 87–96.

Cmiech, H.A., Leedale, G.F. and Reynolds, C.S. (1984) Morphological and ultrastructural variability of planktonic Cyanophyceae in relation to seasonal periodicity. I. *Gloeotrichia echinulata*: vegetative cells, polarity, heterocysts, akinetes. Br. Phycol. J. 19, 259–275.

Darley, J. (1967) Sur quelques résultats de la culture en laboratoire de deux espèces de *Calothrix* Agardh (Myxophycées-Rivulariacées). C.R. Acad. Sci. Ser. D 264, 1013–1015.

Davey, A. (1983) Effects of abiotic factors on nitrogen fixation by blue-green algae in Antarctica. Polar Biol. 2, 95–100.

De Bary, A. (1863a,b) Beitrag zur Kenntniss der Nostocaceen, insbesondere Rivulariaceen. Flora (Jena) 46, 553–560.

De Puymaly, A. (1949) L'algue du *Placynthium nigrum* Gray; son aspect dans le lichen et à l'état d'indépendance. Considérations sur la symbiose lichénique. Botaniste 34, 351–359.

Desikachary, T.V. (1945) Notes on the reactions of *Gloeotrichia raciborskii* Woloscz. to a parasitic attack. Curr. Sci. 14, 207–208.

Desikachary, T.V. (1946) Germination of the heterocyst in two members of the Rivulariaceae *Gloeotrichia raciborskii* Woloscz. and *Rivularia mangini* Frémy. J. Indian Bot. Soc. 25, 3–17.

Desikachary, T.V. (1959) Cyanophyta. Indian Council of Agricultural Research, New Delhi, 686 pp.

Donaldson, A. and Whitton, B.A. (1977) Algal flora of freshwater habitats on Aldabra. Atoll. Res. Bull. 215, 1–26.

Douglas, D.B. (1979) Ultrastructural aspects of tapering in members of the Rivulariaceae (Cyanophyta). Ph.D. Thesis (CNAA), Sunderland Polytechnic, England.

Dromgoole, F.I., Silvester, W.B. and Hicks, B.J. (1978) Nitrogenase activity associated with *Codium* species from New Zealand. N.Z.J. Mar. Freshwat. Res. 12, 17–22.

Edmondson, W.T. (1938) Three new species of Rotatoria. Trans. Am. Microsc. Soc. 57, 153–157.

Fay, P. (1983) The Blue-greens. (Cyanophyta – Cyanobacteria). Edward Arnold, London, 88 pp.

Fay, P., Stewart, W.D.P., Walsby, A.E. and Fogg, G.E. (1968) Is the heterocyst the site of nitrogen fixation in blue-green algae? Nature (Lond.) 220, 810–812.

Feldmann, G. and Guglielmi, G. (1973) Sur la cytologie du *Rivularia mesenterica* Thuret et l'influence du milieu extérieur sur son ultrastructure. C.R. Acad. Sci. Ser. D 276, 1417–1420.

Fliesser, S.B. and Jensen, T.E. (1982) Observations on the fine structure of isolates of the blue-green bacteria *Calothrix*, *Fremyella* and *Gloeotrichia*. Cytobios 33, 203–222.

Fogg, G.E. (1969) The physiology of an algal nuisance. Proc. Roy. Soc. London B 173, 175–189.

Fogg, G.E. (1982) Marine plankton. In N.G. Carr and B.A. Whitton (eds.), The Biology of Cyanobacteria. Blackwell, Oxford, pp. 491–513.

Forest, H. and Khan, K.R. (1972) The blue-green algae – a program of evaluation of Francis Drouet's taxonomy. In T.V. Desikachary (ed.), Taxonomy and Biology of Blue-green Algae. University of Madras, pp. 128–138.

Frémy, P. (1983) Les cyanophycées des côtes d'Europe. Mem. Soc. Nat. Sci. Nat. Math. Cherbourg 41, 1–235.

Friedmann, I. (1956) Über die Blaualga *Gardnerula corymbosa* (Harvey) J. De Toni und ihr Vorkommen im Mittelmeer. Öst Bot. Z. 103, 336–341.

Fritsch, F.E. (1945) The Structure and Reproduction of the Algae. Vol. II. Cambridge University Press, England, 939 pp.

Geitler, L. (1925) Synoptische Darstellung der Cyanophyceen in morphologischer und systematischer Hinsicht. Beih. Bot. Cbl. 41, 163–294.

Geitler, L. (1932) Cyanophyceae. In Rabenhorst's Kryptogamen-Flora, Vol. 14. Akademische Verlagsgesellschaft, Leipzig, 1196 pp.

Geitler, L. (1934) Beiträge zur Kenntnis der Flechtensymbiose, IV–V. Arch. Protistensk. 82, 51–85.

Goering, J.J. and Parker, P.L. (1972) Nitrogen fixation by epiphytes on sea grasses. Limnol. Oceanogr. 17, 320–323.

Goldner, L.L. (1980) Nitrogen fixation (acetylene reduction) in shallow water Bahamian environments. Bull. Mar. Sci. 30, 444–453.

Golubic, S. (1973) The relationship between blue-green algae and carbonate deposits. In N.G. Carr and B.A. Whitton (eds.), The Biology of Blue-green Algae. Blackwell, Oxford, pp. 434–472.

Golubic, S. and Awramik, S.M. (1974) Microbial comparison of stromatolitic environments: Shark Bay, Persian Gulf, and the Bahamas. Geol. Soc. Am. Abstr. Prog., 1974 Annual Meeting 6:7, 759–760.

Golubic, S. and Campbell, S.E. (1981) Biogenically formed aragonite concretions in marine *Rivularia*. In C. Monty (ed.), Phanerozoic Stromatolites. Springer-Verlag, Berlin, Heidelberg, pp. 209–229.

Gomont, M. (1895) Note sur un *Calothrix* sporifère (*Calothrix stagnalis* sp. n.). J. Bot. (Paris) 9, 197–202.

Griffiths, B.M. (1925) Studies in the phytoplankton of the lowland waters of Great Britain. III. The phytoplankton of Shropshire, Cheshire and Staffordshire. J. Linn. Soc. (Bot.) 47, 75–98.

Guglielmi, G. (1975) Étude de quelques espèces marines de Rivulariacées en microscopic électronique. Vie Milieu Ser. A 25, 189–213.

Henssen, A. (1969) Three non-marine species of the genus *Lichina*. Lichenologist 4, 88–98.

Henssen, A. (1973) New or interesting cyanophytic lichens. I. Lichenologist 5, 444–451.

Herdman, M., Janvier, M., Waterbury, J.B., Rippka, R. and Stanier, R. (1979a) Deoxyribonucleic acid base composition of cyanobacteria. J. Gen. Microbiol. 111, 63–71.

Herdman, M., Janvier, M., Rippka, R., Stanier, R. and Mandel, M. (1979b) Genome size of cyanobacteria. J. Gen. Microbiol. 111, 73–85.

Heuff, H. and Horkan, K. (1983) Caragh. In B.A. Whitton (ed.), Ecology of European Rivers. Blackwell, Oxford, pp. 363–384.

Hubel, H. and Hubel, M. (1974) In-situ Messungen der diurnalen Stickstoff Fixierung an Mikrobenthos der Ostseeküste., Arch. Hydrobiol., Suppl. 46, 39–54.

Hudson, J.W., Crompton, K.J. and Whitton, B.A. (1971) Ecology of Hell Kettles: 2. The ponds. Vasculum LVI, 50–58.

Jeeji-Bai, N. (1977) Morphological variation of some species of *Calothrix* and *Fortiea*. Arch. Protistenk. 119, 367–387.

Jensen, T.E. (1984) Cyanobacterial cell inclusions of irregular occurrence: systematic and evolutionary implications. Cytobios 39, 35–62.

Jensen, T.E. and Ayala, R.P. (1976) The fine structure of striated microtubules and sleeve bodies in several species of *Anabaena*. J. Ultrastruct. Res. 57, 185–193.

Johansson, C. (1979) Chlorophyll content and the periphytic algal vegetation in six streams in North Jamtland, Sweden, 1977. Medd. Växtbiol. Inst. Uppsala, 1979, 2, 1–28.

Jones, K. and Stewart, W.D.P. (1969a) Nitrogen turnover in marine and brackish habitats. III. The producton of extracellular nitrogen by *Calothrix scopulorum*. J. Mar. Biol. Assoc. UK 49, 475–488.

Jones, K. and Stewart, W.D.P. (1969b) Nitrogen turnover in marine and brackish habitats. IV. Uptake of the extracellular products of the nitrogen-fixing alga *Calothrix scopulorum*. J. Mar. Biol. Assoc. UK. 49, 701–716.

Kallas, T., Rebière, M., Rippka, R. and Tandeau de Marsac, N. (1983) The structural *nif* genes of the cyanobacteria *Gloeothece* sp. and *Calothrix* sp. share homology with those of *Anabaena* sp., but the *Gloeothece* genes have a different arrangement. J. Bacteriol. 155, 427–431.

Kann, E. (1982) Qualitätive Veränderungen der litoralen Algenbioconose österreichischer Seen (Lunzer Untersee, Traunsee, Attersee) im Laufe der letzen Jahrzehnte. Arch. Hydrobiol./Suppl. 62, 440–490.

Kenyon, C.N., Rippka, R. and Stanier, R.Y. (1972) Fatty acid composition and physiological properties of some filamentous blue-green algae. Arch. Mikrobiol. 83, 216–236.

Khoja, T.M., Livingstone, D. and Whitton, B.A. (1984) Ecology of a marine *Rivularia* population. Hydrobiologia 108, 65–73.

Khoja, T.M. and Whitton, B.A. (1971) Heterotrophic growth of blue-green algae. Arch. Mikrobiol. 79, 280–282.

Kirkby, S.M. and Whitton, B.A. (1976) Uses of coded data in study of *Calothrix* and *Rivularia*. Br. Phycol. J. 11, 407–416.

Kirkby, S.M., Hibberd, D.J. and Whitton, B.A. (1972) *Pleurocladia lacustris* A. Braun (Phaeophyta) – a new British record. Vasculum 57, 51–56.

Komárek, J. and Kann, E. (1973) Zur Taxonomie und Ökologie der Gattung *Homoeothrix*. Arch. Protistenk. 115, 173–283.

Kostyaev, V.Ya. (1984) Intensity of molecular nitrogen fixation by epiphytic complexes of freshwater macrophytes. Akad. Nauk. SSSR Ser. Biol. 1, 117–123, in Russian, with English summary.

Kulasooriya, S.A., Lang, N.J. and Fay, P. (1972) The heterocysts of blue-green algae. III. Differentiation and nitrogenase activity. Proc. Roy. Soc. Lond. B 181, 199–209.

Lami, R. and Meslin, R. (1959) Sur une cyanophycée, *Calothrix Chapmanii* nom. nov., vivant à l'intérieur d'une Entéromorphe limicole. Bull. Lab. Marit. Dinard 44, 47–49.

Lange, W. (1970a) Chelating agents and blue-green algae. Can. J. Microbiol. 20, 1311–1321.

Lange, W. (1970b) Blue-green algae and humic substances. Proc. Conf. Great Lakes Res. 13, 58–70.

Lange, W. (1975) Transformation of nonpolar filaments of the blue-green alga *Gloeotrichia echinulata* U. Wisc. 1052 into double helices. J. Phycol 11, 75–79.

Lannergren, C. (1980) Nitrogen fixation by blue-green algae in a Norwegian land-locked fjord, determined by the acetylene reduction technique. Sarsia 65, 73–77.

Livingstone, D. and Whitton, B.A. (1983) Influence of phosphorus on morphology of *Calothrix parietina* (Cyanophyta) in culture. Br. Phycol. J. 18, 29–38.

Livingstone, D. and Whitton, B.A. (1984) Water chemistry and phosphatase activity of the blue-green alga *Rivularia* in Upper Teesdale streams. J. Ecol. 72, 405–421.

Livingstone, D., Khoja, T.M. and Whitton, B.A. (1983) Influence of phosphorus on physiology of a hair-forming blue-green alga (*Calothrix parietina*) from an upland stream. Phycologia 22, 345–350.

Livingstone, D., Pentecost, A. and Whitton, B.A. (1984) Diel variations in nitrogen and carbon dioxide fixation by the blue-green alga *Rivularia* in an upland stream. Phycologica 23, 125–133.

Macgregor, J.M., Truswell, J.F. and Eriksson, K.A. (1974) Filamentous algae from the 2,300 m.y. old Transvaal Dolomite. Nature, Lond. 247, 538–540.

Maehr, H. and Smallheer, J.M. (1984) Rivularins. Preliminary synthetic studies. J. Org. Chem. 49, 1549–1553.

Mague, T.H. and Holm-Hansen, O. (1975) Nitrogen fixation on a coral reef. Phycologia 14, 87–92.

Mahajan, A.D. (1984) Some Rivulariaceae RABENH. from paddy fields of Kaira District, Gujarat, India. Feddes Repert. 95, 533–539.

Mahajan, A.D. and Patel, R.J. (1984) On *Gloeotrichia* AG. from paddy fields of Kaira District, Gujerat, India. Feddes Repert. 95, 541–547.

Mälmestrom, B. (1972) The genus *Calothrix* in the black zone. Bot. Mar. 15, 87–90.

Mann, J.E. and Schlichting, H.E., Jr. (1967) Benthic algae of selected thermal springs in Yellowstone National Park. Trans. Am. Microsc. Soc. 86, 2–9.

Maxwell, T.F. (1974) Developmental morphology of the blue-green alga *Gloeotrichia echinulata*. J. Phycol. 10, Suppl. 4.

Miller, M.M. and Lang, N.J. (1971) The effect of aging on thylakoid configuration and granular inclusions in *Gloeotrichia*. In B.C. Parker and R.M. Brown (eds.), Contributions in Phycology. Allen Lane Press, Lawrence, USA, pp. 53–58.

Monty, C.L.V. (1967) Distribution and structure of recent stromatolitic algal mats, eastern Andros Island, Bahamas. Ann. Soc. Geol. Belg. Bull. 96, 585–624.

Norton, R.S. and Wells, R.J. (1982) A series of chiralpolybrominated biindoles from the marine blue-green alga *Rivularia firma*. Application of ^{13}C NMR spin-lattice relaxation data and ^{13}C-^{1}H coupling constants to structure elucidation. J. Am. Chem. Soc. 1982, 3628–3635.

Osa-Afiana, L.O. and Alexander, M. (1981) Factors affecting predation by a microcrustacean (*Cypris* sp.) on nitrogen-fixing blue-green algae. Soil Sci. 13, 27–32.

Palla, E. (1893) Beitrag zur Kenntnis des Baues des Cyanophyceen-Protoplasts. Jahrb. Wissensch. Bot. 25, 511–562.

Pankow, H., Festerling, E. and Festerling, H. (1971) Beitrag zur Kenntnis der Algenflora der mecklenburgischen Küste (südliche Ostsee: Lübecker Bucht – Darss). Int. Rev. Ges. Hydrobiol. 56, 241–263.

Pattnaik, U. and Singh, P.K. (1978) Effect of nitrate nitrogen on the growth, heterocyst differentiation and nitrogen fixation in rice field blue-green alga *Gloeotrichia* sp. Arch. Hydrobiol., Suppl. 51, 318–327.

Pearson, H.W. and Taylor, R. (1978) Nitrogen fixation in a polluted canal system. Ecol. Bull. (Stockh.) 26, 69–82.

Pedersen, M. and DaSilva, E.J. (1973) Simple brominated phenols in the blue-green alga *Calothrix brevissima* West. Planta 115, 83–86.

Polyanskiĭ, V.I. (1930) K voprosu o stadiyakh razvitiya *Gloeotrichia natans* (Hedw.) Rabenh. Izv. Glav. Bot. Sada 29, 265–299, in Russian with French summary: On the question of the stages of development in *Gloeotrichia natans* (Hedw.) Rabenh.

Potts, M. and Whitton, B.A. (1977) Nitrogen fixation by blue-green algal communities in the intertidal zone of the lagoon of Aldabra Atoll. Oecologia 27, 275–283.

Prosperi, C.H. (1983) Curiosa asociacion de dos cianofitos. Kurtziana 16, 149–155.

Rai, A.K., Pandey, K.D. and Kashyap, A.K. (1978) Heterocyst differentiation and polarity in *Calothrix*. New Phytol. 81, 647–651.

Rippka, R. and Herdman, M. (1985) Division patterns and cellular differentiation in cyanobacteria. Ann. Inst. Pasteur/Microbiol. 136A, 33–39.

Rippka, R., Deruelles, J., Waterbury, J.B., Herdman, M. and Stanier, R.Y. (1979) Generic assignments, strain histories and properties of pure cultures of cyanobacteria. J. Gen. Microbiol. 111, 1–61.

Roelofs, T.D. and Oglesby, R.T. (1970) Ecological observations on the planktonic cyanophyte *Gloeotrichia echinulata*. Limnol. Oceanogr. 15, 224–229.

Rosenberg, G. and Paerl, H.W. (1981) Nitrogen fixation by blue-green algae associated with the siphonous green seaweed *Codium decorticatum*. Mar. Biol. 61, 151–158.

Savela, K. (1983) Stickstoff-Fixierung der Blaualge *Calothrix scopulorum* in Küstengewässern der Ostsee. Ann. Bot. Fenn. 20, 399–415.

Schafer, A. and Stapf, K.R.G. (1978) Permian Saar-Nahe basin and recent Lake Constance (Germany): two environments of lacustrine algal carbonates. Spec. Publ. Int. Assoc. Sediment. 2, 83–107.

Schwenender, S. (1894) Zur Wachstumgeschichte der Rivularien. Sber. Preuss. Akad. Wissensch. 951–961.

Serbanescu, M. (1966) Sur quelques aspects du cycle de développement du l'algue bleue *Gloeotrichia natans* (Hedw.) Rabenh. Rev. Algol., N.S. 8, 189–195.

Shehata, F.H.A. and Whitton, B.A. (1981) Field and laboratory studies on blue-green algae from aquatic sites with high zinc levels. Verh. Int. Ver. Theor. Angew. Limnol. 21, 1466–1471.

Sinclair, C. (1977) Physiological and morphological studies on tapering blue-green algae. Ph.D. Thesis, University of Durham, England.

Sinclair, C. and Whitton, B.A. (1977a) Influence of nutrient deficiency on hair formation in the Rivulariaceae. Br. Phycol. J. 12, 297–313.

Sinclair, C. and Whitton, B.A. (1977b) Influence of nitrogen source on morphology of Rivulariaceae (Cyanophyta). J. Phycol. 13, 335–340.

Singh, R.N. and Tiwari, D.N. (1970) Frequent heterocyst germination in the blue-green alga *Gloeotrichia ghosei* Singh. J. Phycol. 6, 172–176.

Smith, R.V. and Peat, A. (1967) Comparative structure of the gas vacuoles of blue-green algae. Arch. Mikrobiol. 57, 111–122.

Steinecke, F. (1931) Das Auskeimen alter Heterocysten bei *Calothrix Weberi*. Bot. Arch. 34, 153–160.

Stewart, W.D.P. (1962) Fixation of elemental nitrogen by marine blue-green algae. Ann. Bot. N.S. 26, 439–445.

Stewart, W.D.P. (1963) Liberation of extracellular nitrogen by two nitrogen-fixing blue-green algae. Nature (Lond.) 200, 1020–1021.

Stewart, W.D.P. (1965) Nitrogen turnover in marine and brackish habitats. Ann. Bot. N.S. 29, 229–239.

Stewart, W.D.P. (1967) Nitrogen turnover in marine and brackish habitats. II. Use of ^{15}N in measuring nitrogen fixation in the field. Ann. Bot. N.S. 385–407.

Stewart, W.D.P., Haystead, A. and Pearson, H.W. (1969) Nitrogenase activity in heterocysts of filamentous blue-green algae. Nature (Lond.) 224, 226–228.

Theivendirarajah, K. and Jeyaseelan, K. (1977) The ingestion of blue-green algae by mosquito larvae. Ceylon J. Sci. Biol. Sci. 12, 156.

Tiwari, D.N., Yadava, P.D. and Mishra, A.K. (1979) Developmental behaviour in mutants of *Gloeotrichia ghosei* defective in nitrogen fixation. Microbios. Lett. 12, 121–130.

Tiwari, D.N., Singh, L.J. and Mishra, A.K. (1985) Evidence for a reversible action of methionine sulphoximine during growth of the cyanobacterium, *Gloeotrichia ghosei* Singh. New Phytol. 99, 361–365.

Tsuda, R.T. and Kami, H.T. (1973) Algal succession on artificial reefs in a marine lagoon environment in Guam. J. Phycol. 9, 260–264.

Ueda, K. (1971) Die quantitative Bestimmung des DNS-Gehalts in den Zellen von Cyanophyceen durch Fluorochromierung mit Coriphosphin. Biochem. Physiol. Pflanz. 162, 439–449.

Umezaki, I. (1961) The marine blue-green algae of Japan. Mem. College Agriculture, Kyoto University No. 83 (Fishery Series No. 8).

Van Geel, B., De Lange, L. and Wiegers, J. (1984) Reconstruction and interpretation of the local vegetational succession of a late-glacial deposit from Usselo (The Netherlands), based on the analysis of microfossils and macrofossils. Acta Bot. Neerl. 33, 535–546.

Warmling, P. (1973) Nitrogen fixation on rocks in Oslofjord. Bot. Mar. 16, 237–240.

Webb, K.L., DuPaul, W.D., Wiebe, W., Sottile, W. and Johannes, R.E. (1975) Enewetak (Eniwetok) atoll: aspects of the nitrogen cycle on a coral reef. Limnol. Oceanogr. 20, 198–210.

Weber, R. (1933) Beitrage zur Kenntnis der Gattung *Calothrix*. Arch. Protistenk. 79, 391–415.

Welsh, H. (1965) A contribution to our knowledge of the blue-green algae of South-West Africa and Bechuanaland. Nova Hedwigia 9, 131–162.

Whitton, B.A. (1973) Freshwater plankton. In N.G. Carr and B.A. Whitton (eds.), The Biology of Blue-green Algae. Blackwell, Oxford, pp. 353–367.

Whitton, B.A. (1987) Survival and dormancy of blue-green algae. In Y. Henis (ed.), Survival and Dormancy in Microorganisms. Wiley–Interscience, New York, in press.

Whitton, B.A., Gale, N.L. and Wixson, B.G. (1981) Chemistry and plant ecology of zinc-rich wastes dominated by blue-green algae. Hydrobiologia 83, 331–341.

Whitton, B.A., Khoja, T.M. and Arif, I.A. (1986) Water chemistry and algal vegetation of streams in the Asir Mountains, Saudi Arabia. Hydrobiologia, 133, 97–106.

Wickstrom, C.E. (1980) Distribution and physiological determinants of blue-green algal nitrogen fixation along a thermogradient. J. Phycol. 16, 436–443.

Wickstrom, C.E. and Castenholz, R.W. (1975) Atypical seasonal distribution of thermophilic algae in a hot spring. J. Phycol., Suppl. 11, 7.
Wickstrom, C.E. and Castenholz, R.W. (1978) Association of *Pleurocapsa* and *Calothrix* (Cyanophyta) in a thermal stream. J. Phycol. 14, 84–88.
Wiebe, W.J., Johannes, R.E. and Webb, K.L. (1975) Nitrogen fixation in a coral reef community. Science 188, 257–259.
Womersley, H.B.S. (1946) Studies on the marine algae of southern Australia. I. The genera Isactis and Rivularia (Myxophyceae). Trans. Roy. Soc. South Australia 70, 127–136.
Wood, P. (1984) Structural and physiological studies on the effects of mineral deficiencies on the Rivulariaceae (Cyanobacteria) Vols. I, II. Ph.D. Thesis (CNAA) Sunderland Polytechnic, England.
Zehnder, A. (1963) Kulturversuche mit *Gloeotrichia echinulata* (J.E. Smith) P. Richter. Schweiz. Z. Hydrol. 25, 65–83.
Zhao-liang, X. (1984a,b) Investigation on the procaryotic microfossils from the Gaoyuzhuang Formation, Jixian, North China. Acta Bot. Sin. 26, 216–222, 312–319.

Subject Index